Handbook of Industrial Oil and Fat Products

Volume 2

Preparation, Properties and Uses of Various Oils such as Coconut Oil, Canola Oil, Corn Oil, Cottonseed Oil, Olive Oil, Palm Oil, Castor Oil, Rice Bran Oil, Rapeseed Oil, Sesame Oil, Soyabean Oil, Sunflower Oil, Safflower Oil, Peanut Oil

Handbook of Industrial Oil and Fat Products in 4 Volumes

Volume 1

Chemistry of Fats and Oils, Chemical and Physical Changes in Edible Fats, Fatty Acids and their Derivatives, Toxicity and Safety of Fats and Oils

Volume 2

Preparation, Properties and Uses of Various Oils such as Coconut Oil, Canola Oil, Corn Oil, Cottonseed Oil, Olive Oil, Palm Oil, Castor Oil, Rice Bran Oil, Rapeseed Oil, Sesame Oil, Soyabean Oil, Sunflower Oil, Safflower Oil, Peanut Oil

Volume 3

Butter, Cheese, Casein, Margarine and Spread, Shortening, Cooking Oils, Salad Oils, Marine Oils, and Frying Technology

Volume 4

Rendering, Manufacturing Processes for Emulsifiers, Recovery and Refining of Glycerine, Antioxidants, Fats and Oils in Soap, Surfactants, Paints, Varnishes, Leather Processing, Textile Processing, Pharmaceuticals, Natural Waxes, Cosmetics. Quality Assurance, Storage, Stability and Transport of Fats and Oils

Handbook of Industrial Oil and Fat Products

Volume 2

Preparation, Properties and Uses of Various Oils such as Coconut Oil, Canola Oil, Corn Oil, Cottonseed Oil, Olive Oil, Palm Oil, Castor Oil, Rice Bran Oil, Rapeseed Oil, Sesame Oil, Soyabean Oil, Sunflower Oil, Safflower Oil, Peanut Oil

G. S. BRECK MSc, PhD
S. C. BHATIA BE (Chemical), MBA

CBS

CBS PUBLISHERS & DISTRIBUTORS

New Delhi • Bangalore

Handbook of Industrial Oil and Fat Products • Volume 2

First Edition 2008

ISBN: 978-81-239-1512-8

Published by :
Satish Kumar Jain for CBS Publishers & Distributors,
4819/XI Prahlad Street, 24 Ansari Road, Ansari Road,
Daryaganj, New Delhi 110002, India.
E-mail : cbspubs@vsnl.com
Website : www.cbspd.com

Production Director : Vinod K. Jain

Branch Office :
Seema House, 2975, 17th Cross, K.R. Road,
Bansankari 2nd Stage, Bangalore - 560070
Fax : 080-26771680 • E-mail : cbsbng@vsnl.net

Printed at :
India Binding House, Noida, UP (India)

Preface

This volume covering preparation, properties and uses of various oils such as, coconut oil, canola oil, corn oil, cottonseed oil, olive oil, palm oil, castor oil, rice bran oil, rapeseed oil, sesame oil, soyabean oil, sunflower oil, safflower oil, peanut oil, etc. is the second of the four-volume ***Handbook of Industrial Oil and Fat Products***. It also discusses other oils such as sulphated and sulphonated oil, some of the miscellaneous oils such as grapeseed oil, camelina oil, poppy seed oil, almond and walnut oil, wheat germ oil, tall oil, etc.

This *Handbook* makes an essential reading for all students and teachers, professionals, researchers and industrialists, company executives and managers dealing with fats and oils. It is meant as an effective management tool for them and as a handy aid for their project responsibilities. It is also a valuable source of information for those preparing for or already associated with the oil technology, food processing and foodservice industries. To use these products efficiently, it is necessary to have a good understanding of the properties and behaviour of oil and fats.

The book contains valuable and necessary information related to preparation, properties, applications and analysis of various oils and fats. Oils and fats constitute one of the three major classes of food products. The other two are proteins and carbohydrates. Oils and fats are essential nutrients of the human diet and are concentrated source of energy. In any study of human nutritional requirements, the contribution of each of these major food classes must be understood to set up proper diets for individuals. Knowledge of vitamins and minerals as well as the role of oxygen, water, fibre and trace elements is also essential.

Chapter 1 deals with coconut oil which is the healthiest oil on the earth. Manufacturing aspects along with properties and fatty acids and derivatives from coconut oil are discussed in detail. Chapter 2 concentrates on canola oil which is the only seed crop adopted with temperate regions. Chapter 3 is devoted to corn oil which is a by-product of the cereal industry. Chapter 4 is confined to cottonseed oil along with the manufacturing and biological processes, uses and storage aspects. Chapter 5 focuses on olive oil which enjoys a privileged position amongst the edible vegetable oils. Olive oil is the juice of a fruit and as such it is consumed nearly in its virgin state. Manufacturing aspects of olive oil along with olive oil components are discussed in detail. Chapter 6 deals with palm oil which is derived from fleshy mesocarp of the palm fruit. Manufacturing aspects of palm oil along with applications and treatment of palm mill and palm oil refinery effluent are discussed in detail.

Chapter 7 concentrates on castor oil which is also known as *Ricinus* oil. Preparation, properties, applications and fat-splitting aspects of various types of castor oils are discussed in detail. Chapter 8 is devoted to rice bran oil which is a heart-friendly health oil. Rice bran oil with balanced fatty acid profile and other nutritional compounds can respond to an ever-increasing consumer interest in the health attributes of food. Chapter 9 deals with rapeseed oil which is a vegetable oil derived from rapeseed by expression or solvent extraction. Steps involved in processing like degumming and alkali refining, neutralisation, bleaching, hydrogenation, deodourisation and edible products from rapeseed oil along with applications are discussed in detail.

Chapter 10 is devoted to sesame oil which is markedly different from all other vegetable oils in many chemical, biological and physiological properties. Modern methods of oil extraction, along with sesame cake meal and applications are discussed in detail. Chapters 11 and 12 focus on soyabean oil and sunflower oil; preparation properties and typical uses of these oils are discussed in detail.

Chapter 13 concentrates on safflower oil which was used primarily as a source of dye, in paints, varnishes, alkyd resins, as a food colouring and for medicinal purposes. Steps involved in processing, quality control, storage and transportation are discussed in details. Chapter 14 discusses peanut oil which is nondrying vegetable oil and has a high proportion of unsaturated fatty acids. This oil is used mainly for edible purposes in the preparation of shortening margarine and mayonnaise, as a cooking, frying oil and salad oil.

Chapter 15 focuses on sulphated and sulphonated oils which are the direct derivatives of fats, used for non-edible purposes and are strongly hydrophilic. Chapter 16 is confined to various miscellaneous oils such as grape seed oil, camelina oil, poppy seed oil, oleaginous flax, almond, hazelnut and walnut, evening primrose, borage and black currant oils, wheat germ oil, crambe oil, vernonia oil, cuphea oil, avocado oil, tall oil, etc.

The text in this book is well supplemented throughout with diagrams and tables. The treatment of all topics is in a cogent and lucid style aimed at enabling the reader to grasp the information quickly and easily. The references and index are also given at the end of the book.

While painstaking care has been taken into producing a useful and exhaustive reference book, the authors would welcome any constructive criticism and creative feedbacks from the students, teachers and professionals in the industry. E-mail communications may please be sent at cbspubs@vsnl.com.

We wish to thank all those colleagues who provided figures and photographs and offered us important advice during the preparation of this handbook.

G. S. BRECK
S.C. BHATIA

Contents

Preface — *v*

Chapter 1 Coconut Oil — *1-60*

Chapter 2 Canola Oil — *61-102*

Chapter 3 Corn Oil — *103-116*

Chapter 4 Cottonseed Oil — *117-204*

Chapter 5 Olive Oil — *205-228*

Chapter 6 Palm Oil — *229-331*

Chapter 7 Castor Oil and Its Derivatives — *332-403*

Chapter 8 Rice Bran Oil — *404-436*

Chapter 9 Rapeseed Oil — *437-515*

Chapter 10 Sesame Oil — *516-534*

Chapter 11 Soyabean Oil — *535-591*

Chapter 12 Sunflower Oil — *592-637*

Chapter 13 Safflower Oil — *638-653*

Chapter 14 Peanut Oil — *654-667*

Chapter 15 Sulphated and Sulphonated Oils — *668-683*

Chapter 16 Miscellaneous Oils — *684-714*

References — *715-716*

Index — 717-722

Contents in Detail

Chapter 1 Coconut Oil ***1-60***

Introduction *1*
Coconut palm *1*
Propagation *2*
Seedlings *2*
Tissue culture *2*
Varieties *2*
Fruit *2*
Kernel *3*
Coconut water *3*
Testa *3*
Shell and husk *4*
Copra *4*
Sun drying *4*
Direct fire drying *4*
Hot-air drying *5*
Oil extraction *5*
Dry process *5*
Mechanical extraction *5*
Post-expeller treatment *6*
Solvent extraction *6*
Wet process *7*
Copra meal *7*
Refining *7*
Chemical refining *7*
Advantages of continuous refining process *7*
Batch neutralisation *8*
Continuous neutralisation (short mix) *9*
Bleaching *9*
Deodourisation *9*
Physical refining *9*
Degumming *10*
Bleaching *10*
Steam stripping and deodourisation *10*
Composition of coconut oil *10*
Chemical and physical tests *11*
Fatty acid composition *12*
Acid value/percent free fatty acid *12*
Saponification value *13*
Iodine value *13*
Reichert-meissl value *13*
Unsaponifiable matter *13*
Peroxide value/stability test *14*
Slip/melting points or melting range *14*
Refractive index *14*
Colour *14*
Solid fat content/solid fat index *14*
Titer *15*
Uses of coconut oil *15*
Edible products *15*
Medical and infant food formulations *15*
Non-food products *15*
Virgin coconut oil *16*
Difference between virgin coconut oil and other coconut oils *16*
Coconut health oil *17*
Coconut cooking oil *17*
What coconut oil does not do? *17*
What coconut oil does do? *17*
Why should you use coconut oil? *18*
Storage of coconut oil *18*
Coconut oil in cosmetics *18*
Coconut oil as a body oil *18*
Coconut oil as hair oil *19*
Coconut oil in shampoo *19*
Coconut oil in dental cleaning products *20*
Coconut oil in perfumes and flavours *20*
Fatty acids and derivatives from coconut oil *20*
Types of fatty acids and derivatives from coconut oil and their general applications *21*
Oleochemical derivatives from coconut oil and their general applications *24*
Principles and methods in the manufacture of oleochemicals *25*
Fatty acids *25*
Chemistry of fat splitting *25*
Fat splitting processes *26*
Fatty acid distillation and fractionation operations *28*
Fatty acid distillation *28*
Fatty acid fractionation *29*
Methyl esters *30*
Advantages of methyl esters *31*
Chemistry of esterification and transesterification *32*
Methods of manufacture *33*
Fatty alcohols *35*

Methods of manufacture 36
Fatty alcohols from natural fats and oils 36
Chemistry of hydrogenation 36
High pressure hydrogenation processes 38
Suspension process 39
Lurgi fatty acid hydrogenation method 40
Fractionation of fatty alcohols 41
Glycerine 41
Glycerine from fats and oils 41
Methods of purification 43
Monoalkyl phosphates 47
Chemistry of the reaction 47
Method of manufacture 48
Properties and uses 48
Alkanolamides 48
Chemistry of the reaction 49
Types of alkanolamides 50
Method of manufacture 50
Surfactants 51
Types of surfactants 52
Chemistry of the reaction 53
Method of manufacture 54
Sulphonation of fatty acid methyl esters 57
Other surfactants 58
Tertiary amines 58
Types of tertiary amines 58
Methods of preparation 59
Uses 59

Chapter 2 Canola Oil **61-102**

Introduction 61
Canola oil extraction and processing 61
Extraction 62
Seed cleaning 63
Preconditioning 63
Flaking 63
Cooking 63
Oil extraction by expellers (prepressing) 64
Extruders 64
Extraction of solvent 65
Desolventising of meal and oil 65
Quality assurance: seed, oil and meal 66
Canola oil 66
Alkali and physical (steam) refining 69
Alkali refining 69
Physical refining 69
Adsorptive bleaching 69
Active carbon 70
Synthetic silicas 70
Other methods for removing chlorophylls 70
Hydrogenation 71
Hydrogenated canola oil in edible oil products 72
Interesterification 73
Dewaxing 73
Deodourisation 74
Crystallisation, packaging and tempering 74
Quality assurance 75
Degumming 75
Process control of alkali refining 75
Bleaching 75
Hydrogenation 75
The dewaxing process 76
Deodourisation/physical refining 76
Crystallisation, packaging and tempering 76
Canola oil: physical and chemical properties 76
Physical properties 76
Relative density 77
Crismer value 77
Viscosity 77
Smoke point 78
Flash point 78
Cold test 78
Melting characteristics: polymorphism and crystal properties 78
Solid fat index and dilatation curve 79
Chemical properties 80
Nature of oils and fats 80
Triacylglycerols 80
Fatty acids 80
Geometric isomerism 81
Fatty acid composition of canola oil 81
Minor fatty acids 82
Polar lipids 83
Trace elements 83
Tocopherols and sterols 84

Pigments *85*
Saponification value *86*
Iodine value *86*
Antioxidants *86*
Sedimentation phenomenon in canola oil *86*
Sediment composition *86*
Physical properties of sediment *87*
Nutritional properties of canola oil 87
Nutritional significance of canola oil composition *87*
Saturated fatty acids *88*
Monounsaturated fatty acids *88*
Polyunsaturated fatty acids *88*
Minor constituents: tocopherols and phytosterols *89*
Effect of canola oil on plasma cholesterol and lipoproteins *89*
Hypercholesterolemic subjects *89*
Effect of canola oil on thrombogenesis *90*
Canola oil in infant formulas *91*
Overall properties *92*
Flavour quality of canola oil *92*
Sensory evaluation *92*
Chemical and physical properties *93*
Peroxide values *93*
Thiobarbituric acid *93*
Carbonyl compounds *94*
p-Anisidine values *94*
Conjugated dienes and trienes *94*
Volatile analysis by gas chromatography *94*
Static headspace *94*
Dynamic headspace *95*
Direct injection *95*
Resistance to oxidative rancidity *96*
Active oxygen method *96*
Oxygen absorption tests *96*
Conductivity tests *96*
Flavour stability of canola oil *96*
Oxidative stability *98*
Accelerated storage tests *99*
Ambient storage tests *99*
Stability to photochemical oxidation *99*
Accelerated storage tests *99*
Ambient storage tests *100*
High-temperature stability *100*

Chapter 3 Corn Oil *103-116*

Introduction *103*
Technology *103*
Characteristics of corn oil *104*
Physicochemical characteristics *104*
Triglyceride composition *104*
Fatty acid composition *105*
Composition of unsaponifiable fraction *105*
Processing *108*
Standard recovery methods *108*
Exploratory recovery methods *108*
Refining, bleaching, dewaxing and deodourisation *109*
Alkali refining *109*
Bleaching *110*
Dewaxing *110*
Deodourisation *110*
Physical refining *110*
Effects of processing on mycotoxins and pesticide removal *111*
Mycotoxin removal *111*
Pesticide removal *111*
Effects of processing and handling on oil quality *111*
Packaging *112*
Utilisation and performance *112*
Nutritional considerations *114*
Effects of fatty acids on blood cholesterol and heart disease *114*
Effects of corn oil on cholesterol levels *114*
Effects of corn and other oils on cholesterol levels *114*
Effect of corn oil on other chronic conditions 115
Blood pressure *115*
Diabetes *115*
Immune system *115*
Cancer *115*

Chapter 4 Cottonseed Oil 117-204

Introduction *117*
Production and utilisation of cottonseed and cottonseed products *118*
Cottonseed processing *118*
Composition and characteristics of cottonseed *119*
Physical structure of cottonseed *119*
Development of hull *120*
Nucellus *120*
Cottonseed kernel *120*
Variability in constituents *121*
Fat development in oilseeds *121*
Changes in seeds with maturation *121*
Origin of fat and biosynthesis of fatty acids *123*
Composition of indian cottonseed varieties *125*
Effect of maturity on seed composition *126*
Effect of environment on seed composition *127*
Pigments of cottonseeds *128*
Other minor constituents of cottonseed *129*
Biological processes and storage of cottonseed *129*
Biological and storage changes *130*
Mould activity *130*
Heating and chemical changes *131*
Factors influencing heating *132*
Respiration and lipolysis *132*
Lipolysis *132*
Field and storage damage of cottonseed *133*
Field damage *133*
Storage damage *133*
Handling and practical storage of cottonseed *134*
Handling practices in America *134*
Handling of cottonseed at the mills *134*
Handling practices in India *135*
Processing of cottonseed *135*
Impurities in cottonseed *135*
Seed cleaning *135*
Types of seed cleaning equipments *136*
Construction and working of machines *137*
Cleaning cottonseed at the oil mills *137*
Delinting *138*
Types of delinters and their operation *138*
Dehulling *140*
Types of dehuller and dehulling processes *140*
Processes for meat separation and flaking or rolling of meats *141*
Purpose of flaking and flaking of meats *142*
Cooking of meats *143*
Method of oil extraction *144*
Oil extraction by pressure *144*
Solvent extraction *146*
Cottonseed milling industry in India *149*
Pneumatic seed cleaners *151*
Need for cleaning seed at the oil mill *151*
Test methods of cottonseed-milling industry *151*
Processing and utilisation of cottonseed oil *154*
Processing of cottonseed oil *154*
Pre-treatment and degumming *155*
Refining of cottonseed oil *156*
Refining procedures *156*
Continuous caustic soda refining method *158*
Continuous caustic soda refining process *159*
De Laval caustic soda continuous process *159*
De Laval quick mix process *159*
Continuous refining by sharpels systems *159*
Clayton cottonseed-oil refining process *160*
Modified soda-ash process for refining cottonseed-oil *160*
Caustic soda-ash process for refining crude cottonseed-oil *160*
Refining with ammonia hydroxide *160*
Miscella refining of crude cottonseed-oil *160*
Miscella cottonseed oil refining *161*
De smet-Neumi process *161*
Ranches integrated miscella refining process *161*

Miscella refining by De Laval process 161
Comparison of refining methods 162
Glycerine recovery and bleaching 163
Bleaching of cottonseed-oil 163
Bleaching procedures 164
Hydrogenation 166
Cottonseed oil vanaspati industry in India 172
Deodourisation of cottonseed oil 172
Utilisation of cottonseed oil 174
Properties of cottonseed oil 175
Keeping quality 177
Composition of cottonseed-oil 177
Keeping quality of cottonseed oil 178
Nutritional aspects of cottonseed oil 180
Current uses and new outlet for cottonseed oil 180
Utilisation of by-products of cottonseed milling industry 181
Cottonseed-cake or meal 181
Studies on cottonseed-cake 181
Amino acid content of cottonseed and its meal 181
Protein fraction: gossypol in cottonseed-cake 182
Improving protein quality of cottonseed-cake 183
Uses of cottonseed-cake in rations for livestock and poultry 183
Effect of cyclopropenoid fatty acids (CPFA) on cottonseed-meal 184
Comparative (energy value) nutritive value of cottonseed and cottonseed-cake and meal 184
Aflatoxin in cottonseed and its cake and meal 185
Nutritional quality of glandless cottonseed-meal 185
Edible cottonseed flour for human consumption 186
Nutritive value of cottonseed flour 187
Preparation of edible cottonseed flour and its utilisation 187
Uses of cottonseed protein flour 189
Cottonseed protein concentrates and isolates and uses 189
Preparation of protein concentrates 190
Preparation of protein isolates 190
Uses of cottonseed protein concentrates, isolates 191
Cotton linters 192
Hulls 195
Grading of cottonseed and its primary products 195
Grading of cottonseed 195
Determination of grade 195
Estimation of quality index 196
Advantages of adopting standards of grading 198
Grading systems in India 198
Grading of cottonseed oil 199
Crude cottonseed oil 199
Refined cottonseed oil 200
Washed cottonseed oil and semi-refined oil 200
Types and grades of cottonseed oil 200
Grades of cottonseed oil 201
Grading of cottonseed-cake and meal 201
Cottonseed flour for human consumption 202
Grading of cottonseed linters 203

Chapter 5 Olive Oil 205-228

Introduction 205
Composition and characteristics of olive oil 205
Triglyceride composition 206
Fatty acid distribution in internal and external positions 207
Fatty acid composition 207
Unsaponifiable matters 208
Contaminants 209
Production and utilisation 209
Obtaining virgin oil by pressure 210
Cleaning and washing 210
Grinding 210
Kneading 210
Pressing 211
Separation 211
Other mechanical processes for obtaining virgin oil 211

Other processing methods for olive oil and olive oil meal *212*
Extraction of olive oil *212*
Refining of olive oils *217*
Refining of pomace oil *218*
Olive oil components *219*
Minor glyceridic components (MGC) *219*
Non-glyceridic minor components *219*
Hydrocarbons *219*
Linear short chain alcohols and their esters *219*
Straight long chain alcohols *220*
Cyclic monohydroxy compounds *220*
Cyclic dihydroxy compounds *220*
Triterpenic acids *220*
Chlorophylls *220*
Flavor components *220*
Minor polar components *221*
Analysis of olive oils *221*
Determination of purity *221*
Sterol composition *221*
Total sterol content *221*
Fatty acids composition *222*
Saturated fatty acids in position 2 of the triglycerides *222*
Absolute difference between found and theoretical equivalent carbon number (ECN) 42 (trilinolein) values *222*
Trans fatty acids content *223*
Differentiation between olive oil and olive-pomace oil *223*
Wax content *223*
Dihydroxyterpene alcohol content *224*
Differentiation between virgin and refined olive oil and detection of refined olive oil and seed oils in virgin olive oil *224*
Concentration of stigmasta-3,5-diene *224*
UV absorption at 268 nm *226*
Quality parameters *226*
Organoleptic characteristics *226*
Free fatty acid content *226*
Peroxide value (PV) *227*
Tocopherol content *227*
Impurities *227*
Combined gas chromatography-mass spectrometry (GC/MS) *227*
Quality and specifications: The international COI standard *228*

Chapter 6 Palm Oil *229-331*

Introduction *229*
Oil palm products, their formation and characteristics *230*
Fat formation in the ripening fruit *232*
Palm kernels *233*
Cake-breaking *233*
Separation of fibres and nuts (depericarping) *233*
Preparation of nuts *234*
Nut-cracking *234*
Separation of the kernels from the cracked mixture *234*
Sterilisation and drying of kernels *235*
Palm kernel production controls *235*
Composition of palm oil and palm kernel oil *236*
Palm oil *236*
Palm kernel oil *239*
Oils of Elaeis oleifera *239*
Palm kernel cake *239*
Quality of palm oil and kernels *240*
Palm oil *240*
Formation of free fatty acid *240*
Bleachability of palm oil *243*
Quality standards 245
Palm kernels, palm kernel oil and palm kernel cake *245*
Non-mechanical traditional methods of extraction of oil and kernels *249*
'Soft oil' production *249*
Boiling *249*
Pounding *249*
Separation *249*
'Hard oil' production *250*
Fermentation *250*
Treading and separation *250*
Hand-operated presses *250*
Curb press *250*
Hydraulic hand presses *252*
Kernel extraction *255*
Power-operated palm oil mills *255*
Choice of a plantation mill *256*

Siting of a mill *257*
Sterilisation *258*
Vertical bunch sterilisers *258*
Horizontal sterilisers *258*
Stripping of fruit from bunches *259*
Digestion *260*
Extraction of oil *260*
Centrifuge *261*
Hydraulic press *261*
Screw presses *263*
Comparison of extraction methods *263*
Clarification *265*
Treatment of palm oil mill effluent *267*
Sources of waste-water production *267*
Characteristics of POME *267*
Treatment technology for POME *268*
Ponding system *269*
Open tank digester and extended aeration *270*
Close tank digester with biogas recovery and land application *271*
Kernel extraction *272*
Separation of nuts from fibre *272*
Nut screening and cracking *273*
Kernel and shell separation *275*
Kernel drying *277*
Mill layout *278*
Small-scale mechanical milling *278*
Milling control *279*
Control of quantity *280*
Oil losses *281*
Kernel losses *282*
Oil *283*
Kernels *283*
Other determinations *283*
Milling report forms *283*
Oil palm by-products *286*
Biomass production and availability *286*
Utilisation of by-product *286*
Oil palm trunks and fronds *286*
Empty fruit bunches and fibres and shells *286*
Palm oil mill effluent (POME) *287*
Energy potential from oil palm by-products *288*
Fibre and shell *288*
Empty fruit bunch *288*
Biogas from palm oil mill effluent *289*
Refining and fractionation *289*
Physical and chemical refining *289*
Physical refining *290*
Chemical refining *292*
Refining factor *293*
By-products *293*
Fractionation *295*
Oil composition *295*
Polymorphism *295*
Cooling rate *295*
Process description *295*
Dry fractionation *295*
Detergent fractionation *296*
Solvent fractionation *296*
Double fractionation *296*
Fractionation of palm kernel oil *296*
Quality assurance *296*
Quality chain *297*
Hydrolysis *297*
Oxidation *297*
Quality assurance measures in plantation and milling *297*
Precautions to minimise hydrolysis *298*
Processing of crude palm oil *298*
Raw material *298*
Palm oil refinery effluent treatment *299*
Treatment method *299*
Oils and fats recovery *299*
pH control and chemical treatment *299*
Aerobic treatment process *300*
SBR treatment plant characteristics *301*
Characteristics of palm oil refinery effluent *302*
End uses *303*
Food applications *304*
Shortening *304*
Margarine *304*
Vanaspati *304*
Deep frying fat *305*
Speciality fats *305*
Symmetrical *305*
Lauric *305*
High trans type *306*
Recent food applications *306*
New potential food applications *307*
Non-edible Applications *307*
Direct route *308*
Oleochemical route *311*
Chemical and physical properties of palm oil *317*
Chemical properties of palm oil *317*

Triglyceride and fatty acid composition 317
Minor components 319
Inherent chemical properties of palm oil 321
Physical properties of palm oil 321
Environmental trends: Towards zero waste operation 324
Nutritional effects of palm oil 326
General nutritional properties 326
Effect of palm oil on blood lipids 326
Human studies evaluating the effect of palm oil on blood lipids 326
Possible mechanism for the cholesterol-lowering potential of palm oil 327
Nutritional properties of minor components in palm oil 327
Effects of palm oil on experimental carcinogenesis 328
Palm oil delays ageing 329
Future prospects of palm oil and market requirements 329
Versatility of palm products 330
Technical and economic advantages 331

Chapter 7 Castor Oil and Its Derivatives **332-403**

Introduction 332
Properties of castor oil 332
Composition 333
Adulteration 334
Grades and specifications 334
Applications 336
Direct uses 337
Modified castor oils 339
Dehydrated castor oil (DCO) 339
Dehydrated castor oil (DCO) fatty acids 344
Hydrogenated castor oil 345
Sulphated castor oil 346
Epoxidised castor oil 348
Ethoxylated castor oil 348
Blown castor oil and other products 350
Ricinoleic acid 352
Soaps 353
Methyl ricinoleate 353
Ricinoleyl alcohol 354
Glycerol monoricinoleate 354
12-hydroxy stearic acid 354
Methyl 12-hydroxy stearate 355
Azelaic acid 355
Undecylenic acid 356
Nylon 11 358
Undecylene aldehyde 362
Undecyl alcohol 362
Undecanedioic acid 363
Heptaldehyde 363
Heptanoic acid 364
Sebacic acid 364
Nylon-6,10 371
2-Octanol 371
Urethane foams 372
Dimer acids 373
Miscellaneous 377
Hydrolysis of castor oil 377
Castor fatty acids 378
General considerations for fat-splitting 378
Methods of fat splitting 381
Pressure splitting 381
Atmospheric splitting 383
Twitchell process 384
Enzymatic splitting 385
Kinetics of enzyme catalysed hydrolysis 387
Enzyme catalysed hydrolysis as a technology 387
Commerical potential of enzymatic fat splitting 388
Castor cake 391
Castor hulls 391
Castor stalks 393
Application of castor oil and its derivatives in lubricant industry 393
Chemistry of castor oil 393
Synthetic esters from castor oil 394
Castor oil derivatives in textile industry as lubricants 394
Use of castor oil in metal working operations 395
Applications of castor oil in biodegradable and environmental friendly lubricants 395
Castor oil derivatives in grease industry 395

Future projections: Castor oil/synthetic blends as fuel *397*
Castor oil emerging as vital chemical industry input *397*
Quality aspect *398*
Oil content *398*
Physico-chemical properties of castor oil *398*
Fatty acid profile of castor oil *399*
Castor seed meal *399*
Anti-nutritional factor *399*
Industrial uses *400*
Origin analysis and technology developments of castor oil and its derivatives *400*
Stringent quality requirements for exports *400*
Polycyclic aromatic hydrocarbon (PAC) *401*
Origin of PAH *401*
Analysis of PAHs *401*
Methods/technologies for removal of PAHs *401*
Indian scenario *401*
Trace metal impurities *402*
Methods for isolation and estimation *402*
Indian scenario *402*
Pesticide residue analysis *402*
Indian scenario *402*
Toxic substances *403*
Isolation and detection methods *403*
Techno-commercial Issues *403*

Chapter 8 Rice Bran Oil ***404-436***

Introduction *404*
Nutritional significance of rice bran oil *406*
Basics *406*
Quantity of fat *407*
Quality of fat *407*
Role of fatty acids and micro-nutrients on cardiovascular disease with special reference to rice bran oil *410*
Cardiovascular disease in indians *410*
Cholesterol and other blood fats *411*
Triglycerides: Another important blood fat *411*
Cholesterol and triglycerides as lipids *412*
Cholesterol and triglycerides circulate in blood as lipoproteins *412*
Types of lipoprotein particles *412*
Sources of cholesterol *412*
Sources of triglycerides *412*
Fate of dietary cholesterol and trigycerides *412*
Dietary fat goes to liver *412*
Liver loads triglycerides and cholesterol into VLDL *413*
LDL cholesterol is dumped *413*
HDL cleans out stuck cholesterol *413*
Dietary Fat Recommendations *414*
Value added products from rice bran oil industry *414*
Oryzanol *416*
Lecithin *417*
Upgradation to 65 per cent purity (which is equivalent to commercial soya lecithin) *417*
Wax *418*
Integrated technology for refining rice bran oil and by-products recovery *419*
A novel approach to physical refining of rice bran oil *420*
Recovery of by-products from RBO refining *420*
Process streams *420*
Applying innovations and improvement in rice bran oil processing technology *422*
Rice bran oil—A better choice *425*
Butter/desi ghee *427*
Vanaspati *427*
Kardi oil/sunflower oil *427*
Cottonseed oil *427*
Soyabean oil *427*
Palm oil *427*
Groundnut oil *427*
Rice wax—utilisation *428*
Rice bran oil—Heart-friendly cooking oil *431*
Role of fat in human body *432*
Daily fat intake *432*
Quality of fat *432*
Rice bran oil (RBO) *433*
Health benefits of rice bran oil (RBO) *434*

Cooking advantages of rice bran oil *435*
Edible uses of rice bran oil *436*
Pharmaceuticals and other uses of rice bran oil *436*
Rice bran oil potential in India *436*
Status of RBO in India *436*
Status of RBO in other countries *436*

Chapter 9 Rapeseed Oil 437-515

Introduction *437*
Steps involved in processing rapeseed oil *438*
Seed cleaning *439*
Dehulling *439*
Crushing *439*
Cooking *440*
Screw pressing *440*
Flaking *441*
Solvent extraction *441*
Meal desolventising *444*
Degumming *446*
Storage of crude oil and meal *446*
Influence of the extraction process on the quality of rapeseed oil and meal *447*
Processing of rapeseed oil *448*
Degumming and alkali refining *448*
Neutralisation *449*
Bleaching *452*
Equipment *453*
Hydrogenation *454*
Selectivity and isomerisation *454*
Hydrogen *455*
Oil *456*
Catalyst *456*
Equipment and systems *457*
Practical aspects of the hydrogenation of rapeseed oil *458*
Controlling the hydrogenation process *462*
Post-bleaching *462*
Deodourisation *463*
Edible products from rapeseed oil *465*
Quality factors *465*
Flavour and flavour stability *466*
Physical properties *469*
Specific edible products *470*
Salad and cooking oils *471*
Margarine *472*
Shortenings or compound cooking fats *473*
Other food uses *475*
Future uses *475*
Non-nutritional uses of rapeseed oil and rapeseed fatty acids *476*
Rapeseed oil *476*
Rubber additives (factice) *476*
Lubricants for metal surfaces *477*
Fat-liquoring of leather in tanneries *478*
Varnishes and lacquers *478*
Textile chemicals *478*
Detergent additives *479*
Other reported applications *479*
Rapeseed fatty acids *479*
Erucic acid and its cleavage products *480*
Erucyl amide *480*
Products of ozonolysis *480*
Rapeseed lecithin *481*
Industrial production of rapeseed lecithin *482*
Laboratory preparation *483*
Extraction of lipid material from the seeds *483*
Isolation of the polar lipids from crude extracted oil *483*
Analysis of rapeseed lecithin *484*
Separation and analysis of rapeseed oil phospholipids *484*
Fatty acid composition *485*
Utilisation of rapeseed lecithin *485*
Nutritional value of rapeseed oil *486*
Rapeseed oil/erucic acid in experimental diets *486*
Effect of rapeseed oil on growth of experimental rats *486*
Experiments with partially hydrogenated rapeseed oil *489*
Experiments with heat-treated rapeseed oil *490*
Effect of dietary rapeseed oil on other species than the rat *491*
Digestibility and absorption *492*
Effect of dietary components other than fat on the digestibility of rapeseed oil *493*

Effect of rapeseed oil/erucic acid on 'metabolic fat' *494*
Appetite and palatability related to rapeseed oil diets *494*
Physiological changes induced by rapeseed oil *495*
Effect of rapeseed oil on the suprarenal glands *495*
Effect of rapeseed oil on ovaries, testes, reproduction and milk secretion *496*
Effect of rapeseed oil on the thyroid gland *497*
Effect of rapeseed oil on the liver *497*
Effect of rapeseed oil on blood cholesterol *498*
Pathological changes induced by rapeseed oil *499*
Effect of rapeseed oil on vitamin requirement *499*
Metabolism of erucic acid *500*
Effect of rapeseed oil on longevity *501*
Nutritional value and use of rapeseed meal *501*
Protein content *502*
Amino acid composition *502*
Carbohydrates *503*
Content of minerals and vitamins *504*
Glucosinolates *504*
Chemical properties *504*
Toxicity of various cleavage products *506*
Counteracting the toxicity *507*
Enzyme inactivation *507*
Technical methods of removal or destruction of glucosinolates or cleavage products *508*
Isolation of the protein *510*
Removal of glucosinolates by plant breeding methods *510*
Digestibility *512*
Feeding value *512*
Ruminants *512*
Young ruminants (below 6 months) *513*
Growing and fattening ruminant animals *513*
Milk cows *513*
Swine *513*
Poultry *514*

Chapter 10 Sesame Oil ***516-534***

Introduction *516*
Chemical composition *517*
Lipids *518*
Classification *518*
Composition *518*
Endogenous antioxidants *520*
Properties *521*
Nutritional quality *521*
Proteins *521*
Composition *521*
Solubility *522*
Nutritional quality *522*
Carbohydrates *523*
Minerals and vitamins *523*
Antinutritional factors *524*
Processing *524*
Dehulling *526*
Oil extraction *527*
Traditional methods *527*
Modern methods *527*
Refining *528*
Sesame cake and meal *529*
Protein concentrates and isolates *529*
Roasting *530*
Utilisation *531*
Oil *531*
Sesame seed and other by-products *532*
Future research *533*

Chapter 11 Soyabean Oil ***535-591***

Introduction *535*
Soya chemistry *536*
Food beans and oil beans *536*
Composition *536*
Carbohydrates *537*
Insoluble carbohydrates *538*

Health and soyafoods *538*
Traditional soya foods *539*
Western soya foods *539*
Composition of soyabean oil *540*
Triglyceride structure *542*
Unsaponifiable matter *542*
Physical properties of soyabean oil *543*
Recovery of oil from soyabeans *547*
Storage and preparation of beans *547*
Solvent extraction *548*
Percolation extractors *549*
Immersion extractors *549*
Oil storage *550*
Edible oil processing *550*
Basic processing operations and principal edible oil products *551*
Oil degumming and soyabean lecithin *551*
Refining *553*
Chemical (caustic) refining *553*
Physical (steam) refining *555*
Other novel refining processes *556*
Effects of refining on oil quality *556*
Bleaching *557*
Batch atmospheric bleaching *559*
Batch vacuum bleaching *559*
Continuous-vacuum bleaching *559*
Hydrogenation practices *561*
Selectivity *561*
Conditions *561*
Hydrogenation process *562*
Effect of process conditions *563*
Catalyst activity *565*
Hydrogenation-winterisation of damaged beans *565*
Alternative processes *565*
Alternative catalysts *566*
Deodourisation *566*
Deodourisation of soyabean oil *567*
Effects of Deodourisation on oil quality *568*
New developments *568*
Partially hydrogenated-Winterised soyabean oil *569*
Preparation of high-stability soyabean oils by hydrogenation-winterisation *569*
Shortenings and margarines: base stock preparation and formulation *570*
Formulations for preparing shortenings *570*
Formulation of margarine oils *571*
Soyabean oil quality *573*
Oxidative stability of soyabean oil *574*
Flavour and colour reversion and oxidation rancidity *574*
Colour reversion *574*
Control measure/stabilisation *574*
Evaluation of finished oil quality *575*
Storage and handling *576*
Special processing for off-specification oil *577*
Specifications for soyabean oil *578*
Soyabean oil food products—their preparation and uses *578*
Mayonnaise, prepared salad dressings, salad oils and margarine *578*
Shortenings *579*
Uses of soyabean oils in food products *581*
Consumer products *581*
Baked goods *581*
Fried foods *582*
Canned, packaged and dehydrated foods *582*
Confectionery coatings and candies *583*
Imitation dairy products *583*
Fat replacers *583*
Other uses *583*
Nutritive value of soyabean oil *584*
Metabolic studies *584*
Nutritive value of heated and oxidised oils *585*
Nonfood uses for soyabean oil *585*
Soya—A healthy complex of phytonutrients *586*
Protein *586*
Essential fatty acids *586*
Isoflavones *586*
Oligosaccharides *587*
Fibre *587*
Tocopherols *587*
Saponins *588*
Phytosterols *588*
Lecithin *588*
Inositol-6-phosphate (phytate) *588*
Soya proteins and heart disease 589
Cardiovascular disease *589*
Effect of soya on risk factors *589*
Tocopherols *590*
Essential fatty acids *590*
Oligosaccharides, fibre and others *590*

Chapter 12 Sunflower Oil *592-637*

Introduction *592*
Oil composition *593*
  Oleic sunflower *595*
Sunflower oil *596*
  Crude sunflower oil *598*
Degumming *598*
  Purpose of degumming *598*
  Process description *600*
  Batch degumming *600*
  Continuous degumming *601*
Neutralisation or refining *602*
  Chemical or caustic refining *602*
    Disadvantage of the method *603*
  Use of phosphoric acid *603*
  Phosphoric acid use in alkali refining *604*
  Batch refining *604*
  Continuous refining *605*
  Other neutralisation processes *607*
    Intermix method *608*
    Short-mix method *608*
    Ultra-short-mix methods *608*
  Refining temperature *608*
  Physical and steam refining *609*
  Pretreatment *609*
  Steam refining *609*
Bleaching *610*
  Purpose of bleaching *610*
  Batch bleaching *610*
  Dressing the filter press *611*
  Continuous bleaching *611*
Dewaxing *611*
  Batch dewaxing method *613*
  Continuous dewaxing method *614*
  Continuous dewaxing using a centrifuge *614*
  Use of a surfactant *614*
  Testing for wax *616*
  Preparation of standards *616*
  Refined oil wax testing *616*
  Crude oil wax testing *616*
  Cold test *616*
  Filter aid *617*
Filtration and clarification systems *618*
  Plate-and-frame presses *618*
  Filter precoating *618*
  Vertical tank-type pressure filters *618*
    Advantages *619*
    Disadvantages *619*
  Horizontal leaf-type filter *619*
  Continuous or vacuum filters *619*
  Centrifugal filters *619*
  Bag-type filters *620*
  Cartridge-type filters *620*
Deodourisation *620*
  Operating parameters for deodourisation *621*
  Deodourisation operation *621*
  Deodourisation equipment *621*
  Batch deodourisation *621*
  Continuous deodourisation *622*
  Semicontinuous deodourisation *622*
  Protection and handling of deodourised oil *623*
    Protection from air *623*
    Oil cooling *623*
    Additives *623*
    Filtration *624*
    Nitrogen blanketing *624*
    Stabilisers *624*
Hydrogenation *624*
  Hydrogenation process *625*
  Catalytic hydrogenation *625*
  Nature of the catalyst *625*
  Selectivity in hydrogenation *626*
  Effect of feedstock impurities on hydrogenation *626*
    Carbon monoxide *626*
    Nickel crystallites *626*
    Sulphur *626*
    Phosphorus and phosphatides *627*
    Colour bodies *627*
  Hydrogenation equipment *627*
  Ancillary systems *628*
Continuous hydrogenation *628*
  Buss loop reactor *629*
  Use of hydrogenated sunflower oil *629*
Processing and use of sunflower oil refining by-products *629*
  Purpose of processing the by-products *629*
  Soapstock from alkali refining *631*
  Soapstock as an additive to meal or feed *631*
  Soapstock dehydration *631*
  Batch method *631*
  Continuous method *631*
  Acidulation of the soapstock *631*

Batch method 632
Batch acidulation system 632
Continuous acidulation system 634
Clays from decolourisation 634
Solvent extraction 635
Water extraction 635
Filter aid and dewaxing 637
Continuous dewaxing by-products 637
Regeneration of clay, filter aid and catalyst 637
Commercial uses for fatty materials 637

Chapter 13 Safflower Oil *638-653*

Introduction 638
Technology 639
Physical and chemical properties 639
Safflower seed 639
Safflower oil 641
Physico-chemical characteristics 641
Fatty acid composition 641
Composition of unsaponifiable fraction 642
Safflower meal 644
Processing 645
Extraction 645
Refining, bleaching and deodourising 647
Production of margarine and mayonnaise 647
Industrial processing 647
Safflower oil 647
Safflower meal 648
Quality assessment 649
Safflower seed 649
Safflower oil 650
Safflower meal 650
Storage and transportation 651
Safflower seed 651
Safflower oil 652
Safflower meal 652
Unique uses 652
Bird feed 652
Ornamental 652
Food colouring 652
Dye 653
Medicinal 653
Cosmetic 653

Chapter 14 Peanut Oil *654-667*

Introduction 654
Peanut oil 655
Chemical and physical characteristics 655
Acetyl value 655
Acid value 655
Colour 657
Ester number 657
Free fatty acids 657
Heat of fusion 657
Hehner value 657
Kinematic viscosity 657
Peroxide value 658
Polenske value and Reichert-Meissel Values 658
Thiocyanogen value 658
Melting point and titer value 658
Unsaponifiable matter 658
Gas chromatography 658
Oil content of peanuts 659
Composition of peanuts 659
Triglyceride structure of peanut varieties 662
Major environmental factors that inflence lipid content 663
Modification of fatty acid composition 663
Other peanut products 665
Oil meal 665
Peanuts for direct consumption 665
Generic technology for assured biocontrol of groundnut infections leading to its yield improvement 665
Importance of groundnut crop 665
Heavy pathogen infestation affecting groundnut production 665
Cost-effective biocontrol technology 666
Laboratory demonstration of anti-fungal activity 666
Bioefficacy profile at farm scale trials 666
Complementary interaction (s) with rhizobacteria 666
Salient features of biocontrol package 666

Chapter 15 Sulphated and Sulphonated Oils *668-683*

Introduction *668*
Main reactions for the addition of SO_3 group on fats *669*
Sulphatation *669*
Sulphonation *669*
Sulphated oils *669*
Industrial sulphatation *670*
Main methods of fat sulphatation *670*
Batch sulphatation of fats *670*
Fast sulphatation *672*
Instant sulphatation *673*
Continuous sulphatation *674*
Some sulphated fats *675*
Calculating the amount of combined SO_3 in a sulphated oil *675*
Sulphonated oils *676*
Sulphonation of unsaturated fats by action of sodium bisulphite solutions *676*
α-Sulphonated fats and oils 679
Preparation of sulphonic acids from saturated fats *679*
Direct sulphonation of saturated fats *679*
Preparation of partial a-sulphonates from saturated fats *681*
Preparation of partially sulphonated saturated oils *681*
Industrial uses of sulphated and sulphonated oils and fats *681*
Metallurgy *682*
Textile *683*

Chapter 16 Miscellaneous Oils *684-714*

Introduction *684*
Grapeseed oil *684*
Physico-chemical characteristics *685*
Triglyceride composition *685*
Fatty acid composition *685*
Composition of unsaponifiable fraction *686*
Camelina oil *687*
Physico-chemical characteristics *687*
Fatty acid composition *687*
Unsaponifiable fraction *688*
Poppy seed oil *688*
Physico-chemical characteristics *688*
Triglyceride composition *688*
Fatty acid composition *689*
Unsaponifiable fraction *689*
Oleaginous flax *690*
Oil compositions *690*
Almond, Hazelnut and Walnut oil *692*
Almond oil *692*
Physico-chemical characteristics *692*
Composition of triglycerides *692*
Fatty acid composition *692*
Hazelnut oil *694*
Physico-chemical characteristics *694*
Triglyceride composition *694*
Fatty acid composition *695*
Walnut oil *696*
Physico-chemical characteristics *696*
Fatty acid composition *697*
Evening primrose, Borage and Blackcurrant oils *698*
Evening primrose oil *698*
Upgrading of the evening primrose for production of *g-Ln* *699*
Borage *700*
Main uses of borage oil *702*
Blackcurrant oil *702*
Main uses of blackcurrant seed oil *705*
Wheat germ oil *705*
Conjugated oils *706*
Other oils *707*
Crambe oil *707*
Vernonia oil *708*
Cuphea oil *709*
Avocado oil *710*
Physico-chemical properties *710*
Tall oil *712*
Origin, production, applications *712*
Technology *713*
Chemical composition *713*

References *715-716*

Index *717-722*

Chapter 1

Coconut Oil

INTRODUCTION

Coconut oil is the healthiest oil on the earth. Coconut oil has been used throughout Asia and the Pacific for thousands of years as both a food and a medicine. Even today it holds a highly respected position in the Ayurvedic medicine of India. Only recently has modern medical research confirmed the many health benefits traditionally attributed to this remarkable oil.

Once mistakenly thought to be bad because of its saturated fat content, coconut oil is now known to contain a unique form of saturated fat that actually prevents heart disease, stroke, and hardening of the arteries. The saturated fat in coconut oil is unlike the fat found in meat or other vegetable fats. It is identical to a special group of fats found in human breast milk which have been shown to improve digestion, strengthen the immune system, and protect against bacterial, viral, and fungal infections. These fats, derived from coconut oil, are now routinely used in hospital IV formulations and commercial baby formulas. They're also used in sports drinks to boost energy and enhance athletic performance.

COCONUT PALM

In the tropics, the coconut palm (*Cocos nucifera, L.*) is one of the most useful trees. The palm reaches 30 m or more in height, with a striped, unbranched and relatively smooth trunk measuring approximately 25 cm in diameter and crowned with some 30 fronds, each of which is about 5–6 m long. The roots may extend up to 10 m in sandy soil, serving as anchor in addition to their primary function as nutrient absorbers and breathing organs. As a perennial provider of food, beverage, shelter, animal feed and feed-stock for the oleochemical industries, the palm is reverently described as the Tree of Life, Tree of Heaven and other such metaphors by people of coconut-producing countries.

Palms grown in well-aerated soil are productive and tolerate soil pH ranging from 5 to 8. In determining the suitability of sites for coconut farms, a good supply of soil moisture with adequate drainage are considered the foremost factors. Water supply may come from well-distributed rainfall (1300–2300 mm/year), irrigation, or from accessible water table. The extremes, excessive rainfall and drought, affect the nut yield per palm.

Soil, water and foliar analyses in coconut plantations are conducted to determine the type of fertiliser and micronutrients required by the palm. Long-term studies show that the applications of potassium or sodium chlorides in optimum doses as supplements to the standard nutrients for inland farms increases yields in nuts per tree and copra per nut.

Propagation

The method of propagation for coconut palms is through seednuts. Due to the highly heterogeneous characters of coconut palms, progenies do not necessarily exhibit all the desirable characteristics of the parent palms. Through selection of seeds from elite parents, vigorous and productive palms may be evolved. The long span of six or more years required to evaluate the performance of progenies of coconut palms has underlined the need to explore other methods of propagation; the response appears to be the current interest in the development of tissue culture techniques.

Seedlings

Propagation of the coconut palm starts from the collection of fully matured nuts (11–12 months old) from selected palms. The seednuts are stored in the shade for at least one month. In the nursery, seednuts are set to germinate in loose and friable soil provided with adequate moisture and drainage facilities. Some farmers transfer sprouted seednuts into polybags to allow selection of seedlings showing uniformity in growth and vigour. Further, seedlings in polybags do not suffer damage and shock when transported or when transplanted in the fields.

Six-to nine-month-old seedlings with four to five leaves may be transplanted in fields and spaced to accommodate 115–160 palms per hectare. Coconut palms, on the average, start bearing fruits in the sixth year and continue to provide fruits perennially for more than 50 years, barring any serious damage caused by plant pests, diseases, climatic disasters, or neglect in farm management.

Tissue culture

The most promising vegetative (asexual) method under development for the propagation of elite coconut palms is the tissue culture technique. In this method, the actively dividing tissue from a selected palm is removed and grown asceptically in a specially formulated culture medium containing the prescribed nutrients, hormones and root and shoot growth-promoting factors. Under the right induction conditions, the explant develops into a callus, which is induced to differentiate into shoots and roots forming the plantlet. The plantlet is hardened in a pot kept in a greenhouse before transplanting in the field.

Varieties

There is a wide range of variations that may occur within a coconut palm colony with regard to phenotype characteristics such as colour, size and shape of nuts and shape and symmetry of crown, among others. In some isolated localities, colonies of coconut palm develop well-defined and uniform phenotypic characters arising from generations of natural selection.

Makapuno, a peculiar Philippine coconut variety, has a cavity within the kernel filled with a palatable jellylike substance. These fruits do not germinate although they contain an embryo. However, the propagation of *makapuno* has been successfully developed through *in vitro* culture of its rescued embryo and final establishment of the seedling in the field.

The two most distinguishable varieties of coconut palm are the tall and the dwarf.

FRUIT

The common mature coconut fruit weighs more than 1 kg and is ovoid in shape and green or yellow in colour. The nut has a smooth epidermis over a fibrous mesocarp (husk) that covers the hard endocarp (shell). Within the shell is the endosperm (kernel, meat) approximately 1–2 cm thick. A thin brown layer called testa separates the kernel from the inner surface of the shell. The cavity within the kernel has an average volume of 300 mL and contains the endosperm liquid (coconut water) (Fig. 1.1).

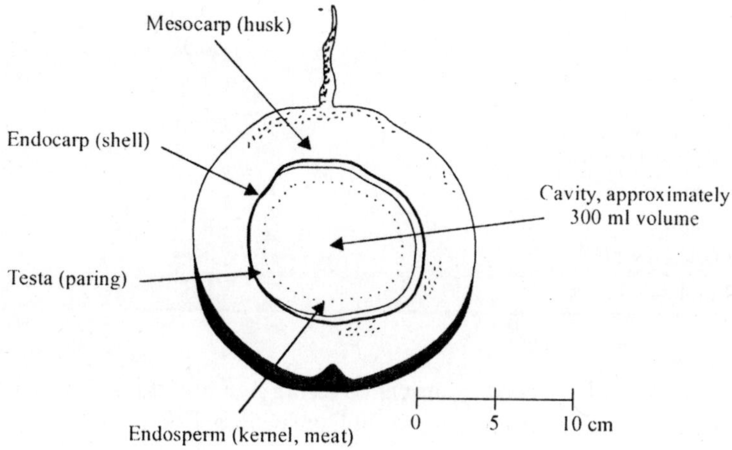

Fig. 1.1. Mature coconut fruit and a longitudinally halved specimen.

Kernel

The kernel is the origin of the following products: coconut oil, desiccated coconut, coconut skim milk, coconut cream, coconut flour, protein powder and copra meal (Table 1.1).

Table 1.1. Typical composition of coconut kernel.

Composition	%
Moisture	50
Oil	34
Ash	2.2
Fibre	3.0
Protein	3.5
Carbohydrates	7.3

Coconut Water

Coconut water is a sterile liquid. From mature nuts, it has the following constituents in percent: total solids 4.7, fat 0.74, protein 0.55, ash 0.46 and reducing sugar 1. The liquid has a pH of 5.6. It is a co-product of desiccated coconut and considered a health beverage. Concentration of coconut water by reverse osmosis may further expand its use in the fermentation and food industries.

Testa

The thin brown layer between the kernel and shell is the testa. This layer is pared off from the kernel's outer surface to eliminate coloured bodies in the production of desiccated coconut. The parings contain oil that has a fair amount of unsaturated fatty acids $C_{18:1}$ and $C_{18:2}$ (Table 1.2).

Table 1.2. Fatty acid composition of paring (testa) oil.

Fatty acid	*Paring oil (%)*
Caproic C6	0.1
Caprylic C8	0.5
Capric C10	1.5
Lauric C12	22.7
Myristic C14	15.8
Palmitic C16	17.2
Stearic C18	1.3
Oleic C18:1	25.4
Linoleic C18/2	15.5

Shell and Husk

The shell that encloses the kernel is a hard spherical covering 3–5 mm thick. It is used mainly for fuel in copra making. Other products derived from coconut shell are charcoal, activated carbon, filler for synthetic resins, glues, components in mosquito-repellent coils and decorative items.

The husk is 5–10 cm thick fibrous cover adhering to the coconut shell. Like the shell, it is also used mainly for fuel in farms. Products derived from the husk are coir, bristle, rubberised fibre, rope, geotextiles and activated carbon.

COPRA

Copra is the dried kernel of coconuts. Fresh kernels contain approximately 50 per cent moisture. Various drying methods are employed to bring down the moisture content ideally to 6–8 per cent (Table 1.3). At this level, mould growth in copra is inhibited. The conversion of kernel to copra is an essential step if the oil is to be drawn by the conventional mechanical extraction method.

Table 1.3. Calculated yields based on an initial 100 g fresh kernel with 50 per cent moisture content.

Moisture	*50*	*15*	*12.5*	*10*	*7.5*	*5*	*0*
Oil	34.5	58.65	60.37	62.1	63.82	65.55	69.0
Nonoils	15.5	26.35	27.13	27.9	28.68	29.45	31.0
Copra wt., g	100[a]	58.82	57.14	55.55	54.05	52.63	50.0

[a] Fresh kernel.

Sun Drying

Dehusked nuts are split into halves and drained. The halves are exposed to the sun and in due time the kernels shrink. The partially dried kernels are separated from the shells for further drying under the sun for 6–8 days. During occasional rains, the kernels are protected with adequate cover, such as plastic sheets or any other suitable material.

Direct Fire Drying

The direct fire dryer consists of a bamboo grill platform where the split nuts are placed. Underneath is a fire hearth where coconut shells and husks are burned to provide heat for the vapourisation of water from the kernels. The kernels shrink and are separated from the shells for further drying. Smoke from the burning fuel imparts a light brown colour on the copra and its oil.

Small farm producers continue to use this traditional method because of the low cost of construction, simplicity of design and ready availability of shell and husk for fuel in coconut farms.

Hot-Air Drying

A low-cost efficient dryer is adaptable for small farms. The dryer uses husks or shells for fuel in a specially designed burner that can operate with an even supply of heat. The dryer is provided with a steel plate bottom to isolate the kernels from the smoke generated by the fuel.

There are a number of hot-air dryer models, devices inspired by modern technology. In principle, these dryers apply hot air to expel moisture from the kernels. Since smoke does not come in contact with the kernels, the product is a cleaner and whiter copra. The relationship of oil, non-oil and moisture contents from fresh kernel to copra as moisture diminishes in stages is given in Table 1.3.

OIL EXTRACTION

On an industrial scale, the dry process is traditional method of extracting oil from the coconut. This is done by crushing copra in an expeller and the meal (or cake) may be further treated with solvents to extract residual oil.

The wet-process feedstock is fresh kernel instead of copra. The extracted oil does not have to be refined, unlike the oil from copra. The co-products of oil from the wet process are edible.

The major steps in the extraction of oil from copra and fresh kernel, are discussed below.

Dry Process

The dry process involves mechanical extraction of oil in crushers or expellers with copra as feedstock (Fig. 1.2). Mechanical extraction may be supplemented with a second extraction, using solvents, to recover residual oil from the meal.

Mechanical extraction

Copra with 10–12 per cent moisture content is conveyed to an automatic scale, passes through a magnetic chamber for the removal of tramp iron and is ground to a particle size of approximately 0.3 cm in diameter. The particles are flaked to facilitate exposure of a large surface area. The flakes are cooked and conditioned for 20 minutes at 115°C in horizontal cookers. Residual moisture is brought down to 3 per cent as fat cells of the conditioned copra flakes are ruptured and phosphatides are precipitated.

The sized, cooked and conditioned flakes are fed to the crushers where a continuous squeezing action extracts the oil. Extraction characteristics follows a smooth diminishing curve with most of the oil drained at the inlet section of the barrel cage and tapers off toward the exit. The resulting copra meal has a typical residual oil content of 7 per cent.

Characteristics of oil and copra mill

The characteristics of oil and copra meal from a well-maintained and properly operated expeller are as follows:

1. Low-coloured oil, 8 Red/50 Yellow in the Lovibond scale.
2. Light brown copra meal with 8 per cent maximum residual oil.
3. Copra meal with a uniform thickness of approximately 0.6 cm and 6 per cent maximum meal fines.

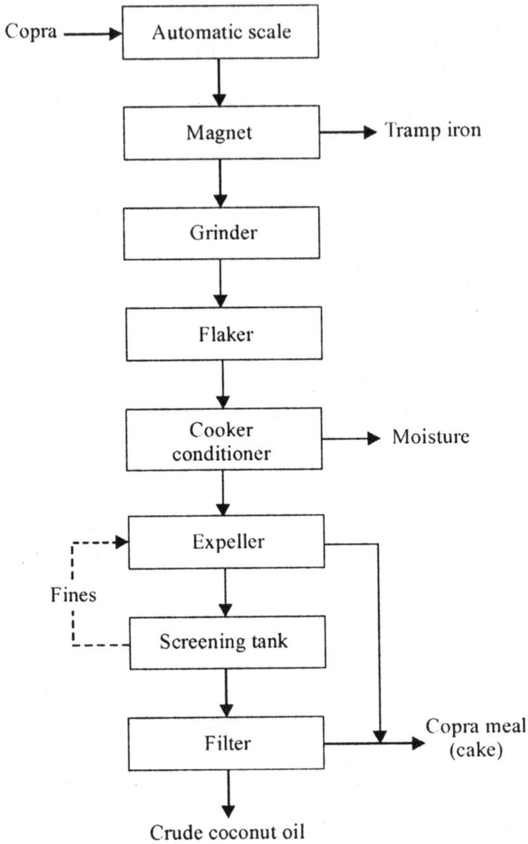

Fig. 1.2. Mechanical extraction flow chart.

A sustained optimum expeller efficiency requires cooking and conditioning of copra flakes to be controlled between 91 and 93°C under an appropriate residence time to reduce moisture content between 3 and 4 per cent. Flakes conditioned below 91°C with residual moisture above 4 per cent have lower oil yield while flakes conditioned above 93°C with residual moisture below 3 per cent would yield dark-coloured oils, charred meal with the consequent disadvantage of poor oil extractability. Screened and settled fines should not exceed 10 per cent of fresh feedstock to prevent the formation of abrasive particles during the cooking/conditioning stage.

Post-expeller treatment

The oil from the expeller passes to a screening and settling tank to initiate the separation of fines, which are recycled with fresh feedstock to the system. To each tonne of supernatant oil 10 kg of bleaching earth is mixed. Passage through a polishing filter gives a clear oil ready for storage or further processing.

Copra meal (cake) from the filter press is pelletised, bagged and despatched to animal feed millers.

Solvent extraction

This operation supplements mechanical extraction and consequently minimises residual oil in the copra meal. Copra undergoes an accelerated preliminary extraction with a controlled residual oil content of

14–18 per cent in the expeller meal; expeller throughput rate is almost doubled. Hexane (bp 68.7°C) is widely used as the solvent for extraction. In an extraction unit that operates on a countercurrent system, the cake is met by oil-rich miscella (hexane + oil) and leaves the extractor as it is rinsed with pure hexane. The solvent-extracted cake has a residual oil content of approximately 3.5 per cent

Hexane in miscella and cake is recovered for reuse in succeeding operations. Escaping hexane vapours are trapped by cold mineral oil spray to preclude hazards and maximise solvent recovery.

Wet Process

Feedstock in the wet process is fresh kernel. In addition to oil, other edible co-products are recovered from the kernel, namely, coconut flour, protein, carbohydrates and vitamins. To encourage wider commercial applications of this process, the following advantages are emphasised: superior quality of the oil product and the recovery of nutrient co-products that would otherwise be lost in copra. The process is briefly described to give a general view on the extraction of oil from fresh kernels.

Dehusked mature nuts are shelled to separate the kernels, followed by paring to remove the testa. The testa is set aside for the extraction of paring oil, a by-product. The pared kernels are finely comminuted through a wedge and die plate mill and through a roller mill. The comminuted mass is passed through a screw press, which expels the coconut milk. The milk is filtered through a screen conveyor. The cream that is separated from the milk by centrifugation is heated to reduce its moisture content. By further centrifugation, oil is separated. Trace moisture in the oil is reduced to 0.1–0.2 per cent level by atmospheric heating. A typical yield of 6.8 tonnes 'natural' coconut oil is extracted from 25 tonnes of fresh kernel.

Copra Meal

The extraction of oil from the dried coconut kernel leaves the residue called copra meal. Copra meal is utilised as a component in animal feeds.

REFINING

Refining of crude fats and oils involves a series of steps for the removal of impurities from the glycerides to make the product suitable for human consumption and improve product shelf life. The impurities are fatty acids, phosphatides, metal ions, colour bodies, oxidation products, solid particles and volatiles that include objectionable odours. Crude coconut oil is refined by any of the following methods:

1. Chemical refining (batch or continuous).
2. Physical refining.

The comparative performance of both methods is summarised in Fig. 1.3.

Chemical Refining

The free fatty acids (FFA) of crude coconut oil are neutralised with dilute sodium hydroxide solution resulting in the formation of soap: $RCOOH + Na^+OH^- \rightarrow RCOO^- Na^+ + H_2O$. The soap and other impurities in the water phase are collectively called soapstock.

In the batch refining process, the separation of the water phase containing soapstock from the oil is by gravity. Some amount of neutral oil is lost by saponification and by occlusion in soapstock.

Advantages of continuous refining process

Continuous refining process, on the other hand, offer the following major advantages over the batch-type operation: (i) saponification of neutral oil is minimised due to a short contact time of 30–45 seconds between oil and sodium hydroxide; and (ii) the time consumed for the separation of the aqueous soapstock and wash water from the oil is reduced considerably by passage through centrifugal separators.

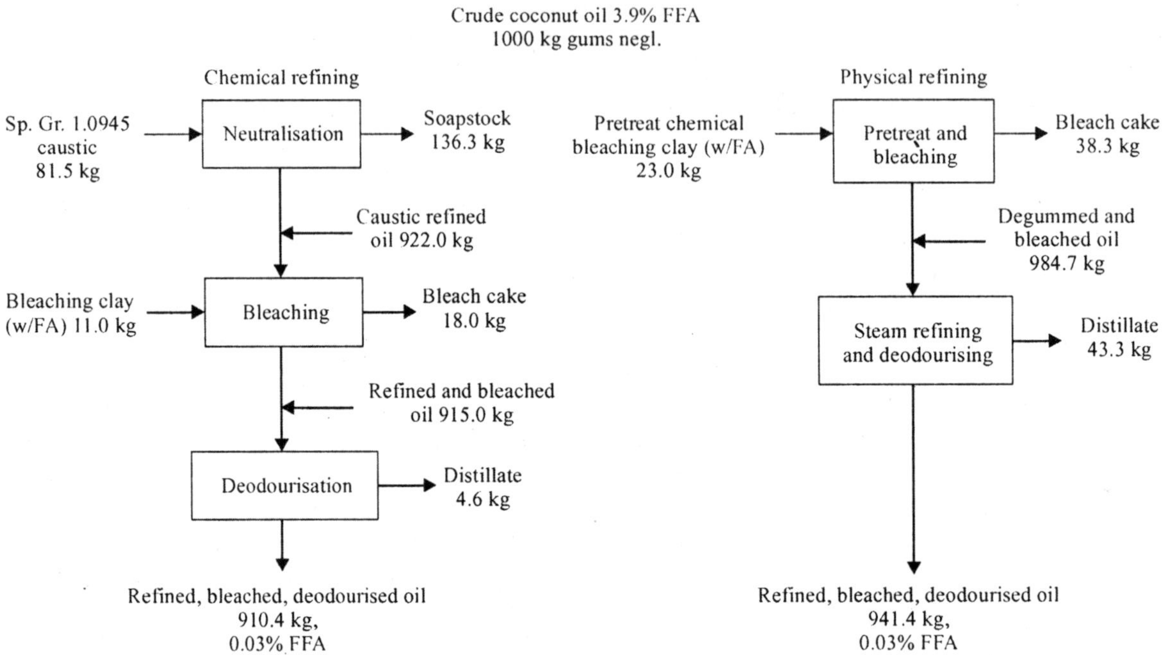

Fig. 1.3. Material balance: chemical and physical refining of crude coconut oil. Units were converted from the original lb and Be to kg and sp. gr.

After refining (neutralisation), the oil is bleached. Colour bodies in the oil are adsorbed on the surface of the bleaching clay and activated carbon particles. Experiments cited by *Brimberg* showed that bleaching process follows the rate formula which is given below:

In $(c/c_0) = -k\sqrt{t}$, where t is the time from the addition of bleaching clay; c is the concentration of pigment at time t, c_0 is the concentration at t_o, and k is the rate constant.

Efficiency in the bleaching process may be improved for certain oils when a small amount of water is present during the mixing steps.

Deodourisation is the last step in chemical refining. Volatile odouriferous substances, including low-molecular-weight fatty acids, are removed by stripping with steam under reduced pressure. The final product is generically called RBD (refined, bleached, deodourised) coconut oil.

A lower grade product called *Cochin* oil is coconut oil that is chemically refined and bleached but not deodourised.

Batch neutralisation

Crude coconut oil from storage is fed into cylindrical tanks equipped with heating coils and stirrers. The oil is heated to 80°C while being stirred and a measured amount of sodium hydroxide solution is introduced by spraying on the surface of the oil. An excess of 5–10 per cent sodium hydroxide over the stoichiometric requirement is added to ensure appropriate neutralisation of free fatty acids. In this stage, hydrated gums migrate to the water phase. Heating and stirring are stopped when soap 'breaks' are formed. A break forms as soap coagulates with some occluded neutral oil, excess sodium hydroxide and other impurities. The aqueous soapstock is allowed to settle and subsequently drawn off for acidulation with sulphuric acid to recover a mixture of fatty acids, occluded neutral oil and other impurities. The mixture is called acid oil.

The neutral oil is washed with soft water at 10–15 per cent dosage to remove traces of soap. The washed neutral oil is dried under vacuum and stored in buffer tanks.

Continuous neutralisation (short mix)

A regulated volume of crude coconut oil is heated through a plate-type heat exchanger to a temperature of 90–95°C, conveyed to a centrifugal mixer where a required volume of 2*N* sodium hydroxide solution is added through a metering pump. The oil–soap-stock mixture is passed through a primary centrifuge for the separation of soapstock from the neutralised oil. Neutralised oil from the primary separator is heated to 90°C and mixed with 10–15 per cent hot water to wash off residual soap. From a second-stage separator, the washed oil cascades over a series of plates in the vacuum dryer maintained under a pressure of 3000 Pa. The dried neutral oil is stored in buffer tanks.

Bleaching

In a cylindrical reactor provided with heating coils and a variable agitator, the neutral oil is brought to a temperature of 90°C under a pressure to 600–1000 Pa and agitated. A predetermined amount of bleaching earth and activated carbon is added via a dosing device installed at the top of the mixer. After 30 minutes, the slurry is dried and later passed through a plate and filter press for the removal of spent earth and carbon. The neutralised and bleached oil is cooled down to 50°C before being transferred to storage tanks. Steam is injected through the spent earth to recover entrained oil.

Deodourisation

The neutralised/bleached oil is pumped into a deaerator operated under a pressure of 500 Pa to evacuate entrained air. From the deaerator, the oil passes through a shell and tube economiser and is heated to a temperature of 240°C by means of a thermal oil heater. The stripper and deodourising column operates under a pressure of 600–1000 Pa; volatile components such as low-molecular-weight fatty acids, ketones, aldehydes and other odouriferous substances are stripped off by live steam. The rising vapours laden with volatile components pass through a cyclone scrubber where fresh fatty acid oil is sprayed on top of the vessel to recover outgoing fatty acids.

The deodourised coconut oil flows down to the drop tank and holding chamber of the distilling column. Residence time in the deodouriser ranges from 1.5 to 2 hrs. The oil from the column is withdrawn by a hermetically sealed and heat-resistant pump and conveyed to the economiser, which preheats the incoming deodouriser feedstock. The oil is cooled down to 50°C, at which temperature citric acid is added. Citric acid enhances the stability of oil by immobilising iron and copper, which are pro-oxidants. (Note: The deodourisation step is omitted in the production of Cochin oil).

Physical Refining

Coconut oil refiners have gained interest in the physical refining system as a substitute for chemical refining for the following reasons:

1. Physical refining has lower oil losses vis-a-vis chemical refining.
2. Pollution problems associated with soapstock acidulation is precluded.

3. Lower installation cost.
4. Lower steam, water and power consumption.
5. Distilled fatty acids are of a higher grade than the acid oil from chemical refining.

Design and technological breakthroughs have improved efficiency of the system. Performance of new packed-column deodourisers for physical refining is predicted by computer calculation, for which a special programme has been developed on the basis of experience with earlier columns.

The main feature in physical refining of crude oils is the application of steam distillation to remove the free fatty acid and volatile components from the oil. The technical feasibility of physical refining depends largely on the pretreatment stages for the removal of phosphatides, colour bodies, metal ions and nonvolatile impurities. Without an effective pretreatment, steam refining may fail to produce an oil of colour and stability characteristics comparable to the classically refined product.

Degumming

Crude coconut oil is continuously heated to 80–90°C and 85 per cent phosphoric acid at a range of 0.05–0.10 per cent by volume of feedstock is dispersed in the oil via an in-line static mixer. The mixture is conveyed into a coagulation tank and agitated for 20–30 minutes.

Bleaching

The oil is pumped to a slurry tank passing through a preheater at a temperature of 90–95°C. The oil, dried and deaerated under vacuum, receives a predetermined dose of bleaching earth–activated carbon (10:1 ratio), agitated for 20–30 minutes for the removal of colour bodies and other adsorbable impurities. The bleached oil is filtered through a plate and frame filter press and further purified in drum-type polishing filters.

Steam stripping and deodourisation

The degummed and bleached oil is deaerated and heated to 240°C as it passes through economisers and thermal oil heating units. The hot oil cascades down the stripping columns operating under a reduced pressure of 600–1000 Pa pulled by four-stage steam ejectors. The oil is met by sparge steam injected at the bottom of the column. Residence time is 1–1.5 hr. for the removal of the odouriferous substances. Rising vapours are sprayed with fresh fatty acids through a nozzle-type sprayer to remove stripped fatty acid distillates. The deodourised coconut oil exits from the column and passes through an economiser for partial cooling by the incoming bleached oil feedstock. Final cooling takes place in a plate-type cooler where the temperature is maintained at 40–50°C. Citric acid is added before passing the product through a polishing filter.

COMPOSITION OF COCONUT OIL

Coconut oil belongs to unique group of vegetable oils called lauric oils. The most abundant fatty acid in this group is lauric acid, $CH_3(CH_2)_{10}COOH$. Other sources of lauric oils are palm kernel, babassu, cohune and cuphea.

More than 90 per cent of the fatty acids of coconut oil are saturated. This accounts for its low iodine value ranging from 7 to 12. The saturated character of the oil imparts a strong resistance to oxidative rancidity. Assessment of the oil by active oxygen method (AOM) yielded results between 30 and 250 hr. Although oxidative stability is reduced in RBD oils, due to losses in the natural antioxidants of crude coconut oils, the addition of citric acid at the end of deodourisation as the oil is cooled to 100°C was effective in regaining considerable oxidative stability in the oil.

Fatty acids with less than 12 carbon atoms are classified as medium-chain fatty acids (MCFA). Esters of MCFA with glycerol, known as medium-chain triglycerides (MCT), are components in medical foods and infant food formulations. With more than 15 per cent C6, C8 and C10 fatty acids, coconut oil is the richest source of MCFA.

Approximately 0.5 per cent of crude coconut oils is not saponified by caustic treatment. The unsaponifiable matter consists mainly of tocopherols, sterols, squalene, colour pigments and carbohydrates. The odour and taste of coconut oil is largely due to δ- and γ- lactones, which are present in trace quantities. Among the unsaponifiables, tocopherols contributes to the oxidative stability of crude coconut oil. A typical sample of crude coconut oil contained 55 ppm total tocopherols of which 40.7 ppm is α-tocopherol. Most of the unsaponifiables are removed in the process of refining, bleaching and deodourising of crude coconut oil.

The various triacylglycerol (TAG) components of coconut oil may be separated and quantified by gas chromatography with the use of stable silicon gum stationery phase under temperature-programmed conditions and identified by reference to standard TAG solutions. The carbon number of a TAG component is the sum of carbon atoms of the fatty acids attached to the glycerol moiety. For example, the carbon numbers of trilaurin and oleodisterin are 36 and 54, respectively. The relative amounts of each TAG in a sample of fat serves to establish its identity. For coconut oil, this test may also serve to distinguish it from other lauric oils (Table 1.4).

Table 1.4. Triacylglycerol number composition (wt %) of coconut oil.

TAG carbon number	*Range (mean)*	%
C28	0.7–1.0	(0.8)
C30	2.8–4.1	(3.4)
C32	11.5–14.4	(12.9)
C34	15.6–17.6	(16.5)
C36	18.3–19.8	(18.8)
C38	15.1–17.1	(16.3)
C40	9.2–11.1	(10.2)
C42	6.5–8.0	(7.3)
C44	3.6–4.6	(4.2)
C46	2.1–3.0	(2.6)
C48	1.6–2.6	(2.3)
C50	0.8–2.0	(1.7)
C52	0.4–2.0	(1.6)
C54	0.1–1.5	(1.2)

CHEMICAL AND PHYSICAL TESTS

For purposes of identifying natural fats and ascertaining their quality, a number of analytical tests are routinely employed. The test results of a sample of fat under assessment should fall within the range of established constants to confirm its identity. For coconut oil the usual tests are fatty acid composition, acid value/percent free fatty acid, saponification value, iodine value, Reichert–Meissl value, Polenske value, unsaponifiable matter, peroxide value/stability test, s.p/m.p, colour and solid fat content (Tables 1.5 and 1.6).

Table 1.5. Range of fatty acid composition of coconut oil cross regional.

Fatty acid	*Range weight (mean)*	*%*
Caproic C6	0.4–0.6	(0.5)
Caprylic C8	6.9–9.4	(7.8)
Capric C10	6.2–7.8	(6.7)
Lauric C12	45.9–50.3	(47.5)
Myristic C14	16.8–19.2	(18.1)
Palmitic C16	7.7–9.7	(8.8)
Stearic C18	2.3–3.2	(2.6)
Oleic C18:1	5.4-7.4	(6.2)
Linoleic C18:2	1.3–2.1	(1.6)
C20	t–0.2	(0.1)
C20:1	t–0.2	(t)

Table 1.6. Coconut oil product specification.

Description	*Crude*	*Cochin*	*RBD*
Moisture and impurities % max.	1.0	0.1	0.03
Free fatty acid (as lauric) % max.	3.0	0.07	0.04
Colour (5¼ inch cell) Lovibond R/Y max.	12/75	1/10	1/10
Saponification value		250–264	250–264
Unsaponifiable matter % max.	0.4	0.1	0.1
Iodine value		7–12	7–12
Peroxide value, max.	2.0	0.5	0.5
Slip/melting point, °C		24–26	24–26
Refractive index 40°C		1.448–1.450	1.448–1.450
Flavour/odour		Coconut flavour and odour	Bland/odourless

The principal criteria presently being used to measure quality of coconut oil are free fatty acids content and colour. In addition, sensory evaluation through taste and odour of oil serves to confirm the acceptability of the product.

Fatty Acid Composition

The individual fatty acid composition of fats and oils are routinely assayed by gas chromatography. Samples are transesterified with methanol to convert the fatty acids into relatively volatile methyl ester derivatives. The esters are volatilised and swept in a steam of inert gas through a stable inert powder support that has been treated with a liquid (stationary phase) that is not volatile under the test conditions. The esters are eluted in succession from the volume in accordance with their individual retention times in the stationary phase. Each emerging fatty acid esters is recorded ideally as an individual peak; the area under the curve is a function of the quantity of eluted component.

Acid Value/Percent Free Fatty Acid

Acid value is the number of milligrams of potassium hydroxide required to neutralise the free acids in one gram of oil.

A measured amount of oil is dissolved in ethyl alcohol and titrated with 0.1 *N* sodium hydroxide. For coconut oil, an equivalent term, 'per cent free fatty acid as lauric' (% FFA), is more commonly used. % FFA as lauric = ml 0.1 *N* NaOH × 20/g oil; acid value (mg KOH/g oil) = % FFA as lauric × 2.8.

Saponification Value

Saponification value is the number of milligrams of potassium hydroxide required to neutralise the free acid and saponify the esters in one gram of fat.

A slight excess of alcoholic potassium hydroxide is reacted with a measured amount of oil sample and boiled gently under reflux with an air condenser. A blank is similarly treated. Upon complete saponification, the treated sample and blank are titrated with 0.5 *N* hydrochloric acid. Saponification value = (B – S) × 0.5 *N* HCl × 56.1/g oil; where B = titer of blank in ml and S = titer of sample in ml.

Iodine Value

Iodine value is the number of grams of iodine absorbed in one gram of oil. The unsaturated carbon atoms of fatty acids absorb iodine according to the following reaction –CH=CH– + I_2 → –CHI–CHI–.

A slight excess of either *Hanus* (IBr) or *Wijs* (ICI) solution is added to a measured amount of oil sample and shaken for 30 minute. A blank is similarly treated. A specified volume of 15 per cent potassium iodide is added to the treated sample and blank, followed by titration with 0.100 *N* sodium thiosulphate.

Iodine value = (B – S) × 0.100 *N* thiosulphate × 12.7/gram oil; where B = titer of blank in ml and S = titer of sample in ml.

Reichert-Meissl Value

The Reichert-Meissl value is the number of millilitres of 0.10 *N* sodium hydroxide required to neutralise volatile fatty acids (mainly C4 and C6) of 5.0 g of oil. *Polenske value* is the number of millilitres of 0.10 *N* sodium hydroxide required to neutralise insoluble fatty acids (mainly C8, C10, C12) of 5.0 g oil. Typical Reichert-Meissl and Polenske values of coconut oil are 8.4 and 11.5, respectively. Both values are significantly high, as expected in coconut oil, which is a rich source of MCFA.

A sample of fat, weighing 5 ± 0.1 g is saponified in glycerol-sodium hydroxide solution. The saponified sample is acidified with dilute sulphuric acid, followed by distillation of the volatile fatty acids. A volume of 110 ml distillate is collected, chilled to 15°C and filtered. The filtrate is titrated with 0.1 *N* sodium hydroxide. A blank determination is carried. Reichert-Meissl value = (B – S) × 1.1; where B = titer of blank in ml and S = titer of sample in ml. The water-insoluble fatty acids on the filter paper are washed, dissolved in ethyl alcohol and titrated with 0.10 *N* sodium hydroxide. Polenske value = B – S; where B = titer of blank in ml and S = titer of insoluble fatty acids in ml.

Unsaponifiable Matter

Unsaponifiable matter in fats in the ether or hexane-soluble components extracted after a fat sample is refluxed with alcoholic potassium hydroxide.

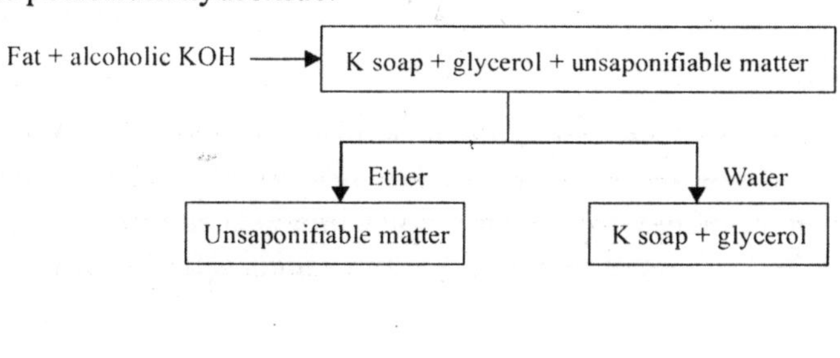

Peroxide Value/Stability Test

Peroxide value is expressed in milliequivalents peroxides oxygen per kilogram oil. Peroxides formed in oil during storage initiate the development of oxidative rancidity.

A measured amount of fat is dissolved in acetic acid–chloroform solvent in the presence of potassium iodide. Peroxides in the fat liberate iodine, which is titrated with 0.10 *N* sodium thiosulphate. A blank run is similarly treated. Peroxide value (meq peroxide oxygen/kg oil = (B – S) × 0.10 *N* thiosulphate × 1000/gram oil; where B = titer of blank in ml and S = titer of sample in ml.

Stability test ('Swift test'/active oxygen method) gives an indication of the oil's resistance to oxidation during storage. The test gives the time in hours required for a sample of oil to reach a peroxide value of 100 when subjected to aeration under specified conditions of temperature and air flow rate. The period of time in hours is determined by interpolation between two peroxide value determinations, which must fall between 75 and 175 meq.

Slip/Melting Points or Melting Range

The slip/melting points or melting range of fat products is the temperature at which the transition from solid to liquid state is observed. Unlike pure substances, which have sharp melting points, fats are mixtures of TAGs exhibiting a range of melting temperatures. Some methods required elaborate equipment while others, such as the capillary tube method are simple and can provide precise results for routine determinations.

By capillary action, molten fat sample is drawn into the tube, which is open at both ends. The fat is solidified by chilling and then melted in water at a regulated rate of increase in temperature. The temperatures at which the fat slips and appears clear in the liquid state are noted as its slip/melting points or melting range.

Refractive Index

The refractive index of a medium is the ratio of the speed of light *in vacuo* at a given wavelength to the speed of light in the medium. Measurement is done by means of a suitable refractometer at a specified temperature for a particular sample. Refractive index is useful for identification of fats and the observation of progress in reactions during hydrogenation.

Colour

The colour of coconut is measured in a Lovibond Tintometer, using a 1 inch or 5¼ inch cell for dark and light coloured oils, respectively. Results are given in red and yellow units describing the combination that matches the sample colour. Alternatively, the optical density of the oil can be measured with the use of a spectrophotometer in a suitable cell at a wavelength of maximum absorbance.

Solid Fat Content/Solid Fat Index

Nuclear magnetic resonance (NMR) is widely used for the determination of solid fat content in fats and oils. The oil dilatometry method, which gives 'solid fat index' values, is time consuming and inadequate as a quick means for control purposes in the processing of fats. The NMR method is quicker, more precise and closer to the absolute solid fat content than dilatometry (Table 1.7)

Table 1.7. Solid fat contents of coconut oil (% by pulse-NMR).

Description	*Fat content (%)*
SP/MP °C	24
Iodine value	8.5
% solids	
20°C	36
30°C	0
35°C	0

Titer

Titer is the solidifying point of the mixed fatty acids derived from a sample of fat. Fat is saponified and subsequently acidulated. The layer of mixed fatty acids is separated, washed, dried and allowed to cool. As the solids begin to separate, the temperature rises due to the liberation of latent heat. The highest temperature reached is taken as the titer.

Uses of Coconut Oil

Coconut oil is used in a wide range of food and non-food products. It is a raw material for the production of medium foods and infant food formulations. In industry, the fatty acids of coconut oil provide a versatile feedstock for an array of products from diesel fuel substitute to hygienic products.

Edible Products

In coconut-producing countries, RBD coconut oil is used extensively as frying oil. Physical blends and interesterified mixtures of coconut oil and hydrogenated palm oil are processed into margarines and shortenings. Coconut jam, a syrupy emulsion derived from coconut cream in cane sugar, is consumed as dessert, bread spread and rice cake topping. Filled milk, in liquid or powder form, contains coconut oil (in lieu of butterfat) and a polyunsaturated oil emulsified in skim milk. As spray oil for crackers and cookies, coconut oil improves shelf life of these products because of its resistance to oxidatives rancidity. Coconut oil is widely used as cream fat and as a component in biscuit cream and confectionery oil.

Medical and Infant Food Formulations

Medium-chain fatty acids (MCFA), mainly C8 and C10, obtained by hydrolysis of coconut oil, is reesterified with glycerol to form a mixture of randomised medium-chain triglyceride (MCT) through a method developed by Babayan. MCTs are absorbed and oxidised rapidly with an energy rated at 34.7 kJ/gram. A number of medical and infant food formulations have MCTs as the principal sources of fat supplemented with polyunsatuates. In a field study by *Intengan* and co-workers a saturated (interesterified) 75 coconut oil–25 corn oil preparation gave better weight gain and nutritional recovery, vis-a-vis a polyunsaturated vegetable oil, when given to malnourished children as supplemental fat source in their diet.

Non-food Products

A commonly used non-food product derived form coconut oil is soap. Laundry bar soaps made by boiled or cold procedures have excellent lathering property even in moderately hard water. A blend of tallow–coconut oil in ratios from 67:33 to 85:15 form an 'ideal' fat change for toilet soaps. Soap from such blends exhibit desirable characteristics related to lather quickness, low mechanical erosion and absence of swelling or cracking of soap bars.

One of the major uses of acid oil from soapstock and the distillates from physical refining is in the manufacture of animal feeds. The fatty components increase the caloric density of the feed.

Derivatives of fatty acids from coconut oil are feedstock for a number of diverse nonfood products. Coconut oil fatty acids and glycerol are released by hydrolysis or alcoholysis of the fat. The fatty acids or their methyl esters, which are subsequently fractionated, constitute the starting materials for the oleochemical industry. The by-product, glycerol is purified by vacuum distillation. The purified product is, among others, a component of pharmaceutical preparations, an important ingredient in toothpastes, a raw material in the manufacture of nitroglycerol and the fluid in hydraulic jacks and shock absorbers.

VIRGIN COCONUT OIL

Virgin Coconut Oil can only be achieved by using fresh coconut meat or what is called non-copra. Chemicals and high heating are not used in further refining. There are currently two main processes of manufacturing virgin coconut oil:

1. Quick drying of fresh coconut meat which is then used to press out the oil. Using this method, minimal heat is used to quick dry the coconut meat, and the oil is then pressed out via mechanical means.
2. Wet-milling. With this method the oil is extracted from fresh coconut meat without drying first. 'Coconut milk' is expressed first by pressing. The oil is then further separated from the water. Methods which can be used to separate the oil from the water include boiling, fermentation, refrigeration, enzymes and mechanical centrifuge.

The method used by Mt. Banahaw Health Products Corporation in the Philippines is fermentation. The coconut milk expressed from the freshly harvested coconuts is fermented for 24–36 hours. During this time, the water separates from the oil. The oil is then slightly heated for a short time to remove moisture, and filtered. The result is a clear coconut oil that retains the distinct scent and taste of coconuts. This is a traditional method of coconut oil extraction that has been used in the Philippines for hundreds of years. Laboratory tests show that this is a very high quality coconut oil, with the lauric acid content being 50 to 53 per cent. This oil is not mass produced, but made by hand just as it has been done for hundreds of years. Most of the suppliers personally guarantee that the best organic coconuts available are used in producing this virgin coconut oil, and that no chemicals whatsoever are used in the growing or processing of the coconuts.

Difference between Virgin Coconut Oil and Other Coconut Oils

Most commercial grade coconut oils are made from copra. Copra is basically the dried kernel (meat) of the coconut. It can be made by smoke drying, sun drying, or kiln drying, or derivatives or a combination of these three. If standard copra is used as a starting material, the unrefined coconut oil extracted from copra is not suitable for consumption and must be purified - that is refined. This is because the way most copra is dried is very unsanitary. Most of the copra is dried under the sun in the open air, where it is exposed to insects and moulds. The standard end product made from copra is RBD coconut oil. RBD stands for refined, bleached and deodourised. Both high heat and chemicals (e.g. solvent extractions) are used in this method.

RBD oil is also often hydrogenated or partially hydrogenated. Hydrogenated oils have been shown to increase serum cholesterol levels which contribute to heart disease.

One of the main differences between Virgin Coconut oil and refined coconut oils is the scent and taste. All virgin coconut oils retain the fresh scent and taste of coconuts, whereas the copra-based refined coconut oils have no taste at all due to the refining process.

Coconut Health Oil

As a 'functional food', coconut oil is now being recognised by the medical community as a powerful tool against immune diseases. Several studies have been done on its effectiveness and much research is currently being done on the incredible nutritional value of pure coconut oil. To add pure virgin coconut oil to your diet, you can purchase from the market.

Coconut Cooking Oil

Coconut oil is a functional skin and health food. You can eat it and use it as a healthy replacement for butter, margarine and cooking oil. Coconut Oil is not stored in the body (liver) as fat, but is converted into pure clean-burn energy. It does not raise cholesterol levels.

What Coconut Oil does not do?

1. Does not increase blood cholesterol level.
2. Does not promote platelet stickiness which leads to blood clot formation.
3. Does not contribute to atherosclerosis or heart disease.
4. Does not promote cancer or any other degenerative disease.
5. Does not contribute to weight problems.

What Coconut Oil does do?

1. Reduces risk of atherosclerosis and related illnesses.
2. Reduces risk of cancer and other degenerative conditions
3. Helps prevent bacterial, viral and fungal (including yeast) infections
4. Supports immune system function
5. Helps prevent osteoporosis.
6. Helps control diabetes.
7. Promotes weight loss.
8. Supports healthy metabolic function.
9. Provides an immediate source of energy.
10. Supplies fewer calories than other fats.
11. Supplies important nutrients necessary for good health.
12. Improves digestion and nutrients absorption.
13. Is highly resistant to spoilage (long shelf life).
14. Is heat resistant (the healthiest oil for cooking).
15. Helps keep skin soft and smooth.
16. Helps prevent premature ageing and wrinkling of the skin.
17. Helps protect against skin cancer and other blemishes.

Why should you Use Coconut Oil?

Coconut oil has a unique role in the diet as an important physiologically functional food. The health and nutritional benefits that can be derived from consuming coconut oil have been recognised in many parts of the world for centuries.

Although the advantage of regular consumption of coconut oil has been under appreciated by the consumer and producer alike for the recent two or three decades, its unique benefits should be compelling for the health minded consumer of today.

A review of the diet/heart disease literature relevant to coconut oil clearly indicates that coconut oil is at worst neutral with respect to atherogenicity of fats and oils and, in fact, is likely to be a beneficial oil for prevention and treatment of some heart disease. Additionally, coconut oil provides a source of antimicrobial lipid for individuals with compromised immune systems and is a nonpromoting fat with respect to chemical carcinogenesis.

STORAGE OF COCONUT OIL

Although coconut oil has a natural resistance to oxidative rancidity, the bulk storage systems should take into account all factors that contribute to the deterioration of the product during storage. These factors are light, prooxidatants, air, heat and moisture. Light may have the least effect because the oil is handled in a closed system.

From the milling of copra for the extraction of crude coconut oil through the refining steps and final storage in tanks, the oil is in continuous contact with iron and possibly copper-containing alloys, both of which are prooxidants. The addition of citric acid or any other appropriate antioxidant in the last stages of deodourisation of RBD oils affords protection to the oil from oxidative rancidity. Crude coconut oil has natural and protective antioxidants.

The introduction of air during handling procedures may be minimised by filling tanks through a subsurface entry point instead of allowing the oil to fall through air into the storage tanks.

Storage tanks are provided with mechanical agitators and heating devices, although the latter may hardly be utilised in most areas of the tropics. During brief spells when temperature falls below slip/melting points of coconut oil, the product is warmed with agitation to prevent localised overheating.

The effect of moisture on fat in storage is well known. In the presence of enzymes, mainly from micro-organisms, hydrolysis of fat is accelerated giving rise to an unpleasant 'soapy' taste peculiar to coconut oil exposed to conditions favouring hydrolytic rancidity.

COCONUT OIL IN COSMETICS

Coconut oil is a popular edible oil in South India. Among the vegetable oils of Indian origin, coconut oil has the largest application in cosmetics. Coconut oil is the oldest and among the most widely used cosmetics vegetable oil because of its low incidence of irritation, sensitisation and for the simple good feel provided on the human skin. The following information highlights the main application of coconut oil in cosmetics.

Coconut Oil as a Body Oil

It is customary in India that a new born baby is massaged by coconut oil before bathing with warm water. The reason of selection of this oil is because of its low viscosity, pleasant aroma and easy washability. Any other vegetable oil could be used, but coconut oil is most suitable. The only draw back is that it freezes during winter and is therefore not popular in North India. In order to keep the pores of the skin open, both men and women particularly in South India take an oil bath once a week. Coconut oil is applied to the entire body even by adults and slowly massaged into the skin for about an hour, the practice is to leave a thin film of oil left on the skin of the body, which keeps the skin soft and moist.

Due to urbanisation the practice of adults taking weekly oil bath has decreased. The trend is to substitute synthetic liquid paraffin oil with stabilised vegetable oil like coconut oil, which is seen as more natural in relation to composition of the skin fats. Liquid paraffins are chemically unrelated to the skin fats, the latter containing only small amount of hydrocarbons and different in structure to those obtained from petroleum. The age-old use of coconut oil as a body oil, baby oil and hair oil is due to the unique unsaponifiable matter (0.1–0.4 per cent) consisting of sterols, tocopherol, squalene and triterpene alcohols present in natural coconut oil, while liquid paraffin oil has no unsaponifiable matter. The thin film of coconut oil is a unique natural vegetable oil suitable for cosmetic uses.

Coconut Oil as Hair Oil

Indians are habituated to use hair oil. Coconut oil is preferred for this due to its low viscosity and pleasant aroma. The practice was to use oil for hair dressing, which prevents dandruff if used daily. Coconut oil on storage develops rancidity due to ambient atmosphere and humidity. This can be prevented by incorporation of antioxidants in small quantities. In order to attract consumers the incorporation of fragrances came into vogue. This increases the cost of the finished hair oil, and some people are allergic to fragrances which is a mixture of several chemicals–both natural and synthetic. Some get headache soon after application of perfumed hair oil, but they are not aware that it is due to the fragrance.

In order to overcome the rancidity drawback of coconut oil, as also its high price, mineral oils are admixtured with coconut oil. These mineral oils are petroleum products with toxic aromatics which are eliminated by chemical treatment giving a new name as 'liquid paraffins', permitted for use in cosmetics. These may be safe if properly purified and tested by sophisticated instruments. It is found that these liquid paraffins on application get absorbed into the skin.

These are called as non-sticking hair oils by TV, and other media. What good or harm it does to human body is to be judged by the user. Industry likes to use liquid paraffins due to its stable price, non-sticky nature and no chance for rancidity formation. Bureau of Indian Standards (BIS) has three types of hair oils, the first containing single or mixture of vegetable oils, the second containing only liquid paraffins and the third type allows mixture of vegetable oils and liquid paraffins. Hair oil manufacturers are taking undue advantage of the lacuna in standards by labelling as 'Coconut Hair Oil', a mixture of Coconut Oil and Liquid Paraffin. It is high time that BIS rectifies such an anomaly.

Coconut Oil in Shampoo

If one needs natural shampoo, coconut oil is the best oil as one can get an excellent shampoo by converting coconut oil to potash soap or mix of potash and sodium soap. The glycerine remaining in the product acts as an emollient. The shampoo lathers well and is liked by the consumer. Instead of synthetic fragrances, one can add extracts of Amla or Shikkakai which gives a herbal touch to the shampoo. The main draw back of coconut oil shampoo is that it does not lather in hard water and one has to switch over to synthetic shampoo. Here again coconut oil is the main raw material as one has to fractionate coconut oil fatty acids to obtain lauric acid. Lauric acid can be converted to lauryl alcohol and then to sodium lauryl ether sulphate which is the main active ingredient of synthetic shampoo. Coconut oil is a rich source of myristic acid which on esterification with isopropyl myristate an excellent binder used in cosmetics.

The higher fatty acids fraction (nearly 20 per cent) of coconut oil consisting of C_{16}–C_{18} fatty acids can be converted to a new surfactant.

Alpha olefin sulphonate is a good surfactant for shampoo as it is tender to the skin and now made in India. Another cosmetic surfactant derived from coconut oil is prepared by the reaction of lauric acid with sucrose to form monoesters which are slightly soluble in water but soluble in many organic solvents. These are neutral and are useful in cosmetic preparations and are non-toxic.

Coconut Oil in Dental Cleaning Products

Both tooth paste and tooth powder contains sodium lauryl sulphate which is the main cleaning detergent. Sodium lauryl sulphate is obtained by the sulphonation of lauryl alcohol and neutralisation by sodium hydroxide. Of course the sodium lauryl sulphate has to meet the BIS requirement of cosmetic grade. The other derivatives of coconut oil used in cosmetics are sorbitan mono laurate, polyoxyethylene sorbitan mono-laurate, sodium lauryl sarcosinate as emulsifiers and solubilisers.

Coconut Oil in Perfumes and Flavours

The lower fatty acids present in coconut oil, namely caproic, caprylic acid capric can be esterified with alcohols like methyl alcohol, ethyl alcohol, isopropyl alcohol, butyl alcohol giving the respective esters, which have characteristic fragrances for use in perfumes and food flavours.

Coconut oil is not only used in cosmetics but also in pharmacy as an ointment base, on account of its ready penetration into the skin and appreciable water absorbing property. Specially prepared coconut oil is used as a good eye lotion. A mixture of coconut oil and lime water is used to apply on burns. Thus, natural coconut oil has several valuable applications in cosmetics.

FATTY ACIDS AND DERIVATIVES FROM COCONUT OIL

Coconut oil and palm kernel oil, a co-product of palm oil, comprise less than 5 per cent of the total natural fats and oils, but they are important feedstocks of the oleochemical industry.

Coconut oil is commercially a major source of lauric acid. Together with palm kernel oil and to a small extent babassu oil, it belongs to the so-called lauric oils, which are characterised by their high lauric oil content of approximately 50 per cent. The lauric oils are highly desirable materials in the oleochemical industry worldwide because of the importance of the lauric fraction especially in the manufacture of soaps and detergents. Coconut oil is well positioned because it has the unique advantage of having its fatty acid composition falling within the carbon chain spectrum highly desired by the oleochemical industry where the C12 and C14 fatty acid fractions are sought after. Table 1.8 shows the fatty acid composition of coconut oil and palm kernel oil.

The caproic to capric (C6–C10) fatty acid fractions comprising approximately 15 per cent are good materials for plasticiser range alcohol and for polyol esters. The latter are used in high performance oil for jet engines and for the new generation of lubricants. These acid fractions are also the basic material for the manufacture of medium chain triglycerides a highly valued dietary fat. The C12–C18 fatty acid fractions, approximately 85 per cent of the coconut oil fatty acid composition, are the primary raw materials for detergent grade fatty alcohols. Coconut oil is a primary source of basic oleochemicals and a host of other oleochemical derivatives. Figure 1.4 indicates some of the major processes by which various oleochemicals and their derivatives can be obtained from fats and oils. These processes are applicable to coconut oil with the exception of epoxidation and direct sulphation since coconut oil lacks the necessary unsaturation to initiate the two reactions. As can be seen from Fig. 1.4, a great number of oleochemicals can be derived from natural fats and oils. These oleochemicals find increasing use in various applications.

Table 1.8. Fatty acid composition of coconut and palm kernel oils.

Fatty acid	*Formula*	*Coconut oil (%)*	*Palm kernel oil (%)*
Caproic	$C_6H_{12}O_2$	0.2–0.8	0–1
Caprylic	$C_8H_{16}O_2$	6–9	3–5
Capric	$C_{10}H_{20}O_2$	6–10	3–5
Lauric	$C_{12}H_{24}O_2$	46–50	44–51
Myristic	$C_{14}H_{28}O_2$	17–19	15–17
Palmitric	$C_{16}H_{32}O_2$	8–10	7–10
Stearic	$C_{18}H_{36}O_2$	2–3	2–3
Oleic	$C_{18}H_{34}O_2$	5–7	12–19
Linoleic	$C_{18}H_{32}O_2$	1–2.5	1–2

Types of Fatty Acids and Derivatives from Coconut Oil and their General Applications

Coconut oil is one of the most important raw materials for the oleochemical industry. The whole range of its fatty acid composition is used as the starting material for a wide variety of oleochemical products. Fatty acids are the building blocks that with proper selection and application of oleochemistry are converted to higher valued products. Coconut oil is considered a saturated oil. From Table 1.8, it can be seen that coconut oil has approximately 92 per cent saturated fatty acid from caproic to stearic and only around 8 per cent unsaturated fatty acid composed of oleic acid and linoleic acid.

Types of fatty acids from coconut oil

Fatty acids obtained by the high pressure splitting of coconut oil, are distilled and can be fractionated into various fractions or individual cuts. Types commercially available are the following.

1. Whole distilled coconut fatty acid. A refined product whose fatty acid composition is identical to that of the original oil.
2. Caprylic, capric acid. The low molecular weight fraction comprising around 55 per cent C8 and 40 per cent C10 fatty acid fractions with small amounts of C6 and C12 fractions.
3. Topped coconut fatty acid. The C12–C18 fraction after topping off the C8–C10 fraction.
4. Lauric, myristic acid. The medium chain fatty acid fraction comprising approximately 72 per cent C12 and 26 per cent C14 fatty acid fractions with traces of C10 and C16 fatty acid fractions.
5. Lauric acid. A pure-cut C12 fatty acid with a purity of 99 per cent minimum with traces of C10 and C14 fatty acid fractions.
6. Myristic acid. A pure-cut C14 fatty acid with a purity of 98 per cent minimum and traces of C12 and C16 fatty acid fractions.

There is a market for individual cuts of fatty acids of high purity. These are highly desired by certain industries such as the cosmetic industry and command higher prices. Table 1.9 shows the typical composition and properties of these types of fatty acids.

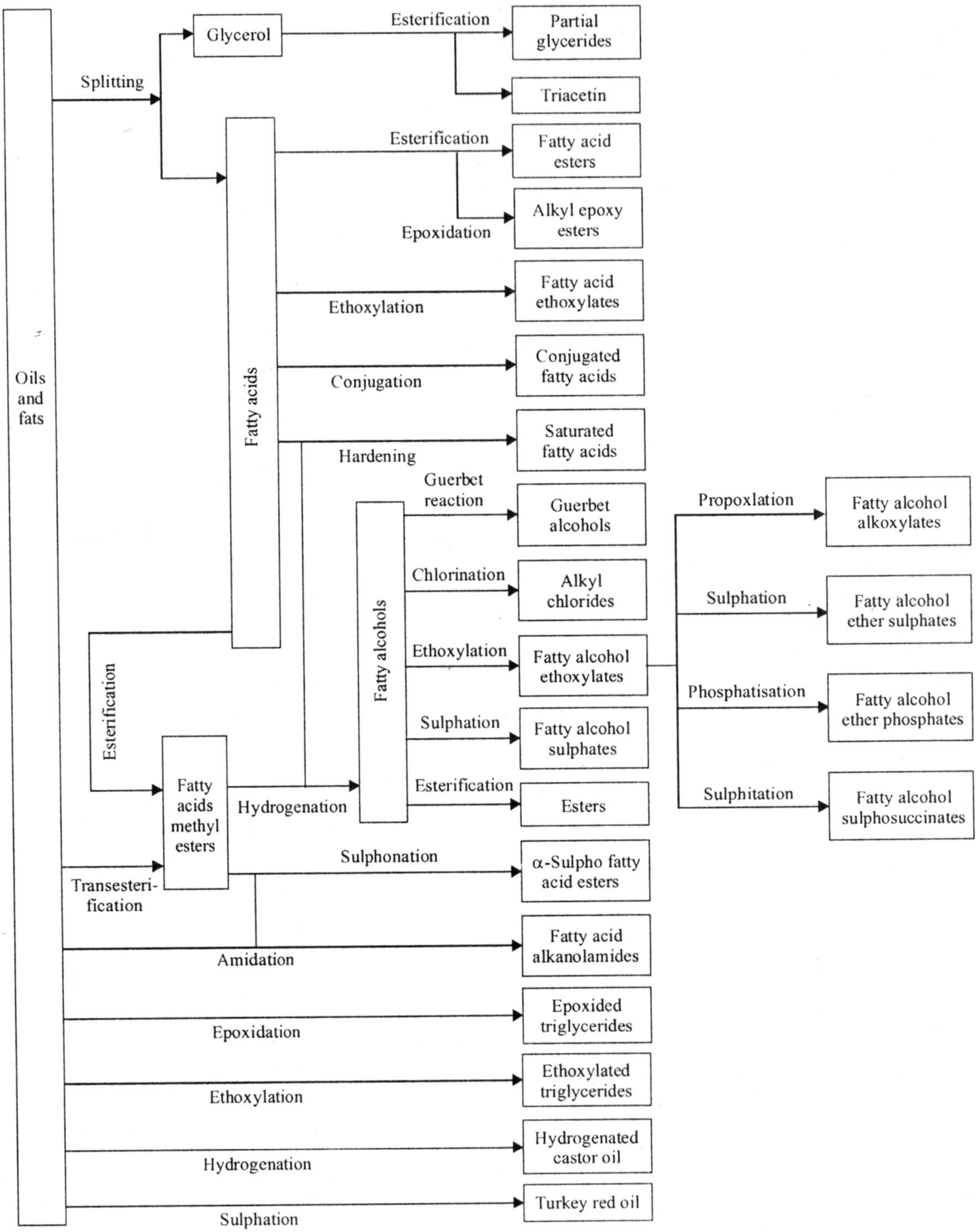

Fig. 1.4. Oleochemical raw materials and their derivatives.

Table 1.9. Fatty acids product specifications.

Fatty acid products	Product code	Iodine value	Acid value	Saponification value	Titer (°C)	Unsapo-nifiable matter (% max.)	Colour lovibond 5.25 inches		Approximate carbon chain distribution								
							Y (max.)	R (max.)	C6	C8	C10	C12	C14	C16	C18	C18:1	C18:2
Whole distilled coconut fatty acid (C8–C18)	Philacid 0818	6–10	268–274	269–275	21–25	0.5	15	1.5	0.5	7.5	6.5	48	18	9	2	7	2
Caprylic–capric acid (C8–C10)	Philacid 0810	0.8 max.	355–365	356–366	1–5	0.5	–	–	4	55	39	3	–	–	–	–	–
Topped coconut fatty acid (C12–18)	Philacid 1218	8–12	254–260	254–260	25–29	0.5	15	1.5	–	–	1.0	55	22	10	2.5	8	2
Lauric acid (C12)	Philacid 1200	0.3 max.	279–281	279–281	42–43	0.5	8	0.8	–	–	0–1	99 min.	0–1	–	–	–	–
Myristic acid (C14)	Philacid 1400	0.3 max.	245–247	245–247	53–54	0.5	8	0.8	–	–	–	0–2	98 min.	0–2	–	–	–
Lauric–myristic (C12–C14)	Philacid 1214	0.5 max.	268–273	268–273	33–35	0.5	10	1.0	–	–	1	72	26	1	–	–	–

Oleochemical derivatives from coconut oil and their general applications

Coconut fatty acids and their various fractions, besides from being used directly, are converted further to other derivatives. Their range of applications covers a broad spectrum in the oleochemical industry. As shown in Fig. 1.4, fatty acids can undergo different processes in the manufacture of various oleochemical derivatives. Among the more common products and applications are the following:

Fatty acids

A large volume of coconut fatty acids are used as major components in toilet soap manufacture. Its high lauric content provides the quick lathering properties of toilet soap.

Fatty acid esters

Different fatty acid fractions can be esterified with a monoalcohol or a polyol to yield various esters. Polyol esters of trimethylol propane or pentaerythritol and C8–C10 fatty acids are the bases for high performance lubricants. Reesterification of C8–C10 fatty acids with glycerol yields medium chain triglyceride, a low viscosity, highly stable oil. Medium chain triglyceride is used as solvent for flavours, in the surface treatment of dried fruits and as a high energy, readily digestible dietary fat.

Esterification with monoalcohol such as isopropanol and myristic acid yields isopropyl myristate, an important cosmetic ingredient. Glyceryl monoesters and wax esters find application as food emulsifiers, mould release agents and lubricants for the plastic industry.

Fatty alcohols

Fatty alcohol is considered a basic oleochemical manufactured by high pressure hydrogenation of fatty acids or fatty acid methyl esters. The majority of the fatty alcohol produced is further subjected to various processes such as sulphation, ethoxylation, amination, phosphatisation, sulphitation and others.

Fatty alcohol can be fractioned to separate the C8–C10 fraction known as plasticiser range alcohol and the C12–C18 known as the detergent range alcohol. The plasticiser range alcohol is a liquid with good dissolving power. It can be used in a limited way as solvent for printing inks and lacquers. Esterification with a polycarboxylic acid such as phthalic anhydride yields an excellent plasticiser especially for PVC. The C12–C14 alcohol finds special application as lubricant additives and in the formulation of bearing and hydraulic oils. The C16–C18 fatty alcohol finds applications as a defoamer, as solubility retarder for syndet bars and as a consistency giving factor in creams, lipstick, pastes and polishes.

By far the greatest application of fatty alcohol is in the manufacture of fatty alcohol sulphate and fatty alcohol ether sulphate. These materials possess good foaming properties and ready biodegradability and are extensively used as base surfactants for laundry detergent products, shampoos, dishwashing liquids and cleaners.

Polyglycol ethers

Polyglycol ethers, produced by the reaction of fatty alcohol with ethylene oxide, constitute the most important class of nonionic surfactants. They possess good wetting properties, produce relatively low foam and are highly effective at low temperature and low concentration. They are used as textile auxiliaries, in dishwashing liquids, degreasing products and liquid cleaners formulations.

Other speciality surfactants

Monoalkyl phosphate, fatty alcohol ether phosphate and fatty alcohol sulphosuccinate are some of the speciality surfactants derived from fatty alcohol with specific applications in the cosmetics and other chemical industries.

Fatty amides

Cocomonoethanolamide and cocodiethanolamide formed by the reaction of fatty acids or esters with monoethanolamine or diethanolamine are popularly used as foam boosters for shampoos and detergent products.

Fatty amines

Fatty amines are the most important nitrogen derivatives of fatty acids. They are produced by the reaction of fatty acids with ammonia and hydrogen. They are the bases for the manufacture of quaternary ammonium compounds used as fabric softeners and biocides. Fatty amine oxides are mild to the skin with good cleaning and foaming properties and find application as a shampoo ingredient.

Principles and methods in the manufacture of oleochemicals

Oleochemicals, by their very name, may be defined as chemicals from oil. These could be natural fats and oils, or oils of petrochemical origin. To have clear distinction, oleochemicals derived from natural oils are termed natural oleochemicals, whereas those derived from petrochemicals are termed synthetic oleochemicals. The natural oleochemicals are obtained from natural oils with the least change in the structure of the carbon chain fraction. In contrast, synthetic oleochemicals are built up from ethylene to the desired carbon chain fraction or from oxidation of petroleum waxes.

Fats and oils are renewable products of nature. One can aptly call them 'oil from the sun' where the sun's energy is biochemically converted to valuable oleochemicals via oleochemistry. Natural oleochemicals derived from natural fats and oils by splitting or transesterification, such as fatty acids, methyl esters and glycerine are termed basic oleochemicals. Fatty alcohols and fatty amines may also be counted as basic oleochemicals, because of their importance in the manufacture of derivatives. Further processing of the basic oleochemicals by different routes such as esterification, ethoxylation, sulphation and amidation (Fig. 1.4) produces other oleochemical products, which are termed oleochemical derivatives.

The detail of the processing methods in the manufacture of the basic oleochemicals and their derivatives are discussed below:

Fatty Acids

Fatty acids and fatty acid methyl esters are probably the most important basic oleochemicals in the oleochemical industry. Fatty acids are used as starting materials for soaps, medium-chain triglycerides, polyol esters, alkanolamides and many more.

Chemistry of fat splitting

Coconut oil, like any other fat or oil, can be hydrolysed or split into its corresponding fatty acids and glycerine. The reaction is represented by the following equation:

$$\begin{array}{ccccccc} \mathrm{RCOOCH_2} & & & & \mathrm{RCOOH} & & \mathrm{CH_2OH} \\ | & & & & & & | \\ \mathrm{R'COOCH} & + & \mathrm{3H_2O} & \rightleftharpoons & \mathrm{R'COOH} & + & \mathrm{CHOH} \\ | & & & & & & | \\ \mathrm{R''COOCH_2} & & & & \mathrm{R''COOH} & & \mathrm{CH_2OH} \\ \text{Triglyceride} & & \text{Water} & & \text{Fatty acids} & & \text{Glycerine} \end{array} \quad \text{... (1.1)}$$

In the case of coconut oil, the fatty acid fractions are C8 to C18 with a trace amount of C6.

Fat splitting is essentially a homogeneous reaction that proceeds in stages. The fatty acid radicals are displaced from the triglyceride one at a time from tri to di to mono. An incomplete splitting will thus contain monoglycerides and diglycerides as well as triglycerides. During the initial stage, the reaction proceeds slowly, limited by the low solubility of the water in the oil phase. In the second stage, the reaction proceeds fairly rapidly brought about by the greater solubility of water in the fatty acids. The final stage is characterised by a diminishing reaction rate as the fatty acids liberated and the glycerine by-product reach equilibrium conditions.

Fat splitting is a reversible reaction. At the point of equilibrium, the rates of hydrolysis and reesterification are equal. The glycerine by-product must be withdrawn continuously to force the reaction to completion.

Increasing the temperature and pressure accelerates the reaction because of the increased solubility of the water in the oil phase and to its higher activation energy. Temperature, in particular, exerts a significant effect. An increase in temperature from 150 to 220°C increases water solubility by two to three times.

The presence of small amounts of mineral acids such as sulphuric acid or certain metal oxides such as zinc or magnesium oxide accelerates the splitting reaction. These metal oxides are true catalysts. They also assist in the formation of emulsions.

Fat splitting processes

There are at least four known methods of fat splitting. These are:

1. Twitchell process.
2. Batch autoclave process.
3. Continuous process.
4. Enzymatic process.

Twitchell process

Twitchell process is one of the earlier processes developed for fat splitting. It is still used in a small way because of its low initial cost and simplicity of installation and operation. However, it is no longer of great commercial importance, due to its high energy consumption and poor product quality. The process makes use of the Twitchell reagent and sulphuric acid to catalyse the hydrolysis. The reagent is a sulphonated mixture of oleic or other fatty acid and naphthalene.

The operation is carried out in a wooden, lead-lined, or acid resistant vat where the fat, water amounting to approximately half the fat, 1–2 per cent sulphuric acid and 0.75–1.25 per cent Twitchell reagent are boiled at atmospheric pressure for 36–48 hr. using open steam. The process is usually repeated two to four times, drawing off the glycerine–water solution after each stage. At the last stage, water is added and the mixture boiled to wash off any remaining acid. The long reaction period, high steam consumption and discolouration of the fatty acids are disadvantages of the process and are the reasons for its limited use today.

Batch autoclave process

The batch autoclave process is the oldest commercial method used for splitting higher grade stock to produce light-coloured fatty acids. It is also more rapid than Twitchell process, taking about 6–10 hr. to complete. Distillation is normally performed to remove glyceride esters.

This process uses a catalyst, usually zinc, magnesium or calcium oxides. Of these, zinc is the most active. About 2–4 per cent catalyst is used and a small amount of zinc dust is added to improve the colour of the fatty acids. The autoclaves are tall clyinders, 1220–1829 mm in diameter and 6–12 m high made of corrosion-resistant alloy and fully insulated. Injection of live steam provides the agitation, although some in addition, use mechanical agitators.

In operation, the autoclave is charged with the fat, water amounting to about half the fat and the catalyst. Steam is blown through to displace any dissolved air and the autoclave is closed. Steam is admitted to raise the pressure to 1135 kPa and is injected continuously at the bottom while venting a small amount to maintain the desired agitation and operating pressure. More than 95 per cent conversion is achieved after 6–10 hr. The contents of the autoclave is transferred to a settling tank where two layers are formed: the fatty acid upper layer and the glycerine (sweet water) lower layer. The fatty acid layer is drawn off, treated with mineral acid to split the soap formed and finally washed to remove traces of the mineral acid.

Continuous process

The continuous countercurrent, high pressure fat splitting process, more popularly known as the Colgate–Emery process is the most efficient of the current methods of fat hydrolysis. The high temperature and pressure used permit short reaction time. Full countercurrent flow of oil and water produce a high degree of splitting without the need of a catalyst. However, catalyst may be used to increase reaction rate further. The splitting tower is the heart of the process. Most splitting towers have the same configuration and basically operate the same way. Depending on the capacity, the tower can be 508–1220 mm in diameter and 18–25 m high and is made of corrosion-resistant material such as stainless steel 316 or inconel alloy designed to operate at a pressure of about 5000 kPa.

Figure 1.5 shows a Lurgi single-stage countercurrent splitting plant. The deaerated fat is introduced by means of a sparge ring, around 1 m from the bottom with a high pressure pump. Water is introduced near the top at a ratio of 40–50 per cent of the weight of the fat.

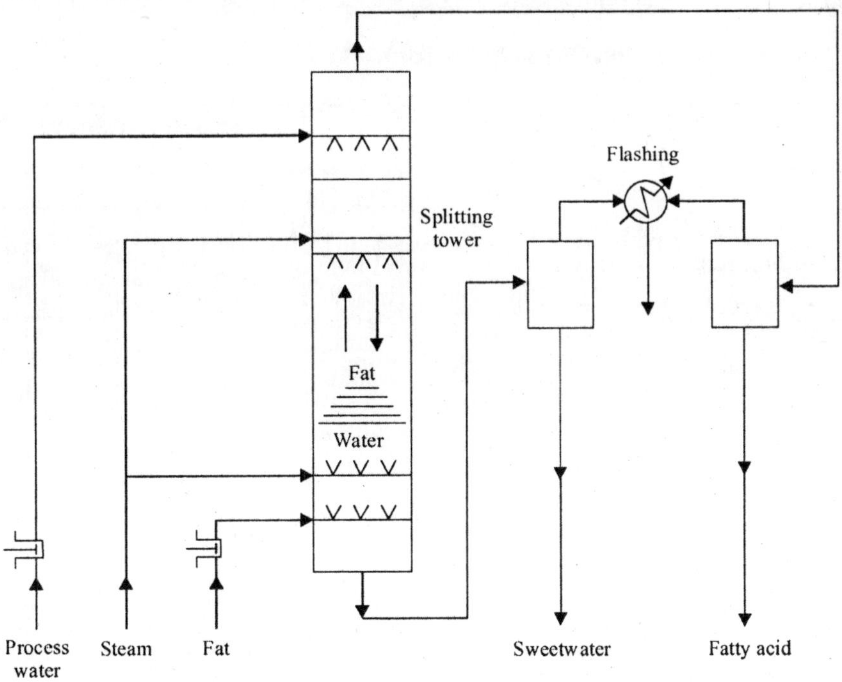

Fig. 1.5. Single-stage countercurrent splitting.

The high splitting temperature (250–260°C) ensures adequate dissolution of the water phase into the fat so that mechanical means for bringing the two phases into contact are not required.

The empty volume of the tower is used as the reaction compartment. The crude fat passes as a coherent phase from the bottom to the top through the tower, while the heavier splitting water travels downward as a dispersed phase through the mixture of fat and fatty acid. Degrees of splitting up to 99 per cent can be reached.

The continuous countercurrent high pressure process splits fats and oils more efficiently than other processes in a reaction time of only 2–3 hr. Little discolouration of the fatty acids occur. Because of the efficient internal heat exchange, this process affords high steam economy.

The utilities consumption per tonne of feed is as follows:

Steam (6000 kPa)	190 kg
Cooling water (20°C)	3 m^3
Electrical energy	10 kWh
Process water	0.6 m^3

Enzymatic splitting

Fats and oils can hydrolyse in the presence of natural enzymes. Fat splitting through the use of lipolytic enzymes had been carried out in experimental trials. However, at present, this process is of doubtful importance because of its high cost and long reaction time.

The enzymatic splitting of fats and oils by lipase from *Candida Rugosa, Aspergillus niger* and *Rhizopus arrhizus* had been studied at a temperature range of 26–46°C for a period of 48–72 hr. About 98 per cent splitting is possible. There are still major problems to be solved in the further development of this process before it becomes commercially viable.

Fatty Acid Distillation and Fractionation Operations

The fatty acids produced from the various fat splitting processes are purified and separated into fractions or even individual fatty acids by distillation and fractionation.

Fatty acid distillation

Distillation of crude fatty acids removes both the low and high boiling impurities as well as odour bodies. Fatty acids are extremely sensitive to heat, oxidation and corrosion effects. This is due to the reactive acid group at the end of the long carbon chain. These factors are taken into consideration in the design of the distillation unit and its operation parameters. Distillation is carried out under high vacuum and lower temperatures and with the shortest residence time allowable.

Technical design of most distillation units features high vacuum with no allowance for air leakage, effective heating to achieve short contact time, good circulation for effective mass transfer between vapour and condensate and steam economy. Internal arrangements of the column vary among suppliers with the ultimate purpose of achieving the design objectives. The basic steps in processing are about the same. Figures 1.6 and 1.7 show Badger's and Lurgi's continuous fatty acid distillation plants.

Crude fatty acids is predried and degassed under vacuum and fed to the distillation unit, which is operated at a vacuum of 1.2 kPa or less and a temperature of approximately 200°C. Modern stills use thermal oil or high pressure steam as the heat source. Stripping steam is provided to improve circulation and reduce partial pressure, thus lowering the temperature and reducing degradation losses. The steam facilitates the removal of low boiling impurities as well as odour and colour bodies from the vapour existing the system and from the light ends. The distilled fatty acids has an almost water white colour and is free of the major impurities. The heavy ends consist of the higher boiling components, usually of lower quality, which can either be withdrawn separately or recycled directly for redistillation. The bottoms or residue are charred viscous polymerised material, which is disposed of by blending with heavy residual oil and used as boiler fuel. It may also find use as an asphalt additive.

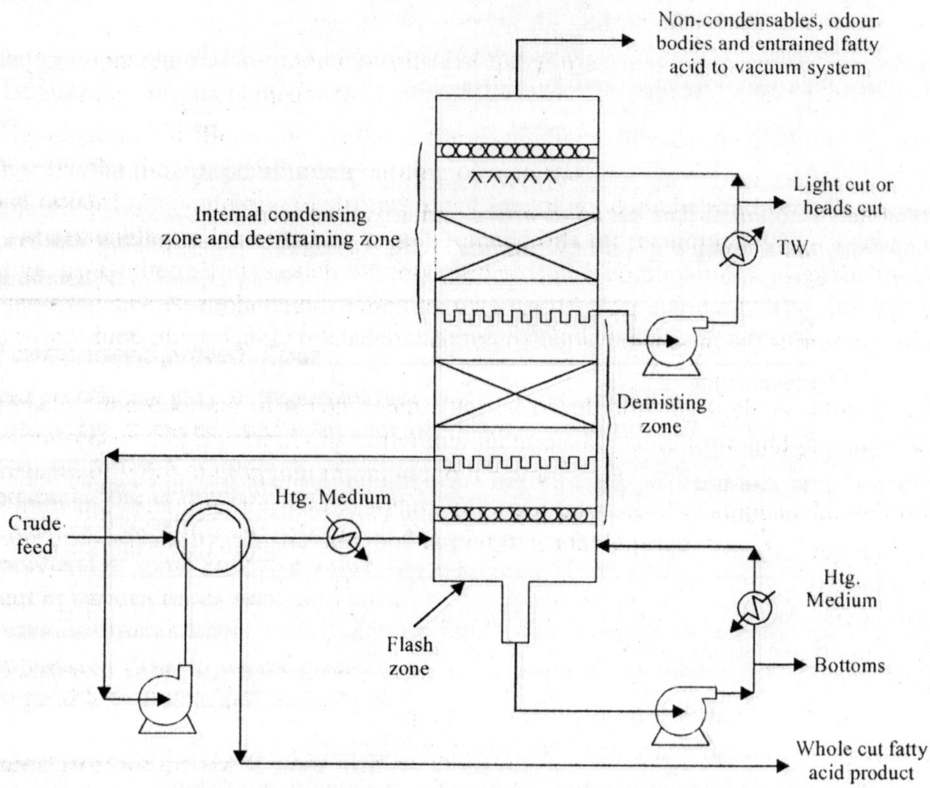

Fig. 1.6. Badger's continuous fatty acid distillation.

Lurgi establishes an approximate utility consumption per tonne of crude fatty acid feed for plant sizes from 50 to 200 tonnes per day as follows:

Heating steam (5000 kPa)	370 kg
Steam (300–1000 kPa)	150 kg
Cooling water (20°C)	15 m^3
Electrical energy	5 kWh
Export steam (300 kPa)	120 kg

Fatty acid fractionation

Today's market is becoming more and more demanding. Specific fractions with purity in excess of 99 per cent are in demand for special products. Fortunately, developments in fractionation technology can now readily meet this challenge. Purities of 99.5 per cent can be achieved for pure cuts of C12 or C14 fraction.

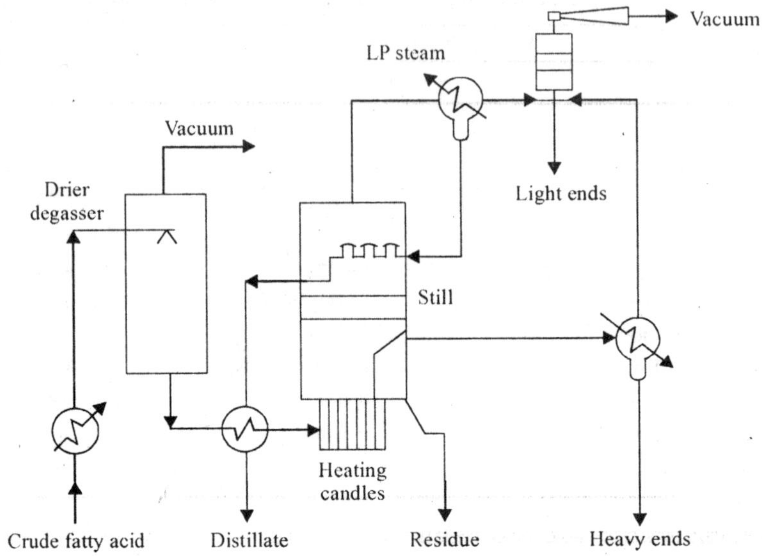

Fig. 1.7. Lurgi's continuous fatty acid distillation.

Fractionation makes it possible to separate the fatty acid mixtures into narrower cuts or even individual components. The detergent-grade feedstock C12–C18 fraction is separated from the whole cut by topping off the C8–C10 fraction. The middle cut, C12–C14, can be further fractionated from the C12–C18 cut through the use of multistage fractionation employing two or more columns. All commercially available fractionation processes can give comparable performances. Basically, each process uses a deaerator, heat source, fractionation column, condensing system and vacuum source. These processes may differ in the internals of the column, in the manner of evaporation and condensation and in the piping arrangement for better heat recovery, but they can give equally good results.

The column internals are designed to provide intimate contact of the vapour and the distillate, together with the lowest pressure drop possible through the volume. Various arrangements are employed such as bubble caps, exchange trays, regular packings or structured packings. Lurgi uses the Thorman tray, which claims high separation efficiency, flexible loading and long-term trouble free operation. Structured packings such as the *Mellapak* or *Sulzer* or the *Glitch* are finding favour in modern columns. These packings are claimed to have high separation efficiencies and minimal pressure drops of 10–50 Pa mbar per stage. Because of their structure, there is little liquid holdup. Slight changes in operating parameters can easily upset the equilibrium conditions and it is thus best to operate the column at steady conditions.

The vacuum system is provided independently for each column and usually consists of a mechanical vacuum pump and a steam ejector to achieve the highest vacuum. A centralised vacuum system is not recommended, since a process upset in one stage can readily affect the other stages. The heat source can be a shell-and-tube heat exchanger using thermal oil. Fig. 1.8 shows a Lurgi fractionation system that has two columns to produce three separate fractions.

Methyl Esters

Fatty acid methyl esters play a major role in the oleochemical industry. Methyl esters have increasingly replaced fatty acids as starting materials for many oleochemicals. They are used as chemical intermediates for a number of oleochemicals such as fatty alcohols, alkanolamides, α-sulphonated methyl esters and many more. One other potential use of methyl esters is as a substitute for diesel oil. Methyl esters are clean-burning with no sulphur dioxide emission. Although the heat of combustion is slightly lower, there is not engine adjustment necessary and there is no loss in efficiency.

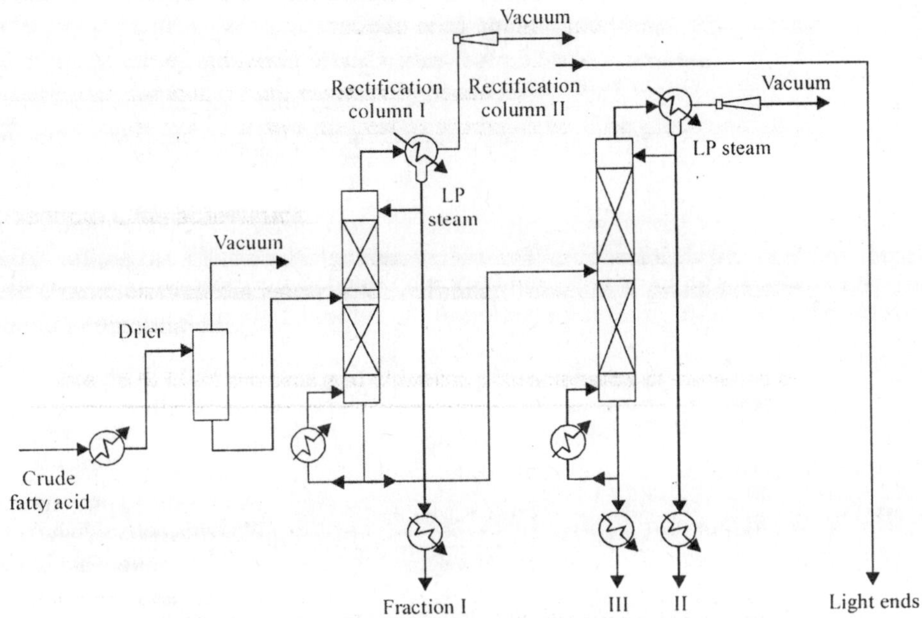

Fig. 1.8. Fatty acid fractionation.

Advantages of methyl esters

The use of methyl esters instead of fatty acids as starting materials for many oleochemicals is rapidly gaining ground because of the following advantages:

1. Lower energy consumption: The production of methyl esters require much lower reactor temperatures and pressures than the splitting of fats and oils to obtain fatty acids.
2. Less expensive equipment: Methyl esters are noncorrosive and are produced at lower operating pressure and temperature conditions, hence they can be processed in carbon steel equipment; fatty acids are corrosive and require heavy-duty stainless steel equipment.
3. More concentrated glycerine by-product: Transesterification is a dry reaction and yields concentrated glycerine, while fat splitting produces glycerine water, which has more than 80 per cent water; thus recovery of the latter uses more energy.
4. Easier to distill–fractionate: Esters are more easily distilled because of their lower boiling points and are more heat stable than the corresponding fatty acids.
5. Superior to fatty acids as chemical intermediates in some applications: In the production of alkanolamides, esters can produce superamides, with more than 90 per cent purity against fatty acids, which can only produce amides with a purity of 65–70 per cent amides.
6. Easier to transport: Because of their chemical stability and noncorrosive property, esters are far easier to transport than fatty acids.

One major consideration is the need to recover and recycle the methanol. Since methanol is a toxic and an explosive material, use of explosion proof equipment and extra safety precautions are mandatory.

Chemistry of esterification and transesterification

Esterification

Esterification is the reaction of an acid with an alcohol in the presence of a catalyst to form an ester. The reaction is expressed by the general equation:

$$\underset{\text{Acid}}{RCOOH} + \underset{\text{Alcohol}}{R'OH} \underset{\text{Catalyst}}{\overset{\text{Acid}}{\rightleftharpoons}} \underset{\text{Ester}}{RCOOR'} + \underset{\text{Water}}{H_2O} \quad \text{... (1.2)}$$

Generally, acid catalysts like sulphuric acid are employed. Esterification is a reversible reaction. Thus, water must be removed to drive the reaction to the right and obtain a high ester yield.

Transesterification

Transesterification, on the other hand, is the displacement of the alcohol from an ester by another alcohol in a process similar to hydrolysis, except that an alcohol is used instead of water. This reaction, cleavage of an ester by an alcohol, is more specifically called alcoholysis and is represented by the general equation:

$$\underset{\text{Ester}}{RCOOR'} + \underset{\text{Alcohol}}{R''OH} \underset{\text{Catalyst}}{\overset{NaOCH_3}{\rightleftharpoons}} \underset{\text{Ester}}{RCOOR''} + \underset{\text{Alcohol}}{R'OH} \quad \text{... (1.3)}$$

In this case, a new ester is formed. Generally, alkaline catalysts are used with sodium methylate said to be the most effective, although sodium hydroxide can also be used.

Transesterification is an equilibrium reaction. To shift the reaction to the right, it is necessary to use a large excess of alcohol or to remove one of the products from the reaction mixture. The second option is preferred where feasible, since in this way, the reaction can be driven to completion.

Transesterification is a general term. More specifically, if methanol is used, the reaction is termed *methanolysis*. Methanol is generally used because it is cheap, but other alcohols can be used.

The reaction with fats and oils and using methanol is represented by the general equation:

$$\underset{\text{Fat or oil}}{\begin{array}{c} RCOOCH_2 \\ | \\ RCOOCH \\ | \\ RCOOCH_2 \end{array}} + \underset{\text{Methanol}}{3CH_3OH} \underset{\text{Catalyst}}{\overset{NaOCH_3}{\rightleftharpoons}} \underset{\text{Methyl ester}}{3RCOOCH_3} + \underset{\text{Glycerine}}{\begin{array}{c} CH_2OH \\ | \\ CHOH \\ | \\ CH_2OH \end{array}} \quad \text{... (1.4)}$$

Although the equation reflects the overall reaction, the reaction usually consists of a series of consecutive reversible steps. The sequence of steps is triglycerides to diglyceride to monoglycerides with 1 mole of methyl esters formed at each cleavage.

The stoichiometry of the reaction requires 3 moles of methanol for each mole of triglyceride. When up to 100 per cent excess methanol is used, the conversion rate is at its highest. The catalyst used are alkaline catalysts. Sodium methylate is commonly used although KOH or NaOH is also used.

The conversion rate is also strongly influenced by the reaction temperature. However, given enough time, the reaction will also proceed to near completion even at room temperature. Generally, the reaction is conducted close to the boiling point of methanol.

Impurities in the oil used also affect conversion rates. Under the same conditions, 67–84 per cent conversion using crude vegetable oils can be obtained, compared with 94–97 per cent when using refined oils. The free fatty acids in the original oils interfere with the catalyst. However, under conditions of high temperature and pressure this problem can be overcome.

Methods of manufacture

Methyl esters of fatty acids can be made either by esterification of the fatty acids or transesterification of the triglycerides using methanol.

Esterification processes

There are two general methods used for esterification: the batch process and the continuous process. Esterification can be done batchwise under pressure at a temperature of 200–250°C. Since it is an equilibrium reaction, the water is removed continuously to obtain a high ester yield.

Henkel has developed a continuous countercurrent esterification using a double-plate reaction column. The technology is based on the principle of an esterification reaction with the simultaneous absorption of the superheated methanol vapour and desorption of the methanol–water mixture.

Figure 1.9 shows the Henkel continuous process of fatty acid esterification. The reaction is carried out at a pressure of approximately 1000 kPa and a temperature of 240°C. One advantage of this process is the excess methanol can be kept significantly lower at 1.5:1 molar ratio of methanol: fatty acids versus the batch process at 3–4:1 molar ratio. The methyl ester, having undergone distillation, does not require further refining. The excess methanol is rectified and reused.

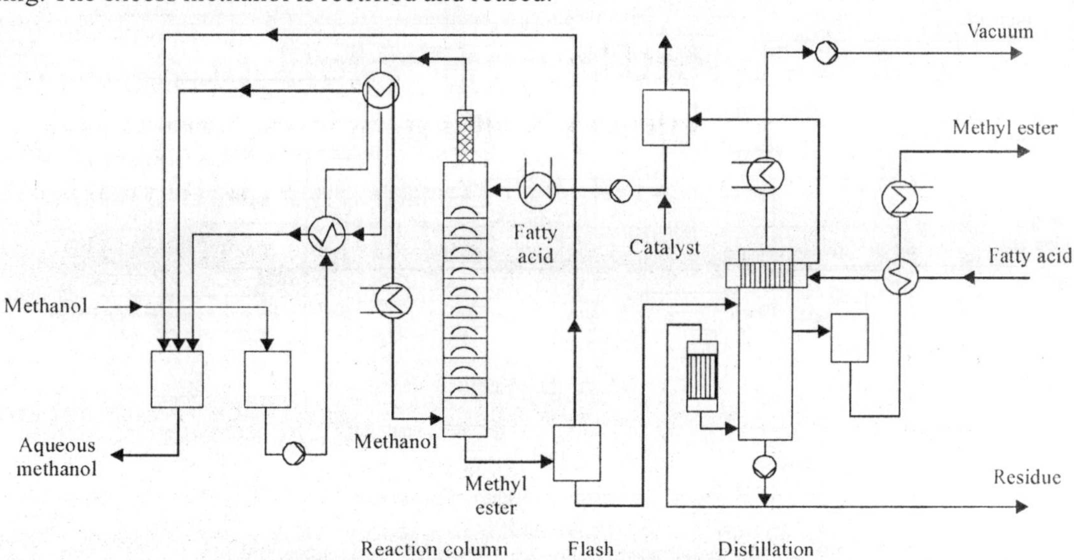

Fig. 1.9. Manufacture of methyl esters by esterification.

The continuous esterification process is superior to the batch process, in that the same high yield can be obtained in a much shorter dwell time and with substantially lower excess of methanol. The esterification process is a preferred method for the production of esters from specific fatty acids.

Transesterification process

Transesterification of fats and oils is the most commonly used process for the manufacture of methyl esters, except in cases where methyl esters of specific fatty acids are needed.

Triglycerides can readily be transesterified batchwise at atmospheric pressure and at a slightly elevated temperature of approximately 60–70°C with an excess of methanol and in the presence of an alkaline catalyst. The mild reaction conditions, however, requires the removal of free fatty acids from the oil by refining or preesterification before transesterification. This pretreatment is not required if the reaction is carried out under high pressure (9000 kPa) and high temperature (240°C). Under these conditions, simultaneous esterification and transesterification take place. The mixture at the end of the reaction is allowed to settle. The lower glycerine layer is drawn off while the upper methyl ester layer is washed to remove entrained glycerine and is then processed further. The excess methanol is recovered in the condenser, sent to a rectifying column for purification and recycled.

Continuous transesterification is well suited for large capacity requirements. Depending on the quality of the feedstock, the unit can be designed to operate at high pressure and high temperature or at atmospheric pressure and slightly elevated temperature.

Figure 1.10 shows a Henkel process flow diagram operated at 9000 kPa pressure and 240°C using unrefined oil as feedstock. Unrefined oil, methanol in excess and catalyst are metered and heated to 240°C before feeding into the reactor. The bulk of the excess methanol is flashed off as it leaves the reactor and is fed to a bubble tray column for purification. The recovered methanol is recycled into the system.

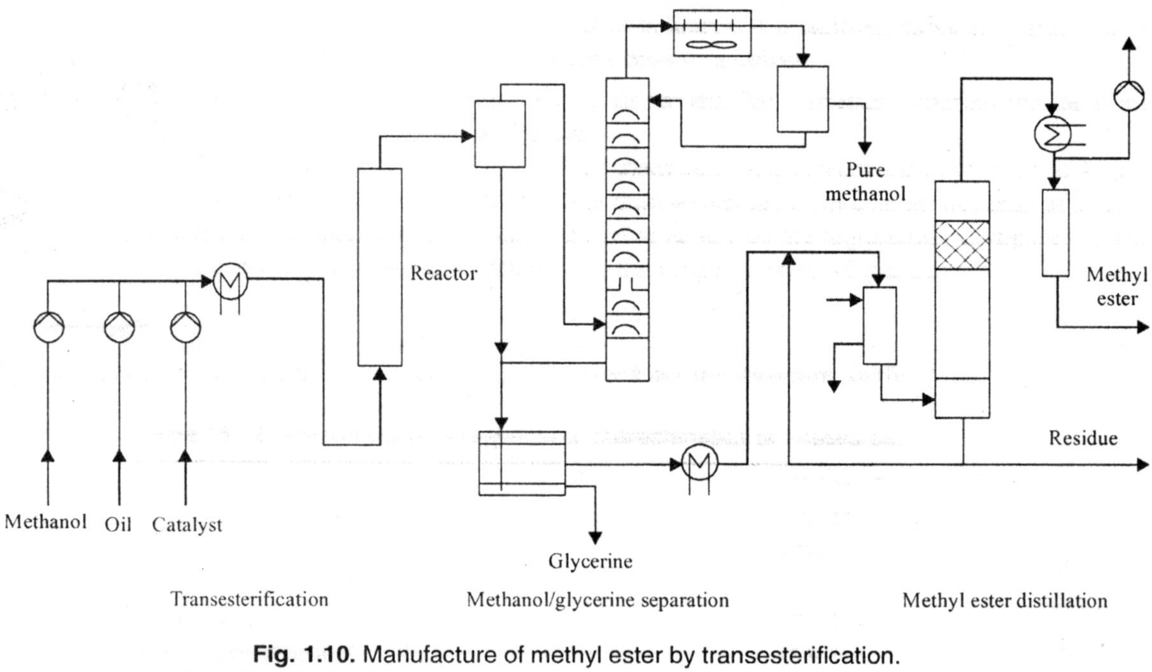

Fig. 1.10. Manufacture of methyl ester by transesterification.

The mixture from the reactor enters a separator where the glycerine in excess of 90 per cent concentration is removed. The methyl ester is subsequently fed to a distillation column for purification. Further fractionation into special cuts may follow if desired.

Figure 1.11 shows a Lurgi process flow diagram operating at normal pressure. The process requires the use of a degummed and deacidified feedstock. The refined vegetable oil and methanol are reacted in a two-stage mixer–settler arrangement in the presence of a catalyst. The glycerine produced in the reaction, dissolved in the surplus methanol, is recovered in the rectification column. Most of the entrained methanol and glycerine are recovered from the methyl ester in the countercurrent scrubber. The methyl ester can be further purified by distillation.

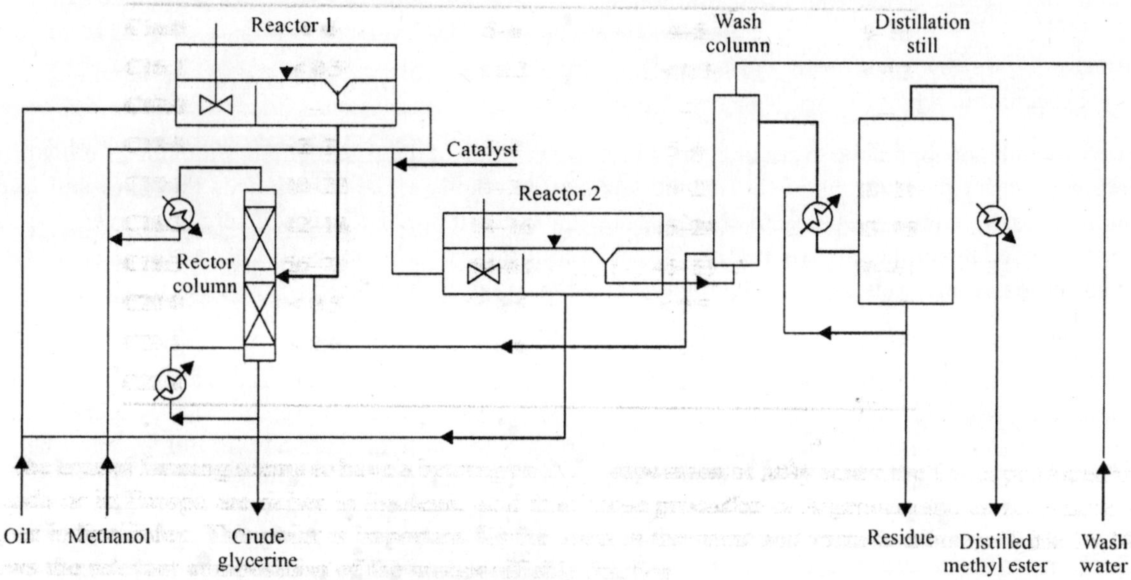

Fig. 1.11. Manufacture of methyl esters by transesterification.

Materials and utilities consumption per tonne of ester

Lurgi's technical data for plant sizes of 30–250 tpd are as follows:

Methanol	142 kg
Refined PKO or CNO	995 kg
Steam (400 and 1000 kPa)	250 kg
Heat energy (thermal oil)	420,000 kJ
Cooling water (20°C)	25 m^3
Electrical energy	10 kWh

Lion corporation of Japan has developed a continuous transesterification process using unrefined feedstock known as the ES process. The free fatty acid in the oil is preesterified by passing the feedstock and methanol through a packed column of a special catalyst resins after which transesterification is conducted through a two-stage reactor. A high conversion rate of more than 99 per cent is claimed.

Fatty Alcohols

Fatty alcohols make up one of the major basic oleochemicals enjoying a continuing growth rate. As a primary raw material for surfactants, its growth parallels increasing economic prosperity and improvement in the standard of living. Fatty alcohols are increasingly gaining favour as the surfactant raw materials of choice because of their biodegradability as well as availability from renewable resources.

Fatty alcohols can be produced from natural oils, or synthesised from petrochemicals.

Methods of manufacture

The different methods used in the production of fatty alcohols from various sources are:

1. Hydrolysis of wax esters using animal fats.
2. Sodium reduction process using fats and oils.
3. Ziegler process using ethylene.
4. Oxo process using olefins.
5. Catalytic hydrogenation of fatty acids and methyl esters from fats and oils.
6. Direct hydrogenation of fats and oils.

Although not applicable to coconut oil, a brief description of the first method is included for historical reference; brief descriptions of the second and sixth methods are also included. The third and fourth methods use raw materials of petrochemicals origin and will not be discussed here. However, it should be mentioned again that approximately 50 per cent of the world's supply of fatty alcohols is produced from these two methods. A more detailed discussion of the fifth method is given later.

Hydrolysis of wax esters

Fatty alcohols was first obtained from the hydrolysis of wax esters of animal origin, primarily spermaceti from sperm whale. With the worldwide ban on whale hunting, however, this source is no longer available. Spermaceti wax is split by heating with concentrated sodium hydroxide above 300°C. The liberated alcohol, approximately 35 per cent yield, is separated by vacuum distillation from the soap and water formed. The product consists essentially of cetyl, oceyl and arachidyl alcohols.

Sodium reduction process

Beauvault and *Blanc* discovered the sodium reduction process of manufacturing fatty alcohols from coconut esters. While the basic process is relatively simple, actual plant operations in handling the reactants and products are much more complex. The hazards of handling metallic sodium is an additional factor that caused this process to gradually lose out to the catalytic hydrogenation process.

Direct hydrogenation

A more recent process, developed and patented by Henkel KGaA, is the direct hydrogenation of natural oils or triglycerides. However, because of the severe conditions employed, a major portion of the valuable glycerine by-product is also hydrogenated to low valued propylene glycol and propyl alcohol. The loss of glycerine, the higher consumption of hydrogen gas and increased use of catalyst do not justify commercial production as yet.

Fatty alcohols from natural fats and oils

The manufacture of fatty alcohols from natural oils can be made either through the methyl ester route or fatty acid route. These two methods are both well established and are strongly competitive with each other. Commercial plants around the world use either route. Figure 1.12 shows the manufacturing routes of fatty alcohols from natural fats and oils.

Chemistry of hydrogenation

The hydrogenation of methyl esters and of fatty acids to form fatty alcohols is given by the following general equations.

$$\underset{\text{Methyl ester}}{RCOOCH_3} + \underset{\text{Hydrogen}}{2H_2} \underset{CuCr}{\overset{Catalyst}{\rightleftharpoons}} \underset{\text{Fatty alcohol}}{RCH_2OH} + \underset{\text{Methanol}}{CH_3OH} \quad ...(1.5)$$

$$\underset{\text{Fatty acid}}{RCOOH} + \underset{\text{Hydrogen}}{2H_2} \underset{CuCr}{\overset{Catalyst}{\rightleftharpoons}} \underset{\text{Fatty alcohol}}{RCH_2OH} + \underset{\text{Water}}{H_2O} \quad ...(1.6)$$

The direct hydrogenation of fatty acids is not used in a large commercial scale due to the need for higher reaction temperatures resulting in lower yields, and to the damaging effects on the catalyst. Traditionally, the fatty acids are first converted to esters before hydrogenation.

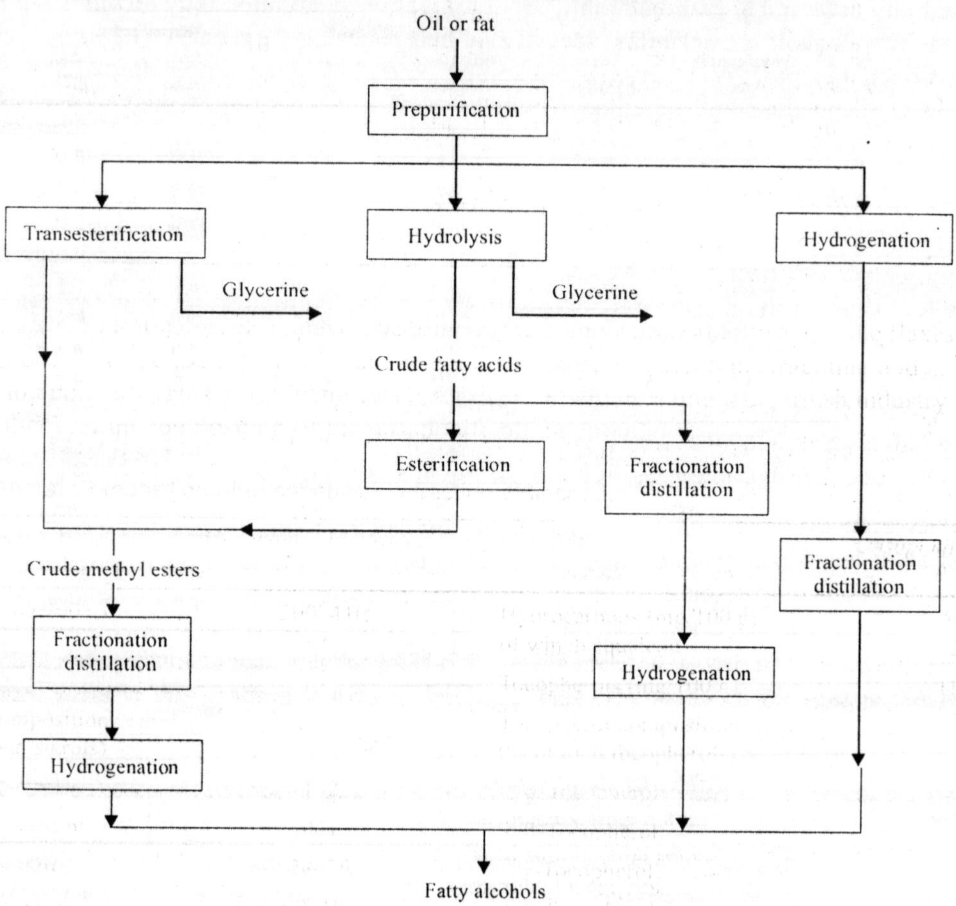

Fig. 1.12. Manufacturing routes of fatty alcohols from natural fats and oils.

Lurgi has managed to overcome this problem by simultaneous esterification of the fatty acids with the alcohol and the hydrogenation of the ester in the same reactor as given by the following equations:

$$\underset{\text{Fatty alcohol}}{RCH_2OH} + \underset{\text{Fatty acid}}{RCOOH} \overset{Catalyst}{\rightleftharpoons} \underset{\text{Ester}}{RCH_2COOR} + \underset{\text{Water}}{H_2O} \quad ...(1.7)$$

$$\underset{\text{Ester}}{RCH_2COOR} + \underset{\text{Hydrogen}}{2H_2} \xrightleftharpoons{\text{Catalyst}} \underset{\text{Fatty alcohol}}{2RCH_2OH} \quad \ldots (1.8)$$

The fatty acids are introduced into a large volume of circulating fatty alcohols, more than 250 times the volume of the acids, so that the esterification is effected rapidly without the damaging effect of the fatty acids on the catalyst.

High pressure hydrogenation processes

The fractionated methyl esters can be converted into fatty alcohols by the high pressure hydrogenation process in the presence of a catalyst. Usually, copper chromite catalyst is used. Copper chromite catalyst also convert any unsaturated carbon double bonds so that only saturated fatty alcohols are formed. If unsaturated fatty alcohols are desired, a special zinc-bearing catalysts is employed.

The hydrogenation process is carried out at 25,000–30,000 kP and a temperature of 250–300°C in a tubular column. Depending on the method by which the catalyst is employed, the hydrogenation can be conducted using the suspension process or the fixed bed process.

Suspension process

In the suspension process, the catalyst is slurried with a small amount of the methyl ester and then fed into the reactor together with the rest of the ester.

Figure 1.13 shows a simplified process flow diagram of the hydrogenation of methyl ester using the suspension process. The methyl ester and hydrogen gas are preheated separately. The copper chromite catalyst is preslurried in a small amount of methyl ester and is introduced simultaneously with the preheated methyl ester and hydrogen gas into the base of a tubular reactor (without packing). The catalyst is metered into the system in an amount of at least 2 per cent. Approximately 20 moles of hydrogen gas per mole of ester feed are used. The hydrogen gas, bubbling through, serves also to agitate the reactants.

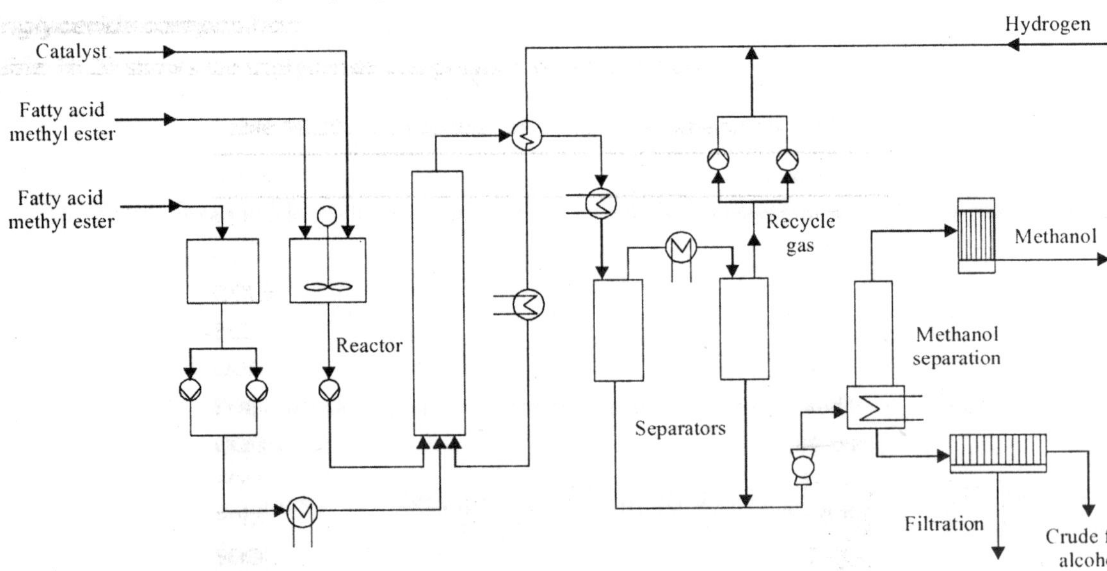

Fig. 1.13. High pressure hydrogenation of fatty acids methyl esters–suspension process.

The reaction is conducted at approximately 25,000–30,000 kPa and 250–300°C. Since hydrogenation is an exothermic reaction, care must be taken to control the reaction temperature to minimise side reactions leading to the formation of undesirable hydrocarbons. From the column, the reaction mixture is cooled, separating the hydrogen gas from the alcohol–methanol mixture. The hydrogen gas is recycled and the alcohol–methanol mixture goes to the methanol stripping unit where, at lower pressure, the methanol is stripped off, recovered and recycled to the esterification or transesterification section. The crude fatty alcohol is filtered to separate the catalyst. A major portion of the catalyst is recycled, so that the consumption average 0.5–0.7 per cent of the alcohol produced.

The filtered alcohol is subsequently treated with caustic soda to form soap with any unreacted esters. The alcohol is finally distilled to strip off any hydrocarbons formed, which amount to about 2–3 per cent. The soap remains in the still bottoms.

Fixed bed process

The fixed bed process, as differentiated from the suspension process, has the catalyst 'fixed' as a bed inside the reactor. The catalyst is either compact pelletised or supported on a carrier. Figure 1.14 is a simplified process flow diagram for the hydrogenation of methyl esters using the fixed bed process.

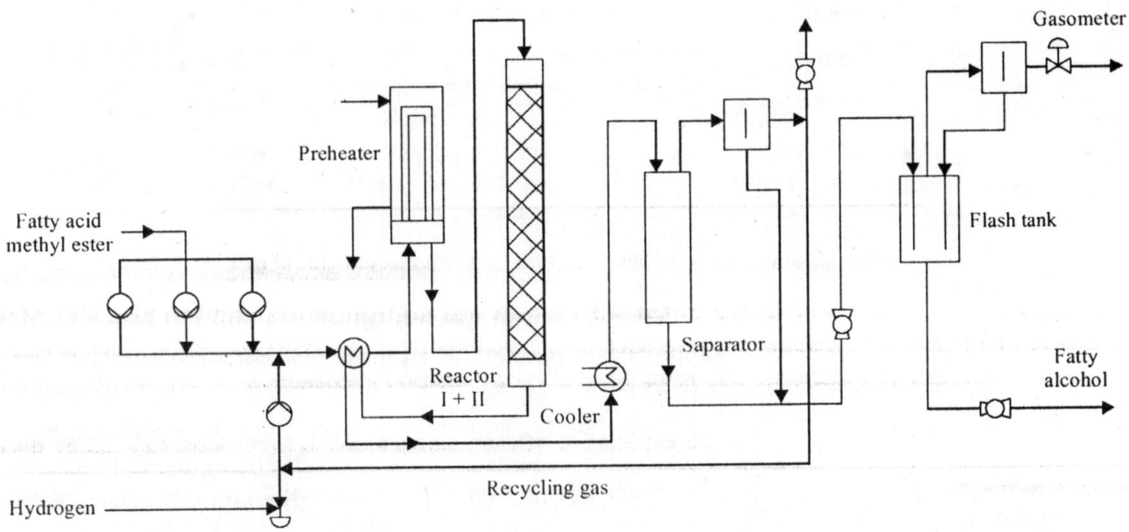

Fig. 1.14. High pressure hydrogenation of fatty acid methyl esters—fixed bed process.

The reaction is conducted in the vapour phase where a part of the organic feed is vapourised in an excess of hydrogen gas (20–25 moles) through a peak heater before passing through the fixed catalyst bed. The hydrogenation is carried out at 20,000–30,000 kPa and 200–250°C. The reaction mixture leaving the reactor is cooled and is separated into the gas and liquid phases. The gas phase, mostly excess hydrogen, is recycled. The liquid phase is expanded into a flash tank to strip off the methanol from the fatty alcohol.

The operating conditions are comparatively mild, so that the fatty alcohol produced does not require further processing. The overall yield is 99 per cent with hydrocarbons and unreacted ester not exceeding 1.0 per cent. Catalyst consumption is claimed to be below 0.3 per cent.

Suspension process versus the fixed bed process

The fixed bed process requires a somewhat higher capital investment, since it needs larger reaction vessels, gas recirculation pumps and pipe work due to the higher volume of hydrogen gas used. The suspension process, on the other hand requires additional equipment for the separation of catalyst, distillation of the crude fatty alcohol and rework of the unreacted methyl ester.

In terms of raw materials consumption, the fixed bed process has a higher yield and the catalyst consumption is only half as much. The fatty alcohol via the fixed bed process has a higher quality. However, the quality of the fatty alcohol via the suspension process can also be brought up to the same level by further distillation.

Lurgi fatty acid hydrogenation method

The Lurgi method, which is a suspension process, makes possible the direct hydrogenation of fatty acids to fatty alcohols overcoming the damaging effects of the fatty acids on the copper-bearing analyst. This is achieved by a two step reaction. The first reaction is the esterification of the fatty acids with the fatty alcohol to yield the ester and water. The second reaction is the hydrogenation of the ester formed to produce 2 moles of the alcohol. Both reactions occur simultaneously in the same reactor. The large volume of the fatty alcohols recirculated, more than 250 times the fatty acid feed, effectively dilutes the feed, thus providing for an optimum conditions for rapid and complete esterification. Figure 1.15 is a simplified process flow diagram of the Lurgi method of converting fatty acids to fatty alcohols.

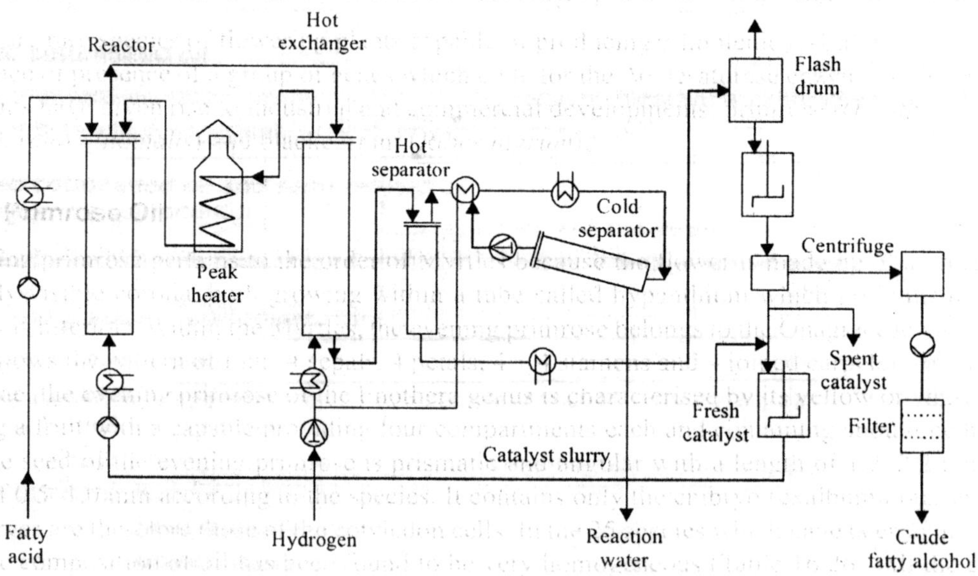

Fig. 1.15. Fatty alcohol synthesis via hydrogenation of fatty acids.

The hydrogenation takes place inside the high pressure reactor where the preheated materials—fatty acid feed, recirculated fatty alcohol with the slurried catalyst and hydrogen gas—are fed continuously. The reaction is carried out at approximately 30,000 kPa and 280°C. The heat of the product mixture leaving the reactor is recovered by the recirculating hydrogen gas through the heat exchanger, after which the products are separated through a two-stage cooling–expansion system.

The gas phase (essentially excess hydrogen gas, some low boiling alcohols and reaction water) is separated from the liquid alcohols in the hot separator. This mixture is cooled further in the cold separator, where the low boiling alcohols and reaction water are condensed and separated. The excess hydrogen gas is recycled into the system.

The liquid alcohols from the hot separator are pumped into a flash drum where the dissolved hydrogen is released and recycled with the make up hydrogen. The catalyst is separated from the crude fatty alcohol using a centrifugal separator. Part of the catalyst is replaced with fresh catalyst to maintain the activity and is recirculated with the fatty alcohol. The clear phase portion from the centrifugal separator is passed through a polishing filter to remove all traces of suspended solids. The resulting crude alcohol undergoes further distillation to remove hydrocarbons and may undergo fractionation if desired.

Materials and utilities consumption per tonne of crude fatty alcohol

Technical data for plant capacities of 50 + T/day are as follows:

Distilled coco fatty acid	1050–1100 kg
Steam (ca. 15 bar)	170 kg
Cooling water (20°C)	27 m^3
Electrical energy	130 kWh
Fuel gas	1.1×10^6 kJ
Catalyst (copper chromite)	5 kg
Hydrogen (0°C, 100 kPa, depending on hydrogen purity and fatty acid specification)	230–300 m^3
Boiler feed water	185 kg
Export steam (ca. 4 bar)	120 kg

Material balance

Figure 1.16 shows the overall material balance for the manufacture of 1000 MT of distilled fatty alcohol starting from distilled coco fatty acids.

Fractionation of fatty alcohols

Crude fatty alcohol resulting from the various hydrogenation processes described previously may undergo fractionation to produce different cuts of alcohol. Figure 1.17 shows a flow diagram for the fractionation of crude fatty alcohols.

Glycerine

Glycerine is a tribasic alcohol occurring in nature in the form of triglycerides, which are glyceryl esters of fatty acids. All fats and oils are made up of triglycerides. Coconut oil contains approximately 13.5 per cent glycerine, highest among commercial fats and oils. The glycerine contents of other fats and oils range from 9 to 12 per cent.

Glycerine is an important high value by-product obtained from the processing of fats and oils. It can also be synthesised from petrochemicals. However, with the increasing output of vegetable fats and oils worldwide, glycerine from this latter source plays a significant role in the world supply of this material.

Glycerine from fats and oils

Glycerine is obtained by cleavage of triglycerides using several methods:

1. Saponification of fats and oils using caustic soda to form soap and glycerine.

2. Splitting or hydrolysis of fats and oils in the presence of a catalyst to produce fatty acids and glycerine; the sweetwater formed contains 16–20 per cent glycerine.
3. Transesterification, which results in the displacement of the glycerine from triglycerides when fats and oils are reacted with methanol in the presence of a catalyst to produce methyl esters; in this process, glycerine concentration in excess of 90 per cent can be obtained since this is basically a dry reaction.

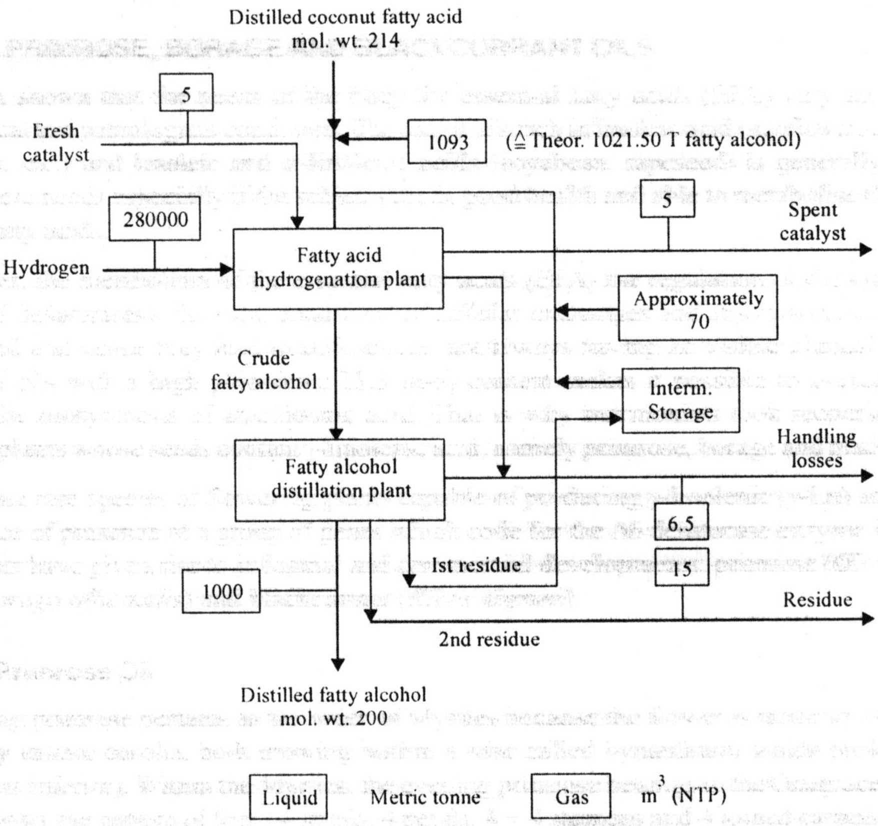

Fig. 1.16. Overall material balance for the manufacture of 1000 MT distilled fatty alcohol from distilled coconut fatty acids.

Both splitting or hydrolysis and transesterification were discussed earlier; therefore, only saponification will be covered briefly in this section.

Saponification

When fats and oils are saponified with caustic soda, the reaction is represented by the following equation:

$$\begin{array}{l} RCOOCH_2 \\ \quad | \\ RCOOCH \\ \quad | \\ RCOOCH_2 \\ \text{Fat or oil} \end{array} + \underset{\text{Caustic soda}}{3NaOH} \longrightarrow \underset{\text{Soap}}{3RCOONa} + \begin{array}{l} CH_2OH \\ \quad | \\ CHOH \\ \quad | \\ CH_2OH \\ \text{Glycerine} \end{array} \qquad \ldots (1.9)$$

Saponification yields soap and spent soap lye containing 8–12 per cent glycerine.

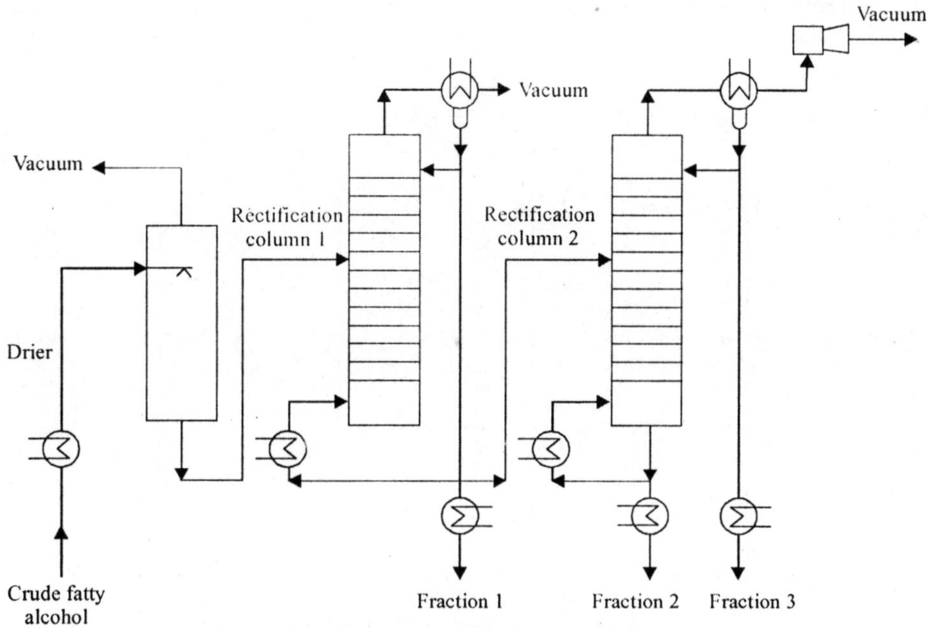

Fig. 1.17. Fractionation of fatty alcohol.

Fats and oils can be saponified via the full boiled process. The saponification process can be briefly described here. A refined blend of fats and oils is charged into a kettle and predetermined amount of soap lye with sufficient concentrations of caustic soda and salt is added. The mixture is boiled vigorously, using closed steam coils, until saponification is almost complete. The amount of caustic soda added is deliberately made less than the stoichiometric requirement, to ensure that the spent soap lye containing the glycerine has minimum alkalinity. The caustic soda in the spent soap lye is neutralised during subsequent treatment of the latter.

The salt in the lye used is necessary to keep the soap in the grained state and to facilitate the separation of the soap and the spent lye. The latter is drawn off after settling and transferred to the glycerine processing section. Meanwhile, the soap undergoes further boiling and countercurrent multiple washing to complete the saponification and recover as much glycerine as possible before being finished to neat soap.

Continuous soap boiling, which is widely practiced, makes use of multiple washing columns or centrifuges. The objective is to effect optimum glycerine recovery with less lye bulk.

Methods of purification

The glycerine obtained as a by-product from the three processes noted above contains impurities and must undergo further processing to purify and concentrate it. Commercially, there are two known processes in use:

1. The conventional methods of purification of the spent soap lye or sweetwater by acid–alum or ferric chloride treatment followed by evaporation, distillation, deodourisation and bleaching.
2. The ion exchange method of purification followed by evaporation and polishing.

Conventional method

The conventional method is widely popular. Figures 1.18 to 1.21 show some glycerine processing flow diagrams.

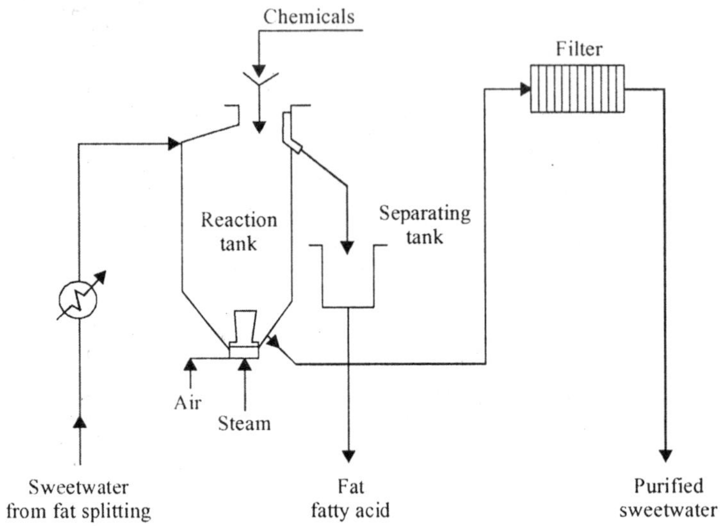

Fig. 1.18. Sweetwater purification (batch).

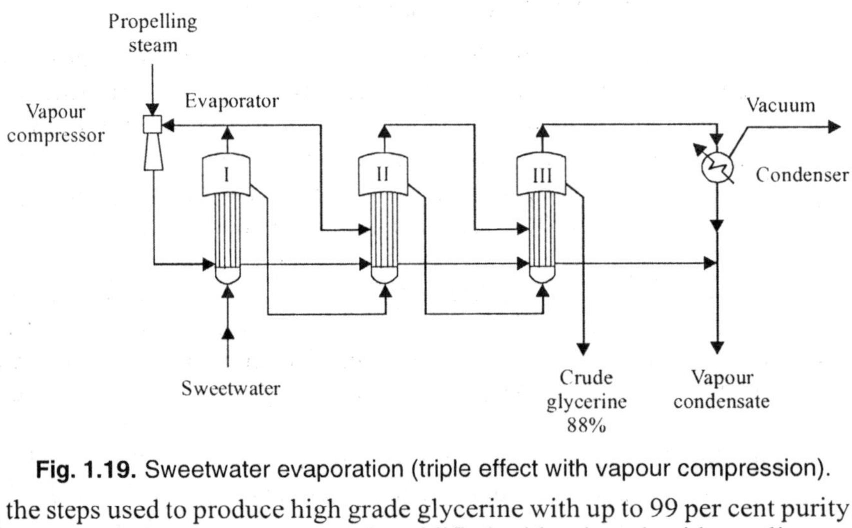

Fig. 1.19. Sweetwater evaporation (triple effect with vapour compression).

Basically, the steps used to produce high grade glycerine with up to 99 per cent purity are essentially the same. The spent soap lye, or sweetwater, is acidified with mineral acid to split any dissolved soap and release it as fatty acid, which is skimmed off. The pH is adjusted and alum or ferric chloride as a flocculant is then added to entrap the impurities, after which the mixture is filtered. The filtered lye is adjusted to a pH of 6.5 or higher before it is fed to the evaporator. The type of evaporators, either single or multiple effect, depend on the volume of materials to be processed. The crude glycerine obtained after evaporation has a concentration of 80–88 per cent and undergoes further processing. Salt, which separates out during the evaporation of treated glycerine soap lye, accumulates in a salt pot positioned underneath the evaporator. The salt is recovered and recycled to the soap-making section.

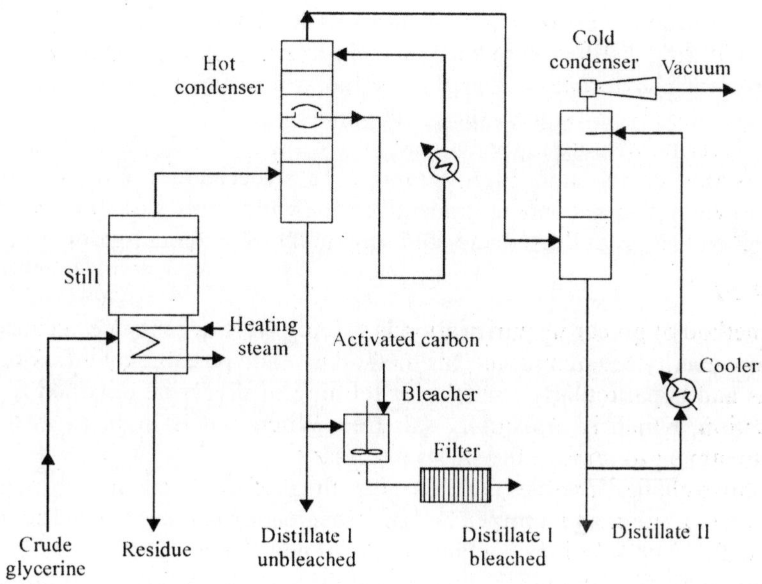

Fig. 1.20. Glycerine distillation and bleaching.

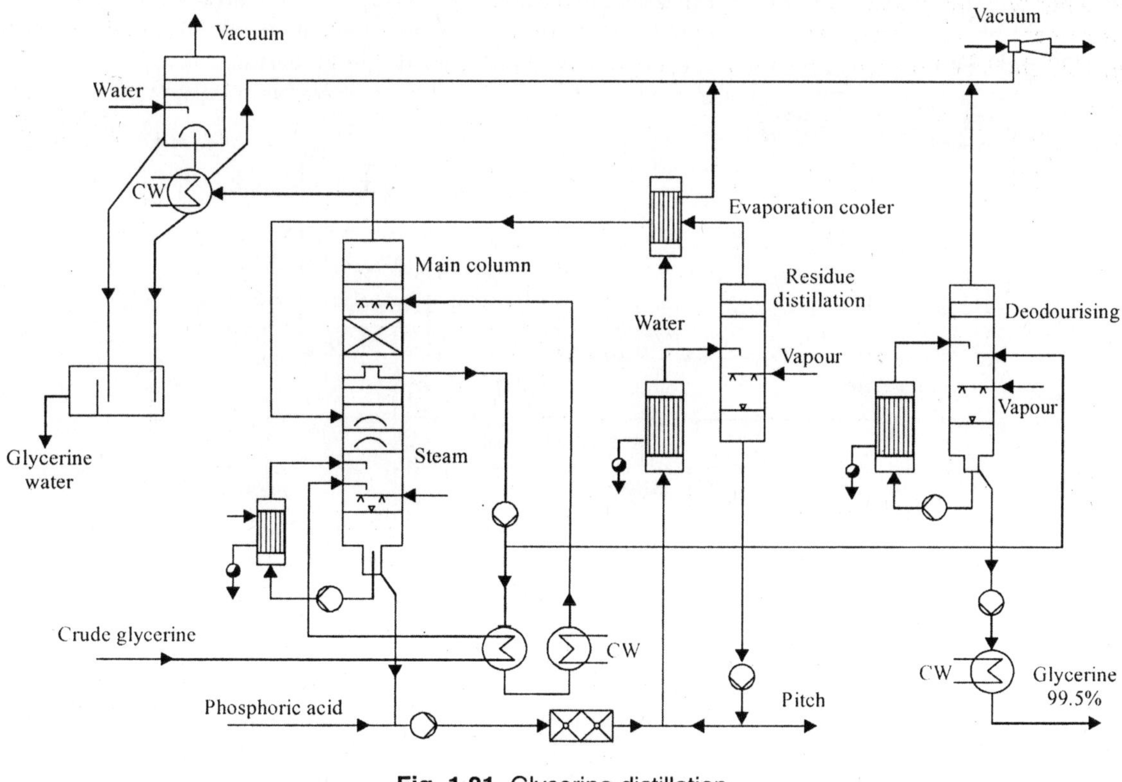

Fig. 1.21. Glycerine distillation.

The crude glycerine from the evaporator is distilled under a high vacuum of 600–1330 Pa absolute. Live steam is injected during distillation to keep the distilling temperature below 200°C. This is done to prevent the polymerisation and decomposition of glycerine, which start at 204°C. Controlled condensation of the vapour separates the glycerine from the water vapour.

The condensed glycerine with up to 99 per cent purity undergoes deodourisation by blowing steam into it in a deodourisation vessel under high vacuum. The glycerine is finally bleached with activated carbon and filtered to yield pharmacopoeia-grade glycerine with a purity of 99+ per cent. Lower quality distillate near the tail end run is collected and sold separately as technical-grade glycerine.

Ion exchange method

The ion exchange method of glycerine purification is gaining wide acceptance because of the simplicity of operations and low energy consumption. This method is made possible by the availability of suitable ion exchange resins and is particularly suitable for refining of glycerine obtained from fat splitting or from transesterification, which is practically salt free. When salt is high, as in the spent lye from saponification, pretreatment to remove the salt is necessary.

Purification by ion exchange involves the passage of the prefiltered material through successive beds of a strong cation, a weak anion and a mixed bed of strong cation-anion resins. The ion exchange units operate efficiently with dilute solutions containing 25–40 per cent glycerine.

Several resins are available commercially. Passage through the resin beds eliminates traces of free fatty acids, colour, odour bodies and other mineral impurities present. Subsequent concentration of the purified glycerine solution is through evaporation using multiple-effect evaporators to produce glycerine with a purity of more than 99 per cent. Final decolourisation by passage through an activated carbon bed or treatment with activated carbon followed by filtration yields pharmaceutical- or CP-grade glycerine. Fig. 1.22 shows a flow sheet for the ion exchange method of purifying glycerine.

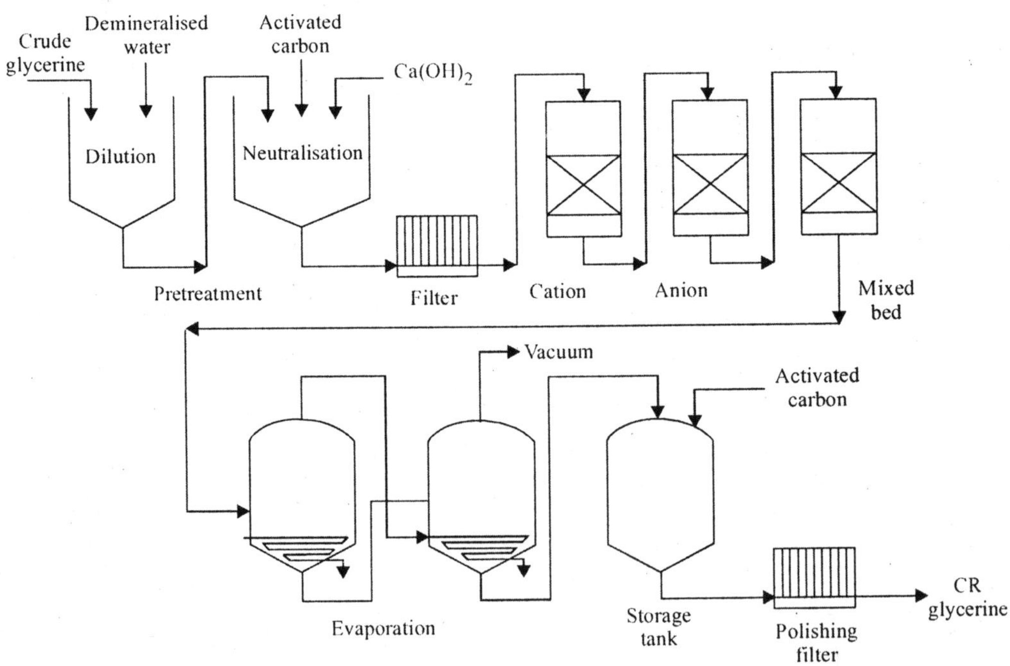

Fig. 1.22. Ion exchange method of glycerine purification.

Conventional method versus ion exchange method

The conventional method offers greater flexibility but uses more energy, considering that much water must be evaporated and that the glycerine is distilled at a higher temperature. The ion exchange method requires less energy but cannot be used for sweetwater containing high chlorides. Chlorides foul up the ion exchange resin.

Monoalkyl Phosphates

Monoalkyl phosphate and phosphate esters are special types of phosphorus-containing anionic surfactants that are of great industrial importance. They are used for flame proofing, as antistatic for textiles, for foam inhibition, as an extreme pressure (EP) lubricant additive, as a surfactant component for alkaline and as acid cleaners and for special cosmetic preparations. The commercially available phosphate ester products are complex mixtures of monoester and diester, free phosphoric acid and free nonionic.

Chemistry of the reaction

Phosphate esters are prepared by the partial esterification of fatty alcohol with phosphorous oxychloride followed by hydrolysis. This method yields a product that is a mixture of monoalkyl, dialkyl and trialkyl phosphates.

Another method of preparing alkyl phosphate is by reacting the fatty alcohol either with phosphorous pentoxide or with polyphosphoric acid. As shown by analysis of experimental results, the composition of alkyl acid phosphates differ significantly, depending on which of the two phosphating agents is used. The use of polyphosphoric acid leads to a higher monoesters: diester ratio than that obtained with phosphorus pentoxide. Typical composition of the reaction products are given in Table 1.10. The difference in composition of the phosphate product also results in different properties and industrial applications.

Table 1.10. Typical composition of the phosphate esters reaction products.

Route	*Components*	*Formula*	*Quantity (%)*
Polyphosphoric acid	Monoalkyl phosphates	$O{=}P(OH)_2$ OR	74
	Dialkyl phosphate	$O{=}P(OR)_2$ OH	2
	Orthophosphoric acid	$O{=}P(OH)_3$	10
	Free alcohol	ROH	14
Phosphorous pentoxide	Monoalkyl phosphate	$O{=}P(OH)_2$ OR	35
	Dialkyl phosphate	$O{=}P(OH)_2$ OH	42
	Orthophosphoric acid	$O{=}P(OH)_3$	1
	Free alcohol	ROH	22

Alkyl phosphates are more easily prepared by the addition of phosphorous pentoxide to fatty alcohols or fatty alcohol ethoxylates.

$$3\,ROH + P_2O_5 \longrightarrow RO{-}\overset{\overset{\displaystyle O}{\|}}{\underset{\underset{\displaystyle OH}{|}}{P}}{-}OH + RO{-}\overset{\overset{\displaystyle O}{\|}}{\underset{\underset{\displaystyle OR}{|}}{P}}{-}OH \quad ...(1.10)$$

The resulting product is a mixture of dialkyl and monoalkyl phosphate esters. These products also contain small amounts of condensed phosphates and phosphoric acid. Neutralisation of the acids with bases like alkali hydroxides, ammonia, or amines produces water soluble anionic surfactants and emulsifiers.

The presence of water in the fatty alcohols or in the pentoxide increases the percent of monoalkyl acid phosphate. The use of higher molecular weight fatty alcohols also increases the monoalkyl acid phosphate content. Products containing higher amounts of monoalkyl acid phosphate can also be derived from the reaction of 2 moles of fatty alcohols with 1 mole of phosphorous pentoxide. This reaction gives a high yielding dialkyl pyrophosphate product, which is readily hydrolysed to give an end product with a higher monoalkyl acid phosphate content than that obtained from the 3:1 ratio of alcohol to pentoxide.

$$2\,ROH + P_2O_5 \longrightarrow RO-\overset{\overset{O}{\|}}{\underset{\underset{OH}{|}}{P}}-O-\overset{\overset{O}{\|}}{\underset{\underset{OH}{|}}{P}}(OR) \xrightarrow{H_2O} 2ROPO(OH)_2 \quad \text{... (1.11)}$$

Method of manufacture

Phosphate esters are prepared at reaction temperatures from 80 to 120°C at essentially atmospheric pressure. It has been claimed that temperature may be maintained at 30–80°C. Lower temperature favours lower colour of the product, whereas at temperature near 100°C, noticeable decomposition occurs. The phosphorous pentoxide is added portion wise to the alcohol at a rate such that the pentoxide dissolves and reacts without lumping. Lumping makes complete dissolution difficult and the P_2O_5 unreactive. The reaction between the alcohol and the P_2O_5 is carried out in the liquid phase and is basically exothermic. No catalyst is required. The addition of a small amount of hypo-phosphorus acid or its salt gives a colourless, colour stable product.

Properties and uses

Monoalkyl phosphates inhibit the foam generation of other anionic and non-ionic surfactants. The composition of the phosphate esters greatly affects the functional properties of the product.

Most phosphate esters are pale yellow to amber in colour, sweet smelling, viscous liquid or pastes. Depending on the type of phosphate esters, they have many important properties, among which are wetting, detergency, solubilising, emulsification, surface tension reduction, foaming, dedusting, lubricity, antistatic, corrosion, inhibition, chelating, dispersing and antisoil redepositing agent. They have been proven to be extremely useful in a variety of applications where alternative anionic, nonionic, or amphoteric surfactants fail to give the desired functional properties.

Phosphate esters possess strong tolerance to electrolytes and alkalis. Their wetting properties are independent of the pH and therefore are used as surfactant components in both alkaline and acid cleaners. Their detergent properties are considered equal to nonionic surfactants and they have excellent temperature stability. In their acid form, phosphate esters are moderately strong acids and when neutralised to pH 5–8 have low order of toxicity. Phosphate esters are considered to be hydrotrope agents and are biodegradable.

Alkanolamides

Alkanolamides are condensation products of the reaction of a primary or a secondary alkanolamine with a fatty acid, a methyl ester, or a triglyceride such as coconut oil. The composition and functional properties of alkanolamides vary considerably, depending on the reactants employed and the reaction conditions. Such properties include superfatting, wetting, foam boosting, foam stabilising, thickening, lubricating, emolliency, skin protection, emulsifying and corrosion inhibition.

Alkanolamides are nonionic surface active agents that find application in a multitude of uses. There are four major groups of fatty alkanolamides: (i) monoethanolamides (MEA); (ii) diethanolamides (DEA); (iii) monoisopropanolamide (MIPA); and (iv) ethoxylated or PEG alkanolamides. Each group has its own specific function and uses in formulations.

The preparation of alkanolamides involves condensation reactions of fatty acids, triglyceride, esters, amides, anhydrides and halides with an alkanolamine. The reaction was carried out at 100–300°C at atmospheric pressure. An important improvement was made by *Meade*, who made use of an alkali metal alkoxide as a catalyst at 100°C at atmospheric and slightly above atmospheric pressure. Further refinement was made by *Tesoro*, who conducted the reaction at 55–75°C and a vacuum of 4–8 kPa. *Schurman* patented a continuous process for making alkanolamide, which makes use of a thin film reactor. It is claimed the short contact time in the reactor produces a high purity alkanolamide.

Chemistry of the reaction

Reaction with fatty acids

When alkanolamines react with fatty acids in the ratio of 1:1 at 140–160°C, *N*-alkylolamides are formed.

$$\underset{\text{MEA}}{HOCH_2CH_2NH_2} + \underset{\text{Fatty acid}}{RCOOH} \longrightarrow \underset{\substack{\text{Fatty acid}\\\text{monoethanolamide}}}{HOCH_2CH_2NHCOR} + \underset{\text{Water}}{H_2O} \quad \text{... (1.12)}$$

The reaction is a two-stage reaction with the initial formation of the alkanolamine soap followed by dehydration to form the alkanolamide. At the same time, significant quantities of amine esters and amide esters are formed by the side reaction. If a dialkanolamine is used as the starting material, smaller amounts of amine diester and amide diester are formed as well as some morpholine and piperazine derivatives.

Reaction of dialkanolamines with fatty acids in a 2:1 ratio at 140–160°C gives a second major type of alkanolamide. This product, in contrast to 1:1 alkanolamide, contains a considerable amount of unreacted dialkanolamine, which accounts for the aqueous solubility of the product.

Reaction with esters

The reaction of an ester such as cocomethyl ester with an alkanolamine at 110°C produces an *N*-alkylolamide.

$$\underset{\text{Ethanolamine}}{HOCH_2CH_2NH_2} + \underset{\text{Ester}}{RCOOR''} \longrightarrow \underset{\text{Ethanolamide}}{R'CONHCH_2CH_2OH} + \underset{\text{Alcohol}}{R''OH} \quad \text{... (1.13)}$$

The reaction of methyl esters of fatty acids with alkanolamines is used commercially to prepare high purity 1:1 alkanolamide. This product, in contrast to the 1:1 alkanolamides prepared from the parent fatty acids, contains only a small quantity of the by-products or unreacted starting material.

There are two competing reactions: amide formation and transesterification. The most desirable reaction is one that gives the highest amide ester by-product. For the yield of the amide to exceed 90 per cent, reduced pressure, lower temperature and a slightly higher catalyst concentration were found to be necessary. Plant operations employ 0.3–0.5 per cent catalyst, a temperature of 70–75°C and a vacuum of 4 kPa or less.

Another variation is the use of triglycerides such as coconut oil as the starting material. This involves the reaction of 1 mole of triglycerides with 6 moles of the alkanolamine. The reaction can be catalysed by sodium methylated at 60°C. Glycerine is a by-product that usually remains with the final product. The typical composition of the product consists of 80–85 per cent alkanolamide, 5–7 per cent or less free amine and traces of fatty acid soap and other by-products.

The reaction is relatively mild and proceeds readily with the use of 0.2–0.3 per cent sodium methylate catalyst at an operating temperature of 70–75°C and at atmospheric pressure.

Types of alkanolamides

There are at least four types of alkanolamides commercially available. The first is formed by reacting monoalkanolamine or dialkanolamine with fatty acids in a 1:1 ratio at elevated temperatures of 140–160°C. Significant quantities of the by-products—amines esters and amide esters—are also formed.

A second major type is the Kritchevsky type made from alkanolamine and fatty acids in the ratio of 2:1. Such product contains 60–70 per cent alkanolamide plus by-products. If diethanolamine is used in addition to the mentioned by-products, some morpholine and piperazine derivatives are also obtained. The excess alkanolamine renders the product water soluble. Both are of commercial importance as detergent and detergent additives.

A third type is the so-called super amide prepared from the methyl esters in a 1:1 mole ratio. They are generally products that have an alkanolamide content in excess of 90 per cent. Some of the same by-products as in the first type are formed but in much smaller quantities. Because of the relatively small amounts of free alkanolamine, superamides have poor water solubility. They are used in conjunction with a small amount of anionic or nonionic surfactant, which acts as a solubiliser to form a clear viscous solution.

The fourth variant is the product of the direct reaction of a triglyceride with an alkanolamine. The glycerine by-product is retained with the other by-products. It is also used as a detergent additive in light-duty liquid detergents.

Other applications of alkanolamides include dry cleaning soaps, fuel oil additives, rust inhibitors, textile scouring and dry leveling agents.

Method of manufacture

Depending on which of the four types of alkanolamide is to be produced, the starting material could be either fatty acid, methyl esters, or triglyceride (coconut oil) and monoethanolamines or diethanolamine. Usually a batch system of manufacturing is employed, as shown in Fig. 1.23.

Alkanolamide from coconut oil

A 2:1 cocodiethanolamide (CDEA) can be produced using 6 moles of diethanolamine and 1 mole of refined and bleached coconut oil. The materials are charged with the reactor and a small amount of catalyst (0.25–0.3 per cent sodium methylate or sodium hydroxide) is added. The temperature of the batch is raised to 70–75°C at normal pressure. After 90 minutes, the reaction is completed. For a 10-T batch, total cycle time from charging the materials, heating them up, allowing the reaction to proceed to completion and transferring the finished product takes at least 4 hr.

If fatty acid is the starting material, a different operating parameter is used to drive off the water formed during the reaction. If superamide is to be produced, cocomethyl esters and diethanolamine can be used as the starting materials in a mole ratio of 1:1. These materials are charged to the reactor with 0.3–0.5 per cent sodium methylate as the catalyst. The reaction is carried at around 100°C and a vacuum of 4–5.3 kPa for a period of 90–120 minutes. The reaction temperature can be lowered to 70–75°C by employing a vacuum of less than 4 kPa. The reaction time takes longer, to enable the maximum vapourisation of the methanol by-product. The methanol is rectified and recycled for use in the transesterification of the coconut oil to produce the methyl ester.

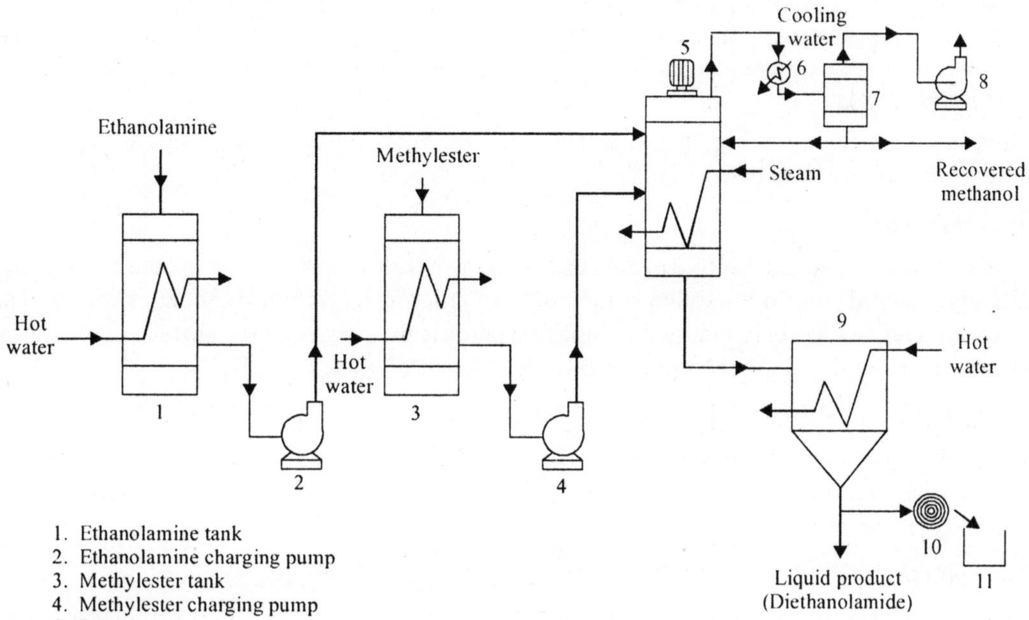

1. Ethanolamine tank
2. Ethanolamine charging pump
3. Methylester tank
4. Methylester charging pump
5. Reactor
6. Condenser
7. Vacuum drum
8. Vacuum pump
9. Surge tank
10. Cooled flaking drum
11. Flaked product (monoethanolaminde)

Fig. 1.23. Manufacture of normal and super alkylolamides.

The final product in the liquid form is transferred into the storage tank. Postreaction can be done by keeping the product in another vessel, preferably at slightly elevated temperature for a few days. This allows further conversion of the ester by-product to amide, thus further increasing the amide concentration.

If monoethanolamine is used as the starting material, the monoethanolamide product (CMEA) is a waxy solid. Accordingly, the finished product after the reaction is passed through a cooled flaking drum to convert it into flakes. It is subsequently packed in bags.

Surfactants

Surfactants or surface active agents make up a special class of chemicals used in practically all industries. Surfactants may be of petrochemical or natural oils origin. This section discusses only those of vegetable oil origin, surfactants, specifically those from coconut oil.

Surfactants are chemical compounds that possess great surface activity. They act so diversely because of the unbalanced molecular structure. A surfactant molecule may be visualised as a tadpole or a mini-racquet. The 'head' is the hydrophilic (water-loving), strongly polar portion and the 'tail' is the hydrophobic (oil-loving) nonpolar portion. The head can be an anion, a cation, or nonion. The tail is a linear or branched hydrocarbon chain. It is this characteristic configuration that makes surfactants perform such diverse function in industry.

Surfactants find broad application in practically all industries as, for example, the main ingredients of detergents and cleaners, foaming agents and emulsifiers in cosmetics and pharmaceuticals, emulsifiers for paints, scouring agents for textiles, flotation agents for the mining industry and emulsifiers and sanitising agents for the food industry.

Types of surfactants

Surfactants are classified into four categories: anionic, cationic, nonionic and amphoteric, each with its own molecular structure and behaviour.

Anionic surfactants

Anionic surfactants are surface active agents in which the hydrophobic portion is connected to an anion or negatively charged ion. In an aqueous medium, an anionic surfactant dissociates into a positively charged cation and a negatively charged anion. The latter is the carrier of the surface active properties. Typical examples are the alcohol sulphates and the ester sulphonates.

Cationic surfactants

Like the anionic surfactants, cationic surfactants also dissociate in an aqueous medium. However, the head (hydrophilic portion) is a cation, which is the carrier of the surface active properties. Examples are the quaternary ammonium compounds.

Nonionic surfactants

The nonionic surfactants do not dissociate in an aqueous medium. Their solubility is provided by the polar group, such as a polyglycol ether or a polyol. The most important type of fatty nonionic surfactants are the polyglycol ethers of fatty alcohols, fatty acids, amines and amides.

Amphoteric surfactants

Amphoteric surfactants in aqueous solution contain both positive and negative charges in the same molecule. Thus, a hydrophobic fatty chain is attached to a hydrophilic group that contains both positive and negative charges. Its behaviour depends on the condition of the medium or its pH value. Examples of this type are the alkyl betaines.

Coco-based surfactants

The most important coconut oil–based surfactants are fatty alcohols sulphate, fatty alcohol ether sulphate and fatty alcohol polyglycol ether. Two relatively new coco-based surfactants are fatty acid methyl esters sulphonate and alkyl polyglycoside, which is produced from fatty alcohol and starch or sugar, both renewable materials.

Fatty alcohol sulphates

Sodium salt of fatty alcohols sulphate of the C12–C18 range is one of the more common active ingredients of choice in the formulation of laundry detergent products, due to its good detergency, wetting and foaming properties and biodegradability. The C12–C14 fatty alcohols sulphate, known in the trade as sodium lauryl sulphate (SLS), provides optimum foam and is used extensively as the foaming agent in most toothpastes. The ammonium and alkanolamine salts of the C12–C14 and C12–C16 alcohol sulphates possess high solubility and are well suited for liquid shampoos and bubble bath preparations. In combination with other surfactants and alkanolamides, they deliver desirable qualities, such as fine structured foam, low irritancy and smooth hair feel.

Fatty alcohol polyglycol ethers

The fatty alcohol polyglycol ethers belong to the most important class of nonionic surfactants. They are characterised by having excellent wetting properties, having low foam and being effective even at low temperature. They are produced by the reaction of fatty alcohol with ethylene oxide (EO) or propylene oxide (PO), using alkaline catalyst such as sodium methylate or potassium hydroxide. The degree of ethoxylation can vary, depending on the chain length of the fatty alcohols and the purpose for which it will be used. The C12–C14 and C12–C16 alcohols used as raw material for the ether sulphates can have 2–3 moles of EO. The C12–C18 alcohol, with 7–10 moles of EO, are used primarily as wetting agents. Higher saturated alcohols, with up to 30 moles of EO, are used as emulsifier for waxes and ointments bases. Fatty alcohol polyglycol ethers have replaced the alkylphenol polyglycol ether in most applications. In the United States, however, the latter is still in use for institutional and industrial products because of cost reasons.

Fatty alcohol ether sulphates

Probably the most important derivatives of fatty alcohol in the C12–C14 and C12–C16 ranges are the fatty alcohol ether sulphates. They are produced by the sulphation of the fatty alcohol, containing 2–3 moles of ethylene oxide, with sulphur trioxide or chlorosulphonic acid and subsequently neutralised with caustic soda, ammonia, or an alkanolamine. The ether sulphates possess superior properties over the fatty alcohol sulphates. They have unlimited water solubility, are unaffected by water hardness and possess superior skin compatibility. Accordingly, they are used in liquid shampoos and bath preparations. One characteristics of this material is its ability to increase its viscosity by the addition of an electrolyte such as salt.

Chemistry of the reaction

Fatty alcohol sulphate of commerce is actually a neutralised salt primarily as sodium coco fatty alcohol sulphate. It is produced by the direct reaction of fatty alcohol with sulphur trioxide and subsequently neutralised with caustic soda, according to the following reactions:

$$\underset{\text{Fatty alcohol}}{RCH_2OH} + \underset{\text{Sulphur trioixde}}{SO_3} \longrightarrow \underset{\text{Fatty alcohol sulphuric acid}}{RCH_2OSO_3H} \quad \text{... (1.14)}$$

$$\underset{\text{Fatty alcohol sulphuric acid}}{RCH_2OSO_3H} + \underset{\text{Caustic soda}}{NaOH} \longrightarrow \underset{\text{Sodium fatty alcohol sulphate}}{RCH_2OSO_3Na} + \underset{\text{Water}}{H_2O} \quad \text{... (1.15)}$$

The intermediate product is a half ester of sulphuric acid. It is quite unstable and must be neutralised immediately. The final product contains approximately 1.5 per cent of sodium sulphate, 1.0–1.5 per cent unreacted alcohol and less than 0.5 per cent of free alkali.

In sulphation, other sulphating agents may be used, such as oleum or chlorosulphonic acid, as the source of SO_3. However, oleum generates a large amount of waste sulphuric acid. It is also not recommended for the sulphation of fatty alcohol ethoxylate.

Chlorosulphonic acid is an excellent sulphating agent, but it generates unwanted hydrochloric acid as a by-product. Because of the corrosive nature of the hydrochloric acid vapour, special equipment must be used.

In neutralisation, other alkali can be used such as ammonia, monoethanolamine, or triethanoalmine. The same chemical reaction and equipment applies to the production of fatty alcohols ether sulphate, except that the raw material is fatty alcohol ethoxylate.

Method of manufacture

The latest technology in the manufacture of fatty alcohol sulphate or fatty alcohol ether sulphate is the direct use of SO_3 gas as the sulphating agent. This process not only produces a high purity product but is also much more economical and generates minimum waste.

The heart of the process is the reactor. Reactors can be batch, cascade, or falling-film type. Industry favours the falling-film type because the reaction is much more controllable and highly efficient. Falling-film reactors can be multitude, monotube, or annular (Fig. 1.24).

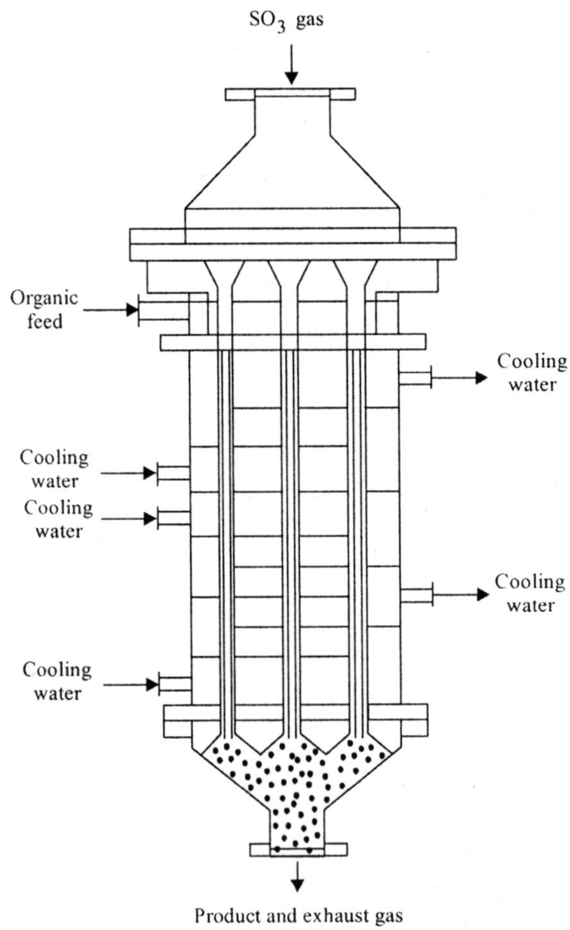

Fig. 1.24. Falling film reactor.

The manufacture of fatty alcohol sulphate or ether sulphate can be divided into five stages: process air preparation, sulphate trioxide generation, sulphation, neutralisation and exhaust gas treatment.

Process air preparation

Process air must be totally dry, preferably with a dewpoint below – 50°C. The presence of moisture not only makes it corrosive (due to its reaction with SO_3 gas) but also increases the colour of the product.

Briefly, air is introduced by a large compressor to a cooling system, where it is cooled down to 3–5°C and most of the moisture is condensed. From there, the air is passed through a dehumidifier such as silica gel where the last traces of moisture are retained. Figure 1.25 shows Ballestra's air drying system.

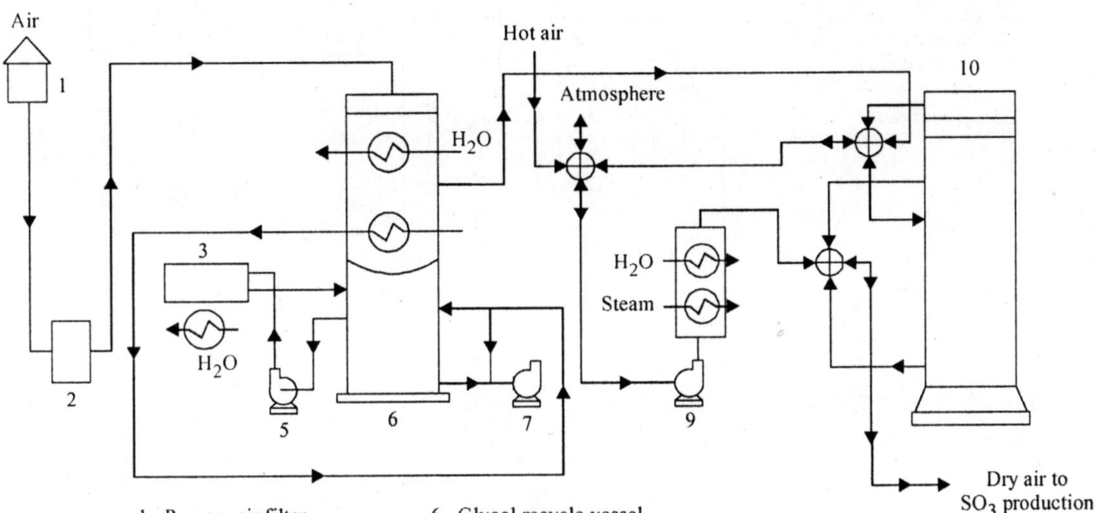

1. Process air filter
2. Process air compressor
3. Chiller
4. Air cooling group
5. Circulating pump
6. Glycol recycle vessel
7. Circulating pump
8. Silicagel regeneration unit
9. Regeneration air fan
10. Silicagel air dehumidifier

Fig. 1.25. Ballestra's air drying system.

Sulphur trioxide generation

For sulphur trioxide generation, high purity sulphur (99.5 per cent pure) is melted in a tank and then kept at 145–150°C to maintain its viscosity at the minimum and constant value. The molten sulphur is introduced into the sulphur burner by a special metering pump and then burned to SO_2, using the dried air. The diluted SO_2 gas (6–7 per cent) leaves the burner at around 650°C and is cooled to 430°C before feeding to the converter.

The catalytic converter with three to four beds of vanadium pentoxide catalyst, converts SO_2 to SO_3 with up to 98 per cent conversion efficiency. The SO_3 gas is cooled to below 60°C, diluted to 4 per cent by volume and is passed through a mist eliminator to remove traces of oleum formed before it is fed to the reactor. Fig. 1.26 shows a typical system for the generation of SO_3 gas.

Sulphation

The sulphation is carried out in a multitude film reactor, especially designed to control accurately the mole ratios between the SO_3 and the organic feed in each tube. The materials enter the top section and flow cocurrently downward inside the tube. Since the reaction is almost instantaneous and exothermic, cooling water at controlled flow is introduced into the jacket to maintain the reaction temperature at 45–50°C maximum. A reaction yield of 97 per cent can easily be achieved (Fig. 1.24).

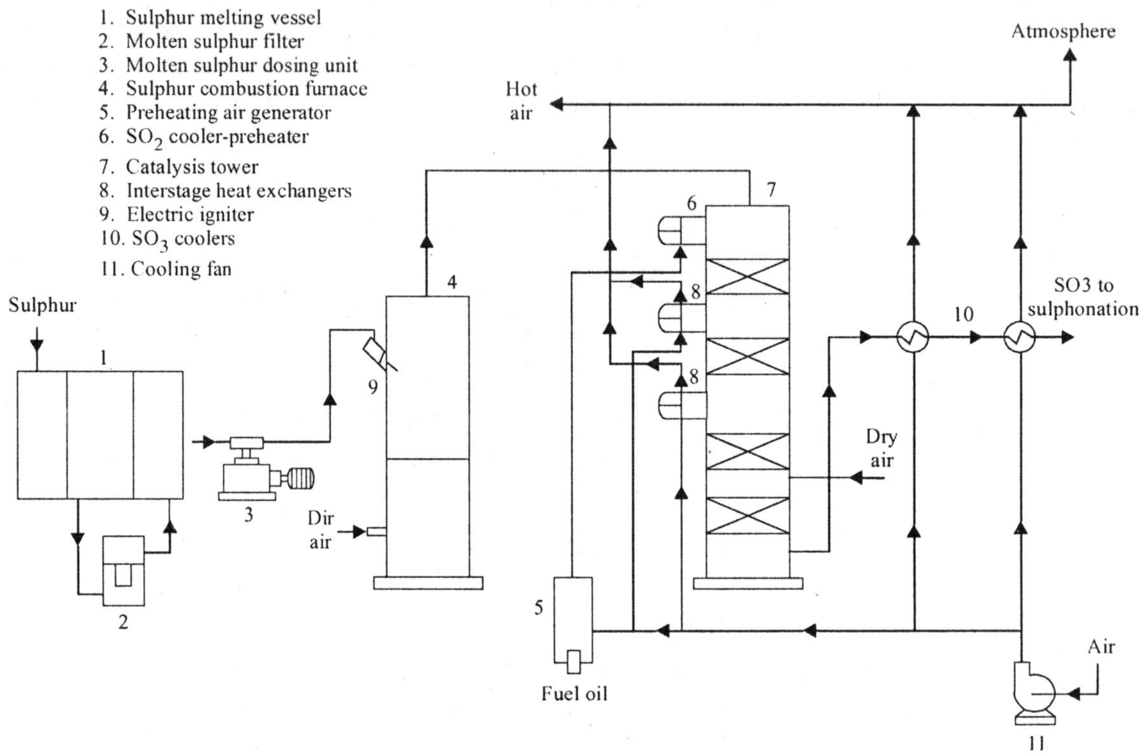

Fig. 1.26. Ballestra's SO_2–SO_3 production.

Neutralisation

The intermediate product from the reactor must be neutralised immediately, as hydrolysis may occur and badly affect the process and the product quality. This is achieved in a double-step continuous neutralisation unit. The multibladed mixer gives exceptionally homogeneous mixing.

It is important that the neutralised paste be kept slightly alkaline to ensure processability and product stability. An average concentration of 72 per cent active matter is readily achievable. A much higher concentration is not recommended, as processing difficulties may be encountered. If a dried product is desired, the paste can be further processed through a wiped film evaporator. Fig. 1.27 shows Ballestra's double-step neutralisation.

Exhaust gas treatment

The gas effluent must be treated to comply with environmental regulations. The exhaust gas from the reactor contains traces of organic matter and unreacted SO_3 and SO_2 gases. The first two impurities are removed by the electrostatic precipitator. The remaining SO_2 gas is removed by the reaction with caustic soda solution flowing countercurrently inside the scrubbing column. The concentration of the residual

SO_2 in the exhaust gas vented into atmosphere is kept at 5 ppm maximum, well below the allowable standards (Fig. 1.28).

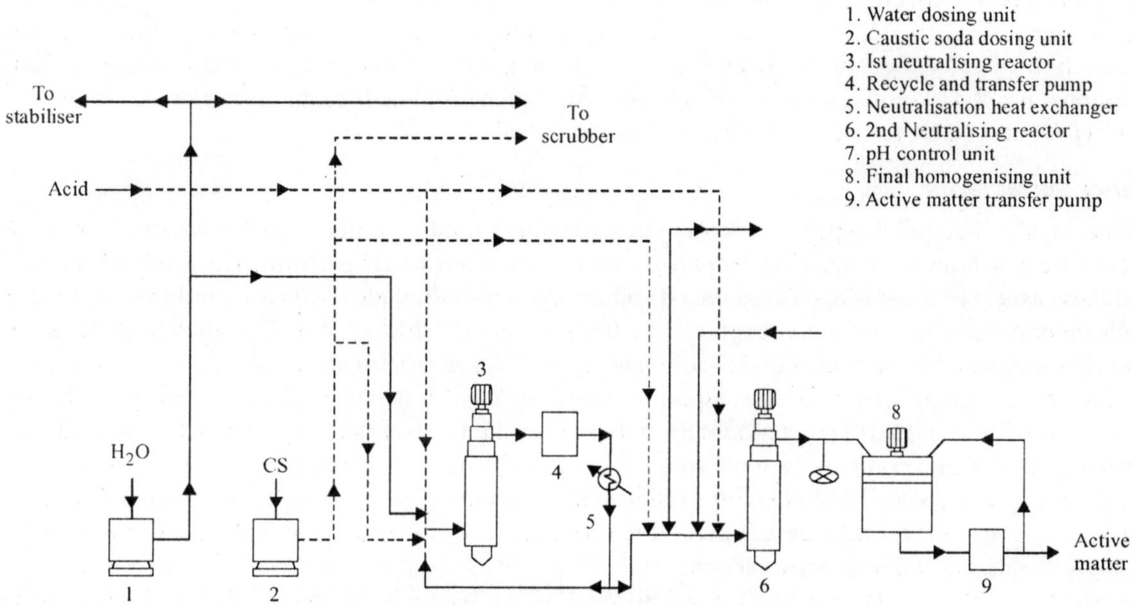

Fig. 1.27. Ballestra's double-step neutralisation.

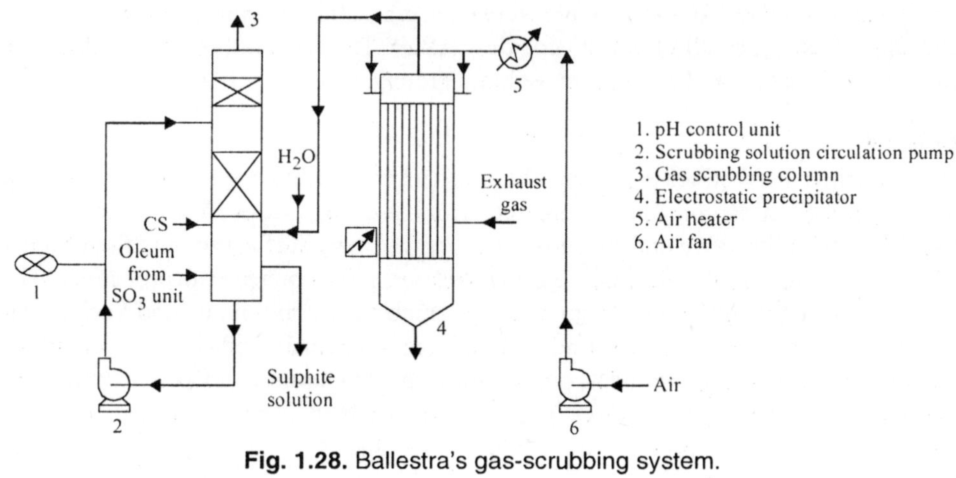

Fig. 1.28. Ballestra's gas-scrubbing system.

Sulphonation of fatty acid methyl esters

Fatty acid methyl esters sulphonate may be considered a material with a great potential. Its superior sequestering effect on water hardness gives it an advantages over fatty alcohol sulphate. The sulphonation of fatty acid methyl ester is different from that of fatty alcohol. The reaction mechanism involves two steps. In the first reaction, the SO_3 gas reacts quickly to form sulphoanhydride. In the second step (which takes 40–90 minute), the sulphoanhydride becomes the sulphonating agent, reacting with the still-unreacted ester.

The reaction requires an excess of SO_3 in the range of 20–30 mol per cent to get initiated. With such an excess, the formation of disalt during neutralisation is unavoidable. This is somehow minimised by subsequent reesterification after the second step.

The neutralisation step can be carried out in much the same way as in the neutralisation step in the production of fatty alcohol sulphate. Because of the nature of the reaction and conditions during sulphonation, a dark-coloured product is produced, which requires bleaching. Postreaction treatment with H_2O_2 and NaOCl yields a product with an acceptable colour.

Other surfactants

Three highly successful types of coco-based surfactants used in toilet soap formulations in United States are worth mentioning. They impart superior hard water lather performance while retaining the mildness aspect of toilet soap. These are (i) sodium cocoyl isethionate (SCI) used in Lever's Dove and their other brands; (ii) sodium cocoglyceryl ether sulphate (LGES) used in Procter & Gamble's Zest; and (iii) sodium coco-monoglycerides sulphate (CMGS) used in Colgate's Vel Bar.

Other coco-based surfactants are sulphosuccinates formed by the reaction of coco fatty alcohol with maleic anhydride and further reaction with sodium sulphite or bisulphite. This product possesses good foaming properties, compatible with soap and is a good lime dispersant. It is used in toilet soap formulation, shampoos, hand cleaning pastes and for scouring raw wool. Its ether variant with 2–4 moles ethylene oxide forms intense, finely structured foam and is used in combination with ether sulphate in baby shampoos and bath preparations.

Alkyl phosphates, also coco based, are formed by the reaction of fatty alcohol with phosphorous pentoxide. The product is a mixture of monoesters and diesters. Its sodium salt (MAP) is reportedly mild and is used in facial wash, Kao's Biore. Other uses are as antistatics in textile auxiliaries, corrosion inhibitors, surfactant additives for extreme pressure (EP) lubricants and surfactant components in alkali and acid cleaners. Coco-based alkyl polyglycosides (APG). The alkyl polyglycoside fully satisfies the demand for a mild surfactant and is completely biodegradable.

Tertiary Amines

Fatty amines and their derivatives represent the most important nitrogen compounds of fatty acids. They possess great ionisation constants compared with other alkyl derivatives of ammonia. They are cationic, basic, biologically active and strongly adsorbed on many surfaces due to their high adsorption potential. They are indispensable in many surface-related physicochemical processes. They are the starting materials for the manufacture of quaternary ammonium compounds and various cationic and amphoteric substances. These derivatives find wide application in many industries as biocides, sanitizing agents for algae control in water treatment, ore flotation agents in mining, effective corrosion inhibitors and lubricants in drilling and formulation. Fatty amines derived from tallow or palm oil fatty acid fractions are widely used in the production of fabric softeners.

The main raw material is the fatty nitrile derived from the reaction of fatty acid with ammonia. Catalytic hydrogenation of the nitrile produces the amines. The amines are available as primary, secondary and tertiary amine, depending on the number of the alkyl groups attached to the nitrogen atom.

Types of tertiary amines

Of the three fatty amines, the tertiary fatty amines, with three alkyl groups, find the most application in industries. They are further classified into three types: (i) symmetrical trifatty amine (R_3N), in which all three alkyl chains are identical; (ii) asymmetrical dimethyl fatty amine [$RN(CH_3)_2$] or methyl difatty amine (R_2NCH_3); and (iii) those derived from primary or secondary amine by the reaction with ethylene oxide. These asymmetrical fatty amines are the bases for the manufacture of organomodified clay for the petroleum industry, biocides and algicides.

Methods of preparation

Several methods are available for the preparation of tertiary amines from fatty acids. One of the oldest is the classical Leuckart reaction, which uses formaldehyde and formic acid for the reductive alkylation of a secondary amine.

$$RNH_2 + 2HCHO + 2HCOOH \longrightarrow RN(CH_3)_2 + CO_2 + H_2O \quad \text{... (1.16)}$$

Modification to this reaction had been reported. *Shapiro* developed the reductive alkylation of primary and secondary amine with formaldehyde in the presence of nickel catalyst.

Of commercial importance is the manufacture of fatty amines via the nitrile route. Fatty acid is first converted to nitrile by the reaction with ammonia at elevated temperatures of 280–360°C at atmospheric pressure.

$$RCOOH + NH_3 \xrightarrow{-H_2O} RCHONH_2 \xrightarrow{-H_2O} RC = N \quad \text{... (1.17)}$$

The reaction that produces the nitrile is essentially a dehydration reaction of the intermediate product. The nitrile is then catalytically hydrogenated to produce the fatty amines.

$$3\ RH = N + 6H_2 \xrightarrow{\text{Catalyst}} (RCH_2)_3N + 2\ NH_3 \quad \text{... (1.18)}$$

Depending on the process conditions, primary, secondary, or tertiary amines can be produced. If more tertiary amines are desired, continuous removal of the NH_3 is necessary to shift the reaction forward. The presence of ammonia favours the formation of primary and secondary amines.

Fatty alcohols are also available raw materials in the commercial preparation of tertiary amines. The fatty alcohol may be converted directly to an alkyl halide by the reaction with a concentrated hydrogen halide. The alkyl halide can then be reacted with dimethylamine to produce the tertiary amine.

$$RCH_2OH + HCl \longrightarrow RCH_2Cl \quad \text{... (1.19)}$$

$$RCH_2Cl + HN(CH_3)_2 + NaOH \longrightarrow RCH_2N(CH_3)_2 + NaCl + H_2O \quad \text{... (1.20)}$$

Long-chain alcohols' can be converted directly to *N*, *N*-dimethylalkyamines by the reaction with dimethylamine at 36°C in the presence of $Th(SO_4)_2$ or in the presence of Cu–Cr catalyst and hydrogen at elevated temperatures and pressure.

$$RCH_2OH + HN(CH_3)_2 \xrightarrow[H_2]{\text{Cu–Cr}} RCH_2N(CH_3)_2 + H_2O \quad \text{... (1.21)}$$

Ethoxylated tertiary amines can be produced by the reaction of primary or secondary amines with ethylene oxide. The water solubility of this amine increases with the increasing degree of ethoxylation.

Uses

The asymmetrical tertiary amines are used exclusively as starting materials for the manufacture of quaternary ammonium compounds, cationic and amphoteric surfactants and amine oxides. Quaternary ammonium compounds used as bactericides and algicides are produced by the reaction of tertiary amines with benzyl chloride, methyl chloride, or dimethyl sulphate. Of these, the benzyl ammonium chloride salt is the most widely used.

Probably the largest volume use of fatty amines is in the manufacture of fabric softeners. The latter is distearyldimethyl ammonium chloride produced by the reaction of distearylamine and methylchloride with continuous addition of alkali. However, as stated earlier, the starting materials for this product are the C16–C18 fatty acid fractions from tallow or palm oil.

Tertiary amines are also used in the manufacture of amine oxides by the oxidation with hydrogen peroxide. The amine oxides are used in cosmetic preparations for its good foaming properties and mildness to the skin. Another raw material used in the cosmetic industry is the betaines produced by the reaction of a tertiary fatty amine with sodium chloroacetate. The betaines exhibit good foam stability over a wide range of pH, are insensitive to water hardness and are mild to the skin.

Chapter 2

Canola Oil

INTRODUCTION

Canola oil produced in Canada is obtained from genetically modified seeds of *Brassica napus* and *Brassica rapa (campestris)*. These cultivars, low in erucic acid and glucosinolates, are quite different in chemical, physical and nutritional characteristics compared to High Erucic Acid Rapeseed (HEAR) oil. Acceptance of canola oil for Generally Recognised As Safe (GRAS) status in the United states allowed it to be sold as an edible oil and expanded canola oil's use into new food products. This chapter provides a comprehensive review of the properties of canola oil that have made it a major edible oil on the world market.

The oilseed rape species that produce canola oil and meal belong to the genus *Brassica* in the Cruciferae family. Other members of the family include the condiment mustards, the *Brassica* vegetables such as cabbage, kale, broccoli, swede, turnip, rutabaga, cauliflower and brussels sprouts, other closely related crops such as radish (*Raphanus sativus*) and weedy species such as wild mustard (*Sinapis arvensis*). The origins of the *Brassica* crops are somewhat uncertain. Several of the species appear to have been domesticated at a number of separate sites and times as the plants became useful to local populations.

Cultivation practices vary throughout the world depending on the species grown, climate and the local agricultural system. In Canada and Europe canola is planted in rows, while in the Indian subcontinent the seed is usually broadcast.

In China, it is still common to transplant canola seedlings into the fields following the rice harvest. In Canada, certified seed is sown not deeper than 2–3 cm into a firm, moist, well-prepared seedbed in soils that do not crust readily so that emergence will not be impaired.

Canola is the only oilseed crop adapted to temperate regions. It is able to germinate and grow at low temperatures, which allows production in cooler regions and at higher elevations than other oilseed crops and also allows it to be grown as a winter crop in areas where the winter is not too severe. Canola is adapted to the temperate and warm temperate zones of Europe and Asia where moisture is not limiting. Yields of canola/rapeseed vary considerably throughout the world due to different species and genotypes grown, climate and production practices.

CANOLA OIL EXTRACTION AND PROCESSING

Because canola was developed in Canada and achieved its greatest importance there, the process and product experience obtained in that country will form the basis of the description.

Extraction

Canola seed contains 38–44 per cent oil (8 per cent moisture basis). This means that the most efficient method of extracting the oil is by mechanical (pressing) expelling of about 60 per cent of the oil, followed by solvent extraction of the remainder, similar to that applied to other seeds of high oil content, such as sunflower. Figure 2.1 gives an overview of the process sequence most commonly applied in canola seed extraction.

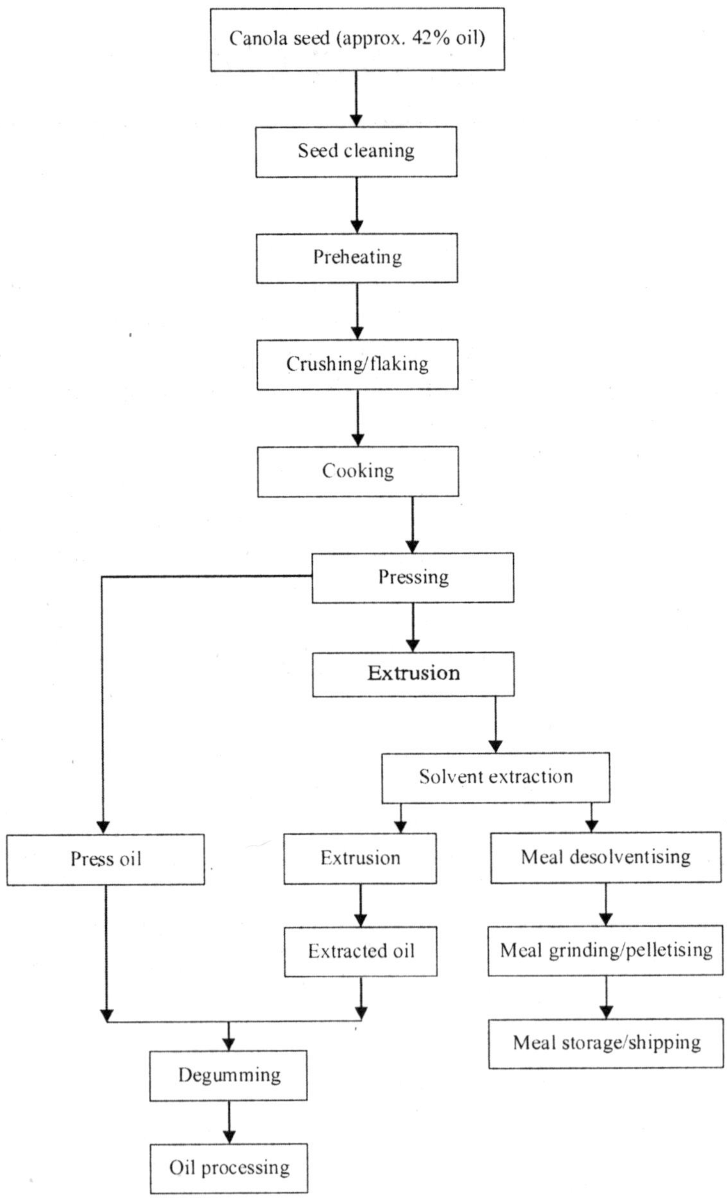

Fig. 2.1. Canola oil extraction from seed.

Seed cleaning

Canola seed is generally received at the extraction plant directly from the farm or a seed gathering station. It therefore includes small amounts of fragments of the canola plant, weed seeds, grain seeds and other, foreign material. This must be removed from the seed prior to extraction of the oil. The seed cleaning process consists of three stages: aspiration to remove dust and very light material, screening to remove oversized particles and rescreening to remove undersized material. The equipment usually comes as one unit. Removal of foreign material is generally to less than 2.5 per cent of it remaining in the seed. Disposal of the separated material is primarily in animal feed. At this time, there are no commercial-scale seed dehulling processes. The subject of dehulling is of interest because canola meal is high in fibre and hence, is limited in its use as a feed material for non-ruminants.

Preconditioning

This is essentially a preheating step prior to flaking of the seed. It is required, first, because of the low ambient temperatures especially during the winter months in growing areas such as Canadian prairies to bring seed temperatures to about 30–40°C before the flaking operation. This makes the seed more pliable and thus prevents shattering of the seed. A second reason is that with the change from low-erucic-acid seed varieties to varieties of low glucosinolate content as well as low-erucic-acid content (canola-quality seed), shattering in the flaking step, even with seed temperatures above freezing was more of a problem. Seed behaviour in flaking and in all subsequent operations was much improved when the seed was preheated. It is speculated that the reason for this different behaviour is in changes in seed coat thickness and fibre content in the seed of these new varieties. The need for preheating is possibly a phenomenon that is limited to the Canadian varietal and climatic situation and to other canola growing areas that are very similar. Preheating equipment may be of the fluid bed type using hot air or steam, direct, or by indirect heat in rotary kilns equipped with steam-heated coils. Moisture content in the seed can be adjusted, if required, during this process step by regulating the air flow through the equipment. The desired moisture content prior to flaking is in the range of 7.0–9.5 per cent. Western Canadian seed is usually in this range without any drying.

Flaking

The preheated canola seed is flaked on smooth-surface rolling mills. Some operators advocate flaking in two stages. In the first stage a flake thickness of about 0.4–0.7 mm is produced. In a second set of rolls a flake of 0.2–0.3 mm thickness is produced. The main reason for two-stage flaking operation is the small size and the deformability of the canola seed, which usually allows some whole seed to pass through in a single-stage operation. Flaking ruptures the cell walls, which allows some of the oil to be separated from the seed residue by simple pressing. Further, the oil retained in the seed residue after pressing can now be more efficiently leached out in subsequent solvent extraction of the press cake with hexane. Properly flaked material is more easily treated in the subsequent cooking operation and requires less mechanical energy in pressing. It is essential for achieving low-residual oil content in the meal and, hence, for an economical operation.

Cooking

The flaked material is subjected to a cooking step. In this operation, the flakes are heated either indirectly on steam-heated surfaces in stack cookers or in rotary cookers equipped with steam coils. Cooking temperatures vary but are usually held to 75–100°C. Ambient air may be admitted to the cooking equipment to adjust the moisture content of the material to the desired level for the expeller operation. This level is about 5–7 per cent.

The cooking operation serves several additional important functions. It serves to coalesce minute oil droplets into larger ones, which can be easily separated. It also changes the properties of the protein so as to make the oil more easily extractable, but temperature and heating time must be carefully controlled to avoid negative effects on colour and sulphur levels in the oil, on protein degradation and on the percolation properties of the flaked material.

There are also effects on enzyme activity. Two enzymes are of particular interest, namely, myrosinase in connection with the hydrolysis of glucosinolates, which produces sulphur containing compounds, and phospholipases in connection with the hydrolysis of phosphatides, resulting in nonhydratable phosphatides. Myrosinase activity has been the subject of much study in order to control the cooking operation so as to minimise hydrolysis of glucosinolates and hence, the sulphur content of the oil. The temperature range of greatest myrosinase activity is from 50 to 70°C. Moisture must be in the 5–7 per cent range.

Phospholipase activity and the effect on it in the cooking operations has also been a subject of interest, but to a lesser extent. *Dahlén* and *Kristofferson* investigated cooking temperatures and found that as cooking temperatures were raised from 85 to 100°C, nonhydratable phosphatides decreased in the oil, but total phosphorus, colour and sulphur content increased. Higher colour and sulphur are undesirable.

The increased sulphur content impairs the catalyst in hydrogenation of the oil. *Unger* showed that not only total phosphorus but also free fatty acids are increased with higher cooking temperatures. In present cooking operations with temperatures in the range of 75–100°C, the sulphur content of canola oil is on the order of 3–13 mg/kg as determined by inductively coupled plasma (ICP) spectroscopy. It is apparent from the above that the selection of cooking temperature in practice is necessarily a compromise between opposing effects on product quality and economics.

Oil extraction by expellers (Prepressing)

The heat-conditioned seed, flaked or whole, is conveyed to continuous screw-presses or expellers. The function of this equipment is to reduce the oil content of the seed from about 42 to 16–20 per cent. Extraction of the remaining oil is then much more efficient and economical. Expellers consist of a rotating screw shaft in a cylindrical barrel. The barrel has flat steel bars mounted edgewise around the inside and spaced to allow oil to flow from between the bars while retaining the solid material within. Pressure and heat is developed within the barrel by the rotating screw shaft working against an adjustable choke at the solids (presscake) discharge end of the barrel. Some discharge of very fine solids (7–15 per cent) with the oil draining from the expeller is unavoidable. These fines are separated from the oil by gravity in settling tanks and by filtration and recycled to the conditioning stage. The presscake, which may be put through a mechanical breaker to produce uniform-sized particles, is now ready for solvent extraction.

Extruders

In some plants, the presscake is subjected to mechanical extrusion to improve its solvent extraction properties. Extruders (expanders) used for this purpose consist of a barrel with a rotating shaft fitted with flights. As steam is added and heating and mixing take place along the length of the barrel. Pressure is developed and the material is then discharged through the small openings of a die plate at the end of the barrel. The pressure release on discharge 'expands' the extruded material, making it very porous.

These small-diameter, porous pieces of presscake (collets) have excellent solvent extraction properties. Throughput through solvent extraction equipment is significantly enhanced. In an extruder, it is possible to achieve very fast, uniform heating to about 100°C. This appears to result in much lower phosphatide (<100 ppm, measured as phosphorus) and lower chlorophyll concentrations in the press oil and in the extracion oil. It, also, results in other beneficial effects, such as lower concentrations of free fatty acids and other polar compounds.

Extraction of solvent

The presscake from the expellers, as is, or as collets from extruders is conveyed to the solvent extraction stage. Some cooling of the 80–100°C material with ambient air is usually done during conveying to minimise vapourisation of hexane when the cake enters the extractor. Extractor capacities are on the order of 600–1200 metric tonnes of seed/day. The solvent is hexane, specially refined for this application and heated to 50–60°C.

In the extractor, the solids, which are about 80°C, are solvent-washed in stages, first with hexane already high in oil content (miscella) and then with progressively leaner miscella and, finally, with pure hexane. The various extractor designs approximate countercurrent extraction to a greater or lesser extent. The oil content in the solid material (meal) is reduced down to about 1 per cent, depending somewhat on equipment design and throughput rate and on how well the cooking, flaking and prepress operations were carried out. The flow rate of hexane to the extractor is adjusted to provide an oil concentration of 25–30 per cent in the oil in hexane solution (miscella) triglyceride oil, with the remainder made up of phosphatides, free fatty acids and other, more sparingly hexane-soluble material.

Desolventising of meal and oil

The meal and the miscella are 'stripped' of the solvent to recover solvent-free meal and oil. The solvent-saturated meal is conveyed to a desolventiser, which is a vertical tank equipped with steam-heated trays and rotating sweep arms just above each tray to agitate the solid material. Reduced pressure and sometimes live steam sparging are used to evaporate the hexane and to dry the meal. The meal is heated to about 105°C. Residence time is about 30–40 minutes. Residual hexane in the meal is in the order of 1500 mg/kg, add moisture in the range of 15–18 per cent. Lower hexane concentrations are quite difficult to achieve. To achieve the best meal quality, the process must be well controlled with respect to temperature (110°C max.) and time. Meal cooling and drying is done in rotary kilns, separate from the desolventising tower or integral with the desolventiser. The temperature is reduced to 30–40°C, moisture to 8–11 per cent and hexane to about 800 mg/kg. Leaving about 800 mg/kg of hexane in the meal corresponds to a hexane loss of approximately 1 litre per metric tonne of seed processed.

The industry has increasingly moved to the use of desolventiser-toaster-dryer-cooler (DTDC) equipment; many plants use it for its energy-saving features. Cooled meal may be ground to a uniform particle size, or pelleted, ready for storage and marketing. The hexane and moisture vapours are vented from the meal desolventising equipment and condensed, the water and hexane are separated, and the hexane is reused. The miscella containing the oil is desolventised in three-stage evaporator equipment. The hexane vapour from this operation is condensed for reuse. A properly sized and operated extraction plant loses no more than about 1–2 litres of hexane/metric tonne of seed processed. Generally desolventising equipment is supplied by the same companies that supply solvent extraction equipment.

Dehulling of the desolventised meal (tail-end dehulling) for nonruminant feed may in the future become commercial practice. Seed source, moisture level in the meal and the type of mill used to grind the desolventised meal were found to be important parameters. This approach to canola dehulling would have the advantage of relatively modest capital investment and would leave percolation properties of the presscake and oil yields unchanged. A negative would be that desolventising of the meal, which is more difficult because of the presence of the hulls, would not benefit from tail-end dehulling.

Quality assurance: seed, oil and meal

The first step in the quality assurance process is the grading of seed deliveries from the farm to the extraction plant. In respect to oil quality, three types of crude oil are distinguished, and these are listed in Table 2.1. Table 2.2 gives the Canadian Oilseed Processors Association (COPA) standard for canola meal. The distinguishing feature of the oil specifications is phosphorus content, depending on the degree of degumming: type I, superdegummed (with acid and water), type II, water degummed and type III, nondegummed.

Table 2.1. Crude canola oil types and specifications.

Specifications	*Type I crude super degummed*	*Type II crude degummed*	*Type III crude*
Free fatty acid (as oleic acid), max. % by mass	1.0	1.0	1.0
Moisture and impurities, max. % by mass	0.3	0.3	0.5
Flash point (typical), min. °C	150	150	150
Chlorophyll content, max. parts per million	30	30	30
Neutral oil (typical), min. % by mass	98.5	98.5	98.0
Loss (typical), min. % by mass	1.5	1.5	2.0
Phosphorus content, max. parts per million	50	200	NA
Erucic acid, max. % by mass	2.0	2.0	2.0
Sulphur (typical), parts per million	10	10	15

Table 2.2. Canola meal specifications.

Characteristics	*Specifications*
Protein, % by mass	min. 36
Fat (oil), % by mass	max. 4
Moisture, % by mass	max. 12
Crude fibre, % by mass	max. 12
Glucosinolates, μm/g	max. 30
Screen analysis	97% by mass of the meal shall pass through a 2.00 mm sieve and 90% by mass shall pass through a 1.70 mm sieve.

Official AOCS methods are used except for erucic acid, for which AOAC method 28.A01–28 A04 is used, and for sulphur, for which the method of *Daun* and *Hougen* is used.

Canola Oil

Processing of crude canola oil to produce edible oil products is, in principle, very similar to that applied to other vegetable oils. Figure 2.2 gives an overview of the process steps that are applied in the industry.

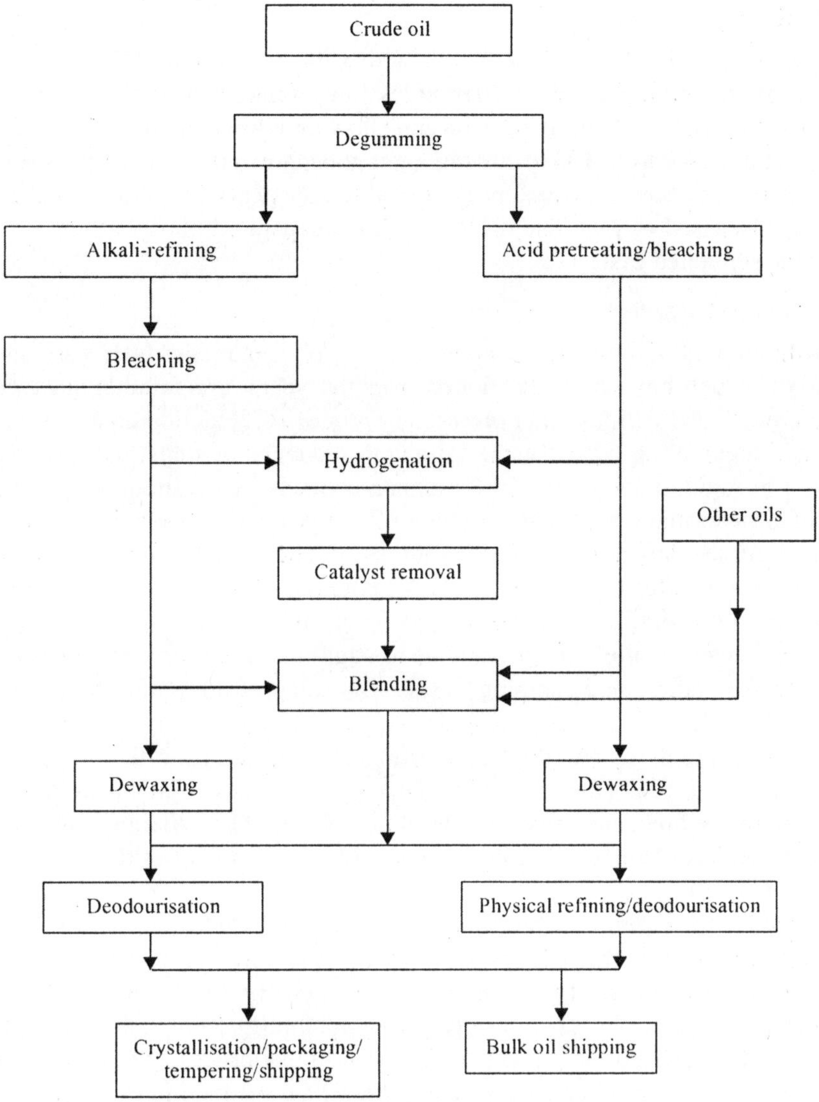

Fig. 2.2. Canola oil processing.

Degumming

The 'crude' oil from prepressing and from solvent extraction is usually blended and then degummed before being stored for sale or further processing. Degumming removes phosphatides coextracted with the oil, some of which tend to separate from the oil as a sludge during storage. The phosphatide content of crude oil varies but is usually on the order of 1.0–1.5 per cent, or, measured as phosphorus, 400–600 mg/kg. Two degumming methods are in use: (i) using water to precipitate phosphatides and (ii) using an acid such as citric, malic or phosphoric and water (superdegumming). After contacting the oil with these reagents, the oil is centrifuged to separate the precipitated material.

Degumming with water

Degumming with water only is usually done by contacting the oil at about 80°C with about 2 per cent water and intensive mixing, giving from 5 to 30 minutes contact time in a gently agitated tank or reactor to keep the precipitated phosphatides in suspension and allow time for agglomeration and then separating in a disk centrifuge. This leaves from 100 to 200 mg/kg of phosphorus (0.25 to 0.5 per cent phosphatides) in the oil, depending on the extent to which nonhydratable phosphatides were formed in the course of seed preparation, as discussed earlier. The oil lost with the separated phosphatides is on the order of 35–40 per cent of the separated material.

Degumming with acid and water

By degumming with an acid and water, the nonhydratable phosphatides (NHP) are also removed. A number of processes, which have been developed over the years, are suitable such as the Unilever Superdegumming process and various other proprietary processes. Reaction conditions are much more critical than in water degumming. Usually, the following sequence of steps, with some variations, are applied: Crude oil is brought to about 60–70°C and then contacted with about 0.1–0.5 per cent of a 50 per cent solution of citric or malic acid with very intensive mixing. Contact may vary from fractions of a second to about 15 minute depending on mixing intensity. The oil is then contacted with about 2 per cent water, either before or after cooling to 25–45°C. About 1–3 hr., of residence time is given in a continuous, stirred reactor. Agitation is designed to avoid shear and keep the precipitated phosphatides uniformly suspended. This is to allow the precipitate to agglomerate to form particles large enough for centrifugal separation. Any subsequent pumping and heat transfer operations must be designed to maintain particle size. The oil-phosphatide mixture is heated very rapidly, immediately ahead of a self-desludging disk centrifuge, in which the precipitated phosphatides are separated from the oil. The oil loss in this process represents from 35 to 50 per cent of the separated phosphatides, with the higher losses occurring with oils that are relatively high in nonhydratable phosphatides. The separated phosphatide phase is added to the meal in the desolventiser. This raises the residual oil content of the meal to about 2–3 per cent and raises its energy content.

Residual phosphatide concentrations are from 5 to 50 mg/kg of phosphorus depending on process details and the level of nonhydratable phosphatides in the crude oil. Nonhydratable phosphatides can vary widely but are usually in the range of 20–45 per cent of total phosphatides. With higher concentrations of nonhydratable phosphatides, longer agglomeration time is required to achieve clean separation in the centrifuge. Phosphoric acid is rarely used in this process, because it tends to produce a precipitate that is very slow to agglomerate. It also presents a potential interference in the analysis for phosphorus to determine process performance whenever traces of the precipitate remain in the centrifuged oil.

In the most recent development, acid and aqueous sodium hydroxide, rather than acid and water, is being used, especially with lower quality oils. This represents an intermediate between acid–water degumming and alkali refining. It obviates the need for cooking, similar to the practice in alkali refining. There are parallels in this development to the processing of soyabean oil, for example. Phosphoric acid rather than an organic acid is preferred in this case. Phosphatides as well as some of the other impurities are removed and, if sufficient alkali is applied to saponify the free fatty acids, fully refined oil is obtained. As seed extraction and oil refining is coming to be more and more integrated at the same plant site, degumming processes with the necessary 'add ons' for flexible 'degumming' treatments with acid and water or acid and alkali up to full refining of the oil may become the preferred process version.

Alkali and Physical (Steam) Refining

Degummed oil is further purified in a process of refining. One of two methods are used, namely, alkali refining, especially with water degummed oil, and physical refining with acid–water degummed, or acid–caustic degummed oil. Alkali refining is the most common process used, even with acid–water degummed oil. Physical refining of canola oil on a plant scale is relatively new. It requires well-degummed oil of moderate chlorophyll and free fatty acid content.

Alkali refining

In alkali refining, the oil is first contacted with about 0.05–0.1 per cent concentrated phosphoric acid in a high-intensity mixer to help precipitate phosphatides. It is then contacted with an approximately 12 per cent aqueous solution of sodium hydroxide to neutralise the free fatty acids present as well as any excess phosphoric acid and to precipitate phosphatides. This requires about 2–3 per cent of the solution. Very intensive mixing is used. Temperatures and contact times may vary from about 90°C and only seconds for both acid and caustic in the 'short-mix' process to about 40°C and about 15 minutes each in the 'long-mix' process. The oil–soap mixture is then heated to about 90°C, if required and centrifuged to separate the aqueous soap phase. This phase also contains the precipitated phosphatides and some triglyceride oil. Usually, self-desludging disk centrifuges are used to avoid frequent shutdowns for cleaning. The centrifuged oil must be further contacted with about 5–10 per cent hot, soft water to reduce soap levels from about 500 to <50 ppm. Solid bowl disk centrifuges are adequate for this service.

Alkali refining reduces free fatty acids to <0.05 per cent and phosphorus to <3 mg/kg. Iron and copper concentrations are reduced to below the detection limit. Coloured compounds (chlorophylloids, carotenoids) are not affected significantly, although chlorophylloids can be reduced by 30–70 per cent of the original concentration under certain process conditions. The concentration of sulphur compounds is reduced slightly. This oil is now ready for bleaching. The soap phase from this operation can be added to the meal, similar to the disposal of the phosphatides from degumming, or it can be acidulated and used as a feed fat. The wash water must be treated to reduce the oil content to acceptable levels for discharge.

Physical refining

In physical refining, acid–water degummed oil with a phosphorus content below 50 ppm is first subjected to a phosphoric acid pretreatment, as in the short-mix alkali refining process. It is then contacted with acid-activated bleaching clay in a standard bleaching process at 95–105°C. The clay, along with precipitated phosphatides and adsorbed chlorophylloid and some carotenoid compounds, is then removed by filtration. Along with acid–water degumming, this is the most important stage of physical refining. It delivers bleached oil ready for physical (steam) refining/deodourising. Except for the free fatty acids in the oil, all other minor constituents are now at the same concentrations as in alkali refined oil, and in addition, chlorophyll is reduced to the concentration required of bleached oil, namely <50 mg/kg. Usually, from 1 to 3 per cent acid-activated bleaching clay is used, depending primarily on the concentration and type of chlorophylloid compounds and the residual phosphatide concentration present in the oil. The removal of the free fatty acids in the oil is done by steam distillation in a deodouriser. This, simultaneously, deodourises the oil. Because deodourisation is also the last process normally carried out on edible oils, this step may be delayed until other processes, such as hydrogenation of the oil, have been done.

Adsorptive Bleaching

Alkali-refined oil requires bleaching. It usually contains traces of soap, and the concentration of chlorophylloid compounds is still essentially that present in the crude oil. Some of the oxidation products and traces of heavy metals, notably iron, are also removed. But, the most important effect to be achieved with canola oil is the removal of chlorophylloid compounds. Chlorophylloid compounds catalyse oil oxidation in light and give an undesirable green colour to the oil. Acid-activated clays are used in bleaching. Their adsorptive properties are especially effective for the removal of 'green' compounds. For especially high chlorophyll concentrations, >30 ppm, 'chlorophyll-specific' clays are used. One feature of these clays is that they are especially highly acid activated. As with other oils, the adsorption carotenoid compounds on bleaching clays is not efficient. Some reduction occurs, partly due to adsorption and partly due to breakdown. In usual bleaching practice, the emphasis is on chlorophyll removal. Heat bleaching of the carotenoids during deodourisation is relied upon to remove most of the red colour present in the oil.

Adsorptive bleaching is carried out under vacuum with the oil at about 100°C. About 5–30 minutes of contact time is given while the oil–clay slurry is progressively dried to about 0.1 per cent moisture content. This appears to give the best adsorption efficiency. As indicated earlier, about 1–3 per cent clay (10–15 per cent moisture content) may be required to achieve chlorophyll removal to <0.050 mg/kg measured as chlorophyll *a*. This level of chlorophylloids is innocuous in respect to oxidation and colour of the oil. *Suzuki* and *Nishioka* have investigated changes in chlorophyll derivatives during seed and oil processing up to the bleaching stage and have determined the ease of adsorption of the various derivatives on activated clay and on active carbon. For acid-activated clay, they found the ease of adsorption to decrease in the following order: pheophytin $a >$ pyropheophytin $a \gg$ pheophytin $b >$ pyropheophytin b.

Chlorophyll *a* and *b* are converted to their respective pheophytins early in seed processing and thus do not play a significant role in the adsorption of chlorophylloids. It is worth noting in this connection that since the various chlorophylloid compounds absorb light at slightly different wavelengths than chlorophyll *a*, which present analysis methods use as the standard, there is a slight inaccuracy in the determination of chlorophylloid concentrations.

Bleaching processes suitable for canola oil do not differ from those used for other oils. Many somewhat different versions of bleaching equipment are in use and most edible oil process equipment suppliers have process designs, which they furnish to clients together with the usual edible oil filter equipment.

Active carbon

Active carbon to remove chlorophylloid compounds is used very little in canola oil bleaching. It presents greater difficulty in handling and it retains more oil than activated clays and is much more expensive. In bleaching efficiency tests, active carbon has been shown to be somewhat more efficient than activated clay at high concentrations of chlorophylloid compounds in canola, but less efficient at very low concentrations.

Synthetic silicas

In recent years, the use of synthetic silicas as a pretreatment in the bleaching of oils has been advocated. In connection with canola oil bleaching, it is claimed that because of the ability of these silicas to adsorb soap and phosphatides more efficiently than acid-activated clays, adsorption of chlorophylloid compounds by the clays in subsequent bleaching is much more efficient. Presently available silicas do not adsorb chlorophylloids or carotenoids.

Other methods for removing chlorophylls

Other approaches to the removal of chlorophylloid compounds from canola oil than adsorption are known and are being practiced. *Bergman* describes the use of concentrated phosphoric acid and vacuum drying of the oil to precipitate impurities from rapeseed oil, especially chlorophylloid compounds. Experience has shown that the precipitation of chlorophyll is much more efficient with superdegummed canola oil. The removal of the precipitated material from the oil can be by filtration on a suitable filter aid or by centrifuging. When the precipitation step is followed by alkali refining of the oil, the acid-precipitated chlorophylloids are simply removed as part of the heavy phase in the soapstock separation. Depending on acid usage and other factors, the concentration of chlorophylloid compounds can be reduced to <5 mg/kg. The extra cost of vacuum drying the acid-contacted oil is more than recovered in bleaching cost savings. The process is proprietary.

Another approach is to make use of the fact that in contact with alkali, chlorophylloid compounds also react to form a precipitate. This can be utilised in the course of alkali refining. The precipitate can be removed together with the aqueous soap phase, but at this time, specific details are proprietary. Experience so far has shown that as much as 70 per cent of the chlorophylloids present can be removed in the course of alkali refining. Typical properties of alkali-refined, bleached canola oil and of acid–water degummed, acid-pretreated, bleached canola oil ready for hydrogenation or steam refining/deodourisation are given in Table 2.3. With the exception of the concentration of free fatty acids, the two process routes produce the same oil quality.

Table 2.3. Bleached canola oil after alkali refining or acid degumming.

Specifications	*Alkali refining/bleaching*	*Acid degumming/bleaching*
Free fatty acids (%)	0.1	As for crude
Phosphorus (ppm)	< 3	< 3
Chlorophyll (ppb)	< 50	50
Iron (ppm)	< 0.1	< 0.1
Peroxide value (meq/kg)	Nil	Nil
Anisidine value	1–3	1–3
Sulphur[a] (ppm)	1–13	1–13
Soap (ppm)	Nil	Nil

[a] Analysis as Raney nickel and ICP sulphur to indicate potential effects on hydrogenation catalysts.

Hydrogenation

Hydrogenation changes the melting behaviour of oils and improves the oxidative stability. It is applied to many oils, including canola. Hydrogen is added to the double bonds of unsaturated fatty acids at temperatures of 160–200°C and pressures of 100–300 kPa in the presence of a nickel catalyst to facilitate the reaction. Chlorophyll phosphatides, soaps and especially sulphur compounds 'poison' the catalyst, raising process costs. The former are removed in refining and bleaching. Sulphur compounds are only reduced somewhat in these treatments.

The determination of sulphur compounds in the range of concentrations that occur in edible oils is difficult. In respect to the effects of these compounds on hydrogenation catalysts, it appears that the Raney nickel method of sulphur determination correlates well. It has the disadvantage of being rather time consuming. The process of hydrogenation of canola oil is usually carried out batchwise. There are various reactor designs in use. Depending on temperature, pressure and to some extent the specific catalyst properties, different melting behaviour in the hydrogenated fat can be obtained.

The oxidative stability improvement also is modified somewhat by choice of process conditions. It is possible to hydrogenate fatty acids with multiple double bonds preferentially. These are the most easily oxidised and their concentration must, therefore, be reduced or eliminated for best effect on stability. Usually, conditions are chosen that provide this selectivity. In canola oil, very low concentrations of linolenic and linoleic acid can be achieved with very little hardening of the oil taking place. Melting behaviour is most usually evaluated by determining the proportion of solid fat in a sample over the temperature range of interest, usually 10–33°C.

Another consequence of hydrogenation is the formation of isomeric fatty acids. Positional isomers as well as geometric, or *trans*, isomers are formed. These compounds contribute importantly to the melting properties and the functional utility of the hydrogenated fat. Recently, nutritional science has again focused on the role of *trans* isomer fatty acids in the diet. This will probably continue to be controversial for some time. Because of the small concentration of sulphur compounds in canola oil, there is a marginally greater tendency of sulphur poisoning of the catalyst, with the result that slightly higher *trans* isomer concentrations may come about compared to, for example, soyabean oil.

Removal of the nickel catalyst from the oil is by filtration of the oil from the reactor after cooling to about 100°C on filter aid precoated filters. Adding filter aid to the oil itself and then filtering is also satisfactory. Nickel concentrations of 3–15 mg/kg are achieved. This is followed by a second filtration on a filter aid precoated filter, possible after contacting the oil with a small amount of citric acid (0.05 per cent as a 50 per cent aqueous solution) as a chelating agent. In stubborn cases it is necessary to bleach the oil with acid-activated bleaching clay. Nickel must be removed to below 0.3–0.5 mg/kg, that is, to Ni-negative by the ammonium sulphide test. Synthetic silicas have also been advocated as a treatment of hydrogenated oil for nickel removal either before or after catalyst filtration.

Hydrogenated canola oil in edible oil products

Hydrogenated fats from canola oil play an important role in the production of the wide range of fat products in the countries in which canola oil is used. They often appear in blends with other oils to combine the advantages of canola oil with those of other oils. They are especially important as lightly hydrogenated oils and are recognised for their improved stability in storage. This is due, in part, to the high content of monounsaturated fatty acids in canola oil. These products are becoming very popular because of their good stability and pourability and their low saturated fatty acid content.

Table 2.4 shows some typical iodine values (IV), fatty acid compositions and solid fat indices ('SFIs') for hydrogenated canola oils used in margarine, shortening and frying fat fomulations. The nonhydrogenated, liquid oil data is given for comparison. It should be noted that the values given may differ somewhat in practice depending on hydrogenated conditions and catalyst used.

Table 2.4. Hydrogenated canola oils for margarines, shortenings and frying fats.

	Fatty acid composition (%)						*SFI at °C*				
IV	*C16:0*	*C18:0*	*C18:1*	*C18:2*	*C18:3*	*Trans*	*10.0*	*21.1*	*26.7*	*33.3*	*40.0*
115 (L)	4.5	2	60	22	10	–	–	–	–	–	–
90	4.5	3	80	9	2	20	2.5	–	–	–	–
85	4.5	5	86	2	1	42	18	5	–	–	–
80	4.5	11	83	–	–	45	30	18	7	< 1	–
75	4.5	14	80	–	–	50	50	31	25	11	4

The hydrogenated canola oil stocks listed in Table 2.4, or similar ones, are used extensively in shortenings and as stable frying fats. Industry makes a wide variety of canola-based hydrogenated oil stocks for tailor-made shortenings.

Lightly hydrogenated canola oil (IV 90), for example, has a special advantage in that even at very low levels of solid fat it has very low polyunsaturation, which makes it an excellent pourable frying fat base of very good stability. Very highly hydrogenated canola oils are rarely used in margarine and baking shortenings because of the tendency to form large β crystals over time. This impairs eating properties and baking performance.

Interesterification

Interesterification is another process for changing the melting properties of fats and oils. This method is used widely in Europe but not in North America. The position of fatty acids in the triglyceride is changed from the naturally occurring to a random distribution. Exceptions are when, for example, high melting material is removed from the reaction mixture, continuously (directed interesterification), or when enzymes are used that are selective with respect to the position of fatty acids on the glycerol moiety. When the process is applied to blends of different oils, triglycerides with fatty acid compositions are formed that can be sufficiently different from the original oils that the melting characteristics of the resulting fat is changed to an extent not achievable by mere blending. The process represents an alternative to the use of partially hydrogenated fats in the manufacture of products of a variety of melting properties.

Thomas and *Desrosier* have investigated interesterification of nonhydrogenated canola oil with fully hydrogenated vegetable oils. If *trans* isomers in fat and oil products must be decreased in future, interesterification with high melting fats, including fully hydrogenated, fats, will assume greater importance. The process has also been proposed for use in controlling the problem with β crystallisation of hydrogenated canola oil. It has not found significant use for this purpose because of cost. *Kurashige* used enzyme-catalysed interesterification of palm oil with canola oil to improve the pourability of the palm oil over that of merely blending with canola oil and used a 1,3 specific lipase. Canola oil was used because of its low content of saturated fatty acids. Industrial-scale processes to carry out enzyme-catalysed interesterification are still rare.

At present, the most usual process in modern plants relies on chemical catalysis to achieve random interesterification. The process is carried out in a heated, agitated tank, equipped with a vacuum system and a water spray. The oil blend must be very dry and low in free fatty acids. Sodium methoxide (a strong base, 0.05 per cent) is added to the heated (100–150°C) oil as a powder under agitation and contacted for about 30 minutes to catalyse the reaction. The oil is then cooled to about 90°C, phosphoric acid is added to 'neutralise' the base. The oil is then washed, or alternatively, bleached and then deodourised.

Dewaxing

Canola oil is a natural salad oil. This means that it remains clear and liquid at refrigerator temperatures and can be used to produce bottled oil and salad dressings without winterisation. In Canada, at present about 50 per cent of the oil is used as salad oil. However, the oil may contain a rather small and variable concentration of compounds (about 20–400 mg/kg) that may over time appear as a sediment in the deodourised oil. For the sake of simplicity, the term waxes is applied to these compounds, but it is known that about 20–40 per cent of these compounds are not wax esters. The wax esters are compounds that are related to triglycerides in composition but are usually crystalline below or at about room temperature. The crystallisation behaviour of the mix of compounds that can crystallise from canola oil is rather unpredictable in that rather low concentrations (<50 mg/kg) can sometimes show up in the oil as a sediment and that the sedimentation can sometimes occur in a few days, or take several months.

It is desirable to dewax the oil to avoid a hazy appearance. The process is carried out on the bleached oil by chilling in a continuous heat exchanger to about 5°C and metering about 0.1 per cent of a filter aid into the chilled oil stream on the way to a filter. Little or no retention time is given before filtration. This reduces the wax content to <50 mg/kg, which then usually no longer produces a visible haze. Usually, in-house dewaxing process designs are used or processes patterned after sunflower oil dewaxing installations. Throughput rates with canola oil are much higher than with sunflower oil, primarily because of the much lower wax content.

Deodourisation

In edible oil processing, this is the final 'refining' step. Its primary function is to remove compound from the oil that impart odour and taste typical of the seed from which it is derived and any odouriferous compounds formed in such processes as bleaching and hydrogenation, Further, at the temperatures required, heat bleaching of the yellow-red carotenoid compounds is an important aspect of the process. As mentioned earlier, deodourising serves to physically refine the oil as a alternative to the removal of free fatty acids by alkali refining.

Deodourisation is essentially a steam distillation of the odour and flavour compounds from the oil, as well as other relatively volatile compounds such as free fatty acids. Tocopherols and sterols are also to some extent removed. The oil, which must be thoroughly purified in the previous process steps, is heated to 225–260°C under very low pressure (0.3–0.6 kPa) to exclude air. This increases the volatility of the compounds to be removed. Steam is blown through the oil (1–3 per cent), which increases the volatility further and allows efficient removal of the volatiles to very low levels.

Retention times under deodourising conditions vary considerably from a few minutes in some processes to over an hour in semicontinuous and batch equipment. Most of the recent emphasis in the design and use of deodourising equipment was on achieving greater efficiency in oil–steam contact, improving energy efficiency and eliminating air and water pollution, which are associated with vacuum generation and the removal of volatiles from the oil.

In the deodourisation of canola oil, it is important that the oil be essentially free of chlorophyll and phosphatidic material and of heavy metals such as iron and nickel. Table 2.3 gives the properties of bleached canola oil suitable for deodourising .It is usual practice to cool the deodourised oil to about 60–80°C after the process and to sparge it with nitrogen gas before storing in tanks or shipping bulk. Often, liquid canola oil is packaged in bottles and larger containers under nitrogen.

Crystallisation, Packaging and Tempering

With the exception of bulk frying fats, margarine and shortening oils are converted into semisolid form for final use. Margarine oils are emulsified with an appropriate water phase first. Shortenings, which do not usually involve a water phase, are crystallised directly. The crystallisation is carried out in systems consisting of ammonia-cooled, scraped-surface heat exchangers (A-units), followed by resting units (B-units) in which further crystallisation with or without agitation takes place. The now partially crystalline, but still pumpable, fat product is then transferred to fillers where it is packaged in a variety of packages and sizes. In the case of shortenings, about 10 per cent by volume of nitrogen gas is added ahead of the B-units. This gives the product its white, even-textured appearance. Essential process conditions are chilling in the A-unit to about 25°C and resting to allow most of the crystallisation to take place before the product enters the package. When done properly, there should be no more than about a 1°C temperature rise in the package. Products high in hydrogenated canola oil require special care to ensure that crystallisation to the very small β′ crystals is achieved. Essentially this means very rapid and complete crystallisation before the product enters the package.

Tempering of crystallised fat products refers to the process of allowing certain equilibrium processes in the crystalline–liquid fat mixture to reach completion. The properly crystallised, packaged product is held for 2–4 days at 25–27°C to improve plasticity, creaming and baking performance. It is a process that is not completely understood, but it is very important for optimum shortening properties. Margarines are usually tempered for a shorter time and then refrigerated.

Quality Assurance

In the processing of canola oil to edible oil products, much the same quality control procedures are applied as with other oils.

Degumming

The in-process quality control commences with the analysis of crude oil from the extraction plant. The main quality parameters are total phosphatides, nonhydratable phosphatides (all measured as phosphorus), free fatty acids and chlorophyll. AOCS methods are used, but for rapid analysis of phosphorus, the oxygen bomb ashing method for mineralisation is often applied. Results correlate well with the AOCS method, but are on the average 10 per cent lower. The level of nonhydratable phosphorus in the crude oil gives a useful indication of the relative amount of acid required in acid–water degumming and how difficult the degumming of the oil will be. Nonhydratable phosphorus is determined by water-degumming an aliquot of crude oil at 80°C with 2 per cent distilled water in a round bottom flask for 30 minutes and then centrifuging or filtering to separate the precipitated phosphatides from the oil.

Process control of alkali refining

This involves the determination of free fatty acids, phosphorus and if desired, chlorophyll, before and after the process. In addition, soap concentration is determined after water washing. Soap determination can present some difficulty in establishing the end point of the titration with oils high in chlorophyll concentration, because of the dark colour of such oils.

Bleaching

The main aspect in control of the bleaching process is determination of chlorophyll. AOCS methodology is the standard, although Lovibond and McCloskey colour meters using colour disks, calibrated to chlorophyll *a* are also used and are very popular. It is accepted practice in the industry that chlorophylloid compounds must be reduced in bleaching to a concentration of 0.05 mg/kg max. Higher concentrations produce a green tinge in the oil and impair the stability of oil when exposed to light. Also, during deodourisation a grayish tinge develops in the oil.

Hydrogenation

Process control in hydrogenation is primarily concerned with degree of hydrogenation, determination of melting behaviour, fatty acid composition, and nickel catalyst removal, all analyses are by AOCS official methods. The solid fat index and where appropriate, the Mettler dropping point are analysed by a method using a hollow stopper in the dilatometer bulb to speed temperatures. It provides results in half the time required by the AOCS method. Determination of the various fatty acid isomers is based on capillary gas chromatography and infrared spectroscopy.

Dewaxing process

This is not controlled by an analytical test that determines the content of waxes in the oil This is too complex as a routine process control or quality control test. Rather, the oil is subjected to the AOCS cold test and this is supplemented by keeping samples at higher temperatures, up to room temperature. These tests are indirect and rather vague as predictors of shelf stability in respect to the oil remaining clear of visible sediment.

Deodourisation/physical refining

This quality control is especially concerned with flavour, free fatty acids, colour and oil stability, and trace contaminates. Secondarily, tests are often performed to provide control over antioxidant levels, both synthetic materials and those naturally present and for iron. Routine flavour testing of the oil is frequently done using a small trained panel. The oil must be essentially bland although very slight beany or grassy notes are tolerated. Free fatty acids, odour and stability testing is done using AOCS official methods. This also applies to qualitative and quantitative testing of antioxidants.

Crystallisation, packaging and tempering

Process control of the crystallisation conditions (temperature of precooling, crystallising, or A-units, temperature rise in the package) and of tempering time and temperatures are especially important with margarines and shortenings containing a high proportion of hydrogenated canola oil. This is because of the greater tendency of such products to recrystallise from the β' to the β form. Polarised light microscopy is sometimes used to provide an immediate assessment of the crystal state of a product. Beyond that, penetrometer readings as well as taste panels for margarine to assess mouth feel and creaming, baking, and icing performance tests in the case of bakery shortenings are the procedures used as a guide to determine tempering time and shelf stability.

CANOLA OIL: PHYSICAL AND CHEMICAL PROPERTIES

Physical Properties

The properties of canola oil are governed by the general standards for vegetable fats and oils. Selected properties are shown in Table 2.5.

Table 2.5. Some physical properties of canola oil.

Parameter	*Value*
Relative density (20°C/water at 20°C)	0.914–0.917
Refractive index (n_D 40°C)	1.465–1.467
Crismer value	67–70
Viscosity (kinematic at 20°C, mm^2/s)	78.2
Smoke point (°C)	220–230
Cold test (15 hr., at 4°C)	Passed
Flash point, open cup (°C)	275–290

Relative density

Typical values for the specific gravity of canola oil are presented in Table 2.5. The relative densities for high-erucic-acid rapeseed oils ranged from 0.906 to 0.914. The higher density of canola oil correlates well with the elimination of erucic acid. This is consistent with fatty acid densities of 0.8706 and 0.8740 determined for methyl erucate and methyl oleate, respectively. The higher specific gravity of 0.9187 observed for soyabean oil can be attributed to the basic variations in composition between these two oils such as the highlinoleic content of soyabean oil compared to a high-oleic-acid content in canola oil.

Crismer value

Crismer value measures the miscibility of an oil in a standard solvent mixture, composed of *t*-amyl alcohol, ethyl alcohol and water in volume proportion 5:5:0.27. The miscibility of an oil is related to the solubility of glycerides and is affected mainly by the unsaturation and chain length of the constituent fatty acids.

Viscosity

Viscosity values estimate the relative thickness or resistance to flow of an oil. Viscosity of refined, bleached and deodourised (RBD) canola oil was higher than for soyabean oil (Table 2.6).

Table 2.6. Viscosity of canola and soyabean oils at different temperatures (mm^2/s).

Temperature (°C)	*RBD canola oil*	*Soyabean oil*
0	–	172.9
4	178.8	–
10	128.8	99.7
25	62.6	50.1
40	36.7	28.9

Figure 2.3 shows the relation between temperature and viscosity for canola oil. Canola and soyabean oils are both more viscous than corn and sunflower oils. This can be directly related to the differences in saturated fatty acids and their higher melting points.

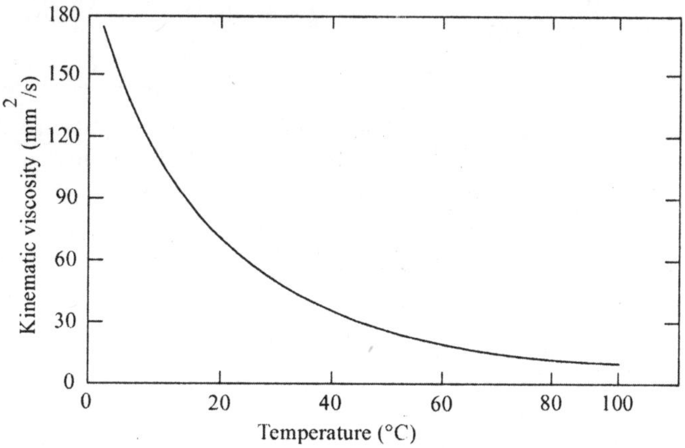

Fig. 2.3. Effect of temperature on canola oil viscosity.

Smoke point

Smoke point is the temperature at which a fat or oil produces a continuous wisp of smoke. This provides a useful characterisation of its suitability for frying. The Canadian government specifications define that frying oil should have a smoke point above 200°C.

Flash point

Flash point defines the temperature at which the decomposition products formed from frying oils can be ignited. This temperature ranges for different oils and fats from 275 to 330°C. Oils with unsaturated fatty acids were reported to have a lower flash point. Canola oil appears to fall well within this range as already mentioned in Table 2.5.

Cold test

The cold test measures the resistance of an oil to formation of a sediment at 0°C. Sediment formation is usually caused by compounds with high melting temperature, mainly waxes. Measurement of this index is generally used to evaluate the effectiveness of the winterisation process.

Melting characteristics: polymorphism and crystal properties

The change to canola oil, as a result of the reduction in erucic acid (C22:1), had a marked effect on the melting characteristics of the oil and the type of crystals formed. This was attributed to the more homogeneous fatty acid composition of canola oil with 95 per cent 18 carbon atoms fatty acids. Hydrogenation of canola oil is used to produce the main products utilised in the formulation of shortenings and margarines with the *trans* isomers formed having higher melting points than the *cis* fatty acids (Table 2.7).

Table 2.7. Melting characteristics of octadecanoic fatty acid family.

Fatty acid	*Melting point (°C)*
Linolenic (*cis*-9, 12, 15)	–11.2
Linoleic (*cis*-9, 12)	–5.1
Oleic (*cis*-9)	13.2
Octadecenoic (*cis*-6)	28.6
Elaidic (*trans*-9-octadecenoic)	43.7
Stearic	69.6

Polymorphism is a well-known phenomenon associated with the crystallisation behaviour of long-chain compounds. Fats can crystallise into a number of subcrystalline forms such as α, β and β', each differing in size and stability. The ability of a fat to exist in a number of different crystalline forms depends on how the molecules arrange themselves in the solid state. It has been established that hydrogenated canola oil has a tendency to crystallise in the β-form, which forms large crystals ranging in size from 5 to 25 μm. The formation of these large crystals causes an increase in graininess, which is directly responsible for gritty and crumbly products. In the manufacturing of margarine the β'-crystal form is the desired form as it has smaller crystals (less than 1 μm in size) giving the formulated product desirable textural characteristics. Energetically, however, the β form is less stable requiring higher amounts of energy for crystals to pack than in the β form. This metastable crystalline form has a tendency to transform into a lower energetical state, the β form.

Trans isomers of fatty acids were found to have a tendency to produce products with higher β′ stability than *cis* acids. This was attributed to the steric effect of these isomers, which hinders the transformation to the β form. To stabilise the β′ form, a blend of three hydrogenated canola oils is often used to increase the heterogeneity of the triglycerides. This prompts the margarine to crystallise in small, needle-shaped crystals, giving the final product a smooth, pleasing mouth feel and good spreadabilty. A more effective approach in avoiding β-crystal formation is to use 10–15 per cent palm oil or 20–25 per cent cottonseed oil to supply triglycerides with the palmitic acid.

Solid fat index and dilatation curve

The solid fat index (SFI) and dilatation curve for a hydrogenated fat describe the amount of solid fat remaining at assigned temperatures. Individual triglycerides differ in physical properties according to their fatty acid composition. Thus, when a fat is kept at a particular temperature, those triglycerides containing unsaturated fatty acids melt first while those containing the saturated and *trans* isomers of fatty acids melt last. The solid fat components expand as the temperature is increased, reaching a maximum prior to melting completely to a liquid. The expansion of the fat or dilatation can be monitored by measurement of the increase in specific volume with temperature and establishing the dilatometric curve. This enables calculation of the percent solid fat at any specific temperature.

The spreadability of margarine can also be predicted from the SFI. To achieve the desired body and melting properties with stick margarines, selectively hydrogenated canola oil are used along with nonhydrogenated oils. The most desirable approach is to use SH (selectively hydrogenated) and SHI hydrogenated canola oils together with a certain amount of liquid canola oil (Table 2.8). The SFIs of the oil phase blend must be in the range of 25–30 per cent at 10°C, 14–18 at 21.1°C, and 2–3.5 at 33.3°C. The low SFI at 10°C in the soft margarines appeared to be responsible for their spreadability at refrigeration temperatures (Table 2.8). Formulation of these margarines requires the use of 70–85 per cent of slightly hydrogenated and/or liquid oils with a minimum of about 11–15 per cent of highly hydrogenated stock usually under nonselective conditions (Table 2.8).

Table 2.8. Solid fat indices of canola hydrogenated stocks and margarines.

	Solid fat index			*Fatty acids*			
Sample	*10.0°C*	*21.1°C*	*33.3°C*	*PUFA*	*MUFA*	*SAT*	*TRANS*
Hydrogenated[b]							
SH[c]	10.8	1.4	0.1	10.6	78.1	9.1	34.0
SH1	41.3	22.5	15.9	3.5	76.7	17.7	51.9
NS[d]	6.2	1.8	1.2	14.4	67.5	14.9	24.6
NS1	24.5	13.4	8.2	7.7	67.0	23.1	31.7
Margarines							
Stick regular	25–30	14–18	2–4	3–10	50–70	16–25	
Stick Hi–Li[e]	16–24	10–15	1.5–4.0	20–40	20–50	13–23	
Soft (70% Liq.)	8–14	5–8	0.5–2.5	30–60	15–42	10–20	

[a] Saturated fatty acids. [b] Hydrogenated stocks. [c] Selective hydrogenation. [d] Nonselective hydrogenation. [e] Margarine with high content of liquid oil.

Using this stock, the formulated product is characterised by good spreadability at refrigerated temperatures and good emulsion stability so there are no oiling-out problems. Typical SFIs of soft margarines are 8–14 per cent at 10°C, 5–8 at 21.1°C, and 0.5–2.5 at 33.3°C. In margarines with good mouth melt characteristics the oil phase in the product melts sharply at body temperature, which results in the total breakdown of the emulsion in the mouth with the release of flavour-laden water phase. If the crystalline structure does not melt rapidly, the margarine feels waxy or thick in the mouth.

Chemical Properties

Nature of oils and fats

Edible oils and fats are composed primarily of triglycerides, esters of one molecule of glycerol and three molecules of fatty acids. Analysis of canola oils showed the triglyceride levels constituted 94.4–99.1 per cent of the total lipid. The typical composition of canola, rapeseed and soyabean oils is presented in Table 2.9.

Table 2.9. Constituents of canola, rapeseed and soyabean oils.

Component	*Canola*	*Rapeseed*	*Soyabean*
Triglycerides (%)	94.4–99.1	91.8–99.0	93.0–99.2
Phospholipids (%)			
Crude oil	Up to 2.5	Up to 3.5	Up to 4.0
Water-degummed	Up to 0.6	Up to 0.8	Up to 0.4
Acid-degummed	Up to 0.1	–	Up to 0.2
Free fatty acids (%)	0.4–1.2	9.5–1.8	0.3–1.0
Unsaponifiables (%)	0.5–1.2	0.5–1.2	0.5–1.6
Tocopherols (ppm)	700–1200	700–1000	1700–2200
Chlorophylls (ppm)	5–35	5–35	Trace
Sulphur (ppm)	3–15	5–25	Nil

Triacylglycerols

Triacylglycerols (TAG) are the most abundant lipid class found in canola oil. The combination of fatty acids on the glycerol moiety are the most complex with n^3 potential molecular species possible, with n being the number of fatty acids. The TAG molecular species profile represents a key to the understanding of the physical characteristics of an oil and also is a unique means of identification which has been used in various instances. Position of fatty acids on the glycerol molecule was originally found for rapeseed oil to be based on saturation. Long-chain (C20:–C24:0) and saturated fatty acids occurring mostly in the 1 and 3 positions, while the octadecanoic (C18) acids, especially the linoleic and linolenic, are integrated in the 2 position.

Fatty acids

Fatty acids are composed of a carboxyl group and a hydrocarbon chain. Individual fatty acids are distinguished from one another by the nature of hydrocarbon chain. This chain can vary in length from 4 to 24 carbon atoms and can be saturated, monounsaturated (on double bond, MUFA) or polyunsaturated (two or more double bonds, PUFA).

The most common fatty acids in edible oils and fats are those containing 18 carbons. These include stearic acid (an 18-carbon saturated fatty acid), oleic acid (18-carbon monounsaturated fatty acid) and linoleic and linolenic acids (both 18-carbon acids containing two and three double bonds, respectively).

Formula	*Common name*	*Abbreviation*
$CH_3(CH_2)_{16}COOH$	Stearic acid	C18:0
$CH_3(CH_2)_7CH = CH (CH_2) COOH$	Oleic acid	C18:1*n*-9
$CH_3(CH_2)_4CH = CH (CH_2$- $CH = CH(CH_2)_7COOH$	Linoleic acid	C18:2*n*-6
$CH_3CH_2CH = CHCH_2$- $CH = CH\ CH_2CH = CH(CH_2)_7COOH$	Linolenic acid	C18:3*n*-3

These fatty acids are abbreviated according to the number of carbon atoms in the molecule and the number of *cis* ethylenic double bonds, on the assumption that they are all methylene-interrupted. The chemical nomenclature requires that carbon atoms be counted from the carboxyl end of the fatty acid. For biological activity, however, carbon atoms are numbered from the terminal methyl group to the first carbon of the ethylenic bond. Such a classification is designated by the symbol *ω-x*, *ωx*, or *n-x*, *nx*, where *x* denotes the position of the double bond closest to the terminal methyl group. For example, linoleic acid with two double bonds, has one that is located six carbon atoms from the methyl group. This fatty acid is thus abbreviated C18:2*n*-6.

Geometric isomerism

In the case of unsaturated fatty acids, several geometric isomers are possible, as a result of the carbon chain being bent into a fixed position at the double bond. When the portions of the chain are bent twoard each other, they are called *cis*; when bent away, *trans*. The natural configuration of fatty acids is the *cis* configuration as shown for oleic acid, where the carbon chain is doubled back on itself. The corresponding *trans* configuration, elaidic acid, results in a straight-chain configuration, except for a slight kink in the chain at the double bond (Fig. 2.4). The importance of *cis* and *trans* configuration is that it is the *cis* form that is considered nutritionally more desirable. Isomerisation from *cis* to *trans* occurs during the hydrogenation of an oil.

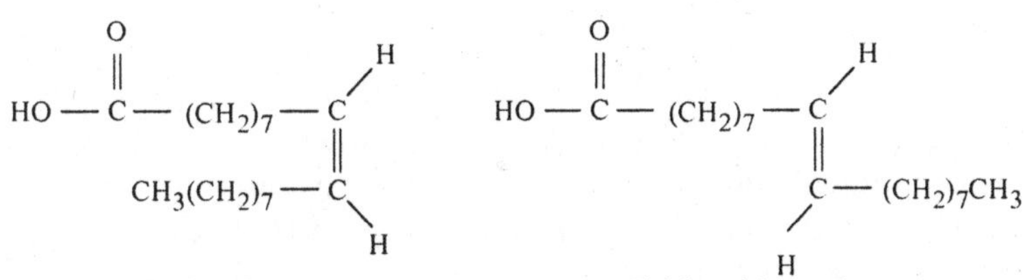

Oleic acid (*cis*) Elaidic acid (*trans*)

Fig. 2.4. Configuration of oleic (*cis*) and elaidic (*trans*) acids.

Fatty acid composition of canola oil

The stigma of the erucic acid (C22:1*n*-9) in rapeseed oil has lingered, despite firm evidence that this fatty acid was more of a threat to rats than to human. It is sufficient to say that the discovery of chain shortening of erucic acid to oleic acid by peroxisomes was one of the most fundamental breakthroughs in our understanding of fatty acid metabolism in the last few decades. Once in the oleic acid form, the erucic acid residue is as readily catabolised by mitochondria as are palmitic and other fatty acids. The reduction of erucic acid (22:1) in rapeseed oil resulted in a marked increase in octadecanoic acids, acids with 18 carbon atoms.

Minor fatty acids

Minor fatty acids present in canola oil often differ from their fatty acid family members by the location of the double bond. Most of these acids are present in the 0.01–0.1 per cent range, except for C16:1*n*-7, which was around 0.3 per cent. Most of these minor fatty acids were *n*-7 to the *n*-9 isomers. A chromatogram of some of these minor fatty acid isomers is shown in Fig. 2.5.

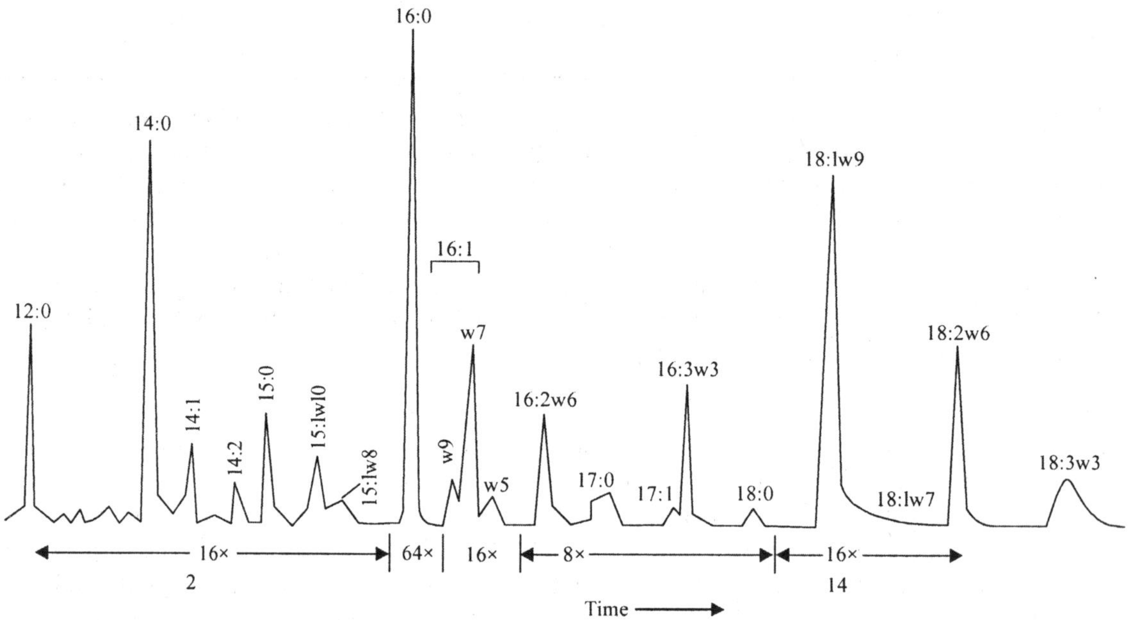

Fig. 2.5. Minor fatty acids found in seeds of *B. Napus* var. Tower.

Conjugated C18:2 fatty acids were also found in canola oils. Some of these compounds are artifacts of oxidation formed during refining. The refining process itself is a source of artifact fatty acids due to the isomerisation of one or more of the double bonds of *cis* linolenic acid. These *trans* isomers can be found in any oil containing linolenic acid after refining, accounting for 1 per cent or more or the parent acid. Canola oil is the only known edible oil that contains one or more fatty acids with sulphur as the integral part of the molecule (Fig. 2.6). The structure of the proposed molecule of this fatty acid suggests the formation or presence of many isomers.

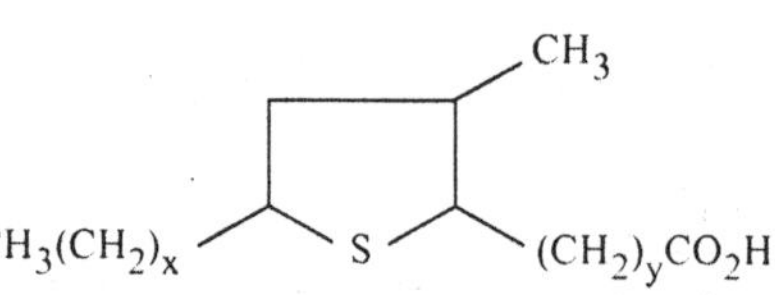

	x	y
1	5	7
2	6	6
3	7	5

Fig. 2.6. Isomers of a sulphur-bearing fatty acid found in canola oil.

In the sediment from industrial winterisation additional minor fatty acids and alcohols with 26–32 carbon atoms in chain have been found. Most of these compounds are extracted from seed coat/hull and can be blamed for occasional haze formation in canola oil.

Polar lipids

It has been found that phospholipids formed the major component (3.6 per cent) of the polar lipids while glycolipids contribute only 0.9 per cent. A more recent study by *Przybylski* and *Esken* reported changes in phospholipids during early stages of conola oil processing (Table 2.10). Significant amounts of phosphatidic acid were formed, which indicates the decomposition of other phospholipids during early stages of processing. This hydrolysis can be attributed to the effect of thermal treatment during the conditioning of flaked seeds.

Table 2.10. Composition of phospholipids in canola oil during processing (%)[a].

Oil sample	*Phosphorus (ppm)*	*PC*	*PE*	*PI*	*PA*	*PS*
Solvent	529.0	31.2	18.8	19.8	21.6	3.1
Expeller	242.3	34.3	16.1	18.7	20.3	4.5
Degummed	12.2	2.8	10.8	28.9	38.4	14.6

[a] Phospholipids: PC, phosphatidyl choline; PE, phosphatidyl ethanolamine; PI, phosphatidyl inositol; PA, phosphatidic acid; PS, phosphatidyl serine.

The presence of highly unsaturated fatty acids in phospholipids is important as they can initiate oxidation of other fatty acids causing accelerated deterioration of the oil. It was reported that phospholipids have a tendency to complex heavy metals. In this form they provide a rather stable catalyst to initiate and stimulate oxidation. The higher amounts of polyunsaturates in soyabean lecithins suggest even greater stability problems compared to canola. For example, linoleic and linolenic acid content in soyabean phosphatidyl choline (PC) was 62 and 7 per cent of the total fatty acids compared to 30 and 1.9 per cent in canola PC, respectively.

Trace elements

The proposed Codex standard for edible Low Erucic Acid Rapeseed (LEAR) provides the maximum levels permitted for iron, copper, lead and arsenic. While these metals are found in other edible oils and are present naturally in the seed, in addition they can be introduced during handling and processing. Mineral element content in canola oils (ppm) is shown in Table 2.11. A very efficient removal of phosphorus and calcium from the oil during processing can be achieved as both of these elements usually formed salts and are removed during degumming. Sulphur in canola oil is in the form of organic compounds as decomposition products formed from glucosinolates. Although these sulphur components occur in trace quantities, they are known to inhibit catalysts used for hydrogenation as well as imparting characteristic odours to the oils. Recent developments in analytical methods for sulphur content evaluation revealed that soyabean, sunflower and even coconut oils contain sulphur at a level of 2–10 ppm. In unrefined canola oil the amount of sulphur present was 25.0 ppm and in RBD oil 9.4 ppm. Sulphur components can not only be destructive to canola oil quality but can also have a positive effect on the stability of this oil. Some sulphur components can also act as antioxidants and protect the oil from autoxidation by complexing hydroperoxy radical with the sulphur present in the compounds.

Table 2.11. Mineral element content in canola oils (ppm).

Oil sample	*Phosphorus*	*Iron*	*Calcium*	*Sulphur*	*Zinc*	*Lead*
Crude oil	1190.0	3.52	296.0	6.5	2.4	0.24
Degummed with water	222.0	1.32	169.0	1.2	2.1	–
Phosphoric acid	117.2	0.63	34.8	1.5	–	–
Bleached						
WDG	0.21	0.23	5.6	–	–	–
PDG	0.19	0.59	4.1	0.87	–	–
Deodourised						
WDG	0.25	–	–	0.25	–	0.07
PDG	0.22	–	–	0.38		

Tocopherols and sterols

The main components of the nonsaponifiables in vegetable oils are tocopherols and sterols, although the amounts vary in different oils. Tocopherols are natural antioxidants and their level in the plant is governed by the level of unsaturated fatty acids, a simple increase in unsaturation results in the formation of higher amounts of antioxidants to protect the oil. The composition of tocopherols in canola oil and some common vegetable oils is summarised in Table 2.12.

Table 2.12. Tocopherol contents in selected vegetable oils (ppm).

Oil	α	β	γ	δ
HEAR	268.0	–	426.0	–
Canola (Westar)[a]	272.1	0.1	423.2	–
Soyabean (refined)	116.0	34.0	737.0	275.0
Soyabean	90.0	–	680.0	230.0
Sunflower	608.0	17.0	11.0	–
Corn	134.0	18.0	412.0	39.0

The distribution of natural tocopherols vary with the different vegetable oils both quantitatively and in the amount of different isomers as shown for soyabean, canola, sunflower and corn oils (Table 2.12). Refining, particularly deodourisation, removes a part of these protective components. Different isomers have different biological and antioxidant activity. The α isomer is thought to be the most biologically active while the γ isomer is perceived as the best antioxidant. Canola oil contains mainly these two isomers of tocopherols, but the amount of the γ isomer is lower and the α isomer is higher than in soyabean oil (Table 2.12).

Sterols are present in canola oil in equal proportions of free sterols and esterified sterols. The total amount of sterols in rapeseed and canola oils range from 0.53 to 0.97 per cent. Composition of major sterols in common vegetable oils is presented in Table 2.13.

Brassicasterol is one of the sterols present in rapeseed and canola oil is used to determine adulteration of other oils with rapeseed oil, even though considerable data indicates the presence of this sterol in other oils. Sterols are also affected by processing. A significant amount is removed from the oil during deodourisation whereas refining causes isomerisation of these compounds and formation of few products.

Table 2.13. Proportions of major sterols in selected vegetable oils (%).

Sterol	*HEAR*	*Canola*	*Soyabean*	*Sunflower*	*Corn*
Cholesterol	0.4	0.1	0.3	0.1	0.1
Brassicasterol	13.2	13.8	–	–	–
Campestrol	34.4	35.2	18.1	7.5	17.2
Stigmaseterol	0.3	0.5	15.2	7.5	6.3
β-Sitosterol	47.9	48.2	54.1	58.2	60.3
Δ^5-Avenasterol	2.1	2.4	2.5	4.0	10.5
Δ^7-Avenasterol	–	–	2.0	4.0	1.1
Δ-Stigmasterol	–	–	3.6	11.5	0.1
Δ^7-Campestrol	–	–	0.5	2.4	–
Total[a] as mg/100 g of oil	881.0	892.0	436.0	496.0	1441.0

[a] Approximate amount in oils.

The amount of total sterols in canola oil is about twice that in soyabean oil. Corn oil, which is produced from corn seed embryo, contains very high amount of sterols, threefold higher than that of canola oil. The chemical structure of phytosterols is similar to that of cholesterol so it is possible that these compounds may be involved in similar type reactions.

Pigments

Pigments present in canola and other oilseeds are important factors as they impart undesirable colour to vegetable oils; they can promote oxidation in the presence of light as well as inhibit catalysts used for hydrogenation. Chlorophylls without phytol such as chlorophyllides and pheophorbides may present a nutritional effect because of their phototoxicity, which may be followed by photosensitive dermatitis. A bleaching step is necessary during oil processing to remove chlorophyll-related pigments and other colour bodies. Changes in chlorophylls during canola oil processing are summarised in Table 2.14.

Table 2.14. Chlorophyll pigments[a] in canola oil during processing (ppm).

Oil after	*Chlor a*	*Pheo a*	*Pheo b*	*Pyro a*	*Pyro b*
Expeller	6.27	4.48	1.79	5.37	0.67
Extraction	1.88	3.31	1.34	16.57	3.13
Expeller + extraction	1.79	5.55	1.34	9.76	1.43
Degumming	0.27	7.16	1.07	9.40	1.84
Alkaline refining	0.22	6.27	1.12	9.13	1.79
Bleaching	–	0.56	0.32	0.21	0.25

[a] Chlor *a*, chlorophyll *a*; pheo *a*, pheophytin *a*; pheo *b*, pheophytin *b*; pyro *a*, pyropheophytin *a*; pyro *b*, pyropheophytin *b*.

During processing, chlorophyll completely decomposes to derivatives that are much harder to remove during bleaching. A bleaching test showed that pheophytin *a* and pyropheophytin *a* are more absorptive than their *b* isomers. Consequently smaller amount of *b* isomers of chlorophyll derivatives are removed from the oil during bleaching compared to *a* isomers. This necessitates much higher amounts of activated bleaching earth to be used in order to achieve similar removal of all chlorophyll derivatives.

In addition to chlorophyll pigments, carotenoids are also found in canola oil. Crude canola oil carotenes are reported to be at a level of 95 ppm, composed of 90 per cent xanthophylls and 10 per cent carotenes.

Saponification value

Saponification value is defined as the weight of potassium hydroxide, in milligrams, needed to saponify one gram of fat. This parameter is inversely proportional to the molecular weight of the triaclyglycerol; in other words, the higher the molecular weight the lower the saponification value. Replacement of long-chain fatty acids such as erucic acid (C22:1) in rapeseed oil by octadecanoid (C18) fatty acids in canola oil increased the saponification value from 168–181 to 181–192.

Iodine value

Iodine value (IV) is an empirical test indicating the degree of unsaturation of a fat or oil. It is defined as the number of grams of iodine absorbed by 100 grams of fat. An iodine value of 97–108 was reported for rapeseed oil with 45 per cent of erucic acid as compared to 110–126 for canola oil. The higher value for canola oil is due in part to the replacement of erucic acid with oleic acid together with an increase in linoleic and linolenic acids. Iodine value can also be calculated from fatty acid composition as proposed by AOCS Method Cd 1c-85 where the fatty acid content as a percentage is multiplied by the characteristic factor for each unsaturated fatty acid.

Antioxidants

The role of antioxidants in retarding rancidity is well established, although the efficacy of some of them has recently been questioned. The best antioxidants are the natural components of canola oil. Tocopherol, particularly its γ isomer, is present in the processed oil in an amount twice that of α isomer (Table 2.12). The most commonly used synthetic antioxidants are butylated hydroxyanisole (BHA), butylated hydroxytoluene (BHT), propyl gallate (PG) and tertiary butylhydroquinone (TBHQ). These phenolic compounds react with free radicals to form relatively stable antioxidant free radicals, which further degrade to produce quinones, thereby terminating the chain reaction of autoxidation.

Sedimentation Phenomenon in Canola Oil

In recent years, the canola oil industry has seen a recurring problem of turbidity (haze, cloudiness) formation during the storage of bottled oil under normal warehouse conditions. This characteristic is detrimental to canola oil quality and may affect consumer preference. Winterisation of canola oil often solves this problem but additional operations and increased costs are involved.

Sediment composition

The major component in sediment isolated from winterisation cake are waxes (Table 2.15). The other major components are the polar compounds, identified as Others in Table 2.15, the nature of which remains unclear. All tests and separation attempts revealed that the polar fraction of sediment was not phospholipid in character, nor did it contain fatty acids as lipids do. Elementary analysis suggested a composition similar to carbohydrates. Lipid components of sediment contained mostly saturated fatty acids made up principally of even numbers ranging from 14:0 to 32:0 carbon atoms in the chain. A small amount of fatty acids with an odd number of carbon atoms and with branched chains, over the range mentioned previously, were also identified. Among wax esters were components with 42–62 carbon atoms. Compounds with long chains can directly initiate crystallisation because of their high melting temperature and require very low concentration to achieve saturation.

Table 2.15. Composition of canola sediment isolated from an industrial filter cake collected after winterisation.

Component	*Content (%) ± s.d.*[a]
Wax esters	78.1 ± 1.0
Free fatty acids	0.2 ± 0.1
Free fatty alcohols	2.0 ± 0.1
Diglycerides	2.7 ± 0.2
Triglycerides	Trace
Others	17.2 ± 0.6

[a] Standard deviation.

Physical properties of sediment

Melting temperature of sediment measured by differential scanning calorimetry (DSC) was found to be affected by the amount of an oil present and by the procedure used to isolate the sediment. A canola oil sediment double-washed with cold petroleum ether gave a sharp DSC peak with melting temperature at 75.5°C. A slightly lower melting temperature, at 74.8°C, was observed for isolated sediment from filtration cake obtained from industrial winterisation. Additional washing with cold petroleum ether resulted in a sediment with a melting point of 77.6°C, indicating removal of residue of oil and/or partial extraction of some shorter chain components. *Ackman* used solvent fractionation for sediment isolation and found that the fraction insoluble in hexane had a melting temperature at 87.3°C. *Ackman* indicated that the major components of this fraction were long-chain waxes. *Liu* and co-workers reported isolation of sediment fractions formed at 5 and 32°C. The melting endotherms for both fractions were similar in shape, although the peak temperatures were 75.2 and 78.7°C, respectively. Further characterisation of these fractions revealed that the one obtained at the higher temperature contained more long-chain fatty acids and alcohols.

The time needed for turbidity appearance in oil containing various fractions was found to be concentration dependent. However, the same concentration of fraction obtained at the higher crystallisation temperature required only 1/25 of the time to form turbidity compared to the fraction obtained at the lower temperature. Turbidity formation is affected by the total concentration of waxes in an oil while the rate is influenced by composition of sediment, particularly components with high melting points. For the fraction isolated at 32°C turbidity was observed at 80 hr. even at a concentration of 20 ppm, while the fraction obtained at the lower temperature did not cause cloudiness. This indicates that high melting components can easily initiate cloudiness at room temperature. The effect of lecithin on crystallisation was evaluated and found only to effect the size of crystals with smaller crystals formed in much larger numbers.

NUTRITIONAL PROPERTIES OF CANOLA OIL

Nutritional Significance of Canola Oil Composition

Dietary fat serves several important nutritional functions. It is the source of essential fatty acids. Members of the *n*-6 and *n*-3 (also known as the ω-6 and ω-3) families of fatty acids are important constituents of cell membranes and serve as precursors of eicosanoids (biologically active compounds such as prostaglandins, thromboxanes, prostacyclins and leukotrienes). Fat also serves as a carrier for the fat-soluble vitamins and it is important source of energy. In addition, it has important culinary properties and contributes to the palatability of food.

The current interest in dietary fat, however, stems primarily from its implication in the etiology of several chronic diseases. Interest has centered on both the amount and type of dietary fat in the development of cardiovascular disease, cancer, hypertension and obesity. As a result, dietary recommendations in many countries call for a reduction in total fat intake to 30 per cent of energy and in saturated fat intake to 10 per cent of energy. In addition, some nutrition recommendations specify recommended levels of *n*-6 and *n*-3 fatty acids in the diets of specific age groups. Interest in the nutritional properties of canola oil arose because of its fatty acid composition. Canola oil is characterised by a low level of saturated fatty acids, a relatively high level of monounsaturated fatty acids and an appreciable amount of the *n*-3 fatty acid α-linolenic acid (18:3*n*-3).

Saturated fatty acids

The adverse effect of saturated fat on blood cholesterol level and its implication in cardiovascular disease has stimulated concern over the level of saturated fatty acids in the diet. Canola oil contains a very low level (<7 per cent) of saturated fatty acids; about half the level present in corn oil, olive oil or soyabean oil and about one-quarter the level present in cottonseed oil. Furthermore, canola oil contains only 4 per cent of the saturated fatty acids (viz., lauric, myristic and palmitic) that have been found to increase blood cholesterol level. Hence, canola oil fits well with the recommendation to reduce the amount of saturated fat in the diet.

Monounsaturated fatty acids

The report that monounsaturated fatty acids (viz., oleic acid) were just as effective as polyunsaturated fatty acids (viz., linoleic acid) in lowering plasma total and low-density lipoprotein (LDL) cholesterol aroused interest in the nutritional properties of canola oil. Canola oil contains 60 per cent oleic acid and is second only to olive oil, among the common vegetable oils in oleic acid content. Although avocado oil and high–oleic sunflower oil also contain high levels of oleic acid (>70 per cent), they are minor constituents in the average diet.

There, also, is interest in dietary monounsaturated fatty acids because of their possible protective effect against oxidation of LDL. There is appreciable evidence that the uptake of LDL and the formation of fatty streaks in the intima of large blood vessels, which is considered an early lesion of atherosclerosis, is enhanced by the oxidation of the LDL. LDL was found to be appreciably more stable to oxidation when subjects were fed diets rich in oleic acid than when fed linoleate-enriched diets.

Polyunsaturated fatty acids

Canola oil is intermediate among the vegetable oils in polyunsaturated fatty acid content; it contains lower levels than corn oil, cottonseed oil, soyabean oil or sunflower oil but appreciably higher levels than olive oil or palm oil. Interest in polyunsaturated fatty acids stems from their role as essential fatty acids and their effectiveness in reducing plasma cholesterol level. Linoleic acid and arachidonic acid, members of the *n*-6 family of fatty acids, have long been recognised as essential fatty acids. Arachidonic acid is an important constituent of cell membranes and is the precursor of eicosanoids, 'hormone like' substances that are involved in a multiplicity of physiological reactions ranging from blood clotting to immune response.

α-Linolenic acid and other members of the *n*-3 polyunsaturated fatty acids likewise are essential. Docosahexaenoic acid (22:6 *n*-3) is a major constituent of lipids of the brain and retina of the eye while eicosapentaenoic acid (20:5*n*-3) is the precursor in the synthesis of an analogous but different series of eicosanoids from those formed from arachidonic acid. Canola oil contains an appreciable amount (10 per cent) of α-linolenic acid. In addition, there is a favourable balance between linolenic acid and linoleic acid (a ratio of approximately 1:2). Soyabean oil is the only other major edible oil that contains a significant amount of linolenic acid (approximately 7 per cent) but the ratio to linoleate is approximately 1:7.

Minor constituents: tocopherols and phytosterols

Vegetable oils are the primary source of tocopherols in the average diet. Canola oil is a relatively rich source of tocopherols (Table 2.12); it is similar in total tocopherol content to corn oil, cottonseed oil, safflower oil and sunflower oil (60–70 mg/100 g). Only soyabean oil contains an appreciably higher level of total tocopherols (100–110 mg/100 g). However, canola oil contains a considerably higher level of α-tocopherol, the main source of vitamin E biopotency among the tocopherols than soyabean oil (27 vs. 12 mg/100 g). Only cottonseed oil, safflower oil and sunflower oil are better sources of α-tocopherol than canola oil (35, 56 and 61 mg/100 g, respectively).

Canola oil contains a relatively high level of phytosterols (892 mg/100 g), about twice the level in soyabean oil or sunflower oil (436 and 496 mg/100 g, respectively) (Table 2.13). β-Sitosterol accounts for about 50 per cent, campesterol 35 per cent and brassicasterol 14 per cent of the total phytosterols in canola oil. Canola oil is the only common vegetable oil that contains brassicasterol. Plant sterols have been reported to lower plasma cholesterol level by inhibiting the absorption of dietary cholesterol and the reabsorption of biliary cholesterol.

Effect of Canola Oil on Plasma Cholesterol and Lipoproteins

Appreciable research on the effect of canola oil on plasma cholesterol and lipoproteins has been reported by *McDonald* and *Ackman.* The primary impetus for this research was the finding that dietary monounsaturated fatty acids were as effective as polyunsaturated fatty acids in lowering plasma total and LDL cholesterol. These findings also provided a possible explanation for the observation that canola oil was as effective as soyabean oil in lowering plasma cholesterol in normolipidemic men. Prevailing theory had held that saturated fatty acids raised plasma cholesterol, polyunsaturated fatty acids lowered plasma cholesterol and monounsaturated fatty acids were neutral, they neither raised nor lowered plasma cholesterol.

Normolipidemic subjects

Studies in Canada, Finland and the United States have shown canola oil to be just as effective in lowering plasma total and LDL cholesterol levels as far sources containing high levels of polyunsaturated fatty acids when each replaced saturated fat sources in the diet. All diets resulted in statistically significant decreases in plasma total (mean of –0.47 to –0.88 mmol/l) and LDL cholesterol (mean of –0.43 to –0.74 mmol/l) levels. Thus, most of the decrease in total cholesterol was due to a decrease in LDL cholesterol, which is consistent with the main objective of dietary intervention programmes aimed at reducing the risk of cardiovascular diseases. Elevated plasma LDL cholesterol level is a primary risk factor in the development of coronary heart disease.

Hypercholesterolemic subjects

Canola oil also was effective in reducing serum total and LDL cholesterol levels in subjects with elevated blood lipid levels. Replacing 50 gram of fat in the regular diet with 50 gram of canola oil mayonnaise resulted in a significant decrease in serum total and LDL cholesterol levels (–9 and –10 per cent, respectively) in a group of men with a mean serum cholesterol level of 7.1 mmol/l. Serum triglyceride and very low density lipoprotein (VLDL) levels also decreased while serum HDL levels increased on the canola oil regimen. Generally, type of fat, at the level of substitution used in this study, has not affected serum HDL levels. *Bierenbaum* also found that substituting canola oil for 30 g of the usual oils and spreads in the diets of hyperlipidemic subjects resulted in a decrease in plasma LDL cholesterol level. Plasma total cholesterol, however, did not differ from baseline levels. Substituting canola oil or canola oil margarine for butter, at much lower levels (approximately one-fifth of the total fat and 8 per cent of total energy) than those cited above, also resulted in a decrease in the plasma cholesterol level of mildly hypercholesterolemic subjects. Substitution of canola oil for butter resulted in a 9 per cent decrease in total cholesterol and 13 per cent decrease in LDL cholesterol level while substitution of canola oil margarine for butter resulted in a 9 per cent decrease in total cholesterol and an 8 per cent decrease in LDL cholesterol level.

Canola oil also was effective in lowering plasma total and LDL cholesterol concentrations of mildly hyperlipidemic subjects fed a low-fat (30 per cent of total energy) diet. The decline in plasma total cholesterol when canola oil or corn oil provided 20 per cent of the total energy was –12 and –13 per cent, respectively, which was significantly greater than the decline when olive oil (–7 per cent) supplied the fat. However, canola oil, corn oil and olive oil were equally effective in lowering plasma LDL cholesterol levels (–16, –17 and –13 per cent, respectively); elevated plasma LDL level is a major risk factor in coronary heart disease.

Effect of Canola Oil on Thrombogenesis

Cardiovascular disease is characterised by two major events: the formation of atherosclerotic plaques on the intima of blood vessels, which reduce the size of the lumen of the vessel; and thrombosis or clot formation, which is the event leading directly to a coronary attack or stroke in many individuals. Clot formation has only recently received attention from researchers even though dietary fat has been implicated in both processes. Marked differences among *Greenland Eskimos* and *Danes* in the incidence of coronary heart disease led to interest in the effect of long-chain n-3 polyunsaturated fatty acids on thrombogenesis. The long-chain n-3 fatty acid, of fish oil, namely eicosapentaenoic acid and docosahexaenoic acid, have been found to inhibit platelet aggregation. A possible mechanism for the effect of long-chain fatty acids on thrombogenesis is thought to be a change in the fatty acid composition of platelet phospholipids and an accompanying change in eicosanoid formation.

Canola oil has been shown to alter several parameters linked to clot formation (Table 2.16). In general, phospholipid fatty acid composition reflects the fatty acid composition of the diet: canola oil resulted in higher levels of oleic acid whereas safflower oil, soyabean oil and sunflower oil resulted in higher levels of linoleic acid. Canola and soyabean oil also resulted in higher levels of α-linolenic acid although very little linolenic acid was incorporated into phospholipids even when flaxseed oil was incorporated into the diet. Dietary linolenic acid, however, has been found to alter the long-chain polyunsaturated fatty acid composition of platelet phospholipids. Several groups reported lower levels of arachidonic acid with canola oil diets. Canola oil diets also resulted in higher levels of eicosapentaenoic acid and docosapentaenoic acid than soyabean oil, sunflower oil or customary mixed fat diets. Not all studies, however, found higher levels of long-chain n-3 polyunsaturated fatty acids with canola oil diets. Part of the explanation for the apparent discrepancy among studies may relate to the balance of fatty acids in the diet. *Chan* and co-workers found that changes in plasma and platelet fatty acid composition varied not only with the level of linolenic in the diet but also with the ratio of linolenic to linoleic acid. Canola oil was found to reduce *in vitro* platelet aggregation although sunflower oil and safflower oil (i.e. low level of linolenic acid and high level of linoleic acid) also reduced *in vitro* platelet aggregation. On the other hand, *Mutanen* and co-workers found that canola oil and sunflower oil enhanced *in vitro* platelet aggregation. Similarly, *McDonald* and co-workers found that both canola oil and sunflower oil increased prostacyclin production (an antiaggregating eicosanoid) and decreased thromboxane production (a proaggregating eicosanoid). Although the effect of canola oil on platelet activity and clot formation is not as well established as its favourable effect on plasma cholesterol and lipoprotein levels, there is evidence that it may impeded clot formation.

Table 2.16. Effect of canola oil on clotting time and factors involved in clot formation.

Parameter	*Effect of canola oil on parameter*
n-3 fatty acid (viz., 20:5 *n*-3 and 22:5 *n*-3) content of platelet phospholipids	Increased No change
Arachidonic acid content of platelet phospholipids	Decreased
In vitro platelet aggregation	Reduced Enhanced
Eicosanoid production	
Thomboxane	Decreased
Prostacyclin	Increased
Clotting time	Increased

Coronary heart disease, however, is a complex, multifactorial disease and the mechanism by which diet affects its development has not been resolved fully. Likewise, there was no difference between the groups in platelet aggregation even though plasma levels of oleic, linolenic and eicosapentaenoic acids were significantly higher and stearic, linoleic and arachidonic acids significantly lower for the experimental group. Nevertheless, the canola oil (α-linolenic acid enriched) diet seemed more effective in the prevention of further coronary events than the diets usually used by patients who had suffered myocardial infarction.

Canola Oil in Infant Formulas

Infant formulas differ appreciably in fatty acid composition from human breast milk. The latter is characterised by relatively high levels of palmitic and oleic acids and contains small but important levels of long-chain polyunsaturated fatty acids (viz., arachidonic acid and docosahexaenoic acid). Infants are capable of desaturating and elongating linoleic acid and linolenic acid to their long-chain homologues (viz., arachidonic and docosahexaenoic acids, respectively) but the efficiency of this process in the infant is only now being clarified. The same pathway is involved in the desaturation–elongation of fatty acids; thus there is competition between the *n*-6 and *n*-3 fatty acids for the enzymes in this pathway. Moreover, there are differences in the rate of incorporation of linoleic and linolenic acids into tissues and in their rate of utilisation for energy. Some of the apparent differences in partition among the various pathways, however, may be due to the linolenate-to-linoleate ratio of the diet. For example, no differences were found in the utilisation of oleic, linoleic and linolenic acids when they were incorporated into the diet at 1:1:1 ratio. There is interest in the possible use of canola oil in infant formulas because of its fatty acid composition, namely the relatively high oleic acid level and the balance between *n*-3 and *n*-6 fatty acids (18:3/18:2 ratio of approx. 1:2).

Overall Properties

Canola oil is characterised by a low level of saturated fatty acids (less than 4 per cent palmitic acid) and relatively high levels of oleic acid (60 per cent) and α-linolenic acid (10 per cent). It is second only to olive oil, among the common fats and oils, in oleic acid level and, except for soyabean oil, the only common dietary fat that contains a significant amount of α-linolenic acid. Furthermore, there is a favourable balance in the levels of linolenic and linoleic acids (viz., 18:3/18:2 ratio of 1:2). Canola oil has been found equally as effective as corn oil, soyabean oil, safflower oil and sunflower oil in reducing plasma total and LDL cholesterol levels in normolipidemic subjects. It was also effective in reducing plasma total and LDL cholesterol levels in hyperlipidemic subjects when it replaced saturated fat in their diets. Canola oil diets also have been shown to affect the fatty acid composition of blood platelet phospholipids and to alter platelet activity and thrombogenesis; although the evidence supporting these observations is not as convincing as its effect on plasma cholesterol levels. The relatively high level of oleic acid and the favourable balance between linolenic and linoleic acids has created interest in the possible use of canola oil in infant formulas.

FLAVOUR QUALITY OF CANOLA OIL

The flavour of fats and oils in one of the most critical factors influencing quality. Fully refined and deodourised canola oil is described as bland and slightly buttery when fresh, but upon storage and heating it is susceptible to the development of off-flavours. These off-flavours have been described as grassy, sulphur, fishy, painty and rancid. Flavour deterioration of canola oil has been attributed to the high content of polyunsaturated fatty acids especially linolenic acid. These fatty acids are particularly susceptible to oxidation and readily form hydroperoxides. Hydroperoxides, the primary products of lipid oxidation, are flavourless but unstable and break down to form a complex mixture of decomposition products including alkanes, alkenes, aldehydes, ketones, alcohols, esters and acids. The volatile compounds formed are mainly responsible for the off-flavours perceived. Many of the secondary products produced are also unstable, particularly unsaturated aldehydes, and decompose to form more stable flavour compounds.

Sensory, chemical and physical methods are used to determine oxidative, photooxidative and thermal stability of canola oils stored under accelerated or ambient conditions. *Frankel* has ranked the various methods on the basis of their usefulness in predicting the stability, shelf life and consumer acceptability of fats and oils. *Frankel* stresses that the degree of oxidation should be determined by more than one method and the methods selected should measure both primary and secondary products of lipid oxidation. Measurement of the primary products of oxidation provide information on precursors of flavour compounds whereas measurement of secondary decomposition products relate directly to flavour compounds that affect flavour quality.

Sensory Evaluation

Flavour and odour assessments by a trained sensory panel provides the most reliable and useful information related to consumer acceptability. In addition, chemical and physical test methods lack the acuity of human senses and the ability to integrate perceptions. Sensory measurements are, however, highly dependent on the preparation and serving of the samples, the physical environment in which the samples are assessed and the quality of training the panel receives. For these reasons, standardised procedures must be followed in order to ensure valid and reliable data. The flavour and odour of canola oil has been assessed by a number of workers using the AOCS flavour intensity scale where 10 = bland

and 1 = extreme. Odour intensity of stored canola oils has also been assessed using an unstructured line scale where 15 = extremely intense/strong and 0 = bland.

Chemical and Physical Properties

Although sensory evaluation is considered the ultimate method for measuring the flavour quality of oils, chemical and physical methods are widely used to monitor lipid oxidation. This is because of the difficulty with sensory measurements in making comparisons of relative flavour scores within samples or to make comparisons with results obtained at different times or locations. Sensory tests are also expensive and time-consuming to carry out. The more common methods for measuring fat and oil quality are given in Table 2.17.

Table 2.17. Common methods for assessing fat and oil quality.

Test procedure	*Parameter measured*
Chemical	
Peroxide value	Hydroperoxide
Thiobarbituric acid	Malonaldehyde
Carbonyl	2,4 DNPH[a] and oximes[b]
p-Anisidine	Unsaturated carbonyls
Physical	
Ultraviolet absorption	Conjugated dienes/trienes
Gas chromatography (GC)	Volatile compounds
Oxidative stability test	
Active oxygen method (AOM)/Swift test	Hydroperoxides
Oxygen absorption test	Oxygen disappearance (GC, pressure reduction, gravimetric)
Conductivity methods (Rancimat, oil stability index)	Conductivity of short-chain fatty acids in water solution

[a] 2,4-dinitrophenylhydrazine.
[b] Oximes, derivatives of carbonyls with hydroxylamine.

Peroxide values

Measurement of peroxide values (PV) is used as an indicator of initial oxidation since hydroperoxides are the primary products of lipid oxidation. There are numerous analytical procedures for the measurement of PV, the most common method being the AOCS iodometric procedure. Because peroxides decompose readily during storage and heating, measurement of PV may not indicate the actual extent of oil oxidation. Despite this, a number of studies have shown a high correlation between PV and sensory odour/flavour scores for stored canola oil.

Thiobarbituric acid

The thiobarbituric acid (TBA) test is based on the reaction between TBA and secondary products of oxidised polyunsaturated fatty acids, namely malonaldehyde. It has been used to measure the stability of canola oil with varying degrees of success. The value of the TBA test has been questioned by a number of workers. *Frankel* discusses its suitability and limitations as a method for assessing oil oxidation and found high correlations between TBA values and PV and sensory scores in his study on stored rapeseed oil. *Ylimaki* found a high correlation for TBA and odour scores of stored canola oil and also reported only fair correlations between TBA values and odour scores of stored canola oil exposed to fluorescent lights.

Carbonyl compounds

The oxidation of fats and oils can also be measured by colorimetric determination of carbonyl compounds formed from the degradation of hydroperoxides. The advantage of these methods is that they measure products that contribute directly to off-flavour development and unlike peroxide determination are not limited to the intital stages of oxidation. It is critical, however, to establish experimental conditions that do not decompose precursors to the carbonyls being assayed. *Przybylski* and *Hougen* developed a method for estimating saturated and unsaturated volatile carbonyl compounds in used frying fat and potato chips. They found the method useful in monitoring the quality of canola and cottonseed frying oils and chips during accelerated storage under heat and light conditions. *Eskin* and co-workers used this method to compare the stability of canola oils heated to frying temperature (185°C). The method was also used by *Przybylski* and co-workers and *Durance* to evaluate the stability of canola oils and canola oil blends, respectively, that had been stored under accelerated conditions. High correlations were found for saturated and unsaturated carbonyls with odour scores in both studies. More saturated and unsaturated carbonyl compounds were formed in canola oil than in sunflower or cottonseed oils.

p-Anisidine values

p-Anisidine values (AV) have been measured to quantify the amount of carbonyl compounds in stored and heated canola oil. The analysis is based on the colour reaction of anisidine and unsaturated aldehydes. *Rossell* recommends AV of less than 10 for fresh fully refined oils.

Conjugated dienes and trienes

Oxidation of oils containing linoleate or more highly unsaturated fatty acids is accompanied by conjugation of double bonds. The conjugated acids formed absorb ultraviolet light and can be used as a relative measure of oxidation. Conjugated dienes show an absorption band at 232 nm whereas secondary oxidation products show an absorption band at 268 nm. Conjugated trienes show a principal peak near 268 nm, together with a peak at 278 nm. Measurement of conjugated dienes (CD) has been used to assess the extent of oxidation in stored and heated canola oils.

Volatile analysis by gas chromatography

Analysis of volatiles by gas chromatography (GC) is closely related to flavour evaluation and is therefore the most suitable method for comparison with sensory results. GC methods also provide useful data on the origin of odour and flavour volatiles and their precursors. While levels of individual and total volatiles can be correlated with sensory scores, the significance of volatiles and their impact on flavour stability are often difficult to determine.

Static headspace

Static headspace involves transferring a set volume of equilibrated gases from the area above a lipid sample kept in a closed container. This gaseous sample is injected directly onto the GC column. Factors governing equilibration include time and temperature and are dependent on the viscosity of the lipid matrix and solubility of volatile compounds in the oil. The volatility of the flavour components is boiling point dependent, which also effects equilibrium. In the case of oil, this requires heating the lipid sample at elevated temperatures up to several hours in order to achieve full equilibration. These extreme conditions can alter the volatiles present in the lipid sample as a result of decomposition, interaction and oxidation. Hydroperoxides, which are always present in oils, can also undergo decomposition and form new volatile compounds that were not originally present in the analysed oil sample. Differences in solubility and partial pressure of individual components can also affect their transfer from the headspace into the vapour phase such that volatiles in the headspace may not accurately reflect the composition of these same components in the lipid sample. Despite these problems, static headspace is easy to use and automated systems are available. The results generated are reproducible when parameters are established for the analysis of a particular type of sample. This technique has been used by *Snyder* and *Warner* to monitor oxidative and photo-oxidative stability of canola oil.

Dynamic headspace

Dynamic headspace or purge and trap methods use a sorbent to collect volatiles from the lipid samples. Volatiles are swept from the heated sample with nitrogen and trapped on a precolumn packed with a porous polymeric sorbent. The trapped volatiles are then transferred from the precolumn onto the GC column by thermal desorption or solvent extraction. With this procedure, lower sample temperatures can be used during trapping to limit alteration of volatiles, but a longer time is necessary to isolate volatile components. The time required for transferring flavour compounds is temperature and concentration dependent, with longer times needed for lower temperatures and lower concentrations of volatiles. Dynamic headspace requires intensive manipulation when transferring volatiles from the sample to adsorbent and from it into the GC system for analysis. Each step requires strict control of the parameters to ensure reproducibility. Alteration of compounds can occur if oxygen is allowed to come into contact with the volatiles. Additionally, the high temperature required for desorption can cause volatile alteration and adsorbent decomposition resulting in formation of artifacts. Another drawback to these procedures is the possibility of contamination. The containers used for purging usually have awkward shapes, making them difficult to clean. This can result in the carry over of volatiles from previous samples. In comparison to static headspace, dynamic headspace is more sensitive and allows the operator to analyse almost all volatile components present in a sample. A high recovery is achieved because there is extensive extraction of the volatiles with the sweeping gas and there is a large area of exposure because small bubbles of sweeping gas are dispersed in the lipid sample. This technique has been used by a number of workers to study the effects of antioxidants on storage stability of canola oil and to monitor the storage stability of potato and tortilla chips fried in canola oil. *McMullen* and co-workers reported high correlations between sensory scores and GC indices of total volatiles, pentane/butane, and 2,4-decadienals, which were identified by *Tokarska* and co-workers as major off-flavour volatiles in stored canola oils.

Direct injection

Unlike static and dynamic headspace techniques, which were developed for analysis of pollutants and later adapted for analysis of lipid samples, direct gas chromatography was developed for analysis of lipid sample. In this method volatiles are directly transferred from the oil onto the GC column. The sample is usually placed on a purified glass wool plug insert. This is placed into the GC injector or external apparatus. The sample is heated and volatiles are transferred directly by the carrier gas onto the column, which is held at ambient temperature when porous packing material is used or cooled with liquid nitrogen or solid carbon dioxide when a capillary column is used. In both cases, lower temperatures are employed than utilised for separation in order to concentrate volatiles and improve separation. Direct gas chromatography permits isolation of volatiles at relatively low temperatures, similar to the purging step of dynamic headspace. Thus, decomposition of hydroperoxides and alteration of volatile compounds is less likely to occur. *Przybylski* used a combination of dynamic headspace and direct injection to examine the stability of canola oils exposed to accelerated heat conditions. Good correlations were found for sensory scores and GC indices for regular canola oil.

Resistance to oxidative rancidity

A number of accelerated methods have been developed to test the resistance of fats and oils to oxidative rancidity. These high-temperature tests involve exposing the lipid sample to elevated temperatures of 100°C or above. According to *Frankel* these tests are unreliable because the mechanism of lipid oxidation changes significantly at the elevated temperatures. This observation has been confirmed by *Warner* and co-workers in their study on stability of vegetable oils including canola oil. They found results from stability tests conducted at 100°C (Rancimat and active oxygen method) did not agree with stability tests performed at 60°C (Schaal oven test).

Active oxygen method

In the active oxygen method (AOM) or Swift test, air is bubbled through oil held at 98°C. Samples of oil are taken at various intervals and PV determined. The PV are plotted against time and the induction period determined as the time at which the slope changes sharply.

Oxygen absorption tests

The oxidative deterioration of an oil can also be monitored by holding the sample at a high temperature in a closed system, which permits a measurement of oxygen absorption over a period of time. Many techniques have been devised that utilise the principle of oxygen absorption by a lipid sample and include the ASTM oxygen bomb, FIRA-Astell Apparatus, Sylvester test and weight-gain or gravimetric procedures.

Conductivity tests

Conductivity methods measure the conductivity of short-chain fatty acids in a water solution. These acids are produced during the autoxidation of fats held at 100°C or above and are mainly formic and acetic acid. Several automated instruments based on this principle have been developed including the Rancimat and the oil stability instrument (OSI).

Flavour stability of canola oil

The flavour stability of a fat, oil or fat-containing food is principally related to the presence of unsaturated fatty acids, proxidants and antioxidants and on the nature of storage and use conditions imposed on the fat. Canola oil has a unique polyunsaturated fatty acid composition with high levels of oleic (55 per cent) and linoleic (26 per cent) acids and a moderate level of linolenic (8–12 per cent) acid. Fatty acids with more than three double bonds have also been found in trace amounts that are presumed to be artifacts formed during processing of canola oil. These fatty acids are important because they oxidise easier and can produce free radicals initiating oxidation of other acids. Volatile compounds formed in canola oil during oxidation belong to four major groups: saturated, monounsaturated and polyunsaturated aldehydes, hydrocarbons, alcohols and miscellaneous. Aldehydes are the most abundantly formed and have a very important impact on the flavour since their threshold levels are extremely low, in the range of a few parts per billion. The volatile compounds formed during oxidation are characteristic for a particular fatty acid.

To find the pathway responsible for the formation of flavour compounds in oxidised canola oil, the oxidised methyl esters of the three major polyunsaturated fatty acids were analysed by *Frankel*. When these results are compared with the volatiles identified in canola oil by *Snyder* and co-workers, it can be seen that not all compounds found in oxidised methyl esters are detected in the oxidised oil and vice versa (Table 2.18).

Table 2.18. Volatile compounds formed during oxidation of fatty acid methyl esters and canola oil (relative %).

	Methyl oleate[a]		*Methyl linoleate*[a]		*Methyl linolenate*[a]		
Component	*Free Rad*[b]	*Photo*[c]	*Free Rad*[b]	*Photo*[c]	*Free Rad*[b]	*Photo*[c]	*Canola oil*[d]
Ethane/ethene	–	–	–	–	17.4	5.7	4.5
Pentane/pentene	–	–	17.1	7.9	–	–	3.1
Heptane	8.4	7.8	–	–	–	–	0.9
Octane/octene	5.2	17.1	–	–	–	–	16.7
Acetaldehyde	–	–	0.5	0.7	1.4	1.1	–
Propanal/propenal	–	–	–	–	13.4	15.9	6.8
Butanal	–	–	–	–	0.2	1.4	–
Pentanal	–	–	1.4	0.5	–	–	2.2
Hexanal	–	–	25.9	31.2	–	–	13.9
Heptanal	1.0	0.8	–	–	–	–	3.6
Octanal	21.0	6.5	–	–	–	–	2.2
Nonanal	28.7	17.1	–	–	–	–	5.4
Decanal	7.5	3.4	–	–	–	–	–
2-Butenal	–	–	–	–	0.9	19.5	2.2
2-Pentenal	–	–	–	–	2.8	2.1	1.3
2/3-Hexenal	–	–	–	–	2.4	6.0	0.9
2-Heptenal	–	–	–	18.2	–	–	4.9
2-Octenal	–	–	4.7	2.8	–	–	0.9
2-Nonenal	1.0	1.2	2.4	2.9	–	–	–
2-Decenal	10.3	20.5	–	–	–	–	0.9
2-Undecenal	3.2	12.1	–	–	–	–	0.4
2,4-Heptadienal	–	–	–	–	16.1	15.6	5.8
2,4-Nonadienal	–	–	0.5	0.5	–	–	–
2,4-Decadienal	–	–	24.2	7.9	–	–	4.9
3,6-Nonadienal	–	–	–	–	0.9	1.9	–
2,4,7-Decatrienal	–	–	–	–	24.3	8.5	–
1-Pentanol	–	–	2.1	0.5	–	–	1.3
1-Heptanol	0.8	0.7	–	–	–	–	–
1-Octanol	0.8	1.7	–	–	–	–	–
2-Decenol	–	–	–	–	–	–	0.4
1-Penten-3-ol	–	–	–	–	–	–	1.8

(Contd...)

Table 2.18. Volatile compounds formed during oxidation of fatty acid methyl esters and canola oil (relative %). *(Contd.)*

	Methyl oleate[a]		*Methyl linoleate*[a]		*Methyl linolenate*[a]		
Component	*Free Rad*[b]	*Photo*[c]	*Free Rad*[b]	*Photo*[c]	*Free Rad*[b]	*Photo*[c]	*Canola oil*[d]
1-Octen-3-ol	–	–	–	3.5	–	–	0.4
2-Pentylfuran	–	–	4.1	1.1	–	–	0.9
2-Butylfuran	–	–	–	–	0.9	0.5	–
Others[e]	12.2	11.4	17.1	22.1	19.5	21.6	8.6

[a] Data adapted from Frankel; [b] Free radical oxidation; [c] Photooxidation; [d] Data adapted from Snyder; [e] Unidentified compounds.

Decenol was detected in oxidised canola oil but not in oxidised methyl esters of fatty acids. In contrast, compounds such as decatrienal were detected in oxidised methyl esters of fatty acids but not in oxidised canola oil probably because this compound can decompose readily to form other components. For many of the volatile components, isomers have also been found, particularly for compounds with double bond(s) present in their structure. These different isomeric forms also have different sensory properties. In addition, it is not known to what extent individual components alter the sensory perception of other components present in the mixture through enhancement, additivity and synergistic effects. Therefore, it is difficult to predict the sensory characteristics of a mixture from either individual or total volatile determinations. The presence of trace elements and pigments in canola oil have exhibited proxidative effects, whereas the presence of naturally occurring tocopherols have demonstrated antioxidant effects. The third element affecting the flavour stability of fat is the nature and stresses of the storage and use conditions that are imposed on the fat. A number of studies have been undertaken to examine the oxidative and photo-oxidative stability of canola oil and the stability of canola oil to thermal oxidation.

Oxidative Stability

The autoxidation of unsaturated fatty acids occurs by way of a free radical mechanism. During the first step a hydrogen atom adjacent to the double bond in the fatty acid is abstracted by exposure to heat, light or metal catalyst to form a free radical (Fig. 2.7).

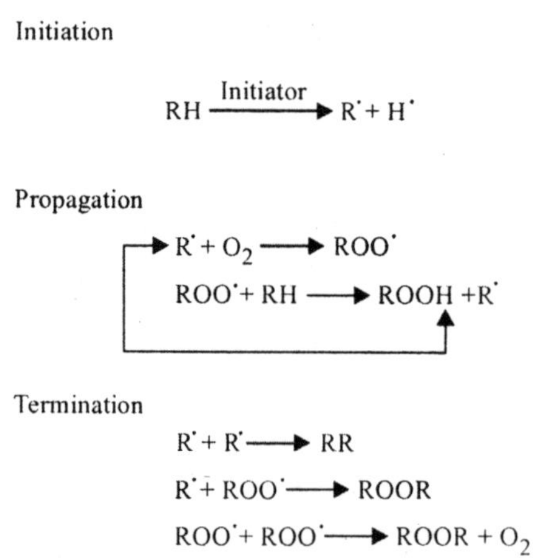

Fig. 2.7. Autoxidation of unsaturated fatty acids by a free radical mechanism.

This free radical reacts with atmospheric oxygen to form an unstable peroxy free radical that may in turn abstract hydrogen from another unsaturated fatty acid to form a hydroperoxide and a new free radical. The new free radical initiates further oxidation. This chain reaction may be terminated by the formation of nonradical products from two free radicals. The rate of oxidation increases in relation to the degree of unsaturation, with linolenic acid (18:3) oxidising twice as fast as linoleic acid (18:2) and 25 times faster than oleic acid (18:1).

Accelerated storage tests

Accelerated storage tests have been used extensively to monitor the stability of canola oil using a modification of the Schaal oven test. In this test, oxidation is accelerated by holding samples of oil at 60–65°C in the dark. Samples may be evaluated at 24 hr. intervals or less frequently. Variations in Schaal oven test conditions exist and include differences in quantity of oil stored, the size of storage containers used and whether or not the containers are covered during storage. Schaal oven tests are more reliable than tests carried out at temperatures of 100°C or higher since they reproduce similar oxidative changes observed under actual shelf-life conditions.

Ambient storage tests

Hawrysh and co-workers evaluated the stability of canola oils stored in clear glass bottles in the absence of light at 24°C for 16 weeks. The oils remained unchanged in flavour and had low PV. From this observation, *Malcolmson* and co-workers speculated that canola oil with a sensory induction period of 2–4 days at 60–65°C may be expected to retain good flavour quality for at least 16 weeks under comparable storage conditions. This was considered consistent with a report by *Evans* and co-workers that found flavour scores for soyabean oil aged 60°C for 4 days equivalent to scores for oils stored 4 months at ambient temperatures. Although testing stability under ambient conditions may approximate real storage conditions, the procedure is too slow to be of practical value. Futhermore under slow oxidation the reproducibility of results may be compromised by many variables that are difficult to control over prolonged storage periods.

Stability to Photochemical Oxidation

Photosensitised oxidation results from the production of singlet oxygen generated by the interaction of light and a photosensitiser, such as chlorophyll or polycyclic aromatic compounds. Singlet oxygen reacts with the double bonds of the unsaturated fatty acids to form hydroperoxides. α-chlorophyll degradation products have been found to be more efficient photosensitisers than chlorophyll itself. Photosensitised oxidation can be inhibited by the presence of naturally occurring substances such as α-carotene and γ-tocopherol, which either deactivate the singlet oxygen or react with it. Since, photosensitised oxidation proceeds by different pathways than free-radical autoxidation, different flavour precursors and volatile products are formed with distinct flavour significance.

Accelerated storage tests

The adverse effects of light on the flavour stability of canola oil has also been examined using accelerated storage conditions. This involves exposing the oil to fluorescent lights often at elevated temperatures. Testing protocols vary in cabinet design, sample 'and' container size, storage temperature and light intensity making it difficult to compare results from one research group to another. *Sattar* and co-workers exposed low erucic acid rapeseed (LEAR), corn, soyabean and coconut oils to fluorescent light with an intensity of 500 fc (5382 lux) at 25°C for 12 hr. As exposure to light increased, PV increased gradually while flavour scores became less acceptable. Good correlations were obtained between PV and flavour scores. Corn oil was found to be the most stable, LEAR was intermediate and soyabean and coconut oil were found to be the most sensitive to light. PV at which the oils were considered unacceptable from the sensory standpoint was unique for each oil: LEAR with a PV up to 10 had an acceptable flavour, whereas soyabean and coconut oils had unacceptable flavours when PV of 2 and 2.5 were achieved, respectively. Corn oil remained acceptable in flavour with a PV of 7.

Durance compared the stability of canola, sunflower and cottonseed oils containing an antioxidant mixture to fluorescent light with an intensity of 250 fc (2691 lux) at 40°C up to 4 days. The sensory data revealed that the cottonseed oil was the most sensitive to light, canola oil was intermediate and sunflower oil was the most stable. In contrast, PV for sunflower oil were found to be twice as high as canola oil and about four times higher than cottonseed oil. *Warner* and co-workers compared the stability of canola, sunflower and soyabean oils, with and without added citrate. Oils were exposed to fluorescent light at 700 fc (7535 lux) at 30°C for 16 hr. Results from sensory evaluation of the oils showed that sunflower oil was the most stable regardless of whether or not citrate was added. In the presence of citrate, canola oil was more stable than soyabean oil, whereas in the absence of citrate, no differences between the two oils was observed. PV did not show differences between the oils. Both soyabean and sunflower oils developed more total volatiles than did canola oil.

Sattar and co-workers examined the effect of light with different wavelengths on stability of oils, including LEAR oil. For all oils PV decreased with increasing wavelengths. The most deleterious wavelengths were those below 455 nm, whereas wavelengths greater than 595 nm produced minimal changes in PV.

Ambient storage tests

Hawrysh and co-workers evaluated the stability of canola oils stored in clear and amber glass bottles exposed to simulated supermarket conditions (1400 lux). Oils were stored for 16 weeks at ambient temperatures. Canola oils stored in amber bottles showed less flavour deterioration than oil stored in clear bottles. Measurement of total volatiles indicated that the oxidative changes for canola oil stored in light-resistant amber bottles for 16 weeks were similar for canola oils stored only 4 weeks in clear glass.

High-Temperature Stability

During the frying of foods, oil is subjected to elevated temperatures of 180–190°C in the presence of air and moisture. Under these conditions a number of reactions occur, including oxidation, hydrolysis and thermal degradation. The breakdown products formed from fat during frying include volatile and nonvolatile decomposition products (VDP and NVDP). The formation of VDP changes the flavour of the frying fat and contributes significantly to the flavour of the fried food. Formation of NVDP is due to primarily thermal oxidation and polymerisation of the unsaturated fatty acids in the frying fat. The presence of these products results in physical changes to the fat such as darkening of the colour, increases in viscosity, and decreases in the smoke point as well as chemical changes such as increases in fatty acids, carbonyl values, hydroxyl content and saponification values.

Stevenson and co-workers examined the effects of extended frying on the chemical and physical properties of hydrogenated solid and liquid canola frying oils. Total frying time for solid and liquid fats was 37.5 and 75 hr. of frying, respectively. After extended frying, only moderate amounts of degradation products as measured by free fatty acids (FFA) and polar compounds (PC) were present in all fats indicating good frying stability. At these levels of degradation products, sensory characteristics of French fries fried in liquid frying fat were not adversely affected. The sensory characteristics of potatoes fried in solid fat were not assessed.

Warner and *Mounts* compared the frying stability of regular canola (10.1 per cent C18:3), LLC (1.7 per cent C18:3), hydrogenated (0.8 per cent C18:3), and brush hydrogenated (0.7 per cent C18:3) LLC oils. Significantly lower FFA were found during frying for the LLC and brush hydrogenated LLC oils than the other two oils. Similar PC were found for all oils during frying with slightly lower PC being reported for the LLC oil. The flavour quality of French fries was significantly better for potatoes fried in the three modified canola oils than regular canola oil.

Heated canola oil has been reported to develop an unpleasant room or heated oil odour that has been described as painty with buttery, sweet, sulphur and fishy notes. This phenomenon has been attributed to the oxidation of unsaturated fatty acids and specifically linolenic acid. Soyabean oil also exhibited this phenomenon, which has been associated with its high linolenic acid content. *Dobbs* found heated rapeseed oil was significantly stronger in odour than heated soyabean oil and both oils were stronger than heated corn, safflower and sunflower oils. The heated room odour reached a plateau after 10 minutes of heating and did not increase with additional heating. Hydrogenation of rapeseed oil and a reduction in linolenic acid content of canola oil decreased heated room odour development when oils were heated up to 40 and 10 minutes respectively. This was confirmed by *Warner* and *Mounts* who reported lower room odour intensity scores for modified canola oils heated for 1 hr. compared to regular canola oil. *Durance* lowered the linolenic acid content of canola oil by blending with either cottonseed or sunflower oil. Acceptability of heated oils increased proportionally with increasing levels of sunflower, whereas the level of cottonseed had to be 75 per cent or higher in the blend before there was a marked increase in acceptability for the heated oil. Lower carbonyl compounds were reported for heated cottonseed and sunflower blends compared to heated canola oil.

Fried foods absorb appreciable amounts of the frying fat, which is then susceptible to oxidative changes during storage. Several studies have examined the stability of snack foods fried in canola oil. *Liu* and *White* evaluated bread cubes that had been fried in laboratory processed canola, soyabean and low linolenic acid soyabean (LLS) oils. Fried bread cubes were evaluated before and after 7 days storage at 60°C. Panel scores for the freshly fried bread cubes showed no significant differences among oils in either flavour quality or flavour intensity. After storage, the bread cubes fried in canola and soyabean oils were judged comparable but had poor flavour quality ratings compared to cubes fried in LLS oil. *Hawrysh* compared the effect of four frying oils–canola, soyabean, cottonseed and partially hydrogenated canola-on the quality of potato chips stored at 60°C for 12 days and at 23°C for 18 weeks. Frying oils did not affect potato chip flavour or off-flavour scores for chips stored at 60°C. After 18 weeks at 23°C, potato chips fried in canola oil had higher potato chip flavour and lower off-flavours than did chips fried in the other oils. PV and total volatiles were highest in fat extracted from chips fried in cottonseed oil. No differences in AV were observed among oil treatments throughout storage.

Hawrysh compared the storage stability of tortilla chips fried in canola (9.9 per cent C18:3), corn, LLC (3.4 per cent C18:3), partially hydrogenated canola (PHC) (1.2 per cent C18:3), and soyabean (PHS) oils. All freshly fried tortilla chips had similar characteristic flavour scores except the chips fried in PHC, which were scored lower in flavour. Flavour scores for chips fried in LLC, PHC and PHS oils did not decrease during the 8-week storage at 60°C but did decrease for chips fried in corn and canola oils. After 24 weeks at 23°C, slightly lower flavour scores for tortilla chips fried in canola, corn and LLC oils were observed. Higher pentane levels and total volatiles were found in chips fried in canola and corn oils and stored under Schaal oven conditions. PV and AV of oils extracted from the stored tortilla chips indicated that PHC and PHS oils were the most stable.

Melton and co-workers examined the stability of potato chips fried in partially hydrogenated canola oil and canola–cottonseed blends stored under ambient dark or light conditions. PV increased with storage time but neither oil type nor storage condition affected PV. Sensory scores of freshly fried chips were similar for all oils. Stored chips were not subjected to sensory evaluation. Greater differences in volatiles were observed among oil types than by storage. However, a modified steam distillation procedure was used to extract volatiles, which may have altered the components present in the oils.

Malcolmson studied the susceptibility of used frying oils to autoxidation and photooxidation. The rationale for storing 'used' oil was that the oil was more homogeneous to store and test than fried foods and reflected the fat absorbed by the snack. A commercial snack food manufacturer heated canola and cottonseed oils for 8 hr. at 190°C and then fried one lot of potato chips before collecting the oil for storage. The used oils were then stored at 40°C in the presence or absence of fluorescent light (250 fc, 2690 lux). Rancid odour scores for used canola oil increased more rapidly over a 63-day storage period than they did for used cottonseed oil. Rancidity of both oils was aggravated by the presence of light and this effect was similar for the two oils. *Hawrysh* also stored used frying oils that had been used for frying potato chips. PV and total volatiles were found to be highest in cottonseed oil and lowest in PHC oil. AV remained low throughout the storage period but were generally higher in soyabean oil. Soyabean and cottonseed oils tended to have the highest levels of 2,4-decadienal and total volatiles after storage. In a subsequent study, *Hawrysh* stored frying oil that had been used to fry tortilla chips. PV, AV, total volatiles, pentane and 2,4-decadienals were lower for PHC stored at 60°C for 16 days compared to canola, corn and LLC oils. Intermediate values for these parameters were observed for PHS.

Chapter 3

Corn Oil

INTRODUCTION

Corn (*Zea mais,* Graminae) is cultivated in all the temperate regions of the world. The principal producer regions are USA, Europe, China and South America (Brazil, Argentina, Mexico).

Oil is a by-product of the cereal industry (corn meal and starch producers). Corn seeds contain from 3 to 6.5 per cent oil (if reffered to the dry key) concentrated in the germ, representing 40–50 per cent of its weight. This content depends on the process chosen for germ removal: dry process (15–25 per cent oil) or wet process (40–50 per cent).

Corn oil is essentially a cooking edible oil, used for frying and seasoning. It is also used in margarine industry. Its richness in linoleic acid makes it interesting for dietetics. The components of its unsaponifiable fraction are employed in pharmaceutical or cosmetic industry. Oil meals made of corn seeds are used as feed (milk cow).

Corn germ, as a rich source of oil, is first separated from the kernel in the dry or wet milling process and provides the raw material for oil recovery. The availability of corn germ for oil recovery is a function of the amount of corn processed by the dry and wet milling industries.

TECHNOLOGY

Two processes are used for the removal of corn germs from the seeds. The dry process (corn meal industry) produces a germ containing 18 to 20 per cent oil and the wet process (starch industry) gives for crushing germs with 40 to 50 per cent oil. However, the oil obtained from the germs treated by wet process is of less good quality than the oil obtained through dry method. This can lead to difficulties during refining.

Not requiring grinding or flatterning in the wet process the corn germs should be first dried to attain a humidity lower than or equal to 12 per cent. The dry germs are then subjected to a new thermal conditioning (105 to 115°C for at least one hour) which reduces the residual humidity to 2 to 3 per cent. Crude raw oil obtained by pressing, passes through a vibrating sieve and then is filtered on a Niagara-type filter. Horizontal decantation may be chosen for this first purification step.

In the dry method, the germs are extracted directly by solvent after shoveling or flaking.

Corn oil is an excellent source of essential fatty acids (EFA).

Characteristics of Corn Oil

Physicochemical characteristics

Table 3.1 shows the principal characteristics of corn oil.

Table 3.1. Principal physical and chemical characteristics of corn oil.

D_{20}	0.917–0.925
V_{20} (cp)	55–56
n_D^{20}	1.474–1.477
Solidification point (°C)	−10a − 18
Iodine value	103–128
Saponification value	187–195

Triglyceride composition

Triglycerides of corn oil are given in Table 3.2 and 3.3. Analysis by gas chromatography separates the triglycerides according to the number of carbon atoms (Table 3.2).

Table 3.2. Triglyceride composition of corn oil (GC).

Nature	*% Total TG*
C42	–
C44	–
C46	0.2
C48	1.2
C50	3.0
C52	25.2
C54	63.4
C56	2.5
C58	0.3

Table 3.3. Triglyceride composition of corn oil (HPLC).

Nature	*% Total TG*	*Nature*	*% Total TG*
LLLn	1.6	PPL	2.1
LLL	26.5	OOO	2.6
OLL	22.9	SOL + POO	4.8
PLL	16.8	PPO	0.3
OOL	9.9		
POL + SLL	12.3		

Table 3.3 shows the content of the main triglycerides identified in corn oil by HPLC.

Table 3.4 shows the fatty acid distribution in positions 1, 2 and 3 of glycerol.

Table 3.4. Distribution of fatty acids on the positions 1, 2, 3 of triglycerides of corn oil.

Nature of fatty acid	*% of total FA in 2*	*% of total FA in 1*	*% of total FA in 3*
C16:0	2	18	13
C18:0	< 1	3	3
C18:1	26	27	31
C18:2	70	50	51
C18:3	< 1	1	1

Fatty acid composition

Table 3.5 summarises the main values which are based on data collected on 788 varieties from all over the world and on more than 200 different samples. But the values usually obtained on commercial samples represent a much narrower range. Corn oil is a linoleic oil. Its linolenic acid content remains lower than or equal to 2 per cent. Its saturated fatty acid content is around 13 per cent, its average P/S ratio is 4.

Table 3.5. Fatty acid composition of corn oil.

	% Total FA	
Nature	*Usual values*	*Literature*
C14:0	< 0.1	
C16:0	8–13	6–22
C16:1	< 1	< 1
C17:tot.	–	–
C18:0	1–4	1–15
C18:1	24–32	14–64
C18:2	55–62	19–71
C18:3	< 2	1–2
C20:0	< 1	< 1
C20:1	< 0.5	< 0.5
C22:0	< 0.5	< 0.3

Composition of unsaponifiable fraction

Table 3.6 gives the contents and the composition of its principal constituents. Sterol content appears to be very variable. *Worthington* has determined for two different oils the distribution between free sterols (0.32–0.37 per cent) and esterified sterol (0.95–1.42 per cent) (with respect to oil).

Among sterols, Δ5-avenasterol generally fluctuates within a more narrow range than that shown in the Table 3.6. Certain references indicate the presence of isofucosterol (< 0.4 per cent). The structure of this sterol is identical to that of Δ5-avenasterol. *Itoh* indicated the presence of a new sterol: 24-methyl (E)-23-dehydrolophenol (4-methyl sterol) in the presence of ferulic acid (phenolic acid) esterified to dihydro β-sitosterol or to triterpenic alcohol, which could contribute to the oxidative stability of corn oil. The tocopherol content is relatively high: one has to note the major percentage in γ-tocopherols.

Table 3.6. Composition of unsaponifiable fraction of corn oil.

Description	*Unsaponifiable fraction: 0.8–2.0%*		
Sterols (mg/100 mg)	830–2550	Hydrocarbon (mg/100 g)	40
		of which squalene	13–24
		Tocopherols (mg/100 g)	113–183
Sterol composition (% total sterols)		Tocopherol composition (% total tocopherols)	
Cholesterol	< 1	α-Tocopherol	8–22
Brassicasterol	–	β-Tocopherol	< 3
Campesterol	16–21	γ-Tocopherol	68–89
Stigmasterol	4–10	δ-Tocopherol	2–7
β-Sitosterol	63–70	Tocotrienols	–
Δ5-Avenasterol	1–9		
Δ7-Stigmasterol	< 1		
Δ7-Avenasterol	< 1		
Ergosterol	1–2		

Triterpenic alcohol contents were determined by *Itoh* on the soapstocks of Mexican corn oil as 13 mg/100 g oil. For the same sample, the *Itoh* reports a hydrocarbon content of 104 mg/100 g and a total sterol content of 1170 mg/100 g.

As shown in Table 3.7, corn oil contains 97.9 per cent unsaturates and 2.1 per cent saturates in the β-position, whereas in the α-position there are 77.5 per cent unsaturates and 22.5 per cent saturates. This is the reason randomised corn oil oxidises three to four times faster than natural corn oil.

Table 3.7. Distribution of fatty acids in α- and β-positions of triglycerides in corn oil, hydrogenated corn oil[a].

	Mol %											
					C	*T*	*C–C*	*C–T*	*T–T*	*C–C–C*		
Position	*14:0*	*16:0*	*16:1*	*18:0*	*18:1*	*18:1*	*18:2*	*18:2*	*18:2*	*18:3*	*20:0*	*20:1*
Corn oil												
α	0.1	18.7	0.2	3.0	29.0	tr.	47.8	–	–	0.4	0.7	0.1
β	0.1	1.9	0.2	0	25.4	tr.	70.0	–	–	2.2	0.1	0.1
Hydrogenated corn oil												
α	0.1	18.6	0.1	9.6	24.9	45.5	0.2	0.2	tr.	0.1	0.6	0.1
β	0.1	1.8	0.1	2.4	7.5	86.3	1.4	0.2	tr.	0.1	0	0.1
Margarine oil												
α	0.1	18.2	0.1	5.7	27.0	21.6	26.3	0.1	tr.	0.2	0.6	0.1
β	0.1	2.6	0.1	1.5	17.4	6.6	40.7	0.1	tr.	0.8	0	0.1

[a] C = *cis*; T = *trans*; tr. = trace.

Another reason for the excellent stability of corn oil is its high level of natural antioxidants such as ferulic acid and tocopherols. Crude corn oil is second only to wheat germ oil with respect to total tocopherols and tocotrienols content. Table 3.8 summarises tocopherols and tocotrienols content in several vegetable oils as reported by different authors. Typically, deodourised corn oil contains 0.080–0.120

per cent of total tocopherol of which 70–80 per cent is γ-, 20–25 per cent α- and 3–5 per cent δ-tocopherol. Consequently, corn oil is a good source of vitamin E, containing from 10 to 20 per cent of the US. Recommended Daily Intake (USRDI) per 14 gram serving. Also, corn oil contains no chlorophylls, which accelerate the oxidative process in the presence of light. An additional contribution to the stability of corn oil may be due to the presence of ubiqui- none (sometime referred to as co-enzyme Q), which is present in corn oil at a level of 200 ppm. This class of compounds has been reported to have biological and antioxidant activity.

Table 3.8. Tocopherols and tocotrienols in vegetable oils.

	Tocopherols (mg/100 g)					
Oil	*α*	*β*	*γ*	*δ*	*Tocopherols*[a] *(mg/100 g)*	*Total (mg/100 g)*
Crude						
Wheat germ	117.9	39.8	49.3	11.8	–	218.8
Corn	19.1	–	94.2	4.2	2.3	119.8
Soyabean	11.6	3.4	73.7	27.5	–	116.2
Olive	9.1	–	1.9	–	–	11.0
Refined						
Corn	13.4	1.8	41.2	3.9	–	60.3
Rapeseed	7.0	1.6	17.8	0.7	–	27.1
Corn	17.1	0.3	70.3	3.4	–	91.1
Soyabean	10.4	2.0	80.9	20.8		114.1
Safflower[b]	31.4	–	2.1	–		33.5
Salad oil[c]	6.6	–	46.2	15.4		68.2
Sesame	–	–	32.5	–		32.5

[a] (α + γ) only; β-tocotrienols are included in γ-tocopherols; γ-tocotrienols is not detected.

[b] High oleic.

[c] Mixture of soyabean and rapeseed oils.

The level of unsaponifiables in corn oil 1.3–2.3 per cent, is higher than in other common oils. Approximately 60 per cent of the unsaponifiables are sterols, the majority of which (about 60 per cent) are esterified with fatty acids. Those predominanting are β-sitosterol, campesterol and stigmasterol. The presence of trace levels of cholesterol (below 10 ppm) has been confirmed by mass spectroscopy in samples of corn oil that were protected from any contamination.

More than 75 per cent of corn oil unsaponifiables originate from components less polar than triglycerides. These nonpolar components have been studied extensively. Hydrocarbons represent 3.3 per cent of the nonpolar fraction, of which squalene is the predominant component (57 per cent). The major part of the nonpolar fraction consists of fatty acid esters of sterols and triterpenols. In addition to β-sitosterol, campesterol and stigmasterol esters, isofucosterol, methylene cycloartanol, Δ5 and Δ7 avenasterol and gernayl–geranol esters were identified. Tocopherol esters, apparently formed during deodourisation, have also been found.

PROCESSING

Standard Recovery Methods

Wet milled germ contains 2-4 per cent moisture after drying and 44–50 per cent oil. This high oil content is the result of the loss of sugars, starches and proteins, which leach into the steep water. When subjected to the full expelling process, wet milled germ yield 89–94 per cent crude oil. The full expelling process is primarily utilised in smaller plants employing a few expellers. Due to high maintenance costs, larger plants utilise expellers for preexpelling at greater throughput rates accepting increased residual oil in the cake (18–22 per cent) as a consequence. The cake is subsequently solvent extracted to recover most of the available oil. Typically, the preexpelled and solvent extracted oils are combined into one crude oil stock, yielding recoveries from 97 to 99 per cent. When the germ is separated from the kernel by the dry milling process, it contains 20–25 per cent oil on a weight basis. This is lower than in germ separated manually (about 33 per cent oil) because of the residual endosperm still attached to the germ. As a low oil material, dry milled germ is most amenable to solvent extraction.

The conventional oil recovery methods described above have several drawbacks. Corn oil is well protected in the germ by external layers of tightly packed, elastic cells that must be broken before the oil is liberated for recovery. The liberation of oil accomplished through conditioning the germ by increasing its moisture and raising the temperature to 90–105°C in order to soften its intracellular structure and provide adequate friction during flaking or expelling. The prolonged temperature exposure may have a detrimental effect on oil quality, is energy consuming and the cost of equipment maintenance is high. One study indicates that increasing the solvent-to-meal ratio from 1 to 1.5 reduces the number of extraction cycles by 50 per cent. During solvent extraction, the percolation of miscella can be hindered by excessive fines (> 5 per cent) and insufficiently gelatinised starch, which absorb more solvent.

To remedy the problems encountered during conventional corn oil recovery processes, extrusion has recently been employed as a means of germ preparation for solvent extraction. During extrusion, the cellular structure of germ is destroyed by internal forces created by sudden evaporation of the germ's moisture as the hot, compressed germ exits the extruder and the pressure drops from 2 to 3 atmosphere to normal. The expanded germ material contains liberated oil, which is loosely absorbed within the porous matrix and can easily be recovered by hexane extraction. Since extrusion subjects the germ to moderate temperatures for a short time, oil abuse is minimised. At an extrusion temperature of 120°C, oxidative changes in the oil are insignificant compared to extrusion at 175°C. Generation of volatile compounds increases three- to fourfold during extrusion at 160°C compared to 120°C.

Exploratory Recovery Methods

Supercritical extraction with a variety of gases has found application in value added products such as flavour extracts. In spite of quality advantages, the supercritical extraction of commodity-type products such as edible oils cannot be justified economically due to high capital expenditures and design constraints for continuous, high-pressure, large-scale equipment capable of achieving high throughput rates.

To eliminate germs drying after wet milling, attempts have been made to recover oil directly from the wet germ (50–55 per cent moisture). In the aqueous extraction process, the wet germs is suspended in water and ground to a particle size below 160 μm to achieve oil release from the cells. Next, the oil is separated from the aqueous phase by decanting or centrifugation. The recovered oil is fully degummed and requires only minimal refining. Since the entire process is performed at temperatures not exceeding 50°C, the oil is of extremely high quality. Comminuting the wet germs in the presence of hexane has also been investigated. The resulting oil–hexane miscella is separated from the solids and the hexane stripped by conventional procedures. The dispersion of wet germs in oil, followed by milling, vacuum drying to remove excess moisture, centrifugation to separate free oil from the ground germ solids and finally, expelling the germ solids by a conventional process to recover the oil. To date none of these experimental processes has gained commercial acceptance due to unfavourable economics.

Refining, Bleaching, Dewaxing and Deodourisation

Crude corn oil from conventional recovery processes contains undesirable components such as free fatty acids (FFA), phospholipids, proteinaceous and acilaginous matter, carbohydrates, pigments, waxes and insolubles. It may also contain oxidation products, mycotoxins and pesticide and insecticide residues. All these impurities must be satisfactorily removed through the refining steps to produce edible, stable and lightly coloured oil. Fig. 3.1 illustrates typical corn oil refining processes.

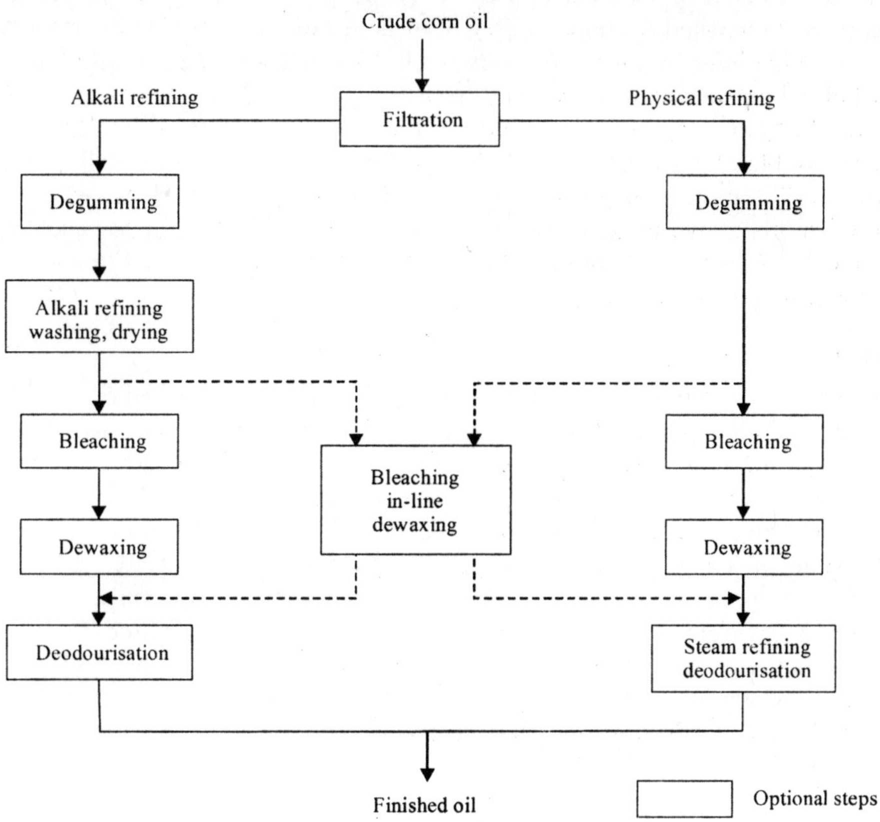

Fig. 3.1. Standard corn oil refining methods.

Alkali refining

Alkali refining is still the most commonly used process in the industry for removing these undesirable components. Usually, crude oil is treated with 12–18 Bé caustic solution using 0.05–0.20 per cent excess to neutralise free fatty acids, precipitate phospholipids and remove insolubles. The advantages of this one-step procedure include simplicity, lower loss of neutral oil and lower capital investment in equipment. After soapstock separation, the neutralised oil is subjected to water washing and vacuum drying.

Bleaching

Bleaching of the refined oil is typically done under vacuum (50 mm Hg absolute pressure), in continuous or batch equipment, using 0.5–1.5 per cent activated bleaching clay at 90–110°C for 20–40 minutes. The oils produced from 'end-of-season' corn may exhibit set colour and require bleaching clay even in excess of 1.5 per cent. Subsequently the oil is filtered to yield a light, clear product.

Dewaxing

Dewaxing is strictly a cosmetic operation to remove components with high melting points, including waxes and trisaturated glycerides present in corn oil in small quantities (less than 0.5 per cent). This ensures clarity and brilliance of the oil at ambient temperatures and even when refrigerated. Dewaxing of corn oil can be accomplished by rapid chilling of bleached oil to 5–10°C in the presence of 0.2–0.5 per cent filter aid and holding at this temperature for 1–2 hr., followed by cold filtration. For US corn oil, a combined bleaching–dewaxing process is also feasible. After bleaching, the oil is chilled directly, without the usual filtration step, to 5–10°C and held at this temperature with mild agitation for 1–2 hours. The suspended bleaching clay acts as a seeding agent for waxes, which crystallise and are removed from the oil by cold filtration. This procedure yields a fully bleached and dewaxed oil with only one filtration step. Corn oils grown in the tropics require a slow, gradual chilling for nucleation of waxes and holding up to 24 hr., at 5–10°C for proper crystallisation of solids before filtration. In such cases, the combined bleaching–dewaxing process is not practical.

Deodourisation

Deodourisation parameters have a significant impact on the quality of finished oil. Temperature, time, stripping steam rate, oil throughput rate and vacuum must be optimised to produce high-quality oil with a long shelf life. Optimisation of deodourisation, based on statistical correlation of process parameters and quality indicators has been demonstrated on model feedstocks and typical commercial oil. Tocopherol retention, FFA level, flavour stability and colour reversion are used as responses for defining optimum deodourisation parameters for batch laboratory and pilot plant deodourisers as well as a plant continuous system. Modern, continuous deodourisers are built of stainless steel and operate at temperature of 240–260°C under absolute pressure of 3–6 mmHg, producing bland or mildly flavoured (slightly nutty, corny) edible oil. Deodourisation parameters must be strictly controlled to achieve complete removal of pesticides and flavour compounds while retaining 85–90 per cent of the tocopherols present in the bleached–dewaxed oil. For the longest shelf life, deodourised oil should always be protected with nitrogen.

Physical refining

The physical refining process is also a viable process for corn oil. It relies on degumming and bleaching for removal of insolubles, phospholipids, pigments, mycotoxins and other nonvolatile components, whereas volatiles such as FFA, pesticides and oxidation products are removed during steam refining–deodourisation. The degumming of corn oil is readily accomplished with 2–5 per cent water alone or through pretreatment with 0.05–0.10 per cent phosphoric or citric acid followed by the addition of water. Degumming must result in at least 95 per cent phospholipids-free oil to ensure economical bleaching. Bleaching consists of pretreatment with 0.05–0.10 per cent phosphoric or citric acid and vacuum bleaching at 90–105°C for 20–40 minute with 1–3 per cent of high-grade activated clay. The dewaxing process is similar to that used in alkali refining and can be done on filtered or unfiltered bleached oil. Steam refining-deodourisation is performed under standard deodourisation parameters but at 20–40 per cent reduced oil throughput rates and increased stripping steam rates compared to alkali refined oil. The steam refiner–deodouriser must be equipped with a properly sized distillate scrubbing and recovery system and be built of 316 stainless steel to resist corrosion. Physical refining is economical when high-quality crude oil from a well-controlled milling/recovery operation is available. Exposure of germ and crude oil to excessively high temperatures and oxidation results in 'fixed' colour that is difficult to remove during refining. In comparing the economics of refining processes, the cost of bleaching clay and its disposal should be carefully assessed for physical refining, whereas the cost of pollution control must be of primary concern for alkali refining.

Effects of Processing on Mycotoxins and Pesticide Removal

Mycotoxin removal

Aflatoxins (AT) are toxic and carcinogenic metabolites produced by the moulds *Aspergillus flavus* and *Aspergillus parasiticus*. Studies have shown that during corn wet milling approximately 40 per cent of AT is removed with the steep water while 34 per cent remains in the fibre, 15 per cent in the gluten, 10 per cent in the germs and 1 per cent in the starch. Hence, crude corn oil produced from mouldy corn contains AT. It has been shown that convention alkali refining, followed by bleaching, completely removes the high content of AT from crude oil obtained from germ intentionally inoculated with *A. flavus*. In laboratory experiments, the degumming of crude oil removes 30–50 per cent of the AT and the bleaching step eliminates the rest. Deodourisation alone is ineffective in removing AT from oil. If AT is present in crude oil, it is completely removed by caustic treatment during alkali refining or by degumming and bleaching steps of physical refining.

Pesticide removal

Control of weeds, diseases, insects and pests is an important part of corn management programmes. Consequently, crude corn oil may contain small amount of herbicides and pesticides. The effect of consecutive refining steps on the removal of pesticides residue from corn oil was studied in a laboratory experiment. Deodourisation is highly effective in removing pesticide residue, whereas the alkali refining and bleaching steps have only margarine impact on pesticide content. This corroborates common knowledge in the industry that deodourisation is the key process in producing pesticide-free finished edible oil.

Effects of Processing and Handling on Oil Quality

The quality of corn oil begins in the field where corn growth, maturation, harvesting, drying and storage conditions have an impact. Harvesting and drying conditions are of great importance for the yield and quality of dry milling products. Corn harvested at 25 per cent moisture or less and dried with 82°C hot air, yields optimum results, including higher oil recovery than corn dried with 150°C air. Corn steeping parameters during wet milling also impact the quality of the oil produced. It has been shown that using high concentrations of SO_2 in steep water (4000 ppm) produces germs of good initial quality, but highly susceptible to oxidation over time as measured by the rapid disappearance of tocopherols and the increase of oil oxidation. Conversely, corn exposed to a lower concentration of SO_2 (1200–1600 ppm) during steeping yields germ more resistant to oxidation, which retains a higher level of tocopherols during storage. These results indicate that attention must be paid to corn steeping, germ separation, washing, drying and cooling to ensure satisfactory stability of the germ's oil prior to its recovery. Ideally, corn germ should be stored at low temperatures (below 20°C) and used for oil recovery within a few days of wet milling. Since quite often the recovery of oil is done at a different location than the wet milling, short transit time and maintaining low to moderate temperatures are beneficial to oil quality.

Storage of crude oil for prolonged periods at elevated temperatures is also detrimental to quality and refining economics. When crude oil was stored at 45–50°C without nitrogen protection, significant oxidation occurred and bleaching clay requirements increased two and fourfold, respectively, for 3- and 6-month-oil crude. A similar experiment with crude oil stored at 5–10°C for 3 months showed no significant increase in oxidation and the bleaching clay demand did not increase. Thus, storage temperature, time and the presence of oxygen are significant factors affecting the quality and refinability of crude corn oil.

Oxidation of oil easily occur during recovery and refining operations. In the modern plants, the level of oxidation present in the germ remains relatively unchanged throughout corn oil recovery, alkali refining, bleaching and dewaxing, with a pronounced reduction of the oxidation after deodourisation. In the outdated plant, there is a significant increase in oxidation during the oil recovery process due to unsatisfactory cooling and prolonged exposure of the oil to air at high temperatures. Alkali refining, removes some oxidation products, but a significant increase in oxidation is noticeable during bleaching. In this case, the introduction of bleaching clay into hot oil (90°C) creates conditions for rapid oxidation (presence of oxygen, large surface area of clay, high temperature). The deodourisation process is effective in reducing the overall oxidation level, but the finished oil is still of inferior quality compared to the product from the modern plant.

The stability of finished oil is affected by the level of oxidation products remaining after processing. Packaged in PET (polyethylene terphatalate) bottles under nitrogen and stored in the dark, oil with an initial Aniside value (AV) of 8.0 oxidises and reverts colour more quickly than oil with an initial AV of 4.3. Colour reversion (darkening) in corn oil is the result of the conversion of tocopherols to tocored in oxidative reactions. The higher the initial content of secondary oxidation products in the finished oil, the lower its stability and the shorter its shelf life.

Packaging

Packaging represents the second most significant portion of the total finished product cost (materials, equipment, labour) after raw materials. Prior to the 1970s, corn and vegetable oils were traditionally packaged in glass bottles and tinplated steel cans. Although these materials are still in use today, plastics have taken control of this market over the last two decades due to mainly consumer preferences for lightweight, 'unbreakable' and easily handled containers.

Initially, extrusion-blown PVC (polyvinyl chloride) bottles were utilised for the smaller sizes to provide the required oxygen barrier characteristics needed for oils and HDPE (high-density polyethylene) was used as a lower cost alternative for the larger sizes with smaller surface-area-to-volume ratios.

Most of today's PET containers are stretch blown, which orients the molecular chains, improving strength and other physical properties. The resulting lightweight containers offer excellent strength, toughness and clarity compared to other plastic resins, while providing adequate protection from oxygen permeation to achieve acceptable shelf life for oils.

UTILISATION AND PERFORMANCE

Corn oil is recognised as a premium quality oil by consumers. Its pleasant, slightly nutty and slightly sweet flavour is often associated with cooked sweet corn. Due to its inherent stability, corn oil has a long shelf life and is resistant to breakdown under adverse conditions, such as those encountered during frying. Originally, most of the corn oil produced was sold as liquid oil for a variety of home uses including salad dressings, pan and deep frying, corn popping and baking. When more corn oil became available at lower prices as a result of steadily increasing corn wet milling, its utilisation extended to margarines, institutional frying and baking, liquid and solid shortenings, mayonnaise, salad dressings, sauces, snacks, baking mixes, frostings, whipped toppings and crumb coatings for meat, poultry and baked goods.

Since corn oil is used extensively for frying, many researchers have studied its performance in this application. In a five-day simulated reuse frying test, corn oil showed remarkable resistance to oxidation and degradation as measured by tocopherols retention. Canola and soyabean oils deteriorated significantly faster during the same test. Apparently, continuous deep frying of potatoes and onion rings is less detrimental to corn oil quality than intermittent frying. Potatoes causes less oil degradation than onion rings. Starch chunks deep fried in corn oil have better flavour stability and oxidise more slowly during storage than those fried in sunflower oil. The same studies reveal that under various storage conditions, corn oil shows a consistently lower rate of oxidation compared to sunflower oil. An extensive study of chemical reactions induced by deep frying was done on corn oil, hydrogenated cottonseed oil, trilinolein and triolein using moist cotton balls. A total of 211 compounds generated during heating were identified. This study found that γ-lactones with unsaturation at the 2- or 3-position are important components of deep-fat-fried flavour. These lactones were present in both corn oil and trilinolein but not in the hydrogenated oil or triolein. This implies that linoleic acid in corn oil generates desirable deep-fried flavour.

In another study of volatiles generated during heating, the sum of polar and nonpolar compounds in corn oil was lower than in soyabean oil but higher than in coconut oil (Table 3.10). An increase of heating temperature from 185 to 250°C results in a 5–15 per cent decrease in volatiles, probably due to a higher rate of decomposition of some compounds. The presence of moisture during heating exerts a protective effect on oil by replacing atmospheric oxygen with steam and by stripping some volatiles (Table 3.11). A comparison of reuse frying on organoleptic qualities of meat croquettes and French fries prepared in corn oil, sunflower oil, partially hydrogenated soyabean oil and hydrogenated palm oil indicates that corn oil produces acceptable food for the longest time; 8 days at a frying rate of 9 kg food per day.

Table 3.10. Volatiles in corn, soyabean and coconut oils, generated by heating (mg/kg).

Description	*Corn oil*		*Soyabean oil*		*Coconut oil*	
	185°C	*250°C*	*185°C*	*250°C*	*185°C*	*250°C*
Polar	869.9	663.1	920.0	819.4	565.4	524.0
Nonpolar	60.3	144.7	41.9	93.3	130.7	281.3
Total	930.2	807.8	961.9	912.7	696.1	805.3

Table 3.11. Effects of heating time and moisture on corn oil volatiles (mg/kg).

Description	*Heating time*			
	1 hr.		*48 hr.*	
	NO H_2O	*With H_2O*	*NO H_2O*	*With H_2O*
Polar	91.5	120.6	2346.3	416.6
Nonpolar	13.0	2.5	158.4	16.2
Total	104.5	123.1	2504.7	432.8

NUTRITIONAL CONSIDERATIONS

Corn oil, a highly palatable food oil, can play a major role in the human diet. It is a concentrated source of energy (calories), is very digestible, provides essential fatty acids and vitamin E and is a rich source of polyunsaturated fatty acids, which help regulate blood cholesterol levels and eicosanoid synthesis and lower elevated blood pressure. Both animal and human studies show that at least 97 per cent of the oil is digested and absorbed. Like all fats and oils, corn oil provides 9 kcal (38 kJ)/g, or about 120 kcal per one tablespoon (14 g) serving. Corn oil is a rich source of linoleic acid, an essential fatty acids that the body cannot synthesise.

The National Research Council and the Food and Agriculture Organisation/World Health Organisation recommend about 2–4 per cent of calories as essential fatty acids with an additional 3 per cent of calories for women who are pregnant or breast feeding. The ratio of linoleic (n–6) to linolenic (n–3) acids should be approximately 5:1. A tablespoon of corn oil will satisfy the daily essential fatty acid requirement for a healthy child or adult.

Effects of Fatty Acids on Blood Cholesterol and Heart Disease

The impact of dietary fat, blood cholesterol and lipoprotein levels on coronary hearth diseases has been extensively studied. It is now widely accepted that a diet high in saturated fat and cholesterol is one of many causative factors in the development of atherosclerosis and coronary heart disease. Saturated fat and cholesterol interfere with the liver's efficiency in removing cholesterol from the body; as blood cholesterol levels increase, the excess of cholesterol accumulates in the arteries, especially the coronary arteries.

It is almost impossible to obtain accurate food intake records for individuals; therefore, several large epidemiological studies have used the linoleic acid content of adipose tissue as a marker for polyunsaturated fat intake since linoleic acid must be obtained through diet. Studies by *Berry* and co-workers in free-living men have shown that those in the upper tertile of adipose tissue linoleic acid had lower concentrations of total and LDL-cholesterol compared to those in the lower tertile, but equivalent HDL-cholesterol levels.

Effects of Corn Oil on Cholesterol Levels

Corn oil has been used extensively in research studying the relationship of dietary fat to blood cholesterol levels. This is due to the fact that corn oil was the only highly polyunsaturated, nonhydrogenated oil readily available to investigators and patients in the mid-1950s. Due to its high stability, pleasant taste and multiple applications, corn oil became the standard against which other oils were compared to assess cholesterol lowering.

Corn oil has shown greater cholesterol lowering than expected from its fatty acid composition alone. This has been attributed to the relatively high level of nontriglyceride compounds, especially the plant sterols found in corn oil.

Effects of Corn and Other Oils on Cholesterol Levels

Table 3.12 summarises the studies in which corn oil was compared to one or more other dietary oils with respect to their cholesterol lowering abilities. The effectiveness of corn oil in lowering serum cholesterol in human compared corn, sunflower and sardine oils with butter. The corn oil group showed greater cholesterol lowering than either the sunflower oil group (by 9.3 mg/dL, $p < 0.01$) or the fish oil group (by 20.2 mg/dL, $p < 0.001$).

Table 3.12. Corn *vs.* other oils in blood cholesterol lowering.

Comparison
Corn > sunflower > sardine > > butterfat
Corn > sunflower > sardine ≥ cottonseed = low fat = olive > > butter fat = 'house'
Corn = safflower = cottonseed = peanut ≥ olive = lard I = chicken > lard III = palm + cocoa butter + beef > > butter ≥ coconut
Corn ≥ sunflower = soyabean
Corn = soyabean ≥ partially hydrogenated soyabean > palm
Corn > olive
Corn > sunflower
Corn > high oleic sunflower > > butter
Corn = sunflower ≥ partially hydrogenated soyabean (4th quartile initial cholesterol levels only: corn > sunflower = partially hydrogenated soyabean)

The work of *Ahrens* and *co-workers* reports extensive human testing in which corn oil was the baseline standard to which all other oils were compared. In this series, no oil was better than corn oil in lowering cholesterol. They concluded: 'The lowest serum-lipid levels were seen when corn oil, safflower-seed oil, or cottonseed oil constituted the sole dietary fats. Higher levels were found with all other fats tested. The differences in these levels are directly related to the degree of saturation of the glyceride fatty acids as measured by the iodine value of the fat'.

Hill and *co-workers* compared corn, sunflower and soyabean oil diets with habitual diets for their abilities to lower cholesterol. They found that all three oils were associated with cholesterol lowering. The diet containing corn oil lowered total plasma cholesterol by 16 per cent, while those containing sunflower or soyabean oils lowered cholesterol by 9 and 11 per cent, respectively.

Laine and co-workers compared the effects of diets containing corn, soyabean and partially hydrogenated soyabean oils on cholesterol levels with diets containing palm oil. Native soyabean and corn oil diets lowered total cholesterol 13 per cent which the partially hydrogenated soyabean oil diet lowered cholesterol by 9 per cent, but these differences were not significant.

Sirtori and co-workers evaluated corn vs. olive oil in 23 middle-aged patients for 8 weeks. They found that the corn oil diets produced significant cholesterol lowering of about 7 per cent ($p < 0.05$) while consumption of the olive oil diets resulted in a nonsignificant 4 per cent cholesterol lowering. *Kohlmeier* and co-workers compared corn and sunflower oils in a crossover design trial. The subjects, all male, ate high-fat diets (46 per cent of calories). Total and LDL-cholesterol were significantly lowered 21.6 and 23.3 per cent from initial results of 221.2 and 165.5 mg/dL, respectively. The corn oil group showed a significantly greater lowering ($p < 0.01$) of both trial and LDL-cholesterol by 6.8 and 12.0 per cent, respectively, compared to the sunflower oil group. *Wardlaw* and *Snook* studied 20 men who ate diets in which 37–43 per cent calories were from fat, the majority of which came from corn oil, high oleic sunflower oil, or butter. Compared to the butter-based diets, substitution of corn oil lowered total cholesterol 21 per cent and high oleic sunflower oil, 16 per cent. These values were significantly different.

In summary, nine human feeding studies involving 350 men and women, carried out over 30 years by different independent investigators in America and Europe, show that no oil results in greater cholesterol lowering than corn oil. In fact, three studies show corn oil to be superior to sunflower oil in overall cholesterol lowering, while a fourth study confirms this observation in 'high' but not 'low' cholesterol subjects. The total composition of an oil, which includes fatty acid composition, triglyceride structure and nontriglyceride components, probably contributes to that oil's ability to lower serum cholesterol. As far as edible, liquid vegetable oils are concerned, no oil is better than corn oil for lowering blood cholesterol levels.

Effect of Corn Oil on Other Chronic Conditions

Blood pressure

Numerous human studies have shown that diets enriched in *n*–6 polyunsaturated fatty acids significantly lower elevated blood pressure. Corn oil was used in many of these studies and it has been shown in animals that polyunsaturates change the production of certain prostaglandins in the kidney so that more sodium is excreted in the urine. Corn oil diets resulted in systolic and diastolic blood pressure reductions of about 12 per cent in men and 5 per cent in women with mild hypertension. No significant effect of polyunsaturates on normotensives has been noted.

Platelet aggregation

Studies in animals and human have shown that *n*–6 and *n*–3 polyunsaturated fats significantly affect blood clotting and platelet aggregation. Again, corn oil was used in many of these experiments as a source of *n*–6 fatty acids. It is believed that on a corn oil diet the increased production of vessel wall prostacyclin (antiaggregatory) is more potent than the increases in platelet thromboxane (proaggregatory).

Diabetes

Enrichment of the diet with linoleic acid has favourably influenced the progression of diabetic angiopathy in adult onset diabetes mellitus. Increasing the linoleic acid from 3 to 13 per cent of calories in the diet over 6 years improved glucose tolerance and micro- and macro-retinopathy.

Immune system

Linoleic acid is needed for normal mitogenic response to occur in animals. Although a diet high in polyunsaturates can be immunosuppressive in animals under certain conditions, diets enriched in corn oil showed no adverse effects on the immune system in human over a 6-month period.

Cancer

There is no evidence that polyunsaturated fats or corn oil are associated with an increased risk of breast or colon cancer, both of which have been associated with a high-fat diet. Although the rat shows a specific increased need for essential fatty acids in the diet for carcinogen-induced tumour growth, especially for chemically included mammary tumours, extensive human epidemiologic studies, whether correlational, case controlled, or prospective, have shown no significant relationship of linoleic acid or polyunsaturated fat to cancer of the breast, colon, prostate, or pancreas. Adipose tissue linoleic acid levels did not correlative with occurrence of breast or colon cancer in two larger studies carried out in Israel, a country with relatively high linoleic acid intake. Analysis of five major diet intervention trials, in which relatively high (up to about 16 per cent of energy) levels of polyunsaturates were fed, showed no increase in cancer or other non-cardiovascular disease mortality.

To sum up corn oil is derived from corn germs. It is inherently stable in spite of its high level of linoleic acid. Corn oil can play a major role in human nutrition not only as a source of essential fatty acids and vitamin E but as an effective dietary oil for lowering blood cholesterol and elevated blood pressure levels and reducing the risk of coronary hearth disease. There are no adverse effects of a diet enriched in polyunsaturated fats such as corn oil.

Chapter 4

Cottonseed Oil

INTRODUCTION

Cottonseed is a by-product of the cotton crop and is a valuable source of edible oil, cake, linter and hull, all of which have several industrial and other applications. Cottonseed is the most widely distributed oilseed in the tropical and sub-tropical areas. Historically, India is the oldest cotton-producing country in the world.

Cottonseed contains two principal components: (i) the hull (outer covering of the seed) from which cotton fibres and cotton linters arise; and (ii) kernel or embryo from which oil and meal are obtained. In addition, the seed contains minor constituents such as lecithins, sterols, some B- and E-group vitamins and minerals. In India, only a small quantity of cottonseed oil (5–10 per cent) is used for manufacturing of soaps, while a much larger quantity is utilised for edible purposes, mostly as edible oil and vanaspati. Cottonseed oil possesses better stability and keeping quality as compared to groundnut- and safflower-oil. It is easily digested with a digestibility coefficient of about 98 per cent. Several non-conventional products can be produced from cottonseed oil. For example, it can be used for production of acetoglycerides for food. Brominated cottonseed oils are used for production of cloudy, flavoured soft drinks. Emulsified cottonseed oil can be used for intravenous administration to patients who require a higher calorie diet than what carbohydrates can provide. Cottonseed-meal or cake (left after extraction of the oil) contains high quality protein and is rich in essential amino acids. It is, therefore, an excellent cattle and animal feed. The metabolisable energy value of cottonseed-meal is similar to that of other oil-meals and animals fed on this diet have shown good growth and performance. A major problem in the use of cottonseed oil and cake is the presence of certain toxic pigments such as gossypol, gossypurpurin and gossyfulvin, in addition to anthocyanins and carotenes. These pigments bind themselves to the amino acids, particularly to lysine and make them unavailable. The presence of cyclopropenoid fatty acids (CPFA) such as malvalic and sterculic acids and of the highly carcinogenic aflatoxins are other negative factors that need to be eliminated.

Cotton linter is similar in composition to cotton fibre and is basically composed of cellulose. Although linters cannot be converted into yarn in view of their extremely short length, they are valuable raw materials for preparation of cellulose derivatives such as cellulose acetate, cellulose nitrate, cellulose xanthate and carboxymethyl cellulose which in turn have several uses in the plastic, textile and other industries. Besides, linters are also used for upholstery, lamp and candle wicks, medical grade cotton and for the preparation of writing papers, currency paper, etc.

Cottonseed hull has poor nutrient value and is generally mixed with cottonseed-meal as roughage to increase the volume of cattle feed. It is also used as soil conditioner and for production of industrially important chemicals such as furfural and active carbon. Besides, particle and fire-retardant boards (to replace costly plywood) can also be made from cottonseed hull powder.

The chapter gives a consolidated account of the utilisation of cottonseed in its entirety—the oil, the meal, the hull and the linters. Important aspects of cottonseed such as production, its physical, chemical and biological characteristics, harvesting and storage practices in vogue, types of processing equipment, methods of oil extraction, refining techniques and procedures, nutritional studies, grading systems, industrial applications and economic criteria have been discussed elaborately.

PRODUCTION AND UTILISATION OF COTTONSEED AND COTTONSEED PRODUCTS

Cotton, as harvested and brought to the gin, consists chiefly of two products, i.e. cotton lint and cottonseed. The ginning process separates the lint from the seed. The proportion of lint to seed in any given quantity of seed-cotton (*kapas*) varies, depending primarily on the variety and also on the location, season, soil and climate. Thus, the proportion of lint to seed from the same variety varies with the locality and the season. The quality of lint obtained in the gins from 100 kg of seed-cotton is termed as the ginning factor or ginning per cent. The variation in the ginning per cent is greater in *desi* cotton varieties (*G. arboreum* and *G. herbaceum*) than in the American cotton (*G. hirsutum*). For instance, it varies from 31 per cent from Bellary to 40 per cent in Bengal *desi*, whereas it varies only from 31 per cent in 'Laxmi' (Phalton) to 36.4 per cent in '320 F' from Sirsa. While the variation in the ginning per cent in the United States is also governed by the above factors, the range of variation of 16 varieties at 16 locations varied from 31.5 per cent to 39 per cent giving an average of 35.5 per cent. In estimating cottonseed production in India, a uniform average ginning percentage of 33-1/3 per cent is being used. The fuzz leftover on the seed coming out of the gins is called the linters and this term is used throughout in the chapter.

Cotton produced in India belongs mainly to four species of *Gossypium*, viz. *G. arboreum, G. herbaceum, G. hirsutum* and *G. barbadense*. The *desi* varieties of cotton come under the first two species and the American varieties under the third. The *barbadense* varieties are of recent origin in India and produce fuzzless seed and superior quality lint. About 50 varieties of cotton from all four species are at present produced on a commercial scale in India.

The cottonseed, in other countries such as China, the USSR and Egypt also is being utilised mainly for production of oil for edible purposes and meal is being used as cattle-feed and to some extent as fertiliser.

Several new varieties have been developed in place of old *desi* varieties. Cottons belonging to American upland types *G. hirsutum* and *G. barbadense* have longer and finer fibres. Also several hybrid varieties have been developed and commercialised. Some of the important hybrid varieties are 'Hybrid-4' and 'Varalaxmi'. 'Hybrid 4' is grown in Gujarat, Maharashtra, Madhya Pradesh, Karnataka and Andhra Pradesh extensively. 'Varalaxmi' is grown in Karnataka, Andhra Pradesh, Tamil Nadu, Maharashtra and Gujarat. Superior long-staple cotton varieties—'Suvin', 'Sujata', 'MC49' 'MC 48'—are grown in irrigated tracts in Tamil Nadu and Andhra Pradesh.

Cottonseed Processing

Seed-cotton (*kapas*) is harvested in India by hand-picking only. The harvesting season varies from state to state, as it differs for different varieties. There are two main harvesting seasons, one from February to April and the other from September to December. The marketing of cotton commences from within a month of harvesting and continues for 3 to 4 months. The seed-cotton is ginned for lint mostly in factories using either power-driven roller gins or saw-gins. The main season for marketing the seed is from March to June in most of the states.

The *kapas* picked from the field is transported to the ginning factories in bullock-carts or lorries. Generally, the *kapas* is heaped in the form of a large conical pile in the open yard of the ginning factory and usually covered by tarpaulins. In 54 per cent of the factories the *kapas* is kept in sheds and in 41 per cent the *kapas* is kept in the open but protected during rains by tarpaulins. It is estimated that 26 per cent of the factories deal with clean *kapas*, 56 per cent with medium clean, 14 per cent with a combination of clean, medium clean and trashy and 3 per cent with clean early in the season and trashy later on. About 64 per cent of factories do not clean the *kapas* before ginning, although it is found that pre-cleaning would be advantageous before ginning.

At the gins, *kapas* is ginned and seeds are separated. Earlier *kapas* used to be mixed before ginning. Therefore cottonseed available after ginning used to be mixed with the seeds of several varieties. However, at present ginning is done more carefully and seeds of individual varieties are collected and sold in the market at different rates. The purchase of the seed depends on its type (*desi* and American) and on its oil content and physical condition. On the other hand, in the USA, where the seed is mainly used for processing after reserving for planting, all mills grade the cottonseed received by them for evaluation, before storage and milling. The factors to determine the quality of cottonseeds are: (i) percentage of oil; (ii) ammonia; (iii) moisture; (iv) foreign matter; (v) linters; and (vi) free fatty acids.

About two decades ago, cottonseed was an important product of food in India for cows and buffaloes in milk and working animals. Before the commencement of the milling of cottonseed for oil, all the cottonseed produced in India was being utilised as cattle feed, after meeting the requirements for planting, for which about 5 per cent of the seed is utilised. The less fuzzy seed of *desi* varieties is preferred by the farmers for feeding their cattle. For this purpose, the prices of American varieties used to be lower than those of *desi* seeds.

However, in recent years, owing to increased demand and scarcity of edible oils in the country, crushing of cottonseed for production of oil has increased tremendously. There has been considerable increase in crushing of cottonseed in the country year after year. At present, nearly 70 per cent of cottonseed produced in the country is crushed for the production of oil.

It is worth mentioning that with the application of scientific cottonseed processing, cottonseed industry and its trade will be able to do much better in future. Therefore, it is suggested that the available cottonseed should be processed by employing delinting, decorticating, crushing and extraction processes.

COMPOSITION AND CHARACTERISTICS OF COTTONSEED

Physical Structure of Cottonseed

The cottonseed is obtained from lint after ginning. The cottonseed belonging to *G. hirsutum* species also known as American type seeds and *desi* varieties belonging to *G. arboreum* and *G. herbaceum* species contain short fibres known as linters. The percentage of linters varies depending on the species and nature of ginning operation. American seeds contain higher percentage of linter compared with *desi* varieties. On the other hand, seeds of *G. barbadense* species are linterless and after ginning the linter content in these seeds is almost negligible. The linters arise from the seed-coat, which is the outer covering of the seed. The seed-coat varies from brown to black in colour. The cottonseed obtained after ginning mainly contains linter and seed. The linters are short fibres, which are removed by delinting process and are used as source of cellulose for production of various cellulose products, pulp and paper and cellulose derivatives. Cottonseed is mainly used for the production of cottonseed oil and a small quantity is used for sowing.

Cottonseed contains two principal components. These include the hull or outer covering of the seed, which is also called spermoderm, from which cotton fibres and cotton linters arise. The other important component is kernel or embryo, from which oil and meal are obtained. Cottonseeds after removal of lint are pointed, ovoid, dark brown or nearly black and vary in length. Both the hilum (point of attachment) and the micropyle (opening to the nucellus) are located at the pointed end and the chalazal is at the other end, towards one side. The broadened oval end of the seed is the chalazal end. The pointed tapering end of the seed is micropylar end near which the hilum and micropyle are located. The cottonseed is a dicotyledon. It indicates two cotyledons in which the radicle, the hypocotyl and epicotyl are located. The cross-section of cotyledon also indicates several dark 'specks'. These specks are also known as 'pigment glands' which contain yellow red colour pigments including gossypol. These glands are easily broken in presence of water and discolour the cotyledon tissue. The cotyledons are kidney shaped.

Development of hull

The hull (spermoderm) is evolved from the external skins of the ovule. It develops rapidly from the time fertilisation takes place. Its surface cells constitute protective tissues in the latter stage of development, the nutrients in spermoderm are utilised, the starch of the parenchaimal is replaced by sugar and oil. Finally all the nutrients are consumed during maturing of embryo. The mature cells of hull indicate the presence of tannin.

The cottonseed hull content is not constant in all the seeds; it varies from 35 to 60 per cent of the weight of whole cottonseed.

Nucellus

It is also know as perisperm. A thin membrane can be seen between the inner wall of the spermoderm and the embryo from a section of cottonseed. This membrane forms the attachment to the cottonseed hull at the chalazal cap. The cell structure of the membrane consists of two distinct structures. The outer structure comprises 'Fring cells' of the perisperm, whereas the inner structure is a tissue of cell structure containing dense fine aleurone grain.

The perisperm and endosperm containing protein, oil and sugar provide nutritive materials for growing embryo. At the end of the development of the embryo these materials fill the area within the hull which develops with great rapidity.

Cottonseed kernel

The kernels of cottonseed consist of two cotyledons and are most important organs of the embryo of cottonseed and these contain oil and meal. The cotyledons are the main source of oil and protein in cottonseed. The parenchyma cells of the cotyledon contain most of the oil present in the seed and bulk of the protein. The pigment glands containing gossypol and related pigments are supported and enclosed by elongated thin walled 'mucilage cells'.

The epicotyl and hypocotyl form the axis of the embryo. The epicotyl is the bud from which the first true leaf develops. The hypocotyl develops into the stem and bears the cotyledons. The main function of the cotyledons is to store and supply the necessary nutrients during germination. The hypocotyl extends to the upper end of the elementary root. The radical is the elementary root. The oil, protein with starch and calcium oxalate are stored in parenchyma cell of the axial organs.

The kernel content in different varieties of cottonseed varies from 49 to 58 per cent depending upon varietal, agronomic and climatic factors. Detailed studies on kernel content of 40 commercially grown varieties showed wide variation for 4 different species and mean values were 53.4, 54.0, 53.9 and 62.0 per cent for *G. arboreum, G. herbaceum, G. hirsutum* and *G. barbadense*, respectively and total mean was 54.4 per cent.

Variability in Constituents

The composition of cottonseeds of different varieties between the species and within the species from different locations and from different crop years shows wide variation in their chemical constituents.

The variation in composition of cottonseed may be due to genetic factors, growth habit, reproduction and environmental factors. However, the studies have indicated that within the same atmospheric condition and agronomic practice the different varieties have shown different results. The seeds of these varieties may be large or small, may have higher percentage of linter or may be lintless. Seeds may also vary in seed index and hull and kernel contents. The seeds may also vary in protein, oil, crude fibre, ash content and other constituents. The cottonseed is considered an important source of edible oil and its high quality protein. The oils are lipids, are mostly in the form of glycerides and present in the liquid state at ambient temperature. The proteins are present in a dissolved state in the cell-sap, whereas the reserve proteins of the seed embryo are found in semi-crystalline or crystalline state. Protein occurs in the form of aleurone grains and as a part of the protoplasm.

Fat Development in Oilseeds

The oils are lipids, present in every living cell and constitute the principal storage along with protein in cottonseed. The reserve lipids of cottonseed are mostly in the form of glycerides. The oil is considered to be the continuous phase of protoplasm which changes during the development of the seed. In mature cottonseed it was observed that oil was embedded in the cytoplasm as minute droplets.

The electron micrographs of fragmented deoiled cottonseed-meal of 35-day-old sample indicated black particles of proteins fragments obtained after 20 minutes fragmentation. These fragments are found to be more often separated. However, in the case of 49 days meal the fragments are seen as clusters. The electron micrographs show fragments of deoiled-meal of cottonseed germinated for varying periods. As the germination progresses the protein fragments which look like individual spots show gradual disintegration.

Changes in seeds with maturation

It is well known that most of the oil in oilseed is produced within the seed while in the early stages of growth and development the seed contains mainly water and starch and no fat. As the development progresses the starch gets reduced and small droplets of fat begin to appear. At the final stage of the seed maturity, the droplets increase in size and fill the oilseed near about. The synthesis of fat in oilseeds has been studied thoroughly. In a recent study, a detailed report on the synthesis of total quantity of fat, carbohydrate, protein, lint percentage in cotton of both 'G. Cot. 10' and 'G. Cot. 11' varieties at different intervals of maturity from 21 to 72 days at weekly intervals from the time of flowering has been published. Initially, there was increase up to 42 days in all components of 'G. Cot. 10' except linter content. It is reported that lipid content showed marked increase from 35 days (0.21 per cent) to 42 days (22.1 per cent). Thereafter, the oil content reached an optimum value after 56 days of growth (22.9 per cent) while there was sharp reduction in linter content (7.5 per cent). The gossypol content free and total, both showed gradual increase with maturation up to 56 days. A similar trend was observed in variety 'G. Cot. 11' during maturation.

Fig. 4.1 shows the changes in seed during maturation in different constituents like moisture content, kernel content, oil content, seed index, linter content, total and free gossypol contents in varieties 'G. Cot. 10' and 'G. Cot. 11'. The figure shows a rapid increase in kernel content, seed index and oil content from 21 days to 42 days maturation time; thereafter all the 3 values level off. The total gossypol and free gossypol contents show gradual increase with seed maturation while linter content remains constant throughout. The changes in kernel composition during seed maturation may be seen in Fig. 4.2. It can be seen that there is a sharp decrease in carbohydrate and protein contents and a sharp increase in kernel and oil contents of both the varieties ('G. Cot. 10' and 'G. Cot. 11') up to 42 days. Ash content shows very little variation. The changes during germination in protein, oil and total and free gossypol contents of 'Hybrid-4' variety may be seen in Fig. 4.3. Figure 4.4 shows the infra red spectra of cottonseed oil during maturation of 'G. Cot. 10' variety after maturation of 35 days, 42 days and 72 days respectively.

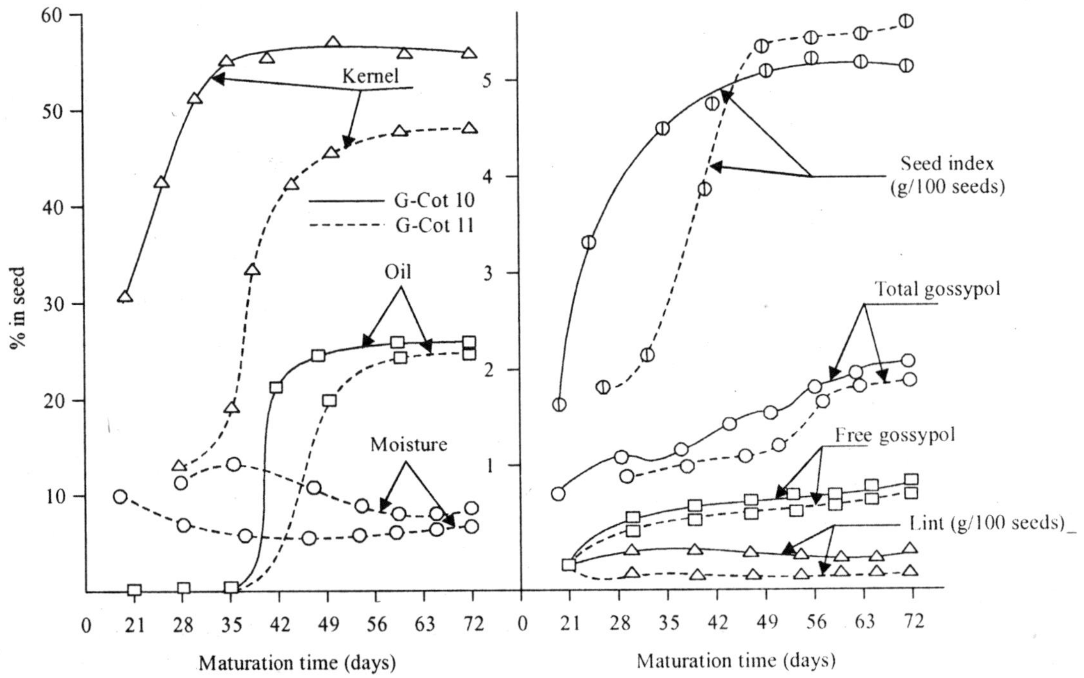

Fig. 4.1. Changes in seed composition during seed maturation.

Formation of all types of fatty acids proceeded throughout the maturation indicating rapid formation of unsaturated acids at later stages. Oil composition of 'G. Cot. 10' during maturation showed marked reduction in total saturated fatty acids from 21 days (33.7 per cent) to 56 days (27.8 per cent), while the unsaturated acids showed increasing trend with maturation. A similar trend was observed in variety 'G. Cot. 11'. These data indicate that various constituents of seed undergo marked changes during maturation. The oil characteristics of cottonseed during maturation of 'G. Cot. 10' and 'G. Cot. 11' varieties showed gradual decrease in free fatty acid (FFA) percentage (as oleic acid) and cyclopropenoic fatty acids percentage (CPFA) (as malvalic acid). FFA percentage for 'G. Cot. 10' decreased from 8.9 per cent for

28 days sample to 2.7 per cent for 72 days sample after maturation. The saponification value (SV) of fat remained unchanged, whereas iodine value (IV) of fat showed gradual increase with maturation. A similar trend was observed in 'G. Cot. 11' cottonseed oil.

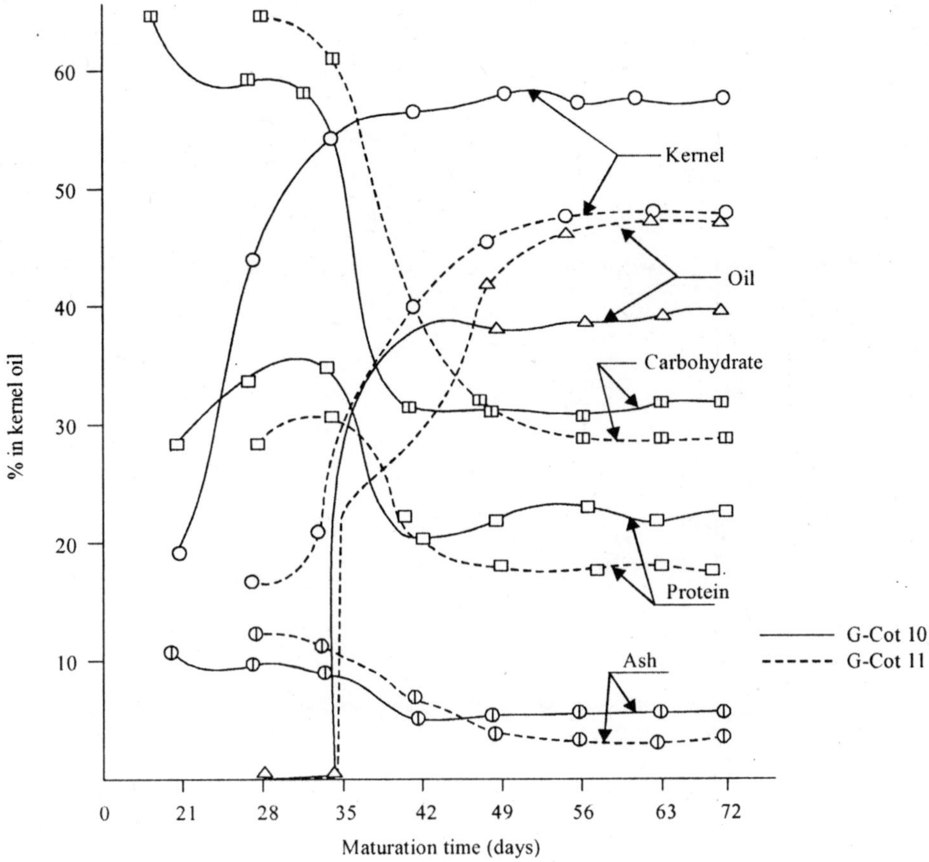

Fig. 4.2. Change in kernel composition during seed maturation.

The changes in fatty acid content during seed maturation may be seen in Fig. 4.5. It is observed that the fatty acid formation start after five weeks of maturation of seed. All the fatty acids initially showed sharp increase up to sixth week; thereafter linoleic acid showed gradual increase up to tenth week. The free fatty acids showed the maximum value after sixth to seventh week of maturity, after which it shows sharp decrease. Saturated fatty acids showed maximum increase at about sixth week of maturation period, beyond which the value levelled off. Figure 4.6 shows the changes in oil and CPFA content, during germination for the period of six days in four cotton varieties, namely, 'Varalaxmi', 'Laxmi', 'Hybrid 4' and 'Suvin'. Figure 4.7 shows changes in fatty acid composition during germination of cottonseed of 'Hybrid 4' variety.

Origin of fat and biosynthesis of fatty acids

The oil is produced at the expense of carbohydrates in the seed. The fats and oils are formed from carbohydrates and fats increased during the germination of oily seeds. The saccharose is detected first and disappears last during germination of oilseeds. Biochemical reactions have shown that intermediate compounds formed from carbohydrate due to oxidation are the real building blocks for fats. The two glycerol molecules, which form the backbone of triglycerides and the fatty acids with which glycerol is attached, originate from carbohydrate-derived products. Glycerol is formed as glycerol-1-phosphate which originates the reduction of dihydroxy acetone phosphate itself derived from fructose 1,6 diphosphate.

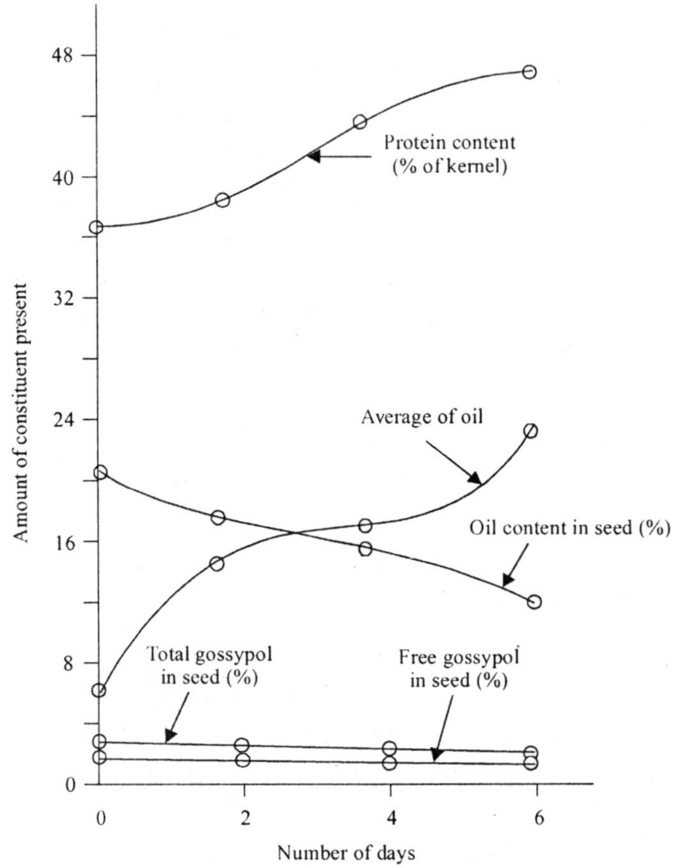

Fig. 4.3. Changes during germination of cottonseed of 'Hybrid 4' variety.

Based on the occurrence of straight-chain fatty acids containing an even number of carbon atoms, it was concluded that two carbon step-wise synthesis is involved. From metabolism and biosynthesis studies it was concluded that acetate was the building block for fatty acid synthesis, which is activated by esterification with co-enzyme A (CoA), the bond involving thioester linkages. It is reported that besides acetyl CoA, malonyl CoA, a dibasic acid derivative from acetyl CoA by the introduction of a carboxyl function, is also a highly reactive substrate in plant tissue. These compounds occur into long chain fatty acids by their initial transfer to a high molecular weight protein, called Acryl Carrier Protein (ACP). These reactions are condensation of the two varieties to give an acetoacetyl derivative which yields the unsaturated crotonyl compound and hydrogenation of this crotonyl compound leads to a butyric acid derivative which is attached to ACP. This butyric acid derivative goes through the cycle of further chain elongation. Once the chain length of acids (saturated) reaches 16 or 18 carbon atoms, they are freed as CoA derivatives from the ACP surface.

Unsaturated fatty acids are formed in plants by dehydrogenation of saturated acyl CoA. Similarly acyl CoA can arise from the stearyl derivative. Similar desaturations of other fatty acid CoA derivatives (C_{12}, C_{14}, C_{16}) are also reported. Desaturation of fatty acids also occurs in seeds and it requires oxygen and NADPH. All these processes are sterospecific and the two hydrogens on the same side of carbon-carbon linkage are removed to give a *cis* and not a *trans* double bond and both hydrogens are removed at time. The enzymes involved in these activities of desaturation appear to become active at different stages of seed maturity in biosynthesis of saturated and unsaturated acids.

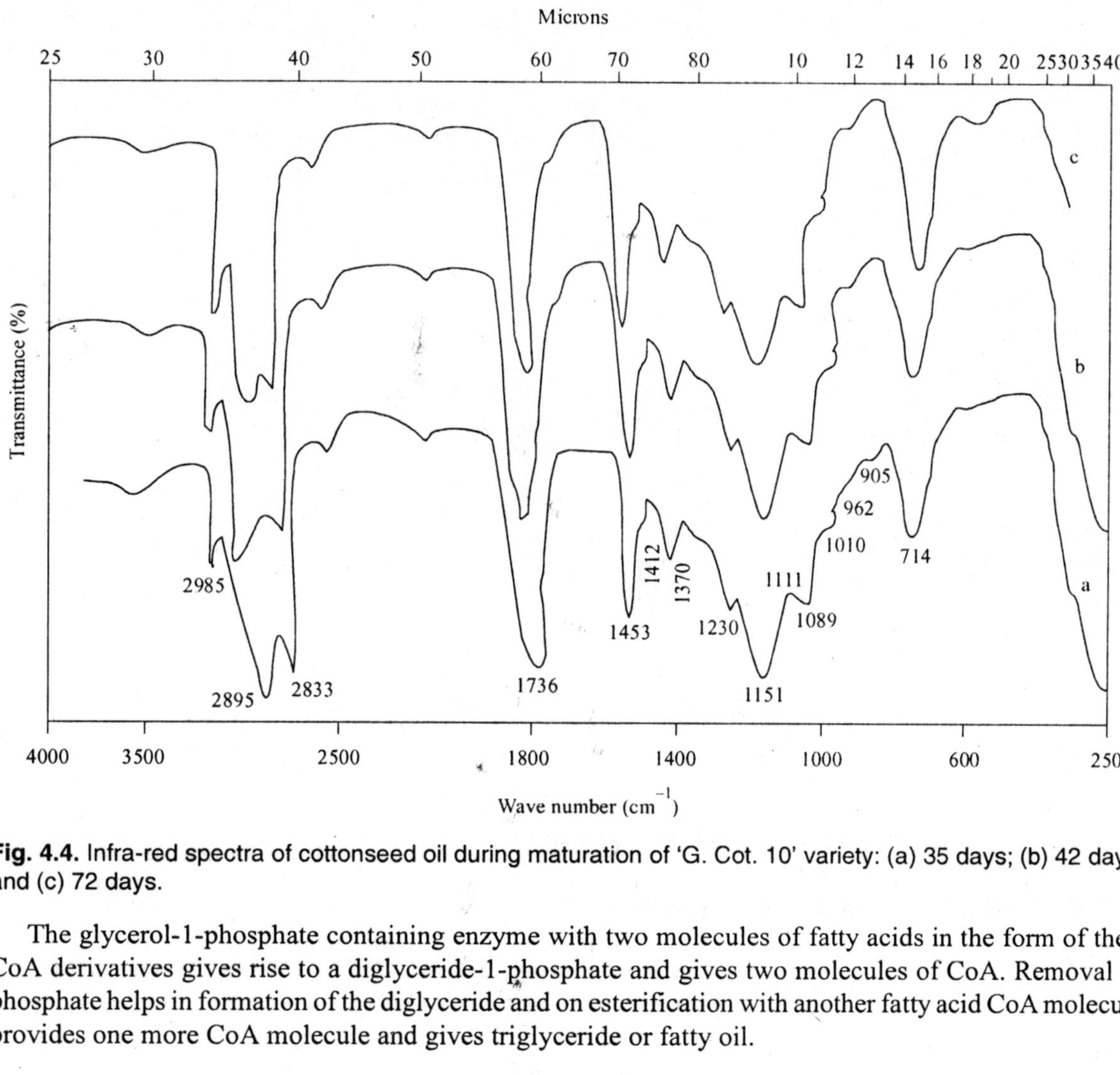

Fig. 4.4. Infra-red spectra of cottonseed oil during maturation of 'G. Cot. 10' variety: (a) 35 days; (b) 42 days; and (c) 72 days.

The glycerol-1-phosphate containing enzyme with two molecules of fatty acids in the form of their CoA derivatives gives rise to a diglyceride-1-phosphate and gives two molecules of CoA. Removal of phosphate helps in formation of the diglyceride and on esterification with another fatty acid CoA molecule provides one more CoA molecule and gives triglyceride or fatty oil.

Composition of Indian Cottonseed Varieties

In India the breeding and production of cotton have been guided by considerations of fibre yield and quality. Income from other seed sources in the oil and feed industries had increased the awareness of the potential importance of cottonseed as food as well as feed all over the world. As a result, National

Cottonseed Producer Association in the USA has suggested intensification of research on cotton with more specific goals relating to cottonseed quality such as: (i) increase in seed yield per hectare; (ii) increased oil percentage; (iii) reduced seed gossypol; (iv) increased protein percentage; (v) increased lysine content; (vi) better micotoxin resistance; (vii) production of non-shattering seeds; and (viii) reduction in the linter percentage.

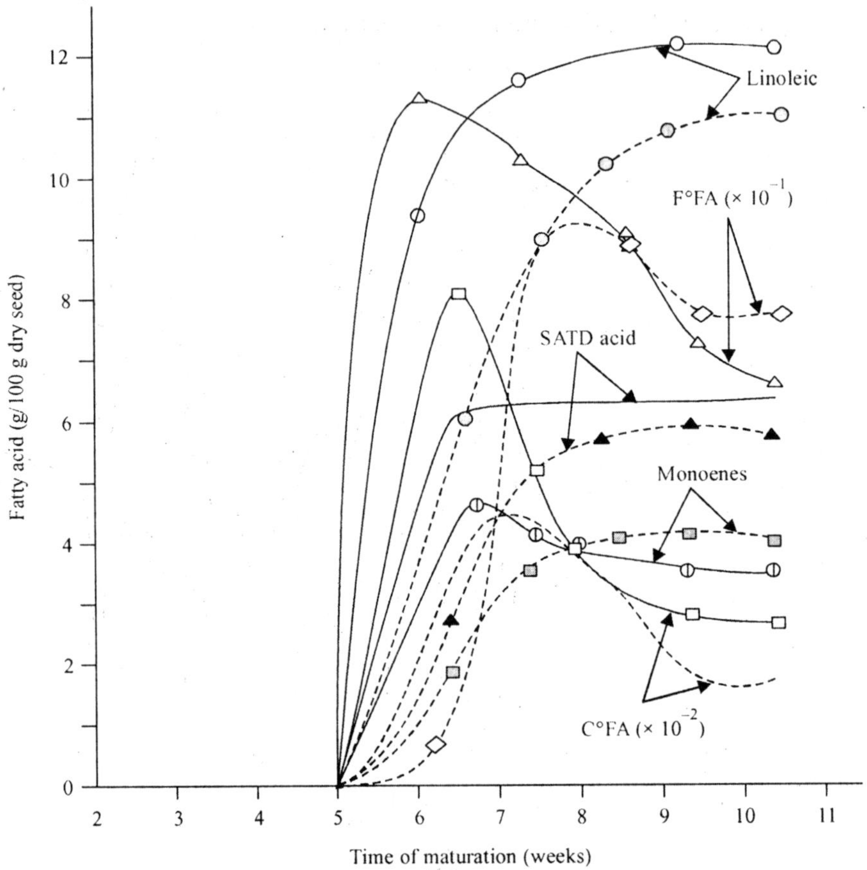

Fig. 4.5. Changes in fatty acid content during seed maturation.

Effect of Maturity On Seed Composition

Detailed observations on maturing of cotton bolls for 8 weeks showed that total weight of the bolls increased progressively with time and so also the average weight of seed and lint content. From seventh day protein, ash and carbohydrate increased progressively up to seventh week. The yields of total lipid from 100 seeds were 0.7, 2.0, 51.8 and 56.6 per cent, respectively, for 21-, 31-, 41-, 51- and 60-day-old seed, whereas the percentage lipid content is 2.2, 2.4, 10.3, 21.7 and 25.3 respectively. The fatty acid composition of the lipid showed a progressive increase in linoleic acid and simultaneous decrease in saturated and oleic acid contents on progressive maturation. The maximum increase in all cases is between 31 and 41 days. It was also observed that crude fibre and fat content increased with the growth process and reached the maximum while the carbohydrate content decreased progressively. It was observed that cyclopropenoic fatty acids (CPFA) content decrease 10 days after flowering and gossypol and tocopherol content increase. Rapid increase in gossypol and oil formation in maturing cottonseed has been reported and it has been postulated that gossypol could be an accelerator in oil formation; protein formation started earlier than oil formation and sugars decreased rapidly during the oil-formation period.

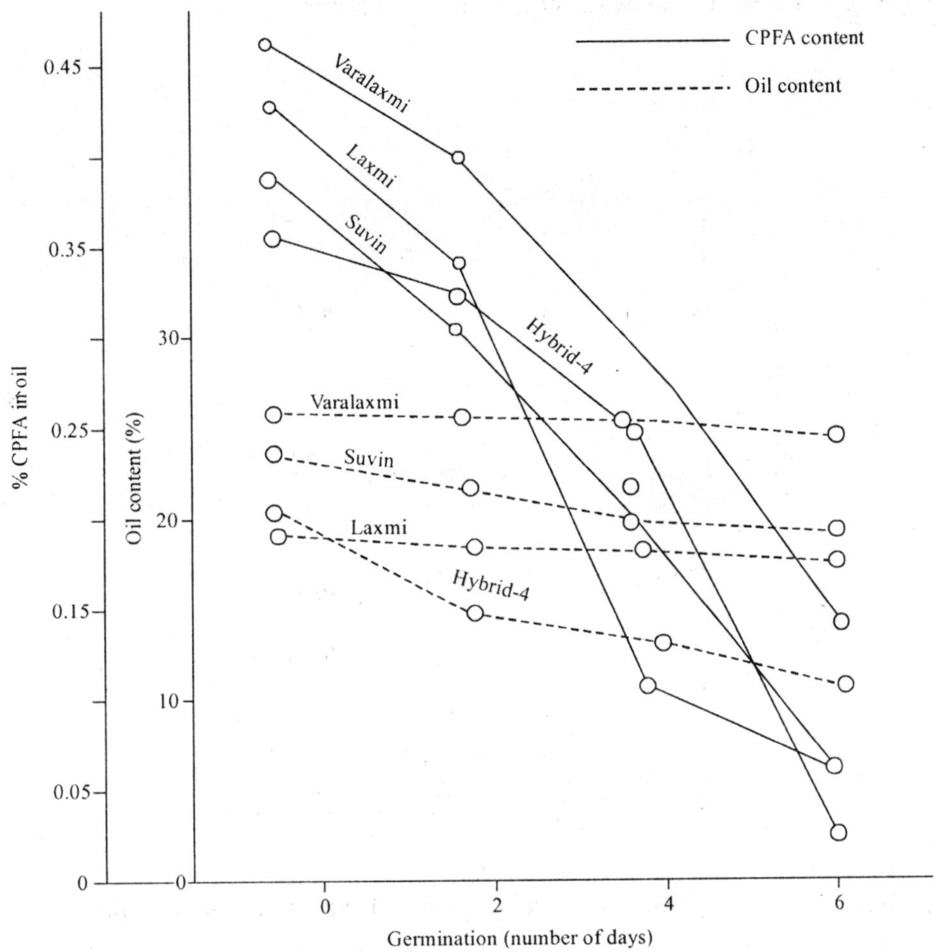

Fig. 4.6. Reduction in oil content of seed and CPFA content of oil during germination.

Effect of Environment on Seed Composition

Cottonseed is the most widely distributed oilseed in tropical and sub-tropical areas and it plays an important role in the economy of agricultural-cum-industrial development. A number of studies have been reported on the influence of place and environment on seed weight and oil content and various other constituents of seed. The effect of place and environment on seed weight and oil content of cottonseed has been studied. Cottonseed of the same variety grown in different localities which showed wide variation of seed index and oil content has been examined. It has been observed that samples grown at Arabhavi showed high seed weight but lower oil content compared to those from Siruguppa samples from Coimbatore. Though the samples at Arabhavi and Coimbatore showed the same seed weight the samples from Coimbatore had higher oil content. Similarly, from another trial the samples from Dharwar and Ranibenur had the same seed weight, whereas the samples from Dharwar had higher oil content. This indicates that factors like soil and rainfall also play an important role in formation of oil content in the seeds. Further, it is observed that the gossypol and iodine values in the oil of the seed are very much governed by the variety, locality, season and temperature.

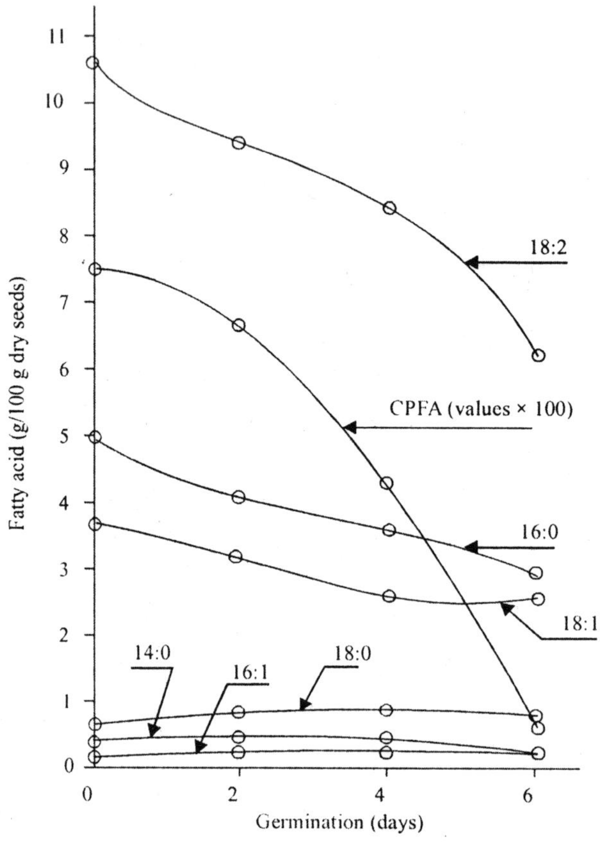

Fig. 4.7. Changes in fatty acid composition during germination of cottonseed of 'Hybrid 4' variety.

Pigments of Cottonseeds

The pigments present in cottonseed play an important role in the processing and utilisation of cottonseed products. One of the most important pigments present in cottonseed is gossypol. The presence of pigments in cottonseed creates enormous problems of seed processing; besides it imparts colour to the oil, causes toxicity to non-ruminant animals and the poultry and prevents the use of cottonseed meal in rations of such animals. The presence of pigments in cottonseed has been known since long and efforts have been made to separate out these pigments for certain specific end uses. No doubt it is difficult and expensive to completely remove the gossypol and related pigments by refining and bleaching of cottonseed oil. Several new processes and solvents have been tried to remove the pigments from cottonseed oil and meal so that this can be used for edible purposes without any adverse effect. Considerable amount of work has been carried out on extraction of pigments and their toxicity at various levels in rations of various animals including pigs, poultry, rabbits and so on. It has been reported that it is free gossypol which is more toxic but at the same time high levels of bound gossypol also reduce the growth of ruminants, since gossypol binds the essential amino acids like lysine. Enormous studies have been carried out in India and abroad on the nature of toxicity and chemical action on gossypol. It is reported that gossypol possesses antioxidant and anti-polymerising properties. Being a polyphenolic compound, gossypol can be used for various medicinal purposes and also for plastics and explosives. It is also reported that gossypol is being used for birth control in China. There are several methods for extraction of gossypol from cottonseed which can be commercially exploited for its use in developing different types of polymers, explosives, etc. Efforts are also being made in India and the USA to develop new varieties free from gossypol pigments, so that cottonseed finds use in utilising its oil and meal for edible purposes by chemical process.

Among the various pigments present in cottonseed gland gossypol is one of the most important. Considerable work has been carried out on studies of gossypol including its synthesis, chemical reactions, etc. Gossypol and other pigments are found in special type of glands which have a rigid wall. The walls of the pigment glands are irregular in shape and are stronger at the centre portion and weaker at the edges. These are composed of pectin, hemicellulose and uronic acid derivatives. The pigment glands are coloured visually showing homogenous contents under the microscope. When the gland is broken the pigments are not discharged unless soaked in solvents in which the pigments are soluble. The pigment glands behave differently with different liquids. The pigment gland on contact with water discharges the contents exhibiting vigorous Brownian movement. When the pigment gland is brought in contact with saturated aqueous solutions of organic and inorganic compounds the gland ruptures. The gland is less affected with low molecular weight water-miscible organic liquids. It is sensitive to anhydrous acetone which ruptures the glands immediately and releases the pigment.

Other Minor Constituents of Cottonseed

Besides linter, hull, oil and cake/meal, cottonseed also contains certain other minor constituents. These have not been given much attention but they play an important role in utilisation of cottonseed products. Some of these are as follows:

1. Phosphorus-containing compounds: They are phospholipids and are present in seed, oil and cake. They contain a mixture of large number of products and are also called lecithins; about 1 per cent of lecithins are reported in crude cottonseed oil.
2. Mineral constituents: They are present in cottonseed, linter and meal. These are estimated from ash content and their presence varies in seeds and linters depending on variety and agricultural practices.
3. Sterols: They are present in cottonseed oils and are estimated in unsaponifiable matter with solvent from the soap solution. They are present in very small amount.
4. Antioxidants: Cottonseed oil contains 0.1 per cent tocopherols which are of 8 types. It also contains trimethyl tocol (vitamin E). It is related with reproductive process and is an important antioxidant. Tocopherol helps in increasing shelf-life of vegetable oils.
5. Vitamins: Cottonseed-meal contains a good amount of vitamin B, including thiamin, riboflavin. The vitamin content of cottonseed-meal is compared with soyabean-meal.

BIOLOGICAL PROCESSES AND STORAGE OF COTTONSEED

It is well known that the storage of cottonseed plays a very important role in determining quality of seed and its biological processes. The moisture content of seed is one of the most important factors in influencing the various biological activities of the seed. It has been clearly shown that the presence of considerable quantity of moisture in seed generates varying amount of heat which ultimately affects the quality of seed and biological activities. Generally it has been observed that the collection of damp cottonseed causes great deterioration in the seed and seed gets heated rapidly after storage, which completely destroys the germination capacity of the seeds and renders them unfit for sowing. Even green unopened bolls, when picked up from the fields and stored for some time, generate considerable amount of heat which ultimately destroys several desirable characteristics of the cottonseed. In view of this there is urgent need to store the cottonseed in a scientific way after taking full precaution to avoid adverse effect on biological actions of cottonseed during storage. Even harvesting and handling of cottons which is very important for its quality is not done by scientific method. The process is totally manual and done by unskilled labour. Many times improper handling and harvesting adversely affects the quality and increases undesirable biological changes in the seed during storage. These biological changes generally take place either in the field or in storage. When seeds get heated in the interior they develop heat and respire vigorously, resulting in loss of their viability and increase in free fatty acids (FFA) contents, which finally produce dark-coloured oil and meal. The effect of biochemical activity of stored seed can be evaluated by measuring heat increase, change in viability, intensity of respiration, lipolysis rate, colour intensity and any other such characteristics.

Biological and Storage Changes

Biological activity in cottonseed causes considerable losses to the cotton and cottonseed oil industries. The main factors causing harmful biological activities are humidity and variable atmospheric temperature. Cottonseeds preserved for sowing in the next season suffer from loss of viability and vigour in germination while seeds stored in mills for extraction of oil suffer from rancidity and loss in colour owing to undesired biological activities. In addition, the yield and quality of oil and cake also suffer.

Damages caused due to biochemical changes in cottonseeds during storage are development of heat in the interior, increase in the rate of respiration and increase of free fatty acids (FFA). All these three factors are very important from the point of view of darkening in the colour of oil and oilcake.

Mould Activity

It has been observed that the seeds with less than 15 per cent moisture content deteriorate solely as a result of enzymatic activity within the seed. However, seeds containing more than 15 per cent moisture are initially damaged by the internal enzymatic activity and as a result viability in seeds reduces and make them susceptible to microbial damage.

Several kinds of micro-organisms attack cottonseed, some of them attack the hulls and others on the fat or cell-walls. Moulds generally grow in moist cottonseed. During growth moulds produce heat and sometimes it may even cause fire. The heat may cause change in colour of the oil or even may break down the oil as a result of which FFA content goes up in the oil. All these harmful activities caused by mould degrade the oil quality to a large extent.

Christensen demonstrated that cottonseed containing 18 per cent moisture became mouldy within 7 days and the rise in temperature was sufficient to kill the moulds after 16 days. Further it is seen that seeds treated with fungistatic agent remained unattacked by moulds initially but after 16 days the seeds had same fats as untreated seeds. In a separate study, *Jensen* observed that during storage of seeds containing 16 per cent moisture, FFA developed due to growth of fungus.

Mayne classified five kinds of micro-organisms responsible for damaging cottonseeds. The first group, all of which were *Xanthomonas*-like bacteria, consumes most of the fat and accumulates large amount of

FFA. The second group, one of which was a fungus of genus *Penicillium*, uses natural fat and accumulates moderate amount of FFA. The third group, one of which was fungus, *Aspergillus flavus*, uses neutral fat but does not accumulate much FFA. It produces aflatoxins which are toxic materials (Fig. 4.8). The fourth and fifth groups were not actively lypolytic. The first and fifth groups were bacteria and hence could grow in extreme high moisture conditions of the moulds. *Penicillia* and *Aspergilli* were most active FFA producer. The high lypolytic A. *flavus* was uncommon in moist cottonseeds.

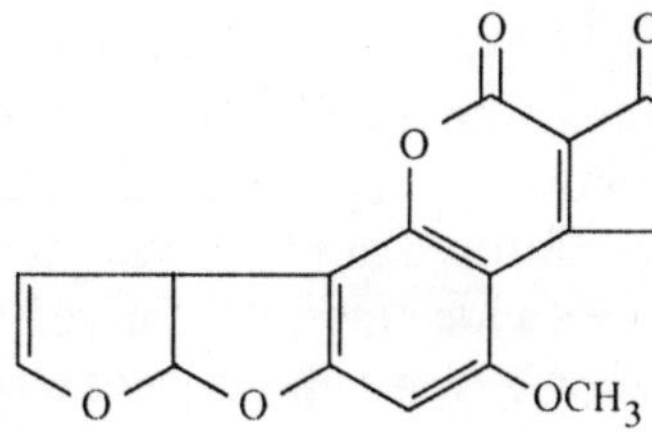

Aflatoxin B_1

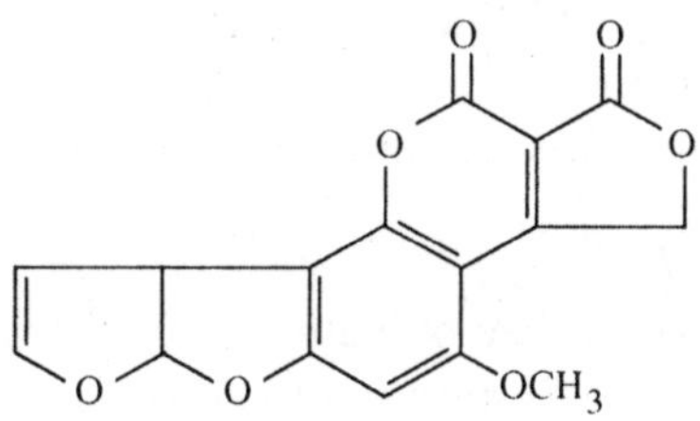

Aflatoxin G_1

Fig. 4.8. Structure of aflatoxin B_1 and aflatoxin G_1.

Heating and Chemical Changes

Production of heat in cottonseeds is a common phenomena when stored in bulk quantity with high moisture content. Generation of excessive heat during storage not only degrades the quality of stored seeds but also may damage the godown. At elevated temperature rate of biochemical changes accelerate resulting in rapid damage to seed in storage. Due to heating, FFA content in seeds increases resulting in degraded quality of refined oil. Excessive heating destroys the germination power, lowers the grade of lint and causes 31–35 per cent loss in fibre strength. Further the rate of germination of seeds decreases with increase in temperature and viability is completely lost when temperature reaches 48°–57°C. It is seen that due to storage of cottonseeds in small piles in sealed tank, the temperature rose very rapidly at the beginning of storage period; it then decreased slowly and unevenly and the seeds showed a several fold tendency to heat again. *Jensen* studied the changes occurring during storage in huge lots and found that seeds of low initial moisture and FFA contents did not heat, whereas seeds with high moisture and FFA showed an increasing tendency to heat up. In addition to that natural, moist seeds heat up faster in comparison to artificially moist seeds.

Cottonseed meats and hulls also heat up when their moisture content is sufficiently high.

From extensive chemical analyses of meat and hull constituents of cottonseed during heating, one can conclude that neither protein nor oil is consumed in heating process, the sole sources causing heating are carbohydrate and pentosans. However, protein is partially hydrolysed and denatured which is evident by the release of ammonia nitrogen. During germination, oil content in seeds reduces whereas it remains same during heating. Loss in oil content in cottonseed during prolonged storage is observed but the exact cause is not known.

Factors influencing heating

The seed contains more than 10 per cent moisture generates heat during storage. The source of excessive moisture in cottonseeds is picking seed cotton before it is fully mature and to the presence of green, unopened bolls and green leaves.

Rate of aeration has a very definite influence on rate and intensity of heating. It is found that even after releasing stored heat from stacks in the godown by forced aeration, the seeds maintain the ability to re-heat again. It was also noted that FFA was formed in seeds despite forced aeration. From a study on hygroscopic equilibrium of cottonseed over a wide range of relative humidity (31–93 per cent), it is seen that humid air may temporarily reduce heating by removing heat of respiration but the increased moisture content stimulates further respiration and heating.

Oxygen is known to be the responsible agent for production of heat in seeds during storage. Inert gas like nitrogen is ineffective, whereas carbon dioxide and hydrogen effectively reduce heat production most probably by cutting oxygen tension. Pentosan-rich materials as impurities in seeds, e.g. burrs and cotton stalks, increase heating effect. Therefore pre-cleaning of seed is necessary before storage.

Respiration and Lipolysis

Respiration is a biological process—controlled oxidation with the help of oxygen—to generate energy necessary for a living cell. The by-product of respiration is carbon dioxide and heat. The ratio of the volume of carbon dioxide liberated to that of oxygen consumed during respiration is termed as respiratory quotient. It should be unity in case of carbohydrate, greater than unity in case of organic acids rich in oxygen and less than one in fats and proteins. The respiratory quotient of cottonseed is unity, indicating that the principal substrate for respiration in cottonseed is carbohydrate, although immature seeds respire at a higher level of intensity than mature seeds of equal moisture content. Respiration increases at higher temperature and is greatly inhibited through the removal of oxygen from the atmosphere surrounding the seeds.

Lipolysis

Lipolysis or hydrolysis of glycerides to form FFA takes place in germinating seeds by the action of lipase enzyme. It also takes place in stored seeds. The rate of lipolysis is influenced by moisture content, temperature of storage, mould activity, disintegration and immaturity of seeds.

Effect of moisture

Increase in moisture content in cottonseed on storage is accompanied by increased rate of lipolysis. Cottonseed having less than 10 per cent moisture does not deteriorate appreciably under ordinary conditions of storage, whereas cottonseed having 14 per cent or more moisture is susceptible to deterioration. It is noteworthy that deterioration or rate of lipolysis increases with the increase in length of storage time and so also the rate of accumulation of FFA. The loss in oil content is proportional to storage interval. Seeds with high initial FFA deteriorate more rapidly than the seeds with low initial FFA content.

Temperature

Increase in temperature favours the conditions for lipolysis. Lipolysis is faster at high temperature and moisture levels. Therefore, lipolysis can be minimised in stored seeds by keeping them at low temperature.

Immature seeds

Immature seeds have a high respiration intensity. Rate and pattern of lipolysis of immature seeds differ from that of normal seed. In matured seeds, lipolysis takes place on prolonged storage whereas in immature seeds lipolysis starts immediately after storage.

Disintegration

Mechanical disintegration of seeds causes production of FFA at a faster rate. Cottonseed meats produce FFA at a higher rate than seed at similar moisture level. At lower moisture levels, the difference of FFA production rate is negligible.

Germination

Germination of seed is adversely affected by lipolysis, especially due to accumulation of FFA in seeds. It is reported that cottonseeds containing more than 1.8 per cent FFA failed to germinate. Low moisture and low temperature which suppress FFA development in cottonseeds prevent loss of viability in seeds on storage.

Field and Storage Damage of Cottonseed

Cottonseeds suffer from two types of damage, viz., field damage and storage damage. Field damage occurs in the pre-harvest stage while storage damage is post-harvest damage during storage, the gin or oil mill. Damaged seeds produce unagreeable colour in extracted oil. Field-damage seeds may have a prime colour, or a greenish background colour, or in extreme cases may have greenish-brown colour.

Storage-damaged seeds are invariably discoloured. Background colour of damaged seeds ranges from light brown to almost black.

Both types of damaged seeds yield oil with higher content of FFA than prime seeds. Damaged seeds due to storage yield more intense dark colour than field-damaged seeds.

Field damage

Field damage is pronounced in harvest seasons in which there is excessive rainfall. There is a relationship between time of harvest and FFA content of cottonseeds while no correlation between oil and moisture content of the seeds was found and the field damage is indicated by FFA content. Humidity of atmosphere at the time of seed maturing and drying in the boll only can influence field damage. Seeds which mature under low humidity do not suffer from field damage.

Storage damage

Unlike field damage, storage damage of cottonseeds occurs mainly due to high temperature evolved owing to respiration in viable seed cells. In field, the heat generated by respiration is dissipated in the open atmosphere, whereas the particle adiabatic condition is set up in closed storehouses, which causes damage in stored seeds. Seeds having higher than average respiration intensity at same moisture level exhibit greater lipolysis resulting in higher FFA value in stored seeds. To avoid such problems controlled aeration or mechanical cooling is often necessary.

Biological changes are thought to be responsible for such damages in cottonseed during storage. Important factors are:

1. Biological activity in stored seeds increases with increasing moisture content of the seed.
2. The intensity of biological activity with increased moisture level depends on various parameters involved during collection of seeds.

So long the immature seed is attached to the plant, it derives energy from the plant in the form of nutrient for its primary synthesis. As the seed matures the attachment gradually loosens and the energy supply from the plant also reduces. However, if the seed matures in such a condition when enzymatic activity is continuing but the energy source is cut off, the seeds tend to autodegradation, i.e. hydrolysis of fat, carbohydrate and protein components causing loss of viability and general deterioration.

The conditions of dehydration during last stages of maturation are extremely important factors for the seeds to field- and storage-damage susceptibility. Freshly opened cottons bolls contain above 50 per cent moisture, which generally reduces to 10 per cent in 10 days under ideal weather. Rainfall during boll-opening stage largely causes field damage. In addition to that if moisture is reduced at a slower rate or faces alternating rains during maturation, it results in field damage. Such seeds should be handled with care to prevent biological activity by quick artificial drying.

Handling and Practical Storage of Cottonseed

Handling and storage of cottonseed is an important phase of cottonseed processing. In cotton-producing countries like the USA, Russia, China and India, it affects the economics of cottonseed processing. Trading regulations are based on quality of seed and rise of FFA in oil. Cottonseeds are highly perishable commodity, especially when the moisture content of seed exceeds 10 per cent and when storage facility is poor. Seeds should therefore be stored properly to avoid economic loss.

Handling practices in America

Cottonseed crushing operation in the USA varies in individual mills from time to time which involve differences in the handling of the seeds. The normal cotton harvesting season in the USA extends from August to December and very little harvesting occurs after January. Normally 80 per cent of the seeds are received at the mills between September and December which is generally transported by trucks.

Handling of cottonseed at the mills

The seeds received in the mill for extraction of oil experience marked variation of weather during harvest. The mills are equipped to handle these seeds properly, analyse them rapidly and segregate according to quality in different storage bins.

Sampling of cottonseeds

When cottonseed is purchased on the standard grades laid down by USDA, samples of the seed for testing are drawn by licensed persons. Each sampler is equipped with (i) suitable sampling forks; (ii) metal container with tight-fitting covers of 2½ bushel capacity; (iii) friction-top cans of 2540 cm^3 capacity; (iv) scales graduated in ounces and half ounces; (v) shaker cleaner provided with a sample-reducing device; (vi) suitable thermometer; (vii) cork screw trier; and (viii) unloading sampler.

Approximately 1 kg sample is drawn per tonne of seed for testing. The samples are placed in moisture-proof bags and transferred immediately to the 2½ bushel metal container. However, foreign matter such as sand, dirt, leaves, sticks, stones, metal pieces and broken and whole carpels or boll shucks introduce heterogeneity in samples. To eliminate such hazards, random sampling from lots is preferable. As much as 25 kg sample taken from a car is first weighed and cleaned over a shaker screen and the clean seed weighed again to give the percentage of foreign matter removed. The percentage may vary from sample to sample. The well-mixed samples are placed promptly in a friction-top tared can and immediately closed. For determining moisture, oil, ammonia and FFA in duplicate, 500 g samples would be necessary.

A sampler's certificate is enclosed and the sample is sent to an official chemist for analysis and grading. A duplicate sample in a friction-top can is retained by the sampler until the results of analysis and grading are received.

Treatment of seed prior to storage

Wet cottonseed having moisture level above 10–11 per cent should be dried and cooled before storage at the mill to prevent heating and deterioration of the seed. The seed is dried at about 104°C in the drying section and cooled in the cooling section to about –9.4°C if needed. In most of the cases heat drying is not needed and seeds are dried by passing dry air over the seeds.

Handling practices in India

Cottonseeds are stored in India in open godown floor or in gunny bags. Several changes have been noticed on storage in laboratory. The important observations are summarised as follows:

1. Higher temperature and humidity increase the rate of FFA formation.
2. Seeds stored below critical moisture level (10 per cent) show little tendency to form FFA.
3. Viability of cottonseeds decreases on storage especially at high humidity and when FFA develops rapidly.
4. Discolouration of seeds depends on various factors and not only on period of storage.
5. Damaged seeds deteriorate at a faster rate than undamaged seeds.
6. Artificially wetted or damaged seeds deteriorate rapidly and turn mouldy.

PROCESSING OF COTTONSEED

Impurities in Cottonseed

Impurities in cottonseed largely affect the quality of extracted oil and cause significant economic loss. Impurities in cottonseed enter from two sources, i.e. from field during harvest and from storage or during processing.

Hand-picked cotton contains very low quantity of foreign materials. The mechanically-picked bolls contain more trash, soil and stone. So hand-picked seeds are easily processed in seed-cleaning machinery with greater efficiency. For getting clean seed, the cultivation field should be kept weed-free as far as possible. Late harvest also may cause large percentage of stalks and stems along with cotton which are difficult to remove by simply ginning operation and trashy cottonseed results.

Some type of foreign matter enters in cottonseed, viz., bolls, cotton fibre or lint, small whole seed, immature and dead seed or seed hulls can be salvaged at crushing mill. However, stones, sand, soil, light trash such as leaves, fragments of bolls and hulls, stalks and sticks, pieces of wood, nails or pieces of metals cause major damage. These impurities largely arise from mechanical processing. Moreover, wet weather during growing or harvesting season produces improperly-opened bolls with immature seed. Such seeds are difficult for ginning and introduce higher quantity of undesirable foreign matters. Insect infestation and unfavourable weather at the time of maturation of fruit result in production of dead and damaged seeds.

Seed Cleaning

In olden days gin operation was mainly aimed at production of clean, high-grade cotton but the maintenance of seed quality was overlooked. But modern gins are equipped with seed-cleaning devices and seed cotton drying facilities for better utilisation of cottonseed.

Proper seed cleaning increases the capacity, reduces maintenance and improves linter and oil quality. Production of high quality linter is not possible without proper cleaning of seed. Other reasons for cleaning cottonseeds are:

1. To minimise wear and tear on delinter saws and grates, huller knives and grinding plates arising from foreign matters like stone, sand and pieces of metals.
2. To reduce the probability of fire which may arise from sparks by friction from stone-like impurities.
3. To maintain hygienic atmosphere in the mill removing dust, fine sand and fly lint residues.
4. To reduce the cause of producing storage heat damage arising from foreign materials attached with cottonseeds.

Types of seed cleaning equipments

Cleaning of seeds before storage is essential in maintaining the quality of the products. Although removal of all types of impurities before storage is not practically feasible, partial removal of sand particles can be done easily by fitting perforated bottoms in the unloading conveyors during unloading. Simply partial removal of sand can increase the efficiency of air-cooling equipment and retard the heating of seed during storage.

There are different types of cottonseed cleaning equipments but most widely used types are: (i) Boll and sand reels; (ii) Bauer/carver pneumatic cleaners; (iii) Rock shakers and air traps for removing stones; and (iv) Electro and permanent magnets.

Boll and sand reels

The device is an inclined, revolving screen of cylindrical or hexagonal cross-section with sufficiently large perforations to allow seed and foreign materials of equipment size to pass through, whereas foreign matter of greater size is discharged at the farther and lower end. The unginned bolls and locks of cotton get separated in this instrument which fall down from the reel when the screen is lodged and reaches the top position by revolution. For good result, both the proper speed and uniformity of feed are very important for successful operation of a reel. The reel should revolve slowly so that its load is not held to the metal by centrifugal force, but fast enough to throw the seed over the shaft passing through the centre of the reel.

Construction of sand reel is very similar to boll reel. In this device sand and other fine particles pass through the opening of the perforated metal and the seeds are discharged from the end of the reel. Some times sand and boll reels are used together.

The reel or barrel is wound with cloth fastened to steel bands and are supported by wood rails. These rails lift the product to the necessary height to give the proper drop. The spiders of the reel are thoroughly truss-rodded to provide rigidity to the reel and reduce the strain on the shaft. The shaft is turned at each end so that the end spiders are mounted against a shoulder which prevents the seed from slipping on the shaft. Special sleeves on the shaft from the master gear to the upper bearing and from the lower spider to the lower bearing eliminate slippage of this gear or the reel assembly. Doors are made removable to give easy access to all parts of the machine. The screen may be clothed for sand or bolls.

The conventional procedure for screening out large trash such as sticks and burrs from cottonseed with reels or shaker screen, removes less than half of this type of foreign matters. Dry seeds are comparatively easier to remove sand and fine trash on the seed screen or the lower tray of the pneumatic cleaner.

Most effective seed cleaners use both shaker and pneumatic principle. Seed cleaning machines of several manufactures employing these two principles differ only in the details of number or location of the shakers. One of the earliest seed cleaners is that of Bauer brothers. This unit along with Carver pneumatic seed cleaner are the most popular seed cleaners used for cleaning cottonseeds. In the beginning these machines were 2-tray units and later on converted into 4-tray pneumatic cottonseed cleaners to increase the capacity and to provide larger shaker area with increased efficiency.

Construction and working of machines

The construction and working of these machines are as follows:

1. Seeds are discharged from the feeder to upper tray and an air current passed in contact with the seeds to remove the dust.
2. Next volume is regulated through the machine by a vane-type feeder. Sometimes a magnetic separator is used below the feeder to remove ferro-magnetic metal particles entered in the seed.
3. Next, the upper shaker tray removes broken bolls and other coarse foreign materials and delivers the seed to the top end of the lower shaker tray.
4. First portion of the lower tray is designed to remove sand, small twigs, chaffs, etc. and the second portion to separate black seed and loose meats.
5. Mechanically-cleaned seeds from lower tray are delivered directly to aspiration chamber where they come in contact with a controlled vertical air stream to carry off fuzzy seed but permitting black seed, stones, metals and other foreign materials to drop into a small screw conveyor.
6. Next light trash, loose lint and shale are carried off by low velocity air currents in the expansion chamber section. The seeds impinge on the baffle of woven wire screen with sufficient impact to loosen both fine sand and dust usually associated with the mat of short fibre and the shale and boll particles. The remainder is picked up by fan and blown to a dust room.
7. Final treatment occurs when the seed falls in a vertical sheet from the discharge roll to a clean seed conveyor. This comes in contact with horizontally flowing air current to remove lighter foreign particles other than seed.

Cleaning cottonseed at the oil mills

Removal of foreign matter from cottonseed received at oil mills and prior to storage benefit the industry in several ways:

1. Decreases the risk of storage because the presence of excess foreign matter in seed tends to cause over-heated during storage.
2. Increases the efficiency of crushing by decreasing loss of linters during cleaning, absorption of less oil by foreign matters and reduction of cost by wear and tear on the expellers.
3. Improves seed quality by checking the formation of FFA level and minimising the oil loss during refining.

In India, cottonseeds are often stored on the ground before ginning. In this way additional sand and dust particles enter with seeds along wtih those already present during harvest. Sand and boll reel as pre-cleaner preceding the pneumatic cleaner helps to reduce the load on the pneumatic cleaner and ensures better cleaning. Use of sand and boll reels are very much helpful in reducing wear on the expeller worms where cottonseeds are crushed directly in expeller without delinting or dehandling. Besides Bauer and Carver seed cleaners, both 2-tray and 4-tray are used in cottonseed oil mills in India.

Another type of efficient seed cleaner unit emerged was named ARS cleaning belt. This consists of moving endless belt covered with a rough-textured rubber surface inclined at an angle of 30°. The operation principle is that trash and seed show differential properties of rolling, bouncing and adhering.

Runs are made with this machine for three principal variables, viz. belt angle, belt speed and capacity rate. It is very efficient in removing stems and shale from linty, trashy cottonseeds. It has some advantage over other cleaning systems such as it completely removes fine sand, dust and linters, consumes less power and ensures additional cleaning on re-passing.

Delinting

Types of delinters and their operation

After cleaning the cottonseeds next essential step is delinting. Cottonseeds received after ginning are covered with short fuzzy fibres. The quantity of short fibres left on seed depends on seed variety and on the efficiency of the gin.

The adhering fibres from ginned cottonseed are removed universally by mechanical machine, known as delinters. The delinter machines are constructed on the same principle as that of saw-type cotton gin with certain modification to remove residual fibres from the seed. The saws are more closely spaced than in the gin. Owing to shortness of the lint to be removed, the delinter requires more refinement in its construction than the gin and greater rigidity to stand high pressure developed by the seed roll.

The main parts of the delinter are steel frame, feed chute and feeders, gratefall, seed board, eccentric adjustment on seed board, i.e. density regulator, grates, float, gratefull lift lever, back board end lip, brush, feeder regulator, roll box, permanent magnetic separator, mote board and mote chamber and flue and condenser. Two motors of 7.5 hp are used, one to drive the saw cylinder and the other to drive both the float and brush.

The seed from the cleaners is conveyed and falls through a chute to hopper or intake opening of feeder. The feeder is constantly filled with seed and then fed to roll box of the delinter. The seed slides over the permanent magnet located between the feeder and the roller to remove tramp iron. The seed roll is revolved by a winged shaft or float in opposite direction to that of the saw and at a lower peripheral speed than the saws. The seed roll is confined between a curved metal sheet—the seed board and the ribs of gratefall. The saws project through these ribs and cut through the seed roll which carries linter fibre between ribs not to allow seeds to pass through. Bottom of the seed board is adjustable towards the ribs.

The amount of linters to be removed from the seed is governed by opening and closing discharge area under seed board. The capacity of the delinter depends on the density or hardness of the roll. Roll density is adjustable. During operation, the feed or seed is continuously controlled by the amount of seed discharge from delinter. Multiple cuts, often 3–4 cuts, can remove linters from seed with very high efficiency (less than 1 per cent remained on seeds).

Delinting in India

Not much emphasis has been given for delinting. Generally first or second-cut linters are only taken. Reasons for not taking third or fourth cuts are low price offered for them. All the mills even do not have proper lint-cleaning machinery because of high cost.

Only recently the need for delinting have been realised. After first and second cuts the delinted seeds are sent to cottonseed-crushing mills for extraction of oil or sold to agriculture department for sowing purposes, since delinter seeds germinate better than undelinted seeds.

Lint cleaning

Most common impurities in linters are trash, stick and shale. Lint cleaning is essential as it removes the impurities and improves the cellulose content of the linter and the quality. The principles involved in cleaning are beating of linters and pneumatic separation of impurities. There are several lint-cleaning devices such as air separator, centrifugal separators, supper-jet cleaner, whirligig pneumatic attachment and so on. The secret of making good clean linter is to have an excess of both cleaning capacity and linter room capacity. It is recommended that a mill handling 120 tonnes seed per day should use a large bool reel, fans and cleaners to handle all the fractions without any manual handling. One suggestion to make better second-cut linters is to take more first cut using 15 delinters of 141 saws on the first cut, 20 on the second-cut and 5 on the third-cut. First-cut linters should be smooth and bright in colour, second-cut linters should be free from large hull particles and low in shade content with proper cellulose content for chemical purposes. When cottonseed is closely delinted by a heavy single cut or by both first and second cuts, the lint should be cleaned with lint-cleaning machinery. For using such machinery, the delinters should be provided with a flue-collecting system to deliver the lint to a cyclone collector.

Cleaning of first-cut linters

Cellulose-control machine is used to clean first-cut linters which is designed to handle the lint properly over perforated metal in two sections each having three chambers. First, the linters pass over the fine perforated metal section of three chambers where contaminant hull bran passes through the time perforations and get discharged at the bottom. The linters then pass to the next section, which consists of three chambers of bigger perforated metal and again the objectionable hull particles and shells are removed. In this operation, some linters pass through the larger perforations which are recovered in the tailings, beater and shale shaker. The clean lint is delivered from the end of the machine to bailing press.

Cleaning of second-cut linters

Linters are collected from the delinters with a flue system and discharged from a cyclone into a feeder at the far end of the top unit. Linters are cleaned by passing through the top unit and are picked up with air by a large suction nozzle at opposite end. The linters entered through this nozzle which pass over a moving gap where large and heavier particles drop out. Impurities present pass through the bottom of the top unit but carry some linters with them which is recovered in two beaters below. During cleaning of second-cut linters, the nozzle on the bottom beater is not used. However, the recovered linters are discharged from the end of the beater over a lint shale shaker before returning to the flue system. The cleaned second-cut linters are then delivered to a bailing press.

Prospects for cotton linters in India

In Indian cottonseed industry the average recovery of linters is only about 2 per cent. This is because a substantial number of the mills do not have delinters and many of them who have delinting machine do not delint seed as they find it uneconomical. They simply dehull to separate the meat from the lint-bearing hulls and extract oil from the meat. The main reason for this low recovery is due to insufficient number of delinters and lack of technical knowledge to maintain and operate them.

The main consumers of cotton linters in India are the rayon and carboxymethyl cellulose industries. These mills are sufficient to consume the present production and the incoming pulp mills increase the demand for cotton linters. The Bureau of Indian Standards Institution (BIS) has formulated specifications for first-cut, second-cut and mill-run linters and careful processing of cottonseed which can be easily followed. By using cotton linters the dependence of rayon industry on imported wood pulp can be reduced.

Moreover, the hulls which are consumed as cattle feed may gain higher price if delinted properly. By delinting the seeds higher recovery of oil is possible. Mills which crush cottonseed directly without decortication can fetch higher quantity of quality oil and better oil cake.

Dehulling

The next step after delinting is to remove hulls which surround the kernels or meat of the seed. The hulls contain around 0.5 per cent oil whereas the meat contains 30–40 per cent oil. So if whole seed is crushed for extraction of oil after delinting, the hull portion absorbs the extracted oil and thus reduces the yield of oil and the oil cake becomes inferior in quality. Oil extracted from dehulled seeds gives higher value of protein in oil cake. In addition to that crude fibre content of hull is about 35 per cent. Moreover, oil extracted from undehulled seed will be inferior and darker in colour owing to the presence of some pigmented materials in hull.

Dehulling equipments consist of huller, shaker-separator of the purifying type with huller supporting to mount the huller at the rear end, purifying nozzle at the other end, hull and seed separator double-drum hull beater, meat purifier, rotating drum tailing beater and cyclones separation. Number and arrangement of these equipments depend upon capacity of the plant. For a 50–70 tonnes seeds capacity plant per 24 hr., a single hulling and separation unit will be sufficient. For handling 40–150 tonnes per 24 hr. all units should be doubled except 1.4 metre safety shaker with feeder, hull beater, meat purifier and failings beater.

Safety shaker preceeds the huller and shaker separator. Function of this unit is to remove loose linters from the delinter seed and the residual foreign matter and tramp iron. Seed coming out from this unit is sent to the top of the huller machine placed at the rear end of the shaker separator.

Types of dehuller and dehulling processes

Cottonseed hullers are of two types: (i) Bar-type huller; and (ii) Disc-type huller. The Bar huller consists of a frame which holds stationary knives running the full width of the huller. The knives are individually adjustable towards the cylinder. The cylinder is a heavy drum which runs at high speed with knives along wtih full width of the drum. These knives are usually I-shaped cross-section and are held by plates or clumps.

Another popular machine is Disc huller. This machine is similar to an attrition mill; the difference is the plate on one side is stationary. The disc huller functions by the operation of one rotating disc. Upon the face of each, bars or ridges are cast with cutting edges radiating from the centre. The face of each plate is concave. This permits free entry of seed near the centre. The seeds then travel centrifugally outward through a gradually narrowing space between discs which are the minimum distance apart at the periphery. The seeds are cleanly cut between the stationary and rotating sharp edges.

Two different makes of knife-type bar hullers are popular. The two makes differ mainly in details of their knife-edge settings. The Carver perfection huller has a single adjustment. By turning eccentric shaft located on the concave, the distance between the fixed knives in the concave and the rotating knives in the cylinder could be adjusted to regulate seed cut in the huller. The concave is held by safety studs, designed to break if any non-ferrous foreign materials enter the huller.

The chandler bar huller permits the use of screens to free the meat from dust and lint during the process of hulling. The screens are interchangeable with the breast knives and may be alternated. The cylinder is cast in one piece, consisting of a heavy solid rim with no deep grooves to weaken it and 8 supporting spokes.

The speed of Carver perfection huller for normal cottonseed is 750 rpm. For very dry seed the speed may be reduced to avoid production of very fine meats and hulls. At proper speed a bar type huller (1.15 metre) can process about 75 tonnes of seed per 24 hr., the uncut seeds are returned to the huller.

The delinted seed after passing through the safety shaker is conveyed and dropped on the huller fixed on the rear end of shaker separator. In the huller, seeds cut and hulls are loosened from kernels of meat by the series of knives. The mixture of meat, hulls and uncut seeds coming out of the huller falls on the shaker separator.

Separation of cottonseed meats from hulls is an essential step for obtaining products like oil, cake and hulls at optimum proportions. Separation process depends upon several factors like (i) moisture content of seed; (ii) amount of linters left on seed; (iii) degree of decortication of seed; (iv) amount of oil absorbed in the hull of seed before it reaches hulling and separation operations; (v) quantity of loose fibre, hull bran and dirt on the seed; and (vi) set-up and operation conditions of machinery. Separation practices vary considerably, the main processes are single hulling, double hulling and universal hulling.

In single-hulling process, seed is passed through a huller mounted on a shaker separator. The seeds are cut coarsely to avoid oil absorption by the hulls. The meat is shaken out of the hulls as the materials passes over the shaker. The meats from the shaker are collected and passed over a meat purifier consisting of several shaking trays equipped with adjustable suction nozzles to remove the floating hull particles, carrying residual lint from the meat layer by suction. In single-hulling process, the hulls and uncut seeds are separated before the hulls are allowed to go for storage.

In double-hulling process, the seeds are passed through a huller mounted on a shaker separator. The seeds are cut coarsely to avoid oil absorption by the hulls and free meat is separated by passing over shaker separator. The mixture of hulls and uncut seeds are passed through a hull beater to remove fine powdery meat. The entire process is then repeated by passing the hull and uncut seeds through a second hull shaker and beater. The second huller is set close enough to hull the remaining seed. Meats from the shakers are collected and passed over a meat purifier consisting of several shaking traps equipped with adjustable suction nozzle to suck out the floating hull particles from the meats layer. The unit is designed to efficiently beat cottonseed hulls and thus remove the fine floury meats adhering to them to increase the oil content.

This system has some disadvantages also. The entire hulls passing over the first-cut shaker and hull beaters is again passed through the second-cut hullers. As this second-cut huller is set very close, the hulls are exposed to oil absorption a second time in hulling the uncut seed. Moreover, a large amount of fine hulls produced during second hulling tends to pass through the preforation of the shaker and are difficult to be separated from meat.

Universal hulling process is a combination of single- and double-hulling techniques. In this process, the hull fraction from the single-hulling operation is passed through a second more closely set huller shaker and beater before going for storage. This modification is used where (i) large amount of linters is left on the seed and a relatively low protein in cake is required; and (ii) where the seed contains a large amount of immature seed.

Processes for Meat Separation and Flaking or Rolling of Meats

During dehulling operation cut meats and hulls are discharged from the huller on an upper tray. This tray has a single deck of corrugated perforated metal which allows the meat and some hulls to drop through on the lower tray. Coarse hulls and uncut seeds are discharged to the sides through a spout at the end of the tray. The lower tray has a single deck of perforated metal which allows very fine meats and very fine particles to pass through on the solid bottom. A mixture of small hulls and coarse meats is discharged over the end of tray from the perforated metal deck. These coarse meats can be purified at this point by installing a purifying nozzle at the end of tray. This nozzle is adjustable and able to remove any remaining hull particles by air suction which are floating on the coarse meat. Removal of excess hulls from cottonseed meats is done by passing through at meat purifier. The equipment is a combination of screening and separation by air. This separation can be carefully regulated to control the hull content of meat delivered to oil extraction system. This also regulates the protein content of the extracted cake or meal.

During separation of meat the following points should be considered:

1. Separating arrangement should be simple.
2. In operation of the separation equipment uniformity of feed should be maintained.
3. Materials should be handled in small quantity and gently.
4. Seeds should be delinted perfectly, so that separation of hull from meat could be more efficiently done.
5. Rotation of drum and direction of spikes of hull beater should be same.

Meat purifier is a machine designed to purify meat by removal of lint and linty hulls and thereby enabling to increase nitrogen level in the cake and extract more oil from pure meat. Meat purification is essential because when a mill is delinting seed closely, it is difficult to separate hulls completely and prevent some hull from going to meat. On the other hand, when the mill is not delinting closely because the huller is set coarsely and coarse perforations are placed on the shaker, both lint and linty hull will separate with meat. In both the cases meats are not pure. Meat purifier consists of 2 shaker separator trays which receive the hulls through a feeder fed from the purifier nozzles of hullers shaker where the fine meats, fine hulls and coarse hulls get separated.

In earlier days rolling of cottonseed meats was believed to result in actual rupture of high percentage of oil-bearing cells. But from recent studies it is evidenced that only a small percentage of oil-bearing cells are disrupted fully or partially during rolling of cottonseed meats. Irrespective of the extent of total rupturing of cells, cottonseed should be flaked thin for efficient extraction of oil. In this way the meats are reduced to a physical state which will be conducive to most efficient contact between the meats and hot water vapour under the conditions later encountered in cooking process.

Five high rolls are usually used for flaking cottonseed meats. A pair of horizontal rolls of sufficient dimensions can also do the job satisfactorily. The five high rolls of vertical type are placed one over the other. Cottonseed-meats are fed into the feed hopper at the top of the rolls in which a grooved feed roller rotates. The outlet from the hopper is regulated by means of an adjustable slide. The feed roller distributes the meats evenly on to a guide plate which conducts them between the first and second crushing rolls, besides preventing them from further falling. Crushed meats adhering to the back face of top roll are removed by a stationary scraper and fall into the space between the second and third rolls. Partially crushed seeds then fall on to a second guide plate on the other side of the rolls and are deflected between second and third rollers, where they are further crushed and flaked. In this way the seed particles are passed between third and fourth and then fourth and fifth rollers. The bottom 3 rollers have smooth surface, so the meats finally emerge as flakes of adjustable thickness.

Purpose of Flaking and Flaking of Meats

Flaking of purified meat is an essential step in extraction of good quality oil. Optimum thickness of flakes is thought to be about 0.016 cm by many workers which is able to provide large surface area of the meat for rupturing the cells. Experienced mill operators consider the following points for poor flaking: (i) use of a single wornout roll; (ii) higher quantity of uncut seeds in the feed to the roll; (iii) unrolled meats escaping through the chipped ends of roll; (iv) presence of excess hulls in the meat; and (v) poor distribution of feed over a roll.

Cooking of meats

It is important that rolling must produce uniformly thin flakes, otherwise cooking and subsequent pressing will not be effective.

Methods of extraction of oil from cottonseed meats

Cooking of seed meat

Before the cottonseed meats go to the rolls, their moisture level should be about 10–11 per cent. There are several methods to humidify dry seeds to a desired level but a very simple method is to add controlled quantity of water to the meats just before they fall into the hopper at the top of the roll before flaking. Next to rolling or flaking of cottonseed-meats is the operation known as cooking. Objectives of cooking are as follows:

1. To complete the rupture of oil cell needed for extraction of oil.
2. To increase the fluidity of oil by increasing temperature of meat and oil.
3. To coagulate the protein fraction of the meat so that oil should be free of proteinaceous matter.
4. To coagulate phosphatides in the meat.
5. To detoxify the free gossypol by rupturing gossypol glands and thereby causing them to be absorbed into proteinaceous materials.
6. To destroy microbes and free enzymes present in the meats which may cause increase of FFA level in oil and meat.
7. To dry meats to a suitable moisture content for pressing operation.

Purpose, equipments and cooking

The equipments used for cooking of cottonseed meats are as follows:

1. Single-batch cooking kettles which deliver cooked material to a central charging kettle or single or double compartment kettles which deliver directly to the screw presser or expellers.
2. Stack cookers with 3 to 6 stacks which normally operate as continuous cookers.
3. Pressure cookers.
4. Horizontal cookers, usually combined cooker-dryer.

In India, most of the cookers are mounted above the expellers and form part of the expellers equipment. Pressure cookers are practically not used. In India, 1 or 2 stack cookers commonly used are mounted on the expeller top. They vary in size and have provision for heating with open steam and a jacket at the bottom. Provision exists for heating with closed and open steam. These cookers are not suitable for furnishing the heat treatment required to cook meats to required degree and much less to dry excess moisture after cooking.

For proper extraction of cottonseed oil from meat to obtain oil cake with low level of gossypol, it is necessary to cook the meat with about 10 per cent moisture level and to reduce the moisture content of cooker meat below 5 per cent level. The cooking temperature and time of cooking are also very important for extraction of good quality oil.

Drying operations in India

For extraction of oil from cooked meat 3–4 per cent moisture level is considered optimum. The moisture level in cooked meats, the power applied, the amount of oil left in the cake and capacity of pressing machine are all interrelated. If cooking conditions are not properly controlled, oil content in cakes will be more and quality of extracted oil will be inferior. In such oils refining and bleaching cost will be higher. In one standardised method when meats were cooked at 85–95°C for 15–20 minutes and moisture content during cooking the meat was 11–12.5 per cent, the oil content in cake was 4.2 per cent. In this case the moisture level of cooked meat before pressing was kept approximately at 4 per cent level.

Cooking conditions are the major factors influencing the distribution of gossypol between the meal and oil. Reduction in free gossypol during cooking was due to binding with meal components. Nitrogen solubility is checked during cooking of meat. Moisture content during cooking is generally not allowed to drop below 11.5 per cent and when the moisture level of cottonseed-meat stands at 10–11 per cent, direct use of water is avoided and steam heating from the top of the kettle is preferred. Relatively lower temperature of cooking i.e. 65.6–77°C instead of normally used 88–96°C produce slightly higher yield of crude oil, although this is not gainful economically as the refining cost will be more. As a matter of fact higher top kettle temperature above 93.8°C lowers the extraction but produces better quality of oil. The drying temperature should be 104°C–116°C and this should also be the temperature of meat at first stage of pressing.

The pre-treatments like cooking and drying before crushing of cottonseed is normally followed in India. After 1930, imported machine was installed for cooking and drying of cottonseed for the purpose of extraction of oil. After 1950, Indian initiation was started for manufacturing of standard expellers. The expellers have only small 1 or 2 stack cookers seated on them. They vary in diameter from about 45 cm to 100 cm and have an open steam heating arrangement in the top kettle and a jacketed steam heating system in the bottom kettle. The space of the kettle was not sufficient to cook or dry oilseed or cottonseed-meat for proper degree and time. All these expellers were low-powered motor-driven units. As a result 7–8 per cent oil was left in the cakes even after two crushings. Only recently few firms have started manufacturing 4–5 stack cooker with foreign collaboration. The cookers have 100 tonnes capacity for cooking and drying of cottonseed meats per day.

It is thus not possible for all Indian mills to provide proper pre-treatment, viz., cooking and drying in processing cottonseed-meats. The prime need in Indian cottonseed oil and cottonseed-cake industry is to reduce gossypol content and increase protein value in cottonseed-cakes and produce oil of high quality. An important step in this direction is adequate equipment for cooking and drying flaked cottonseed-meats under controlled conditions, prior to oil extraction by screw pressing or by direct solvent extraction.

Method of Oil Extraction

Oil extraction by pressure

Hydraulic pressing

Hydraulic pressing system for extraction of oil is now not a popular method for extraction of oil and as such is not used in India. So it is now of historic importance only. The hydraulic press base contains a hollow cylinder of about 37 cm in diameter, carrying a ram which rises the press boxes against the top head supported by four steel column. Four heavy nuts hold it against the hydraulic pressure exerted by the ram. A typical press contains 15 boxes or space for the formed cake. The press is filled with 15 formed cakes covered in press cloth and pressure is applied for liberation of oil from the cakes.

The presses operate by means of automatic hydraulic pumping system in two stages, first a low pressure and then high pressure. At first stage a pressure of 2000 psi is attained in 5 to 7 minutes and allowed to drain out maximum amount of oil coming out and then gradually pressure is increased up to 4500 psi to drain for sufficient time. An average pressing cycle is of 30 minutes and the oil content in cake is 5.5 per cent.

Hydraulic pressure system has certain advantages over expellers, viz., the initial capital investment is lower, less power is required for operation, maintenance cost is low and oils are lighter in colour. However, in screw-pressing system overall running cost is low due to requirement of less labour cost, higher yield of oil and low free gossypol content in cakes.

Expelling or screw pressing

Modern trends are towards larger capacities, higher screw speeds and development of greater pressure within the expeller barrel.

The mechanical expellers has five essential elements: (i) main worm shaft and worms; (ii) choke mechanism; (iii) drainage barrel; (iv) motors, transmission and thrust bearings; and (v) cooling machine. The main shaft and worms can exert a pressure of 10,000 to 30,000 psi on the seed. The choke or cone mechanism permits a final adjustment of this pressure. The thickness of the emerging cake can be varied by opening out or closing in the cone mechanism. The drainage barrel or case is made of rectangular bars of variable size and length. There are spacings in the bars which not only permit the drainage of oil from material being pressed but also acts as a coarse filter medium for the solids. The motor, transmission and bearings are essential elements of any motor-driven machines. Main worm shafts of some of the screw press now-a-days are operated with 30–100 hp motors. The maximum torque on a main worm shaft may be 16,000 ft-lb and maximum thrust may be as high as 110 tonnes. A significant amount of heat is produced due to friction between seed being pressed and the encasing element of the machine. To control the heat, cooling devices are necessary in an expeller. The heat generated in mechanical screw press also creates an acute erosion problem. Some mills use water-cooled shaft for use and others add water cooled ribs on the bar case.

The capacity of an expeller mainly depends on (i) preparation of the raw materials to be pressed; (ii) speed (rpm) of the shaft; and (iii) arrangement of worms on the shafts.

As the speed of the shaft of expeller is increased, the oil content in oil cake will also increase. In typical comparative study it was found that the hydraulic meal contained 0.07 per cent gossypol and about 31 per cent protein soluble in 0.5 M NaCl, while the expeller meal contained 0.02 per cent gossypol and only 11.3 per cent NaCl soluble protein.

Skipin processing

A new concept was developed by *Skipin* for extraction of oil from various oil-bearing seeds. The concept assumes that water by a type of selective wetting can remove the oil from seed interior. *Skipin* observed that free discharge of the major part of the oil from the seed meal is possible when the water content in the material remains about 14–18 per cent and oil separation stops beyond the moisture level. At a fixed moisture content and at fixed temperature, depending upon the type of seed, a violet oil discharge takes place. At the oil discharging stage the seed materials change into an intimate, coherent granular mass and the volume reduces considerably. The process is based on fixed limits of moisture and temperature. For cottonseed 14 to 20 per cent moisture has to be maintained and 70–72°C temperatures are optimum. The oil begins to discharge freely at 60°C and stops at about 80°C.

Skipin process consists of three successive operations: (i) moistening the meats with the requisite amount of water to release the oil, followed by heating the meat to free oil from binding forces; (ii) allowing the freed oils to draw through the screened bottom of the heating kettle; and (iii) drying or cooling the partially defatted meat to a moisture content which permits subsequent solvent extraction. Skipin oil contains less gossypol and the recovery of oil is also much higher.

Cottonseed crushing in India

In India, presently most of the cottonseed is crushed in expellers by screw pressing and solvent extraction. Hydraulic pressing is totally an obsolete method. Pre-pressing and solvent extraction of the pre-pressed cake is also practised. Although 90 per cent of cottonseed oil is extracted by screw pressing and solvent extraction the remaining 10 per cent is only extracted by various other methods.

In the decentralised sector of cottonseed-milling industry, the entire undelinted seed is crushed directly by expellers. The process is primitive and yields only 12–13 per cent crude oil which is inferior in quality and dark in colour. This process produces 80–85 per cent cake which has low nitrogen and high fibre content. However, in Indian trade condition the process is economical.

The cottonseed-crushing mills in India face innumerable problems owing to ununiformity in seed collection and preservation. Moisture is one such uncontrollable factor. To overcome this problem seeds are cooked in top kettle or in horizontal cooker provided with good agitator which moves the seeds thoroughly forward and backward at 100–110°C for 15–20 minutes. The cooked seeds are then further dried in the bottom kettle to reduce moisture as much as possible. The prepared materials are then expelled in high-powered expeller. The entire crude oil coming out of the expellers is screened.

Another modification used is to dehull the whole cottonseed with or without seed cleaning but without delinting the seeds. The meats are then cooked and pressed in expellers to obtain oil. Such processing is also profitable in India, since the capital investment is reduced due to omission of steps like delinting and lint cleaning. However, in the process considerable amount of linter and oil is lost, which can fetch better remunerative price of processed scientifically.

Solvent extraction

Extraction of oil from oilseeds is the most efficient method for recovery of oil Applications of solvent extraction can increase oil yield by about 11 per cent from cottonseed of 17.5 per cent oil content.

Cottonseed flakes disintegrate readily and cause trouble through the production of fines. In general, mechanical strength of oilseed flakes is directly proportional to the content of solid materials and inversely proportional to its oil content. So solvent extraction of cottonseed of low oil content is mechanically simpler as well as more economical. However, by pre-pressing the seed under low pressure expeller to remove a portion of oil makes cottonseed more suitable for solvent extraction.

Solvent extraction involves very little heat treatment. Oil produced by this method is of high quality and meal contains unaffected protein by subjecting to low heat treatment. However, solvent extraction system has the following disadvantages.

Solvent extraction equipments are costly, danger of fire and explosion from inflammable solvents and meals feed from oil tend to be dusty. The complete solvent-extraction process involves materials handling, size reduction, size separation, drying, flow of fluids heat transfer, extraction, evaporation, stripping, filtration and absorption.

Complete extraction of oil can be achieved only by the use of a large volume of solvent and this solvent must be recovered from the oil. In practice, solvent extraction on large scale is carried out only in continuous system where operation is entirely automatic. In these systems, highest economy of steam power, labour and materials is achieved.

Solvent for oil extraction

The most common solvent used in India is a commercial hexane (bp 67–70°C). For extraction of edible oil food-grade hexane is used.

In the USA, hexane type naptha (bp 63–69°C) is generally preferred. In Europe, cyclic hydrocarbons are more commonly available and a product consisting predominantly of cyclohexane (54 per cent) (bp 71–85°C) is widely used. Commercial hexanes in the USA contain mainly *n*-hexane, methylcyclopentane, 3-methylpentane and 2-methylpentane in varying proportions. The main disadvantage of most of these solvents is their extreme inflammability. Non-hydrocarbon solvents like trichloroethylene or carbon disulphide have been used for extraction of oil from oilseeds other than cottonseed. Five commercial solvents, viz, hexane, benzene, ethyl ether, acetone and butanone, have been studied for extraction of oil from cottonseed. However, hexane has been the best of the lot. Solvent extraction of cottonseed using azeotropic mixture of solvents hexane, acetone and water was found to remove gossypol almost completely from the meal.

Types of extractors

There are two types of extractors in practice: (i) batch type; and (ii) continuous type. Batch type extractors are commonly used in India. They are indigenously made and are thus economical and popular. In advanced countries batch process is used for extraction of costly oils. Batch type with single extractor is not economical. In Germany, oilseeds of high oil content including cottonseed are re-pressed once to thrice in the expeller after usual flaking, humidification and cooking-drying. In single, double and triple pressing, 24 per cent, 20 per cent and 13 per cent oil, respectively, remains in the oil-cake. Oil produced in all stages of pressing and in solvent extraction operation are combined together for refining. Generally 90–95 per cent of the solvent is distilled in the apparatus.

The guiding principle of a continuous solvent extraction system is to pass the solvent through the travelling mass of oil-bearing materials continuously. Continuous extractors are of three types: (i) percolation type, where solvent flows by gravity through a stationary or moving bed of solids; (ii) immersion type where the solids are immersed in the solvent or miscella; and (iii) combined system of both types. Percolation type is now replacing the immersion type rapidly because of the inherent advantages of small size and high efficiency. Through recirculation of miscella within percolation-type extractor, both increased miscella concentrations and relatively clear miscella can be obtained. In addition, the percolation type usually incorporates much better drainage of liquid from the spent solid which further decreases steam requirements for solvent recovery. There are several extraction systems in use. But recently high-capacity modern extractors have been introduced. These capacities refer to extraction of either soyabean or pre-pressed cakes of other oilseeds. Such high capacities are achieved by (i) increasing the extractor speed; (ii) adding additional baskets and lengthening extractor shell; and (iii) increasing height of the existing basket and arranging a second level of baskets. The main aim for replacing lower-capacity extractor by higher-capacity extractor is to reduce the overall cost of processing per tonne of seed materials.

Solvent extraction of cottonseed

Solvent extraction of cottonseed is not convenient for various reasons. The unique features are highest percentage of oil present in cottonseed-meals, the existence of troublesome linter coat and the presence of undesirable pigments and phenolic compounds like gossypol. However, all these difficulties have been overcome and solvent extraction of cottonseed oil is practical on commercial scale. All prior processing steps for cottonseed, such as seed cleaning, delinting, dehulling and hull separation are more or less same. Factors like temperature, time and moisture level in flaking operation are more closely controlled in solvent extraction than in mechanical or hydraulic pressing to achieve satisfactory extraction.

Cooking for pre-press solvent extraction does not greatly differ from that for straight expeller operation. Cooking for direct solvent will, however, differ to some extent. It has been reported that the rate at which oil is extracted by solvents is the function of the surface area and thickness of the solid extracted. Two general objectives are considered in selecting cooking conditions: (i) to avoid heat damage to protein as far as possible; and (ii) to suppress the development of phospholipid material.

In case of cooking for direct solvent extraction, the conditioning temperature will be 49–66°C while raising the moisture from 9.5 to 11.5 per cent. These meats are then flaked quite thin and sent to the extractor. Cooking conditions for pre-press solvent extraction are less drastic than those for other extraction processes.

A few precautions are required to be taken for extraction of oil from cottonseed which are:

1. A uniform flow of material to the extractor should be maintained and the solvent going to the extractor should be heated to 60°C.
2. It must be ensured that all solvent is recovered from extracted cake and moisture is reduced (7–10 per cent) for safe storage.
3. Sparge steam should be added uniformly to obtain maximum benefits.
4. Vacuum in stripping column should be preferably 58.4–63.5 cm of mercury.
5. Adequate supplies of cooling water should be maintained.
6. All leaks should be stopped immediately and seals and packing glands should be kept tight.
7. A combustible gas alarm should be provided to show if any solvent is present in meal, oil, vent and waste-water.

Recovery of phospholipids-lecithins from cottonseed or cottonseed oil by solvent extraction is not economical because the dark colour of its lecithins make them unsuitable except for non-food uses.

New trends in solvent extraction

The outcome of recent research has introduced new extractant solvents like isopropyl alcohol, ethanol, acetone and azeotropic mixtures in place of hexane.

The extraction of cottonseed-meat using isopropanol has two advantages over *n*-hexane extraction: (i) it reduces the formation of fine particles; and (ii) it reduces the free gossypol level to a safe level (less than 0.03 per cent). Moreover isopropanol-extracted oil is found to be richer in nutrient values than hexane-extracted oil and it also has less fire-hazard problem.

However, isopropanol extraction has some limitations also. The main disadvantages are its higher cost, need for rectification of the alcohol, lower solvent power in comparison with hexane, need for separation of phosphatides, carbohydrates, etc. Because of these problems isopropanol is not used to extract oil from cottonseed.

The extraction of oil from cottonseed using ethanol as solvent extractor has been exercised. The moisture content in ethanol has some bearing in extraction of oil from cottonseed. It is found that complete solubility of oil is attained when moisture level in ethanol is less than 1.5 per cent. Higher moisture content results in effective separation of oil from the solvent when miscella is cooled after extraction. It has been found that critical solution temperature increases steadily with moisture content of alcohol. Pressure in the system also varies directly with temperature. Ethanol is found to produce meal having higher nutritive value and improved flavour. Moreover, alcohol-extracted oil gives lighter colour and lower refining loss than expeller oil. However, it suffers from defects as mentioned in case of isopropanol. For these reasons, ethanol could not be introduced even as an alternative solvent for cottonseed oil.

Acetone has been used as a solvent to extract oil from cottonseed in place of hexane. It has the merit of removing gossypol almost completely with very low refining loss and high neutral oil yield. Physico-chemical characteristics of acetone, viz. boiling point, specific gravity and viscosity, are almost like hexane but the major difference between the two solvents is that acetone dissolves water present in the seed slowly and it gets diluted and its solubilising power for oil decreases rapidly. Nonetheless, acetone extraction gives an exceedingly light-coloured meal and also provides superior quality oil than all other commercially-extracted meal. Another plus point of acetone as solvent is extraction cost, which is almost on par with hexane. The extraction of cottonseed oil using acetone as a solvent in place of hexane is reported to be carried out commercially in Italy.

Besides all these solvents, mixture of acetone, commercial hexane and water in proportion of 55:44:3 has been employed to extract raw cotton meats at ambient temperature. This mixture could extract 97 per cent of oil in 4 hr., whereas acetone alone could extract only 78 per cent oil and petroleum ether 71 per cent in this period. The mixture is also able to extract practically all free gossypol within 1 hr. whereas only about 60 per cent could be removed wtih acetone alone in 4 hr. Moreover, lysine was left entirely in the extracted flakes. The meals extracted by this mixture were superior to commercial cottonseed-meal as feed for poultry birds. The mixed solvent extraction process is promising because it is commercially feasible. Moreover, the meal and oil products are also of very high quality.

COTTONSEED MILLING INDUSTRY IN INDIA

The milling industry in India in its present form is not well organised. It is seen that basic principles involved in quality products are not taken into consideration. The major important operation of seed cleaning and proper storage are neglected, which are in fact absolutely essential for good quality and better yield of seed oil. The proper cleaning of cottonseed reduces maintenance and provides better quality seed by-products, such as linter, hull, oil and seed-meal. The presence of foreign matter in seed like dirt, trash, etc. deteriorates the product yield and introduces several additional operations to remove them and increases complication in seed-milling operation. In fact, it increases huge losses by increasing processing cost.

The impurities and foreign matter present in cottonseed are of two types. One can be salvaged at the time of crushing and others cannot. Those that can be salvaged are bolls, cotton fibre or lint, cottonseed meats, cut, dead and immature seeds and seed hulls. Other class of impurities are stalks, leaves, bolls, stone, sand and dirt. It is also seen that wet weather during harvest affects the maturity of boll and seeds and such cotton cannot be ginned easily and contains foreign matter. Also it affects the seed maturity and produces immature and damaged seeds. In general, there is awareness about the presence of impurities in the seeds and its effect on quality of seed and its by-products.

Further, in general, ginning process is essentially carried out to get clean lint. The ginning units, few in number, are equipped with seed-cleaning and drying machinery but in several factories such facilities are not satisfactory. Many ginners in the country are using obsolete roller ginner, hence it is not possible to produce quality seeds. This is because many of them are doing job work for others and their interest is maximum earning and not the quality of the seed. It is also found that sometimes American and *desi* seeds are mixed and ginned to get acceptable fibre length. This results in mixture of fuzzy type of American seeds and *desi* seeds. The broken seeds found during ginning operation cause damage during storage. It is necessary to clean the cottonseed before storage. Even if partial cleaning is attempted, the foreign matter like sand can be removed through seed-cleaning equipments with perforated bottom in the unloading conveyors. The removal of sand increases the efficiency of air-cooling and reduces the heating of cottonseed during storage. Major operations involved in cottonseed-milling industry are shown in Fig. 4.9.

In general, several types of seed-cleaning equipments are in use in the USA. Some of these are also in use in India. They are: (i) Bauer/Carver pneumatic cleaners; (ii) boll and sand reels; and (iii) rock shakers and air traps for separating stones.

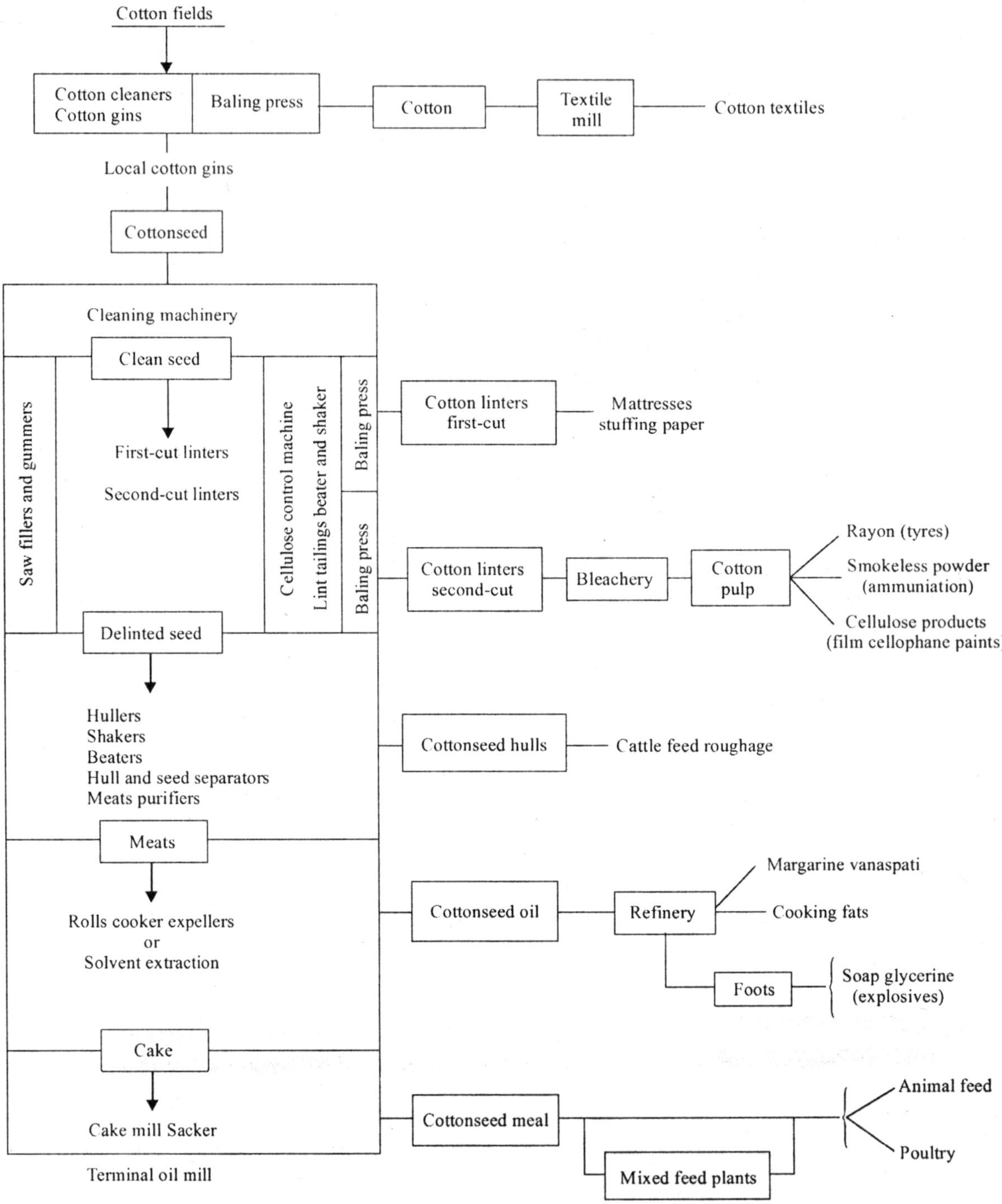

Fig. 4.9. Major operations involved in cottonseed oil industry.

Pneumatic Seed Cleaners

These machines have both shakers and pneumatic principles. Oldest machines in use are of Bauer Brothers. Similarly, Carver seed cleaner are also equally old and these are most popular machines for seed cleaning in India as well as in the USA. These machines have similar construction with varying size and operating capacity. In recent years both Bauer and Carver have introduced 4-tray or double-decker machine with a larger screening capacity with several multipurpose cleaning arrangements where various foreign matters and objectional materials in lint and seed are separated.

Need for cleaning seed at the oil mill

It is advantageous to clean the seed and remove the foreign matter from cottonseed as it is received at the oil mills and before keeping in godown. This will benefits the seed-crushing industry to a great extent and to users of cotton linters. The practices followed in the USA are quite sufficient to remove the foreign matter from the seed before storage; however, such practices are lacking in India. In view of this, it will be more economical to have seed cleaners at the oil mills instead of gins.

There are several advantages in cleaning seed at the oil mills and prior to storage. The removal of foreign matter in seed will reduce risk of heating during storage. It will also increase the efficiency of crushing industry. Seed quality will remain intact after storage as the absence of foreign matter during storage of the seed will reduce free fatty acid content and minimise loss of oil during refining.

Test methods of cottonseed-milling industry

From experience it has been seen that methods of test are an effective and accurate means of sampling and analysis of cottonseed before and after cleaning to find out the trash content. In a study it has been shown that larger samples should be taken to get more reliable results. While sampling the seed samples a sufficient quantity of sample from different lot should be taken. The sample should be screened to determine the content of seed and find trash and hand-picked trash. It is easier to remove trash sand on screen or the lower tray of the pneumatic cleaner. If the samples are properly cleaned and not over-loaded, sufficient quantity of broken sticks and burrs can be removed. In this way 90 per cent of fine trash can be removed.

Busket-type seed cleaners

The busket cleaners are used for two manual applications, one as a shaker pneumatic-aspirator seed cleaning unit and the other is first-cut delinting machine. The speed of the cleaner is varied, average from 90–100 rpm on No. 18 or 100–150 rpm on No. 20 unit. These units are most commonly used in the USA. These seed-cleaning units take out extra-small percentage of seed trash from the seed. However, it is worth mentioning that in cottonseed high percentage of trash is difficult to clean.

In India, cottonseed is generally stored on the ground at the gins. In this way, the seed contains large amount of sand, bolls, leaves and twigs in the heap from the fari. Cleaning machines are not properly used for seed cleaning. It is necessary to install sand and boll reel as a pre-cleaner prior to use on pneumatic cleaners to get better cleaning. Few mills have installed sand and boll reels and these are found useful to the mills which are directly crushing the seeds in expellers without delinting. Bauer-and-Carvers seed cleaner, both 2-tray and 4-tray are in use in a number of cottonseed oil mills. A few factories have built their own seed cleaner indigenously.

Delinting of seeds

After cleaning of cottonseed and ginning of the seed, short fibres known as linter or fuzz, are leftover on the seed. The quantity of short fibres 'fuzz' left on the seed is very small and varies from 5 to 15 per cent of the total weight of the seed of *G. hirsutum*. In India, the linter content in *desi* varieties ranges from 1 per cent to 10 per cent or more depending on the varieties. These fibres are removed from the seed by delinting machines which are fabricated on the same principle as the short-type cotton gins. During the seed linting operation the delinting process is sometimes repeated. When the seed is delinted only once the linter is known as 'mill-run' linter. When delinting is done twice then the two lots of linters are called as first-cut and second cut linters.

The delinting machines are basically called saw-type cotton gin. The main parts of the delinters are still frame, feed chute and feeder, gratefall, seed board, eccentric adjustment on seed board, density regulator, grates, float, gratefall lift lever, brush, feeder regulator, rollbox, permanent magnetic separator, mote board and mote chamber and condenser. It contains two motors of 7.5 hp. The seed from cleaners is conveyed to the hopper which is kept constantly filled with seed. The seed roll revolves by a winged shaft in the opposite direction of the saw and at the lower speed. The amount of linters to be removed from the seed is adjusted by opening and closing discharge area under the seed board. The capacity of the delinters depends on the density or hardness of the roll. The feed of the seed is continuously controlled by the amount of the seed discharged from the delinter. The linters are removed by a cylindrical brush revolving with the saws at somewhat higher speed than the latter. The diameter of the standard delinter cylinder is about 30 cm and it can recover about 20 kg linter per hour. The delinters are built originally with 106 saws. However, it is now increased to 176 or even more. The yield of linters is increased with the number of saws to the cylinder. The moting of the delinters can be adjusted by the draft shield. The saws of the cylinders should always be kept cleaned. It is necessary to adjust the peripheral speed of the saws.

For second-cut delinters, about 7 tonnes of seeds are required, from which first-cut linters have been removed. The speed of the saw should be between 700 and 750 rpm, whereas it should be between 600 and 650 rpm for thin-hulled seed. It is necessary to leave some linters on the seed which helps in the hulling operation. In the USA, about 4 per cent linter will pose problem with existing hulling and separating equipments.

The number of mills of cottonseed-crushing industries are inadequately provided with delinter machines. However, the process of delinting in India includes mill-run which is more popular. Some mills also delint seeds in two-cuts employing twing as many delinting for the second cut as for the first. The delinted seed is normally sent to the cotton-crushing mills and is utilised partly for sowing purposes. In India, the indigenous delinter machines are available. Recently an additional process to recover the residual linter in the seed is also followed and such linters are known as defibrator and the machine is called defibrator unit. This linter is suitable for cellulose nitrate and chemical cotton. It is heartening to see that production and utilisation of linter in our country has not improved on the same line as the seed production has gone up. This shows that a number of mills are not following proper removal of linters and crushing the seed with linter which may be very uneconomical.

Cleaning of linter

The common impurities present in the linters are trash, shale, leaves and sticks. The presence of such foreign materials is responsible for poor quality of linter and lower percentage of cellulose. In view of this, cleaning of linter by employing proper lint-cleaning device is a must for good quality linter. There is special type of linter-cleaning machinery. The principles involved in cleaning are beating of linter and pneumatic separation of impurities. This produced better-quality seed and linter in one process.

It has been seen that various types of lint-cleaning units are in use. There are air separators, centrifugal separators, the super-jet cleaners and whirling pneumatic cleaners. These are commonly used in the USA and very few find application in India. It is essential to have sufficient facility for linter cleaning by way of creating enough capacity of machinery as well as storage of linter. However, the methods and technologies used in advanced countries have to be carefully examined before implementing in India.

The significance of good quality cleaned linter suitable for various end-uses cannot be ignored in view of the present situation of cottonseed-milling industry in India. Many of Indian cottonseed-crushing mills are not properly equipped with modern linter-cleaning machineries for proper use and realisation of suitable price. Linter has to be of good quality. First-cut linter should be clean and free from impurities with bright colour. Similarly, second-cut linter should be free from hull and shale contents and have high percentage of cellulose content.

First-cut linter

The first-cut linters are cleaned with cellulose-control machine. The machine handles the lint over perforated metal in two sections, each of which has three compartments. When it enters the machine, the linter passes over the fine perforated metal section of three chambers; the fine impurities pass through the fine perforations and get discharged at the bottom. Again it moves to the next section which is provided with three compartments of bigger perforated metal, where objectionable hull particles and shell are removed. In this process some loss of lintes takes place due to large perforations, which can be recovered later on in the tailings, beaters and shaker. Therefore the clean linter is passed on to the baling press.

Second-cut linter

The cleaning of second-cut linter can be done in similar way with minor modifications. Normally second-cut linters are collected from the delinters by flue system and passed on into feeder at the far end of the top unit. The linters are cleaned by passing through top unit and are picked up with air by a large suction nozzle at the opposite end and then it passes over moting gap to drop out large, heavier foreign matter. The cleaned second-cut linters are then passed on for baling machines.

In smaller sized mills, cellulose-control machines are preferred over three-high lint cleaner. In this case, lint tailing beater is used to separate hull bran, after which the linters pass on to the lint shale shaker. After being freed of shale in the shale box, the materials from the first-cut linters are cleaned, second-cut linters go to the flue system for treatment.

Critical examination of commercial delinting

If delinting process is reviewed critically, it will be seen that the linters are a valuable by-product of cottonseed. It will be in the interest of cottonseed-proceesing industry to produce quality linters to meet the requirements of various industries utilising the linters.

In this context linters must be removed from the seed to facilitate hulling and separation. It is observed that anything in excess of 4 per cent lint left on the seed would make the seed difficult to process during dehulling and separation of hull. Also, if linter is cleaned and sold in the market, it will fetch higher price, making cottonseed milling more profitable. In the USA, normally delinted seed contains 1.5–2.0 per cent linters. It is necessary to work out more precisely the cost and profitability of linter in each mill when the moderate market price is prevailing. For working out the cost one has to include various operations including hull costs. Even interest and depreciation on machinery cost, yield of linter, cellulose content and colour of the linter are to be taken into account. It may be profitable to make slight reductions in linter yields in order to produce the export grade.

If adequate facility is provided for proper delinting and lint cleaning permitting minimum of 1.5–2 per cent residual linter on the seed, there is enough scope for reasonable profitability. The mill should keep their target for production and also keep watch on the process for better yield and market.

Dehulling of cottonseed and separation of hulls and meats

After removing the linter cottonseed is subjected to dehulling. For dehulling cottonseed is directly fed to the dehuller. There are two types of cottonseed hullers. One is the bar type and the other is the disc huller. The bar huller consists of a frame which holds stationary knives running the full width of the huller and one near the cylinder. Cylinder is the drum running at high speed with knives along the full width of the drum. The other type of machine is disc huller. In this plate on one side is stationary. The two types of bar hullers are popular—the Carver perfection huller (central concave model) and the Chandler bar huller. This huller permits the use of screen to free the meats from the dust and lint during the hulling. The separation of seed meats from hull is an essential process to obtain further products like oil, cake and hulls in maximum yield. The separation should ensure production of hulls from meats and whole seed. There are three types of dehulling processes: (i) single hulling; (ii) double hulling; and (iii) universal hulling process. It is seen that flaking and rolling of meats are another important process for good oil recovery both by hydraulic and screw press methods. Possible causes for poor flaking are: (i) use of a single worn-out roll; (ii) too many uncut seeds in the feed to roll; and (iii) excess of hulls in the meats.

Thus looking at the increasing population of the country and future need of oil requirement, there will be pressing need to improve the production of oil in the country and also by-products of the cottonseed. Production of cottonseed is likely to go up in near future. Similarly, there will be more pressure to find better utilisation and milling practices for cottonseed. Newer application for linter, hull oil and seed meat has to be investigated.

PROCESSING AND UTILISATION OF COTTONSEED OIL

Processing of Cottonseed oil

Cottonseed oil is obtained from cottonseed by different methods. The quality of oil depends on the processing methods and variety of seeds. Crude cottonseed oil is dark coloured complex material and contains several impurities. Some of the impurities present in the cottonseed oil are free fatty acids, free gossypol and related pigments, phospholipids, phytosterols, inositol, phosphates, xanthophyll, chlorophyll, roffinose, pentosans, peptones and similar other products. The presence of gossypol and a like substances is well known in cottonseed oil and these are rarely found in other oils with a few exceptions. Presence of these substances in crude cottonseed oils necessitates its refining. The impurities present in the cottonseed oil are generally fat insoluble and fat soluble.

The insoluble impurities dispersed in the cottonseed oil including those present in colloidal form can be removed by filtration, centrifugation, steam, water and electrolyte treatment. The method of processing will depend on the nature of impurities present. Some of such impurities are seed particles, mucilages, dust, mineral matters which can be normally removed by filtration. The other materials present in solution or colloidal form are phosphatides, carbohydrates and proteinaceous substances which can be removed by centrifugation or by filtration with absorbants. The fat soluble substances are free fatty acids, colouring matters like gossypol, carotenoids, chlorophylls, sterols, resins and similar other products.

Crude cottonseed oil needs refining, which is very important process. Dark, black-coloured cottonseed is refined to remove the above given impurities and undesirable substances to get fats with high yield of purified glycerides. It is essential that refining should be done with utmost care to get high quality finished products. Refining process affects the processes too. If product is refined well, subsequent processes like bleaching, hydrogenation and deodourisation will also give good results and produce high quality oil and higher yield of oil which gives positive return to the producer.

Pre-treatment and degumming

It is essential that crude oil, before storage, should be treated to remove fat insoluble impurities. This will help in maintaining good quality of oil. Even presence of moisture in oil causes fat splitting and hydrolysis of glycerides and increases the free fatty acids. Also presence of moisture will help the growth of certain micro-organisms. These may deteriorate the quality of oil and increase rancidity and undesirable odours and tastes. It has been reported that after hot pressing, cottonseed oil stored without filtration gave higher acid value compared to control whereas filtered oil under similar conditions showed insignificant change. The methods used for removing fat-insoluble impurities are settling, filtration and centrifugation.

The method based on settling principle is slow, inefficient and time consuming for removing suspended impurities. If moisture is present in the oil, this method provides opportunity to micro-organisms to develop and fatty acids are formed at fast speed. It produces poor quality oil and the process is expensive because it is time consuming. Also because it does not produce very clear oil, further purification by employing other methods is necessary. The filtration method by filter process is very effective for removing suspended impurities from cottonseed oil. In this method, suspended and solid particles are separated and only clear oil or is obtained through filters and centrifugation.

The filters such as cotton cloth (special grade), wool, synthetic fabrics (woven and non-woven) retain suspended and solid materials during filtration process. The filter cell and dicalite, if mixed with oil prior to filtration, absorb most of the moisture and help in producing clear and better quality oil after filtration. Centrifuge method for removing suspended particles in oils is done by employing high-speed centrifuge of cylindrical rotor types or the bowl and disc type. To prevent oxidation hermetically sealed equipments are found more efficient.

Degumming

It is necessary to remove gums, resins, protein and related substances from crude cottonseed oil. This process is known as degumming. Several methods are in use to remove such products. Use of alkali treatment with crude cottonseed oil facilitates the removal of such substances. Gums and colouring matter are absorbed by the soaps formed by reaction of free acid with alkali. At present following two main methods are employed for degumming of edible oils; (i) specific treatment before neutralisation with removal and at times isolation of the gums; and (ii) degumming without separation of the gums and alkali treatment.

The second method is preferred for degumming of crude cottonseed oil as it removes the gum which is later on combined with soap stock and added to the meal to improve its quality for use as cattle-feed. The gum content in crude cottonseed oil is almost equal to soyabean-oil, i.e. about 1.8 per cent is higher than other vegetable oils. The removal of gums and phosphatides facilities the alkali-refining process, reduces loss, improves storage stability, taste and flavour of the oil.

Degumming by hydration

The phosphides and protein when wetted or hydrated swell to form gels which separate from the oil and settle at the bottom. The hydration treatment for most of the oils containing above impurities is similar. It requires 2.5 per cent of hot water which is added with vigorous stirring and hydrated colloids are separated from the oil. More simple hydration treatment is to heat the oil to about 80°C, pass direct steam through perforated coils, warm the oil to 90°C for 10–15 minutes and then allow the residue to settle. Use of aqueous solution containing salt, starch and alkali and degumming at 60°C is also suggested. It is suggested that the optimum condition for degumming of crude cottonseed oil is addition of 4.5 per cent of water, heating to 75–80°C with good stirring for about 15 minutes and separating the gums from the oil.

A popular method is used to degum oil in Germany by two successive water-washings, followed by centrifugal separation and third washing with a mixed solution of aluminium sulphate, sodium sulphate and sulphuric acid to effect a very thorough removal of phosphatides. The cottonseed oil is generally degummed along with refining and neutralisation process with a caustic soda solution.

Refining of cottonseed oil

The refining of cottonseed oil is neutralisation by alkali treatment. There are several methods employed in the industry. Following are the popular methods; (i) neutralisation with caustic alkali; (ii) neutralisation with weak alkalies; and (iii) neutralisation in solvent solution (miscella refining).

The sodium hydroxide is mainly used for refining cottonseed oil. It reduces free fatty acids content to 0.01–0.03 per cent while it is not possible by weaker alkali to reduce FFA below 0.1 per cent. It is reported that FFA can be completely removed from oils by chromatographic absorption on alumina or silica gels but the process is uneconomical. Alkali refining is capable of removing phosphatides completely from oils. Also used for bleaching earth which removes impurities like phosphatides, gums, slimers, mucilaginous materials. Similarly, nickel catalysts are better absorbents than bleaching earths and these are effective even after use for hydrogenation. During alkali refining, colouring pigments like gossypol react with alkali and are removed completely. Acid refining reduces the red and yellow colour of oils to greater extent than alkali refining. Cottonseed oil darkens when heated to a high temperature, colour gets fixed which cannot be removed later by any method. Oxidation has an important effect on the colour of oils. Main attempt in refining of cottonseed oil is on minimising the refining loss of oil rather than complete refining.

In refining, a few important principal practices are to be kept in mind. During neutralisation at the time of soap formation, the temperature needs to be kept high enough to allow soap to separate from the oil and settle quickly. Also the formation of scum due to presence of carbon dioxide and high temperature during neutralisation should be avoided by deaerating the crude oil. If excessive agitation is given, there is risk of an emulsion-forming which may spoil the product. The quantity of lye required will depend on the type of impurities present in oil. In cottonseed oil after removing the impurities and FFA by dilute alkali wash before the oil is sent for storage or for further processing, the acidity is 0.01–0.02 per cent. Such oil is called washed oil. Finally it is neutralised with strong lye before use as an edible oil or product.

Refining procedures

(Open) kettle method

Refining of cottonseed oil with alkaline lyes when subjected to various types of vessels (open, closed, vertical and cylindrical) is called kettle method. The most common kettle in use is cylindrical-type vessel provided with a sweep agitator having varied speed of 10–40 rpm. The capacity of kettles may vary depending on need from 20 tonnes to any suitable size. Kettle is normally 3 m in diameter and 4 m deep. The agitator shaft (1 metre long, 10 cm wide and about 1 cm thick) are made of steel and are fitted at 45° angle. The kettle is fitted with steam coil and operated at steam pressure at about 150 lb/sq. inch. This can be taken away through a swivel pipe fitted above the cone to empty the kettle. When agitator works, the swivel and skin pipeline are pulled out and cables are stretched to clear the paddles. It is also fitted with spray nozzles at the top for spraying caustic lyes, brine and water for efficient refining. For efficient refining vessels are fitted both with internal coils and steam jackets for steam heating.

Vessel can also be closed one for neutralisation. It should be operated under vacuum. In this case grid stirrer is provided which moves at 30 rpm and mixes oil and lye without heating up the soap.

Dry refining of cottonseed oil

It is called dry refining method as oil is refined by treating it with strong lye of caustic soda and recovery of soap stock in solid or semi-solid form from the cooled oil. It differs from the wet method of refining where weaker lyes are used and the soap stock is washed with plenty of water and recovered in the form of liquid. The dry refining has advantages over wet refining as it is rapid, convenient and produces concentrated soap stock and lighter colour refined oil free from soap and moisture. Dry method of refining is less suitable for other oils such as linseed and degummed soyabean.

The strength of caustic lye used in refining depends on the end-use of the oil. If the oil is for production of vanaspati, where very light colour is essential, larger quantity of soda lye is required, for darker colour oil for cooking, lesser quantity of soda lye can be used. The quantity of excess soda needed for efficient refining is worked out from refining loss obtained for the standard cup refining test. For carrying out refining, crude cottonseed oil is charged in the kettle after deaeration adjusting the temperature of oil around 25–30°C. Agitator is operated at the highest speed and caustic lye is added through spray nozzle. Stirring at 40 rpm is continued for 15–20 minutes to disperse the soda in oil. During this period the mixed solution becomes slightly granular due to soap particles. This stage is called 'pin break', and indicates to reduce the speed of agitator and increases heating at the rate of 20°C per minute. The pin break gets replaced by true break forming large soap flakes. The oil temperature should be kept at 51.5°–55°C. When the foot settles fast at the bottom agitation is stopped and contents are allowed to settle overnight or minimum for 8 hr.

After that refined oil is taken out through swivel pipes into another tank. Refining loss is worked out from the differences between the total weight at the crude oil before refining and the weight of refined oil. The remaining oil is separated from the soap-stock by heating and sprinkling salt. This oil is dark in colour and it further refined with the fresh crude oil.

Washing of refined oil

Refined oil is washed in kettles with 5–10 per cent of hot water to remove the traces of soap from it. The washed refined oil is dried under vacuum or oil is filtered through a press filled with spent cake or to treat it with 0.25 per cent of bleaching clay before storage. The dry-refining method is used both for batch and continuous methods of refining cottonseed oil.

Wet method of refining

The wet refining method is employed where oils are high in FFA and produce soft soap stock. Wet refining can be used more successfully with washed or once refined cottonseed oil. In general, the method consists of heating of the oil to higher temperature 65.5°C, mixing of the lye and washing of the precipitated soap stock with a spray of hot water. In some cases salt or sodium carbonate or any other electrolyte is used to separate the emulsion of soap stock and oil. Several water washes are needed to remove soap from the oil.

A modified wet process for cottonseed oil consists of heating the oil charged in the kettle to 50°C and strong caustic lye of 18°C Be in excess of 100 per cent is added to the oil under agitation for about ½ hour till the break occurs. Then 10 per cent of water at 50°C is added and settling takes place in about an hour. Thereafter 2–3 washes with 10 per cent of lye (0.4 per cent) at 100°C are given followed by draining. After the wash oil is heated to 95°C and washed 2–3 times with boiling water, 8–10 per cent each time. Entire processing takes about 10 hour. It is reported that removal of pigments from cottonseed oil in the presence of aqueous caustic soda depends on several variables such as speed of agitation, contact time of oil and alkali during agitation, quality and strength of caustic soda solution used, initial temperature of oil alkali dispersion and shearing effect of agitator. At high shear agitation sodium hydroxide solution of 30°–40° Be gave desired lovibond colour. Most of the colour removal took place within 5 minutes.

Refining of cottonseed oil in India

Refining of crude cottonseed oil in India is done both by batch-kettle process and continuous process. Refining is done in crushing mills. The small crushing mill do not have adequate facilities to carry out proper refining. In view of this Bureau of Indian Standards has developed standard for determination of bleachability (IS, 1968), so that refined oil can be checked for proper bleachability. It is observed that it is very difficult to remove the residual colour from oils once treated by sodium hydroxide. Therefore, utmost care should be taken to carry out proper refining at the first instance. The FFA of crude cottonseed oil varies between 1 and 3.5 per cent and refining loss is found to be 2:5 times of the FFA. Refining of cottonseed oil has been studied extensively in India. In these studies various aspects of refining such as use of additive, degumming prior to alkali refining, dry and wet methods of refining and colour fixation and refining loss have been covered.

Few of the important observations are mentioned here:

1. On storage at room temperature crude cottonseed-oil showed increased colour fixation with time of storage. Hence, it becomes difficult to remcve the colour of oil after storage. There was no change in colour of refined cottonseed oil. Therefore oil should be stored after refining.
2. The addition of 0.10 per cent salicylic acid, tetrasodium pyrophosphate, oxalic and citric acids on the oil weight lowered the refining loss of crude cottonseed oil.
3. Degumming of cottonseed oils prior to refining followed by separation of the gums by setting or by removal gave lower refining losses than straight refining of crude cottonseed oils.
4. Dry method of refining produced lower refining losses and lighter coloured refined oils than the wet refining. Pilot plant trials indicated that dry method of refining was superior to wet method both in refining loss and colour of the oil.
5. Storage studies in temperature and humidity controlled conditioned chamber indicated that temperature was mainly responsible for colour fixation of cottonseed oil without increase in free acidity. Free acid increase was due to humidity effect.

Continuous Caustic Soda Refining Method

Earlier work on continuous refining method of cottonseed oil was carried out by a number of workers. However, these processes were not very successful. Later on processes based on centrifugal method were developed in place of employing gravity forces, which were found useful as continuous refining process. The contact of oil with alkali was reduced to about 3 minutes and this process has almost replaced the batch process in advanced countries.

The continuous process has several advantages over the batch process. A few of them are as follows:

1. Owing to short-time contact of alkali with oil, saponification is reduced and the loss of neutral oil is less.
2. Loss of oil in soap stock due to high efficiency is reduced.
3. The batch-kettle process requires high skill for selection of lye strength and excess control to effect a good break and separation of oil and soap stock.
4. Continuous process reduces the cost of processing and gives higher yield (20–30 per cent).
5. The process being continuous is much faster and more efficient than the batch-refining process.

6. Moisture content in refined oil by continuous process is reduced considerably from 0.5 per cent (in batch process) to 0.05 per cent.

There are number of continuous processes in applications. Some of these are as follows: (i) continuous caustic soda-refining process; (ii) two stage soda-ash process; (iii) modified soda-ash process; and (iv) miscella-refining process.

Continuous caustic soda refining process

There are several processes developed by different workers. They differ in industrial applications with each other. Some of these systems are dealt in brief below:

A process is developed using disc bowl in De Laval equipment for continuous alkali refining of oils. In this process the centrifugal separator is the bowl, which rotates on vertical shaft. In this process the contents of one of the vessels are centrifugally separated, the other vessel is used for neutralisation, which keeps the separators in continuous use. About 20 per cent of hot water is used in the washing tank and after centrifugal washing, oil is free from soap. The residual moisture in the oil is removed in the vacuum evaporator.

De Laval caustic soda continuous process

This process refines and combines various operations like degumming, neutralisation and deodourisation in one process. After separation of soap stock in disc-bowl centrifuge, the neutral oil is mixed with hot water and washed through secondary water-wash separators prior to drying oil in the vacuum vessel. Other advantages of this process are ease of cleaning of hermetically sealed bowls without dismantling, ready adoption to solvent operation, absence of troublesome release of carbon dioxide in some variations of refining and discharge of a drier soap.

De Laval quick mix process

In this process, phosphoric acid is used to degum the oil in centrifuge and then it is mixed regularly with caustic soda solution, oil and soda and kept in contact for a few seconds. Soap stock is removed at once in another separator. Washing and drying of the refined oil is done along the conventional lines for continuous processes. The refining of crude cottonseed oil includes five stages—degumming, first alkali treatment, i.e. neutralisation, second alkali treatment, water wash and vacuum drying. In this process, the hydration of the gums can therefore be kept to a minimum thereby reducing the loss of lecithin in a shorter evaporation time.

Continuous refining by sharpels systems

Disc or hollow-bowl centrifuges are used for refining of crude cottonseed oil in this system. Crude cottonseed oil from feed tank is flowed to mixer where a ratio flow controller maintains constant flow of alkali in relation to flow of oil. This lye passes to low-steam heater. The 'break' is obtained by raising the temperature to 60–71°C and then refined oil and soap stock are separated. Soap stock is passed on to wash tank. The entire operation between introduction of lye and separation of the soap stock is about 3 minutes. In this process oil is mixed continuously with 10–20 per cent hot water keeping the temperature of mixture at about 76.5°C. The time of contact between the oil and water is about 10 minutes. In some cases oil is double washed. The water then is allowed to pass to the sewer and oil is passed to vacuum dryer. From the dryer the refined oil is washed, dried and passed for storage.

Another two-stage process known as sharpels low-loss process with two-stage caustic treatment of cottonseed oil has been in use. In the first stage cottonseed oil is conditioned with 0.1 per cent citric acid. Mixture is then heated to 66°C and FFA is neutralised by caustic addition and oil is separated after centrifuging at high speed of 15,000 rpm.

The oil separated from the soap stock is passed to mixer for the second refining treatment. This improves the colour of the oil and removes the traces of free fatty acids. The refined oil is separated by second super centrifuge. The remaining treatments is similar as in single-stage continuous refining. This is the most efficient and popular process among various processes of sharpels.

Clayton cottonseed-oil refining process

In this process, degumming of crude cottonseed oil is done in first stage using soda ash solution, followed by additional process of dehydration and re-dehydration of the gums. Sodium carbonate being non-saponifying alkali, no loss of oil in saponification takes place. After degumming, oil is decolourised by refining in the next step with strong caustic soda solution which neutralises the free fatty acids (FFA) and removes colour. Then it is double washed to free the oil from soap and dried under vacuum. However, this process has now been replaced by more recent methods.

Modified soda-ash process for refining cottonseed-oil

It is somewhat improved and modified version of soda-ash process. In this process, water or carbon dioxide is not removed by dehydration. Elimination of dehydration reduces the operating costs. This process is applicable to cottonseed oil which contains FFA up to 3 per cent. It has several advantages over soda-ash process.

Caustic soda-ash process for refining crude cottonseed-oil

This process uses strong caustic soda to neutralise FFA keeping the soap stock value to a minimum. The soda ash further functions to complete the neutralisation without any side effect. The amount of soda ash varies from 1.5 to 4 per cent. The oil is refined wtih caustic-soda solution and is washed to remove the colour as far as possible.

The above process has few advantages: (i) no saponification of neutral oil takes place in the main treatment and losses of oils are limited; and (ii) there is no possibility of loss of oil by oven heating with soda ash in the first step. Now caustic soda-soda ash process has taken over the soda-ash process.

Refining with ammonia hydroxide

Efforts have been made to use ammonia for refining cottonseed oil. The process is known as 'Clayton ammonia refining process' for crude cottonseed oil. A 14 per cent aqueous ammonia solution is used as the neutralising agent and it can be evaporated from the soap stock and re-used for neutralisation. Ammonia refining removes neutralisation of soap stock by acid. Storage test revealed that ammonia-refined oil is comparable in quality to caustic-refined oil. The savings in neutral oil compared to cup loss are appreciable say about 40 per cent.

Miscella refining of crude cottonseed-oil

This is simple refining process of crude cottonseed oil. Miscella obtained from the extraction plant is mixed with sufficient caustic soda solution to neutralise the free fatty acids and remove the bulk of the colouring matter. The mixture is then passed through centrifugal separator where soap stock is separated from the refined miscella. The refined miscella is then washed with water and finally freed from solvent to give refined yellow oil. Miscella refining has number of plus points, such as: (i) over saponification of glycerides oil is reduced considerably; (ii) in this process, soluble carbohydrates and protein fines are removed; (iii) colour of the final oil is superior to that of other methods and it can be varied by altering the strength of caustic soda without appreciable refining loss; and (iv) there is no possibility of colour fixation of oils on storage before refining.

Miscella Cottonseed Oil Refining

Some of the miscella cottonseed oil refining processes are discussed below:

De smet-Neumi process

In this process, hexane-isopropanol solvent mixture is used to separate oil and soap stocks. The reaction between soda and FFA takes place instantaneously even at room temperature and the soap stock is dissolved as soon as it is formed. The products separated after reaction into three distinct phases in a decanter—the top layer contains neutral oil in hexane, the bottom layer holding the soap stock in alcohol and the middle layer contains other materials besides oil and soap. The bottom layer is regularly withdrawn and washed with hexane to recover the traces of oil in it and then acidified with sulphuric acid in a special mixture. Finally acids are collected through thermostatically controlled column where the acids and solvents are separated and stored.

Miscella from the main decanter is also washed with solvent to remove traces of soap from the miscella. With a small quantity of water even traces of soap is removed from miscella and neutral oil is recovered. Thereafter if necessary miscella can be bleached. The salient feature of the process is automatic regulation of the processing conditions which enable to achieve a ternary heterogenous azeotrope of hexane, isopropanol and water distilling at a very low (56.5°C) temperature.

Ranches integrated miscella refining process

It is an integrated process, starts from seed and ends with the finished refined oil. The speciality of the process is interaction of a non-toxic alkali metal salt or sodium hydroxide with colour bodies, waxes, gums, free fatty acids, gossypol and other impurities present in the seed. Generally soda ash crystals are added at a rate of 0.01 to 10.0 per cent before the cooked meats come out to the screw press. The filter miscella is pumped to a miscella head tank and it is conditioned. The miscella concentration (50–60 per cent) is used for refining. The concentration of the alkali used for refining varies from 10–15 per cent. The mixing is accomplished in the homogeniser at pressure of 500–2000 psi. The mixture is passed through an economiser of heat exchanger at about 60°C device and cooling unit. The mixture is heated for 5–10 minutes in the soap stock formed in the homogeniser and is melted and agglomerated. Then the mixture is cooled to about 27°C and stirred which helps in settling of soap stock. From cooling section soap stock is passed to a centrifuge and then passed on to the drier. The residue can be even used to feed non-ruminant animals.

Miscella refining by De Laval process

The crude miscella (50–60 per cent oil in solvent) is passed to plate heat exchanger 55–65°C, then to filter to remove any meal scale or insoluble impurities. Then it passes through a flowmeter such as an indicating rotameter with ratio controller for the reagent treatment system. The miscella is then treated with additive like glacial acetic acid (accurately measured quantity), then passed to reaction section where caustic soda is added until impurities from oil separate as soap stock. Then it is heated to 57–60°C separation, where refined miscella is separated from soap stock. The refined oil passes at 104.5°C through filter where last traces of soap are removed, then it is passed for storage or for bleaching.

The main part of De Laval miscella refining process is the hermitic separator, which is operated to meet requirement of varying specific gravity of oil, hexane and soap stock to avoid the possibility of solvent loss. As an integral part of completely closed system, the hermitic centrifuge operates under pressure of 55–120 psi at the feed inlet and 50–90 psi at miscella discharge. The refined miscella contains 10–50 ppm soap on oil basis which does not require water washing.

Comparison of Refining Methods

Refining losses for crude cottonseed oil by various methods have been compared earlier. The values for refining losses may be seen in Table 4.1. Comparative study of plant refining of cottonseed oil of high and low FFA of four methods indicated marked difference. The comparative values are given in Table 4.2.

Table 4.1. Refining losses by various methods of processing crude cottonseed oil in the USA.

Seed-processing procedure	*Refining process*	*Number of samples refined*	*Refining loss % mean values*
Hydraulic press	Batch	21	15.4
	Centrifugal	9	5.1
Screw-press (expeller)	Centrifugal	9	8.7
High-speed screw press	Centrifugal	5	15.3
Pre-press solvent extraction	Centrifugal	12	5.6
Direct solvent extraction	Miscella	10	4.3
	centrifugal	3	4.9
Blends of oils	Miscella	9	5.3
	centrifugal	13	9.5

Table 4.2. Plant scale processing of cottonseed.

Processing methods	*FFA of oil*	*Refined oil obtained (kg/tonne)*	*Refining factor*	*Soap in refined (ppm)*	*Refined oil colour red units*	*Bleached oil colour red units*	*Bleaching earth used (%)*
Cottonseed with oil of high FFA							
Ranchers	8.8	120.2	1.72	388	13.5	5.9	0.30
Sambra	1.1	119	1.95	1710	12.8	9.2	0.20
Low-loss (Sharpels)	7.7	112.1	2.08	3388	12.6	6.5	0.61
Cottonseed with oil of low FFA							
Ranchers	0.47	157.8	4.9	2.0	4.3	1.2	–
Modified soda Ash (Clayton)	0.55	146.9	7.2	6.0	7.8	2.1	–

Table 4.2 indicates comparative yield of refined cottonseed oil and various other characteristics of different methods. It is seen that among these methods Ranchers method produces high yield as well as better quality oil compared to other processes. It will be advantageous to use continuous refining process as against conventional method for better quality and higher yield of refined oil.

Glycerine recovery and bleaching

The soap stock obtained after neutralisation of oil contains small amount of neutral oil, i.e. glycerine. The split water which contains glycerine is pre-purified with mineral and aluminium salts, then concentrated to about 90 per cent in vacuum evaporators and then distilled under high vacuum. After giving treatment with fuller's earth, the distillate contains pure glycerine suitable for pharmaceutical uses. Even an ion exchanger will purify glycerine fully by removing all minerals and pigments present from split water, producing pure and colourless glycerine.

Bleaching of soap stock or acid oil

Among the various agents used for bleaching soap stock, hydroperoxide, sodium chlorite and sodium hypochlorite are prominent. Among these, sodium hypochlorite is the cheapest. However, sodium chlorite is widely used for bleaching of oils and soaps. For bleaching soap oil with sodium chlorite, stannous chlorite is reported to be better activator than orthophosphoric acid/persulphate. Use of chlorite (0.1–10 per cent) and hypochlorite (0.002–3.0 per cent) together is reported for bleaching cottonseed oil. Similarly solutions of alkali metal or alkaline earth chlorite have been used together with phosphoric acid or phosphates. Sodium chloride with stannous chloride as activator has been used for bleaching Indian cottonseed oil soap stock. The yield of soap stock in laboratory scale experiment was 60–70 per cent.

Composition of cottonseed soap stock indicates wide variations. The moisture content varied from 17–65 per cent, neutral oil content from 3–32 per cent, the total fatty matter from 20–60 per cent and the unsaponifiable matter from 1–4 per cent. Average values are around 40 per cent moisture, 45–50 per cent total fatty matter and about 20 per cent neutral oil. Bleaching of refined cottonseed oil is an important process in which various undesirable colouring matters are completely separated before oil is used for various purposes. The common colouring matters present in cottonseed oil are gossypol and related pigments and carotenes. Decolourisation is caused by a coupled oxidation with fat involving a free radical mechanism which destroys the conjugated system and is catalysed by heat and light. Gossypol, a yellow phenolic pigment, is easily soluble in alkali but it can react with the glycerides, phospholipids, fatty acids, proteins and metals to give oil soluble colours which are much more resistant to alkali and need special bleaching. Free fatty acids react with gossypol and carotenoids.

Bleaching of cottonseed-oil

Bleaching process

The common absorbents used for bleaching refined cottonseed oil are earths and carbon. Earths can be natural and activated. The carbons are made with high capacity colour absorption and are known as active carbons. The natural earth (clays) have been mostly replaced with acid-activated clays but are still in use being cost effective. They perform better in removing residual soap/gum from oil. Neutral activated clays are used for bleaching edible oils while acidic clays are used for bleaching edible oils and oils which are difficult to bleach. The fixed colour of refined cottonseed oil can be easily removed by acid-activated earths. Earths reduce peroxide value of the oil.

Active carbons (5–20 per cent) are generally used along with bleaching earths mixtures. Carbon is better in removing soap from oils and bloom produced in mineral oil.

Among the bleaching processes, open-kettle bleaching is quite effective if earth is properly selected. Mixing of earth with oil should be done at lower temperature and a whole lot of mixture should be warmed to final temperature to get desired result. Vacuum bleaching removes the danger of darkening of the oil by oxidation. In vacuum bleaching absorbent can be added either as in oil slurry drawn from a separate clay slurry tank, or when bleaching is completed the oil is cooled to avoid oxidation and filtration is carried out in normal way. Bureau of Indian Standards has developed suitable standards for preparation of bleaching earth of different types (IS 1965–61). Methods of sampling and test for bleaching earths are described in IS 1035–1964 and bleaching procedure for cottonseed-oil is described in IS 543–1968 using two grades of washed and one grade of semi-refined oils. Indian standard specifications for bleaching earth are shown in Table 4.3

Table 4.3. Indian standards specification IS: 1965–1961 for bleaching earths of Indian origin.

Characteristics	*Acidic*		*Neutral*
	Grade 1	*Grade 2*	
Moisture % by wt., max.	15.00	10.00	10.00
Bulk density, g per ml, max.	0.9	0.9	1.2
Screen analysis: retained on IS seive 12% by wt., max.	15.0	20.0	25.0
Passing through IS seive 5% by wt., max.	40.0	40.0	40.0
Activity as H_2SO_4, % by wt., max.	0.2	0.2	Nil
Alkalinity as NaOH, % by wt., max.	Nil	Nil	0.4
Decolourising power, %, min.	80	60	30
Oil retention, % by wt., max.	40	35	30
Test for filterability, time in min., max.	15	10	8

Active carbons and bleaching earths

There are absorbents prepared from a range of carbonaceous materials by treatment at elevated temperature with air, steam or metal salts. Low-grade coals, lignites, woods, rice hulls and coconut shell, bitumines, peat, paper-mills waste and bones are used for its manufacture. The process of making active carbon is quite simple. The carbonaceous material is carbonised in absence of air, then crushed and screened to desired particle size of about 30 mm and activated at high temperature in presence of air or steam.

The bleaching earths and active carbons remove the colouring matter from the oil in similar way. The important factors are presence of large surface area of irregular structure and of surfaces which retain certain types of molecules. The colour is removed in various ways. It may be mechanical filtration, absorption of pigments by bleaching agent. Montmorillonite absorbs the colours by flexible swelling layer while altaputgite colour is absorbed by empty gutters and channels of which it is made. Colour can also be removed by oxidation. In acid-activated earths, the metal ions are partly replaced by hydrogen ions which re-exchange for metal ions. Active carbons have strong acidity and base exchange capacity. The most important takes place by physical absorption or chemisorption by residual surface or Van der Waal's forces.

Bleaching procedures

Batch bleaching and vacuum bleaching

Bleaching of refined cottonseed oil is carried out by treating oil with bleaching clay, using activated carbon as an adjunct to the clay at high temperature. In India, normally bleaching is done in a tank or vacuum bleacher. The tanks vary in size and are equipped with steam coils and mechanical agitators. The tanks are cylindrical, with shallow cone bottoms through which oil-clay mixture is pumped to the filter processes.

The vacuum bleachers can bleach higher quantity of oil and are equipped with steam coils and mechanical agitators. A vacuum of 27–30 mm is maintained by two stage ejectors or vacuum pump. Before bleaching, refined oil is dried and it is heated to around 104.5°C under continuous stirring to mix up the oil with earths (0.5 to 2 per cent or even more). After completion of the process, oil with clay is passed through filter.

After filtration, the residue contains sufficient quantity of oil (20–40 per cent), which should be allowed to cool to avoid oxidation and then oil is recovered.

In vacuum bleaching, vessels should be strong enough and completely air tight. Some absorbent is added in stages instead of adding all at a time. After vacuum bleaching, oil is cooled and vacuum is discontinued and oil mixture-clay is filtered through filterpress. Filterpress or leaf filters are used under pressure and bleaching oil is stored in suitable tanks.

Bleaching of cottonseed oil in solvent

Bleaching of cottonseed oil is also reported in hexane. It is reported that this process gives better oil colour than that obtained in absence of solvent. The cottonseed oil dissolved in hexane gives 30 per cent miscella, which is bleached by activated clay and carbon. At about 25°C, the process improved the absorption capacity of pigments to three-fold. The colour reduction is dependent upon the concentration of oil. With the decrease of cone of oil, colour decreased. Counter-current bleaching of cottonseed oil hexane in packed columns was found very efficient regarding utilisation of the absorptive capacity of a given quality of clay and loss of oil by absorption. Such columns of spent bleached clay could be regenerated easily with acetone. The colour of bleaching oil varied from 2.8 to 3.8 lovibond red.

Continuous bleaching process

There are number of modifications in this process. In one of the processes, the absorbent is mixed in desired proportion with the oil at 55°C to form slurry, which is sprayed into upper portion of a two-compartment tower kept under a vacuum of 30 mm. It removes air and moisture. The slurry is passed through heat exchanger, where it is brought to the desired bleaching temperature (100–110°C) and sprayed in lower section of the tower to release bound moisture. After some time, slurry is pumped through filter process to remove earth and from there oil is passed to storage tank. The vacuum-bleached oils are found better in stability than open-kettle bleached oils. It has also been found that besides 10–20 per cent higher stability, peroxide value is lower and flavour stability is higher than that of open-kettle bleached oils.

Other advantages claimed for some of the continuous processes are: lower operation cost, improved oil quality, lower space requirement, automatic operation and cleanliness of plant.

Bleaching of oil by alumina

The oil with red coloured bodies produced by oxidation cannot be bleached to desired colour by activated earths. It is observed the activated alumina has been found superior absorbent for removing such pigments (red colour) from cottonseed oil. The conditions required for optimum bleaching with alumina are: (i) about 225°C temperature; (ii) a moisture content of 10 per cent; and (iii) contact time of 5 minutes. Bleaching temperature (100–120°C) with earth was also found effective in removing cyclopropenoic acid and undersirable pigments from off-coloured cottonseed oils, which are specific to Halphen test. It is also found that the mixture of equal parts of activated and sulphurous acid treated alumina is excellent to obtain all the desired effects ofbleaching, viz., removal of red, yellow and green pigments and reduction in the intensity of the Halphen test.

Recovery of oil from bleaching residue

Around 35 per cent of oil remains in filterpress residue. The loss of oil is a costly affair in bleaching process. In view of this maximum recovery of the oil is essential. The normal method of recovery of oil is to mix the residue with ground oilseed and pass it through extraction or crushing process. Avoid long storage of oil, pass it immediately for recovery of oil, after removal from filterpress. The following methods are employed for the recovery of oil.

Deoiling with surface-active agents

In this method residues are treated with surface-active agents mostly alkali solutions at boiling temperature to recover about 80 per cent of the oil. The residue is mixed with an equal quantity of hot water in an open tank and sufficient caustic soda is added to yield a lye of 2° Be strength. It is steam heated, then 10 per cent salt is added and the contents are boiled for 30 minutes, then hot water (three times of the material) is added to the tank, it is then allowed to settle and oil is collected from the surface of the mixture. Sometimes it is advantageous to use a mixture of alkali, sodium silicate and soda ash.

Solvent extraction of bleaching residues

This method gives higher recovery of oil (about 95 per cent) than other methods. The stationary as well as rotating cylindrical types of plants are used for solvent extraction. These are similar to oil extraction plants. It is preferable to use two stage extraction process. In first stage, higher quality of oil may be removed followed by remaining darker oil. Benzene is the best solvent. To avoid the extraction of colouring pigments, extraction is carried out at lower temperature. The residue is allowed to settle after extraction and miscella is decanted and filtered. Flow sheet of cottonseed oil mill is shown in Fig. 4.10, and flow chart of oil refinery in Fig. 4.11.

Hydrogenation

The purpose of hydrogenation is to modify liquid oils to solid or semi-solid products to increase the stability of the products to oxidative deterioration or flavour reversion. The hydrogenation reduces the contents of the most unsaturated acids, dience or triene and improves the stability. It improves the colour of fat. It finds applications in the edible fat, soap and industrial oils. It has provided better substitute 'margarine' or better products with some desired properties. The addition of one molecule of hydrogen to saturate a double bond causes a drop of one unit in iodine value. To reduce the iodine value by one unit, 100 kg of oil will require about 0.8 per cent gram of hydrogen. During hydrogenation to vanaspati, the iodine value of cottonseed oil is lowered to around 30 units. The normal hydrogenation catalyst is nickel.

Characteristics of reaction

The oil is hydrogenated at about 180°C under low agitation with hydrogen at low pressure (30–50 lb/m^2) using very small amounts of nickel catalysts (0.02–0.1 per cent), in the form of metal precipitated from nickel compound on to a carrier. The fat is allowed to melting print of 37°C. The main changes occur during the process are the conversion of 20 per cent of linoleic acid to *trans* monoenoic acids and transformation of 15 per cent of *cis* monoenoic (oleicacid) to *trans* monoenoic acids. A *trans* double bond causes slight change in the linear disposition of fatty acids, thus the molecule can pack tightly like saturated acids increasing the melting point; a *cis* double bond causes a big change in the molecule, leading to low melting point.

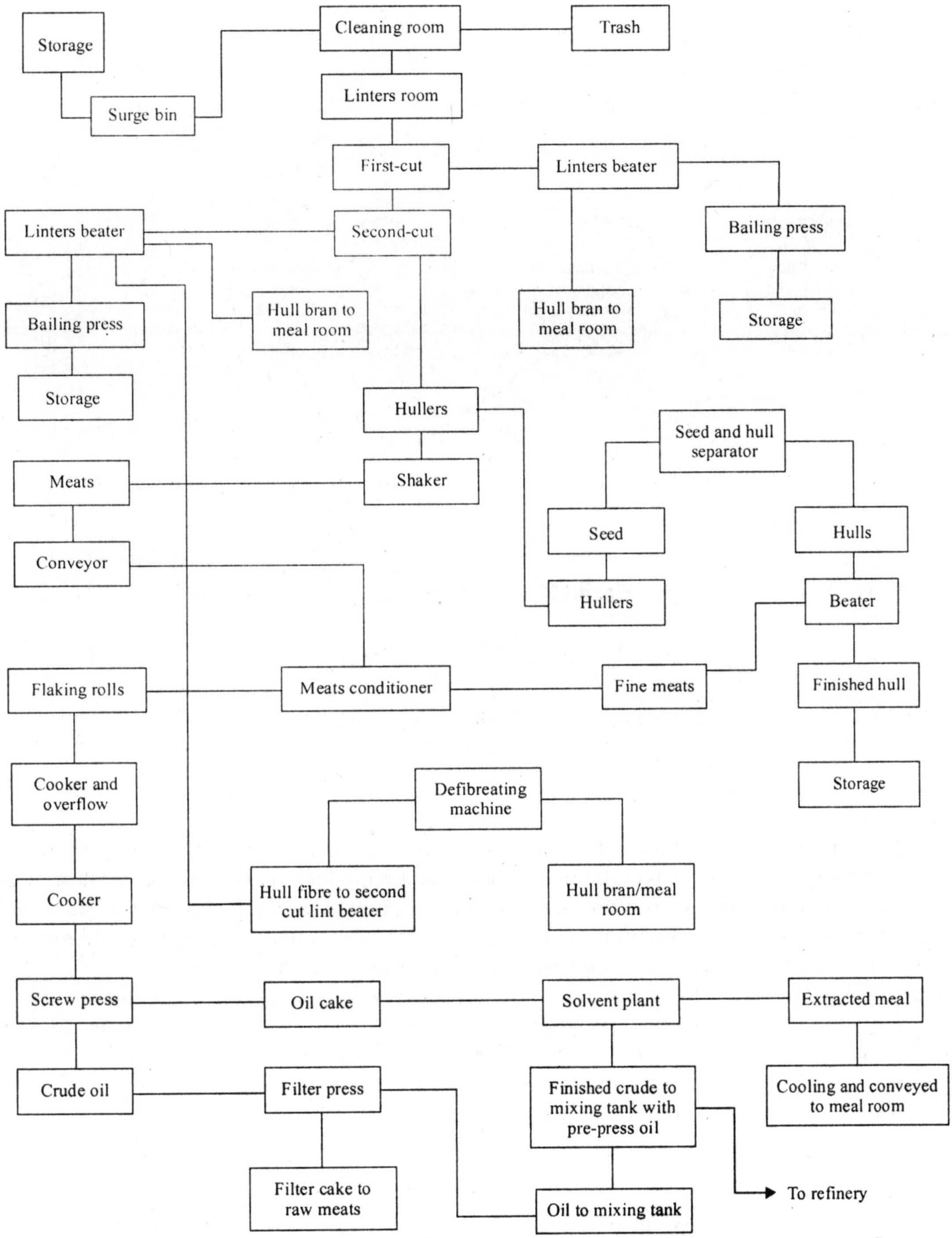

Fig. 4.10. Flow sheet of cottonseed oil mill.

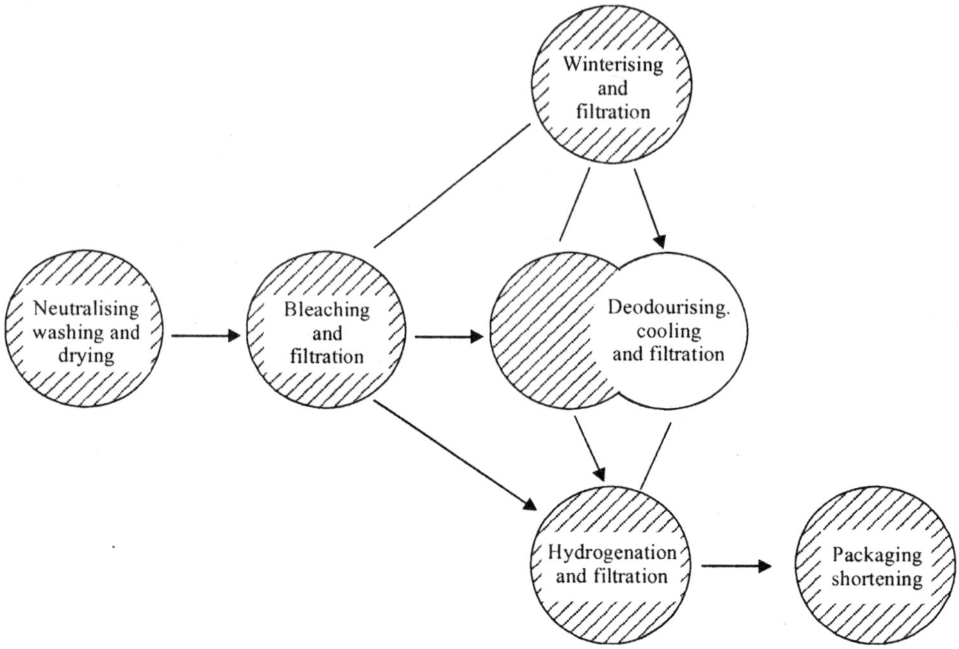

Fig. 4.11. Flow chart of oil refining.

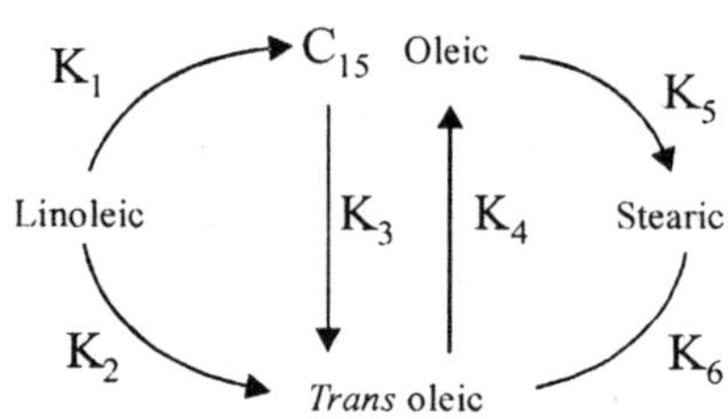

In vanaspati, the content of diene falls markedly, whereas the monoenoic acids do get saturated. In view of this, the process is selective. Hydrogenation saturates linoleate and not oleate. In other processes, both types of selective and *trans* monoene take place. In non-selective hydrogenation both linoleate and oleate are reduced in unsaturation. The hydrogenation of oil containing linoleic and oleic can be shown in six reactions. These *trans* reactions may be of different rates.

It is seen that several reactions take place during hydrogenation process of unsaturated fats. Saturation of oleate to stearate or semi-saturation of linoleate to yield 9 or 12 *cis* octadecanoate is one reaction. The migration of double bonds from oleate in linoleate is accompanied by change in configuration of the double bonds from *cis* to *trans*, which indicates lower energy. Also the *cis* 9-octadecanoate can give rise to *trans* 8–10 octadecanoate and these may give rise to various forms for carrying double bonds in place of 9 or 12 positions.

Catalytic hydrogenation

The catalyst which is normally employed in very small quantity alters the rate of reaction without any change in energy factors of reactions and without any change in its weight and will not take part in the reaction. The metallic catalysts used in such reactions are nickel, palladium and platinum. A common feature of these catalysts is that they possess high electron affinities and can accommodate hydrogen atoms to be used for hydrogenation. The heterogeneous process reaction takes place on the surface of catalyst. In view of this very finely divided catalyst with large surface area are found more effective. In hydrogenation process it is believed that heterogeneous reactions occurs as follows:

$$H{:}H + M \longrightarrow HM + H^+$$

where, hydrogen atom is hydrogenating. The nickel surface dissociates the molecule of hydrogen. The molecule of methyloleate left chemisorbed on the surface of catalyst can react with a hydrogen atom to give free radicals which are partially saturated with free valancies:

$$CH_2CH_2{}^{10}\ CH^9CH_2-$$

$$-CHCH^{10} = CH^9CH^{+H} \longrightarrow CH_2CH_2^{10}CH^9CH_2$$

or

$$-CH_2CH^{10}CH_2^9CH_2 \longrightarrow$$

Similar changes occur with linoleate. In final stage of hydrogenation process all the groups are transferred from the surface of catalyst to the oil.

Batch hydrogenation

This is most common process in India as well as in other countries. The hydrogenation is carried out in batch in convertor or autoclave to obtain a thorough mixing of the contents—the oil, catalysts and hydrogen—to obtain good dispersion of hydrogen within the oil. Batch hydrogenation plants are of two types: (i) recirculation system; and (ii) dead-end system (Fig. 4.12). Among the two systems, the dead-end system is more commonly used.

Recirculation system for hydrogenation

The reactor is tall, cylindrical of large capacity, to hold 20 tonnes oil. It can stand 10.54–14.06 kg/cm^2 pressure, provided with agitator suitable devices of heating and cooling and distribution of hydrogen in oil. It is also equipped with external blower or compressor of large capacity for withdrawing and circulating a large volume of hydrogen from the head space of the reactor. The reaction is carried out slightly above the atmospheric pressure. The converter is filled with oil and catalyst is added to it in sufficient quantity. On this hydrogenation takes place during the heating period and continues till the oil is reduced to the desired iodine value. To maintain the temperature within limits, water is admitted to cooling coils. Hydrogenation is stopped on completion of the reaction. The finished product is pumped out to filterpress and catalyst is removed.

Dead-end batch hydrogenation system

In this system, hydrogen gas is stored in convertor in the head space above the oil. The convertor is a closed, vertical, cylindrical pressure vessel of high quality stainless steel. Size of vessels varies from 5–20 tonnes and pressure 7.03–10.54 kg/cm^2. It is also provided with cooling and heating arrangements and agitator. In this process, refined oil from measuring tank is added to the converter under vacuum. The agitation is started and catalyst is added in required quantity to the oil. Oil is regularly deaerated. After oil is filled in the converter, it is heated with steam to remove the moisture and volatile impurities. When temperature reaches 135°C, vacuum is closed and hydrogen is added in the vessel till required pressure is formed. The hydrogenation process is exothermic and takes place rapidly.

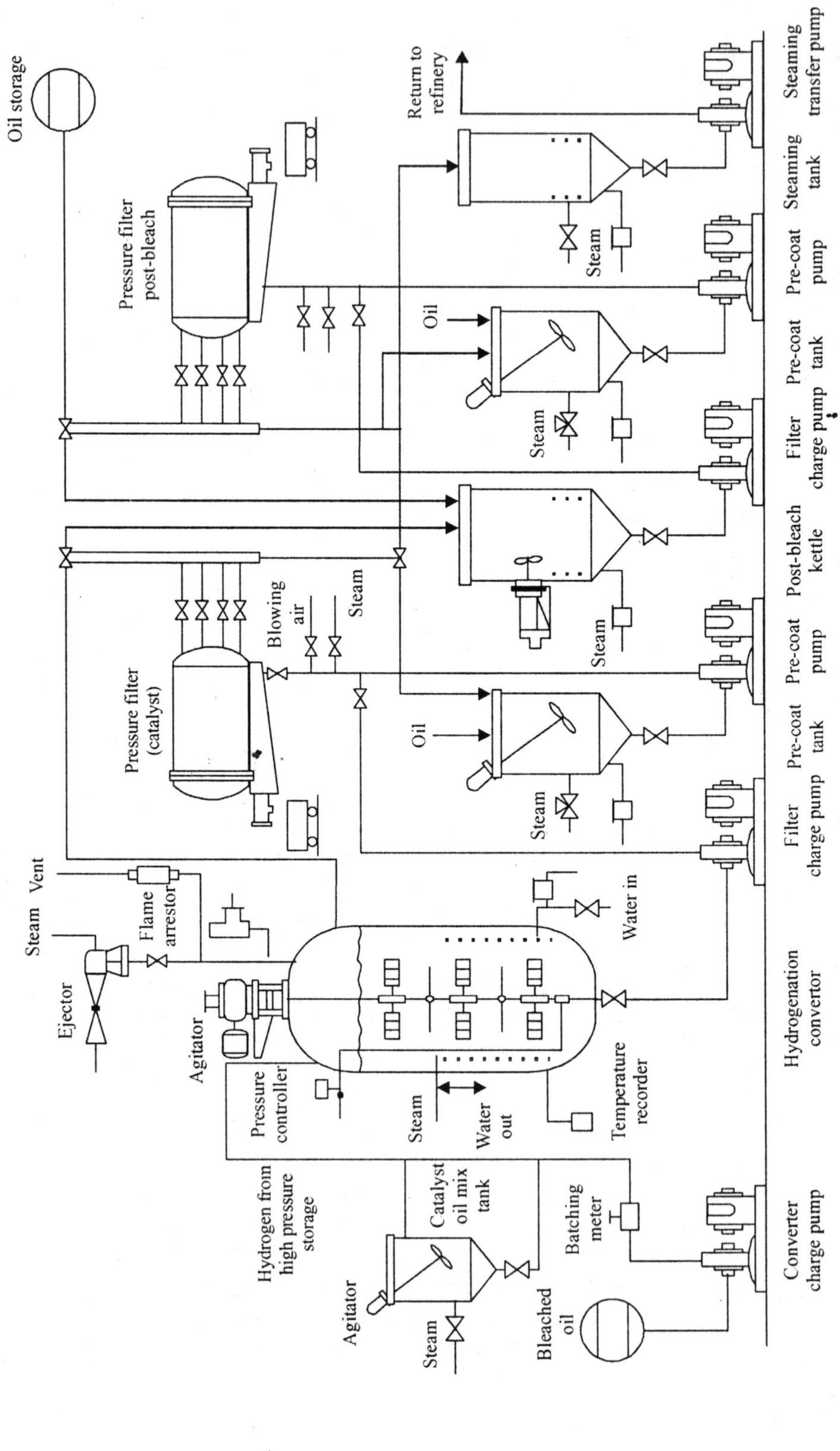

Fig. 4.12. Batch system hydrogenation plant, dead-end type.

Hence, temperature of oil rises due to nature of reaction, steam is stopped and temperature is controlled by circulating cold water in the coil. After desired hardness is achieved, the vacuum is removed, the cloudy oil is passed to the converter back.

The material processed is cooled to about 60°C, filtered to remove catalysts, filterpress remains is sent to the mix-tank for re-use.

Sometimes it may be necessary to post-refine, post-bleach or post-deodourise the hydrogenated product. In case FFA is too high, then it is necessary to remove it with alkali treatment afresh. Sometimes even post-bleaching may be required to remove nickel which might be present over there.

Continuous hydrogenation

The process has not been very popular due to its inherent problems. However, number of processes are in use in different countries. In one of the processes the oil and hydrogen are passed through heated vertical tubes in which catalysts are kept in series. The oil is kept in closed hydrogenated vessel and falls over catalyst at 170–180°C, where hydrogen is added at the base of the vessel. In the tube, oil gets partially hydrolysed. Hydrogenation is controlled by the speed of passage of the oil through the tubes and the gas pressure of 10.5 kg/cm^2. The process can be operated by drip method in which oil drips down the catalyst or by overflow method.

Solvent hydrogenation of cottonseed oil

Miscella refine has been found advantageous in many respects. Hence, it will be economical to hydrogenate the refined oil while solvent is present. This will save considerable amount of energy as the reaction can be carried out at lower temperature. Other advantages of this process are less mass transfer, resistance of the reactants and products, lower oil loss, and easiness in filtration. A process using solvent has been patented which indicated higher selectivity and less isomerisation. In case of solvent hydrogenation at low temperature, solvent should be inert and volatile for effective process. Such solvent are methanol, ethanol, isopropyl alcohol, cyclohexane, acetone and ethyl ether. Studies on kinetics of the solvent hydrogenation of cottonseed oil indicated most rapid rates at hydrogen pressures of about 600 mm of mercury.

Evaluation of hydrogenated products

Generally samples are evaluated for their iodine value and refractive indices, besides some other test such as congeal point and the cloud point (micropenetration). These tests indicate that the end point has reached. Based on this hydrogenation is considered complete and further processing is stopped.

Test methods for determining hydrogenation

Various test methods for determining the hydrogenation are brief discused here:

1. Iodine values and refractive index: A refractometer with thermostatic temperature controlled bath at 40°C is used for refractive index tests. Refractive index of oil is reduced on hydrogenation. Hence it is convenient method of measurement. Each oil has characteristics iodine value-refractive index curve.
2. Melting point: Melting point through capillary tubes are determined. These also give some indication of the completion of the process.
3. Congeal point: This method indicates the solidification temperature of the fat. In this method, melted sample is cooled with stirring until the fat becomes cloudy. The sample is then allowed to be at 20°C. Then temperature rises and maximum temperature is congeal point of the fat.

Modified hydrogenation for specific fats

Depending upon end use of product and its requirement by the consumers, the hydrogenation process has to be modified to suit the test for individual products. Cottonseed oil is straightaway converted to various products like margarines, shortenings, stearins (hard oils), cocoa, butter and so on. In case of margarine, oil is hydrogenated to produce product similar to butter. The margarine, therefore, has to be made as soft as butter and firm at room temperature. In view of this, margarine oils are hydrogenated at lower pressure 0.035 kg/cm^2, temperature up to 205°C and special catalyst to form acid from 0.10–0.15 per cent which helps in formation of large amount of *trans* acid and lower percentage of high melting fat. This imparts required firmness.

Similarly shortenings are used for vegetable oils in place of products like lard and tallow. The shortening for cakes and similar products requires two iodine values of about 65–80. These should be soft and plastic at ambient temperature. In view of this the hydrogenation process has to be selective. These require two-stage operation. One to minimise *trans* acids at low temperature of 93–149°C and low catalyst concentration of 0.05–0.10 per cent using selective catalyst, which will not form *trans* acids. It is also possible to add hard stearin hydrogenated product.

Characteristics of hydrogenated fats

In recent years various products of hydrogenated fats like margarine have become very popular throughout the world. Studies on various make of samples on fatty acids composition showed *trans*-acid content of 18–55 per cent. The soft margarine and blends with non-hydrogenated oils showed lower *trans* fatty acids content than all hydrogenated margarine.

Cottonseed oil vanaspati industry in India

The cottonseed oil or other vegetable oils after refining, bleaching, hydrogenation and deodourisation are called vanaspati in India. The use of cottonseed oil in vanaspati was made mandatory by the government. This prompted the industry to use large quantity of cottonseed refined oil for production of vanaspati. This also helped the cottonseed crushing industry to boost its production and make available sufficient quantity of cottonseed oil for the manufacture of vanaspati. The All-India Cottonseed Crusher's Association has been actively involved in promoting the activities of production of cottonseed and its products. Complete specification of vanaspati are to be strictly followed in maintaining the quality and composition of vanaspati produced in India. Cottonseed oil in certain percentage along wtih other vegetable oils is permitted for production of vanaspati. Addition of vitamin A is an essential requirement.

The vanaspati produced in India contains large amount of groundnut-oil (75 per cent), as much as 20 per cent or more of cottonseed oil and 5 per cent of sesamum-oil. Percentage of cottonseed oil can vary from 15 to 50 per cent. Both dead-end and the re-circulation systems of hydrogenation are practised in the Indian industry. In vanaspati manufacture, groundnut-, cottonseed and soyabean-oil are hydrogenated to some degree by straight hardening processes.

Vanaspati in India is mainly used for edible purpose for cooking. It is carefully processed to produce the products to look similar to 'ghee'. Vanaspati is used for the production of bakery products and same oils as started above are used in these products.

Deodourisation of cottonseed oil

The process used to remove products causing undesirable odours in fats and oils is known as deodourisation. The process of deodourisation is followed after refining and bleaching of the oil or hydrogenated product (vanaspati or margarine). Deodourisation removes odour-producing volatile products, destroys peroxides, carotenoid pigments, removes aldehydes from the oils, reduces the FFA of the oil and removes the earthy odour left after bleaching.

The colour of vegetable oils is appreciably reduced by deodourisation as the carotenoid pigments present in oils are unstable to heating. In cottonseed oil 10–30 per cent of the red colour is lost on deodourisation, which depends on the temperature as well as time of deodourisation. Maximum deodourisation may take about an hour at temperature 227–238°C.

Several methods besides steam deodourisation are reported for edible oil. The free fatty acids of an oil are completely removed by deodourisation. The oil is charged into the vessel and vacuum is created within a few millimetres of that required and the charge is heated to about 232°C. The temperature, vacuum and flow of steam for 2–6 hr. are maintained. The deodourisation depends on the types of oil to be processed. After the completion of the process, the fat charge is cooled and filtered.

Basis and fundamentals of deodourisation

The deodourisation of oil is a distillation process and is based on the laws covering a steam-stripping operation. In this process, concentration of the undesirable materials found in most fat-oils is small (0.2–0.4 per cent) which include FFA, aldehydes, ketones and peroxides. The aim of this process is to reduce 1 per cent or lower than the original content. Removal of such constituents by deodourisation depends on the following:

1. Vapour pressure should be very low.
2. Closely mixing of the steam and the fat, where volume of steam is important.
3. The absolute pressure should be low under which deodourisation is done.
4. Temperature of deodourisation should be high.
5. Time of deodourisation, which is dependent upon the absolute pressure and temperature.

Different processes of deodourisation

There are number of processes used for deodourisation of cottonseed oil. Out of these the main processes are: (i) standard and modified batch process—most important in India; and (ii) semi-continuous and continuous process—common in advanced countries.

The deodourisation plant comprises deodourisation vessel along with auxiliary plants including vapour traps and coolers plant for production of vacuum and condensation of vapours.

Batch deodourisers

This is the most commonly used process in India for deodourisation of cottonseed oil. Vertical cylindrical type deodourisers are most commonly used. About 10 tonnes of vessel, with height being twice of the diameter is employed. Oil should be filled only half the capacity. Cold oil is introduced into the deodouriser. It is heated to get required temperature and then cooled before passing to storage tank. Cycle of operation for vanaspati in India is as follows:

Filling and heating	0.25 hr
Heating, steam stripping	1.25 hr
Holding steam stripping	5.00 hr
Employing, external cooling and filtering	0.75 hr
Internal cooling, steam agitation	0.50 hr
Total	8.00 hr

Time cycle may be adjusted on needs. The principal component part consists of deodouriser with heating coils, a vacuum system to provide low absolute pressure, means to provide high temperature heat, internal water cooling coils for reduction of oil temperature. Steam should be saturated while being passed through deodouriser.

Other system uses a drop tank below the deodouriser. In this, hot air is drawn from the deodouriser into a separate vessel which can be utilised as a cooler. The main advantage of batch-refining system is its easy operation. It can be utilised for long and short durations.

Semi-continuous deodourisers

This type consists of tall cylindrical shell, which is fitted with 5 pans of stainless steel. In the top pan oil is heated by steam which deactivates the feed. In the second pan, oil heating is continued by means of condensing dowtherm. The third and fourth pans are not heated and fifth provides cooling under the same vacuum used for deodourisation. Heating coils are provided to pans to compensate for the heat loss due to radiation and other causes. Holding time in each pan is about 30 minutes. Stripping steam is passed to all the pans. The main advantage of this process is that all the pans of deodouriser are under the same but high vacuum and the oil gets uniform treatment. The oil after deodourisation is collected continuously through filter. Such plants have flexibility for different oils.

Continuous deodourisers

There are several modifications in continuous plants used by different companies. One of the plant is based on bubble cap column where oil and stripping steam are brought into contact. In this, vacuum of 6 mm is provided at top of the tower, which drops between the trays and finally increases to about 20 mm at the bottom. Owing to short retention time in the tower which is about 20 minutes, the system is operated at higher temperature 230–250°C. Similarly in another system series of vessels are used under high vacuum and oil is allowed to pass from them controlling temperature, vacuum and steam in each vessel. In another plant, oil is continuously charged to a deaerator under high vacuum, where stripping steam, is passed at desired level. Volatile matter and moisture are removed at low temperature.

The continuous deodourisers have advantage of having most of the heat needed to heat the oil for deodourisation purpose. Also steam keeps continuous removal of undesirable volatile substances. Similarly super deodourisers are also in use where better economy is claimed. In this system, 550 lbs of steam is required for super heating every tonne of oil. The advantage of the plant is that less stripping steam is required and its quantity is controlled. The comparative study of different systems indicate clear cost saving in continuous process over the batch process.

Finishing treatments

Care should be taken to prevent oxidation of oil, as unsaturated bonds in fats and oils are susceptible to oxidation. In fatty acids like linolenic and linoleic, the presence of methylene groups between the two double bonds continuously increases suceptibility to oxidation, which gives rise to rancidity and it becomes a chain reaction. The prevention of oxidative rancidity is controlled by altering the nature of oil or fat by processes like hydrogenation. Also excess heating should be avoided by keeping temperature as low as possible. Oxidised oil should not be allowed to collect on the sides of the tanks. Use of copper, brass, copper alloys and iron vessels should be avoided in handling the fats. Also synthetic antioxidants can be added to enhance the shelf life of fat products. Some of these products are butylated hydroxyanisole (BHA) and propylgallate, which should be used very carefully.

Utilisation of Cottonseed oil

Cottonseed oil is one of the most important constituents of cottonseed. Owing to shortage of edible oil in the country, cottonseed oil has become very important commodity. The cottonseed oil production has gradually improved and at present most of the cottonseed oil available in the country is fully utilised for edible purposes and other industrial applications. The world production as well as production in India of cottonseed oil has increased significantly in recent years. Most of the refined cottonseed oil finds uses in Indian vanaspati industry. The properties of cottonseed oil are standardised and the Bureau of Indian Standards has developed specification for various grades of oils.

Properties of cottonseed oil

The properties of cottonseed oil are useful to trade, producers and consumers. There are four grades of oil among express oil as well as solvent-extracted oils. In case of expressed oils, there are refined, washed grade 1A, washed grade 1B and raw grade 2. The refractive index is 1.4630–1.4660 at 40°C, specific gravity is 0.910–0.920 at 30°C, saponification value 190–198, iodine value 98–110. Moisture and insoluble impurities for the four grades of oil are 0.10–0.25 per cent and unsaponifiable matter 1.5–2.0 per cent. Colour except for raw oil, other three grades are: refined washed grade 1A and washed grade 1B should be colourless expressed as Y + IOR unit on lovibond.

For the four grades of refined, semi-refined, raw grade 1 and raw grade 2, the values of refractive index, specific gravity, saponification and iodine values are the same as those for expressed oils. The levels of moisture and insoluble impurities in order are 0.10, 0.25, 0.75 and 1.5 per cent maximum and unsaponifiable are 1.5, 2.0 and 2.5 per cent maximum. The acid values are 0.3, 0.75, 10.0 and 25.0 maximum. The maximum colours are 14 and 35 for refined and semi-refined grades.

Physical properties

Some of the important properties are discussed below:

Density and solidification

Due to presence of certain percentage of saturated fatty acids in cottonseed oil, it solidifies partially on storage below 10°C. The density of cottonseed oil at 15°C is between 0.912 and 0.922. Hydrogenation reduces the density of cottonseed oil.

Melting point, congeal point and titre

Melting point of cottonseed oil is reported to be 5–10°C. However, for oil having different iodine values after hydrogenation give different values. Titre is the useful property to find the solidification point of the fat. Cottonseed oils due to high percentage of saturated fatty acids have higher titre than most of the oil of some iodine value. The normal range of titre is 33.5–35.5°C.

The congeal (setting) point which is the solidification point of the fat determined under specified conditions is not useful or easily determined characteristic of the unhydrogenated oil. The value will depend on the method of estimation. Margarine oils are hydrogenated to a congeal point of 26–32°C.

Cloud and pour points, cold test

Cloud point of cottonseed oil as per ASTM method varies from 0.5–4.5°C. In AOCS cold test, the oil sample is immersed in ice and water at 0°C. Cottonseed oil clouds within an hour. Winterisation of the oil and stearin reduce the cold point to –5.5°C to –3.3°C and the pour point to –7.8°C to –2.8°C and increases the cold test to 8–20 hours.

Solubility and miscibility

Cottonseed oil is miscible with most of the organic solvents at ordinary temperature. At low temperature, it is immiscible in some polar solvents such as acetone and methyl acetate. Cottonseed oils readily dissolve 4–10 per cent of their own volume of air at room temperature.

Thermal properties

The heat of combustion of raw or hydrogenated cottonseed oil is similar to other oils (9.5 kcal/g). The values of specific heat vary at different temperatures. Specific heat of hydrogenated cottonseed oil varies from 0.270 (solid at 69.8°C) to 0.524 (liquid at 70°C).

Smoke, fire and flash points

The properties of the oil depends on the FFA content of oil or fat. A refined oil with very low FFA content (0.01–0.03 per cent) has a smoke point of 221–233°C, a flash point of 324–329°C and a fire point of 357–363°C. A crude oil with FFA content of 1.8 per cent had a smoke point of 145°C, a flash point of 293°C and a fire point of 356°C. IS:543–1968 has indicated flash point of 250°, 125°, 100° and 90°C (minimum) for the four grades of solvent-extracted oils, i.e. refined, semi-refined, raw grade 1 and raw grade 2 respectively.

Refractive index

Refractive index of cottonseed oil with an iodine value (IV) of 108 is 1.4724 at 20°C, 1.4648 at 40°C and 1.4572 at 60°C. A difference of about 0.00038 in refractive index is found by a difference of 1°C temperature. The normal range of refractive index is 1.4630–1.4660 at 30°C.

Chemical properties

Chemical properties of cottonseed oil are very important. These determine the main value of the fat and its utility.

Iodine value (IV)

The (IV) is the number of grams of iodine absorbed by 100 g of oil, and is a well known measure of the unsaturation (double bond) in the oil. Several studies have been reported on (IV) of Indian cottonseed oil. The (IV) of crude cottonseed oils (82 samples) and washed cottonseed oils (170 samples) indicated the range of 100–110).

Saponification and related values

The saponification value (SV) is the number of milligrams of potassium hydroxide required to saponify one gram of oil, and is measure of average molecular weight of glycerides in the oil. The average SV and saponification equivalents are 180–198 and 283–295.

Acetyl and hydroxyl values

These values measure the free hydroxyl groups present in oils. The acetyl value is the number of milligrams of potassium hydroxide required to neutralise the acetic acid obtained by the saponification of one gram of acetylated oil. The hydroxyl value is calculated from the weight of the unacetylated oil and is defined as the number of milligram of potassium hydroxide equivalent to the hydroxyl content of a one gram sample. Most oils should show 'O' value if these are ordinary triglycerides. The acetyl value or hydroxyl value of refined cottonseed oil varies from 8 to 14.

Neutral oil content

This is an important property and determines both refining capacity and the yield of neutral oil from refining. The American Oil Chemists Society (AOCS) has prescribed a chromatographic method as its official procedure for determining neutral oil content.

Glycerol content

In refined or neutral cottonseed oil, glycerol content varies from 10.4 to 11.0 per cent (10.7 per cent average) on the weight of the oil.

Unsaponifiable matter

Unsaponifiable matter of oil is determined by selected analytical methods which include sterols, tocopherols, aliphatic alcohols, hydrocarbons, terpenic alcohols and other products which after treatment with alcoholic potash are insoluble in water and soluble in petroleum ether. The unsaponifiable matter in crude and refined Indian cottonseed oils are found within 2 per cent and 1.5 per cent respectively, whereas in American cottonseed oil its percentage is lower (0.5–0.7 per cent).

Keeping quality

The most important test for quality of an oil is the keeping quality. This measure the degree of hydrolytic attack to which the oil has been subjected when present in the oilseed. This is oxidative rancidity for which peroxide value is measured. This is expressed in milli-equivalents per kg of oil. Peroxide value is determined by liberation of iodine from KI, which is estimated by titration. It has been found that refined and deodourised cottonseed oil of good quality takes 10–12 hr. to develop a peroxide value of 30 in the swift stability procedure. Studies on Indian cottonseed oil indicated good stability and very low rancidity. In detailed study following peroxide values for crude alkali washed, refined bleached and deodourised oil was reported. Also peroxide values of 165 and 175 were reported for 90 days refined and bleached oils when stored in glass containers.

Halphen tests

This test is used to find quantitatively the presence of cottonseed oil in other oils. The test is carried out as per Indian Standard method. For Halphen tests a reagent solution is prepared using 1 per cent (W/V) sulphur in carbon disulphide and then an equal volume of amyl alcohol is added. About 5 ml of the oil sample is mixed in a test tube with an equal volume of the above reagent, shaken and heated gently in hot water (70–80°C) for a few minutes with shaking until the carbon disulphide has boiled off and the sample stops foaming. The tube is placed in an oil bath or saturated brine bath at 110° for 1–2 hr. A red colour at the end indicates the presence of cottonseed oil. If sufficient quantities of cottonseed oil are present a positive reaction will be obtained within 1 hr. in water-bath or near the boiling temperature or smaller quantities longer period (2 hr.) at higher temperature of 110–115°C may be required. The Halphen test is dependent on the presence of cyclopropenoic fatty acids (CPFA) which is found in the range of 0.04 per cent to 2 per cent. Even slight hydrogenation can reduce the intensity of Halphen tests. It is an extremely sensitive test and is capable of detecting 0.25 per cent or less cottonseed oil in admixtures with other oils.

Composition of cottonseed-oil

Cottonseed-oil is similar to other fatty oils and contains triglycerides. These contain triesters of glycerol. The fatty acids present in oil determine the physico-chemical properties of the oil. The fatty acids differ in properties. Saturated fatty acids have high melting point are unreactive and without double bond, whereas the unsaturated fatty acids have low melting point, are reactive and have double bonds. The fatty acids, which are major components, are analysed by gas chromatography or high pressure liquid chromatography. The iodine values of oils are found in the range of 98–110 as reported in Indian

Standards. The average values for fatty acids are: myristic (0.5), palmitic (2.58), stearic (2.3), oleic (19.1) and linoleic (51.1) and average (IV) of 105.9. Besides, presence of other fatty acids like arachidic (0.3 per cent), linolenic (0.2 per cent), capric (0.5 per cent), lauric (0.4 per cent), myristic (0.8 per cent), pentadecanoic (0.6 per cent) and hexadecanoic (0.4 per cent) have also been found in cottonseed oils. Cottonseed oil also contains unusual fatty acids with cyclopropane ring. These are know as cyclopropenoic fatty acids (CPFA), malvalic and sterculic acids. Malvalic acid is present in larger quantity.

$$\text{Malvalic acid: } CH_3\,(CH_2)_7\,\underset{\displaystyle CH_2}{\underset{|}{C}} = C(CH_2)_6\,COOH$$

$$\text{Sterculic acid: } CH_3\,(CH_2)_7\,\underset{\displaystyle CH_2}{\underset{|}{C}} = C(CH_2)_7\,COOH$$

Sterculic acid is present in sterculia seeds in large quantity (70 per cent) and its name is derived from *Sterculia foetida* and malvalic acid from family Malvaceae. These acids give Halphen test colours which indicate presence of cottonseed oil. The detailed study on CPFA has been reported on various cottonseed varieties using hydrobromic spectrophotometric method. These acids have specific effect if cottonseed or its oil is fed to laying hens. The eggs on storage show a brown or pink dicolouration in the yolk.

Gossypol content and related pigments of cottonseed oil

Several studies have been reported on pigments content of cottonseed and cottonseed oil, which are already discussed.

Glycerides of cottonseed oil

These are present in varying percentage. The main glycerides together are present in oil upto 90 per cent, out of which are palmitodilinolein (PLL) and palmitoleolinolein (POL); about 17 per cent of oleodilinolein (OLL), and about 9–10 per cent each of dipalmitolinolein (PPL) and dioleolein (OOL).

Phospholipids of cottonseed oil

The phospholipids are mixtures containing number of products. These are lecithins, which are 1,2 diglycerides with the third hydroxyl group esterified to phosphoric acid and then to amino acid, choline and cepholins. As most of the phospholipids contain phosphorus, phosphorus is estimated and multiplied by 25 to get phospholipids. Phospholipids are present in the oil, seed and cake. In crude cottonseed oils 1 per cent of phospholipids are reported. It is reported that cooking affects phospholipids content of oil. Cottonseed oil contains about 1.90 per cent of phospholipids. Solvent-extracted oil contains higher quantity of phospholipids than the expressed oils. From hexane extract by repeated precipitation with acetone, 35 per cent phospholipids yield could be obtained.

Unsaponifiable matter

On saponification of oil or fat with alkali, it can be extracted with ether or other solvent from the diluted soap solution to find unsaponifiable matter. It consists of several colouring matters, fat soluble vitamin, sterols and their esters, hydrocarbons, aliphatic alcohols, triterpene alcohols and other constituents. The unsaponifiable matter in washed cottonseed oil is found to be in the range of 1.5–11 per cent, whereas in crude cottonseed oil it ranges from 2 to 3 per cent.

Tocopherols

Tocopherols are mixed products of aliphatic aromatic nature. There are eight types, out of which four are based on tocol skeleton and carry 3, 2 and 1 methyl group, while four are based on a tocotrienol skeleton. α-tocopherols, which is a trimethyl tocol, is also called vitamin E. It is related with reproductive processes and is important biological antioxidant which permits the utilisation of vitamin A. Tocopherols help vegetable oil, better shelf life than the animal fats. Tocopherol content of crude or refined cottonseed oil is reported to be 50–100 mg/100 g of oil.

Keeping quality of cottonseed oil

Most of the oils give rancid flavour which may be attributed to their oxidation due to dissolved air or oxygen. It is reported that high energy singlet form of oxygen at the *cis* double bond of an unsaturated fatty acids is main cause for it. This leads to loosening of the adjacent hydrogen atom, followed by migration of the 2 oxygen atoms to form a hydroperoxy free radical in which the double bond is not only shifted but is changed to a *trans* configuration. Thus in oleate:

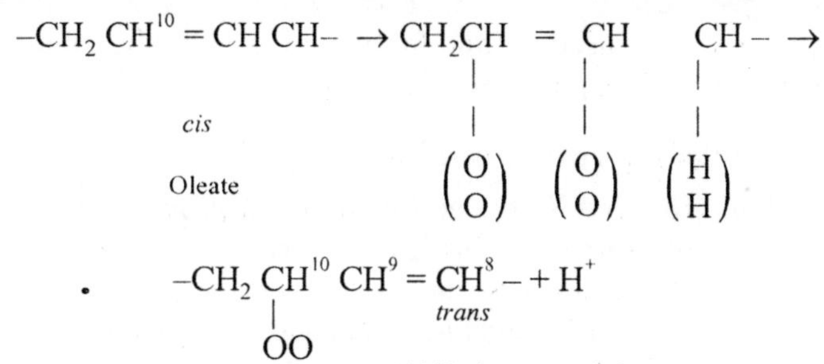

(10 Hydroperoxy 8, 9 *trans*)

Linoleate has two *cis* double bonds separated by a central methylene group.

$$\overset{13}{-CH} = \overset{12}{CH}\ \overset{11}{CH_2}\ \overset{10}{CH} = \overset{9}{CH}-$$

These groups are very susceptible to attack by oxygen than the two double bonds. Linoleate gives rise on oxidation to 2 hydroperoxy free radicals. Hydroperoxides have no odour, but their degraded products such as short-chain unsaturated aldehydes, give very strong and unpleasant odours when present in traces and they are responsible for rancidity of fats.

The rancidity of fats can be checked by introducing antioxidant. These are substances like gossypol in cottonseed oil, or can be added as synthetic compounds. These antioxidants function by liberating hydrogen atom to free radicals which terminates their activity. Synergists enhance the effectiveness of antioxidants. They often function by chelating with trace amount of metals that catalyse oxidation. The most important synergists in oil are the phospholipids. Sometimes citric or phosphoric acids are added for this purpose.

The cottonseed oil has very good stability. It contains unsaturated fatty acid: linoleic (50 per cent) and oleic (20 per cent). These give increased stability to cottonseed oil. Besides, cottonseed oil also contains small amount of tocopherols, gossypol which help in maintaining better stability and shelf-life.

Several studies have been reported on stability and keeping quality of cottonseed oil. It is observed that crude cottonseed oil has better shelf-life than refined and deodourised oil. However, comparative studies with other oils indicated that hydrogenated products gave improved and better shelf-life than other products. In recent study on comparative stability of cottonseed oil, safflower-oil, sesamum-oil, groundnut-oil and vanaspati, it was seen that cottonseed oils showed better stability in terms of peroxide values and organo-leptic properties compared to groundnut- and safflower-oil, while sesamum-oil and vanaspati showed still better shelf-life.

The stability of cottonseed oil can be further improved by adding approved antioxidants like gallates (ethyl, propyl, octyl and dodecyl), BHA, BHT, citric, tartaric and gallic acids.

Nutritional aspects of cottonseed oil

Like other edible vegetable oils, refined cottonseed oil can be easily digested by the normal people and its digestibility coefficient is about 98 per cent. The triglycerides are far more digestible than their respective fatty acids. Cottonseed oil has very high percentage of linoleic fatty acid (50 per cent), which is very good for nutrition. Several studies are reported on utilisation of cottonseed oil including different kinds of oil which indicate that refined, deodourised cottonseed oil/vanaspati compares well with other oils in this regard. Mean fatty acid content of the cottonseed oil of all the species together for 43 varieties for myristic, palmitic, palmiloeic, stearic, oleic and linoleic acids were 0.7, 24.4, 1.2, 2.3, 20.2 and 50.6 per cent respectively.

Besides cottonseed oil is widely used in food for human consumption but the possibility exists that the cyclopropenoic fatty acids (CPFA) may produce deleterious effects.

Current uses and new outlet for cottonseed oil

In India entire cottonseed oil produced is utilised for edible purposes, mostly for vanaspati, only small quantity (5–10 per cent) is used for manufacturing of soaps.

Several new products can be produced from cottonseed oil. It can be used for production of acetoglycerides for food. It is produced by substituting acetic acid for a portion of fatty acids in ordinary fats and oils to give diaceto- or monoaceto-triglycerides. These can be prepared by acetylating oil with monoglycerides or combinations of these. The reaction takes place at 110°C and excess acetic anhydride is removed by boiling with water and washing. Also the acetoglycerides can be produced which will be stable to oxidation and will solidify at temperature far below those of other fats of the same degree of saturation. These can be used for food and cosmetics.

Brominated oils

Cottonseed oil is also used for bromination. Brominated-oils are used as wetting agents and for production of cloudy flavoured soft drinks. This product is made by using a solution of the oils to which bromine is added at a low temperature with regular mixing to enable addition at the double bonds.

Fat for intravenous nutrition

Emulsified cottonseed oil can be used for intravenous administration to patients, who require a high calorie diet than carbohydrate can provide.

Tallow substitutes from cottonseed oil

Sufficient amount of tallow is imported into India for soap making and textile sizing. Cottonseed oil has been hydrogenated to obtain tallow like products.

Preparation of palmitic acid and by-product plasticiser from cottonseed acids

From cottonseed oil, purified palmitic acid can be prepared by crystallisation of its cyclohexylamine salt from acetone and subsequent regeneration of acid. Cyclohexylamine can be modified to morpholides, which can be used as vinyl plasticisers.

UTILISATION OF BY-PRODUCTS OF COTTONSEED MILLING INDUSTRY

Cottonseed-Cake or Meal

Cottonseed-cake is one of the product of cottonseed milling industry. Here cottonseed-cake is referred as expeller oil cake including both decorticated as well as undecorticated. The solvent-extracted product is referred as extraction (meal).

In the beginning of the cottonseed crushing industry production of cake was very much restricted to whole cottonseed. Even seeds were not delinted. The production of oil was only 4 per cent of the total seed produced, and gradually importance of cottonseed by-products was realised and more and more quantity of seed was available for crushing. This helped in increased production of cottonseed-cake and meal. The demand and supply position of oil in the country has brought marked change in processing and utilisation of cottonseed. In view of acute shortage and increased demand, the cottonseed crushing has increased and the practice of feeding of whole seed to cattle has been reduced considerably. In view of the realisation of feed value of cottonseed-cake, farmers prefer to use cake as animal feed. They also prefer undecorticated cake to decorticated in view of its lower price. However, there is need to motivate the farmers by convincing them that the decorticated cake is superior in protein content than the undecorticated seed cake.

Studies on cottonseed-cake

Cottonseed-meal has been found as one of the best protein-concentrate feeds for milch cattle. Cottonseed-meal is very economical ration and it can be fed to cattle as dry feed. Animals fed cottonseed-meal ration show better growth and performance than many other feeds. Normally 2.5 kg cottonseed-meal is fed to milch cattle per 450 kg live weight.

Cottonseed-hull is used as roughage in the animal feed. It can be fed up to 8 kg per day per adult animal. The best way of feeding hulls to cattle is to mix it with salt and other grains or meal in the form of slop or gruel. Cows and buffaloes eat it well. Comparative feed studies on cottonseed, sarson and tora cakes showed higher digestibility coefficients for fat and fibre for cottonseed-cake than for the others. Similarly, there are few more studies reported on feeding of whole cottonseed and cottonseed-meal to cattle and these studies give conflicting information.

Amino acid content of cottonseed and its meal

Proteins are made-up of well defined peptide unit, which consists 20 or more different amino acids arranged in several combinations. These structural arrangements modify the properties of protein products. Proteins also contain non-amino acid materials or prosthetic groups. These may be sugar, lipids or similar products which affect various properties of proteins. These proteins are classified in two major groups—amdox fibrous (wool, silk, muscle, etc.), and globular proteins (globulin, pepsin, insulin, pepsin and egg protein).

The amino acids present in cottonseed protein belong to l-system, which indicates the special structural arrangement with alpha-carbon atoms. Out of 21 amino acids, 8 are essential for the development of human body. These are threonine, valine, leucine, isoleucine, phenylalanine, lysine, tryptophan and methionine in l-form. Besides, histidine and arginine are also considered very important for the growth of the body. Data on amino acid content of Indian cottonseed-meal, amino acid analysis based on auto-analyser are given in Table 4.4.

Table 4.4. Ranges of essential amino acid composition of Indian cottonseed protein (g/16 g of nitrogen).

Amino acid	*Range*	*Mean value*
Valine	0.81–2.55	1.86
Isoleucine	1.67–2.71	2.14
Leucine	4.66–5.63	5.16
Threonine	1.67–2.27	1.99
Methionine	0.72–2.82	1.13
Phenylalanine	2.16–3.34	2.70
Lysine	2.57–3.04	2.83
Histidine	1.61–2.60	2.23

Detailed analysis of data on amino acid content of American cottonseed has shown marked variation in varieties. The results include 10 nutritionally essential amino acids. The amino acids like methionine, lysine and tryptophan are mostly deficient in feeds of animal, poultry and swine. The cottonseed-meal contains slightly higher methionine but lower lysine than soyabean-meal. It is observed that a good quantity of lysine in cottonseed-meal binds with gossypol and thus, becomes unavailable to non-ruminant animals with simple stomachs.

The cottonseed protein contains globulin 12 per cent, β-globulin 67 per cent, gluteline 3 per cent and pentose 6 per cent. This indicates that β-globulin is present in much large quantity. Amino acid content of some oil cakes is given in Table 4.6.

The solubility of protein in water and other solvents furnishes useful information. The cottonseed protein solubility in water indicates the effect on the non-protein nitrogen for constituents and solubility in 3 per cent sodium chloride solution indicates globulin fractions containing 80 per cent of total protein nitrogen. The solubility in 3 per cent sodium chloride solution is found in the range of 10 to 47 per cent (35 per cent average) compared to 75–80 per cent for solvent-extracted meal. Cooking at lower temperature improves the quality of meal protein and it has its influence on digestibility of cake. In view of this, temperature, time and moisture for manufacturing of cakes should be carefully selected. It is reported that the apparent digestibility of the proteins of cottonseed eliminating oil and gossypol by ethylether was higher than the digestibility of the proteins of the commercial cottonseed-meal.

Application of cottonseed protein also depends on various other factors such as desirable solubility properties and its colour, viscosity, adhesive strength and so on. These properties play important role on properties of artificial protein fibres, films, glues, water paints, paper and plywood adhesives. The solubility of cottonseed protein is less at acidic pH and gelation occurs at higher viscosity. The industrial application of cottonseed meal can be grouped as: (i) meal with unchanged protein molecules; (ii) with increase molecule size due to chemical treatments; (iii) with poultry degraded protein; and (iv) hydrolysed protein in amino acids.

The conditions of processing in various manufacturing units vary, which alter the protein properties. Proteins are widely used for production of paints, plastics, fibres coatings, emulsifiers, binders, adhesives, inks, coatings and films.

Protein fraction: gossypol in cottonseed-cake

Earlier work indicates that the presence of pigments gossypol causes deleterious effect on hen, swine, guinea pigs, etc. It is found that screw pressing reduces gossypol content (both free and total), lower proportion of soluble nitrogen which gives poor nutritional value. The solvent-extracted meal showed the best nutritive value although per cent of gossypol remains high.

Improving protein quality of cottonseed-cake

The most important factors for nutritive value of cottonseed-cake are the percentage of protein, its quality and its free gossypol content. Numerous studies have been conducted for improving the protein quality and nutritive value of cottonseed-cake. The major objectives of such studies have been to maintain the quality of protein and keep the free gossypol and bound gossypol within accepted limits without adversely affecting the quality of cake protein and yield of oil. The free and total gossypol of these meals varied from 0.02 to 0.04 per cent and from 0.08 to 0.95 per cent respectively. At constant moisture level and after raising boiling temperature from 93–110°C, free gossypol content and soluble nitrogen both showed lower value. Higher temperature cooking up to 137°C reduced free gossypol and also reduced soluble nitrogen. In view of this cooking of meats at lower temperature (93°C) is advocated to give better quality products, which can be fed to chicken and ruminants.

It may be mentioned that small amount of free gossypol present in cottonseed-cake can be removed by boiling with water. In view of this cottonseed protein can be used for edible purpose without any fear. As regards bound gossypol in the meal, if percentage in the original cake is low, meal can be used for human consumption after careful processing.

Physiological studies on animals did not reveal any adverse effect on dogs. It was seen that the quantity of gossypol fed was the same in excreta although free gossypol was much higher in excreta than it was fed. It is possible to inactivate the gossypol by use of stearylamine calcium hydroxide and ferrous sulphate. Further it is seen that presence of gossypol is responsible for discolouration of egg-yolk on storage.

Many other ways are suggested to eliminate gossypol from cottonseed-meal. One of the method is to develop glandless cottonseed varieties which will completely avoid the presence of pigment glands and cottonseed of such varieties will be free from gossypol. The other method is to remove mechanically pigment glands of cottonseeds so that gossypol and related pigments will not come in contact with oil and meal while processing. In other method, use of pure solvent or mixture of solvent is advocated, which can dissolve gossypol in oil such as acetone or azeotropic solvents like hexane-acetone-water.

To achieve satisfactory lysine level of about 4.2 per cent, sometimes it is necessary to supplement the meal with lysine-rich soyabean-meal or fish-meal. It is seen that low gossypol meal supplemented with lysine has produced marked improvement in growth in swine and poultry.

Uses of cottonseed-cake in rations for livestock and poultry

Cottonseed-cake or meal is an excellent cattle-feed and is very rich source of protein. It is fed to cattle as concentrate along with fodder or roughage in India and in other countries. Cottonseed-meal/cake contains about 60 per cent high quality protein and is rich in essential amino acids like lysine, methionine, trytophan, histidine, arginine and other amino acids compared to other oilseeds. Cottonseed-cake is one of the best feeds to cattle and milch cows.

The major problem with the use of cottonseed-meal/cake in the diet of non-ruminants (single stomach animals—poultry, swine, rabbits) is the presence of a number of toxic pigments such as gossypol, gossypurpurin, gossyfulvin, in addition to anthracynins and carotenes. These pigments cause darkening in the meal and also react with epsilon amino group of lysine thereby making these amino acids unavailable. However, several processes have been developed where gossypol and related pigments can be reduced to a very low level without adversely affecting the quality of protein.

Development of glandless cottonseed varieties may help further in producing gossypol-free oil and cake. This cake containing negligible gossypol can be fed to non-ruminants without any problem. Cottonseed-cake/meal is being used for feeding livestock, milch cattle from generations. In rural areas it is generally believed that feeding of cottonseed-cake/meal to milch cattle increases the production of milk and its fat content. Thus, some quantity of cake/meal is fed to cattles in India. Feeding of whole cottonseed needs to be discouraged as it contains valuable oil (20 per cent) which cannot be digested by the cattle. Recently, the National Dairy Development Board (NDDB) has formulated cattle-feed in which cottonseed extraction has been essential ingredient as a source of proteins and fat, to improve the general condition of the cattle and milk production and the technology is known as by-pass protein technology.

In USA, most of the cottonseed-meal in different forms is used for feeding cattle. Its use also depends on the market price. When it is cheap it is mostly used alone as cattle-feed in the form of cracked cake and pellets. Otherwise it is used along with soyabean-meal. Degossypolised cottonseed-meal is used for poultry and swine rations. High protein meal (50 per cent protein) is used for feeding chickens and swine. It is found that laying hens when fed cottonseed-meals give eggs, which show discolouration in yolks.

Effect of cyclopropenoid fatty acids (CPFA) on cottonseed-meal

It is seen that the presence of malvalic and sterculic CPFA results in certain biological effects in stored eggs, such as the development of pink whites, decreased pH of whites, increased pH of yolk and higher rations of stearic to oleic acid in yolk lipids and in tissue lipids. Besides, other disorders are decreased egg production, retardation of ovary, oviduct and comb development, enlargement of liver and gall bladder, increased chick embryos mortality and growth retardation of chicks. It has been found that hen fed with the ration containing 2 per cent cottonseed oil gives positive results when eggs are stored at 2°C for periods of even less than a month. The effect of cottonseed oil on eggs in storage is as follows:

0.5 per cent oil gives a positive response after a storage period of 3 months and 0.1 per cent oil gives a negative response even after the egg has been stored for 6 months. In other case, when hens fed with 2–4 mg per day of sterculic acid produced eggs, which show pink colour when kept for 3 months. Further, it is reported that if malvalic or sterculic acid is fed 250 mg/hen/day, it depressed the laying of egg within 2 days. In view of this, screw press meals are not given to hens. Similarly, solvent-extracted meals are suitable for poultry feeding.

Comparative (energy value) nutritive value of cottonseed and cottonseed-cake and meal

The metabolisable energy value of cottonseed-meal is similar to other oil meals. The energy value of 41 per cent cottonseed-meal is about 56 therms per cwt. This compares with 72 therms per cwt of 44 per cent protein of soyabean-meal. The energy value of cottonseed hulls is nil. In view of this only 41 per cent protein cottonseed-meals are used in non-ruminant rations, which has fibre value of 14 per cent. If the fibre value can be reduced to 8 per cent or lower then the energy value of cottonseed-meal will be 72 therms per cwt as compared to a value of 56 therms per cwt for 44 per cent soyabean-meal. Cottonseed-meal along with grain like corn will be suitable substitute to soyabean-meal.

Cottonseed-meal contains minerals and vitamins. These are compared with soyabean (Table 4.5). It may be seen that cottonseed-meal is fairly rich in vitamin B including thiamine and riboflavin. The thiamine content in 43 per cent cottonseed-meal is 6.13 mg/lb (14 PAM).

Aflatoxin in cottonseed and its cake and meal

The presence of highly carcinogenic aflatoxins in cottonseed-meal or cake may cause liver cancers to trouts. However, cottonseed is free from aflatoxins but it gets contaminated during harvest or storage with aflatoxins. Studies indicate the aflatoxin contamination is higher in Indian seeds, meals and cakes compared to samples of meals from USA.

Table 4.5. Vitamin and mineral contents of cottonseed and soyabean-meal.

Vitamins (mg/lb)	*Cottonseed-meal*		*Soyabean-meal*	
	41%	*43%*	*44%*	*44%*
Biotin		0.50		0.15
Choline	1301	1300	1247	935
Folic acid	0.46	0.34	1.70	0.44
Niacine	20.70	20.70	12.20	12.20
Pantothenic acid	8.10	8.10	6.35	6.60
Pyridoxine		2.50		2.20
Riboflavin	2.10	2.10	1.50	1.00
Thiamine			3.00	
Minerals				
Calcium (g/lb)	0.68	0.68	1.45	2.71
Magnesium (g/lb)		0.62		0.27
Phosphorus (g/lb)	1.63	1.36	0.99	1.08
Potassium (g/lb)		6.56		8.60
Sodium (g/lb)				
Zinc (g/lb)	26.0			11.0
Manganese (mg/lb)	9.1		12.5	

Nutritional quality of glandless cottonseed-meal

The development of glandless cottonseed varieties and glandless cottonseed-meal will go a long way in expanding the use of cottonseed-meal for feeding ruminants. Feeding trials of glandless meal, ordinary cottonseed-meals produced with and without heat treatment by extraction with hexane and hexane-acetone-water azeotrope and commercial pre-press solvent meal from ordinary cottonseed fed to chicks with 21 per cent protein diet with and without 0.3 per cent lysine indicated best results with commercial glanded types and least with glandless meals. The glandless meals were found equal to soyabean for chickens. Feeding value of glandless and glanded-seed meal was further studied in broiler diets. Glandless cotton variety was almost free from gossypol while samples contained 0.008 to 0.020 per cent total gossypol. When some glanded cottonseed with gossypol content of 0.5–1.3 per cent were processed by solvent extraction, the meal was found to contain 46 per cent protein, 11.3 per cent fibre, 0.009 per cent free gossypol, 0.044 per cent total gossypol, 2.19 per cent available lysine and 85.6 per cent nitrogen solubility. The result indicated that the diets containing mixed ratio of soyabean and glandless cottonseed-meal showed better results than either fed alone. The studies on storage protein of glandless cottonseed flour by gel permeation column chromatography indicated six fractions of protein with molecular weight in the range of 2000 to 6,00,000. Each fraction showed some similarity in amino acid content. Fraction VI was rich in lysine and arginine.

Edible cottonseed flour for human consumption

Malnutrition is one of the most serious problem in the world today. The importance of protein on development and growth of human being particularly for young children is well known. There is shortage of protein in food all over the world but it is more particularly in developing countries. Deficiency of protein in infant food retards mental and physical development. To meet the deficiency of protein in diet of children and adults the food chosen must be economical, locally available and acceptable to the people. Cottonseed protein is one of the valuable dietary supplements which has been so far neglected. Cottonseed is the most widely distributed oilseed in tropical and sub-tropical areas. Detailed studies on storage and quality of seed protein has been reported wherein wide variation has been found in seed constituents. Cotton protein concentrates/flours can be very well utilised, if it is processed carefully. Cottonseed-meal contains about 60 per cent of high-quality protein. It is rich in essential amino acids like lysine, methionine, tryptophan, histidine, ariginine and other amino acids compared to other oilseeds (Table 4.6).

Table 4.6. Amino acids in cottonseed and some other oilcakes (g/16 g of nitrogen).

Amino acids	*Cottonseed*	*Soyabean*	*Groundnut*	*Safflower*	*Sesame*
Arginine	9.02	6.04	12.38	7.78	10.04
Histidine	2.50	2.40	2.23	1.99	2.36
Isoleucine	4.32	6.32	3.89	3.85	3.56
Leucine	6.06	7.80	7.11	5.52	6.12
Lysine	4.65	6.32	4.53	2.71	2.66
Methionine	0.96	0.68	0.71	1.54	2.03
Phenylalanine	5.15	8.77	5.41	5.25	4.48
Threonine	2.86	3.79	2.84	2.94	3.49
Tryptophan	1.30	1.33	0.92	1.18	1.22
Tyrosine	3.04	3.18	3.28	–	4.70
Valine	4.58	4.86	5.50	4.93	4.82

From Table 4.6 it may be seen that the cottonseed-meal is comparable in amino acid contents with soyabean and it contains higher percentage of essential amino acids like lysine, histidine, methionine and tryptophan compared to other meals. Cottonseed protein being rich in essential amino acids and vitamins, its flour can be utilised for supplementing protein and vitamin with wheat flour and other grains which are deficient in essential amino acids and vitamins as food for millions of needy people.

For making cottonseed flour for edible grade, it is necessary that seed should be mature, free from aflatoxin and foreign matter. The seed should be delinted, cleaned and dehulled. Then it should be carefully screened and all hull including traces should be separated from meat. The cleaned meat should be rolled to a small size in thickness (0.010–0.020 cm) and cooked at about 107°C for 1–1½ hr. Then the meat is subjected to hydraulic pressing for oil removal and press cake is stored for about a month before processing it in grinding mills. After grinding the cake, flour is screened through 200-mesh sieve. In this way about 400 kg of flour containing 43 per cent of protein can be produced from 1 tonne of good-quality cottonseed. The flour thus obtained is light in colour and give nutty smell. It also contains about 2 per cent of fibre, 6.5 per cent oil and 6.2 per cent ash. These types of products find use in bakeries and confectionaries. The flour has high percentage of protein of very high biological quality and rich in vitamin B. It has good stability on storage.

There are certain undesirable inherent products present in cottonseed which needs to be eliminated carefully while processing the cottonseed protein-rich flour. Such products produce deliterious effect in the long run and can be completely eliminated or their quantity can be brought down within permissible limits prescribed. The main substances are pigment gossypol and cyclopropenoid fatty acids CPFA. Besides the seeds get contaminated with mycotoxins particularly aflatoxin.

The effect of gossypol and its adverse effect on human beings has been discussed earlier. Sometimes its presence may cause serious problem. Besides, gossypol chemically reacts with protein and reduces the availability of lysine and other amino acids like glutamic acid and aspartic acid. To avoid such problems it is advisable to select cottonseed which contains less than 1 per cent of gossypol. Further, presence of gossypol and its effects can be suppressed by feeding it along with other good-quality protein. Also if cottonseed protein is boiled in water free gossypol is eliminated. Aflatoxins are other undesirable products in contaminated seeds as discussed earlier. *Aspergillus chevaleri* and *A. flavus*, var. *rubrum* are two of the moulds which are different from the *Aspergillus flavus* link Ex-Fries. *A. chevaleri* infests hay and *A. flavus var. rubrum* infests grain and corn. Cattle, mice and swine are susceptible to it. *Aspergillus falvus* link ex-Fries infests peanuts, cottonseed and grains. Its toxins are aflatoxins, which are harmful to ducks, swine, calves and rats. Aflatoxins are found in cottonseed and its meal. Conventional alkali refining, washing and bleaching removes aflatoxin from the oil.

Chemical treatment of cottonseed-meal with ammonia, methylamine, sodium hydroxide and formaldehyde decreased its aflatoxin content and the meal was found suitable for practical use. Treatment with hydrogen peroxide has also been found effective to destroy aflatoxin from edible protein products.

Nutritive value of cottonseed flour

Vegetable protein mixtures containing 9 per cent of cottonseed flour were recommended for the supplementary feeding of young children. The mixture gave satisfactory results when fed as protein-rich diet to chicks, rat and children. Another mixture with 38 per cent of cottonseed flour was used for feeding tests. The mixture gave very good results showing satisfactory growth and feeding efficiencies, without any adverse or toxic effect. Free gossypol decreased on cooking and the mixture was considered good quality and suitable for edible use. Cottonseed protein-rich flours developed in mixture for edible purpose having 50 per cent protein of high biological value contained 1 per cent of total and above 0.05 per cent of free gossypol. These mixtures furnished about 80 per cent of the dietary protein for periods from few weeks to three months. The mixtures were called ‘Incaparina’. On the basis of extensive biological and chemical experiments, the products were recommended for supplementary and mixed feeding to young children and adults. Feeding studies on rats and children at the National Institute of Nutrition, Hyderabad and Central Food Technological Research Institute, Mysore were made. In meal-nourished children, a solvent-extracted product with low gossypol per cent gave excellent results in reducing leg oedema and raising serum albumin levels in children.

Preparation of edible cottonseed flour and its utilisation

There are several processes for production of edible cottonseed protein-rich flour. Few of them are described briefly:

Conventional process

Salient features of the process are—the seed used for making flour should be of very good quality based on gossypols and FFA contents and well cleaned free from foreign matter. Care should be taken to remove the hulls and linters completely. Rolling of flakes is desirable to thickness of about 0.02 cm.

This helps in rapid and uniform cooking by exposing large surface and better heat penetration. A moisture content of about 10 per cent, improves the rolling and helps to break some of the pigment glands. Cooking process of meats should be uniform to reduce the free gossypol to an acceptable level and avoid loss to protein quality. A satisfactory process of cooking may be done with 12 per cent moisture cooking for about 60 minutes at temperature range of 101–107°C depending on type of extraction needed. Oil extraction can be followed by any one of the standard processes such as continuous screw-press operation, pre-press solvent extraction and direct solvent extraction to provide quality products cake meal sufficient ageing for cooling before processing.

Grinding should employ suitable heavy impact type hammers to provide meal free from hulls. Separation of the fine flour is best done by air flotation method. It is possible to get flour yield up to 50 per cent without loss of protein and its lysine. Such process has been used for UNICEF projects.

The flour quality should be relatively bland in flavour. Normally low percentage of fat is desirable for better keeping quality with high percentage of protein. Gossypol content both free and total should be within acceptable limit. Processing condition in plant should take care of sanitary practices to maintain the standards prescribed for production of food for edible purpose. Crude fibre content should be carefully controlled to keep it at minimum. A process for production of edible flour from cottonseed has been described by Regional Research Centre, Hyderabad, wherein the yield of quality edible flour was 25–28 tonnes per 1000 tonnes of cottonseed processed. Better quality flour is obtained even by mild pre-pressing followed by extraction with hexane.

Processes employing unconventional extraction solvents

A process has been developed by RRL, New Orleans, USA, wherein use of azeotropic mixture of hexane-acetone-water has been indicated. Another process using acetone has been developed in Italy. Both the processes give edible protein products of excellent quality and low in gossypol.

Process by isolating gossypol

Number of processes based on separation of gossypol-rich material from cottonseed kernels are described earlier. Among these the cyclone-hexane process (Table 4.7) appears to be the best. The composition of cottonseed protein concentrate is given in Table 4.8. Modified version of direct hexane extraction process for glanded cottonseed has been described and the process indicates that uncooked flakes can be extracted to yield low residual lipids by using a low-solvent to flake ratio and ambient temperature hexane solvent.

Table 4.7. Composition of defatted cottonseed flours.

Composition %	*PAG guidelines*	*Glandless direct solvent*	*Glanded liquid cyclone process*
Moisture	10.0	9.5	3.8
Crude fat	6.0	2.4	1.2
Crude fibre	5.0	3.3	2.3
Total gossypol	1.2	0.06	0.06
Free gossypol	0.06	0.04	0.02
Free fatty acids (of oil)	1.8	–	–
Protein (N × 6.25)	58.0	54.7	66.2
Available lysine (g/16 g N)	3.6	3.9	4.0
Nitrogen solubility	–	95.7	99.6

Table 4.8. Composition and yield of cottonseed protein concentrates: method of preparation.

Method	*Air classification*	*Aqueous extraction*	
		90% ethanol	*0.008 M $CaCl_2$ H_2O*
% Composition			
Protein (N × 6.25)	73.7	71.9	75.9
Lipid	0.7	0.1	2.6
Ash (N × 6.25)	9.4	8.7	8.5
Crude fibre	1.7	2.7	3.7
Total sugar (as invert)	4.7	2.1	0.3
Yields (as in basis)			
Percentage of total weight	56	84	60
Percentage of total N_2	63	93	77

Uses of cottonseed protein flour

Cottonseed protein can play an important role in fulfilling the requirement of protein for edible purposes in the world today. It can be used as an additive in bakery products, as a filler in meat-like products and as a protein supplement to diets of cereal grains or of vegetable protein mixtures. When mixed with bakery products, the cottonseed protein flour reduces dough stickiness, increases the water-binding capacity, increases shelf-life and reduces fat absorption. If used as filler in the meat type ground products, it gives products of good nutritive value and supplements cereal diets. Also it can be used as protein supplement in various foods such as to enrich rich preparation with protein, bakery preparation, chapati made of various flours such as wheat, ragi and barley. It has been reported that the addition of 8–10 per cent of cottonseed flour to wheat flour improved its nutritive value and weight of rats. In mixture of various vegetable proteins, cottonseed protein can provide required quality of essential amino acids.

Cottonseed protein concentrates and isolates and uses

Protein content in cottonseed-meal is found to vary from 45 to 70 per cent or still higher depending on the quality of seed, variety, as well as technique of processing. The cottonseed protein is a rich source of protein, it has been converted in various products to provide high percentage of protein in diet to meet the protein deficiency for human consumption. Cottonseed protein isolate is one of such products which is prepared by removing non-protein constituents to provide product rich in protein say about 90 per cent or above. Presently there are three items used by the industry to describe protein products: flour, concentrates and isolates. Such products differ based on their protein content. Flour containing 50 per cent protein or more concentrates contain a minimum of 70 per cent protein on dry weight basis. Concentrates can be prepared either by dry or aqueous fractionation process. With carefully ground defatted meal, the intact protein bodies can be concentrated at the expenses of the cell wall fragments by the dry operation of air classification. This process can provide two products—one that accepts or concentrate containing 70 per cent protein the other rejects. This closed dry process of air classification (Fig. 4.13) has advantages from the pollution aspects. The classical procedures described give protein isolate with maximum recovery of nitrogen from cottonseed flours after extraction with dilute alkali followed by acidification.

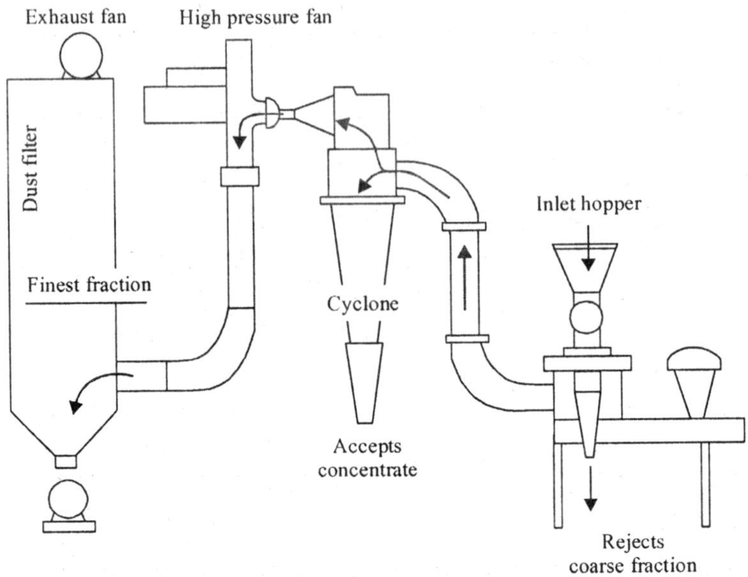

Fig. 4.13. Diagram of air classification process for dry preparation of protein concentrates.

Preparation of protein concentrates

Cotton protein concentrates can also be prepared from cottonseed by two extraction methods. Extraction with aqueous ethanol removes the extractable sugars, free amino acids and residual lipids, as well as certain colour and flavour components. Because of the insolubility of the protein, a concentrate can also be produced by a simple aqueous extraction with or without 0.008 M calcium chloride. The successive extraction with calcium chloride solution and water removes the low molecular weight, water soluble proteins, in addition to sugars, free amino acids and certain colour and flavour compounds. However, these methods including aqueous alcohol have problems of by-products recovery and water pollution. Composition and yield of concentrates of these methods are given in Table 4.8. These concentrates differ in crude fibre, total sugar contents, etc.

Preparation of protein isolates

Various types of protein isolates can be prepared from different cottonseed flour employing different methods. In the normal or classical method, the maximum amount of nitrogen is extracted with dilute alkali followed by acidification of the centrifugal protein liquor at a pH of 5, which provides maximum precipitation of protein. The separated protein is then washed and dried, either at the pH of minimum solubility, pH 5 or at pH 7 as the sodium proteinate. It is mentioned that these isolates contain both the functional and storage proteins. Similarly based on extraction characteristics of protein, a two-step process is described, wherein the low molecular weight functional proteins are extracted first with water and then with dilute alkali.

It is observed that storage protein isolate functions very well in bread dough up to 10 to 15 per cent substitution. It is found that the lower the moisture content of the flour, lower the consistency. This indicates that the hydration characteristics of a flour can be controlled by desolventising operation.

Uses of cottonseed protein concentrates, isolates

There is protein deficiency in the diet of 50 per cent children of the world. According to Food and Agricultural Organisation of United Nations, at least 69 g of protein per day should be made available in the diet of every person even in undeveloped countries. On account of shortage of proteins, the cottonseed protein isolates and concentrates offer the immediate opportunity for increasing the supply of good grade protein for human consumption.

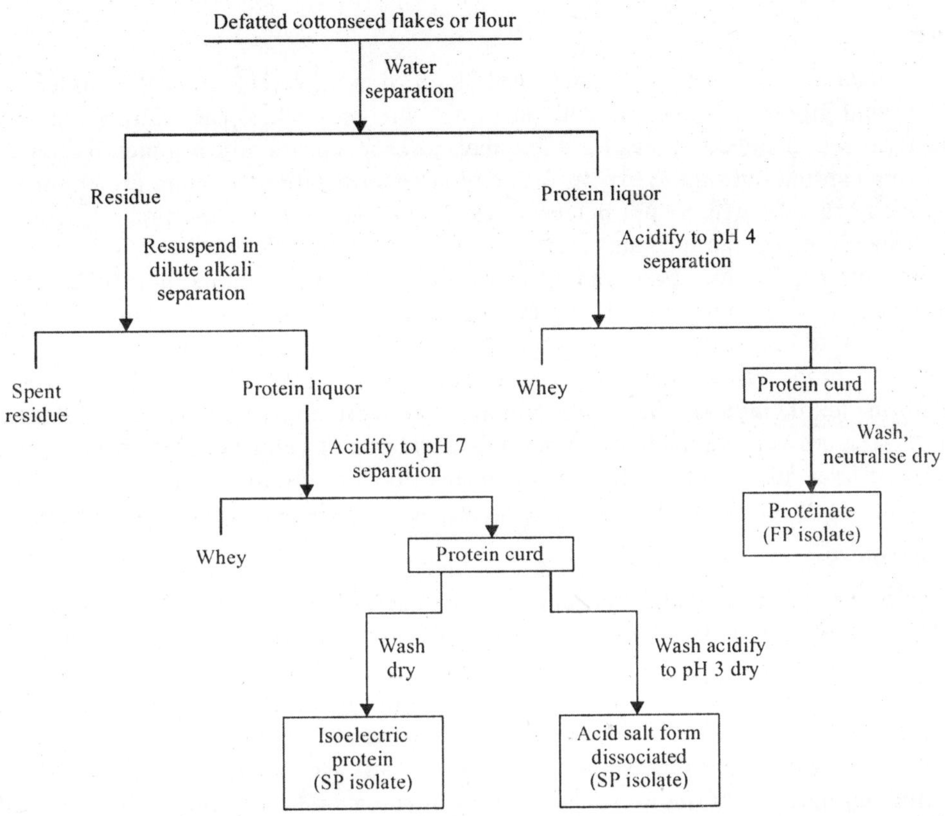

Fig. 4.14. Flow diagram of classical procedure for preparation of cottonseed protein isolate.

In Central and South America, cottonseed protein, flours, concentrates and isolates are being used in food formulations. 'Incaparina' sponsored by the Institute of Nutrition of Central America and Panama, contains 38 per cent of cottonseed flour. It is being used in children food for beverages, soaps and guels in Gautamala, Panama and Venezuela. Pervita, a high protein food product is used in Peru, 'Profolo' a food grade cottonseed products is being used for several years. It is used in the USA in bakery products. Texturised protein can be blended with ground meat to make loaf or pattice. Cottonseed protein products can also be used as a protein source in frozen desserts. Protein isolates can be used in beverages for protein fortification and as an excellent whipping agent. The major isolates will be useful in acidic systems as citric acid base beverages.

Cottonseed protein isolates, concentrates can be used to meet protein deficiency in human food particularly in developing countries where protein meal nutrition is prevalent. These can be used in bakery products, rice preparation, vegetable and drink. It can be supplied to school children, infants as drink along with other proteins. There are several other medicinal and biological applications of cottonseed protein concentrates and isolates which can play a vital role in meeting the requirement of protein and which will solve the protein malnutrition in various countries, facing acute shortage of edible protein. Cottonseed protein isolates can also be used as substitutes to casein and soya protein. These can be used for industrial purposes also such as for production of fibres, films, adhesives, in paper and textiles sizes or coatings and in paints as thickener.

Cotton linters

Cotton fibre is a seed hair, formed by external protrusion of a single cell from the epidermis of the cottonseed. The rapid growth in the epidermal cells into fibre cells takes place during the opening of the flower and fertilisation and continues for a few days after fertilisation. It is generally believed that the fibre cells which sprout during the first to fifth day after fertilisation develop into commercial lint fibres while those fibre cells which appear later develop into fuzzy, fibres or 'linters', which form a dense mat adhering closely to the seed surface.

After ginning, commercial lint fibres are removed, while seed still contains short fibres adhering to it, which cannot be removed from the seed by ginning machines. These fibres are known as 'linters' or fuzz and can be separated from the seed by delinting machines. As such cotton fibres and linters are similar in chemical composition and basically composed of cellulose. While cotton fibres form valuable raw materials for the textile industry, the linters cannot be converted into yarn in view of their extreme short length but these are very valuable raw materials as source of cellulose of high industrial value. These can be used as a high-grade cellulose for production of cellulose acetates, viscose rayons, carboxymethyl celluloses, cellulose ethers and esters, nitrocellulose lacquers and various other products.

Types of cotton linters

Basically linter fibres are similar in chemical composition but based on the delinting process, these are classified as: (i) first-cut linters; (ii) second-cut linters; (iii) mill-run linters; and (iv) difibrated linters.

Delinting of cottonseed has two-fold advantage. It gives valuable cotton linter required to feed the chemical pulp industry and it also helps in a more scientific crushing of cottonseed. In two-stage delinting process first-cut linters and longer fibres left on seed are removed in first operation by delinting machine. The second-cut linters which are shorter in length but more uniform and clean are removed in second stage of delinting. Similarly when delinting is done only in one operation all linters are removed in one stage of operation and such linters are known as mill-run linters. Mill-run linters contain more heterogenous linter fibres compared to second-cut linters.

There are standard machineries used for delinting of linters. Linter fibre contains considerable amount of impurities and foreign matters. Presence of such matters in linters creates problem in application of linters and reduces their industrial value. Therefore, it is essential that cotton linters should be cleaned before their industrial application. Some industries also follow a process that they dehull the seed before delinting and hull is passed to defibrating machine. In this process fibres are separated from hull particle by air flow method through blower. Hulls being heavier than the fibres are easily separated. Such linter fibres are called defibrated linters. These fibres contain larger foreign material and impurities compared to first or second-cut linters.

The production of linters in India is not very encouraging. The cottonseed delinting-installed capacity of industry is not fully utilised. The main problem is high cost of delinting process owing to high power requirement and low prices and poor demand of linters in the country.

There is urgent need that such an important commodity should be utilised for making valuable cellulose derivatives and various products the Indian industry needs. In this regard it will be advisable that import of pulp should be stopped wherever linter can be utilised. Further, whole crushing of cottonseed as well as feeding of cottonseed to cattle should be discouraged. This effort will pave the way for industrial application of this valuable commodity in the country.

Quality and specification of cotton linters

It is shown that cellulose content in linters collected from ginning factories during mill-run, first-cut and second-cut of different varieties after purification found the cellulose content was over 98 per cent in all the cases whereas ash content in the mill-run and first-cut linters were higher than the permissible limit but the values were with in permissible limit in second-cut linters.

Further, it is reported that the Indian linters are found to be comparable in quality to the linters from other countries, although few samples of Indian linters might have higher ash content or iron content. However, it is observed that second-cut lintes are better in quality than the first-cut while the mill-run linters are better in quality than the second-cut linters.

Trash content of linter can be analysed by Shirley analyser. The values are considerably high ranging from 9.2 to 14.3 per cent. Another important factor about linters quality is length of linter fibre. It is seen that linters show wide variation in length of individual fibres. Some of the fibres are very short and get lost during chemical purification processes. Also, in Shirely analyser such fibres are lost and are considered as 'invisible loss' in this test. It is reported that invisible loss varied from 4 to 22.5 per cent in linter. In other study it is shown to vary from 10 to 49.6 per cent.

Utilisation of cotton linters

Cotton linters as by-product of cottonseed are very useful industrial raw material for production of various types of cellulose derivatives such as cellulose nitrate, ethers, esters, acetates, which are used for production of various kinds of chemicals, plastics and fibres. Besides linters can be used as a source of chemical cotton for speciality paper and filters, non-woven fabrics, filler aids, for films, plastics etc. The details of utilisation of cotton linters for industrial use are described below. The linters are converted for various products by different routes. Such as it is converted into dissolving pulp which is converted to cellulose nitrate, viscose, cellulose esters and ethers, cellulose acetates which are used for production of different industrial products. The linters are classified on the basis of their utility are as follows:

1. Dissolving pulp—Cellulose nitrate, viscose, cellulose esters and ethers and cellulose acetate.
2. Felts—For automotive upholstery, pads, cushions, furniture upholstery, mattresses.
3. Yarns—Lamp and candle wicks, twine, rugs, mops.
4. Absorbent cotton medical grade—Cotton swabs, cottons bails, gauze, pads, papers.
5. Fibre pulp papers—Fine writing and bond papers, filter-paper, currency, laminating papers, sanitary products, battery separators.

Cellulose nitrate is used for production of plastics, dynamite, lacquers, finger nail polish, smokeless gun powder and solid rocket propellants. Viscose pulp is used for food casings such sas bologna, sausages and frank furthers and rayon for air nose and industrial fabrics. Similarly other uses of cellulose esters and ethers and cellulose acetate are indicated in Fig. 4.15.

During recent years Bureau of Indian Standards (BIS) has developed several specifications for linters and chemical cotton required for various end uses which are given in Appendix-I.

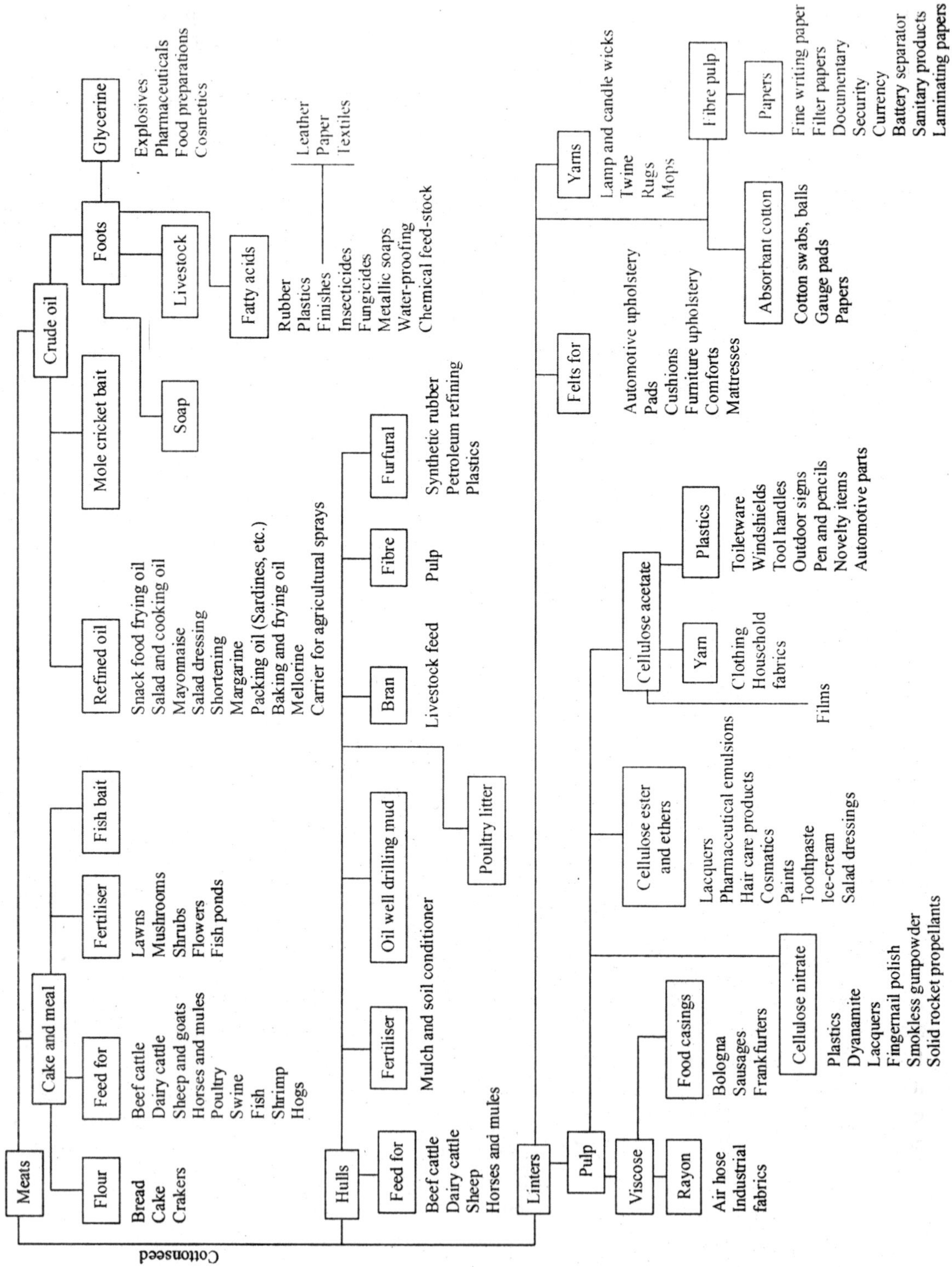

Fig. 4.15. Cottonseed products and their uses.

These BIS standards specify various characteristics of Indian cotton linters required for various purposes. Limit of different constituents of various types of linters are given in first specification. Similarly detailed specification of test methods and other products of linters are described in these standards. These specifications clearly indicate that the yield of chemical cotton and limits of ash, silica and iron are important for end uses of linters for various end products.

Hulls

Nature and composition

The hull is outer covering of the seed and is about 37 to 60 per cent of the seed weight depending on the variety, species and various agronomic factors. The average value of hull content of Indian varieties is found to be 47 per cent in *G. arboreum* and *G. herbaceum*, both *desi* cotton varieties, 45 per cent in *G. hirsutum* and 38 per cent in *G. barbadense*. Hull contains about 35–47 per cent cellulose, 19–27 per cent pentosans, 15–20 per cent lignin and 5 per cent ash, protein, fat, etc.

Uses of hulls

The hull is mostly used for cattle feed. It is mixed with other feed as roughage to increase the volume of the feed. Its nutrient value is very poor. It can also be used for more important industrial applications. The hull has been used for the production of furfural, plastics, soil conditioner, fertilisers, insulating materials and for α-cellulose, for viscose rayon and active carbon. Very recently hull powder has been used for production of various kinds of particle boards, laminated particle boards and fire-retardant boards to replace the use of plywood. These boards can be used for furniture, doors, windows, partitioning, false ceiling and for various kinds of packaging. Various products of cottonseed and their several uses have been shown in Fig. 4.15. It may be seen that cottonseed oil, cake, linter and hull have several important industrial applications and can play vital role in uplifting the economy of the country.

GRADING OF COTTONSEED AND ITS PRIMARY PRODUCTS

Grading of Cottonseed

Grading plays an important role in proper utilisation and its placement among the producers and consumers. Same is true about cottonseed. Cottonseed being by-product of cotton crop needs suitable classification and grading or its trading hence it generates additional income to the farming community and those engaged in trading business of cottonseed. It is disappointing to see that in spite of acute shortage of edible oil in the country not much attention has been given to this important commodity as regards grading is concerned. The importance of grading of cottonseed has been felt since long by the cottonseed-crushing industry in India. Similarly, in the USA the grading of cottonseed and its by-product was well adopted in the beginning of this century. It is reported that cottonseed was considered an important commodity and more than 50 per cent of the seed was available for oil crushing, wherein quality of the product and standardisation of the processes were given lot of importance. Records reveal that initial grading of cottonseed in the USA was started in 1899 where cottonseed was classified into prime and non-prime. Later on grading was based on mature and damaged seeds. Thereafter more importance was given to quality and quantity of oil and cake in the seed.

Determination of grade

At present, grades of cottonseed in the USA are based on scientific analysis of various characters and these are certified by the US Department of Agriculture. The grades of cottonseed include quantity index and quality index in the USA. These index values are divided by 100 and are expressed in whole number of the nearest half numbers. These are determined as: (i) the basis grade of cottonseed shall be grade 100; (ii) high grades of cottonseed shall be those grades above 100; and (iii) low grades of cottonseed shall be those grades below 100.

The quantity index of cottonseed is determined as follows:

1. The quantity index shall equal the result of four times the per cent of oil plus six times the per cent of ammonia plus the applicable linters premium for discount obtained on the basis of the formula given below plus 5.
2. Formulae for determining linters premium and discounts (total linter content to the nearest (0.1 per cent) are as follows:

Linter content in cottonseed per cent	Premium or discount factor
10.6 and above	Premium (per cent linters minus 10.5) × 1.0
10.5	None
10.4–9.0	Discount (10.5 minus per cent linters) × 1.0
8.9–4.0	Discount (9.0 minus per cent linters) × 2 × 1.5
3.9–0	Discount (4.0 minus per cent linters) × 2.5 × 11.5

Estimation of quality index

Quality index of cottonseed shall be an index of purity and soundness and will be estimated as follows:

Prime quality cottonseed

Seed with more than 1 per cent of foreign matter, not more than 12 per cent moisture and not more than 1.8 per cent of fatty acids in the oil will be designated as prime quality cottonseed and will have quality index of 100.

Below prime quality cottonseed

The quality index of cottonseed by analysis that contains foreign matter, moisture of free fatty acid in the oil or seed in excess of the per cent indicated for:

1. Above shall be found by reducing the quality index of prime quality cottonseed as given below:
   (a) 0.4 of a unit for each 0.1 per cent of free fatty acids in the oil in the seed in excess of 1.8 per cent.
   (b) 0.1 of a unit for each 0.1 per cent of foreign matter in excess of 1.0 per cent.
   (c) 0.1 of a unit for each 0.1 per cent of moisture in excess of 12.0 per cent.

Off-quality cottonseed

The cottonseed which has been subjected to mechanical or chemical process other than the usual cleaning, drying and ginning (except sterilisation required by the US Department of Agriculture (USDA) for quarantine purposes), or that is fermented or hot, or that upon analysis is found to contain 12.5 per cent or more of free fatty acids in the oil in the seed, or more than 10.0 per cent of foreign matter, or more than 20.2 per cent moisture and or more than 25.0 per cent of moisture and foreign matter combined, shall be called as 'off-quality cottonseed'.

Below grade cottonseed

The cottonseed when analysed as per the above guidelines gives below grade 40.0 shall be designated as 'below grade cottonseed'.

Although grading system has been described but in commercial practice most of the companies or factories grade the seed according to their own convenience, which plays an important role in grading. Guidelines are given by the USDA to facilitate the sampling of large bulk of cottonseed. It also indicates the details of tools and quantity of samples of cottonseed per tonne of cart load. After doing sampling, each lot is combined and foreign matter is obtained by cleaning the samples. A representative sample of about 1 kg is drawn from the composite sample for analysis by the USDA and certificates are issued by the chemists after analysis. The qualities determined are oil content, ammonia, moisture, foreign matter, linters and free fatty acids related to quality and quantity index are given in Tables 4.9 and 4.10 respectively.

Table 4.9. Computation of cottonseed quality indexes and grade by qualities as per official standards in the USA.

	Quality factors							
Example No	*FFA of oil*		*Foreign matter*		*Moisture*		*Total reduction units*	*Quality index*
	Total (%)	*Reduction units*	*Total (%)*	*Reduction units*	*Total (%)*	*Reduction units*		
Prime quality seeds								
1	0.5	0.0	0.3	0.0	10.0	0.0	0.0	100.0
2	1.8	0.0	1.0	0.0	12.0	0.0	0.0	100.0
3	0.2	0.0	0.8	0.0	9.6	0.0	0.0	100.0
Below prime quality seeds								
4	1.9	0.4	1.0	0.0	12.0	0.0	0.4	99.6
5	1.8	0.0	1.1	0.1	12.0	0.0	0.1	99.4
6	1.8	0.0	1.0	0.0	12.1	0.1	0.1	99.9
7	2.5	2.8	3.2	2.2	7.4	0.0	5.0	95.0
Off-quality seeds								
8	12.5	42.8	0.7	0.0	12.0	0.0	42.8	57.2
9	1.8	0.0	10.1	9.1	12.0	0.0	9.1	90.9
10	1.8	0.0	1.0	0.0	20.1	8.1	8.1	91.9
Below grade seeds								
11	10.5	34.8	20.8	19.8	15.6	3.6	58.2	41.8
12	17.5	62.8	0.5	0.0	14.1	2.1	64.9	35.1

These data indicate that in practice, the yield of products are close to the data obtained from grading. An important point in this grading system is that the actual oil content of the seed no longer finds a place.

Table 4.10. Computation of cottonseed quantity index and grade by quantities as per official standards in the USA.

Example No.	*Oil Total product (%)*		*Ammonia Total product (%)*	*Linters Total product*		*Sum products (%)*	*Adjustment factors*	*Quantity index (%)*	*Grade*
Prime quality seeds									
1	19.0	76.0	3.6	21.60	12.0 + 1.5	99.10	+ 5	104.10	104.0
2	17.8	71.2	4.10	24.60	10.0 – 0.5	95.30	+ 5	100.30	100.5
3	16.2	64.8	3.97	23.82	9.5 – 1.0	87.62	+ 5	92.62	92.5

(Contd...)

Table 4.10. Computation of cottonseed quantity index and grade by quantities as per official standards in the USA. *(Contd.)*

Example No.	*Oil Total product (%)*		*Ammonia Total product (%)*	*Linters Total product*		*Sum products (%)*	*Adjustment factors*	*Quantity index (%)*	*Grade*
Below prime quality seeds									
4	18.5	74.0	3.50	21.0	12.0 + 1.5	96.50	+ 5	101.50	101.0
5	19.7	78.8	3.75	22.5	11.0 + 0.5	101.80	+ 5	106.80	106.5
6	17.3	69.2	4.23	25.38	8.7 – 2.1	92.48	+ 5	97.48	97.5
7	24.7	98.8	4.15	24.50	1.7 – 17.85	106.45	+ 5	111.45	106.0
Off-quality seeds									
8	20.1	80.4	3.67	22.02	11.5 + 1.0	103.42	+ 5	108.42	62.0
9	16.9	67.6	4.13	24.78	8.9 – 1.7	90.68	+ 5	95.68	87.0
10	18.8	75.2	3.80	22.80	13.2 + 2.7	100.70	+ 5	105.70	97.0
Below grade seeds									
11	17.0	68.0	3.41	20.46	11.5 + 1.0	89.46	+ 5	94.46	–
12	19.5	78.0	3.94	23.64	13.9 + 3.4	105.04	+ 5	110.04	–

Advantages of Adopting Standards of Grading

The data based on these grading systems are of immense benefit to the producers and sellers and those engaged in dealing with cottonseed processing. People would take proper care in maintaining the quality of the product during its production, harvest, transportation and storage. Grading could also be used as check against mill-processing practices. Mixing of properly graded seed would enhance the productivity of oil and prevent the losses. It may also be noted that grading increases the efficiency of milling industry. It helps in recovery of protein and oil on the basis of whole seed. With regard to the loss of oil cake and protein, the method of preparing the meats for pressing or expelling, pressure used and various conditions affect the quality and quantity of these products.

Based on accepted grading systems, concerned personnel are supplied with such certificate from approved chemists based on statistical details from which variation and losses in oil and cake can be derived. The USDA-treated cottonseed is on par with wheat and other agricultural commodities as regards grading and marketing.

Grading Systems in India

It is astonishing that cottonseed which has been recognised as a valuable agricultural commodity for the production of edible grade oil, cottonseed flour and as source of linter in developed country like the USA, has been used as cattle feed for long time in India. Only after acute shortage of edible oil in the country and need for import of large quantity of oil, some realisation came during last two decades and then the seed was recognised as a source of edible oil. However, not much emphasis was given for the grading of cottonseed oil and its various products.

It became amply clear to the industry that the seed with higher oil content will be more useful and beneficial for crushing to get higher production of oil per tonne of seed crushed. Seed of cotton varieties containing lower oil content paid lower price to the cultivator. In view of this, considerable amount of seed was used for cattle feeding. However, this gave awareness of quality and grade of cottonseed. The Bureau of Indian Standards (BIS) with its various committees tried to cover different aspects of cottonseed and its by-products by developing suitable grades and specifications with the help of various research organisations and the seed processing industries in the country.

The specifications have been laid down by the BIS for cottonseed, where requirements for specific grades are indicated in IS Specification (IS 4620, 1968).

Grading of Cottonseed oil

Initially the cottonseed oil was not subjected to complete refining and normally it used to be alkali washed just to remove free fatty acids and colour and it was mostly consumed by vanaspati industry. Gradually with the realisation of importance of this important commodity in various uses, standards for Cottonseed oil were laid down by the BIS with the help of research organisations, consumers and producers.

Crude cottonseed oil

Cottonseed oil is obtained from cottonseed by mechanical crushing and also by solvent extraction. The oil thus obtained from the seed without subjecting to any further treatment is termed crude cottonseed oil. This oil is dark black or brown and normally not traded in the country. The colour of this oil depends on various factors including colouring matter content. It is seen that presence of certain pigments like gossypol and related pigments produce more dark coloured oil. The various factors determine the quality of this oil, such as colour, flavour and odour, sampling and refining loss methods. These aspects need considerable attention and suitable grading system in view of the variability in quality of crude cottonseed oil as has been reported in recent studies. It has been found that crude cottonseed oil for the same variety varies in colour, free fatty acid (FFA), gossypol and odour when extracted by different processes. Marked variation in composition of oil for the same variety has been seen when processed by different mills by the similar method. Storage and shelf-life of oil has been affected to some extent when packed or stored in different ways. There is an urgent need to have suitable grades including all for separating crude cottonseed oil which will be helpful to identify the crude oil based on its grade. The first Indian Standard for cottonseed oil was brought out in 1954. It was indicated as IS: 543–1954. This was revised in 1966. In earlier grade (IS: 543–1954) only one grade of refined oil and one grade of washed oil was included. On revision in 1966 it included one grade of raw oil, one grade of refined oil and two grades of washed oil. This was further revised in 1968, IS 543–1968 which includes four grades of expressed oil and four of solvent extracted oil (Table 4.11). In this standard refining loss for crude cottonseed oil was included. The standard also includes moisture content, refractive index, specific gravity, acid value, iodine value, saponification value and percentage of unsaponifiation value.

Table 4.11. BIS Specifications: Requirement of cottonseed oil IS: 543–1968.

Characteristic	*Expressed*				*Solvent extracted*			
	Refined grade	*Grade 1A (washed)*	*Grade 1B (washed)*	*Grade 2 (raw)*	*Refined grade*	*Semi-refined grade*	*Grade 1B (raw)*	*Grade 2 (raw)*
Moisture and insoluble impurities % by wt. max.	0.10	0.10	0.10	0.25	0.10	0.25	0.75	1.5
Lovibond, colour, 1-cell as (Y-IOR), not deeper than								

(Contd...)

Table 4.11. BIS Specifications: Requirement of cottonseed oil IS: 543–1968. *(Contd.)*

Characteristic	*Expressed*				*Solvent extracted*			
	Refined grade	*Grade 1A (washed)*	*Grade 1B (washed)*	*Grade 2 (raw)*	*Refined grade*	*Semi-refined grade*	*Grade 1B (raw)*	*Grade 2 (raw)*
Original oil	10	25	35	–	14	35	–	–
Bleached oil	–	8	15	–	–	15	–	–
Refractive index at 40°C	1.4630 to 1.4660							
Sp. gravity at 30°C	0.910 to 0.920							
Saponification value	190 to 198							
Iodine value (Wijs)	98 to 110							
Acid value, max.	0.3	0.5	0.5	10.0	0.3	0.75	10.0	25.0
Unsaponifiable matter, % by wt. max	1.5	1.5	1.5	2.0	1.5	1.5	2.0	2.5
Refining loss, % by wt. max.	–	–	–	15.0	–	–	15.0	40.0
Flash-point Pensky-Martens (closed, 0°C, min.)	–	–	–	–	250	125	100	90

Refined cottonseed oil

Crude cottonseed oil, refined (neutralised) by alkali treatment, bleached with bleaching earth or activated carbon or both and deodourised is called 'refined cottonseed oil'.

Washed cottonseed oil and semi-refined oil

Crude cottonseed oil, neutralised by alkali treatment and subsequently washed and dried. Washed cottonseed bleached with bleaching earth or active carbon or both is semi-refined oil.

Types and grades of cottonseed oil

Expressed: Refined grade
Grade 1A (washed)
Grade 1B (washed)
Grade 2 (raw)
Solvent extracted: Refined grade
Semi-refined grade
Grade 1 (raw)
Grade 2 (raw)

Refined grade of oils are suitable for direct edible consumption. All other grades of expressed type and the solvent extracted type except grade 2 (raw) are suitable for production of vanaspati and refined oil and not for direct edible consumption. Grade 2 (raw) is for industrial uses other than the manufacture of vanaspati and refined oil. In the USA, trading in cottonseed and its products is governed by the trading rules of the National Cottonseed Products Association (NCPA). They have various grades for cottonseed and cottonseed products. There are following eight grades of crude cottonseed oil.

Grades of cottonseed oil

1. Prime crude cottonseed oil: It is prime in flavour and odour, in which FFA does not exceed 3.25 per cent and which when refined is sweet in flavour and odour, shall be graded prime in flavour and odour.
2. Basis prime crude cottonseed oil: Oil must refine to a colour not higher than AOCS 12 red (matched against 70 Y in 51/4″ cell) or with loss in weight not exceeding 20 per cent.
3. Off-crude cottonseed oil: Must refine to a colour not higher than AOCS 20 red (matched against 70 Y) with loss in weight not exceeding 25 per cent.
4. Reddish off-crude cottonseed oil: Must refine to a colour not higher than AOCS 30 red, with a loss in weight not exceeding 40 per cent.
5. Low-grade crude cottonseed oil: This oil is not up to the two specifications immediately above and shall be sold either as sample or as 'low grade' cottonseed oil.
6. Screw-press or expeller cottonseed oil: Cottonseed oil produce by screw-press or expeller is tenderable on contracts for the same grades oil previously defined, if such oil will refine within the requirements of the foregoing rules.
7. Solvent-extracted cottonseed oil: Crude cottonseed oil produced by solvent extraction with or without pressing shall be tenderable on contracts for the grade of oil previously defined.
8. Slow-breaking cottonseed oil: Should be notified by seller that the oil is slow-breaking type, such oil shall be analysed by the special slow-break procedure provided for.

Grading of Cottonseed-Cake and Meal

Grading of cottonseed oil cake is as important as for cottonseed oil. The BIS has framed few Indian Standards Specifications of Indian cottonseed-cake and meal. These standards are as follows.

1. IS: 1712–1960 (has been revised). It covers decorticated and undecorticated cottonseed-cake.
2. IS: 3592–1968. Solvent-extracted oil cake (meal) for use as livestock feed.
3. IS: 1714–1960. Sampling and test procedures required for use for the above standards.

Details of these standards are shown in Table 4.12.

Table 4.12. The BIS specifications for cottonseed-cake and solvent-extracted meal as livestock feed.

	IS: 1712–1960			*IS: 3592–1968*			
	Decorticated cottonseed oilcake			*Undecorticated cottonseed oil cake*		*Solvent-extracted oilcake (meal) as livestock feed*	
Description	*Grade HF*	*Grade 1 (% wt.)*	*Grade 2 (% wt.)*	*Grade 1 (% wt.)*	*Grade 2 (% wt.)*	*Grade 1 (% wt.)*	*Grade 2 (% wt.)*
Moisture, max.	8.0	8.0	8.0	10.0	10.0	10.0	10.0
Crude protein (N × 6.26), min.	40.0	41.0	37.0	24.0	20.0	42.0	40.0
Crude fat or other extract, min.	8.0	5.0	5.0	7.0	7.0	1.5	2.0
Crude fibre, max.	10.0	12.0	14.0	24.0	28.0	16	18
Total ash, max.	7.0	8.0	9.0	9.0	9.0	–	–
Acid insoluble ash, max.	1.5	2.0	2.5	2.5	2.5	2.0	2.5
Castor husk, max. on moisture-free basis	Nil	Nil	Nil	Nil	Nil	Nil	0.005

Note: Test procedures are described in IS: 1714–1960. Method of scanning and test for oil cakes as livestock feed.

The more detailed information on the BIS standards may be referred from the original specification. These specifications may be seen from the table which covers three types of cake (meal)—undecorticated cottonseed-cake, decorticated cottonseed-cake, and solvent-extracted cottonseed-cake as cattle feed.

Undecorticated cake will be of two grades—grade 1 and grade 2. Decorticated oil cakes are classified further in two grades depending on presence of oil, such as grade HF (high fat), grade 1 and grade 2. Similarly, the cake sample should be free from foreign matter and impurities. Also, it should be free from rancidity, adulterants and other objectionable foreign matter. The protein and fibre contents are considered the most important segments as indicated in the Table 4.12. The protein content shall be 40, 41 and 37 per cent and crude fibre content 10, 12 and 14 per cent in decorticated oil cake.

Grade 1 and 2 indicate 5 per cent original oil while grade HF (high fat) indicates 8 per cent of residual oil indicating poor oil pressing. Undecorticated oil cake, which is not desirable to be produced for various reasons, contains very high fibre content (24–28 per cent) and low protein content (20–24 per cent). The solvent-extracted cake indicates very high percentage of protein (40–42 per cent) and low residual fat content (2 per cent). Grading of cottonseed-cake in the USA is done on similar basis but more scientifically by the National Cottonseed Products Association. These grades cover cottonseed-cake, sized cottonseed-cake, cottonseed-meal and whole pressed (undecorticated) cake. All these cakes should contain 36 per cent of minimum protein, should have acceptable colour and flavour and be free from fungus. Meal is defined as a ground product from cake. Also gossypol content should be very low (0.04 per cent free gossypol) in special meals.

Cottonseed Flour for Human Consumption

Special products developed from cottonseed-meal for edible purposes have been standardised and specification after careful testing and evaluations have been given. This has been done to encourage the use of cottonseed flour to meet protein deficiency and remove the protein meal nutrition to some extent. The need to process quality meal and to preserve the nutritive value of protein in meal has been specially indicated. Extensive investigation on the use of cottonseed protein for human consumption has been carried out. In view of this, grading of cottonseed flour for edible use has been done in the USA as well as in India. It has been often reported that gossypol during processing binds the lysine epsilon amino acid in presence of moisture and heat and makes it deficient in this essential amino acid if proper care is not taken during processing of cottonseed flour.

It is therefore emphasised that maximum care should be taken during screw-pressing of cottonseed to reduce the binding of the lysine. The lysine content in natural cottonseed protein is 4.2 g/100 g of crude protein. Any reduction in this quantity indicates the loss of this amino acid. Besides free gossypol itself is toxic pigment for non-ruminants as has been indicated by several feeding trials. Efforts should be made to keep the gossypol quantity as minimum as possible by carefully removing it with cottonseed oil, where it can be easily washed out with alkali refining. This will help in maintaining the quality of cottonseed protein.

The standards were laid down by the World Health Organisation and by the UNICEF for screw-pressed cottonseed protein concentrates. Composition of cottonseed concentrate is given in Table 4.13. This envisaged a minimum percentage of protein in concentrates and available lysine of 3.6 g/100 g protein and nitrogen solubility of 70 per cent.

These specifications were developed at the Institute for Nutrition in Central America and Panama, Guatlmala.

Table 4.13. Specifications for edible cottonseed flour.

Description	*Percentage*	*Aflatoxin*
Maximum moisture content	10	0.03–0.05 ppm
Protein (minimum)	50	
Lysine (available)	3.6	
Nitrogen solubility	70	
Free gossypol	0.050	
Total gossypol	1.0	

Similarly, the BIS has formulated standards IS: 4874–1968 for edible cottonseed flour (expeller pressed) and IS: 4876–1968 for edible cottonseed flour (solvent-extracted). The detailed specifications are given in Table 4.14. The protein content is 45 per cent for expeller-pressed and 47 per cent for solvent-extracted meal and free and total gossypol are 0.065 and 1.1 and available lysine is 3.6 g/100 g protein. Besides other important factors like microbes, moisture and fat contents are also indicated. The standard compares well with the specification laid down by the WHO. There is need for further work on revision of these specifications based on nutritional experiments for edible purpose.

Table 4.14. Indian Standards Specifications for edible cottonseed flour (expeller-pressed).

Description	*Percentage by weight*
Moisture, max.	8.0
Crude protein (N × 6.25), min.	45.0
Total ash, max.	5.0
Acid-insoluble ash, max.	0.35
Fat, max.	6.0
Crude fibre, max.	5.0
Free gossypol, max.	0.065
Total gossypol, max.	1.10
Available lysine, g/100 g crude protein, min.	3.6
Acid value of extracted fat, max.	4.0
Total bacterial count/g, max.	50.0
Coliform bacteria/g, max. 10	
Salmonella bacteria	Nil

Note: Test procedures are described in the specification itself and also in IS:4684–1968, Specification for edible groundnut flour.

On dry basis, IS: 4876–1968 for edible cottonseed flour (solvent-extracted) is identical, except for (i) crude protein, minimum, 47 per cent; (ii) fat, maximum, 1.5 per cent; and (iii) residual solvent, maximum 170 ppm.

Grading of Cottonseed Linters

Linter being short fibre leftover on cottonseed after ginning and by-product of cottonseed is valuable product for manufacturing cellulose and cellulose derivatives, for pulp and paper (writing, printing, filter, currency paper and other speciality paper), and for filling material in textiles, crushing, non-woven products, plastics and films. There has been considerable improvement in the production of cottonseed linters, yet many cottonseed-crushing factories do not delint the seed and prefer to crush the seed without delinting. The recent facts regarding delinting and production of linters have been stated. It is stated that most of the linters produced in the country is used in the manufacturing of ordinary paper. The development of sophisticated uses of linters will not only mean better utilisation of linters but will also help scientific cottonseed-processing industry by increasing its realisation from linters to some extent. It is stated that production of linters takes place in India essentially during production of cottonseed extraction and not for production of linters themselves.

The BIS has formulated at least 16 specifications on cotton linters so far. These include the grading of linter and its various derivatives.

Appendix-I specifications for linters and chemical cotton required for various end uses.

1. IS: 3517–1979 Cotton linters (Ist revision) Re-affirm June, 1990.
2. IS: 3518–1966 Chemical cotton for viscose and cellulose acetate, Re-affirm January, 1989.
3. IS: 3519–1976 Methods of sampling and test for chemical cotton Re-affirm January, 1989.
4. IS: 3520–1979 Water soluble sodium carboxymethyl cellulose (Ist revision), Re-affirm December, 1984.
5. IS: 3532–1987 Chemical cotton for nitrocellulose (IInd revision).
6. IS: 1091–1985 Nitrocellulose (IIIrd revision), Re-affirm June, 1990.
7. IS: 8782–1980 Chemical cotton for paper manufacture, Re-affirm April 1981.
8. IS: 9594–1980 Cellulose powder, Re-affirm December, 1986.
9. IS: 9191–1979 Code of practice for storing cotton linters, Re-affirm June, 1990.
10. IS: 9596–1980 Method of determining of cellulose in chemical cotton, Re-affirm December 1986.
11. IS: 9598–1980 Micro-crystalline cellulose powder, Re-affirm April, 1987.
12. IS: 11213–1985 Defibriated linters, Re-affirm June, 1990.
13. IS: 12179–1987 Method of determination of sulphuric acid luminosity and bottle acelycation of chemical cotton.
14. IS: 12253–1987 Glossary of terms relating to linters and its allied products.

Chapter 5

Olive Oil

INTRODUCTION

Olive oil enjoys a privileged position amongst edible vegetable oils. There are many reasons for this status. Historically, it is the oldest oil known to man, it was used in antiquity and is therefore several thousand years old.

Its production zone is identical to the territory once covered by the Greco-Roman civilisation and it thus continues to thrive in our regions; it is this civilisation which promoted olive groves and production and consumption of olive oil throughout the Mediterranean basin, which has remained the olive tree's natural area.

Olive oil is the juice of a fruit and as such it is consumed nearly entirely in its virgin state. Today this fact makes it a distinctive product. Production of this oil requires special crushing equipment unsuitable for crushing of any other oil seed. Olive oil is a relatively exclusive and expensive oil (compared to other dietary oils and fats) and therefore undeniably plays an important social role in its production zones.

For all these reasons, the olive and its oil have become the object of special attention on the part of various national, European and International organisations.

It is well-known that there are two types of olive trees, distinguished by the destination of their fruits: direct consumption and olive oil production. The former are larger fruits which are picked before reaching maturity while the latter have smaller and are picked on reaching optimum maturity, when their oil content is measured at 18 to 22 per cent.

Depending on the climate and the region, olives are harvested between mid-autumn and the beginning of winter. In spite of research with the aim of mechanising and automatising this operation, olives are still generally picked by hand.

The water content of mature olives is high. This results in the disadvantages of enymatic changes leading to acidification of the oil. Another distinctive feature of the olive, in comparison to other oleaginous fruits and seeds is its oil-rich kernel.

COMPOSITION AND CHARACTERISTICS OF OLIVE OIL

As we have seen, olive oil occupies a special position among edible oils because of its image, consumption as a virgin oil, price, etc. A considerable amount of research has been conducted in order to ensure its purity, authenticity and quality. Many new techniques were first tested on olive oils: UV spectrophotometry, GPC, HPLC, enzyme analysis, etc. Olive oil was also the first product for which man observed the influence of variety, region of cultivation and climate on its composition and characteristics.

There are three types of olive oil: virgin oil, refined oil and crude or refined oil produced by extracting the olive oil meal with the help of a solvent. In this section we shall deal with virgin olive oil. The main physical and chemical characteristics of oilve oil are shown in Table 5.1.

Table 5.1. Main physical and chemical characteristics of olive oil.

D_{20}	0.910–0.916
V_{20} (cp)	75–79
n_D^{20}	1.468–1.470
Solidification point (°C)	
Iodine value	75–94
Saponification value	184–196

Triglyceride Composition

Triglyceride composition is determined by HPLC. For example, Table 5.2 shows the average values measured in Spanish and Tunisian olive oils which are rich in linoleic acid.

Table 5.2. HPLC analysis of olive oil triglyceride composition.

Nature	*% Total TG spain*	*Nature*	*% Total TG tunisia*
LLL	–	LLL	0.8
OLL	0.3	OLL	5.8
OLLn	0.9	OLLn	0.6
OOLn	1.0	OOLn	1.5
PLL	0.5	PLL	2.8
POLn	0.3	POLn	1.1
OOL	10.4	OOL + POOO	18.2
PoOO	1.1	–	–
POL	4.5	POL	12.3
PPoO	0.4	PPoO	1.2
PPL	0.7	PPL	2.1
OOO	43.1	OOO	21.8
POO	23.1	POO	20.0
PPO	2.9	PPO	5.3
PSPo	0.8	PSPo	–
PPP	0.5	PPP	–
SOO	3.6	SOO	3.7
PSO	0.4	PSO	1.2
PPS	0.6	PPS	0.5

Samples of known origin have been tested for trinolein (LLL) content. These tests showed that LLL content is always very low and often below the measurable threshold of 0.1 per cent. Following a recent ring analysis, the standardised IUPAC (2.324) method of triglyceride analysis was adopted. For the time being, the upper limit of LLL has been set at 0.5 per cent. For example, this allows detection of the presence of small quantities of sunflower oil (2 per cent). However, this limit is not satisfactory for the analysis of some Tunisian olive oils.

Fatty Acid Distribution in Internal and External Positions

This distribution was determined with the help of the enzymatic method. Originally, the method was used to detect the presence of esterified oil in olive oil and it still helps to detect this fraud. Table 5.3 shows a typical analysis of this distribution. In esterified olive oil, the content of saturated acids (C16:0 + C18:0) in position 2 is higher, with values of approximately 13–15 per cent compared to the normal 1.5–2 per cent.

Table 5.3. Fatty acid distribution in position 1, 2 and 3 of olive oil triglycerides.

Nature of FA	*% Total FA in 2*	*% Total FA in 1 + 3*
C 14:0		
C 16:0	1.4	15.0
C 18:0	–	3.4
C 18:1	82.9	72.8
C 18:2	14.0	7.4
C 18:2	0.8	0.9

Fatty Acid Composition

Gas chromatography has been used for this analysis. There are two known types of olive oil: one with a relatively high C18:1 (oleic) content and a relatively low C18:2 (linoleic) content and another with just the opposite proportions. The first type corresponds to oils of European origin and the second to oils of North African origin. Table 5.4 shows fatty acid values for oils marketed in France, as well as Portuguese (low linoleic acid content) and Tunisian oils (high linoleic acid content). The table shows the variations of composition, according to the place of production.

Table 5.4. Olive oil fatty acid composition.

Nature	*% Total FA*		
	French market	*Portugal*	*Tunisia*
C14 :0	< 0.1	< 0.1	< 0.1
C16 :0	8–14	9–18	8–21
C16 :1	< 1	1–3	< 4
C 17:0	< 0.2	< 0.3	< 0.2
C 18:0	3–6	1–4	1–4
C 18:1	61–80	66–79	53–75
C 18:2	3–14	3–12	9–24
C 18:3	< 1	< 1.5	< 2
C 20:0	< 0.5	< 0.4	< 0.4
C 20:1	< 0.4	< 0.4	–
C 20:0	<0.9	< 0.7	–

Portuguese oils never contain more than 12 per cent linoleic acid whereas only 1 per cent of the Tunisian production contains less than 12 per cent linoleic acid. The presence of linolenic acid was a subject of controversy, but we now know that all olive oils contain a small percentage of this acid.

Unsaponifiable Matters

The unsaponifiable content and the nature and content of its constituents have similarly been studied and discussed. The main reason for this interest is their importance for the virgin oil's purity and sensorial properties. Table 5.5 provides the information on the main unsaponifiable components. β-sitosterol and Δ7–avenasterol contents in total sterols vary considerably, depending particularly on the maturity of olives. On the other hand, the sum of β–sitosterol and Δ7–avenasterol remains more or less constant and for virgin olive oils it is about 93–95 per cent.

Table 5.5. Composition of olive oil unsaponifiable.

Unsaponifiable content		*Virgin oil: 0.4–0.8%* *Oil from cakes: 1.0–2.0 %*	
Sterols (mg/100 g)	98–184	Hydrocarbons (mg/100 g),	310–730
		including squalene	300–700
		Tocopherols (mg/100 g)	5–15
Sterol composition (% of total sterols)		Tocopherol composition (% of total tocopherols)	
Cholesterol	< 0.3	α-Tocopherol	52–87
Brassicasterol	–	β-tocopherol	10–25
Campesterol	2–4	γ-Tocopherol	7–23
Stigmasterol	1–2	δ-Tocopherol	–
β-Sitosterol	75–90	Tocotrienols	–
Δ5–Avenasterol	3–14		
Δ7–Stigmasterol	< 0.8	Triterpenic alcohols	
Δ7–Avenasterol	< 0.7	(mg/100 g)	100–300
		Erythrodiol (mg/100g)	
		Virgin olive oil	1–7
		Oil from cakes	40–80

Olive oil unsaponifiable is made up of a number of minor compounds whose functions vary: hydrocarbons, alcohols, aldehydes, ketones, esters, etc. These compounds play an important role in determining the aroma of virgin oil and in the analysis of purity or quality. Identification of these molecules calls for a certain degree of competence and the use of tools such as mass spectrometry.

The following molecules have been identified:

1. Short-or long-chain aliphatic hydrocarbons (such as octane or dodecane), aromatic hydrocarbons (such as naphthalene) and terpenic hydrocarbons (such as squalene) which the unsaponifiable fraction contains in large quantities.
2. More or less long-chain, saturated or unsaturated aliphatic alcohols, (from methanol to pentanol, octadecanol, etc.) and triterpenic alcohols (such as cycloartenol, butyrospermol, α-and β-amyrine, erythrodiol).

3. Aldehydes, from ethanal to decadenal.
4. Ketones, from acetone to nonanone.
5. Furanic and thiophenic derivatives, various esters, etc.

The following pigments have been identified: chlorophylls (0.1 to 1 mg/100 g) xanthophylls and caretenoids (0.5 to 1 mg/100 gram).

A scientific team has provided the centesimal pigment composition, identifying a dozen constituents, including: (i) in the carotenoid fraction: 30–60 per cent luteine, 5–15 per cent carotene and various xanthines; (ii) in the chlorophyll fraction: 20–40 per cent pheophytine 'a' and several per cent of pheophytine 'b' and chlorophyll 'a' and 'b'.

Finally, as all oils, olive oil contains some phospholipids, but in lesser quantities than those found in seed oils (5–15 mg/100 g), waxes and alteration products: free fatty acids, partial glycerides, oxidised acids, etc.

Contaminants

Like all industrially processed vegetable products, olive oil can be contaminated by foreign substances during handling, processing, filling and packaging, etc. Certain substances such as water, volatile matters and common insoluble impurities have no major effects; others can cause problems for quality and even safety of foodstuffs.

Amongst the contaminants to be detected are:

1. Trace metals: Fe, Cu, but mostly Pb, Cd, Hg, As; the content of the latter should not exceed agreed levels, i.e. 0.05 to 0.1 ppm.
2. Polycyclic aromatic hydrocarbons (PAH), limited to the level of a few ppb.
3. Pesticides, particularly organo-chlorinated or-phosphorated, some of which are used in fighting the olive tree fly, a kind of parasite capable of causing much damage.

Presence of these and other contaminants such as aflatoxins (they have never actually been detected but there is always a risk, particularly for non-refined oils) is regularly controlled by analytical methods.

Finally, one cannot ignore the occasional 'atypical case', which creates a stir in both professional and media ranks following some 'sensational revelation'.

The latest substance to make the headlines is perchlorethylene following accidental contamination of several batches of olive oil. Traces of up to 100–200 ppb were found and in some exceptional cases, even higher levels. Today, systematic controls ensure levels of less than 100 ppb.

PRODUCTION AND UTILISATION

Olive oil, due to its properties and qualities, is used almost entirely in dietary consumption:

1. In home where a major part of it is used for dressing, cooking, baking, etc.
2. In food processing industries such as fish canneries, pre-cooked food industry, seasoning products industry, etc. For these purposes, the International Standard authorises the addition of antioxidants such as α-tocopherol in refined oils.

A small amount of different types of olive oil goes to the pharmaceutical and cosmetic industry; and finally the non-food grade matter is used in lipochemistry since olive oil is extremely rich in oleic acid.

The fruit of the olive tree is an egg-shaped drupe, consisting of a pericarp and an endocarp. The pericarp includes an epicarp (skin) of varible thickness according to the variety and a mesocarp (pulp) surrounding the endocarp (woody pit) in which the seed is enclosed.

In addition to oil, the pulp and epicarp contain a variety of natural components soluble in the oil. As will be seen later, the oil is obtained from the olive by a variety of techniques, always physical, leaving a residue (pomace) that contains up to 8 per cent oil, which is then extracted by solvent (usually hexane) and named pomace oil.

Because of the behaviour of the solvent, solvent-extracted oil contains more minor components at higher levels than those found in physically extracted oil. This provides the basis for designating pomace oil as a commercial product distinct from virgin oil (obtained only be mechanical means) or refined (lower grade) virgin oil mixed with virgin oil (olive oil, Riviera type).

Olive oil is obtained from the oleaginous matter of a small fruit containing a small quantity of oil (about 20 per cent), a large quantity of water loaded with soluble carbohydrate substances (about 70 per cent) and a kernel. This oil is most often consumed in its virgin state, the preservation of its organoleptic qualities is a *sine qua non* since it is essentially on these criteria that it is termed as 'extra virgin' or 'fine' and purchased by consumers.

The process by which olive oil is obtained is therefore characterised by the following concerns, which create a number of constraints for manufacturers:

1. Specific equipment good for extraction only of this oil, in good working condition.
2. Risk of degradation of bruised olives by acidification with subsequent loss of quality and rejection.
3. Short olive-growing season (a few months at the most) and hence heavy investments, low productivity and long depreciation.

Mature plucked olives have a short period of conservation and suffer if transported long distances. They must therefore be processed quickly. Due to this necessity, grinding centres are quite numerous and spread out over the production region so as to keep delivery and conservation times short. As a result, grinding centres have low production capacities and the character of a cottage industry.

Obtaining Virgin Oil by Pressure

Immediately after plucking, the olives are stored in the oil mill. Storage rarely lasts more than a few days, except in case of bumper crops which call for special precautions. As for obtaining of oil, the 'pressure process' is the most traditional and common one. It includes a series of mechanical operations described below.

Cleaning and washing

The objective of this operation is to eliminate leaves, mechanical impurities, soil, etc. endowed with the risk of adultration of the oil and of premature wearing out of the equipment. Cleaning is done mechanically and magnetically; washing in water is often repeated twice.

Grinding

Clean olives are submitted, without heating or boiling, to a thorough grinding during which the pulpy drupe replete with oil bursts and the stone and kernel inside are crushed. Oil exudes slowly from the paste thus obtained. Grinding is done by a horizontal grindstone, the quintessential traditional image of the olive oil mill. In more modern mills, hammer grinders or disc grinders are used.

Kneading

This operation complements grinding and gives the paste perfect regularity and homogeneity. This is also an intermediate step prior to the next operation. It is carried out in a double-walled kneading machine with a water circuit-heated, if necessary–for maintaining the paste at an appropriate temperature.

Pressing

The paste which has been well kneaded and homogenised is spread out in pans (sort of berets made of vegetable fibres) subsequently stacked on the plate of a hydraulic press (or super-press). To maintain the correct vertical line of the stacks, the press has often a needle. By means of pumps, pressure is built up gradually upto 300/400 kg/cm^2. During this process, the liquid phase made up of oil and water is expressed and collected. The residual solid phase is gradually used up to form the olive oil meal.

Separation

The two components of the liquid phase are separated by a mechanical process, either simple decantation or, more frequently, centrifugation. The latter is carried out through centrifugal cleansers which produce olive oil, on the one hand and the aqueous phase called vegetation water or margine containing the hydrosoluble olive components, on the other hand.

For 100 kg olives, this traditional pressure process, by far the most practical, produces: 20 kg oil and 40 kg olive oil meal, with 8–12 per cent of oil unextracted by pressure and 40 kg margine with low oil content.

Other Mechanical Processes for Obtaining Virgin Oil

Over the years numerous alternatives to the traditional pressure process have been proposed. Each of the large producer countries has set up experimental or model workshops where the new processes developed by the equipment industry are tested.

In the alternative technique which has received the most attention, pressure is replaced by direct centrifugation of the paste. A significant gain in productivity has been obtained by this new technique.

The initial steps of the new process is similar to those of the pressure process. It begins with washing and grinding by hammer crushers. The mixing operation is improved in order to enhance the paste's consistency, viscosity and homogeneity for the centrifuging process. Often water is added.

Some processes also include the removal of the kernel in these pretreatment operations. Carried out in continuous kernel removers, this process results in a more homogeneous and fluid paste, without hard fragments from the crushed kernel. The quality of the final olive oil meal is improved, but at a higher price.

The core of the process is the centrifugation which enables separation of the liquid and solid phases. The horizontal centrifugal separators used are quite impressive due to their size and speed of rotation. Two models are used, depending on whether they have two or three discharge points. The model with three discharge points enables the separation of:

1. A solid phase corresponding to the olive oil meal, still quite wet and fatty.
2. An aqueous phase corresponding to the margines, still containing some oil.
3. An oily phase constituted by virgin oil, still containing some impurities.

The margine and oil phases are then treated by centrifugation, the former for extraction of residual oil and the latter for clarification and purification.

As can be understood from this simple description, the direct centrifugation process is quite complex and requires qualified operation. Well-known equipment manufacturers-CentriOlive for example – offer turnkey installations. Their main advantage is greater productivity for virgin oil of quality comparable to that of pressured oil. The inconveniences are mainly the production of more humid olive oil meal (up to 50 per cent), on the one hand and the production of a clearly higher volume of margines due to addition of water, on the other hand. This raises the problem of their purification and upgrading. This is one of the major difficulties faced by olive oil mills. Several studies are being conducted, particularly in France, to find practical and economically viable solutions to this problem.

Another type of process known as the 'percolation' or 'selective filtration' process. It is based on the difference of tension at the oil/metal and water/metal interfaces, a characteristic already discovered during the first decade of this century. This process seems to be interesting and competitive. It has not, however, met with the expected success.

Other processing Methods for Olive Oil and Olive Oil Meal

Oil obtained by one of the mechanical processes described above can, if necessary, undergo one of the following additional treatment:

1. Super or fine virgin oil is simply filtered through paper or inert earth to eliminate the last traces of impurities and water, giving it its typical brilliance. This treatment does not in any way alter the oil's properties and characteristics nor its denomination.
2. Crude virgin oil, which has higher acidity, may be consumed in some producing regions. But generally it is refined by the normal refining processes including alkaline neutralisation, discolouration and deodourisation. Oil obtained by this method is known as refined olive oil.

As for the olive oil meal obtained by the traditional pressure process or by the centrifuging process, which still contains 8–12 per cent oil, it is generally submitted to hexane extraction in order to recover the residual fat. The process is the same as that used for cakes of oleaginous plants. As cakes are fairly moist, initial drying is indispensable. This is a comparatively costly operation and therefore inflates production costs.

The oil obtained is known as olive meal oil. It is either used as such or refined to obtain refined olive meal oil (Fig. 5.1).

EXTRACTION OF OLIVE OIL

Ripe olives contain a variety of components, including water, oil, sugars, proteins, organic acids and cellulose. Olive cultivars with medium-size fruits generally provide the best oil yields. The pulp-to-kernel ratio of olives for oil production ranges from 4:1 to 8:1.

The epicarp contains a number of components of relatively high polarity that are not removed by mechanical extraction and remain in the pomace. Removal of these components along with the oil by solvent extraction accounts for the higher unsaponifiable content of olive-pomace oil.

Most of the oil (96–98 per cent) is in the pulp along with most of the water 'vegetation water' (VW), which accounts for 40–60 per cent of the weight of the fruit. The woody pit inside the mesocarp holds a seed whose oil is more unsaturated than the mesocarp (pulp) oil because of a higher content of linoleic acid. The ratio of fruit oil to seed oil is 50:1.

The approximate chemical composition of olive fruit is as follows: water 52.4 per cent; oil 19.6 per cent; proteins 1.6 per cent; sugars 19.1 per cent cellulose 6.8 per cent; and ash 1.5 per cent. Oil yield and quality depend on the cultivar of olive tree, ratio of the various anatomical parts and levels of minor components as well as growing conditions and health of the trees. Soil moisture is very important during fruit development.

Harvesting of fruit for oil production begins in the middle of autumn and lasts until the end of February. In some regions it begins earlier and in other locales it lasts until March. Accordingly, differences in oil quality and composition can be expected along with variations caused by climatic and soil conditions. Variations in quality are chiefly related to the levels of minor components and flavour compounds, acidity and the presence of mono- and diglycerides.

Analytical and organoleptic data show that oil content is lower at the beginning than at the end of the harvesting period, but is of higher quality. Harvesting technology is very important for production of high quality oil. Olives should be collected as soon as they reach optimal maturity; however, it is difficult to have mechanical collection devices available where and when needed. In addition, because of the conformation of the tree branches, strong adherence of the fruit to the tree and limited accessibility, most olives are picked by hand.

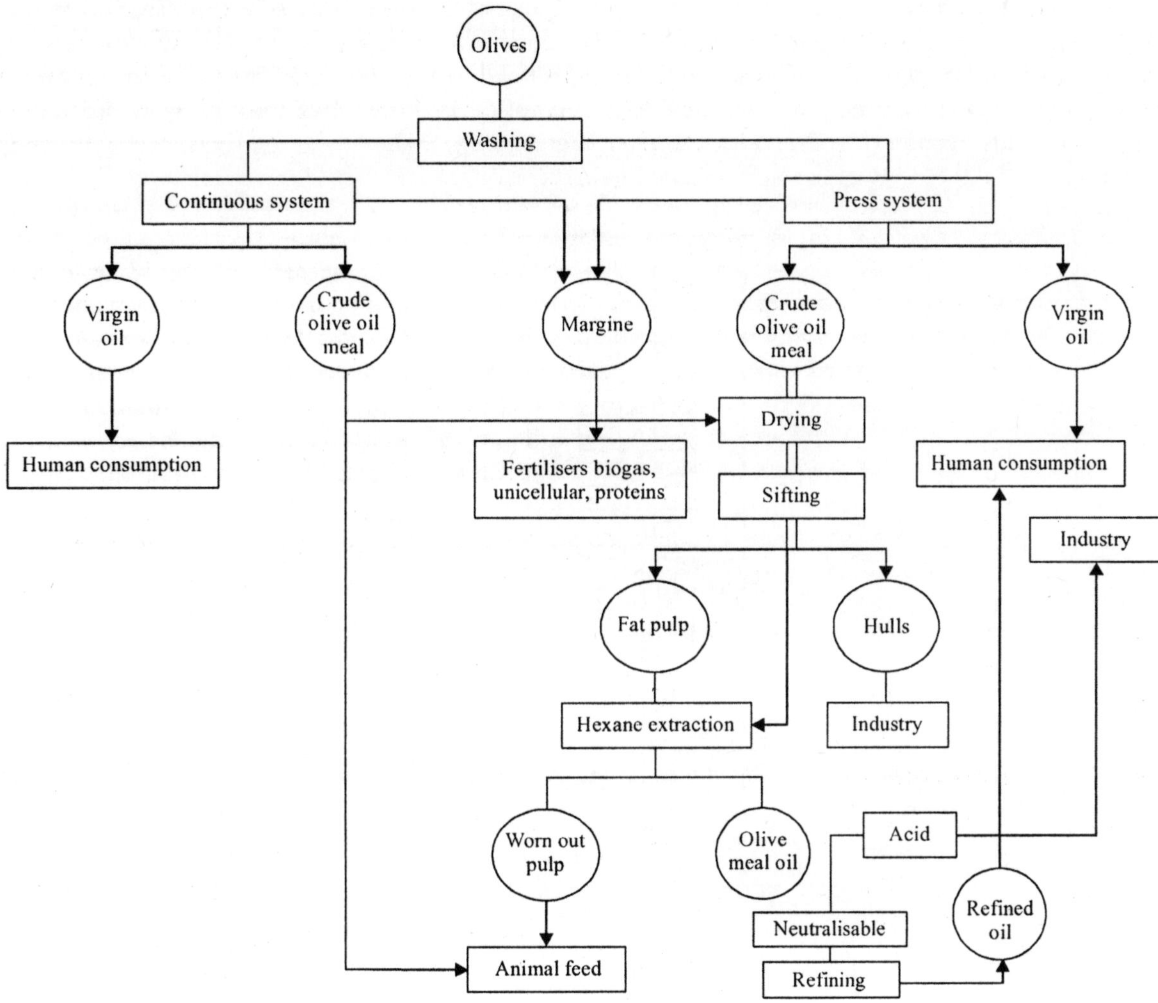

Fig. 5.1. Optimum olive processing.

Another harvesting procedure is to wait until the olives drop naturally and then collect the fruit with a system of nets. When the ripening period is delayed, both this procedure and handpicking are used. Although attempts have been made in the past to use chemicals to influence dropping time, however, chemicals are seldom used. Mechanical devices must be used with caution so that neither the tree nor the branches are damaged. When mechanical devices are used, the olive fruits are caught in nets to avoid and damage to the fruit contact with the ground.

Under optimum conditions, the olives are transferred from the nets to cages (usually plastic), forming layers not higher than 30 cm each and the olives are sent promptly to the extraction plant. In most regions of Italy and Greece, cages are stored no more than 3 to 5 days before extraction. This procedure ensures high quality oil if climatic conditions are good, the trees receives proper care and the fruit are not damaged by pests. If proper precautions are not taken and the olives are collected in large batches and held in piles several meters high, the fruit may be damaged. The enzymes released will cause hydrolytic and oxidative transformations resulting in off-flavours that affect the quality of the oil. Even with low acidity, such oils will have an unpleasant taste not acceptable for virgin oil and will have to be refined. Because of the difference in price between virgin and refined oils, economic losses to the farmer can be high. Three systems are used for mechanical extraction of oil from the olive fruit: pressure processing (Fig. 5.2); centrifugation (Fig. 5.3); and adhesion filtering (Fig. 5.4).

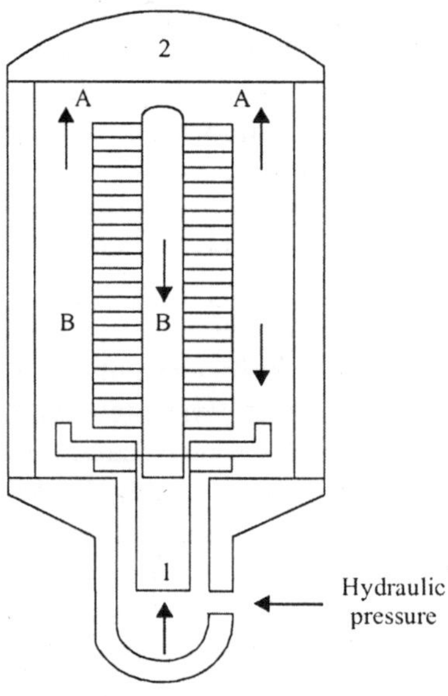

Fig. 5.2. Pressure extraction of oil. A, movement of the rack; B, movement of the oil; 1, mobile head; 2, fixed head.

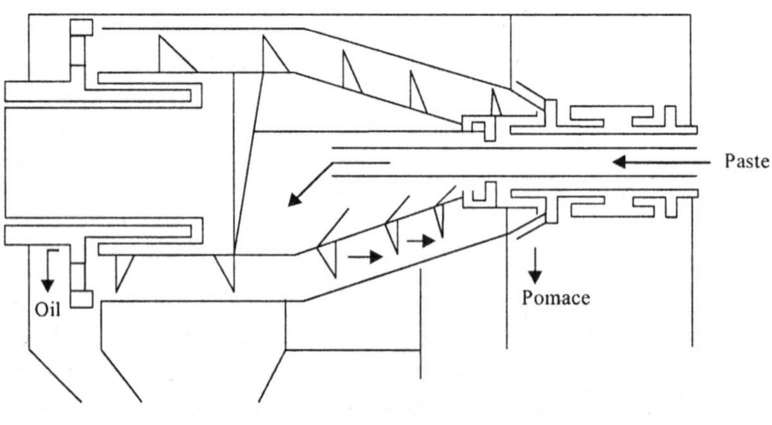

Fig. 5.3. Centrifuge for oil extraction from olive-pomace.

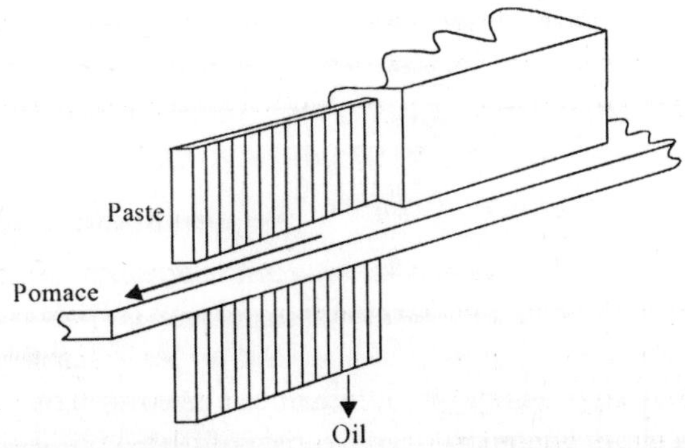

Fig. 5.4. Diagram of adhesion extraction of oil.

Pressing is the oldest and most often used method for olive oil extraction. High speed rotating machines are used for centrifugation extraction. With adhesion filtration, a series of steel plates or blades are dipped into olive paste; when withdrawn, the oil drips off the blades.

Several processing steps are required before extraction. The fruit must first be cleaned to eliminate branches and leaves and any extraneous materials that might damage plant equipment. The fruit is then washed to remove dirt and agricultural contaminants and finally crushed and milled to a coarse paste (Fig. 5.5). During the last step, enzymatic action breaks up the bitter components and reduces the level of peppery constituents while increasing the amount of minor polar components and tocopherols in the oil. If enzymatic action is prolonged, the minor polar components break down into water-soluble compounds that are removed from the oil, causing the loss of much of the antioxidant strength of the oil. Milling releases the oil from the oil-bearing cells and helps smaller droplets of oil to merge into larger drops, thus preparing the fruit for the following extraction step. A solid residue and vegetation water are produced during extraction in addition to oil (Fig. 5.6).

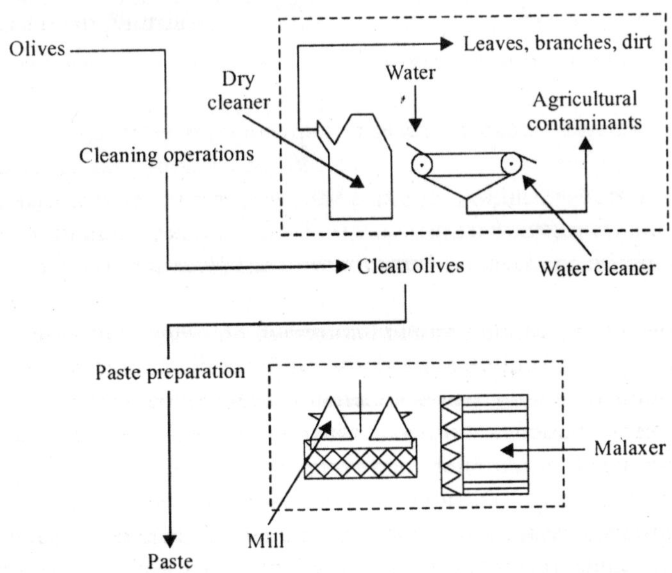

Fig. 5.5. Flow diagram of steps to prepare olives for extraction of oil.

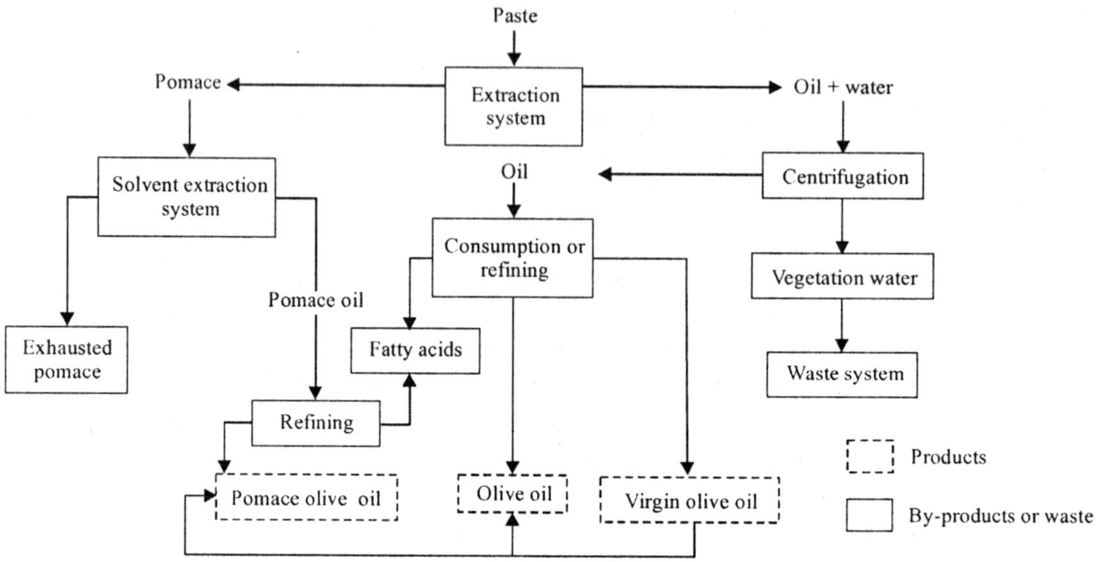

Fig. 5.6. Flow diagram of olive oil extraction and processing to yield olive oil products and by-products.

The vegetation water must be purified before discharge into a municipal sewer. A number of alternative technologies are available for waste water purification; however, they are costly and difficult to apply. If suitable for consumption, the oil is centrifuged after extraction to eliminate solid impurities and residual water. If the free fatty acid content is too high or organoleptic properties are unsatisfactory, the oil is refined. At the solvent extraction plant, the cake (pomace) containing up to 8 per cent residual oil is dried in a rotary kiln before proceeding to the solvent extraction unit, usually a semicontinuous system (Fig. 5.7). The extracted pomace oil is always refined. Spent cake is used as fuel or is separated into two fractions, the pulp (including skin) and the pit. In addition to use as fuel, the pit is occasionally used to produce fibreboard.

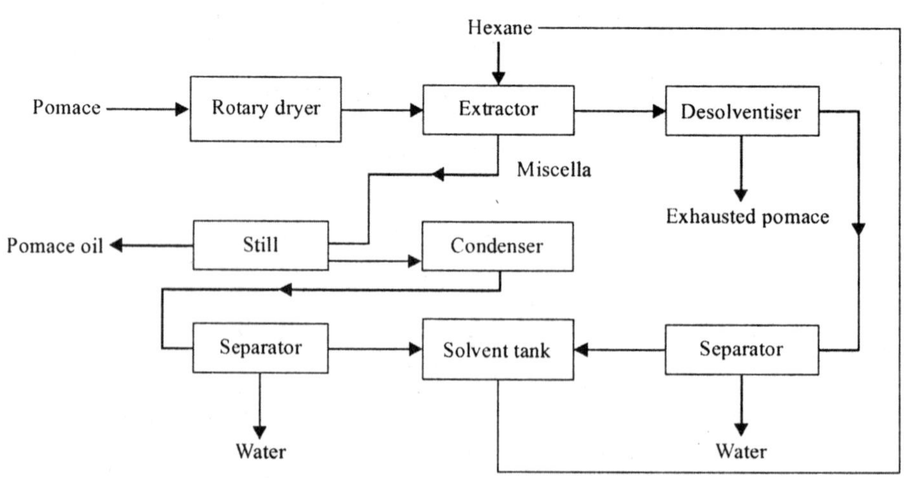

Fig. 5.7. Flow diagram of solvent extraction of pomace.

REFINING OF OLIVE OILS

Olive oil refining is carried out in either of the two ways: by alkali refining, generally used for animal and vegetable oils and fats; or physical refining, a technology not usually used for seed oils. Flow diagrams of the two procedures are shown in Figs 5.8 and 5.9. In the first procedure the oil is treated with dilute acid to precipitate the gums (phosphatides and proteinaceous material), which are separated by settling or centrifugation. Phosphoric acid and citric acid are the two most common degumming agents. After degumming, the oil is neutralised (alkali refined) either in a batched or continuous system.

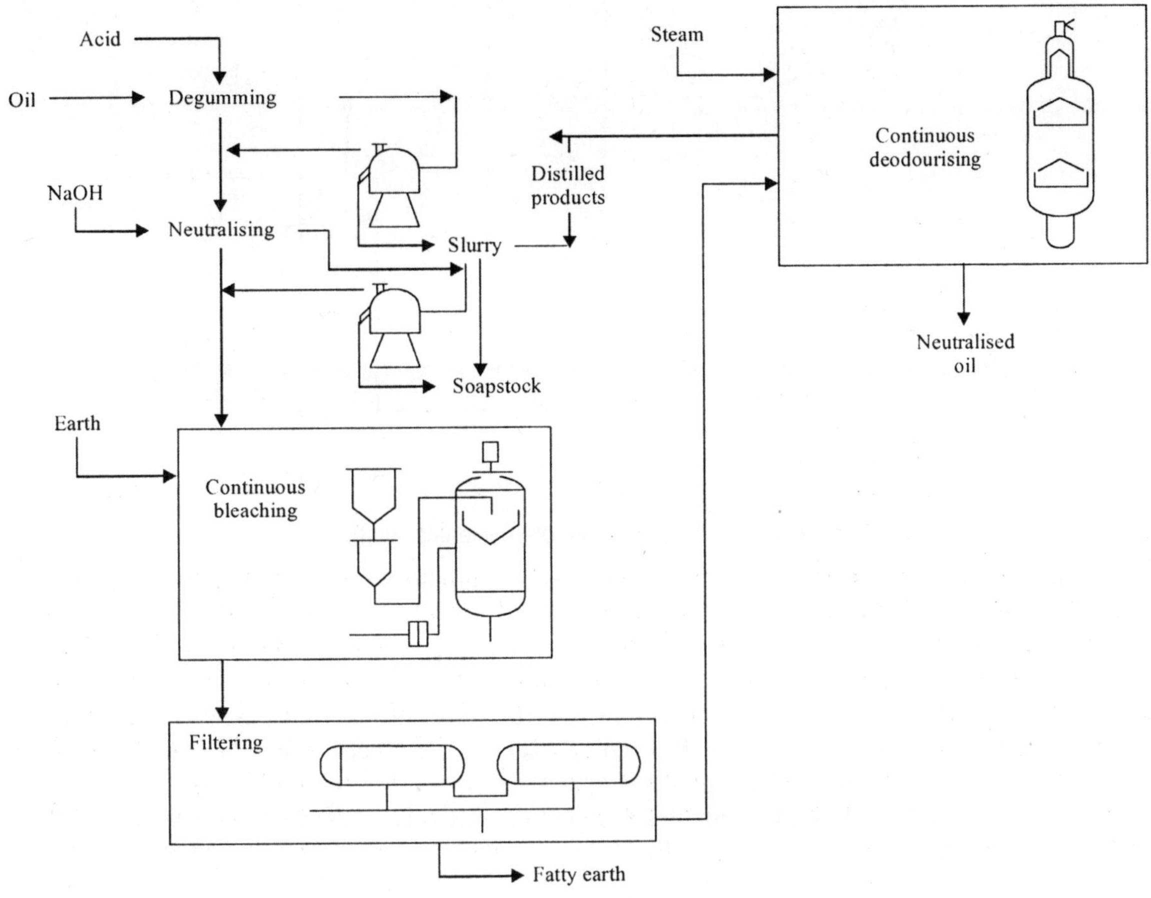

Fig. 5.8. Flow diagram of alkali refining.

Batch neutralisation is currently preferred because centrifuging of only the settled soap fraction lowers the neutralisation coefficient values, thereby shortening the washing time of the oil. The separated soap solution is acidified with sulphuric acid to recover the free fatty acids (containing 30–40 per cent triglycerides) for industrial applications. The alkali-refined oil is then bleached under vacuum with mixtures of various adsorbents (bleaching earth or clay and sometimes small amounts of activated carbon) and filtered by any of a number of available filter presses occasionally equipped with a solvent system for recovering oil entrained in the bleaching earth.

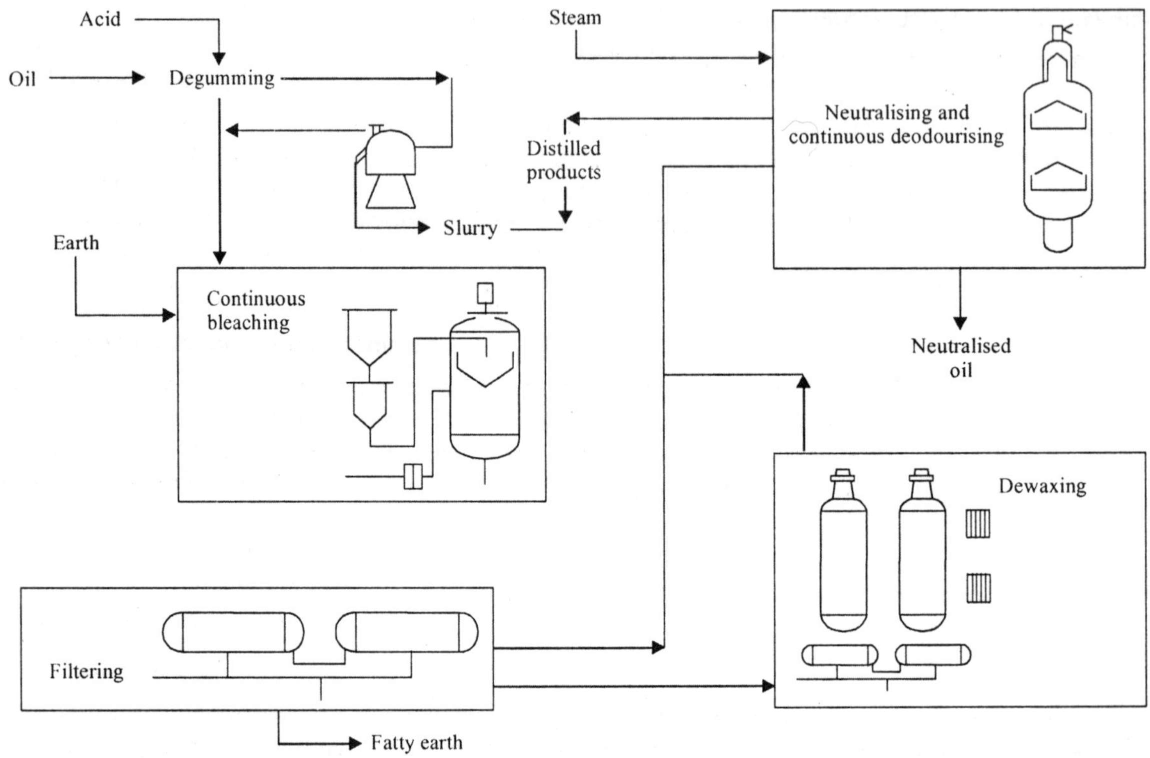

Fig. 5.9. Flow diagram of physical refining.

The bleached oil is deodourised in a semicontinuous or continuous deodouriser operation at a vacuum of less than 2 mm Hg. The final step involves mixing refined oil with virgin oil to improve the organoleptic and keeping properties of the oil. A good olive oil will contain at least 20 per cent virgin oil, but the product must, of course, meet consumer preference which sometimes requires a very light flavour and taste. With physical refining, the oil is first degummed and bleached and then fed to a continuous distillation (deodourisation) unit which removes the free fatty acids (92–95 per cent) and volatiles. The refined oil is blended as above. Frequently, distillation is stopped before removal of all of the free fatty acids and the oil is alkali refined to remove the remainder of the free fatty acids. This procedure has the advantage of eliminating oxidation by-products and pro-oxidant metals, thus improving product stability.

REFINING OF POMACE OIL

The technologies adopted for refining pomace oil are based primarily on physical refining because the acidity of pomace oils is about 10 per cent (expressed as oleic acid). Because degumming of pomace oil requires more drastic conditions than those for pulp oil, larger amounts of acidulant are used (phosphoric acid is preferred) and occasionally, the precipitate (gum) which entrains a high proportion of oil, is centrifuged to recover the oil. Larger amounts of bleaching earths are required to remove the intense green colour of the oil. Additional processing of the bleached oil usually follows the same procedures described for physical refining of olive oil, including incomplete distillation (deodourisation) followed by alkali refining of the partially deodourised oil.

Dewaxing (winterisation) of pomace oil is mandatory because of its high content of waxes (olive oil may also be winterised, especially if it is used to produce margarine or mayonnaise). Higher melting point triglycerides are also removed. Winterisation can be carried out after bleaching or following partial deodourisation and alkali neutralisation (alkali refining). If alkali neutralisation is performed at a low temperature, winterisation can be carried out simultaneously. A continuous apparatus is generally used for winterisation (Fig. 5.9) coupled with continuous filtering units. The winterised oil is then blended with virgin oil to restore the oil's antioxidant properties.

OLIVE OIL COMPONENTS

Glycerides account for at least 97 per cent of a virgin oil if the acidity is disregarded. The free fatty acid content is used to distinguish the various classes of virgin oil, from extra virgin to lampant. It must be emphasised that virgin olive oil is a natural product and therefore subject to variations in composition, both qualitative and quantitative. The origin, cultivar, extraction technology, state of ripening of the fruit, climatic conditions and rainfall all influence biosynthesis within the fruit and, therefore, the composition and quality of the oil.

Several classes of minor components are present in virgin olive oil. The structure, concentration and number of these substances are characteristic of virgin oils. Some are minor glyceridic components (MGC); other fall into other categories. (i) hydrocarbons; (ii) tocopherols; (iii) linear short chain alcohols and their esters; (iv) linear long chain alcohols and their esters; (v) sterols and their esters; (vi) α-methyl sterols; (vii) monohydroxytriterpenes; (viii) dihydroxytriterpenes; (ix) triterpenic acids; (x) phytol; (xi) geranylgeraniol; (xii) phenols and related compounds; (xiii) flavour components; (xiv) methyl and ethyl esters; and (xv) other components.

Minor Glyceridic Components (MGC)

Monoglycerides (MG) and diglycerides (DG) in the olive fruit are due to enzymatic hydrolysis of the triglycerides and incomplete triglyceride biosynthesis. In general, DGs are more abundant than MGs. Determination of DG concentration is useful for evaluating oil freshness and time of fruit harvesting because the DG level is strongly related to climatic influences. DG concentration can even be used to determine the source of an oil, even a refined oil, since the DG content of edible virgin olive oils differ from that of high acidity oils or solvent-extracted oils. Phospholipids are essentially absent from olive oil.

Non-Glyceridic Minor Components

Hydrocarbons

Both even and odd chain *n*-paraffins, including branched chain (*iso* and *anteiso*) compounds, which are minor components of the hydrocarbon fraction, are present in virgin olive oil. The polyunsaturated triterpenic hydrocarbon squalene and biochemical precursor of sterols, is the main component of the hydrocarbon fraction. The squalene content of olive oil ranges from 150 to 700 mg per 100 grams. β-Carotene is also present in olive oil as are aromatic hydrocarbons including benzenoid, napthalenic and more complex aromatic hydrocarbons.

Linear short chain alcohols and their esters

Methanol and ethanol esters of the fatty acids present in olive and in the same proportions as in the olive are present among the volatile compounds in virgin olive oil.

Straight long chain alcohols

Linear long chain alcohols with carbon numbers between C22 and C32 are present in olive oil both free and esterified (waxes). The components are abundant in the epicarp of the fruit and concentrate in solvent extracted oil. Phytol, probably derived from biodegradation of chlorophyll, is also present along with geranyl.

Cyclic monohydroxy compounds

Triterpenic tetra and pentacyclic monohydroxy compounds are quite characteristic of olive oils. The following compounds have been shown to be present, accompanied by small amounts of lanosterol and obtusifoliol:

Tetracyclic: Cycloartenol 24–methylene cycloartanol.

Pentacyclic: α-amyrin β-amyrin.

Methylsterols (4–desmethyl triterpenes) and sterols (4,4-di-desmethyl triterpenes) present in olive oils are derived from the tetracyclic alcohols. The following methyl sterols (4α-methyl–7–cholesten–3β-ol compounds) are present: 24–methylene, 24–methyl-, 24–ethylidene and 24–ethyl.

The main sterols of olive oil are: campesterol, stigmasterol, clerosterol, β-sitosterol, sitostanol and δ–5–avenasterol.

These are accompanied by small amounts of: cholesterol (max. 0.5 per cent), brassicasterol (max. 0.1 per cent), 24–methylenecholesterol (max. 0.5 per cent), campestanol (max. 0.5 per cent), δ –5,24,–stigmastadienol (max. 1 per cent), δ–7–stigmastenol (max. 0.5 per cent) and δ–7-avenasterol (max. 1.1 per cent).

Analysis of the sterol fraction isolated from the unsaponifiable fraction is very important, for determining the authenticity of the oil. The triterpenes and sterols are present both as free alcohols and as fatty acid esters.

Cyclic dihydroxy compounds

Pentacyclic triterpenes in olive oil include 3β, 17β-dihydroxy-12-oleanene (erythrodiol) and its parent compound uvaol, obtained largely from the epicarp and therefore characteristic of solvent extracted oils.

Triterpenic acids

The following pentacyclic mono-and dihydroxy triterpenic acids are present in virgin olive oil 3β-hydroxy-17-carboxy-δ-12-oleanene (oleanolic acid); 3β,2α-dihydroxy-17-carboxy-δ-12-oleanene (maslinic acid); 3β-hydroxy-17-carboxy-δ-12-ursene (ursolic acid); 2α, 3β-dihydroxy-17-carboxy-δ-12-uresene (2α -hydroxyursolic acids); and deoxyurasolic acid (structure not fully elucidated).

Chlorophylls

Both chlorophyll α and chlorophyll β are present in olives and are partially extracted into the oils.

Flavor components

Olive oil volatiles contain at least 100 compounds in several categories: hydrocarbons (5 compounds), aliphatic alcohols (13 compounds), terpenic alcohols (4 compounds), aldehydes (27 compounds), ketones (8 compounds), ethers (2 compounds), furans (3 compounds), thiophenes (6 compounds) and esters (29 compounds).

Minor Polar Components

The olive mesocarp contains a number of phenolic and polyphenolic compounds and their esters, small amounts of which are present in olive oil. These include monohydroxy-and dihydroxy-phenylethanol, including tyrosol and other phenols and a series of carboxy-phenols, including caffeic, *o*-coumaric, *p*-coumaric, cinnamic, ferulic, gallic, *p*-hydroxybenzoic, protocatechuic, sinapic, syringic and vanillic acids. Benzoic and cinnamic acids are produced by hydrolysis of flavonoids. The hydroxyphenyl-ethanols arise from hydrolysis of oleoeuropein. Their esters are responsible for the bitterness and pepperlike sensation occasionally dominant in the taste of olive oils. Olive oil contains α-tocopherol in the range of 12–190 mg/kg. According to one report, olive oil tocopherols were found to consist of 88.5 per cent α-tocopherol, 9.9 per cent β- + γ-tocopherol and 1.6 per cent δ-tocopherol. Tocopherol content can be used to detect adulteration of olive oil with seed oils.

ANALYSIS OF OLIVE OILS

Olive oil is initially examined to determine purity, then to place it in the proper category and finally to establish its quality.

Determination of Purity

Sterol composition

Sterol analysis involves preparation of the unsaponifiable fraction, fractionation by thin layer chromatography (TLC) and gas chromatographic analysis of the TMS derivatives. The following limits apply to all types of olive oils (Table 5.6).

Table 5.6. Sterol composition of olive oils.

Sterol	*Sterol fraction (%)*
Cholesterol	Max. 0.5
Brassicasterol	Max. 0.1
Campesterol	Max. 4.0
Stigmasterol	Less than 4.0
δ–7–stigmastenol	Max 0.5
The sum of the following sterols must be more than 93.0% of the sterol fraction:	
β-Sitosterol	
δ–5–Avenasterol	
δ–5,23–Stigmastadienol	
Clerosterol	
Sitostanol	
δ–5,24–Stigmastadienol	

Total sterol content

The gas liquid chromatographic method for sterol determination using an internal standard (cholestanol) is used to calulate the absolute (total) sterol content of an oil. Limits (mg/100 g) are as follows: virgin olive oil, refined olive oil and olive oil (mixture of refined and virgin) > 100; crude olive-pomace oil > 250; and refined olive-pomace oil, olive oil and olive-pomace oil (mixture) > 180.

Fatty acids composition

Olive oil triglycerides are converted into methyl esters and the methyl esters are analysed by gas liquid chromatography (GLC). The limits of fatty acids in genuine olive oils are as follows (per cent m/m) (Table 5.7).

Table 5.7. Minimum and maximum limits of fatty acids genuine olive oils.

Acid	*CAN*[a]	*Minimum*	*Maximum*
Myristic	14:0	–	0.05
Palmitic	16:0	7.50	20.00
Palmitoleic	16:1	0.30	3.50
Heptadecanoic	17:0	–	0.30
Heptadecenoic	17:1	–	0.30
Stearic	18:0	0.50	5.00
Oleic	18:1	55.00	83.00
Linoleic	18:2	3.50	21.00
Linolenic	18:3	–	0.90
Arachidic	20:0	–	0.60
Eicosenoic	20:1	–	0.40
Behenic	22:1	–	0.20
Lignoceric	24:0	–	0.20

[a] CAN = carbon atom number.

Saturated fatty acids in position 2 of the triglycerides

Hydrolysis with pancreatic lipase is followed by thin layer chromatographic isolation of the monoglyceride fraction, which is converted to methyl esters. The methyl esters are analysed by GLC. Maximum acceptable level is the sum of palmitic and stearic acid (per cent m/m):

Virgin olive oil	≤ 1.5
Refined olive oil	≤ 1.8
Olive oil (mixture of refined and virgin)	≤ 1.8
Crude olive-pomace oil	≤ 2.2
Refined olive-pomace oil	≤ 2.2

Absolute difference between found and theoretical equivalent carbon number (ECN) 42 (trilinolein) values

The triglyceride composition of the oil is determined by high performance liquid chromatography (HPLC). (A chromatogram of an olive oil sample (ECN 42, 0.8 per cent) is shown in Fig. 5.10.

The theoretical triglyceride composition is calculated with a Lotus 123 programme provided by the IOOC. The maximum difference of theoretical ECN 42 *vs* ECN 42 found is calculated. (ECN = CN–2*n*, where CN is the carbon number and *n* is the number of double bonds.) The maximum difference between the real and theoretical ECN content of olive oils and olive-Pomace oils should be 0.3 and 0.5, respectively. This procedure avoids errors due to miscalculation of trilinolein alone.

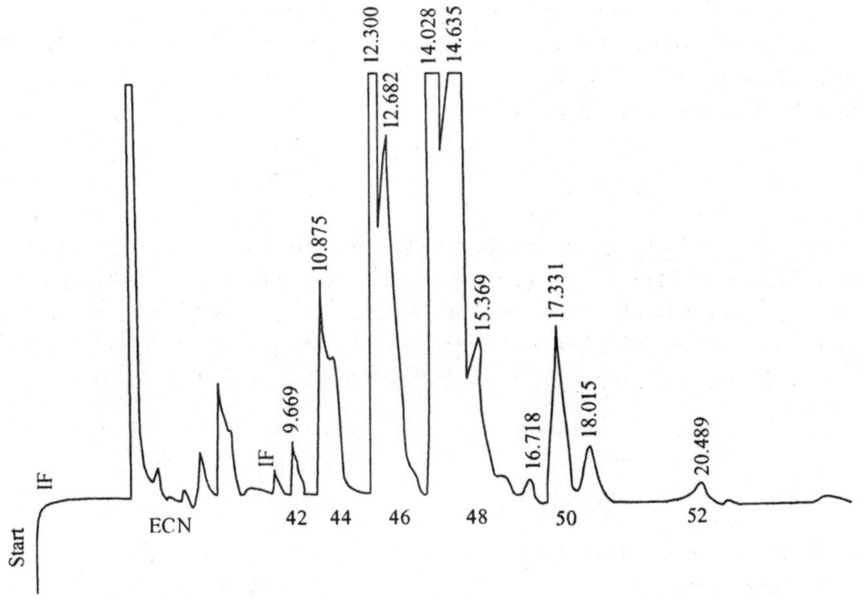

Fig. 5.10. HPLC chromatogram of olive oil triglycerides. Column: LC–18, 200 × 4.6 mm i.d.; mobile phase: acetone: acetonitrile (60 : 40, v/v); flow rate: 0.75 mL/min; refractive index detector; oven and detector temperature: 40°C. IUPAC method 2.324 (72) with injection of 10 mL test sample diluted 1: 20 with acetone. ECN 42, 0.8 per cent of total triglycerides.

Trans fatty acids content

Trans fatty acids arise during refining of vegetable oils as well as during hydrogenation, or from attempts to eliminate the sterol fraction of seed oils with a fatty acid composition similar to that of olive oil. Methyl esters are analysed by capillary column GLC. The following limits (per cent m/m) are mandatory (Table 5.8).

Table 5.8. Limits of methyl esters analysed by GLC.

Oil	*18:1 trans*	*18:2 trans + 18:3 trans*
Virgin olive oil	< 0.03	< 0.03
Lampant olive oil	≤ 0.10	≤ 0.10
Refined olive oil	≤ 0.20	≤ 0.30
Olive oil (mixture of refined and virgin)	≤ 0.20	≤ 0.30
Crude olive-pomace oil	≤ 0.20	≤ 0.10
Refined olive-pomace oil	≤ 0.40	≤ 0.35
Olive-pomace oil and olive oil mixture	≤ 0.40	≤ 0.35

Differentiation between Olive Oil and Olive-Pomace Oil

Wax content

Olive oil fatty acid esters of straight chain alcohols (wax esters present in solvent extracted olive-pomace oil are isolated by column chromatography on silica gel (LC) and quantitated by GLC to determine if olive-pomace oil is present in olive oil. LC separation of the wax esters can be replaced with HPLC to automate the separation step and improve reliability and repeatability. Limits for content of C40 + C42 + C44 + C46 wax esters (mg/kg) are as follows:

Virgin olive oil	≤ 250
Lampant olive oil	≤ 350
Refined olive oil	≤ 350
Olive oil (mixture of refined and virgin)	≤ 350

Dihydroxyterpene alcohol content

Olive-pomace oil contains relatively high levels of erythrodiol, uvaol and wax esters. Erythrodiol and uvaol (total diol) content is determined by the same procedure as that used for sterol analysis. Limits for total diol content (as per cent of total sterols) are as follows:

Virgin olive oil	≤ 4.5
Lampant olive oil	≤ 4.5
Refined olive oil	≤ 4.5
Olive oil (mixture of refined and virgin)	≤ 4.5

Differentiation between Virgin and Refined Olive oil and Detection of Refined Olive Oil and Seed Oils in Virgin Olive Oil

Concentration of stigmasta-3,5-diene

When olive oil and seed oils are refined, stigmasta-3,5-diene is produced by dehydration of β-sitosterol, the parent sterol. Refined olive oils contain significant amounts of stigmasta-3,5-diene (3–100 mg/kg) not present in any significant amount in virgin olive oils. Refined seed oils also contain significant amounts of steroidal hydrocarbons, including campesta-3,5-diene and stigmasta-3,5,22-triene in addition to stigmasta-3,5-diene. The relative amounts of these steroidal hydrocarbons can be used to detect refined seed oils or seed oils desterolised for the purpose of adulterating olive oil. Isolation of the hydrocarbon fraction from the unsaponifiables by column chromatography on silica gel followed by GLC is used to determine the concentration of stigmasta-3,5-diene and accompanying hydrocarbons. A chromatogram of the hydrocarbon fraction from an olive oil is shown in Fig. 5.11. A chromatogram of the hydrocarbon fraction from an olive oil admixed with desterolised, refined seed oil is shown in Fig. 5.12.

Ratios of stigmasta-3,5-diene to campesta-3,5-diene (R1) and stigmasta-3,5-diene to stigmasta-3,5,22–triene (R2) are determined when the level of stigmasta-3,5-diene exceeds 4 ppm (Table 5.9).

Table 5.9. Ratios of stigmasta-3,5-diene (R_1) and stigmasta-3,5-diene to stigmasta-3,5,22-triene (R_2).

Oil	*Stigmasta–3,5–diene (ppm)*	*R1*[a]	*R2*[b,c]
Virgin olive oil	≤ 0.1	–	–
Lampant olive oil	≤ 0.5	–	–
Refined olive oil	≤ 50.0	≥ 15	≥15
Olive oil	≤ 50.0	≥ 15	≥ 15
Crude olive-pomace oil	≤0.5	≥ 15	≥ 15
Refined olive-pomace oil	≤120.0	≥ 15	≥ 15
Pomace and olive oil mixture	≤120.0	≥ 15	≥ 15

[a] R1 = ratio of stigmasta–3,5–diene to campesta–3,5–diene.
[b] R2 = ratio of stigmasta–3,5–diene to campesta–3,5,22–triene.
[c] Provisional limits.

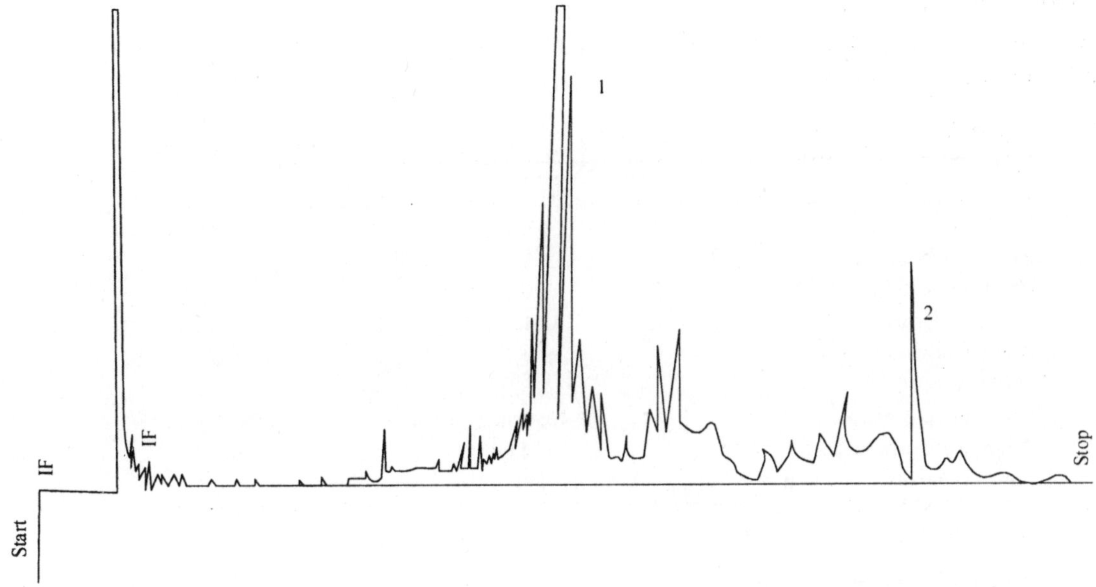

Fig. 5.11. Capillary GLC of the hydrocarbon fraction of olive oil (blend of refined and virgin olive oil). Column: DB-5,25 m × 0.25 mm i.d., 0.2 μ m film thickness; split ratio; 1 : 15; temperature programme: 235°C, 6 min; 20°C/min; 285°C final temperature; injector: 300°C; detector: 320°C; 1, cholesta–3,5–diene (internal standard); 2, stigmasta-3,5-diene.

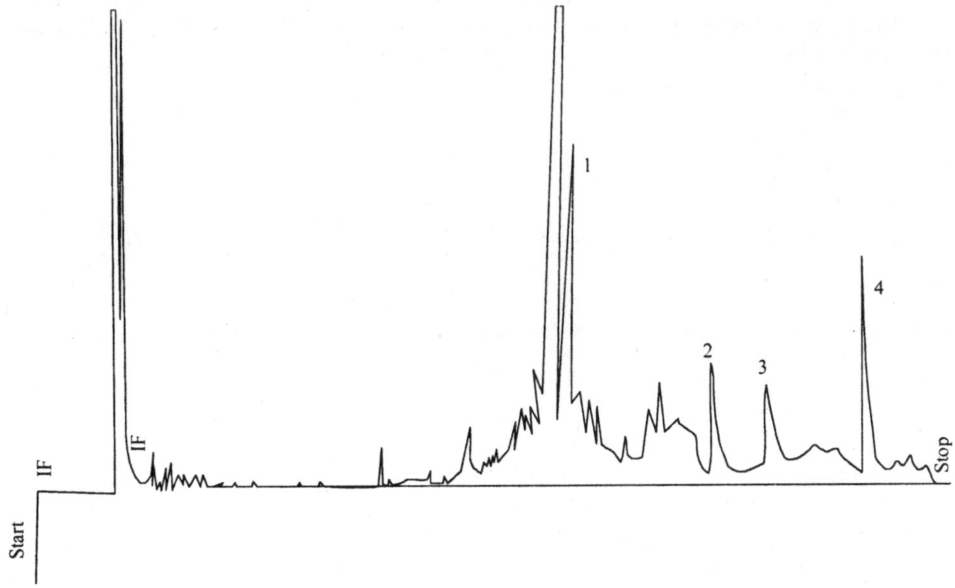

Fig. 5.12. Capillary GLC of the hydrocarbon fraction of olive oil admixed with a desterolised seed oil (GLC column and operating conditions as described for Fig. 5.11). 1, cholesta-3,5-diene (internal standard); 2, campesta-3,5-diene; 3, stigmasta-3,5,22-triene; 4, stigmasta-3,5-diene.

UV absorption at 268 nm

K (1 per cent, 1 cm) and related value, δ-K, are useful for readily classifying olive oil quality according to the following values (Table 5.10).

Table 5.10. K (1 per cent, 1cm) and related value, δ -K for classifying olive oil quality.

Oil	*K270 nm*	*δ-K*[a]
Extra virgin olive oil	≤ 0.25	≤ 0.01
Virgin olive oil (fine)	≤ 0.25	≤ 0.01
Virgin olive oil (semifine)	≤ 0.30	≤ 0.01
Lampant olive oil	No limits	No limits
Refined olive oil	≤ 1.10	≤ 0.16
Olive oil	≤ 0.90	≤ 0.15
Crude olive-pomace oil	No limits	No limits
Refined olive-pomace oil	≤ 2.00	≤ 0.20
Pomace and olive oil mixture	≤ 1.70	≤ 0.18

[a] δ-K = K268 – [(K262 + K274)/2].

Both K and δ-K are altered when oxidation products are present. In this case, the oil is dissolved in hexane and passed through an alumina column before measurement of K and δ-K.

Quality Parameters

Organoleptic characteristics

Organoleptic properties of virgin oil can be determined by a 'panel test' which gives results that are often controversial. Organoleptic testing is currently undergoing revision. The method is based on examination of virgin oil by a panel of 8 to 12 trained personnel who grade various characteristics and defects which are then converted into a number score. The following scores apply to various grades of virgin olive oil:

Extra virgin olive oil	> 6.5
Fine virgin olive oil	> 5.5
Semifine virgin olive oil	> 3.5
Lampant virgin olive oil	> 3.5

Free fatty acid content

Free fatty acid content (expressed as per cent oleic acid) is used to define the various grades of virgin olive oil:

Extra virgin olive oil	< 1.0
Fine virgin olive oil	< 1.5
Semifine virgin olive oil	< 3.3
Lampant virgin olive oil	< 3.3

Refined olive oil and mixtures have the following limits:

Refined olive oil	≤ 0.3
Olive oil	≤ 1.5
Refined olive-pomace oil	≤ 0.3
Olive-pomace and olive oil	≤ 1.5

Olive oil and mixtures of olive-pomace and olive oil have higher free fatty acid contents because they are generally mixed with virgin olive oils of high acidity.

Peroxide value (PV)

PV (expressed in meq per kg oil) allowed for various grades of olive oil is as follows:

Extra virgin, fine and semifine virgin olive oil	≤ 20
Refined olive oil	≤ 10
Olive oil	≤ 15
Refined olive-pomace oil	≤ 10
Pomace oil and olive oil mixture	≤ 15

Virgin olive oil contains components that interfere with conventional PV determination. Even freshly expressed olive oil have PV values of about 10 and under certain climatic conditions (dry weather) the PV value can be higher than 10.

Tocopherol content

Tocopherol can be determined by colourimetry or GLC, or by HPLC. Added tocopherols are not permitted in virgin olive oils and crude olive-pomace oils. Added α-tocopherol is allowed in refined olive oil, olive oil, refined olive-pomace oil and olive-pomace oil to restore natural tocopherol lost during refining with a maximum level of 200 mg/kg of total α-tocopherol in the final product.

Impurities

Water content of virgin olive oil should not exceed 0.2 per cent (m/m); for refined oil and mixtures (olive oil, olive-pomace and olive oil), the maximum value is 0.1 per cent; for lampant olive oil, 0.3 per cent; for crude olive-pomace oil, 1.5 per cent. Allowable hydrocarbon (hexane, petroleum ether) residues are as follows (per cent m/m).

Extra virgin, fine and semifine virgin olive oil	≤ 0.10
Refined olive oil, olive oil	≤ 0.05
Refined olive-pomace oil, olive-pomace and olive oil	≤ 0.05

The occurrence of μg/kg to mg/kg amounts of tetrachloroethylene in some olive oils led to an EEC regulation limiting the tetrachloroethylene content of olive oil and products containing olive oil to not more than 0.1 mg/kg, as determined by a head space/electron capture GLC method. Maximum allowable contents of iron and copper are 3 and 0.1 ppm, respectively. Smoke point is a function of acidity level in the oil. The smoke point for olive oil generally ranges from 150 to 163°C.

Combined Gas Chromatography-Mass Spectrometry (GC/MS)

GC/MS has become a powerful tool for identification and confirmation of the various components of olive oil. With GC/MS in the selective ion mode, unresolved GC peaks can be identified and accurately quantitated. For example, an apparent β-sitosterol peak in the sterol fraction was resolved into clerosterol (m/z 218) and Δ-5-avenasterol (m/z 314) and both sterols measured regardless of inadequate GC resolution. Italian and Spanish olive oil contained a very high level of 9,19–cyclolanosterol (>400 mg/kg) which was not found with the standard method for sterol analysis. Two isomers of this sterol were identified by GC/MS of the unsaponifiable fraction and their levels were found to be inversely proportional to the levels of β-sitosterol in the oils. GC/MS of the unsaponifiable fraction with high resolution GC capillary columns provides a relatively rapid means of checking product purity and the identity of individual components.

QUALITY AND SPECIFICATIONS: THE INTERNATIONAL COI STANDARD

The consumer's interest in olive oil is evidently based on its nutritional and sensorial qualities and its purity. Quality and purity should be guaranteed by certain labelling and specifications which must be:

1. Measurable by tested analytical techniques.
2. Agreed by a consensus of scientists and experts.
3. Published and officially recognised.

It is the goal of official organisations, particularly COI, to study these questions and define concrete quality and safety measures. This Standard includes several classes of oil and provides their denominations and definitions:

1. Virgin olive oil, which is subdivided into extra, fine, semi-fine, 'lampante', according to their acidity and other analytical criteria.
2. Refined olive oil.
3. Crude and refined oil from olive oil meals.
4. Different types of blends or mixtures of these oil.

The Standard dwells on purity criteria which include the composition of fatty acids, saturated acids in position 2 of the TG and sterols, on the one hand and values such as saponification, bellier, etc. on the other hand.

The Standard also expands on the quality critieria which include free fatty acids, peroxide value (less than 20 meq/kg), E absorbance in UV at 270 nm less than 0.25 for virgin oil, extra and fine and the E232/E270 ratio. The other series of quality criteria include organoleptic criteria such as odour, taste, colour, appearance. Sensorial analysis which helps in determining these properties of olive oil has made a lot of progress.

Chapter 6

Palm Oil

INTRODUCTION

Palm oil is an edible oil referred to by the FAO/WHO Codex Alimentarius as being derived from the fleshy mesocarp of the oil palm fruit. In the unprocessed form palm oil is reddish brown in colour and it has a semi-solid consistency at ambient temperature. One should not confuse palm oil with palm kernel oil, which is another product obtained from the kernel of the oil palm fruit while palm oil is derived from the mesocarp or fruit flesh. The two oils have different chemical composition and physical characteristics and they are used and marketed separately according to their own supply and demand situations.

Before the extraction of oil palm products can be considered, attention must be paid to the nature and characteristics of these products and the manner in which they are held in the harvested organ, the bunch. When the harvesting bunch arrives at the place of extraction, a portion of the fruit will already be loose; but the remainder will be held on the bunch for some time. Complete loosening of fruit from a bunch may take more than a week and if natural loosening is awaited the quality of oil, as indicated by the free fatty acid (FFA) content, will deteriorate, since a proportion of both the loose fruit and fruit on the bunch will inevitably be bruised and lipolysis, which is the splitting of fat molecules by hydrolysis into glycerol and fatty acids through enzyme action, will proceed.

It is recognised, therefore, that the rapid extraction of fruit from the bunch* is an important part of the product extraction process and that provision must be made to arrest the deterioration of palm oil quality even before the fruit has been removed from the bunch. This is to some degree recognised in non-mechanical processes through the careful cutting up of the bunches into sections and the hastening of natural loosening by sprinkling heaps of these sections with water and covering them up with leaves. The value of even this crude expedient is shown by the fact that low FFA oil can be produced, after hand picking, by non-mechanical extraction methods, or by mechanical methods not including equipment for dealing with bunches.

* As some mill engineers use the word bunch to indicate that part of the bunch which is left over after stripping, it is necessary to define the terms used in this chapter as follows:

Bunch = the whole ripe bunch with loose fruit as cut from the palm.
Fruit = the palm fruit as detached from the bunch before or during stripping and sometimes containing a proportion of perianth segments ('the calyx leaves').
Bunch refuse = that part of the bunch which is discarded in the stripper.

Apart from the overall retention of the fruit in the bunches, the palm oil is held in the cells of the fibrous mesocarp of the fruit and the kernel is firmly encased in the hard shell of the nut.

Palm oil is solid at ambient temperatures in a temperate climate. At tropical temperatures it is a fluid with certain fractions held in crystalline form. On settling, there is a clear liquid section and crystalline fluid base. The pre-requisites for the release of palm oil from the fruit are a physical breakdown of the mesocarp sufficient to rupture the cells, and a temperature sufficient to aid in this rupturing and fully to homogenise the fat constituents.

For the satisfactory release of the kernels from the fruit the requirements are that the oil-bearing mesocarp shall be removed and the shells cracked without damage to the kernels.

Palm kernel oil is not usually extracted on the plantations, though occasionally mills contain presses designed for this purpose. The conditions for the release of palm kernel oil, which is liquid at tropical day temperatures, are different from those of palm oil, but similar to those of copra and hard oil-bearing seeds. Very small quantities are extracted in producing countries by primitive means, but the greater bulk of the palm kernels produced are subjected to industrial processes. As the matrix is solid and hard it must be crushed to a meal before the oil can be extracted under pressure. Though hammer or attrition mills can be used, roller mills are usually employed to grind the material so fine that a high proportion of the oil-containing cells are ruptured. 'Cooking' in stack cookers under steam pressure releases the oil still further and expression of the oil is then undertaken in various types of hydraulic batch presses or in continuous screw-press expellers, or the oil may be extracted by solvents.

OIL PALM PRODUCTS, THEIR FORMATION AND CHARACTERISTICS

The three commercial products of oil palm fruit are palm oil, palm kernel oil and palm kernel cake. Chemically, the word 'fat' is coming to be used to cover vegetable oils and fats whether they are in the solid or liquid state, though in normal parlance the oil is applied to a fat when it is in the liquid state. Fats have been defined as the esters of fatty acids with the trihydric alcohol glycerol and they must be distinguished from other simple lipids like the waxes which are esters of fatty acids with high molecular weight, straight chain alcohols. The triglyceride fats, which predominate in plant and animal fats, have the following general formula:

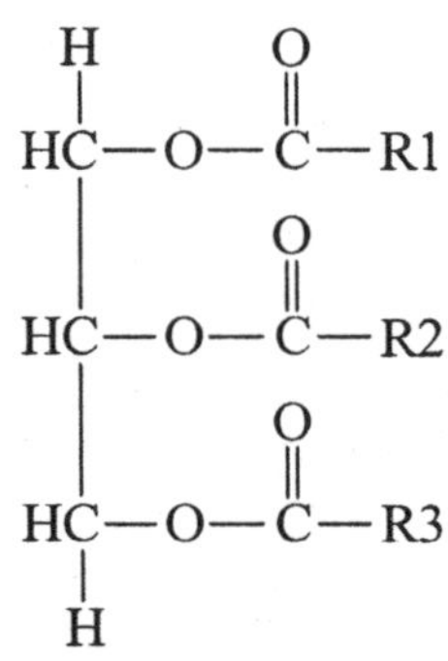

where R1, R2 and R3 represent the hydrocarbon chains of fatty acid radicals.

Fatty acids may also combine with glycerol to form mono- or diglycerides when only one or two of the hydroxyl groups of the glycerol will be in fatty acid combination, but these occur naturally only in fats which have become partly hydrolysed. When more than one fatty acid radical is involved, as in tri- or diglycerides, these may be alike or different. There being many different naturally occurring fatty acids, there will therefore be a multiplicity of fats formed from them.

Naturally occurring vegetable fats are mixtures of fats and their characters are taken largely from the fatty acids which predominate in them and from the arrangement of these fatty acids in the triglycerides. The fatty acids are hydrocarbon chains in which two hydrogen atoms are attached to all or the majority of carbon atoms within the chain. The carbon atom at one end of the chain has three hydrogen atoms attached to it and the one at the other end is attached to a carboxyl group to give the general structure:

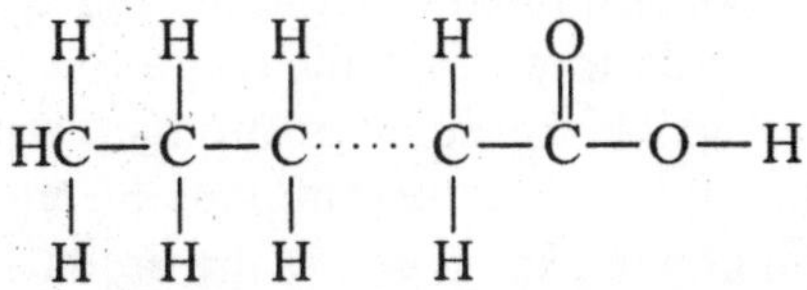

Fatty acids of this general formula are saturated fatty acids as they have the full number of hydrogen atoms attached to the carbon atoms of the chain. In unsaturated fatty acids there are one, two or three double bonds between carbon atoms which they have only single hydrogen atoms attached to them, in the following manner:

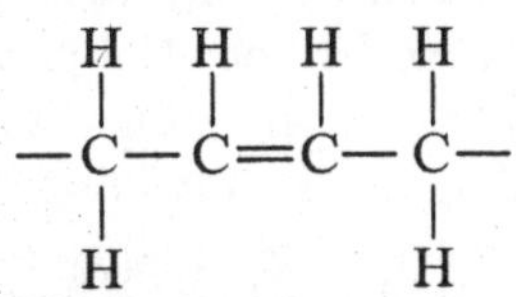

Double bonds can occupy different positions in the chain, thus giving rise to different isomers. Furthermore, there are also geometric isomers (the *cis* or the *trans* forms) according to whether portions of a molecule joined by a double bond extend in the same or opposite directions.

Chemically, the most important reaction of the fats from the producer's point of view is hydrolysis, i.e. the formation of free glycerol and free fatty acid through a splitting of the fat molecule and the addition of the elements of water which may be partly represented by the equation

$$CH_2OOCR + H_2O = CH_2OH + HOOCR$$

where CH_2 represents one-third part of the glycerol radical and R the hydrocarbon chain of the fatty acid radical. Hydrolysis can be either autocatalytic in the presence of water, be catalysed by metals, or be brought about by the action of the enzyme lipase. The latter is, of course, the fat-splitting enzyme of animal digestion, but it is also found in palm fruit and in fungi and other organisms which gain access to fats. One of the most important tasks of the palm oil producer is to prevent hydrolysis by reducing to a minimum the amount of water and impurities present in the oil and by the destruction of the enzyme. Hydrolysis in alkalies is distinguished as saponification and gives rise to soaps and glycerol.

The second important reaction of fats is oxidation. Unsaturated fats are commonly oxidised at the double bonds and the oxidation products, the first of which are hydroperoxides, lead to rancidity with the loss of palatability due to obnoxious flavours and odours and may effect the bleachability of the oil. In oil production the substances most likely to promote oxidation (pro-oxidants) are free atmospheric oxygen and traces of metals; the process is accelerated by light. Oxidation and consequent rancidity does not, however, proceed so fast in vegetable as in animal fats owing to the presence of naturally occurring protective materials or antioxidants.

Oxidation to hydroperoxides is measured as the peroxide value of a fat; this represents the reactive oxygen content and is estimated through the liberation of iodine from potassium iodide in glacial acetic acid and recorded in terms of milli-equivalents of peroxide-oxygen per 100 gram fat.

The third important reaction of the fat is hydrogenation, which is a process of manufacture. Hydrogenation or 'hardening' of fats has contributed more to the interchangeability of fats and fatty oils than any other process and therefore is a factor in the maintenance of stable economic conditions in the production of all fats. Broadly, hydrogenation processes add hydrogen atoms at the double bonds of unsaturated fats converting these into the higher melting point saturated fats. Thus the oleic ester triolein (the fatty acid of which is a primary constituent of palm oil and is present in palm kernel oil) can be directly hydrogenated to the stearic ester tristearine, the respective melting points being –32° to 5.5°C and 54° to 73°C. Fatty substances which are liquid at ordinary temperatures can therefore be made suitable for products which must be sold in the solid state. Apart from this main feature of hydrogenation processes, selective hydrogenation now makes possible the production of products containing almost any desired range of fats. These processes are particularly applied to marine oils and to subsidiary parts of liquid oils and palm oil.

Fourthly, mention may be made of the reaction which is used as a measure of the proportion of unsaturated constituents present in a fat. This is halogen addition to the double bonds of the unsaturated fatty acids and the quantity of halogen taken up is expressed in terms of iodine as the iodine value, which is the number of grams of iodine absorbed per 100 gram fat.

Lastly, with much larger quantities of palm oil coming on the market and fears, so far unfounded, that there might be difficulties in disposing of these quantities, there has been an increasing interest not only in providing a standard product of very high quality but also in the possibilities of improving overall income through fractionation of the oil in the countries of production. The liquid fraction would then be marketed as cooking oil for local consumption while the solid fraction would be sold as a component of frying fats, margarine and other products. Various fractionation methods have been compared and with solvent fractionation a production of 75 to 85 per cent liquid oil and 15 to 25 per cent solid fat is achieved. In a fractionation process by transesterification a liquid fraction can be produced with 59 per cent *tri* unsaturated and 34 per cent *mono* or *di* unsaturated glycerides giving a composition not far removed from those of soyabean or corn oils. This oil has a high stability and can be used for superior quality salad oil. It has an iodine value of 80; the residual solid fraction has an iodine value of 5 and a melting point of 62°–63°C.

Much use has been made of the physical properties of fats, particularly optical properties, in identification of constituents. Solid fats were found to have several melting points. This was shown to be due to polymorphism, i.e. the triglycerides can crystallise to give different structural forms which in turn have different melting points as well as differing in other physical properties. The solid forms all revert to the same liquid when heated to the melting point of the highest melting form. Vegetable fats such as palm oil therefore melt gradually over a range of temperatures not only because they contain a mixture of fats, but also because any solid fats of the same chemical composition may have several crystalline forms or *cis* and *trans* isomers.

Fat Formation in the Ripening Fruit

The crude physical changes accompanying ripening and development of fats in the kernel precedes that of the mesocarp. At 8 weeks from pollination the content of the seed is really hard until the fifteen week. At 10 weeks it becomes semi-gelatinous and it does not become liquid. At 10 weeks from pollination the amount of fat is very small and the composition is different from that of mature kernels. It is believed that at this stage the oil is present only as basal protoplasmic fat. Unsaturated fatty acids preponderate and this is indicated by high iodine values. As much as 67 per cent oleic and 14 per cent linoleic acids have been found at this stage with comparatively small quantities of saturated acids.

From this stage there is a slow accumulation of fats until about the twelfth to the thirteenth week when fat formation becomes more rapid; the fats laid down are largely saturated, principally lauric (C12) which rises to 46 to 50 per cent and myristic (C14) which reaches 18 to 20 per cent by the twentieth week. The biggest accumulation occurs around the fourteenth to sixteenth week. The fact that considerable quantities of oleic acid are also formed has led to the belief that the laying down of the highly saturated fats is a separate physiological development. There is no evidence of conversions of one fat to another or of free fatty acid accumulation prior to esterification. Fat formation in the mesocarp takes place very late in fruit development. From the eighth to the sixteenth week after pollination fats constitute less than 2 per cent of the dry weight. There is in fact very little addition of any kind to the dry weight of the mesocarp from the eighth to the nineteenth week when, just prior to ripening, dry weight increases by 300 to 500 per cent and fats rather suddenly come to constitute 70 to 75 per cent of dry matter. During the long period of low oil content palmitic and linoleic acid esters predominate. Oleic esters are present only in very small quantities. During the final week of ripening all the fatty acids in combination increases in quantity and there is no evidence of conversion. Oleic acid, however, increases in greatest proportion and becomes second only to palmitic acid in quantity.

PALM KERNELS

Palm kernels are the kernels contained in palm nuts. Dried to less than 7 per cent moisture content, their palm kernel oil (PKO) content is about 50 per cent. In wild palm trees, the palm kernels are more important than palm oil as their extraction rate often exceeds 12 per cent, whereas that of oil does not always exceed 10 per cent. For this reason, early on oil mills included a unit for the recovery of palm kernels in their palm layout.

This custom has been continued down to this day, even though the proportion of palm kernels in the new selections of palm trees is smaller and their extraction cost sometimes excessive.

During the production process, the palm nuts begin to appear in the cake leaving the presses. The following operations are used to recover them:

1. Cake-breaking.
2. Separation of fibres and nuts.
3. Preparation of nuts.
4. Nut-cracking.
5. Separation of kernels and shells from the cracked mixture.
6. Sterilisation-drying of kernels.

Cake-Breaking

This is an operation during which a relatively compact press-cake is made to swell. It is carried out in a cake-breaking conveyer, a kind of screw conveyer where the continuous spirals are replaced by blades. The cake is carried along while being beaten, which makes it disintegrated. Many of the cake breaking conveyers are fitted with steam-jacket, the heat of which reduces the moisture of the products and thus makes the following operations easier.

Separation of Fibres and Nuts (Depericarping)

A rotary feeder feeds the cake into a column with an ascending air current. Its speed is adjusted so that the air carries away lighter fibres and lets the nuts and other heavier impurities like small stones fall to the bottom. Its setting is between 12 and 18 m/s.

The fibres carried away by the air current are then removed by cyclone separation. Nuts usually are collected in an appliance known as a polishing drum, where fibres still adhering to the nuts are separated by a rubbing motion.

This drum has a small slope which makes the nuts move down slowly with the help of low-speed rotation and the friction between nuts due to the motion. The end of the drum is fitted with a screen which holds back larger foreign bodies still present.

Preparation of Nuts

In current installations, the nuts coming leaving the polishing drum are transferred into silo-driers. One of the purposes of this storage is to control the downstream feeding of the palm kernel section. Its main role is to bring moisture down to a threshold value beyond which the kernel contained in the nut detaches itself from the shell. This value is about 1 per cent.

It is obtained by drying for a minimum of 20 hr. with the help of an air current whose temperature is not higher than 65°C. This limit should not be exceeded in order to prevent the oil from oozing from the loose kernels due to the pressure resulting from the storage height.

Silos are therefore large, designed to prevent any preferential flow and equipped with devices for heating and blowing air through the storage mass. Heating is provided by steam radiators.

The preliminary drying of nuts before cracking is not required with the use of new nut-crackers called 'ripple mills' and in this case the role of the nut silo is reduced to a small storage with downstream feeding control.

Nut-Cracking

The dried nuts are feeding the nut-crackers which break their shells and as far as possible without breaking the kernels. The common nut-crackers are fitted with a turbine which throws the nuts against the wall of their cylindrical bowl by a centrifugal force. Their speed of rotation can be set up to break a maximum of nuts with a minimum of broken kernels. This speed usually is between 700 to 1500 rpm.

Small nuts require higher speed than larger ones for cracking. That is why most of the efficient installations are supplied with an upstream size-classification of the nuts, which is followed by two or even three nut-crackers, each one working at an optimum speed which corresponds to the nuts size it is dealing with.

Never the nut-cracking is a perfect operation, whatever could be the adopted precautions and the process refinements. It always gives some unbroken nuts and some broken kernels and the amount of those can change significantly without any setting modification.

These problems are simplified to a larger extent with the new 'ripple mills' technology. The mode of functioning of this equipment is far from being as clear as that of the traditional nut-crackers. But paradoxically it has proved to be more efficient by bringing the proportions of unbroken nuts and broken kernels to low levels never achieved before.

Separation of the Kernels from the Cracked Mixture

This separation usually is done in two stages. The first stage is designed for the elimination of residual fibres and dust present in the mixture, because this presence could disturb the following operations. The second stage is the separation of kernels from shells still present in the cleaned mixture.

The first operation, called 'cleaning', is carried out in the same manner as the depericarping, by the introduction of the cracked mixture into an ascending air current. The speed is adjusted to remove fibres and dust but some shells and broken kernels are often carried away with them.

In South East Asia this simple system has been improved by placing different sections of passage over the air circuit thus giving different air speeds in each section and therefore different carry over forces. This design allows to have one extraction of products for each section of the cleaning column, depending on their properties of suspension in the air. This system is relatively simple but voluminous, consumes few energy and has an acceptable yield.

Some people still prefer hydrocyclones to separate the kernels from the shells in the cleaned mixture. The shells and kernels mixture is placed into water which serves as the carrier fluid for pumping and two consecutive cyclonings are given to it.

The first one is designed and set for the separation of the pure kernels from the shells. The second hydrocyclone takes from the first one the rejected shells still containing some kernels and separates the pure shells which are sent to the boilers while the residues with kernels are being recycled. This system does limit the kernel losses.

Hydrocyclones are still the most widely used in kernel recovery, even though their operating cost appears to be high in terms of consumed energy and of wear and tear.

Since the beginning of the palm oil story, salt and clay-bath separators had been used for a long time before falling into oblivion. Now-a-days they have again found supporters because of the emergence of new technologies which overcome the problems met during their operation, i.e. bath density control and corrosion resistant construction materials.

The bath, whose density is set between 1.10 and 1.12, lets the kernels float with a density of about 1.08 while the shells sink with a density of about 1.18.

None of the above-mentioned technologies is perfect and the obligation to limit the amount of impurities for marketing palm kernels, commonly below 4 per cent, often forces the mills to supplement these mechanical separations by manual picking.

Sterilisation and Drying of Kernels

Currently, following nut-cracking the kernels still have a residual moisture content of about 10–12 per cent, or 16–18 per cent when the nuts have not been dried before cracking. In both cases, they are unsuitable for long storage, because the moisture will favour fungal growth and oil acidification.

They should therefore be dried to less than 7 per cent, a level at which these phenomena stop. Some drying installations are also supplemented by live steam sterilisation, in order to destroy the enzymes which catalyse hydrolysis.

Drying is done inside silos with hot air. The dimensions of the silos are calculated for continuous operation, with kernels being fed in at the top and extraction of dried kernels at the bottom. Drying time is a minimum of 20 hr. with an air current whose temperature is not higher than 65°C.

Dried kernels are sent to oil mills, where they are crushed, laminated and pressed. More and more palm oil mills are adding a palm kernel oil mill to their plant to take the most of available energy.

Palm Kernel Production Controls

The same as in the case of palm oil, palm kernel production is subjected to quantity and quality controls, which are usually carried out by the laboratory.

Quantity controls

Their aim is to evaluate daily all the elements which combine to achieve the maximum possible extraction rate. The amount of kernels contained in the fresh fruit bunches cannot be determined and since this figure is missing (as for palm oil), it is not possible to calculate the extraction efficiency directly.

The process control identifies all the sources of kernel losses and it is always possible, like for the oil, to calculate the ratio between kernel output and the same kernel output plus kernel losses. But the figures resulting from this operation have no accurate significance if moisture losses at each stage of kernel processing are not taken into consideration.

The following main controls are carried out:

1. Analysis of the respective amounts of nuts and fibres in the cake leaving the presses.
2. Analysis of the nuts in the cake and determination of the respective amounts of nuts, broken nuts, loose kernels, broken kernels, shells.
3. Analysis of the fibres going to the boiler and determination of the amounts of kernels and broken kernels carried over with the fibres.
4. Determination of the amount of kernels and broken kernels lost in the polishing drum.
5. Determination of the nut-cracking efficiency by the comparative analysis between the uncracked nuts going to the nut-cracker and the cracked mixture. This efficiency is estimated by the amount of residual unbroken nuts in relation with the increase in kernel breakage.
6. Analysis of the product leaving the crusher: unbroken nuts, shells, sticking shells, full kernels, fragments, fibres and debris.
7. Weighing of kernels and kernel fragments carried over during the cracked mixture cleaning operation.
8. Weighing and percentage of impurities contained in kernels at the outlet of hydrocyclone.
9. Weighing and percentage of kernels contained in shells at the outlet of the hydrocyclone.

The final kernel output is metered with their bagging and kernel extraction rate is calculated with it.

Quality controls

These controls are so named because their specific aim is to make sure that the kernel quality corresponds to the commercial requirements. They are led on representative samples during the time of bagging or bulk despatch.

Here are some standards commonly used in kernel trading:

1. An impurity ratio of less than 4 per cent.
2. Less than 10 per cent broken kernels.
3. Less than 7 per cent moisture.

Sometimes stricter conditions will includes:

1. Minimum oil content.
2. Maximum acidity.
3. Colour sorting.

Composition of Palm Oil and Palm Kernel Oil

Palm oil

Although palm oil has a higher proportion of the saturated palmitic acid (C16) it also contains a higher quantity of unsaturated fats, principally those derived from oleic acid. About three-quarters of the glycerides are mixed saturated and unsaturated triglycerides. The oil melts over a range of temperatures from 25°–50°C.

Eckey quoted the fatty acid composition of twenty-one samples of commercial palm oils, seven from plantations and fourteen from countries producing oil mainly from the groves. The ranges and the means of the percentages of the constituent acids are given in Table 6.1 together with analyses from other sources. *Loncin* found traces of linolenic and lauric acids in his Zaire oils and stated that oil from grove palms had almost the same composition as plantation oil.

Table 6.1. Fatty acid composition of palm oil (*E. guineensis*) (per cent: means and ranges).

Description	*Lauric C12*	*Myristic C14*	*Palmitic C16*	*Stearic C18*	*Palmitoleic C16:1*	*Oleic C18:1*	*Linoleic C18:2*	*Linolenic C18:3*
Eckey								
Range	–	0.6–5.9	32.3–45.1	2.2–6.4	0.8–1.4*	38.6–52.4	5.0–11.3	–
Sierra Leone, Liberia and Ivory Coast samples	–	1.8	34.8	5.4	–	50.4	7.4	–
Far East, Zaire, Nigeria, Cameroon and other samples	–	2.3	41.2	4.3	–	42.5	9.6	–
Loncin and Jacobsberg								
Zaire plantations	0.2	1.2	43.7	6.2	–	36.5	11.8	–
Malaysia								
Plantations†	–	0.6	49.9	2.8	–	40.6	6.2	–
Range	–	0.1–1.0	45.2–58.5	0.6–5.4	–	34.6–44.0	4.3–9.2	–

* In three samples only.

† Oil from four *dura* and four *tenera* bunches.

With the considerable range in fatty acid composition it is obvious that progenies will differ significantly. Some significant differences were found in Malaysia between fruit forms and types, but as the range of composition overlapped the differences were probably ones of progeny within the forms and types.

Loncin and *Jacobsberg* showed that in Zaire the triglycerides were to be found in the following proportions:

Triglycerides S-S-S about 6 per cent
Triglycerides S-S-U about 48 per cent
Triglycerides S-U-U about 43 per cent
Triglycerides U-U-U about 3 per cent
where S = saturated
U = unsaturated fatty acids.

The low quantity of fully unsaturated triglycerides of high melting point is said to make the oil particularly suitable for margarine manufacture.

Thus palm oil is firstly a fat containing a very high proportion of palmitic acid to which may be attributed its value in soap-making. Secondly, the high quantities of oleic and linoleic acids give the fat a much higher unsaturated acid content than that of coconut or palm kernel oils which are essentially lauric oils giving a hard soap with greater lather.

The most important minor constituents of palm oil are the carotenoids; among these the carotenes are so conspicuous that the fat was termed 'red palm oil' in the Far East. Both the total carotenoid content of the oil and the proportions of the constituents vary. The carotenoids includes α, β, γ and ζ carotene together with lycopene and lutein (xanthophyll). Vitamin A is derived from the carotenes, β-carotene having twice the 'vitamin A activity' of either α-carotene or γ-carotene. The carotenes, which are among the most highly pigmented substances in nature, are hydrocarbons with long chains of conjugated double bonds; they have the formula $C_{40}H_{56}$. When the carotene molecule is split with the addition of the elements of water to both halves, vitamin A is produced. In β-carotene the two halves are identical, but in the other carotenes they are not, only one-half being of the constitution necessary for vitamin A formation.

The carotenoids are easily separated chromatographically as they are absorbed from solution in light petroleum by several powders, e.g. aluminium oxide. In chromatographic columns the different affinities are made use of, lycopene being absorbed preferentially to the carotenes and therefore appearing at the top of the column. β-carotene appears above α-carotene.

Differences in fruit colour have already been noted and apart from the *albescens* type which contains very little carotene, there are big differences in the carotene content of the mesocarp. For instance, two fruit samples which were 'red' and 'orange' when ripe and gave almost identical quantities of oil to dried mesocarp had the following carotene contents:

Fruit	*Red*	*Orange*
Carotene in dry mesocarp (mg/100 gram)	207	89
Total carotene in oil (ppm)	2560	1100

The carotene content of the red sample is exceptionally high; oils taken direct from individual palms rarely contain more than 2000 ppm. The carotene content of the orange sample is quite usual for palms in West Africa. Samples of Sumatran oil, however, were found to have carotene contents of between 320 and 475 ppm and commercial Malaysian oils are reported as having between 300 and 500 ppm. There is thus evidence that, on average, Deli palms produce a lighter coloured oil than the general run of African palms. In the examination of progenies in Nigeria it has been shown that carotene content may vary widely from bunch to bunch. In one progeny examined, thirty-eight out of fifty-one bunches had carotene contents in the low range 250–500 ppm and ten were in the range 501–700. In another progeny, in which sixty-two bunches were examined, there was a much greater scatter, viz:

Carotene range (ppm)	250–500	501–600	601–700	701–800	801–900	901–1000	Over 1000
Number of bunches	12	7	3	9	11	9	11

Bienaymé and *Servant* have suggested that carotenoids are in greater concentration in oils from Togo, Dahomey and Guinea than elsewhere in Africa and that in the palms of Angola carotene contents are as low as in the Deli palm.

The proportions in which the different carotenoids are found are said to be variable. In the 'red' and 'orange' samples referred to above the proportions were as follows:

Different carotenoids	*Red (%)*	*Orange (%)*
α-Carotene	28	34
β-Carotene	54	50
Lycopene	<3	<3
Lutein and ζ-Carotene	<1	<1

The red fruit thus had a higher proportion of the higher 'vitamin A activity' β-carotene. The above examples are typical of many other analyses.

Of the other minor constituents of palm oil, the tocopherols are found in a number of forms. They are antioxidants and may be found in quantities as high as 800 ppm in well-prepared plantation oil, though in oil coming from the groves the quantity is usually around 500 ppm. Palm oil contains only very small quantities of phospholipids and sterols.

Palm kernel oil

Palm kernel oil resembles coconut oil with which it is readily interchangeable. Both fats have a preponderance of saturated fatty acids, but palm kernel oil has a lower quantity of the low molecular weight acids, caprylic and capric. The usual ranges of fatty acid constituents, per cent, are as follows:

Description	*Saturated*						*Unsaturated*	
	Caprylic C8	*Capric C10*	*Lauric C12*	*Myristic C14*	*Palmitic C16*	*Stearic C18*	*Oleic C18:1*	*Linoleic C18:2*
Range (%)	3–4	3–7	46–52	14–17	6–9	1–2.5	13–19	0.5–2
Typical sample (%)	3.0	6.0	50.0	16.0	6.5	1.0	16.5	1.0

Traces of the saturated caproic and arachidic acids, C6 and C20 and of the unsaturated palmitoleic and linolenic acids, C16:1 and C18:3 are also found.

With such a high proportion of saturated acids it is not surprising that saturated triglycerides constitute over 60 per cent and monooleic disaturated triglycerides more than 25 per cent of the total glycerides. The iodine value is about 17 and the melting point is 25–30°C.

Oils of Elaeis oleifera

The mesocarp oil of the American oil palm, *Elaeis oleifera*, has an iodine value of 78 to 88, much higher than that of the oil of *E. guineensis*, showing that it contains higher percentages of oleic and linoleic acids and lower percentages of palmitic and other saturated acids. Similarly, the kernels oil of *Elaeis oleifera* has an iodine value of 25 to 32, indicating that saturated acids do not predominate to such an extent as in palm kernel oil. In these respects the American oil palm is similar to quite a number of other American palms which can yield both pulp and kernel oils. (In these respects the American oil palm is similar to quite a number of other American palms which can yield both pulp and kernel oils. For instance, *Jessenia batana*, the Patana or Seje palm of Brazil and Venezuela, has a pulp oil with 84 per cent of unsaturated fatty acids.

In analyses of the mesocarp and kernel oils of *E. oleifera* and of its hybrids with *E. guineensis*. The unsaturated fatty acid content of the mesocarp oil of the hybrid is intermediate between those of the parent species. It has been postulated that if the hybrid is back-crossed with either of the parents then half the progeny will have hybrid-type fatty acid analysis and half will have the analysis of the parent; support for this was obtained from two back-cross analyses on Malaysian material which showed both a hybrid-type and an *E. guineensis*-type analysis. There is, however, considerable overlapping in analyses and it will be noted that *Eckey's* data in Table 6.1 for *E. guineensis* oils from the western part of West Africa are not unlike those of some hybrid oils.

The fatty acid content of the hybrid kernel oil approximates more to that of the *E. guineensis* parent.

Palm kernel cake

Palm kernel cake contains around 18 to 19 per cent protein and is thus the lowest of the oil cakes in protein value. The amino acid composition has been given by *Babatunde* and shows high proportions of arginine and glutamic acid. The cake or meal is more suitable for use in the rations of ruminants than of pigs; for the latter its gritty and rather fibrous nature, together with its low level of lysine, makes it less effective than other available protein sources. An average analysis is as follows:

	Carbohydrates	*Oil*	*Proteins*	*Fibre*	*Ash*	*Water*
Per cent	48	5	19	13	4	11

The protein content of *E. oleifera* and hybrid meals is lower than that of *E. guineensis* though the amino acid composition is similar.

Producing countries with developing livestock industries are increasing their palm kernel processing plants and much larger quantities of cake and meal are now available in Africa and Asia.

QUALITY OF PALM OIL AND KERNELS

The processing of oil palm bunches has as its object not only the extraction of the maximum quantity of the products, but also the production of palm oil and kernels of high quality. Before considering extraction methods, therefore, it is important to know the meaning of high quality and what factors will promote or prevent its attainment.

Palm Oil

A poor quality palm oil may be defined as having any or all of the following characteristics:

1. A high free fatty acid (FFA) content.
2. Contamination with water and other impurities.
3. Poor bleachability.

These characteristics are not entirely independent, high water and dirt contents being contributory causes of high FFA; factors causing high FFA, may also contribute to poor bleachability. It should also be noted that, for local consumption in Africa, oil of a low FFA content is not always desired. Dishes made with oil of 10 to 20 per cent FFA are often preferred.

Formation of free fatty acid

The fatty acids of palm oil may be formed by autocatalytic action, by the action of the lipolytic enzyme lipase from the palm fruit, or by microbial lipases. In practice the main 'set-up' of FFA is caused by the action of lipase before processing.

Palm fruit contain a very active lipase which can only effect the breakdown of the fats into fatty acids and glycerol when the cellular structure of the fruit has been disturbed. The lipase is active even when the temperature has been reduced below 15°C. It is believed that the fat is protected from the lipase in the fruit by the membranes of the vacuoles and that these may be ruptured either mechanically or by low temperature. The enzyme is completely inactivated by high temperature.

Oil from fresh ripe fruit contains as little as 0.1 per cent fatty acid (estimated as palmitic acid), but in bruised and crushed fruit the FFA may increase to 50 per cent in a few hours. *Fickendey*, who was the first to draw attention to enzyme lipolysis, showed that oil from mesocarp peeled from the fruit attained a free fatty acid content of 67 per cent after 24 hours, but that if the fruit was heated to 90°–100°C before depulping, little rise in FFA took place. Later it was shown that a temperature of 55°C was sufficient to prevent enzyme action, that with crushed mesocarp the speed of fatty acid formation was remarkably fast (Fig. 6.1) and that partial bruising of fruit prior to sterilisation may be followed by only a small overall rise, to around 5 per cent, in the FFA of the oil of the fruit. In all this work the fat splitting process, as indicated by the FFA content, reached its peak in 30 to 60 minutes. With careful harvesting and carriage, the bruising of fruit need not be very extensive and these results show how it is that on well-regulated estates low FFA oil can often be obtained from bunches kept overnight.

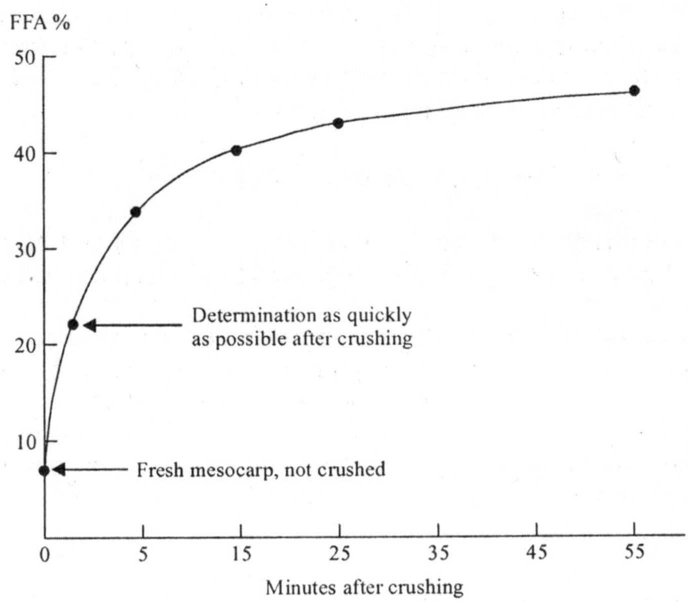

Fig. 6.1. Free fatty acid set up in mesocarp following crushing.

Fruit which has been kept for several days before processing, or which has been allowed to become over-ripe on the palms, may be covered and invaded by a number of moulds. Usually these fungi invade the base of detached fruit or wounds on the fruit surface. Many years ago an *Oospora* was claimed to have been responsible in the Congo for causing lipolysis but more recently lipolytic species of *Rhizopus, Aspergillus, Penicillium, Trichoderma, Circinella, Cunninghamella, Fusarium* and *Phoma* have been identified in Nigeria. Under ordinary plantation practice, however, this cause of FFA rise is likely to be insignificant in comparison with lipase activity originating in the fruit itself.

In Malaysia *Turner* examined the microbial flora which causes lipolysis prior to processing. He drew attention to the invasion of fruit by *Marasmius palmivorus* and species of *Sclerotium, Diplodia* and *Glomerella* before the bunch is removed from the palm and to invasion of loose fruit by seventeen different fungal species, fourteen of which showed lipolytic activity, together with yeasts, bacteria and nematodes. Of the fungi, *Aspergillus* species were particularly abundant. However, oil degradation through these agents was only thought to be very slight on plantations with good harvesting standards.

Turner has also examined the possibilities of microbial degradation of oil during processing and has shown that lipolytic fungi can be found in all parts of a mill both in the atmosphere and on the floors, walls and apparatus. Sterilised fruit if left for more than 24 hrs. before further processing is invaded by micro-organism, particularly *Neurospora sitophila* which grows rapidly over both bunches and fruit. However, temperatures are normally so high at all stages of processing that the potentially lipolytic organisms have little chance of significant activity except when a forced close-down of a mill takes place.

From the above it will be clear that for the production of low FFA oil the major requirements are: (i) minimal bruising of the fruit during harvesting, carriage and movement at the mill side and; (ii) minimal time between harvesting and sterilisation. To these may be added; (iii) the processing system must be such that the fruit or extracted oil does not cool down and come into contact with apparatus or materials which could cause a recommencement of lipolysis.

While enzyme action is the major cause of rises in FFA between harvesting and processing, the work of Loncin has shown that, after destruction of the enzymes, oil in storage can deteriorate through autocatalytic hydrolysis. The fatty acids already present in small quantity act as catalysts in the reaction between the triglycerides and water according to the formula

$$kt = 2.3 \log \frac{A}{A_0}, \text{ or } \log A = \log\left(A_0 + \frac{kt}{2.3}\right)$$

where k = the velocity coefficient, t = the acidification time, A_0 = the initial FFA content, A = the FFA content after time t. If t is given in 10-days unit, k = 0.12 at 60°C and is approximately doubled for every 10°C temperature increase.

The reaction velocity is independent of the water concentration so long as enough water is available to saturate the oil. Below the water saturation level water concentration has to be taken into account, but in practice hydrolysis stops almost completely when the moisture content is kept below 0.1 per cent; increases in FFA between purification centres and destination should therefore be negligible. Figure 6.2 is *Loncin* and *Jacobsberg's* representation of spontaneous hydrolysis at 70°C, yielding diglycerides and monoglycerides and glycerol.

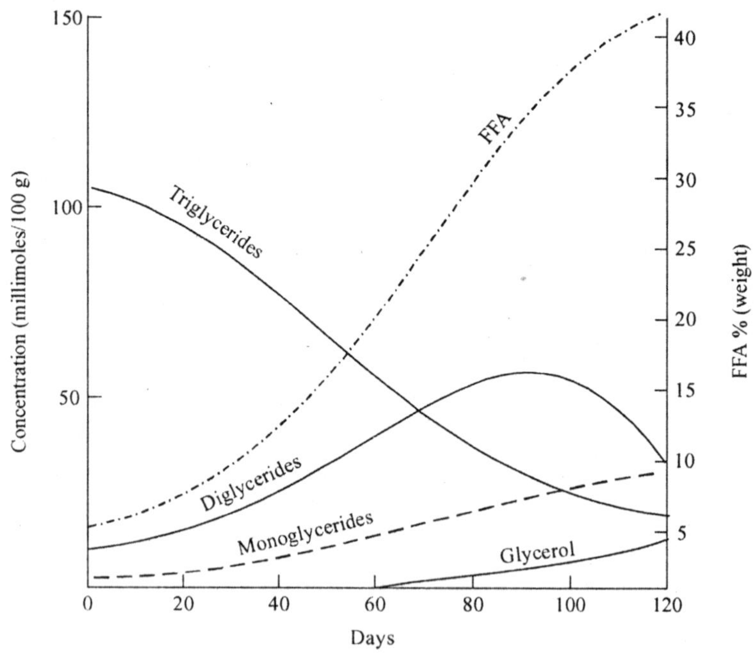

Fig. 6.2. Spontaneous hydrolysis of water-saturated palm oil at 70°C.

There appears, however, to be two modifying factors to *Loncin's* formula. Firstly, some experiments have shown that at high temperatures the water concentration has an effect on hydrolysis, i.e. it increases the rate of FFA formation.

Vanneck and *Loncin* obtained the following figures:

Time and temperature	*FFA percentage (as palmitic acid)*	
	Oil with 0.25% water (%)	*Same oil with addition of 5% water (%)*
At the beginning of experiment	2.8	2.8
After 62 days at 18°C	3.3	3.3
After 55 days at 55°C	3.5	4.1
After 36 days at 75°C	4.6	9.7

Secondly, there has been increasing evidence of the action of lipolytic micro-organisms in stored oil. *Desassis* showed that with oil stored in drums in Dahomey the FFA rise was greater than could be expected from autocatalytic hydrolysis. *Loncin* reported that *Geotrichium candidum*, possibly the same organism reported many years previously as a kind of *Oospora* causing acidification on palm fruit, has been found in many mills in Zaire to be responsible for lipolysis. More recently, *Coursey* has clearly demonstrated that increased hydrolysis can be brought about by infection with lipolytic micro-organisms. Examination of samples of oil in Nigeria showed that the more important identified lipolytic fungi were species of *Paecilomyces, Aspergillus, Rhizopus* and *Torula*. Inoculation experiments showed that additional rises of FFA of the order of 1 to 3 per cent could be induced during 8 week's storage. The sources of infection are palm fruit, bunch refuse, oily films on drums and other receptacles. Rises in FFA due to lipolytic fungi are likely to occur wherever oil is produced under generally dirty conditions and where the means of drum sterilisation are inadequate. With drum storage, variations from the autocatalytic reaction formula have been most common at the lower temperatures and it has been suggested that the biochemical factor is unlikely to be important where storage at high temperatures in bulk tanks is the rule.

It has further been suggested that the effect of impurities, including water, on the rates of hydrolysis may also be explained by supposing that they contain, or provide nutrients for, micro-organisms. When sterile conditions do not obtain, therefore, the rate of FFA formation will become dependent on the amount of dirt and water present in the oil. The blending of oils containing lipolytic organisms is likely to increase further the FFA production in storage.

It should now be clear that in order to provide an oil which will maintain a low FFA content, the final milling operations, must be directed primarily to attaining a water content of less than 0.1 per cent, since below this figure little autocatalytic hydrolysis is likely to take place. Secondly, the dirt content must be reduced to a minimum and clean and sterile conditions must be maintained in order to avoid the invasion of the oil by lipolytic micro-organisms.

Bleachability of Palm Oil

Although some of the materials for which palm oil is used, e.g. margarine, are coloured, in manufacturing practice oils must be bleached to definite specifications for the various uses to which they are put. They are blended following the refining process and the quantities of each oil used depend on availability and price. The composition, soft texture and plastic range of palm oil make it satisfactory for blending in fair quantities in margarine and shortening fats and for use in commercial baking and biscuit manufactures. But the oil must first be transformed into a bland, colourless, stable, edible product and colours in the range 5.0 red to less than 1.0 red with the 5¼ inch Lovibond cell are needed.

Any difficulty in bleaching palm oil and any additional expenditure involved must, therefore, militate against its use in manufacture; the bleachability of the shipped product has thus become an important factor. Nigerian oils became in other respects suitable for use in edible fat manufacture.

Manufacturers found that the bleachability of oil received from plantations in the Far East was superior to oil received from the west coast of Africa, but later it became apparent that easily bleached oil could also be obtained from plantations in Zaire and other parts of Africa. It was at first thought that the carotene content of the oil would be the primary factor in bleachability, but later there was evidence that oxidation of the oil and of the carotenoids was of more importance than the absolute carotene content.

Oil from the Deli palm estates of Malaysia and Indonesia is generally found to have a fairly uniform carotene content of around 500 ppm. Wide variations exist in Africa, however, both in carotene contents and bleachability. One of the first surveys carried out of oils from fruit and from small scale and industrial extraction plants gave some remarkably diverse figures. Certain fruits in the extreme west of Africa were found to contain over 3000 ppm carotene while Deli palms growing in the same region gave, as expected, oil with only 400–600 ppm. Some African fruits, however, also gave low carotene contents and mills provided oils varying in carotene content from 600 to 1600 ppm.

Of particular interest was the discovery that oil from the residual fibre was much higher in carotene content than oil as normally expressed, suggesting that the oil first squeezed out from digested fruit is unlikely to show the full carotene content of all the available oil. Subsequent work showed that hand-squeezed oil has a lower carotene content than press-extracted oil. It is thought that the exocarp fibres, which lie in that part of the fruit having the highest pigmentation, are less completely separated in the course of pounding or digestion than those of the remainder of the pulp (i.e. the true mesocarp).

Studies carried out at the Tropical Products Institute, London, suggested that poor bleachability of West African oil; which was largely produced by small-scale processing methods, might be due to (i) oxidation of lipoxidases in bruised fruit stored for various periods before processing; (ii) oxidation, catalysed by iron, during processing and bulking; and (iii) mixing of oils of good and poor bleachability. It was also suggested that high carotene oils are more likely to deteriorate than oils of low carotene content. More recent work has helped to determine the relative importance of these factors.

Fruit storage for a few days is quite common in West Africa but provided care is taken no rise in peroxide values or deterioration of the bleachability of the oil occurs. If fermentation is allowed to take place, as in one of the traditional extraction processes, bleachability will be affected, since oxidation of the unsaturated fatty acids gives rise to compounds responsible for colour fixation in fats. However, these oils, which are also characterised by a high FFA content, do not normally enter into international trade and fruit storage cannot, therefore be considered as an important cause of poor bleachability.

No significant deterioration in bleachability occurs during the milling process in a well-regulated mill of standard design; oxidation can be reduced to a minimum by the avoidance of prolonged heating in the presence of air at any stage, by the careful regulation of the sterilisation process and by the elimination of copper parts. However, in palm grove areas where crude or improperly regulated extraction methods are used, bleachability may be impaired even though the oil is relatively resistant to oxidation through its low polyunsaturated fatty acids content and the presence of tocopherols.

The handling of oil in metallic containers at various stages of village processes is likely to be of some importance. Metals and metallic soaps formed from the fatty acids have pro-oxidation effects. Copper is a strong oxidation catalyst and zinc, tin and aluminium form metallic soaps. Vessels of these metals are commonly used.

Of similar or perhaps greater importance is the storage, transport and blending of crudely extracted palm oils in metal drums. Surface contact in this case is increased by rolling the drums during handling. Moreover drums in continual use, even though steamed out at bulk oil establishments, become encrusted with a chocolate-coloured, tarry, rancid oil. Eight samples of this oil gave a mean FFA of 4.4 per cent, a mean peroxide value of 22.3 and a mean carotene content of 232 ppm, indicating considerable carotene breakdown. This drum residue was unbleachable.

If drums are not properly cleaned out and steam-sterilised before filling, considerable deterioration takes place. The manner in which mill processes are operated has also been shown to effect bleachability. In the sterilisation process both failure to eliminate air and the application of too high a pressure impair bleachability. Digestion with steam injection and clarification with centrifugal separation help to maintain the blechability of the oil.

Quality standards

Palm oil is traditionally bought on a 5 per cent FFA basis by importing countries, with penalties for exceeding this figure. To keep within these standards producers must achieve an oil of about 3½ per cent FFA and most mills in the Far East and elsewhere have been satisfied with the production of oils with FFA ranging from 2.5 to 4.0 per cent. Although palm oil is still bought on the basis of its FFA, water and dirt contents alone (with the exception of SPB oil from Zaire) consumers are becoming much more quality conscious of and have been demanding, good bleaching qualities.

Water and dirt levels have been variable. The maximum dirt content has been 0.1 per cent which again compares unfavourably with normal mill performance. Most mills in the Far East, Zaire and elsewhere achieve moisture contents of 0.1 per cent or below, though occasionally samples with as much as 0.3 per cent are to be found. Dirt contents are often as low as 0.005 per cent.

Under efficient management the mill would inevitably produce oil of around 2 per cent FFA and dirt contents well below requirement, but there was no incentive to reduce the moisture content below the legal limit.

Arnott has given the results of some 1500 analyses done by the Department of Agriculture in Malaysia. Histograms obtained showed considerable variability in FFA. There were two peaks, at 2.6 per cent and 3.6 per cent, while the single peaks for moisture and dirt were 0.15 per cent and 0.01 per cent respectively. Quality was categorised as follows:

Factor	*Very low*	*Low*	*Medium*	*High*	*Very high*
FFA	< 2.0	2.0 to 2.7	2.8 to 3.7	3.8 to 5.0	> 5.0
Moisture	< 0.1	0.1 to 0.19	0.2 to 0.39	0.4 to 0.6	> 0.6
Dirt	< 0.005	0.005 to 0.01	0.011 to 0.025	0.026 to 0.05	> 0.05

The characteristics of this oil (SPB) in comparison with what may be termed 'ordinary' plantation oil are given below:

Description	*SPB*	*Ordinary*
FFA, as palmitic acid (%)	1–2	3–5
Moisture (%)	< 0.1	> 0.1
Dirt (%)	< 0.002	0.01

(Contd...)

(Contd.)

Description	*SPB*	*Ordinary*
Iron (ppm)	< 10	> 10
Copper (ppm)	< 0.5	> 0.5
Iodine value	53 ± 1.5	45–56
Carotene (ppm)	500	500–700
Tocopherol (ppm)	800	400–600
Bleachability (Bleaching standard: Lovibond 5¼ inch)	2.0R, 20Y	3.5R, 35Y

While commending the above standards, there is an inherent danger in producing oil of around 1.5 per cent FFA which should not go unnoticed. The handling of fruit before milling and in particular the loose fruit, plays an important part in the FFA 'set-up' achieved. If a very low FFA oil is insisted upon, managers, and harvesters may be forced into the cutting of underripe bunches instead of concentrating on the careful handling of fully-ripe bunches and the loss of production may then be far greater than any advantage to be gained by a 1 per cent lower FFA content. This tendency has, indeed, often been noted in the field.

Although users of palm oil have been increasingly demanding oil which is easily bleached, the measurement of bleachability and the devising of standard bleachability tests still presents problems. This is because the damage which palm oil may have suffered can be due to the formation of oxidation products or to a decrease in the natural antioxidants or to the presence of oxidation promoters. No single test has been found satisfactory. Until recently the peroxide value was commonly determined, but a 'totox value' (2 × peroxide value (milli-eq/kg) + 1 × anisidine value (E_{350}) has been used by Swedish importers. A recent study has shown a good correlation between various analytical oxidation figures and bleachability. Peroxide value alone gives insufficient information since for highly oxidised oils this value alone gives insufficient information since for highly oxidised oils this value is not well correlated with the oxidation process. UV 233 nm and UV 269 nm absorbance have also been tested. Various direct bleaching tests with active earth or a combination of active earth and heat bleaching have been pur forward.

Palm Kernels, Palm Kernel Oil and Palm Kernel Cake

The quality of the eventual products palm kernel oil and cake depend primarily on the quality of the kernels. It is therefore important to know what qualities are required in the products and what kernel characteristics are needed to provide these qualities. Palm kernel oil is required to be of a low FFA content and a light yellow colour, easily bleached.

Palm kernel cake is required to be relatively light-coloured and its nutritive value, particularly in respect of the constituent amino acids of the proteins, must not be impaired.

It is not surprising, therefore, that palm kernels are themselves judged mainly on the FFA content of the oil they contain, on the amount of external and internal discolouration and on factors such as moisture content and presence or absence of mould which are likely to effect the eventual FFA content and colour. An average sample of palm kernels in a producing country has a composition as follows:

Description	*Oil*	*Moisture*	*Protein*	*Extractable*	*Cellulose non-nitrogen*	*Ash*
Per cent	47–52	6–8	7.5–9.0	23–24	5	2

There appears to be some variation in the composition of the non-oily and non-protein solids. The portion given as 'extractable non-nitrogen' has been found to contain variable quantities of sucrose, reducing sugar and starch, but in some samples no starch has been detected though mannose has been present. As to their moisture content, kernels eventually come to an equilibrium according to the prevailing relative humidity. This being, on average, in the region of at least 80 per cent in producing countries, the equilibrium moisture content is around 7 per cent, though in very wet periods this might rise to 8 to 9 per cent. It is believed that a safer moisture content for seed storage is 14 per cent of the non-oily portion of the seed and it would therefore appear that the equilibrium moisture content is on the border line of what has been termed 'safe'. Therefore, until the moisture has been reduced to its equilibrium level kernels are liable to reactions or activities for which moisture is needed.

These reactions are similar to those of palm oil, namely autocatalytic hydrolysis and lipolysis by fat-splitting enzymes in the kernels and from lipolytic moulds. At ordinary tropical temperatures the more important of the latter are four species of *Aspergillus* and species of *Paecilomyces*, *Syncephalastrum* and *Penicillium* identified in Nigeria and a species of *Rhizopus* and a lypolytic yeast, *Petasospora rhodanensis*, found in Zaire. In stacks of palm kernels which heat up to 50°–60°C a different fungal flora, which is thermophilic, appears. The most prominent species are a *Thermomyces*, *Chaetomium thermophile* and a *Penicillium*, probably *P. duponti*. All are actively lipolytic and their optimum temperature for growth lies between 42°C and 52°C. Only C. *thermophile* grows below 30°C. *Turner* has detected lipolytic micro-organisms at various stages in the milling process but particularly in hydrocyclones and clay baths during shell and kernel separation. Disturbing features of infection with these organisms are their ready carriage by insects and the continued activity of their lipases even when the moisture content of the kernels has been brought well below the equilibrium value after several hours heating at 100°C. Lypolysis may be reduced by sterilisation and a reduction in the number of cracked and broken kernels; but, as will be seen later, overheating produces discolouration of the kernels. It is for these reasons that the highest quality kernels are to be obtained by skilled hand-cracking of nuts which under village processes, have not been subjected to high temperatures for long periods. West African kernels produced by these means and properly dried and bagged are usually of superior colour to mill-produced kernels. It has been reported that oil from the testa of the kernel has an FFA content three times as high as that of oil from the enclosed endosperm, indicating that access to the oil-bearing tissue is all-important and that the amount of breakage will inevitably effect quality.

For the provision of kernels giving a low FFA oil, therefore, low moisture contents and low breakage are required. This has been well illustrated in Sumatra for whole and broken kernels of average and high moisture content and the results obtained are shown in Fig. 6.3. Even though precautions may be taken to reduce FFA production by proper drying and minimum breakage, there still remains, however, the problem of the heat-resistant enzymes already mentioned. In Zaire kernels are found to have in their oil an FFA content, as lauric acid, of 0.5 per cent after extraction from the nut and this rises to 1.0 to 1.5 per cent after drying.

During storage the FFA may rise to any point between 2 and 10 per cent and this has been shown to be due to the variable action of microbial lipases, undestroyed by drying or high temperature. Free fatty acid content of 3 to 7 per cent or more are common in the oil of kernels arriving in importing countries and it is not possible to prevent this except by preventing the development of micro-organisms before and during storage. This has been effected by successive injections of steam and cold air, thus sterilising and drying the surfaces of the kernels before introducing them into the drier. By this method the FFA of the oil in the kernels has been maintained below 1.5 per cent for 6 months of storage.

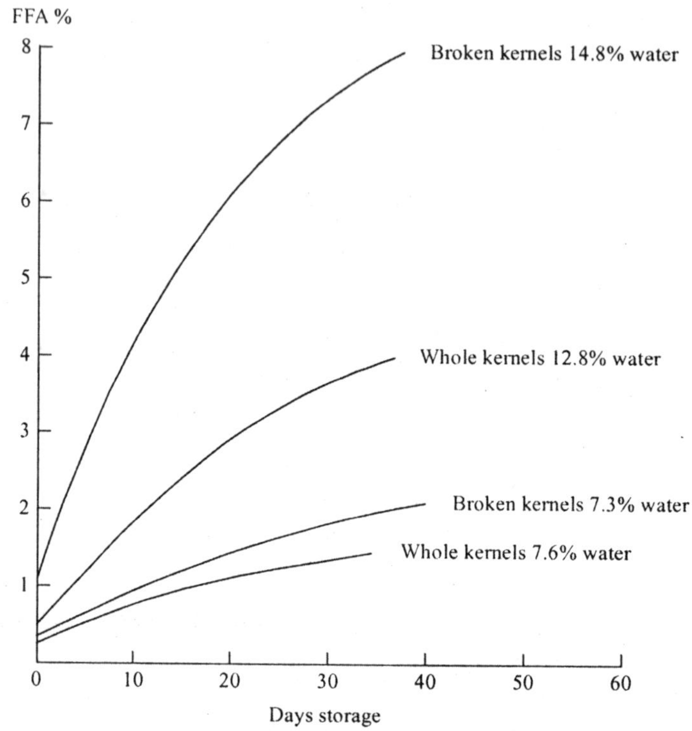

Fig. 6.3. Rises in FFA of oil of old kernels of average and high moisture content stored at about 28°C.

In Malaysia steam sterilisation at temperatures exceeding 90°C for 5 to 6 minutes has been shown to inhibit rises in FFA and to give rise to kernels which are stable during storage.

It has long been realised that the browning of the endosperm and darkening of the testa of palm kernels extracted in mills fitted with steam pressure sterilisers was to some degree inevitable. Further discolouration in the digester is normally slight and in the nut bin and kernel drier negligible. Kernels which show only moderate browning do not provide a more coloured oil than white kernels, but when severe browning has occurred bleaching is impaired or may even become impossible. Any browning effects the colour of the cake.

A correlation which was found in Malaysia between the percentage of kernels classified as 'off-colour plus mouldy' and the FFA of the kernel oil was no doubt due mainly to the lipolytic activity of the moulds. A somewhat similar relationship shown for kernels of differing colour and mouldiness arriving in the United Kingdom seemed to have a similar cause since the appreciable rises were nearly always associated with the presence of mould. However, small FFA rises are also obtained on browning kernels in the laboratory. Heating, with consequent discolouration, can also take place in heaps of uncracked nuts and in kernels stored under unsuitable conditions.

The chemical reasons for browning are not fully understood but are likely to be (i) the reaction of free amino groups of proteins with aldoses to give brown polymers and co-polymers; and (ii) the reaction of mannose (among the carbohydrates) with amino groups to form brown compounds.

Some attention has been given by mill manufacturers to the colouring of kernels since this is so intimately bound up with the bunch sterilisation process. In Indonesia with sterilisation at 130°C for

2 hr., it was observed that by the end of the milling process the kernels were coloured and it was concluded that at 120°C the risk of discolouration was much smaller.

NON-MECHANICAL TRADITIONAL METHODS OF EXTRACTION OF OIL AND KERNELS

The satisfactory extraction of palm oil requires specially designed machinery, whether hand or machine operated and the provision of ancillary equipment of correctly calculated capacity for the prior preparation of the fruit and for the subsequent preparation of the products for sale.

Before the advent of machinery, oil was extracted in Africa by crude means to give a product of generally poor quality; and kernels were extracted from the nuts one by one by hand cracking. These processes still continue to be used in many parts of Africa, owing partly to lack of capital and partly to the unchanging organisation of the rural communities. Though they vary considerably in their details, their essentials are those briefly described below.

'Soft Oil' Production

In the great areas of Eastern Nigeria from which the bulk of the early export suppliers were brought, the process employed gave rise to the so-called 'soft oil', thus named because the greater part of the oil was liquid at tropical temperatures. The harvested bunches are cut up into sections and kept in heaps for 2 to 4 days. The heaps are sprinkled with water and covered with leaves.

Boiling

The fruit is picked from the bunch sections and boiled in larger pots for about 4 hours; 44 gallon petrol drums are commonly used. The fruit may be left in the vessel for up to 3 days.

Pounding

The boiled fruit is pounded in a wooden mortar with a wooden pestle until a mixture of nuts and crushed pulp of more or less even consistency is obtained.

Separation

The oil is separated from this mass of pulp by immersing the latter in water. The initial stage may be carried out either in a special pit with its sides coated with cement or mud or in some larger vessel. The whole mass is stirred and first the crude oil which has risen to the surface is skimmed off into another vessel, then the fibre is shifted out of the water and finally the nuts, now largely free of fibre, are picked out and laid out to dry. Later they are hand-cracked. The crude oil thus obtained is boiled in smaller vessels where any fibre it contains sinks to the bottom. Now the purer oil is again skimmed off and is then 'fried' in a shallow pot to get rid of the last traces of water. When this condition is reached, drops of water sprinkled on the oil will rapidly evaporate with a crackling noise—hence the use of the term 'frying'. The amount of oil extracted depends largely on how far heat has been maintained throughout the process and how assiduous the women are in skimming and in teasing out the fibre. Usually the mass of pounded pulp is allowed to cool off and extraction rates are low. Such figures as exist for extraction rates in the soft oil process, i.e. 6 to 10 per cent oil to fruit of low mesocarp content, suggest that a normal efficiency (extracted oil to total oil in the fruit) would be 40 to 45 per cent, occasionally rising to 50 per cent, but sometimes falling to as low as 30 per cent. Average FFA content used to be about 7 to 12 per cent, but lower FFA oils can be produced by this method.

'Hard Oil' Production

In some parts of Africa the people are not accustomed to pounding and a less efficient method, in which the fruit is trodden, is the rule.

This is the typical method of the people of the Niger Delta. The sequence of events is:

Fermentation

After hand picking from the stored, chopped up bunches the fruit is placed in a pit or in a long wooden canoe and covered over with leaves. Fermentation caused by microbial and enzyme action takes place with the generation of heat and the fruit thus becomes softened.

Treading and separation

After some days the fruit is in a condition where it can be vigorously trodden in a canoe. After the first treading the oil is allowed to drain for 3 days from the lower end of the canoe. Water is then added and a second treading is done. When the work is considered to have proceeded far enough and the remaining separated oil has risen to the surface it is skimmed off and boiled in vessels for final preparation as in the soft oil process, though usually with less care. The oil thus prepared solidifies rapidly. Considerable hydrolysis and oxidation takes place during the process and the proportion of low melting point constituents is reduced. As would be expected, the extraction rate is very low, about 4 to 6 per cent oil to low mesocarp *dura* fruit, with a correspondingly low efficiency of 20 to 30 per cent; and the FFA content, owing to the high initial set-up during fermentation, is usually between 30 to 50 per cent. The changes taking place in the fruit and oil during the process of fermentation in the canoe have not been fully studied. The hard oil process continues in use because it makes low demands on labour and firewood and much carrying and boiling of water is avoided. In some areas it is claimed that there is still a taste for high FFA oil in cooking.

HAND-OPERATED PRESSES

With an increasing demand for palm oil from the west coast of Africa it was clear even by the turn of the century that traditional methods of extraction could not satisfy either demand or quality requirements and that they gave a very meagre return to the producer. Consideration was therefore given to providing oil extraction methods which could be easily adopted by the small or large producer.

Curb Press

The press which was eventually found most satisfactory and captured the market was a curb type of press similar to those used for the extraction of juice from soft pulp fruit in the wine and cider industries.

The press consists of a screwed steel shaft, fixed in the centre of a base plate and a cage, composed of strips of stout wood set vertically about 3 mm apart and looped externally with two iron bands (Fig. 6.4). The cage, which is in two halves, can be opened and lifted off so that the pressed fibre and nut mixture can be easily removed. Pressure is applied by a ram which is worked downwards on the shaft by a cross-head turned manually by two long iron bars on which two to four men work. The bars, which are detachable, are fitted into the screw cross-head which lies above the ram. The base of the press is wooden and is surrounded by a metal trough fitted with one spout. The oil is squeezed out of the digested material between the wooden strips into the trough and runs through the spout into any collecting vessel. The first term is unobjectionable, but the later may cause confusion with other types of screw press.

In Africa, sterilisation and pounding proceeds as described for the soft-oil process and inferior extraction rates are often attributable to the cooling of the pounded mixture before and during transfer to the press, as well as to variations in the length of the sterilisation period and the efficiency of pounding. In a trial in Malaysia, Deli fruit was boiled in a 40 gallon (181 litres) drum for 5 to 6 hours and immediately pounded with a wooden pestle in smaller drums for 10 to 15 minutes and transferred hot to the press. A recovery of 72 per cent of the oil contained in the fruit was obtained by double pressing which involved removing the whole charge after the first pressing, separating the nuts and re-pressing the fibre from the first charges; this fibre had been kept warm and partially dried by placing it on iron sheets over a small fire. It was noted that with mixed fruit there was difficulty in obtaining fully mashed material from the pounding drums if small undeveloped fruit were present. It seems probable, therefore, that a better extraction rate can always be expected from good Deli fruit than from the mixed lots of fruit usually obtained from the African groves.

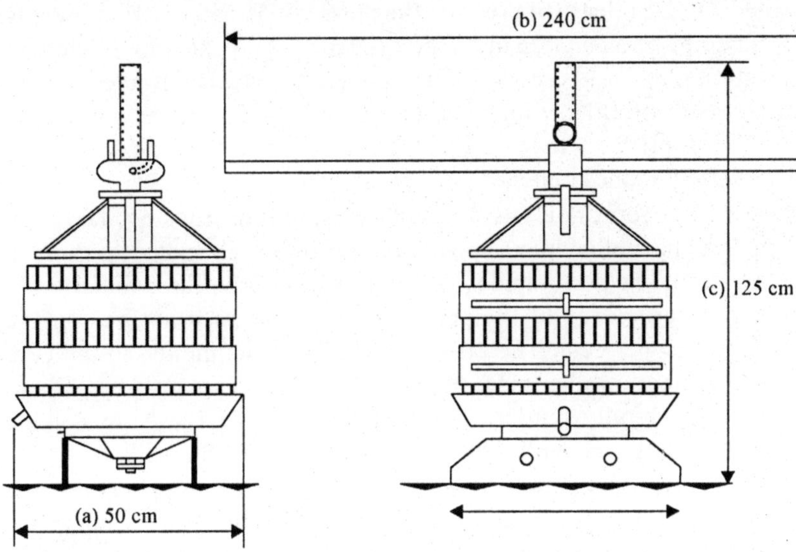

Fig. 6.4. The Duchscher curb press.

The Malaysian trials suggested that of the 72 per cent oil extracted 5 per cent was attributable to the second pressing, a maximum of 67 per cent efficiency being obtained from one pressing. In Africa double pressing is done by adding fresh material to a charge already pressed.

Presses of various sizes have been used, but an average charge of the Duchscher NGI press is about 70 kg fruit, equivalent to about 115 kg of bunches.

The quantity of fruit treated would be around 450 kg, equivalent to 770 kg of bunches. It was pointed out that the press was not the limiting factor and that if more men were available at least twice the number of pressings could be undertaken, giving the equivalent of 1½ tonnes bunches per day.

The fruit was hand stripped from the bunches and sterilised by boiling for half an hour in water. It was then pressed in hand-screw presses consisting of perforated steel cylinders and a screw ram. Pounding of the 14 kg charge was undertaken in iron mortars after this pressing and the nuts were separated by hand. The oil-bearing fibre was then heated over boiling water and pressed, wrapped in sacking, in the same screw press, heating and pressing sometimes being repeated. Clarification was done, as in

West Africa, by heating and skimming. An extraction efficiency of between 60 and 70 per cent was obtained and the quality was surprisingly good, moisture being 0.3 to 0.4 per cent and FFA 3 to 4 per cent.

Hydraulic Hand Presses

While the use of the curb press was a substantial advance on traditional methods of extraction and it was soon shown that curb-press operators were able to produce low FFA oil, the loss of extractable oil was still considerable. Recently a hand-operated press of much greater efficiency has been developed.

Hydraulic hand press

It consists of a ram, which is the cylinder of the hydraulic mechanism, moves downward into a perforated press cage when hydraulic fluid pressure is increased by the hand operation of a two-piston pump; both pistons are operated until considerable force is required and thereafter the small piston only is operated until full pressure is reached. On the release of pressure the press ram is withdrawm upwards by springs.

The cages are filled and emptied on a table in front of the press ram and with skilled operation one pressing, including the insertion and withdrawal of charges, takes 6 to 10 minutes. As two cages are provided and one can be emptied and filled with hot digested material during the pressing of the charge in the other, it is possible to complete six to ten pressings per hour and thus to press 270–450 kg fruit per hour, equivalent to 0.45–0.75 tonnes bunches per hour.

The cages are surrounded by covers or oil guards while pressing is proceeding. The crude oil, which is heavily laden with sludge, is channelled into a spout and runs off into buckets. As there is seepage of oil on to the press table this should be perforated with weep holes and be fitted underneath with an oil catch. *Nwanze* showed that the efficiency of the press alone, which must not be confused with full process efficiency, can exceed 95 per cent. The efficiency of this machine led to the construction of hydraulic hand presses by a number of other manufacturers. In some cases pressure was exerted from below, but the mechanism was generally similar. Later, small motors were used by some operators to activate the press and presses of this type with motor drive can now be obtained.

A hand press of the type described is suitable for use by a small holder or an operator buying in fruit from groves or small holdings; such an operator will not normally have much capital for expensive ancillary equipment. Nevertheless, it is desirable on economic grounds that the press should be kept in continuous operation at least throughout the day. Most early operators of hydraulic hand presses were surprised at its rate of operation and were unable to supply sufficient equipment and men to deal with the amount of fruit required by the press. This situation, as already mentioned, was not uncommon even with the curb press.

Suitable cheap ancillary equipment which can be made locally has been designed for use with hydraulic hand presses and this, with the quantities needed, is described below.

Sterilisation

If loose fruit is being dealt with (some operators buy in loose fruit) boiling is best carried out in 44 gallon (200 litres) sterilising drums set in a special tipping frame (Fig. 6.5). If sufficient loose fruit is being supplied to keep the press in operation during the daytime then about four drums will be needed. Fruit boiling is carried out for 1 hour after steam appears at the top of the drum. While the drums can also be used for sterilising cut up bunches, even with as many as seven of these it is difficult to supply sufficient fruit for the press and fuel consumption (wood, fibre and shell) and labour usage is excessive.

Large bunch sterilisers for use with hand presses have been designed to hold about 1 tonne of quartered bunches. These are locally constructed from cut-up oil drums welded together to form a cylindrical vessel 2.5 m tall by 1.1 m diameter and fitted below with two boiler compartments surrounding a fire space. Two or three steam pipes rise from the boiler compartments to distribute the heat.

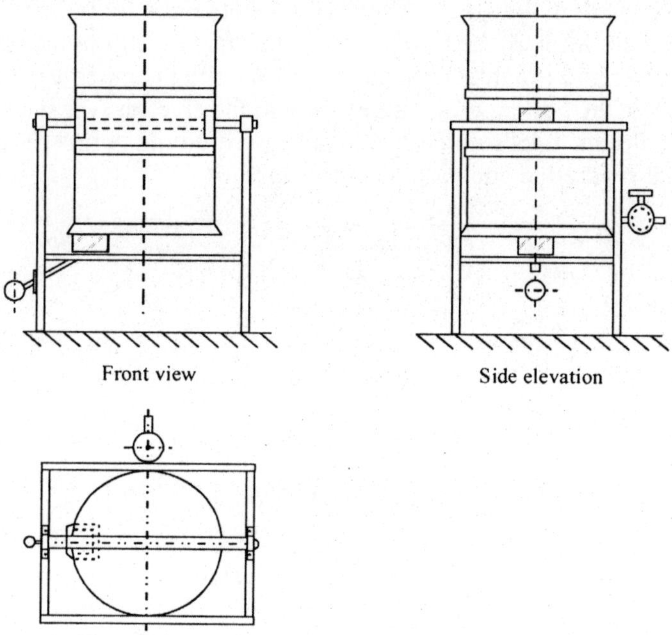

Fig. 6.5. Fruit steriliser for use with the hydraulic hand press.

Stripping

Stripping in a hand-operated mill can be by beating out the fruit from the sterilised bunches by means of wooden batons or pronged forks, or by the use of a hand-turned octagonal slatted drum. The latter has proved very efficient.

Digestion

Digestion is by pounding stripped fruit but, in the hand press system, the fruit tends to get cold while waiting its turn for pounding. For this reason two fruit re-heating drums of similar design to the bunch steriliser are used for storing the stripped fruit until ready for pounding. The pounding is best done in a large concrete mortar of 90 cm diameter and 50 cm deep; six to eight men using wooden pestles pound the fruit until a 'mash' of uniform consistency is achieved.

The pounded material is then placed in special pounded-fruit re-heating drums which are fed at the top and from which the material can be continuously removed through a door near the bottom for transfer to the press. This drum has a water compartment with perforations for steam to rise into the drum. At least two pre-heating drums are needed.

Clarification

With occasional operation of the press, 44 gallon (200 litres) drums may be used for clarification. For continuous milling, however, two large gallon drums measuring 90 × 100 cm are employed. These are fitted with a 44 gallon (200 litres) drum as an inner compartment and a funnel of at least 25 cm diameter attached to a tube of 5 cm or more which forms a feed pipe and extend downwards near the side of the large drum to within 7.5 cm of the bottom. The inner 44 gallon drum is supported in the centre of the large drum by bars or an attachment to the feed pipe. After the drum has been one-quarter filled with water which is brought to the boil, the oil and-water mixture obtained from the press is poured through the funnel. The clean oil rises through the water and the dirt tends to fall to the bottom. As more oil or water is added the level rises and eventually the oil overflows into the inner drum and can be drawn off through a pipe leading from the bottom of the inner drum to the exterior. Various arrangements may be made for linking the two drums together so that oil partly purified in the first drum may pass over into the second drum for final separation and clarification (Fig. 6.6).

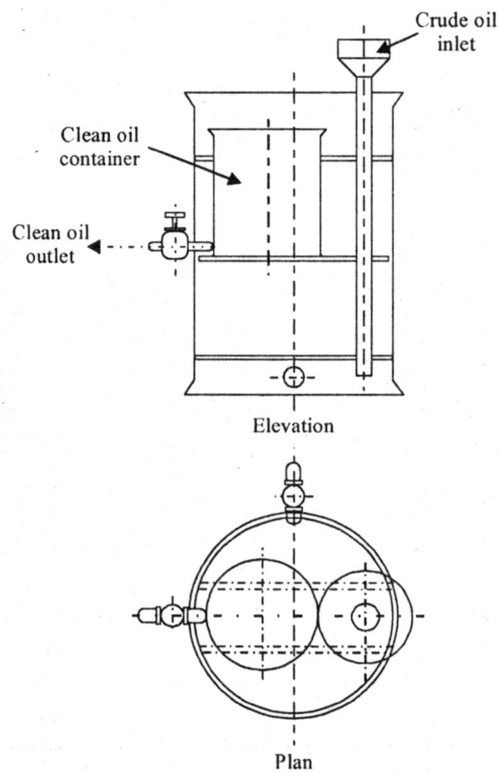

Fig. 6.6. Clarification drum for use with the hydraulic hand press.

Oil exudations at various stages of the process, e.g. re-heating, pre-heating, may be caught and transferred to the clarification drums and conversely, drainings of heavy unclarified sludge from the first clarification drum may be used for wetting the pounded material in the pre-heating drum. The final extraction rate may thus be kept at a maximum.

An efficient system of hand milling requires a substantial opensided building for housing the equipment and giving space for the cutting up of the bunches.

The operation of this equipment and the number of vessels required for continuous operation have been extensively studied and described by *Nwanze*, who showed that with one press, three large sterilising drums, one stripper, two fruit pre-heaters, two fruit re-heating drums and two large clarifiers, throughput could be raised to nearly 100 tonnes of bunches per month on double shift working (3.8 tonnes per working day) and over 130 tonnes if a three-shift system was introduced. The pressing rate was about 0.5 tonnes per hour pressing time, but when working with two shifts in weekly stretches (i.e. starting operations at the beginning of a week and closing down at the end) overall working time was still two or three times the pressing time through daily throughout could be gradually raised to 4 tonnes, The extraction efficiency of the whole process varied from 75 to 85 per cent; it is believed that this could be raised to 93 per cent if pounding were as efficient as conventional digestion. Over a period of 7 months the following results were obtained:

No. of working days	*Weight of bunches milled (tonnes)*	*Weight of bunches per milling day (tonnes)*	*Weight of oil produced (tonnes)*	*Extraction rate (%)*	*Extraction efficiency (%)*
183	648.4	3.54	88.9	13.7	80.4

Theoretically the press could deal with 12–16 tonnes of bunches per 24 hours, but in a hand-press mill the practical operation time of the press itself is unlikely to exceed 8 or 10 hours per day, or a throughput of 4–7 tonnes of bunches. Only with continuous operation is this likely to be achieved or pressing time to approach overall working time.

Nwanze has demonstrated the considerable gains to be had from a change from the curb press to the hydraulic hand-press and has shown how, in countries where the bulk of the rural population are not wage earners, the economics of hand-press operation cannot be satisfactorily assessed by reference to the high statutory or large-enterprise wave rates which may exist alongside the very much lower income level of the farmer. With one press and the equipment described it is clearly possible to achieve annual throughputs of well over 1000 and perhaps nearly 2000 tonnes of bunches. Thus, depending on locality, areas of 80 to 200 hectares might be served. Areas of this magnitude might be able to command the capital for small mechanised mills and thus the hydraulic hand-press is best suited to the grove areas or to groups of very small holdings agreeing to mill in co-operation.

In spite of this potential and its relatively low costs and high efficiency the hydraulic hand-press system has not taken on in the West African grove areas. It has even been reported that the number of the curb-press operators has decreased while employers of primitive soft oil production methods have increased. The reasons why primitive methods have persisted in spite of the availability of relatively inexpensive and efficient apparatus has been the subject of some controversy, it being maintained on the one hand that small operators have only family, unpaid labour and a better choice of fruit and on the other hand that these factors are spurious or insufficient explanations and that much more social and economic research is required.

Kernel extraction

In Africa it has been common practice in hand-operated mills to give back the nuts to the women folk for hand-cracking. There are on the market, however, a number of nut-crackers driven by small petrol engines and these can with advantage be introduced into an otherwise hand-operated mill. Motors of around 3 hp are required to turn the cracker at between 2000 and 2500 rpm and about 0.75 tonne of nuts may be cracked per hour.

POWER-OPERATED PALM OIL MILLS

The mechanical milling of oil palm fruit started in Africa before the First World War. A French firm, Fournier, erected a small plant at Cotonou was the first to develop the idea of pressing the fruit whole.

Thereafter a number of rather crude bunch strippers was invented and the central problem of how to extract the oil from the pulp began to be tackled in earnest.

Two pressing methods were tried. In the first, the fruit was sterilised, after removal from the bunch, by steam-heating for 15 to 30 minutes and was then transferred to hydraulic presses. This was the principle of the Cotonou mill of French design. Two pressings were carried out, with heating in between and the nuts were then separated from the fibre in a rotating drum and cracked in a centrifugal cracker. The process sounds thoroughly modern except for the omission of digestion. Digestion was first undertaken in a stamping mill in a plant of German design in Togoland. The entire spikelets seem to have been cut- off the bunches and subjected to digestion, so that fruit-loosening and pulp-crushing were simultaneous. The whole mass was treated in hydraulic presses, the nuts were separated in a rotating drum and cracked in a Haake centrifugal cracker.

Centrifugal extraction was first used in Zaire and was the main feature of the Lever Brothers mills. Centrifuges proved very suited to the African grove fruit with its thin layer of mesocarp, but they also became the standard extracting equipment in Malaysia during the first 10 to 15 years of the industry's life in that country. It should be realised that the centrifuge was an entirely new modern method of extracting oil from fruit or seeds and as such it was highly successful. Although the most modern mills now employ hydraulic or screw presses there are centrifuge mills still operating in Asia and America.

Centrifuges and hydraulic presses became efficient oil extractors in combination with rotary-arm digestors. Oil mills are normally powered by steam engines and the boilers use fibre and shell in their furnaces. Transmission in small and medium sized mills was, from early days, by transmission shaft and belting and this method is now being superseded by the generation of electricity for driving electric motors of suitable sizes.

Choice of a Plantation Mill

In their essential features oil mills are standardised in that they consist of sections, sometimes grouped as 'stations', for (i) sterilisation of bunches; (ii) stripping of bunches; (iii) digestion and mashing of fruit; (iv) extraction of mesocarp oil; (v) clarifying the oil; (vi) separation of fibre from the nuts; (vii) nut drying, (viii) nut grading and cracking; (ix) kernel separation and discarding of the shell; and (ix) kernel drying and bagging.

The central machines of the mill are the extracting machines. It is the rate of operation of these machines, in terms of weight of bunches entering the mill, that determines the capacity of the mill. All other machinery is, or should be, adjusted to the rate of operation of the extracting machines. This to some extent simplifies the procedure of choosing a suitable mill, but there are many other considerations to the taken into account and these are discussed below.

In the first place it is necessary or desirable to know the area to be planted, the rate of planting per annum and the expected yield at maturity. From this data estimates can be made of the probable crop for each year until the plantation is fully mature. If there are to be possible extensions, then it must be decided whether to allow for these extensions or to adjust the mill to the original area and to erect further mills for any extensions which may eventuate.

Secondly, it is necessary to estimate the probable distribution of the mature crop throughout the year. In countries with an even climate it is usually reckoned that in the peak month about 12.5 per cent of the total annual crop will be harvested. In certain countries, however, as much as 18 to 19 per cent of the crop may be harvested in 1 month. The mill's capacity must allow for this.

Thirdly, it must be known what form of fruit is to be produced. This will effect the extraction and kernel sections and will be discussed under these headings. The fuel supply will also be affected. In addition, it may be that the mill will lie in an area where it is profitable to include palm kernel oil extracting machinery and to sell kernel oil and cake in addition to palm oil.

The milling capacity is based on the assumption that three shifts will be worked in the peak month with a total operation of 20 hours per day giving 500 hours per 25 working-day month. Obviously if conditions are such that three-shift work cannot be done then the calculations must be based on a lesser number of hours (including overtime) but this will lead to a larger capital outlay.

Usually, after calculations of this kind have been done, decisions have to be taken on (i) whether to have one mill or several smaller mills; (ii) the stages of instalment of the equipment, particularly the extracting equipment, so as gradually to expand the mill or mills to their maximum capacity; and (iii) the method of dealing with the small crop of the first few years.

There are various methods of dealing with the latter problem. In the first place harvest may be delayed and a larger first harvest obtained by the practice of ablation, i.e. the removal of young inflorescences during the first few months of flowering. In this way the need for a small plant may be obviated and early crops may be accommodated in the mill.

Secondly, an hydraulic hand-press or presses may be used and these may be mechanised. This must, of course, be combined with the provision of sterilising and digesting equipment and if considered worth-while, a simple nut-cracker. With a press capacity of over half a tonne of bunches per hour it will be seen that four mechanised hand-presses could deal with production in the scheme for 2 to 3 years and that the heavy outlay of capital on the large mill could be delayed. Such a scheme is very suitable in areas which are expected to come slowly into production.

Milling machinery can be supplied and installed in several stages. The stages will depend on the rate of plantation development, on the machinery chosen and on the number of lines of machinery to be installed. Six or seven hand-operated hydraulic presses of 3 tonnes, four automatic presses of 4½–6 tonnes or two screw presses of 10 tonnes bunches per hour rating could handle this crop. Thus, the choice of stages of mill development will be bound up with the type and individual capacity of the units of extracting equipment chosen. When units of 3 tonnes capacity are being used, three stages might be adopted; sterilisers, strippers, digesters and much other equipment would then be provided in stages to serve the extracting units as they are supplied. It will be seen that the choice of the extracting unit is in more ways than one the most important exercise in the choice of mill.

Siting of a Mill

To reduce transport costs to a minimum a mill should be in the centre of a plantation. Other considerations may prevent this, however. These are:

1. The mill must be close to a suitable water supply sufficient for its needs. As a rough guide a mill of about 18–20 tonnes/hr. capacity will require about 40,000 litres or 40 m^3 per hour. The exact water requirements will be specified by the manufacturers. The water, or at least the part required for boilers, must be of specified purity and if necessary a filtering and purifying plant must be installed.
2. The mill must be as near as possible to the point of despatch of the palm oil, particularly if stored in large tanks. The latter are with advantage placed on a slight rise.
3. The site must not be subject to flooding, must have sufficient space for overnight storage of bunches and for the necessary workshops and other subsidiary buildings.
4. It must be possible to discharge sludge into a fast flowing river or stream or to effect disposal in some other manner.

Sterilisation

Vertical bunch sterilisers

These are only suitable for small mills where the cages on rails for horizontal sterilisers cannot be afforded. They are usually of 2 to 3 tonnes capacity, but can be constructed to take up to 6 tonnes. In one type the bunches are forked at ground level into a bucket elevator which drops them into the steriliser through a circular door at the top. After sterilisation at pressure the bunches are discharged through a rectangular hinged door at the foot of the steriliser. The discharge, though manual, is aided by the presence of a sloping perforated plate at the bottom of the steriliser. Nevertheless, discharge may take 30 to 40 minutes.

An alternative type of vertical steriliser containing an immersion basket involves less labour and time in charging and discharging. The bunches are tipped from lorries into the basket which is then lifted by monorail and lowered into the steriliser which is closed by a specially locked lid. After sterilisation, the basket is hoisted out of the steriliser and moved to the platform above the stripper; there the load is discharged automatically on lowering.

Horizontal sterilisers

The immediate advantage of the horizontal steriliser is that it does not require hand emptying and that the bunches are sterilised in cages in which they have been packed and are therefore not subject to further bruising before sterilisation. It is also claimed that, largely owing to the way the bunches pack in a vertical steriliser, there is less loss of oil in the condensed steam and in the bunch stalks coming from the horizontal type of sterilisers.

Horizontal sterilisers are long cylinders placed at ground level into which cages of standard design can be pushed on rails through a door which then closes the whole of one end of the cylinder. Their capacity depends, therefore, mainly on their length. A large steriliser would hold about 15 tonnes of bunches in six cages. Several of these would be required for mills of high capacity.

Mention has already been made of the importance of sterilisation in reducing the FFA of the oil and the danger, during sterilisation, of causing the kernels to become brown. One of the primary purposes of bunch sterilisation, however, is to loosen the fruit on the bunch so that stripping is not difficult. 'Hard bunches' are often a problem and much attention has been given to methods of injecting and 'blowing off' the steam and raising and lowering the pressure. Few of the important considerations in practical sterilisation are mentioned below:

1. The air has to be swept out of the steriliser to attain the temperature value corresponding to a given pressure and to avoid oxidation. This takes about 5 minutes.
2. Increasing the steam pressure above 2 kg/cm^2 (30 lb/inch^2), which corresponds to a temperature of approximately 130°C, leads to kernel discolouration and may affect the colour of the oil.
3. To avoid zones of very high temperature due to the presence of superheated steam, particularly near the inlets, it is important to work up to full pressure slowly.

With regard to 1, it is common practice to introduce steam slowly from a high point and to allow the air to flow out from the bottom, continuing this procedure for a short time after steam is seen to be coming from the outlet. Venting systems have also been used. A single venting system entails bringing the pressure to 2 kg/cm^2 for 5 minutes and then venting-off the steam and re-charging to the same pressure. Double venting entails a second 'blow-off' after a further 5–10 minutes. These systems employ a greater amount of steam.

Sterilisation of bunches is carried out for 60–75 minutes of which 5 minutes is taken up by replacing the air and 15–20 minutes in working up to full pressure which is maintained for 40–55 minutes. The higher figure is usually employed as reduction of sterilising time has been shown to increase the number of hard bunches. With horizontal steriliser 5 minutes is occupied in blowing off and 10–12 minutes in discharging and re-loading so that the whole cycle occupies about 96 minutes. If the mill is large enough to have four sterilisers some steam may be saved by a system of blowing over from one steriliser to another. The system must be organised in such a way that there is no overall delay in the sterilising process.

The required sterilising time is affected by the size of bunches and the degree of ripeness. If a harvest consists solely of bunches from 4 or 5 year-old palms, then sterilisation for half an hour may be quite sufficient to loosen the fruit. Similarly if the harvest consists of only very ripe bunches with a high proportion of loose fruit sterilising time can be reduced.

Sterilisation usually results in a loss of weight of about 10 per cent of the weight of fruit, due to evaporation. This dehydration is important, as, particularly with *tenera* fruit, it reduces the volume of digested fruit in the presses and improves oil extraction. In small mills, such as the Pioneer, in which fruit, not bunches, are sterilised, the time taken is much shorter; a sterilising time of 12 minutes has been found sufficient in an autoclave designed to hold 200 kg of fruit.

Stripping of Fruit from Bunches

There are two kinds of stripper (or thresher), the beater-arm type and the rotary-drum type. The former is smaller and cheaper and is suitable for small mills with a capacity of upto 5 tonnes bunches per hours. It consists of an inclined cradle of curved bars between which the beater arms attached to the stripper shaft can pass. Bunches falling into the top of the cradle incline are knocked and turned by the tips of the beater arms and pass gradually down the cradle. The empty bunches pass out of the cradle at the end and the fruit and 'calyx leaves' which have been knocked out of the bunches fall through the bars to be picked up by a bucket conveyer and raised to the top of the digester.

The rotary drum stripper has a diameter of about 1.8 metres and a length of 2.7 metres to 5 metres. The longer the drum the more complete the stripping should be. Even feeding is very important. Bunches falling on the platform above the stripper are usually pushed manually into a shute carrying them into the drum and loose fruit fall through a grill on the platform to be carried straight to the digester without entering the stripper. Automatic feeders can, however, be employed. These consist of large hoppers into which the sterilised bunches are tipped. At the bottom of the hopper is a moving belt with slots which feeds the bunches at a uniform rate into the stripper. The stripper drum is composed of horizontal metal bars with sufficient space between them to allow the stripped fruit to fall through onto a conveyer; this carries the fruit to the bucket conveyer which takes it to the digester.

The centrifugal force exerted on the bunches is sufficient to raise them nearly to the top of the revolving drum whence they fall back onto the bottom and scatter their fruit. The bunches, while repeating this motion many times, gradually work their way along the drum aided by guide strips attached to the inside of the drum and they fall out at the end and may be carried by a conveyer to a special bunch refuse incinerator outside the mill. It will be obvious that the speed of the drum is important; if too fast the bunches will be simply carried round and round, if too slow they will remain rolling gently at the bottom. Speeds suitable for larger heavy bunches will be too high for small light ones. Speeds of between 21 and 23 rpm are usual with drums of 1.8 metres diameter. The maximum capacity of a rotary drum stripper is about 20 tonnes of bunches per hour.

The rotary drum stripper is favoured not only for its higher capacity, but also for its smoother and more efficient running. Bunches occasionally get jammed in beater-arm strippers and this may increase loss of oil in the bunch stalk.

In the use of any stripper a check has to be kept visually on the completeness of fruit removal from the bunches and, in the laboratory, on both the fruit retention and the oil loss in the bunch refuse. Bunches seen to be 'hard' i.e. to have retained their fruit through the stripping process, must be returned for re-sterilisation. Both losses of fruit and losses of oil in bunch stalk are much reduced by making sure that large heaps of sterilised bunches do not accumulate on the platform and that the bunches are fed into the strippers at a regular pace.

Incinerators are designed to operate by autocombustion, after initial firing, with a natural draught. The resulting ash is a useful fertiliser containing 30 to 35 per cent K_2O and 3 to 5 per cent MgO.

Digestion

The treatment of fruit in digesters prior to oil extraction is a very important process and it is partly through the examination of the action of digesters over a relatively long period of time that the proportion of oil extracted from the digested material is now so high. The purpose of digestion, as with pounding, is to break up the pulp physically and further, to liberate the oil from the cells in which it is contained. The fault of simple pounding and of some small-mill digesters, is that heat is not also provided, for heat assists in the loosening of the cells from the fibre.

Modern digesters are steam-jacketed cylindrical vessels with a vertical rotating shaft in the centre driven from the top or, less commonly, from the bottom. The fruit is mashed by pairs of stirring arms attached to the shaft. Various devices are used for preventing the mash going round with the arms. Baffles may be set vertically between linear plates on the wall of the digester or bars may be inserted across the vessel. Sufficient steam pressure is maintained in the jacket to ensure that the mash leaves the digester at about 90°C. There are perforations at the bottom to allow seepage of up to 50 per cent of the oil direct to the crude oil tank. For poor *dura* fruit there should be no perforations. Digesters are sometimes provided with a steam-jacketed hopper in which a quantity of fruit is kept hot before passing into the main cylinder where the stirring arms are revolving.

The digester should usually be of sufficient size to provide fruit for an hour's pressing and the size must therefore be adjusted to the rate of work of the press or centrifuge to the be used. Fruit will be fed at the top as charges are removed from the bottom through a special discharging hopper.

The important points to be checked in the digestion process are: (i) the temperature must be maintained at around 95°C but the water in the mash must not be allowed to boil; (ii) the mash as produced must be homogeneous without any undigested fruit; (iii) the perforations must be kept cleared so that crude oil may escape at the bottom; (iv) the arms must be replaced when they show wear; and (v) the digester must be kept at least three-quarters full at all times.

If digesters are being used with centrifugal extractors they are slightly modified.

Extraction of Oil

Although early trials were made with hydraulic presses, the earliest efficient machines for extracting oil from the digested fruit were centrifuges. It was mainly in Nigeria and Zaire that pressing difficulties were encountered and they were attributable to the fact that the bulk of the fruit treated had a very thin layer of pulp. For a long time it was considered that presses would inevitably crack the nuts and that therefore a depulper must first be used. However, it was later found that hydraulic presses working at relatively low pressures could be used. Pressing was still in two stages; in the first the digested fruit was pressed and in the second stage, after the nuts had been removed and the fibre dried, the latter was pressed alone. This practice was later abandoned in favour of a single pressing or the 'pre-pressing' of the digested material followed by the addition of further material at the top of the press for final pressing. In post-war years screw presses were developed, the larger models of which have a much higher throughput than any single centrifuge or hydraulic press. The three types of extraction are briefly described below.

Centrifuge

Centrifuges can be driven from the bottom or top but the latter type has become standard in oil palm factories. A basket is rotated at high speed, between 950 and 1250 rpm, by a belt-driven shaft. The charge of digested material depends on the size of the basket which is usually between 91 and 125 cm in diameter with capacities, in terms of bunches, of 1–2 tonnes per hours. Charging, spinning and discharging takes from 20 to 25 minutes and three spinnings should normally be completed in the hour. The oil is expelled through the mass by centrifugal force and passes through the perforations of the basket into the space between the basket and the stationary casing from where it is pumped to the clarification section. A typical schedule for centrifuging in a small centrifuge would be:

Filling	3 minutes
Accelerating	2 minutes
Spinning	6 minutes at full speed
Spinning	2 minutes at full speed with steam injection
Spinning	2 minutes at full speed
Stopping	2 minutes
Discharging	3 minutes

The following characteristics of centrifugal extraction should be noted. The oil obtained is comparatively sludge free and clarification losses are small. Steam injection is important, but should not be used until towards the end of spinning. At the very end of the spinning it is desirable to cease steam injection and then dry out the centrifugal matte to assist subsequent fibre separation. The centrifuge should be covered with a metal cover during spinning. Discharging the matte which has caked around the circumference of the cage is an arduous job which must be done manually through the bottom of the cage after the steel ring around the shaft, which closes the bottom, has been raised. Though covered digesters with steam inlets to the interior were used with centrifugals, open-type digesters, as already described, are now usually employed, but they are fitted with steam injection tubes at the bottom to ensure that the digested material has not dried out too far for satisfactory centrifuging.

Hydraulic press

The hydraulic hand-press has already been described. Standard presses for large mills may be hand-operated or automatic. The former have capacities from 1.5 to 3.0 tonnes bunches per hour. The material is introduced into the perforated press cage from the digester shute through a hole (of the same diameter as the press) in a table-like sliding door which is operated by a hand-wheel. The door is closed on the interior of the press by sliding it forward so that the hole no longer lies above the press cage. This type of press works at a cake pressure of 75 kg/cm^2 and six or seven pressings can be done in an hour. Hydraulic pressure is applied at the base of the press to a ram which moves upwards to press the mash and returns by its own weight. The mash is introduced in portions with up to six circular steel plates inserted between the portions. This assists the operator by enabling him to push off the press-cake in manageable slices after pressing. If pre-pressing or double pressing is adopted a plate is held back to cover the material already in the press before material is added for the second pressing. A typical schedule for pressing at the rate of six pressings per hour would be:

Filling	2 minutes—from digester.
Pressing	6 minutes, including double pressing if employed.
Emptying	2 minutes—pushing press-cake into conveyer.

The time can be controlled either from a clock or by a system of lights a and the pressing cycles are recorded from a pressure gauge onto a drum recorder chart.

Hydraulic presses may be of the revolving type comprising two cylinders swivelling on a central column. Discharging and filling is carried out in one press cage while pressing is proceeding in the other. Ejection of the press-cake, loading the digested fruit, insertion of the plates and some pre-pressing is carried out with the aid of an auxiliary piston. These presses may have capacities up to 5 tonnes of bunches per hour.

An automatic hydraulic press has been evolved in which the digester forms one unit with the press. In this unit the pressure is applied by a ram from above the press cage. When the ram is at its highest position mash passes freely into the cage in which are retained the press-cakes of four or five previous pressings. Thus, by a succession of pressings the mash is moved slowly downwards through the press-cage cylinder. This is made possible by the presence of a cone at the bottom of the cylinder which closes on the cylinder when the ram moves downwards and presses the mash, but opens after pressure has been applied for a certain time. When the cone moves downwards to open the cylinder the ram continues on its way pushing out a portion of the press-cake. The ram then moves swiftly upwards, opening the digester door and closing the cone which in its motion breaks up the press-cake exuding from the cylinder. At the same time the cone pushes back the press cake in the cylinder causing some friction which is considered to assist oil expulsion. The speed of action of this automatic press can be controlled and its capacity thus increased from the usual 4.5 tonnes bunches per hour to 6.5 tonnes. The oil to dry fibre will of course vary with the pressing rate.

It has been shown that the performance of an hydraulic press depends to a great extent on the composition of the press-cake. Trials conducted by press manufacturers suggest that there is an optimum proportion of nuts to fruit for maximum oil-extraction, i.e. there is an increased loss of oil in the fibre when the proportion of nuts to fruit exceeds or falls short of a certain percentage. Where there is an excess of nuts, they bear on each other and full pressure is not exerted on the oil-bearing fibre. No very clear reason has emerged for the low extraction obtained when there is a shortage of the nuts but it is thought that the addition of nuts to *tenera* digested fruit may reduce friction resistance within the press-cake and help to ensure a better distribution of pressure through the press-cake.

Another measure which improves distribution of pressure is the insertion of the metal plates between sections of the introduced digested material. If this is not done the press-cake will tend to be more compact and drier in the middle than at the circumference and more fully extracted at the bottom than at the top. The insertion of plates reduces this tendency by re-distributing the forces bearing on the cake.

It has also been shown that the presence in the cake of small pieces of trash such as the 'calyx leaves' (perianth segments) from the bunch far from hindering pressing, actually improves oil extraction from the cake. To obtain the most satisfactory expression of oil, therefore, (i) fibre should be added to the sterilised fruit if the fruit has a high nut/mesocarp ratio, e.g. above 50/50; (ii) nuts should be added if the nut/mesocarp ratio is low, e.g. below 25/75; and (iii) no attempts should be made to remove calyx leaves. The actual quantities of nuts or fibres to be added usually need to be found by trial.

Screw presses

Two types of continuous screw press have been used for the expression of oil from digested material and it is claimed that with these presses the digestion period may be shorter. Both types of press allow for the gradual passage of the mashed fruit through a cylinder while under compression by screws against the wall of the cylinder and against cones at the end of the cylinder.

The single-shaft screw press has a gear box and a pressing section consisting of two screws—a feed screw and a press screw. The screws are placed end to end on the same shaft but are of opposite thread and rotate in opposite directions. The feed screw in large screw presses of this type rotates at 10 rpm while the opposing press screw rotates at 6 rpm. The digested fruit is fed to the press screw by the action of the feed screw and the tapering design of the former gradually reduces the volume available around the screw, thus increasing the pressure until the matte is expelled around the end cone, which is adjustable. Oil is expelled outwards through the perforated cage around the screws and inwards through a short cage around the shaft at the end of the press screw.

This type of screw press can be obtained in small sizes (about 3 tonnes bunches per hour), but very large presses capable of dealing with 13 tonnes of bunches per hour are also in operation.

The double-shafted screw press also consists of a gear box and pressing section, but the latter contains twin shafts with screws turning at the same speed but in opposite directions. The digested fruit enters a hopper below which lie the screws and a strainer through which some of the oil will pass. The material then passes into the perforated press cylinder through which oil is expelled; finally the matte is screwed further along the perforated cylinder where the shaft tubes are also provided with holes through which further expelled oil passes. The outlet at the end of the cylinder is regulated by adjusting the cones by means of an hydraulic pressure control system.

Another make of double-shafted screw press expels the oil only through the press cage and not through the shaft tubes. It is equipped with a feed screw between the digester and press and incorporates a variable speed drive which regulates the throughput and an automatic cone adjustment system which depends on the power absorbed by the main screws.

Comparison of extraction methods

It is rarely possible to make satisfactory comparisons of milling processes or individual milling machines. This is because two parallel plants of different construction are hardly ever found on the plantation to enable crops from the same fields, treated in all other respects identically, to be fed to the machines to be compared. This is, however, the only way in which a valid comparison can be made.

Centrifuges provide a clearer crude oil, but give a higher oil content in the residual cake. Oil from the centrifuge is more easily clarified and losses in sludge are not so high.

A valuable comparative study of the performance of centrifugals, hydraulic presses and screw presses showed that the type of extraction unit chosen will influence not only the oil losses, but also kernel breakage and quality, iron absorption during processing and wear and tear of the extraction machinery, including the digesters (Table 6.1). *Nelson* studies confirmed that the loss of oil in the fibre was higher with centrifugals than with presses, though the loss in sludge was lower. It was also shown that losses, in the fibre were lower with screw presses than with hydraulic presses, but the losses in sludge were higher. Controversy about the relative merits of the two types of press at one time centred on the non-fatty pressing quotient (NFPQ) which is the ratio of non-oily solids (NOS) in the sludge to the total of NOS in press-cake and sludge. This quotient tended to be lower with hydraulic presses. It will be seen from Table 6.2 that the overall extraction efficiencies are only marginally in favour of screw presses;

and the view has been expressed that efficiency has now reached a level which, with any kind of pressing, is unlikely to be exceeded.

Table 6.2. Loss of oil, analysis of residues, nut breakage and iron contamination when using different extraction equipment (material: 85 per cent Deli *dura*, 15 per cent *tenera*).

Description	*Centrifuge*	*Hydraulic press*	*Automatic hydraulic press*	*Screw press*
Oil loss, kg per 100 kg bunches, in:				
Steriliser condensate	0.040	0.040	0.040	0.040
Empty bunches	0.450	0.450	0.450	0.450
Press residue	1.400	0.874	0.701	0.606
Nuts	0.148	0.124	0.150	0.152
Sludge	0.054	0.107	0.192	0.231
Total	2.092	1.595	1.533	1.479
Oil to dry fibre (%)	14.91	9.82	8.32	7.27
Oil to total dry matter in sludge (%)	3.35	5.09	7.48	8.67
NFPQ*	14.6	19.1	23.3	4.2
Iron loss, kg per tonne bunches†		0.034	0.027	0.013
Nuts in press residue (%)				
Unbroken	99.54	94.48	84.66	76.11
Smashed	0.28	2.95	5.61	8.33
Split	0.00	1.62	4.43	5.29
Free shells	0.12	0.81	4.11	8.14
Free kernels	0.05	0.14	1.19	2.14
Extraction efficiency				
Deli *dura* material	89.3	91.8	92.2	92.4
Deli with 20% *tenera*			93.5	93.8

* Non-fatty pressing quotient = non-oily solids in sludge/NOS in press-cake + NOS in sludge.
† From recorded losses in weight of press plus digester after 1000 hours' pressing.

The advantages and disadvantages of screw presses in comparison with hydraulic presses may be summarised as follows:

Advantages

1. Higher throughput.
2. Lower capital outlay per tonne of throughput.
3. Lower power consumption.
4. Lower labour requirement.

Disadvantages

1. Higher kernel breakage and loss of kernel in fibre.
2. More difficult and costly clarification.

In the past maintenance costs have been higher with screw presses, but it is now claimed that this is no longer the case. It should also be mentioned that although screw presses have been successfully employed with mixtures of *dura* and *tenera* material, they are certainly best adapted to the plantations of pure *tenera* material which are now usual, particularly in new areas. Another advantage claimed for screw presses is flexibility of throughput. Presses with a nominal throughput of 10 tonnes bunches per hour can be run down to 6 tonnes or up to 14 tonnes when necessary.

Solvent extraction has often been suggested and tried on a small scale, but the technical and economic difficulties have not been entirely overcome. In the early days of the Sumatran industry solvent extraction of the residual fibre after the first pressing (containing over 20 per cent oil to dry matter) was put into practice on a number of estates. The solvent used was naphtha (boiling point 85°–105°C) which was a product of local refineries. The solvent extraction of residual oils in fibre after normal pressing or centrifuging has also been suggested and much discussed. Various objections have been responsible for this idea not being pursued. The quantity of solvent required would be large and the loss, under tropical conditions, considerable. Oil extracted from the residual fibre is known to contain a higher proportion of carotenoids and is contaminated with an unacceptable quantity of non-oil components such as waxes and this would introduce bleaching problems.

The method of impulse rendering (the Chayen process) used for the extraction of oil and proteins from some seeds has also been suggested for the extraction of oil and small-scale tests have been carried out. The principle of the process is as follows. If any cellular matter is subjected, in a fluid such as water, to impulses or shock waves of sufficient frequency and intensity, the cells are burst and their contents mix freely with the fluid. Oil produced from oil seeds by this method is of good quality and light colour, though complete separation from the protein fraction is not obtained. If the process could be adapted for the extraction of oil from oil palm fruit, the product might be expected to be of both low FFA content and good bleachability.

Clarification

Oil coming from the presses or centrifuges is crude oil, that is to say it is accompanied by water, dirt and cellular matter from the mesocarp. Oil from centrifuges will contain only about 40 to 50 per cent water and comparatively small quantities of dirt and other matter. Oil from hydraulic presses contains about 55 per cent water and from screw presses about 60 per cent with considerably more impurities, particularly cellular matter. Crude oil is largely an oil-water mixture in various phases of oil dispersion in the water. The majority of the oil settles out on top of the water easily enough but, if care is not taken, emulsions, may be formed. Impurities tend to prolong emulsification and temperatures of over 100°C tend to produce emulsions.

Crude oil flows by gravity straight to the crude oil tank where it passes through screens to get rid of the larger impurities. In a small plant the screens may be stationary and set in the crude oil tank, but in larger plants vibrating screens of stretched gauge are used above the crude oil tank and the material which does not pass through is automatically returned to the digester. Sand sinks to the bottom of the tank. Crude oil pumps direct the screened crude oil to the clarification tanks.

In small mills the oil is separated from the water and dirt (the sludge) in simple settling tanks containing water and heated to about 95°C by steam coils. Pure oil is drawn off from the top and sludge passes to another settling tank for further separation after addition of more water. The quantity and kind of clarification equipment depends on the size of the mill and the market requirements.

What are often termed continuous clarifiers are really semi-continuous since although crude oil may be supplied at a regular pace to them and oil and sludgy water will emerge from two separate pipes, the flow and sometimes the internal piping, need regulating and, moreover the really heavy particles must be specially drawn off from the bottom from time to time. In these clarifying tanks (Fig. 6.7) the crude oil is passed via a heat exchanger (A) at 85–90°C to near the bottom of the tank and the oil collects and rises in the thick layer until it reaches the top of a pipe (B) through which it will be automatically or intermittently drawn off. The sludge water syphons off at a slightly lower level since the specific gravity of the oil plus sludge is lower than the specific gravity of sludge alone which rises in the side pipe (C). The liquid fractions X + Y in the tank and Z in the pipe are in balance. Temperature is maintained by a steam coil in the tank of a stream jacket. The quantity of crude oil being fed and the volume of the oil layer together determine the residence time of the oil in the tank and if this becomes too short for satisfactory clarification then an adjustment must be made to the height of the oil exit pipe. A slow supply leads to too long residence and tends to increases the FFA set up. When supply ceases, the separated oil must be drawn off by allowing entry of water into the sludge layer through pipe D and the sludge water and the heavy material must be run off by manipulation of the various external taps. It will be seen, therefore, that control of clarification, though continuous for much of the time, needs careful handling by an experienced operator.

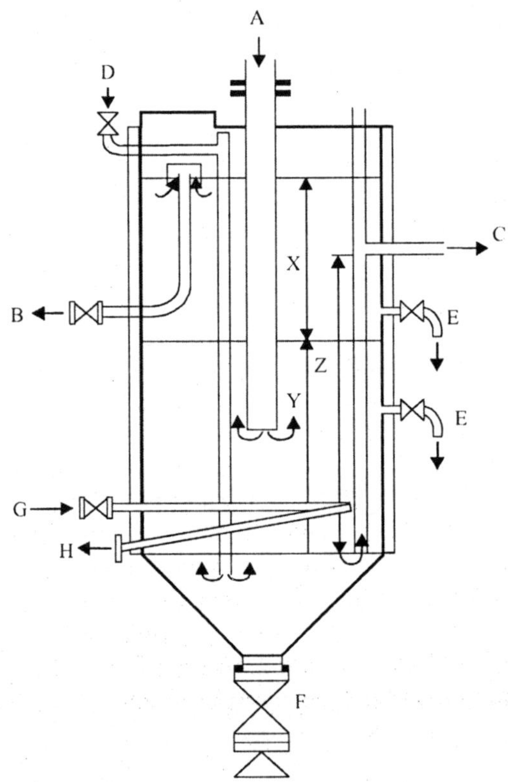

Fig. 6.7. Stock continuous clarification tank.

The remaining clarifying equipment depends on the scale of operations and the results required. Further settling tanks may be employed and vacuum driers or centrifugal separators. A number of alternative systems have been advocated. In one system a new type of centrifugal separator, designed to prevent air entry with consequent oxidisation, is used and the oil is then passed to a vacuum drier which reduces moisture to below 0.05 per cent. In another system a nozzle-type centrifugal is used and a method of recycling the water is included. The oil is sent direct, or through a drier, to the storage tanks.

The old and usual method of dealing with the sludge is to pass it through a series of fat pits. Sludge is objectionable, malodourous material and it should be the object of every mill manager to discharge from his care the very minimum of solids and water contaminated with rancid oil. This task will be much assisted if the bunches enter the mill in as clean a condition as possible. The fat pits are in series: the heavier water and dirt go to the bottom of the first and succeeding pits and pass to the next pit in order under a partition which itself holds back the lighter oil. The partitions can usually be let into the pits in slots and are easily removed. In Africa it is often possible to dispose of the sludge oil for local soap-making rather than risk contaminating the pure oil with it. Otherwise it must be returned to the clarifiers.

An alternative method of treating sludge is to conduct it to a tank for boiling and then to a special sludge centrifuge of which several types are available. In one type a star-shaped rotor has four nipples with 2 mm orifices. Sludge water enters through the hollow half-shaft and is thrown by centrifugal forces towards the four nipples at the extremities of the star. The heavier particles and water pass out, but the oil builds up in the centre and is forced back through a small diameter outlet shaft inside the inlet shaft and returns to the continuous clarifier.

TREATMENT OF PALM OIL MILL EFFLUENT

Sources of Waste-Water Production

Large quantities of water are required in the palm oil milling operations. It is estimated that about 1 tonne of water (including boiler feedwater) is required to process 1 tonne of fresh fruit bunches. Obviously, a great proportion of the water will be discharged as waste-water, commonly known as palm oil mill effluent (POME). Some water is lost as vapour (steam).

POME is mainly generated from sterilisation and oil clarification processes in which quantities of steam/hot water are used. Another waste stream originates from hydrocyclone operation where water (with clay or salt) is used as a medium to separate shell and kernel. For a well-operated mill with good housekeeping practices, the amount of waste-water generated from the sterilisation process (steriliser condensate), oil clarification process (separator sludge) and hydrocyclone are 0.9, 1.5 and 0.1 tonne per tonne of oil produced, respectively. Thus about 2.5 tonnes of POME are generated for every tonne of palm oil produced. Obviously water consumption and waste-water productions can be minimised through good housekeeping and process control.

The steriliser condensate and separator sludge are segregated into separate oil pits for residual oil recovery before they are mixed again for treatment. The hydrocyclone waste contains very little residual oil and is discharged directly into the treatment plant. The final mixed waste-waters are commonly known as palm oil mill effluent.

The residual oil recovered from the oil pits are of poor quality. It is drummed and sold as technical oil for non-edible applications. It is not advisable to recycle or mix this poor-quality oil with the normal production oil as such practice will negatively affect the normal production oil quality.

Characteristics of POME

POME, when fresh, is a thick brownish colloidal slurry of water, oil and fine suspended solids. It is hot (80–90°C) and acidic (pH 4–5) and contains very high organic matter as indicated by its high biochemical oxygen demand (BOD_3, sample incubated at 30°C for 3 days) (Table 6.3). In terms of BOD, POME is 100 times as polluting as domestic sewage. It also contains very high suspended solids (SS), which are mainly oil-bearing cellulosic materials from the fruits. The POME is non-toxic as no chemical is added to the oil extraction process. The characteristics of the steriliser condensate, separator sludge and hydrocyclone waste-water are also shown in Table 6.3.

Table 6.3. Characteristics of palm oil mill effluent.

Parameter[a]	*Steriliser condensate*	*Separator sludge*	*Hydrocyclone water*	*POME*
pH	5.0	4.5	300	4.7
Oil and grease	4000	7000	300	6000
BOD	23,000	29,000	5000	22,000
COD	47,000	64,000	15,000	61,000
SS	5000	23,000	7100	13,000
DS	34,000	22,000	100	21,000
TN	500	1200	100	800
AN	20	40		35

[a] All in mg/l except pH: BOD, biochemical oxygen demand; COD, chemical oxygen demand; SS, suspended solid; DS, dissolved solid; TN, total nitrogen; and AN, ammoniacal nitrogen.

In addition, POME also contains high metal content, which can be of importance for other application like recycling as plant nutrients. Some of the essential metal contents are given in Table 6.4.

Table 6.4. Metal content in palm oil mill effluent.

Metal	*Conc. (mg/l)*
P	180
K	2270
Mg	615
Ca	439
C	25,440
B	7.6
Fe	46.5
Mn	1.98
Cu	0.89
Zn	2.30

Treatment Technology for POME

As palm oil milling processes require large quantities of water, it is common to find palm oil mills located near rivers where (free) water is readily available. Because of its high organic content (high BOD), if discharged untreated into the watercourse, POME will soon undergo biological oxidation that depletes the dissolved oxygen in the water system. Oxygen depletion destroys aquatic life in the water and consequently the surrounding environment. Therefore the industry has both the social and ethical obligations to reduce the environmental impact caused by POME.

There are several options available to reduce the pollution problem created by palm oil mills. These include complete treatment and disposal of POME or systematic utilisation of POME for beneficial purposes. The choice depends very much on the local environment.

POME, because of its high organic content (BOD) is easily amenable to biodegradation. Therefore biological oxidation is the most suitable process to breakdown the organic pollutants in POME. Biological treatment of waste-water is a process in which a mixed population of micro-organisms utilise as nutrients substrates contaminating the water. Waste-water containing polluting substances is brought into intimate contact with a dense population of microorganisms for a duration sufficient for the microbes to breakdown and remove the pollutants to the desired level.

The oxygen required for the microbial activities is supplied through dissolved oxygen in water. Invariably the biological treatment system consists of a train of anaerobic (absence of oxygen), aerobic (presence of oxygen) and/or facultative (anaerobic and aerobic) processes to meet the required treatment efficiency. The end products for anaerobic process are biogas [a mixture of methane (60–70 per cent), carbon dioxide (30–40 per cent) and a small amount of hydrogen sulphide] and biosolids. The end products for aerobic process are mainly carbon dioxide and some amount of biosolids that need to be disposed of separately. The three most common and effective treatment systems developed for the POME are shown in Figs. 6.8, 6.9 and 6.10. It has been shown over the last decade that all these systems if operated according to design and maintained properly could meet the local discharge standards for POME as shown in Table 6.5.

Table 6.5. Discharge standards for palm oil industrial effluents for Malaysia.

		PORE and OIE standard	
Parameter[a]	*POME*	*A*	*B*
Temperature (°C)	45	40	40
pH	5.0–9.0	6.0–9.0	5.5–9.0
BOD_3 (mg/l) (3 days 30°C)	100	20	50
COD (mg/l)[b]	–	250	250
SS (mg/l)	400	50	100
O and G (mg/l)	50	–	–
AN (mg/l)	150	–	–
TN (mg/l)	200	–	–

[a] BOD_3, biochemical oxygen demand; COD, chemical oxygen demand; SS, suspended solid; O and G, oil and grease; AN, ammoniacal nitrogen; and TN, total nitrogen.
[b] Ungazetted. POME, palm oil mill effluent; PORE, palm oil refinery effluent; OIE, oleochemical industrial effluent. Standard A or B depends on locations.

Ponding System

This is by far the most popular treatment system used by the palm oil mills in Malaysia. It is the most economic system provided suitable land is available at very low or no cost to the palm oil mill. Large land area is required for a ponding system. Various design and configurations of ponding systems are used. Fig. 6.8 shows a typical system used in Malaysia. It consists of essentially a number of ponds for different functions. Ponding systems are reliable, stable and capable of producing good-quality final discharge with a BOD of less than 100 mg/l.

The anaerobic ponds are usually 5–7 metre deep each and the facultative ponds are 1–1.5 metre each. The hydraulic retention times (HRT) for the deoiling tank, acidification, anaerobic and facultative ponds are 1, 4, 45 and 16 days, respectively.

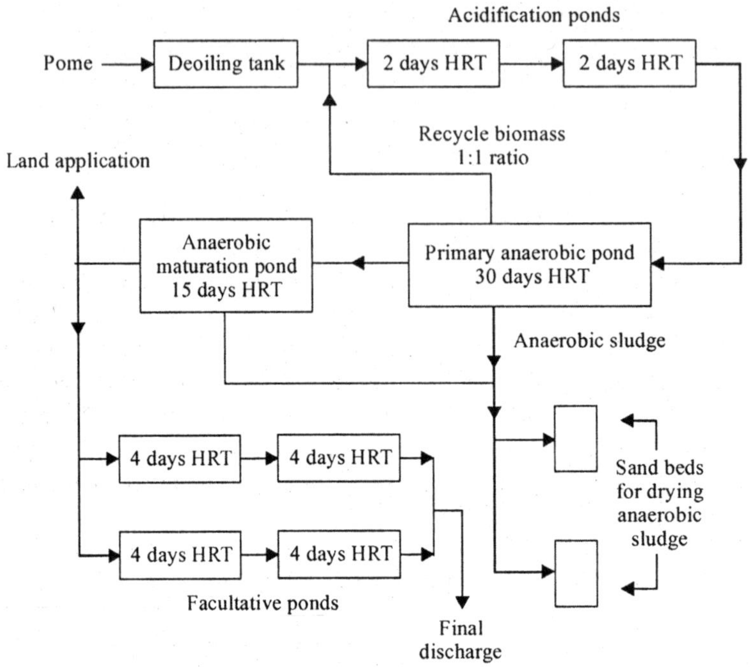

Fig. 6.8. Schematic flow diagram for ponding system.

A ponding system is normally operated at low rate, with organic loadings ranging from 0.2 to 0.35 kg per BOD per cubic metre per day. Because of the size and configuration of the ponds, they are quite difficult to control and monitor. Furthermore, there is very little mixing. Mixing is only achieved through bubbling of the biogas generated by the anaerobic process. It is hardly adequate to mix the digester content. Also the rising biogas brings along very fine solids to the top of the ponds. Therefore it is very common to find islands of solids floating in the anaerobic ponds. This often results in dead spots and short-circuiting in the ponds, which reduces the treatment efficiency of the system. Obviously it is very labour intensive and expensive to maintain the ponding system in very satisfactory conditions. It is also imperative to ensure that as little oil as possible is allowed to get into the anaerobic pond. Otherwise the oil will agglomerate with the rising solids brought up by the biogas and form a sticky scum that is difficult to remove. It is not advisable to allow excessive accumulation of the scum so that the effectiveness of the system is not adversely affected.

Due to inadequate mixing by biogas, solid build-up at the bottom of the anaerobic pond poses another maintenance problem to the palm oil mills. Excessive solid build-up at the bottom of the ponds will reduce the effective digester capacity and consequently shorten the hydraulic retention time. Thus, the treatment efficiency will be adversely affected.

Regular desludging of the ponds is recommended. Submersible slurry pumps can be used for this purpose. The solids are removed at regular intervals to a series of drying beds constructed besides the ponds. The dried solids, which contain high plant nutrients, are used as fertilisers in the estates.

Open Tank Digester and Extended Aeration

In this system, after the oil recovery pit, POME is treated in a two-phase anaerobic digestion process followed by extended aeration in a pond. The digesters are open-top and unstirred. Figure 6.9 shows a schematic flow diagram of the system. The HRT for the acidification, anaerobic and aerobic process are 1, 20 and 20 days, respectively. The organic loading of the anaerobic digester is in the range of 0.8–1.0 kg BOD per cubic metre per day.

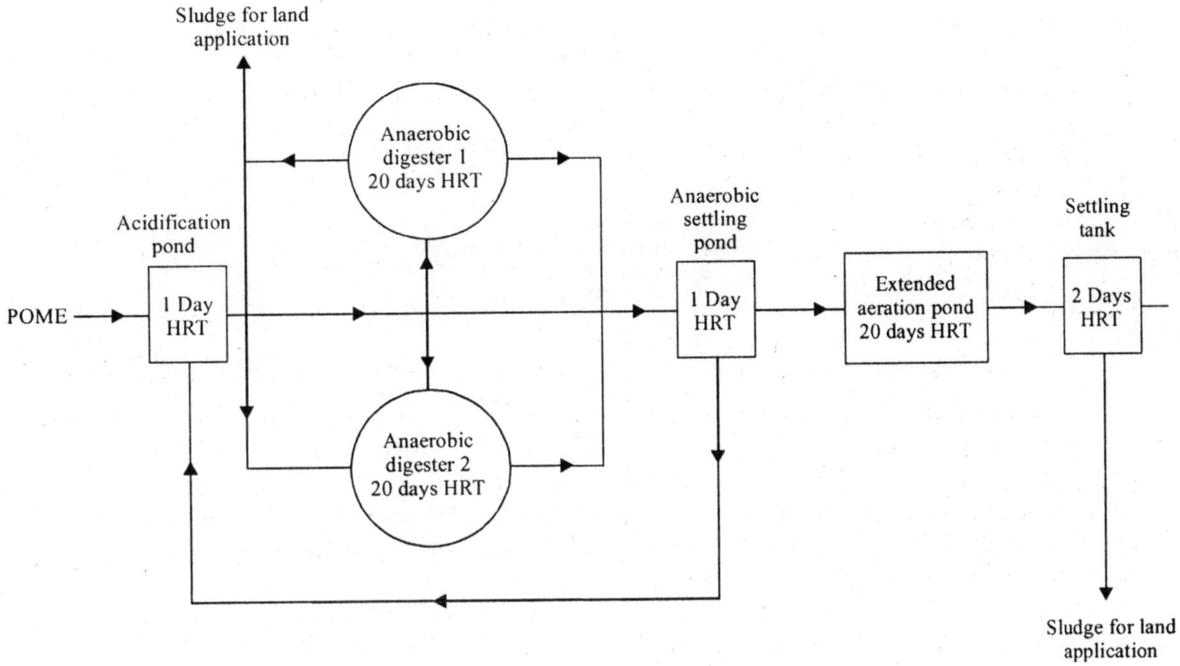

Fig. 6.9. Schematic flow diagram of opened tank digester and extended aeration system.

Similar to the ponding system, limited mixing is provided by the biogas generated. Hence such a system faces the same problem of solid build-up at the bottom of the digester. In order to maintain sufficient HRT for effective digestion, the solids have to be removed at regular intervals as in the ponding system. This can be easily done by means of slurry pumps. The solids are carted away for land application in the estates. As compared to the ponding system, the scum and solid build-up in the digester can be readily monitored and controlled.

Mechanical surface aerator is used to supply air/oxygen to the aerobic pond. In aerobic systems it is important to ensure that enough oxygen is supplied to the aerobic micro-organisms to do the job.

Close Tank Digester with Biogas Recovery and Land Application

The digesters (Fig. 6.10) are operated as conventional high rate systems with organic loading of 4.8 kg volatile solids (VS) per cubic metre per day. The HRT for anaerobic digestion is about 10 days operating at a slightly elevated temperature of 42–50°C.

Good mixing is ensured by recycling the compressed biogas through an emitted and a draught tube. From the emitter, the biogas rises through the draught tube in large bubbles. Thus the digester liquor is drawn into the bottom of the draught tube and discharged from the top, causing effective circulation and hence mixing of the digester content. As the content of the digester is well mixed, there is no problem of solid build-up at the bottom of the digester.

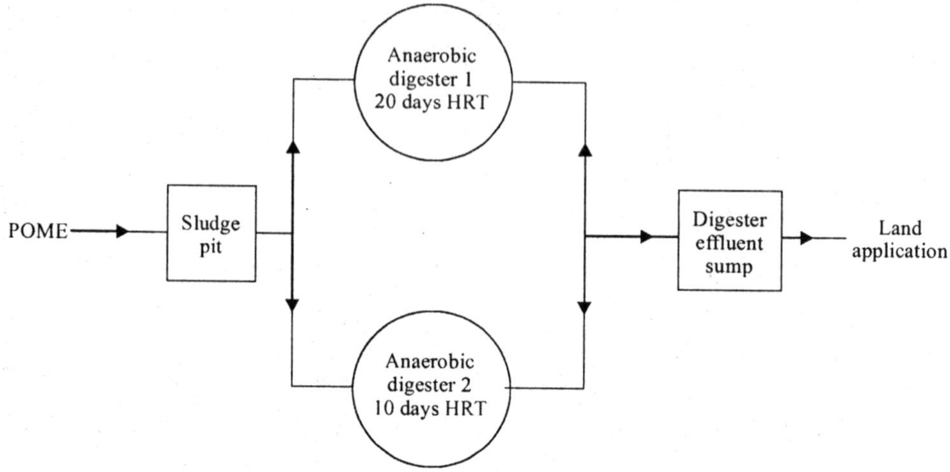

Fig. 6.10. Schematic flow diagram of closed anaerobic digester and land application system.

In this system, the digester liquor, having a BOD of about 200 mg/l, is applied to the plantation nearby as fertiliser. Several systems have been developed for land applications of POME. The biogas generated can be harnessed for heat and electricity generation. Excess biogas is flared off. About 0.59 m^3 of biogas is produced per kilogram of VS added. Thus for a 60 tonne FFB per hour mill operating 20 hr. per day, about 20,000 m^3 of biogas is obtainable. The biogas contains about 65 per cent methane, 35 per cent carbon dioxide and less than 2000 ppm of hydrogen sulphide. It has a calorific value of about 5300 kcal/m^3. It can be used as an energy source to supply heat or electricity to supplement the energy requirements if necessary. Substantial saving in energy bill can be realised.

Kernel Extraction

The matte coming from the presses consists of nuts and moist fibre with some residual oil. To extract the kernels it is necessary to (i) extract the nuts from the fibre; (ii) crack the nuts; and (iii) separate the kernels from the cracked shells. In large mills the kernel extraction plant occupied a great deal of space and consisted of a large fibre separator, nut drying silo, nut screens, hydrocyclones or mechanical clay baths, with at least four elevators and numerous screw conveyers for moving nuts, shell or kernels. In comparison with the oil extraction plant the kernels recovery equipment was very expensive, it being estimated that while the investment and operating costs of the oil extraction equipment was only twice as great as that of the kernel equipment, the value produced in the oil extraction section of the mill was nearly nine times as great as the kernel value produced by the kernel extraction plant. This costly and space-occupying equipment has been simplified by using air only to separate the fibre and to blow the nuts first to the silo and later to the hydrocyclones. A pneumatic transport system is also used for conveying kernels to the kernel silo and shell to the boiler platform.

Separation of Nuts from Fibre

This process has unfortunately been called depericarping a term already used for removal of the fresh mesocarp prior to oil extraction and the machines used are often called depericarpers. The terms 'fibre separator' has also been used and is more accurate. Fibre separation may be hydraulic, pneumatic or mechanical. The hydraulic separator performs a washing process in which the fibre floats and the nuts fall to the bottom; this process is not used today.

Pneumatic fibre separators which are partly mechanical have been the rule in nearly all large modern mills. The press-cake has a high temperature as it comes from the press and if it is immediately pushed into a 'breaker conveyer'—i.e. an open-topped, but steam-jacketed conveyer with a shaft carrying blades which cut up the press-cake as well as forcing it along the conveyer—it will dry out appreciably under its own heat. This assists the loosening of the fibre from the nuts. The effect may be enhanced by arranging for the air which is to pass through the separator to be heated by a steam-operated air-heater.

In the most commonly used type of fibre separator the fibre-nut mixture passes into a large drum rotating at about 15 rpm. This drum has baffles mounted on the inside which carry the mixture upward and allow it to drop. The current of air passing through the drum is sufficient to carry the partially dried fibre to the exit tube and blow it to the boiler platform. The nut and fibre mixture is gradually moved by the rotating motion towards the end of the drum and in the course of this motion is further dried by the hot air current. The nuts will fall into a smaller, lower rotating drum where they are polished by friction. Some air passes through this drum also, carrying any light particles upwards to join the main flow. Various releases can be arranged under the second drum to allow the exit of polished nuts and the passing over to the end of the drum of heavy pieces of bunch stalk and other debris.

The currents of air are provided by a fan placed above the separator and the air-fibre mixture is blown to a cyclone which is situated above the boiler house and serves the purpose of separating the fibre from the air, the former falling onto the boiler platform or some convenient place whence it can be easily employed as fuel. The separator is efficient but, besides being noisy, it makes comparatively heavy calls on the power supply of the mill.

Purely mechanical fibre separators have been used for a long time, particularly in small mills or in mills, such as the Pioneer, where a low capital cost was imperative. One type consists of a screened drum which is rotated and allows the separated fibre to fall through the screen. A second type is a modification of a cotton ginning machine; a revolving shaft is fitted with studs which tease off the fibres from the fibre-nut mixture fed into the machine. A third type has a rotary cage bounded by rollers revolving in opposite directions in pairs which remove the fibre to the outside of the cage but retain the nuts inside.

Rotating drum separators are now being replaced by stationary direct air separation columns in which the velocity of the upward current of unheated air which removes the fibre is adjustable. Separators of this type occupy less space and reduce power consumption. They can be combined with simplified cracking and kernel separating systems in which pneumatic transport replaces chain elevators and screw conveyers.

Nut Screening and Cracking

The clean nuts may be dried in a nut silo or, if the drying in the breaker-conveyer and in the fibre separator has been sufficient for cracking, they may be conveyed straight to screens for grading according to size before cracking. It is generally considered that in a mill with a large throughput the additional drying in the silo is necessary and the opinion has been voiced that the moisture content of the kernels must be less than 16 per cent if they are to be sufficiently shrunk away from the shells for easy cracking. Nevertheless, quite satisfactory cracking is often obtained in mills without nut silos and if heating in the latter is too high kernel quality will suffer. It should be remembered, however, that small nuts are more difficult to crack than large ones and that proper drying is therefore likely to be of greater importance an all *tenera* plantations. Too prolonged drying leads to a high percentage of broken kernels.

For satisfactory nut cracking it has been usual to send manufacturers samples of the nuts produced or to be produced on the plantation. The calculation of the correct, or most nearly satisfactory, screens can then be carried out and a decision taken on how many nut screens, crackers and cracked-mixture screens are required and of what size and type they should be. The cracking section usually consists of revolving screens to grade the nuts, the crackers themselves and screens to separate the uncracked nuts from the mixture. In old mills this section is often difficult to improve especially when nuts of widely different sizes are being produced and if an incorrect installation has been made at the start it may be difficult ever to rectify it completely. More modern systems avoid these difficulties.

The larger the number of nut grades used, the less will be the number of kernels returned to the nut-cracker. This is because the frequency curve or histogram of nut size usually overlaps the frequency curve of kernel size. If no grading is done, therefore, some large kernels will pass over the cracked mixture screen and be returned to the cracker where they may be broken and partially lost as dust. If the perforations of the cracked mixture screen are made wider, then the return of kernels to the cracker will be less, but more small nuts will pass through the screen and appear in the cracked mixture for separation into kernels and shell. Cracking in at least two fractions is usual and in three fractions if the size of the plant warrants it. This reduces the overlap of kernel and nut size; with large nuts there will be no overlap, though with small nuts a small overlap remains. In grading into three fractions calculations have to be made as to the proportions of each grade passing through the cracking section of the mill. If, then, five crackers are being used there may need to be one small, one medium and three large graders or any other suitable combination of five screens of three sizes.

The object here is to distribute the work as evenly as possible over the five crackers while bearing in mind that, there being the greatest overlap of nut and kernel size in small nuts, more than one small nut screen will be undesirable. With *tenera* nuts, which are often small, the overlap of nut and kernel size is inevitably larger than with *dura*, particularly Deli, nuts and it is partly for this reason that kernel recovery is less satisfactory with the former fruit.

Nut screening is not as simple as might be supposed. If hole size is correct for the desired screening, many small nuts may still pass over the screen if it is operated too fast, fed too much or is too small. There is thus a relationship between capacity and hole size and if high efficiency is required more screens of lower capacity must be installed. Design of the screen is also important and *dura* grading is always easier than *tenera* grading; it is therefore not surprising that methods of avoiding screening altogether have now been sought.

Modern nut-crackers are centrifugal, the nuts fed into the inlet falling into and then being thrown out of slots on the face of a shaft rotating at high speed and being hurled against a cracking ring. The vertical crackers (horizontal shaft) previously employed are suitable for large *dura* or Deli nuts, but in predominantly *tenera* plantations cracking is best done with horizontal crackers (vertical shaft) which have a larger diameter. The theory behind the use of the latter crackers is that, firstly, *tenera* nuts, having a distinct rounded head and tapering fibre-covered tail are enabled to take up a 'head-foremost' position in the extra distance covered and so be cleanly cracked on hitting the plate. If the fibre-covered tail hits the plate the nut may not be cracked. Secondly, in the confined space of a vertical nut-cracker the *tenera* nuts may be deflected by re-bounding shell fragments. Thirdly, the larger the distance between the cracker rotor and the cracking ring the more obtuse will be the angle at which the nut strikes the ring and there will then be less likelihood of the nut glancing off the ring surface. The speed of nut-crackers varies from 800–2500 rpm according to the diameter of the rotor.

Not all the nuts will be cracked when they are first fed into the cracker. A 90 per cent cracking is usually achieved and this may rise to 95 per cent with efficient crackers. The object of the cracked mixture screens is to allow dust and small shell particles to be first released and then fed by conveyer to the boiler house, to allow the cracked mixture of shell and kernels to pass to the kernel separators and to return the uncracked nuts to the cracker. The cracked mixture screens have in some modern mills been replaced altogether by vertical separating columns (below) in which dust and small pieces of shell are blown away; uncracked nuts are recovered after kernel and shell separation.

Kernel and Shell Separation

The task of kernel and shell separation can be lightened, particularly with *tenera* nuts or with *tenera* and *dura* mixtures, if a winnowing system is placed between the cracked mixture screens and the kernel separating machine. Fragments of light shell and small nuts with long fibres often have a specific gravity differing little from that of kernels and if they can be first blown away a more perfect separation will be effected. Normally the winnowing device consists of a rectangular section vertical tube in which the cracked mixture meets an upward current of air. The fragments and dust carried up by the air current are separated in a cyclone.

In the early days of milling the mixed shell and kernels were placed in a salt bath of such a specific gravity that the shells sank and the kernels floated. The kernels were then skimmed off and dried for bagging. Later a mixture of water and clay to give a specific gravity of 1.17 was found more suitable. The shells have a specific gravity of 1.3 to 1.4 and sink to the bottom; the kernels, which have a specific gravity of about 1.07, are skimmed off, washed and dried. However, shell of *tenera* material has a lower specific gravity than *dura* shell. This makes separation based on specific gravity differences more difficult.

A fine clay must be used which does not contain any appreciable quantities of sand or organic matter. It has been estimated that about 2 cubic yards of clay is required per 100 tonnes of bunches. Simple perforated tanks have been devised which are suspended in clay bath tanks and assist in separation and in clearing away the shell after removal of the kernels. The cracked mixture is placed in the perforated tanks and after the kernels have been skimmed off the tank can be raised manually or on hinges and the shell tipped out.

In one type of mechanical clay bath there is a shaft running longitudinally above or at the surface of the clay-water mixture and carrying two discs or wheels with lifting buckets and on either end, washing drums (Fig. 6.11). The bath tank consists of two compartments separated by an inclined baffle. From one compartment (B) clay water is continuously scooped by the buckets of one wheel. This clay water is tipped out of the buckets and runs into the other compartment (A) so that the latter is continually overflowing back into the first compartment (B). The cracked mixture is introduced into the overflowing compartment and the floating kernels pass out with the overflow into one of the washing drums while the clay water is running back into compartment B. At the other end of the bath the shells fall along the inclined baffle to the bottom of the overflowing compartment A and are picked up by the second set of lifting buckets and dropped into the other drum where they are washed and pass out of the machine. The wear and tear on these machines is naturally high and efficiency is limited owing to the tendency for floating kernels to get picked up by the shell buckets.

In another type of mechanical clay bath the clay-water mixture is circulated by a pump from a ground-level tank into a large cone shaped funnel above it. The mixture is introduced into this funnel. The floating kernels pass from the surface through a pipe on to one side of a vibrating screen while the shells sink to the neck of the funnel and are syphoned over with the clay water on to the other section of the screen. The clay water from both pipes falls through the screen back into the ground-level tanks and is recirculated. Both kernels and shell are washed by a sprayer on the screen and are discharged. Kernel separators of this type have capacities of 2–4 tonnes of nuts per hour.

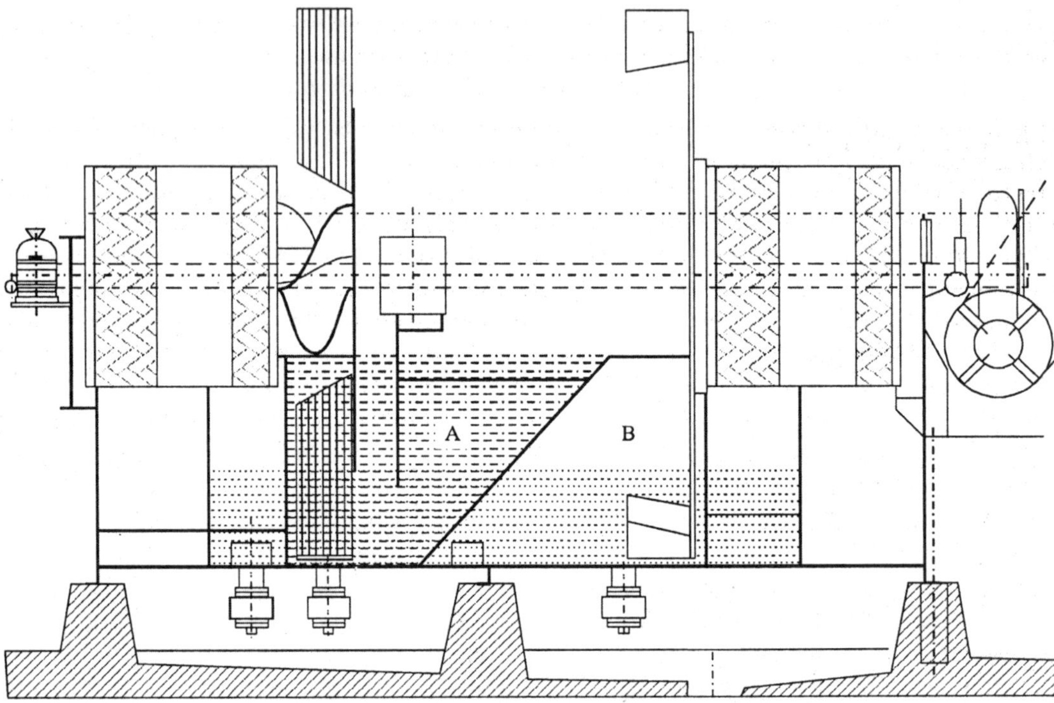

Fig. 6.11. Mechanical clay-bath kernel and shell separator.

The advent of the hydrocyclone has reduced the use of clay baths. The hydrocyclone's main advantage is that it is compact and requires only water. Clay and the manual labour that goes with its collection is now eliminated from the mill; however, the hydrocyclone has a higher capital cost than the clay bath and its efficiency may not be higher.

A hydrocyclone unit consists of two cyclones, two tanks, two washing drums and to pumps with connecting pipes. The mixture is introduced into the tank (Fig. 6.12) which is filled with water and is forced with the water by a pump (C′) to a cyclone (D). Here the water is subjected to two forces; firstly, a centrifugal force and secondly a non-tangential force directing the water to the vortex in the centre of the cyclone through which the majority of the water flows (Fig. 6.13). A particle lighter than water and/or with high flow resistance tends to leave the cyclone by the vortex with large quantities of water. Heavy particles and/or particles with a low flow resistance follow the centrifugal force and are discharged at the apex at the lower end of the cyclone with the smaller quantity of water. It appears that with *dura* nuts flow resistance plays little part and thus kernels are discharged from the vortex and shell at the apex; but with *tenera* material, separation, as in the clay bath, is less efficient. In practice the greater part of the kernels pass out through the washing drum (F) from which the water is allowed to flow back into the tank (A); the shell, with some kernels, passes into the other water-filled tank (B) and is pumped to another cyclone (E) where separation continues. In this case the kernels going out at the vortex re-enter the first tank (A) while the shell from the apex goes to the other washing drum (G) and so leaves the separator.

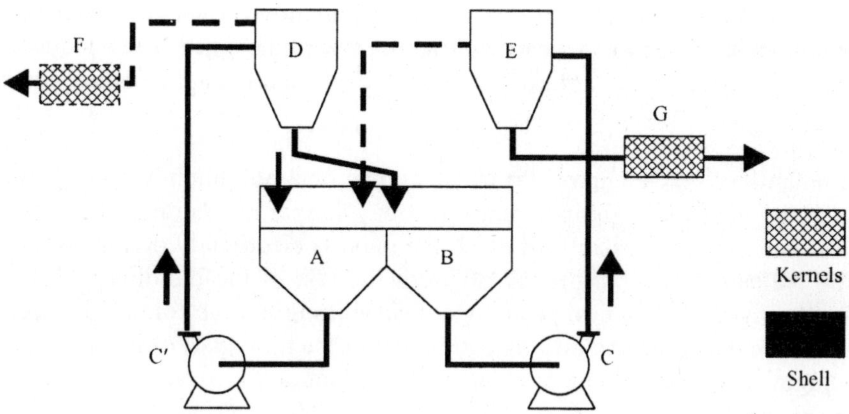

Fig. 6.12. A representation of the operation of the hydrocyclone.

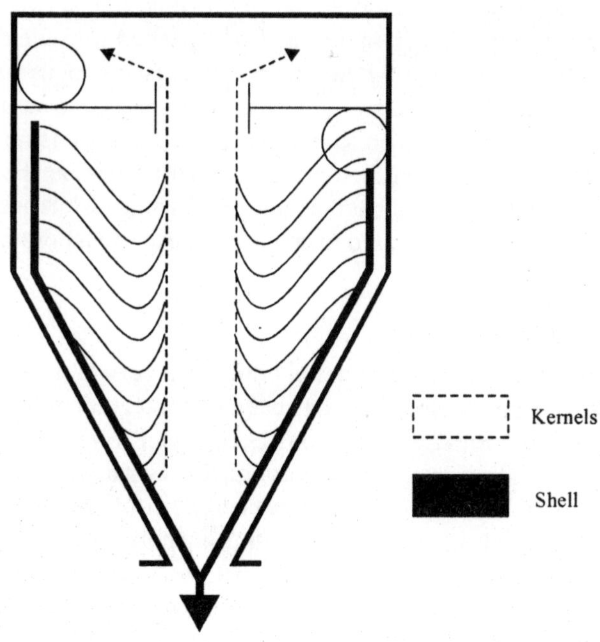

Fig. 6.13. The movement of kernels and shell in the hydrocyclone.

Kernel Drying

The washed shell passing out of the clay bath or hydrocyclone is carried by conveyer to the boiler house where it serves as fuel. The kernels have to be dried before they are fit for bagging and if the separation is not sufficiently complete they must be picked over by hand. This is usually done by passing the kernels slowly in front of a picker on a conveyer belt either before or after the kernels have been dried.

Kernels can be dried on trays passing slowly over hot air, but the kernel silo is now the most common method of drying. The wet kernels are admitted at the top and move down the silo in a continuous flow as kernels at the bottom are discharged through a shaking grid and bagged. The silo incorporates a drier which blows heated air through the kernels at different levels and is thermostatically controlled.

Mill Layout

In small or medium mills all the equipment is usually under one roof, but in very large mills the boilers, engines and electricity generating equipment may occupy a separate building. A separate power house assists cleanliness and is made possible when all the plant is electrically driven. When shaft and belt drive is used the boiler room is partitioned off from the rest of the building and the furnaces face outwards; the steam engine is then best positioned in an adjoining room for belt drive on to the middle of the long shaft which drives all the milling equipment in the main part of the building.

Next to the engine room will come the clarification plant and next to that a small laboratory and office. Bunches are brought to one end of the building for loading into vertical or horizontal sterilisers while kernels are bagged at the other end. Oil may be drummed at the side entrance to the clarification room or pumped to storage tanks.

In a large installation the reception area must be arranged so that lorries may tip bunches on to a loading ramp for temporary storage. The bunches are then pushed into the cages for sterilisation. An electric hoisting crane is required for hoisting the cages to sterilised bunches onto the stripper platform.

The bunch refuse usually leaves the mill by conveyer, but it may go to an incinerator or to a cutter so that it may be used in the boiler, or it may go to the fields. Incineration is now the usual practice.

A layout for a large electrically driven factory of over 20 tonnes per hour capacity usually has a separate power house with boilers, steam engines and alternators. Clarification is carried out in the main building of the mill though well separated from the presses. The laboratory, offices and workshops are separated from the main building.

Small-Scale Mechanical Milling

For a long time palm oil production was considered either as a very large-scale business or as a small-scale peasant undertaking. For this reason there has been much technical progress in the installation of very large mills and at the same time, attention has been given to providing for the non-mechanical needs of the small producer. However, in some parts of the world, notably America, the planting of medium-size holdings of a few hundred hectares has become common and is increasing. Under these circumstances it might be thought that several plantation owners would combine to build a medium-sized mill of high efficiency; such a course is not always possible however, owing to the isolation of small estates and other causes. For this reason there has been an increased interest in small plants of high efficiency. Such plants need to have a capacity of between 1 and 3 tonnes of bunches per hour and a low installation cost. Re-organisation of the Pioneer mill machinery and the use of pulleys and gravity feeds much reduced the labour costs of operation and a number of minor technical improvements such as steam-jacketing the digester and covering the centrifuge increased the efficiency.

Furthermore, small mills will not be operated economically unless there is skilled supervision and spare parts are readily available. A plan of a small mill ¾ to 1 tonne bunches per hour and incorporating the hydraulic hand-press with added mechanical drive is shown in Fig. 6.14.

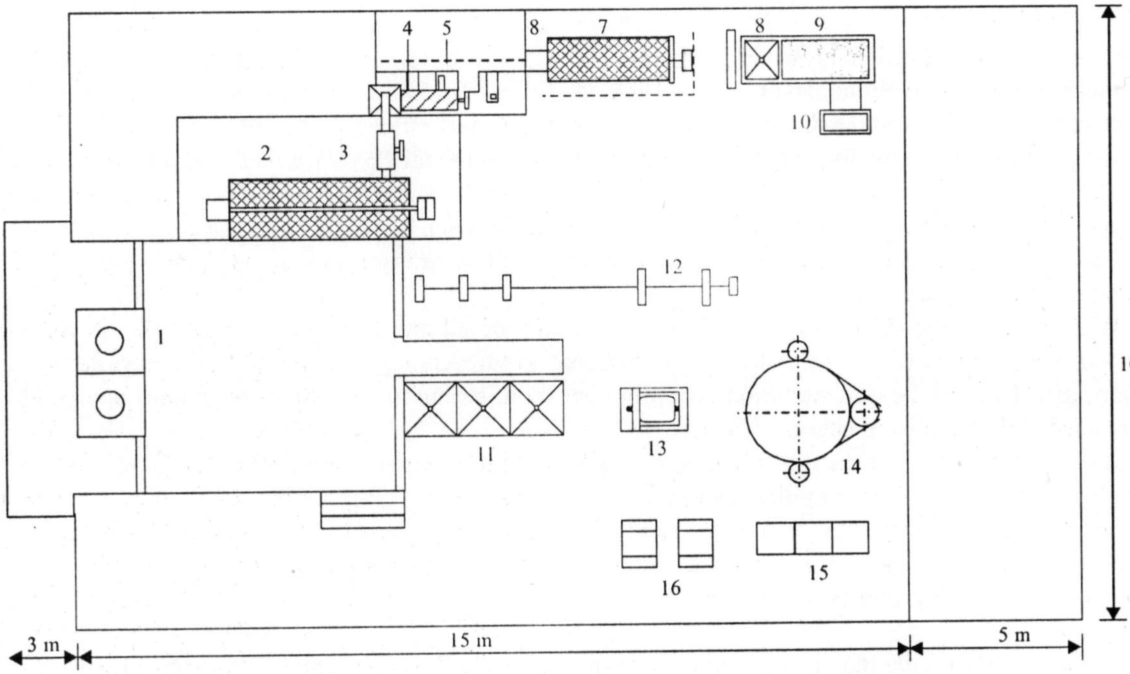

1. Steriliser
2. Bunch stripper
3. Fruit elevator
4. Digester
5. Hydraulic press
6. Cake breaker
7. Fibre separator
8. Nut-cracker
9. Cracked mixture screen
10. Clay bath
11. Clarifier
12. Transmission
13. Diesel engine
14. Boiler and pump
15. Feed water tank
16. Fuel oil tanks

Fig. 6.14. Plan of a small capacity mill incorporating the hydraulic hand-press, but with mechanical drive.

Mills of ¾ to 3 tonnes capacity using hydraulic presses have been installed in several parts of South America. These mills have been locally designed and incorporate short horizontal bunch steriliser with a compact rail system which enables the sterilised bunches to be moved easily from the sterilisers to the stripper. In many cases the mills have been constructed on the side of a hillslope so that gravity can be utilised for moving the bunches to the sterilisers and to the strippers and the fruit to the digesters.

MILLING CONTROL

The processing of bunches for the extraction of oil and kernels cannot be properly controlled unless what is going out of the mill is known in relation to what is entering it:

Coming into the mill are	*Going out of the mill are*
Bunches	Bunches refuse
Water	Palm oil, containing varying quantities of free fatty acids, water and dirt
Steam	Palm kernels, containing varying quantities of shell, dirt and moisture
Dirt	Sludge water
	Steriliser condensate

In addition press fibre and shell is going to the boiler to be used as fuel.

Control of Quantity

The calculation of the efficiency of oil and kernel extraction would be much simplified if the composition of bunches was relatively constant or if sufficient of them could be sampled and analysed to give a representative analysis of the daily or weekly intake. Figures of efficiency for oil and kernel extraction could then be obtained by dividing the quantity going out of the mill by the calculated quantity coming in. Such efficiency calculations can be made in a mill such as the Pioneer, which has an intake of stripped fruit, not bunches, since samples of unsterilised fruit can be readily analysed at intervals and a very fair estimate can be made of the average quantity of oil and kernels contained in the weight of fruit which has been milled.

With bunches, the difficulties of a direct calculation of oil and kernel content are (i) the varying fruit-to-bunch ratio; and (ii) the difficulty of obtaining a representative sample of fruit before stripping; a sample of fruit can be obtained after stripping, but by this time its moisture content has appreciably decreased and losses have begun to occur.

A direct method of estimating the oil and kernel content of batches of bunches has, however, been developed. This entails clearing the stripper before the test begins, determining the total bunch weight of the test bunches and the weight of sterilised fruit obtained from them. The oil and kernel contents are then determined by a system of sampling. Such a method requires the separation of different lots of bunches for sampling and treatment and it will not in itself show the cause of any sudden drop in efficiency.

For these reasons the majority of large mills are 'controlled' by a system of analyses for losses. All possible sources of loss are examined and analyses for oil on the one hand and kernel on the other are done. The system is admittedly imperfect and the total of oil produced plus oil losses are rightly recorded in some mill control forms as 'total accounted for' but wrongly recorded in others as 'total oil present in bunch'. With oil there is inevitably some loss on machinery and some leakage or spillage which cannot or does not get recorded. Some of the kernel present may disappear as dust. The real total of oil and of kernels coming into the mill is in fact never known and never fully estimated under the loss system, but an approximate figure is known and in a well-managed mill, the various losses will be fairly constant or improving and any aberrent figure will soon point to a cause of inefficiency. The impossibility of obtaining a true and complete balance of incomings and outgoings has its accounting disadvantages, however and may give rise to a certain complacency. The method does not constitute a complete check on events and the need for supervision of all possible points of leakage remains.

In the quantitative control of the mill the following absolute weights are obtained:

1. The weight of bunches entering the mill.
2. The weight of clarified oil produced.
3. The weight of kernels bagged.

Extraction rates depend on type of fruit, age of palms, seasons, etc. so although they are exact and very useful for estimation of returns, they do not in themselves give any information on how the mill is operating.

Most of the other data relating to quantity analysis are obtained by sampling; attention must therefore be given to the adequacy of sampling. There is a great need to submit mill sampling methods to statistical examination. Most sampling methods, though sensible and possibly adequate, are empirical and if a proper statistical examination of sampling could be undertaken it might be possible to reduce some of the work while at the same time judging the limits of each determination.

Loss measurements are made as follows:

Oil losses

Steriliser waste or 'condensate'

The quantity of steriliser waste is measured and samples taken at regular intervals from the discharge line are analysed for oil and hence the total quantity of oil lost over a period is estimated.

Bunch refuse

Bunch refuse (sometimes termed empty bunches, though they are not entirely empty of fruit) must be weighed *in toto* or sampled and analysed for (i) unreleased fruit; and (ii) oil absorbed in the stalk and empty spikelets.

The weight of bunch refuse is commonly obtained by weighing daily about 10 per cent of the cages of refuse going to the incinerator or field. Average weight × total number of cages then gives the total weight of bunch refuse. One method of analysis is to check over one random truck per shift and weigh the quantity of unreleased fruit, analysing a sample for oil and kernel content. Fixed percentages may be used for unreleased fruit if the crop is reasonably uniform, but they must be checked from time to time. For oil absorbed in the refuse one stripped bunch can be taken hourly at random from the stripper outlet conveyer and weighed. All the samples from one shift are then re-weighed, quartered, one-quarter of each being taken after removal of loose fruit; the bulk is then mixed and quartered down to two samples of 100 gram. The total dry weight of the bunch refuse coming from the mill is calculated from the dry weight obtained in the course of these analyses. Percentage oil and kernel losses on dry weight can then be used to calculate total losses. This avoids inaccuracies due to different rates of evaporation.

Chan has studied the variation of oil content of bunch refuse. He found there was more oil in the stalks of overripe than underripe bunches and that more oil was absorbed by the empty spikelets than by the stalk, though there was a positive and significant correlation between the oil contents of stalk and spikelets. As a result of his studies he recommended the sampling of three bunches every half-hour and quartering longitudinally to take one-quarter of each sampled bunch.

Hard bunches are usually returned to the steriliser and stripper for a second treatment, so bunch refuse weight estimations must be made during the final exodus of stripped bunches from the mill.

Fibre

The oil loss in fibre leaving the mill must be obtained from estimates of the mesocarp fibre entering the mill in the bunches and the oil-to-dry fibre leaving the extraction plant. The oil-to-dry fibre estimation is simple and may be done at any stage after the matte has left the centrifuge or press, but in view of the difference in oil, content of fibre from different parts of the press-cake the sample is perhaps best taken at the fibre cyclone. Then the quantity of nuts and kernels being loose in the fibre may also be analysed and estimated. Samples of about 200 gram are taken throughout the day or shift and mixed and quartered down to duplicate 100 gram samples.

Two methods of estimating the amount of mesocarp fibre entering the mill have been commonly employed. In one the quantity of fibre is estimated by sampling the sterilised fruit travelling in the fruit elevator. Through a knowledge of the ratio of sterilised fruit (calculated as the difference between the weight of sterilised bunches and bunch refuse) to bunches entering the mill and of the quantity of nuts entering the nut drier an estimation of fibre in the total weight of bunches may be made.

An alternative method, in which the weight of sterilised bunches is not required, is to carry out trial runs during a period of, say, 4 hours and measure the total quantity of all products including the press-cake in its component parts and hence through sample analysis to obtain an average figure of dry fibre to bunches entering the mill.

In sampling for oil-to-dry fibre, it was found that variations between samples of a bulk sample were small compared with hourly variations and that duplicate tests from a bulk sample made of hourly samples gave a result within 0.6 per cent oil-to-dry matter of the true figure in 95 out of 100 cases. Thus bulking hourly samples were recommended which give a daily sample and duplicating the analysis.

The figure of oil-to-dry fibre being the largest among the accounted losses and this loss obvious to the eye and touch, a great deal of notice has been taken of it. Moreover gradual improvements in digestion and extraction methods have tended to reduce the average figures over the years from a little under 20 per cent to well under 6 per cent. There has been a tendency even for the oil-to-dry fibre figure to be taken as a kind of measure of efficiency and this may lead to a neglect of other and more easily rectified sources of loss. Two points should be borne in mind concerning the oil-to-dry fibre estimations. In the first place they will depend on the solvent used in the Soxhelt apparatus and will tend to decrease if the fibre is stored for any length of time before analysis. It is necessary, therefore, to adhere to the use of one solvent and to carry out the analysis as soon as possible. Secondly, it should be realised that a difference of 1 per cent in the oil-to-dry fibre results in a very small difference in estimated milling efficiency. For instance, in a mill where the total accounted losses were estimated to be 1.3 per cent of the weight of bunches entering the mill and the total oil produced by the mill was 17.2 per cent of the weight of bunches, an increase of oil-to-dry fibre from 9 per cent to 10.5 per cent only led to a reduction of estimated efficiency from 93.0 per cent to 92.4 per cent. In general, with other losses unchanged, a decrease of 1 per cent in oil-to-dry fibre will only increase efficiency by about 0.4 per cent. Thus, while this loss should clearly not be treated as of no account, it is important that those in charge of mills should not concentrate their attention on oil-to-dry fibre to the exclusion of other losses which may be occurring, though less obviously, in other parts of the mill.

Sludge

The number of sludge tanks filled and discharged is counted or some other method is adopted for measuring liquids. Analysis of samples for oil and for solids not fat are carried out. Regular samples are taken and mixed and the analysis is by day or by shift.

Nuts

There are various ways of estimating the small amount of oil lost on nuts. In some mills the nuts are weighed automatically entering the nut drier so that samples are taken at the point and analysed for shell, kernel, moisture and oil on shell and these estimations are related to the input of bunches. If the nuts are not weighed, however, then samples must be taken of nuts coming out of the fibre separator and, after cracking, the oil and moisture content of the shell must be estimated in the laboratory. The calculation of the oil loss will then depend on measurement of the quantity of shells produced and determination of the quantity of shell per unit of bunches entering the mill and on estimation of the percentage of oil and solids not oil in the shell.

Kernel losses

Losses of kernels are incurred (i) in bunch refuse; (ii) among the shells going to the boiler; (iii) sticking to shell; (iv) in uncracked nuts among the shell; and (v) in nuts in fibre.

(i) and (v) have already been mentioned when oil losses in bunches and fibre were discussed; (ii), (iii) and (iv) can be estimated by sampling the shell going to the boiler house and separately they give an indication of the working of the cracking section of the mill.

In relating these losses to the total weight of bunches entering the mill, account is taken of the quantity of nuts, if this is measured and of the quantity of shell, fibre and bunch refuse.

Efficiencies

The efficiency of oil extraction is obtained from the estimations described, as follows:

$$\frac{\text{Oil produced} \times 100}{\text{Oil produced} + \text{oil losses accounted for}} = \text{Estimated oil extraction efficiency}$$

The efficiency of kernel extraction is similarly obtained:

$$\frac{\text{Kernels produced} \times 100}{\text{Kernel produced} + \text{losses accounted for}} = \frac{\text{Estimated kernel}}{\text{Extraction efficiency}}$$

or, if nuts are automatically weighed entering the nut drier and the dry weight of kernels estimated in nut samples, the efficiency of cracking and separating may be estimated as follows:

$$\frac{\text{Dry weight of kernels produced} \times 100}{\text{Dry weight of kernels entering nut drier}}$$

This, however, leaves out of account losses in bunch refuse and in fibre which take place before the nut drier stage.

Control of Quality

A great deal has already been said in this chapter about oil and kernel quality and, for completeness, it remains only to record here that the quality analyses required to be done in the mill laboratory are:

Oil

1. Moisture content: The sample is weighed to constant weight in a drying oven at 105°C.
2. Dirt content: This is determined by processes of filtration.
3. Free fatty acid content: This is normally estimated by titration against caustic soda.

Kernels

This is entirely a physical process, pieces of shell, dirt and broken kernels being separated from weighed samples of bagged kernels and weighed separately. Moisture content is determined by grinding and drying to constant weight in an oven.

Determination of the amount of discolouration is a somewhat subjective process. Usually it is considered sufficient to divide the sampled kernels, which have been bisected with a sharp knife, into white, slightly discoloured, badly discoloured or mouldy.

Other determinations

The majority of other determinations in the mill laboratory are either moisture or oil determinations. The former are carried out by drying in ovens to constant weight while the latter, e.g. of fibre, bunch refuse, is usually done in the Soxhlet apparatus.

Milling Report Forms

Milling report forms or control sheets will vary according to the measurements actually made in the mill and the analyses carried out in the mill laboratory. They will be made up from the many books which have to be entered up daily.

The forms should have the sections shown in Table 6.5 where imaginary figures are entered for a mill processing predominantly Deli *dura* bunches. In this scheme there is no repetition of figures and the form can thus be kept in manageable proportions. The tables can be extended to the right to allow for the recording of past or cumulative data for comparison.

Table 6.5. Milling report forms.

A. Throughput, output and intermediate weights

Throughput and output	*Weight (tonnes)*	*Extraction rate (%)*
1. Bunches milled	4462	–
2. Palm oil produced	794	17.8
3. Palm kernels produced	187	4.2
Intermediate + other products	*Weight (tonnes)*	*As per cent of bunches milled*
4. Sterilised bunches	3998	89.6
5. Loss of weight on sterilisation	464	10.4
6. Bunch refuse, wet	1258	28.2
7. Bunch refuse, dry	415	9.3
8. Sterilised fruit (4–6)	2740	61.4
9. Fruit in bunch refuse	12	0.27
10. Fibre, dry	339	7.6
11. Nuts	897	20.1
12. Kernel in nuts, dry	170	3.8
13. Shells	678	15.2
14. Sludge	830	18.6
15. Steriliser condensate	647	14.5

B. Oil and kernel quantitative analyses – losses and efficiency.

Oil in	*Per cent oil*	*Loss tonnes*	*As per cent of bunches milled*	*Notes*
16. Steriliser condensate	0.32	1.7	0.04	Calculated from No. 15 above
17. Bunch refuse:				
Fruit	33.3	4.0	0.09	Calculated from No. 9 above
Dry trash	2.10	8.7	0.19	Calculated from Nos. 6 and 7 above
18. Dry fibre	9.92	33.6	0.75	Calculated from No. 10 above
19. Nuts	0.33	3.0	0.07	Calculated from No. 11 above
20. Sludge	0.25	2.1	0.05	Calculated from No. 14 above
21. Total accounted losses	–	53.1	1.19	
22. Mill efficiency	93.7%			(No. 2 × 100) ÷ (No. 2 + No. 21)
Kernels in	*Per cent kernels*			
23. Fruit in bunch refuse	5.00	0.6	0.01	Calculated from No. 9 above

(Contd...)

Table 6.5. Milling report forms. *(Contd.)*

Kernels in	*Per cent kernels*			
24. Losses among shell	1.11	7.5	0.17	Calculated from No. 13 above
25. Sticking to shell	0.31	2.1	0.05	Calculated from No. 13 above
26. In uncracked nuts (to shell)	0.36	2.4	0.05	Calculated from No. 13 above
27. In fibre, dry	0.02	0.8	0.02	Calculated from No. 10 above
28. Total accounted losses	–	13.4	0.30	
29. Mill efficiency	93.3%			(No. 3 × 100) ÷ (No. 3 + No. 28)

C. Laboratory qualitative analysis

Oil	*Per cent*	*Previous periods*
30. FFA	2.1	
31. Moisture	0.072	
32. Dirt	0.006	
Kernels		
33. Moisture	6.4	
34. Shells	0.68	
35. Dirt	0.25	
36. Broken	4.2	
37. White	85.0	
38. Slightly discoloured	12.0	
39. Badly discoloured	3.0	
40. Mouldy		

Operating data are also recorded for the proper physical and mechanical control of a mill. The following records (among others) are usually kept:

Sterilising
- Maximum steam pressure
- Average sterilising time
- Maximum temperature
- Number of sterilising cycles
- Average steriliser load, tonnes
- Capacity per milling hour, tonnes
- Condensate, in kg per tonne of bunches

Presses
- Number of pressings per pressing hour
- Number of presses in use
- Quantity pressed per hour, in terms of bunches

Kernel cracking
- Cracking hours
- Quantity of nuts cracked per hour

Clarification
- Data on throughput and sludge/oil ratio

In some mills measurements of the degree of ripeness of the bunches are attempted. This is in the nature of an overall check on the harvesting. The figures are open to misinterpretation since the amount of fruit loosening will depend, apart from ripeness, on the number of hours between harvest and the recording of the number of detached fruit and on whether the bunches are dry or have become wet in the rain. In some mills the bunches are sampled on receipt and are classified according to the number of detached exterior fruit. In other mills the quantity of loose fruit delivered is measured and related to the total bunch plus loose fruit weight. In spite of the difficulties and admitted inaccuracies of these measurements it is certainly advantageous from the point of view of the mill engineer to obtain some picture of the state of the produce he is receiving from the field so that he may be able to note any variations from his usual supply. In one scheme devised for a group of estates bunches are sampled each day and divided into three ripeness classes: (i) ripe—more than 1 loose fruit per pound of bunch weight (2 per kg); (ii) over-ripe—less than one-eight of the fruit retained in the bunch; and (iii) under-ripe—bunch with less than 1 loose fruit per pound weight.

Bunches are further classified and enumerated if found unsatisfactory in the following respects: (i) empty bunches; (ii) poor fruit set; (iii) diseased; (iv) damaged by pests; (v) dirty; (vi) bruised; and (vii) long stalks.

OIL PALM BY-PRODUCTS

Apart from the production of CPO and palm kernel oil (PKO), the oil palm industry also generates large quantities of by-products in the form of biomass. The bulk of the by-products derived from the palm oil industry are basically lignocellulosic and organic in nature and with a high plant nutrient content. With proper handling and management, these by-products could be utilised and converted into value-added products.

Biomass Production and Availability

The biomass production from the palm oil industry is derived mainly from two sources, i.e. the plantations and the palm oil mills. From the plantation the biomass produced per hectare of oil palm has been estimated and shown in Fig. 6.15. On an annual basis, about 0.4 tonnes of palm fronds per hectare are available through routine maintenance pruning and harvesting. However, at the end of its economic life span, another 14.4 tonnes of palm fronds per hectare are available during re-planting. In addition a total of 74.4 tonnes of palm trunks per hectare are also available during the re-planting programme.

At the palm oil mills, about 20 per cent CPO and 1.6 per cent of PKO are recovered from the FFB thus leaving about 78.4 per cent biomass, including palm kernel meal.

As shown in Fig. 6.15 the processing of fresh fruit bunches generated about 1.5 tonnes of empty fruit bunches (EFB), 1.6 tonnes of palm press fibres, 0.9 tonne of palm kernel shell, 2.4 tonnes steriliser condensate and 0.7 tonne of dry mill effluent per hectare of oil palm annually.

Utilisation of By-Product

Oil palm trunks and fronds

Under normal plantation practices, the pruned fronds are placed along the palm interrows and act as mulch. Besides conserving soil moisture and reducing solid surface erosion, the fronds on decomposition return organic matter and slowly release plant nutrients to the soil.

During the re-planting programme, the oil palm trunks and fronds are chipped into small pieces and pulverised using a special pulveriser. The biomass residues are left in the field to allow for decomposition processing, which could then yield organic matter and release of plant nutrients. The placement of trunk residues on field terraces could also reduce soil erosion.

Empty Fruit Bunches and Fibres and Shells

Traditionally, the empty fruit bunches generated at the palm oil mill are mostly incinerated to produce bunch ash. Bunch ash is considered a good source of potasshic fertiliser and is also useful as liming materials because of its high alkalinity (pH 12). However, incineration of EFB could cause air pollution and this practice is not encouraged by the Department of Environment.

Alternative uses of EFB have been investigated and results have shown that they are suitable as a mulching material for oil palm. The EFB are fibrous in nature and have a high moisture content (about 60 per cent). The application of EFB in the interrows of palm avenues has been shown to improve oil palm growth and yield performance. When applied onto the soil surface, the EFB undergoes a degradation process that will yield organic matter and slowly release plant nutrients for crop uptake.

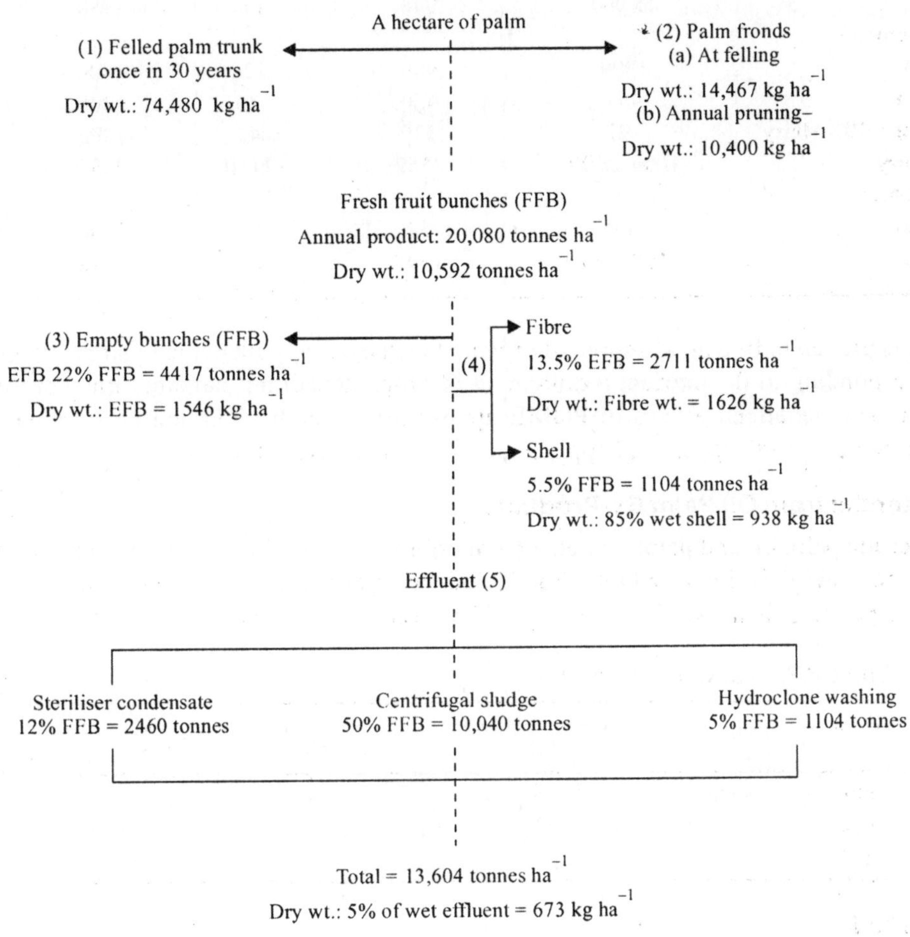

Fig. 6.15. Diagrammatic representations of actual quantity of oil palm wastes.

Some of the oil palm fibres and shells are usually used as surface mulching material for oil palm seedlings at the nursery stage. These materials are beneficial in conserving soil moisture and reducing fertiliser leaching in the polybags and thus they enhance the growth of the seedlings.

Palm Oil Mill Effluent (POME)

Palm oil mill effluent is essentially organic in nature and non-toxic but has a higher polluting potential. In its raw state, POME has an extremely high concentration of biochemical and chemical oxygen demand

(BOD and COD) and high in plant nutrient contents, particularly in nitrogen and potassium. After treatment processes various types of POME are available and their chemical composition is shown in Table 6.6.

Table 6.6. Types of POME available and their chemical compositions.

	Chemical composition (mg/l)				
			Total		
Types of POME	*BOD*	*N*	*P*	*K*	*Mg*
Raw	25,000	948	154	1958	345
Digested (anaerobic)					
Stirred tank	1300	900	120	1800	300
Supernatant	450	450	70	1200	280
Supernatant + 10% slurry	191	320	42	1495	258
Bottom slurry	1000–3000	3552	1180	2387	1509
Digested (aerobic)					
Supernatant	100	52	12	2300	539
Bottom slurry	150–300	1495	461	2378	1004

POME has proven to be a good source of organic fertiliser and is available in larger volume. Applied at rates corresponding to the nutrient requirement of crops, it will not have detrimental impact on the environment. The beneficial effects of POME application on crop yield and performance have been investigated. Several methods of land application systems are available.

Energy Potential from Oil Palm By-Products

Apart from crude palm oil and palm kernel, a palm oil mill also produces a large quantity of biomass as by-products. In general, an FFB contains about 20 per cent palm oil, 6–7 per cent palm kernel, 14–15 per cent fibre, 6–7 per cent shell and 23 per cent EFB. The heat value of each biomass is shown in Table 6.7.

Table 6.7. Heat value of biomass.

Biomass	*Moisture content (%)*	*Oil content (%)*	*Heat value kcal/kg (dry)*
Empty fruit bunches	65	5	3700
Fibre	42	5	4420
Shell	7	1	4950

Fibre and shell

The palm oil mill uses fibres and shell as boiler fuel to produce steam for electricity generation and palm oil and kernel production processes. The fibre and shell alone can supply more than enough electricity to meet the energy demand of a palm oil mill. It is estimated that about 20 kWh (lower for higher-capacity mill) of electrical energy is required to process 1 tonne of FFB.

Empty fruit bunch

Apart from fibre and shell, EFB is another biomass that can be readily converted into energy. However, this material has only been utilised to a limited extent. This is because there is already enough energy available from fibre and shell. Also due to its physical nature and higher moisture content (50–65 per cent), the EFB has to be pre-treated to reduce its bulkiness and moisture content to below 50 per cent in order to render it more easily combustible.

Biogas from palm oil mill effluent

Biogas is generated from anaerobic treatment of POME. It contains about 65 per cent methane (CH_4), 35 per cent carbon dioxide (CO_2) and trace amounts of hydrogen sulphide (H_2S). It has a calorific value of about 4740–6150 kcal/Nm^3. About 28 m^3 of biogas are generated for every cubic metre of POME digested. In a gas engine, about 1.8 kWh of electricity can be generated from every cubic metre of biogas.

REFINING AND FRACTIONATION

Physical and Chemical Refining

Crude palm oil extracted commercially from the fresh fruit bunches contains a small but variable amount of undesirable components and impurities. These include some mesocarp fibres, moisture and insolubles, free fatty acids, phospholipids, trace metals, oxidation products and odouriferous substances. As a result, palm oil is normally refined to a bland, stable product before it is used for direct consumption or for formulation of edible product.

Two methods, namely physical refining and chemical refining, are available for refining crude palm oil. They differ basically in the manner in which the free fatty acids are removed. Physical refining has become the major processing route because of its cost effectiveness, efficiency and simple effluent treatment. Both processes are able to produce refined, bleached and deodourised (RBD) palm oil of desirable quality and stability suitable for edible purposes. The unit operations involved in these two processes and the components removed are shown in Fig. 6.16 and Table 6.8, respectively.

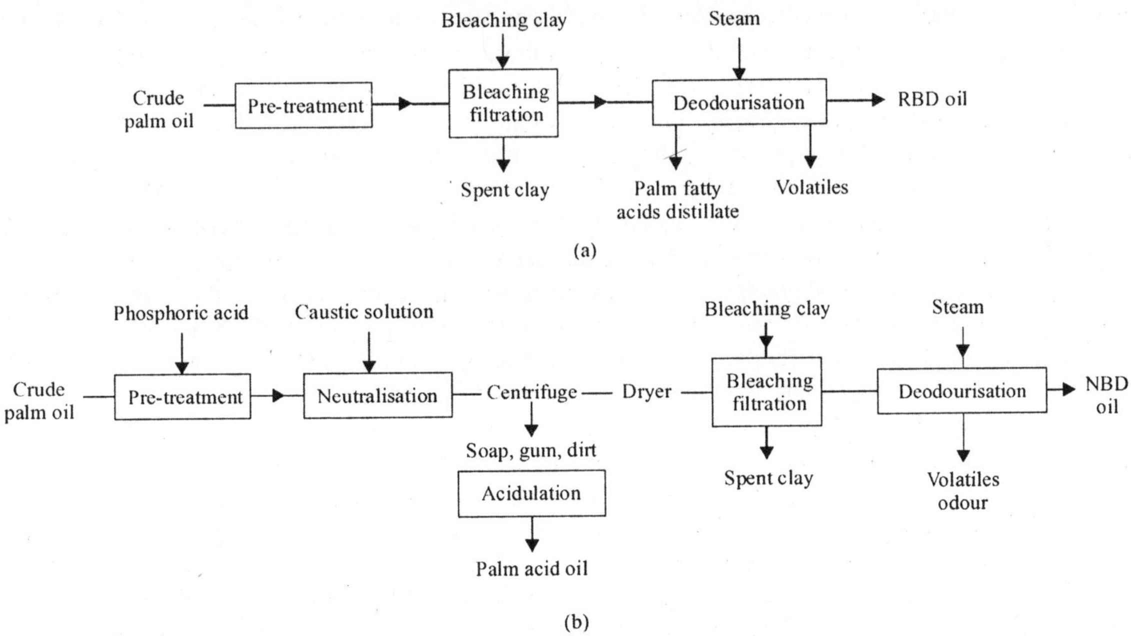

Fig. 6.16. Flow diagrams of: (a) physical refining; and (b) chemical refining of crude palm oil.

Table 6.8. Refining crude palm oil: unit processes.

Stage	*Principal impurities reduced or removed*
Degumming	Phospholipids, trace metals, pigments
Neutralisation	Fatty acids, phospholipids, pigments, oil insolubles, water solubles
Washing	Soap
Drying	Water
Bleaching	Pigments, oxidation products, trace metal, traces of soap
Filtration	Spent bleaching earth
Deodourisation	Fatty acids, mono- and diglycerides, oxidation products, pigment decomposition products
Physical refining	Fatty acids, mono- and diglycerides, oxidation products, pigment decomposition products
Polishing	Removal of trace oil insolubles

Physical refining

Its unique feature is that the deacidification, deodourisation and thermal decomposition of carotenoids are accomplished in one process in a stainless steel deodouriser. It is a continuous processing consisting of a two-step operation of pre-treatment followed by steam distillation.

Pre-treatment

Pre-treatment refers to the initial degumming of crude palm oil with concentrated phosphoric acid and the subsequent adsorptive cleansing with bleaching clay. Crude palm oil is dosed with phosphoric acid (80–85 per cent concentration) at a rate of 0.05–0.2 per cent (of the feed oil), heated to 90–110°C and given a residence time of 15–30 minutes before passing to the bleacher where bleaching earth is added as a slurry. The earth required ranges from 0.8 to 2.0 per cent, depending on the quality of the crude oil.

The purpose of the phosphoric acid is to precipitate the non-hydratable phosphatides while the function of the earth is four-fold: (i) to adsorb the undesirable impurities such as trace metals, moisture, insolubles and part of the carotenoids and other pigments; (ii) to reduce the oxidation products; (iii) to adsorb the phospholipids precipitated by the phosphoric acid; and (iv) to remove any excess phosphoric acid present in the oil after degumming.

The final residual colour of the pre-treated oil alone is unimportant as the role of the bleaching earth is not so much of colour removal but more critically in its ability to act as an adsorptive cleansing agent. Complete removal of residual phosphoric acid in the bleaching stage is also critical as any 'slip through' can result in the rapid rise of free fatty acid content and colour of the final RBD oil. As a further assurance, a suitable quantity of calcium carbonate is often added after dosing of the bleaching earth to the degummed oil, to help neutralise the residual phosphoric acid.

Bleaching is carried out under a vacuum of 20–25 mm Hg and at a temperature of 95–110°C with retention time of 30–45 minutes. The slurry containing the oil and earth is then filtered to recover a clear, light orange colour pre-treated oil. Usually a small amount of diatomaceous earth is used to pre-coat the filter leaves to improve the filtration process. As a quality precaution, the filtered oil is polished through another security filter bag in series, to trap any earth particles that escape through the first filter. This is essential as the presence of spent earth particles in the pre-treated oil reduces the oxidative stability of the final RBD oil. The spent earth from the filter normally contains about 20–40 per cent oil and this is the major source of oil loss in the refining process.

The pre-treatment process can be carried out in batch, semi-continuous or continuous equipment and the filters used are either plate and frame presses or verticle or horizontal pressure filters with verticle stainless steel filter screens.

Deodourisation

The pre-treated oil is then ready for deacidification and deodourisation. The pre-treated oil is first deaerated followed by heating to 240–270°C in an external heat exchanger before pumping into the deodouriser, which is kept under a vacuum of 2–5 mm Hg. Traditionally thermal fluids are commonly used as the heating medium. However, to eliminate the risk of possible contamination of refined oil with thermal fluid, superheated higher-pressure steam is now commonly being used, especially in new plants. Temperatures above 270°C are to be avoided to minimise loss of neutral oil, tocopherols/tocotrienols and also the possibilities of isomerisation and undesirable thermochemical reactions. Under such conditions and with the help of stripping steam, the free fatty acids, which were still present in the pre-treated oil, are distilled together with the more volatile odouriferous and oxidation products such as aldehydes and ketones, which otherwise would impart undesirable odour and taste to the oil. At the same time, the residual carotenoids present are also thermally decomposed and the end result is the production of a light-coloured, bland RBD palm oil. To maximise the recovery of thermal energy, the hot deodourised oil is heat exchanged against incoming pre-treated oil to be cooled down to a temperature of 120–150°C. Further cooling is effected by water down to 55–65°C prior to storage. Antioxidant and citric acid, if required, are dosed into the RBD palm oil at this stage.

The desirable qualities of the pre-treated and RBD palm oil are given in Table 6.9.

Table 6.9. Desirable quality of pre-treated and RBD palm oil from the factory.

Parameter	*Pre-treat palm oil (Degummed/bleached)*	*Refined bleached and deodourised palm oil*
Free fatty acids (as C16:0) (%)	Same as crude feed	0.10% max.
Peroxide value (meq/kg)	0	0
Moisture and impurities (wt. %), max.	0.1	0.1
Iron (mg/kg), max.	0.12	0.12
Copper (mg/kg), max.	0.05	0.05
Phosphorus, (mg/kg), max.	4	4
Lovibond colour, max. (5¼-inch cell)	–	3.0 R

Development in palm oil deodourisation

The main operation in the deodouriser is the stripping of volatile materials and thermal action due to the combined effects of superheated steam, high temperature and efficient vacuum. The older deodourisers use bubble caps or sieve-tray designs to effect the countercurrent mixing action between the stripping steam and oil flow. However, technological innovations have resulted in many plants changing over to new deodourisers of packed column or the falling film types.

These new features reduce pressure drop and improve the contact between the oil film and stripping steam, thus enhancing mass-transfer efficiency. Incorporation of this new design has resulted in deodourisers of larger capacities, faster throughput, lower loss of neutral oil and thus lower steam consumption on a per tonne basis.

Chemical refining

Also called caustic refining, chemical refining involves three stages; (i) gum conditioning and neutralisation; (ii) bleaching and filtration; and (iii) deodourisation.

Gum conditioning and neutralisation

The crude oil is heated to a temperature of 80–90°C. Phosphoric acid of 80–85 per cent concentration is then dosed in at a rate of 0.05–0.2 per cent (of the feed oil). This serves to precipitate the phospholipids. After this, the degummed oil is further treated with a caustic soda solution of about 4 N (or 20 Bé) concentration with a calculated excess (based on free fatty acid content of the crude oil) of about 20 per cent. The reaction between caustic soda and the free fatty acids in the degummed oil results in the formation of sodium soap, which is readily removed by a centrifugal separator. The lighter phase discharged consists mainly of neutralised oil containing 500–1000 mg/kg of soap and moisture while the heavy phase is mainly soap, insoluble impurities, gums, phosphatides, excess alkali and a small quantity of oil loss through emulsification. As an excess of alkali is used, it is unavoidable that a slight loss of neutral oil through saponification also occurs.

The neutralised palm oil (NPO) is then washed with 10–20 per cent hot water to remove traces of soap still present. After another stage of centrifugal separation, the washed oil is then dried under vacuum to a moisture level below 0.05 per cent.

Bleaching and filtration

The neutralised palm oil is treated with bleaching earth in a similar manner as that described in physical refining. However, in this case, the earth also removes traces of soap that are present.

Deodourisation

The neutralised and bleached oil is then channeled to the deodouriser in a similar manner to that in the physical refinery. The oil is subjected to distillation at a temperature of 240–260°C and a vacuum of 2–5 mm Hg with direct steam injection. Under such conditions, residual free fatty acids, volatile oxidation products and odouriferous materials are removed together with thermal decomposition of carotenoids. The final product, called neutralised, bleached and deodourised (NBD) palm oil is then cooled down to 60°C and passed through polishing filter bags before pumping to the storage tanks. The desirable quality characteristics of intermediate and final products are given in Table 6.10.

Table 6.10. Desirable quality of freshly produced intermediate and final products in alkaline refining of palm oil.

Parameter	*Neutralised palm oil*	*Neutralised and bleached palm oil*	*Neutralised and bleached and deodourised palm oil*
Free fatty acids (as C16:0) (%) max.	0.15	0.15	0.10
Peroxide values (meq/kg)	–	0	0
Moisture and impurities (wt %) max.	0.1	0.1	0.1
Iron (mg/kg)	–	0.15	0.12
Copper (mg/kg)	–	0.06	0.05
Phosphorus (mg/kg)	–	4	4
Soap content (mg/kg)	20	0	0

Refining Factor

The efficiency of the refining process is estimated by the use of a refining factor (RF):

$$RF = \frac{\text{Per cent total oil loss}}{\text{Per cent FFA in crude oil}}$$

Both figures are corrected for moisture and insoluble impurities in the crude oil. In alkaline refining, the term acid oil factor (AOF) is sometimes used:

$$AOF = \frac{\text{Per cent acid oil}}{\text{Per cent FFA}}$$

The AOF is used for monitoring losses of neutral oil in the neutralisation process.

Values of RF for chemical refining range from 1.5 to 2.0, while lower figures of 1.2–1.4 are usually recorded in physical refining.

Refining of other palm products

Beside crude palm oil, crude palm olein, crude palm stearine, crude kernel oil, crude palm kernel olein and crude palm kernel stearine can also be refined by either chemical or physical processes described before. The basic unit operations and processing conditions for crude palm olein and stearine are similar to those of palm oil.

However, in refining palm kernel products, due to the virtual absence of carotenoids, the earth dosage required in the bleaching stage is lower, usually less than 1.0 per cent. Furthermore, due to the presence of shorter chain (C8–C14) fatty acids, the deodourisation temperature required is about 230–250°C. Typical achievable quality of RBD/NBD palm kernel oil is given in Table 6.11.

Table 6.11. Desirable quality of RBD/NBD palm kernel oil.

Parameter	*Value*
Free fatty acid (as % C12:0), max.	0.05
Moisture and impurities (wt %), max.	0.1
Lovibond colour (5¼ inch cell), max.	1.5R
Iron (mg/kg), max.	0.5
Copper (mg/kg), max.	0.1
Phosphorus (mg/kg), max.	1.0

By-products

Chemical refining

The neutralisation of free fatty acid in the crude palm oil with caustic soda results in the formation of soapstock, which is treated with dilute sulphuric acid of pH 2.0–3.5 at 110–130°C for 30 minutes. A by-product called palm acid oil is then separated from the aqueous phase by centrifugation followed by hot-water washing. It consists mainly of free fatty acids, neutral oil and partial glycerides. A small amount of unsaponifiable matter is also present.

Characteristics and properties of palm acid oil (derived from chemical refining of crude palm oil, stearine and olein) are given in Table 6.12.

Table 6.12. Characteristics of Malaysian palm acid oil.

Parameter	*Range*	*Mean*
Free fatty acids (% as C16)	66.9–88.7	72.8
Moisture content (wt. %)	0.10–0.68	0.28
Iodine value (Wijs)	41.8–64.4	53.2
Titer (°C)	38.8–47.1	44.6
Unsaponifiable matter (wt. %)	0.40–1.95	0.79
Fatty acid composition (wt. %)		
C12:0	0.1–0.5	0.1
C14:0	1.0–1.6	1.2
C16:0	31.8–56.0	47.1
C18:0	4.1–5.2	4.6
C18:1	29.9–48.9	36.3
C18:2	6.3–12.0	9.6
C18:3	0.3–0.8	0.7

Physical refining

The by-product in the physical refining of crude palm oil is the palm fatty acid distillate (PFAD). It is obtained as a condensate of the volatile matters carried over from the deodouriser by the action of the stripping steam. It consists of 80–90 per cent of free fatty acid. It has often been used as a raw material for soap making, feed compounding and oleochemical feedstock. An important and valuable constituent of PFAD is vitamin E in the form of tocopherols and tocotrienols. A process for the economical recovery of vitamin E from PFAD has been developed by the Palm Oil Research Institute of Malaysia (PORIM). Characteristics and properties of PFAD from the physical refining of palm oil are given in Table 6.13.

Table 6.13. Characteristics of palm fatty distillates from the physical refining of palm oil products.

	PFAD from palm oil		*PFAD from palm olein*		*PFAD from palm stearine*	
Parameter	*Range*	*Mean*	*Range*	*Mean*	*Range*	*Mean*
Free fatty acid (% as C16:0)	72.3–89.4	83.3	71.8–98.6	85.5	77.7–89.5	85.9
Moisture content (wt %)	0.03–0.15	0.08	0.03–0.12	0.07	0.04–0.16	0.09
Iodine value (Wijs)	51.2–57.4	55.3	45.6–59.1	57.4	44.3–52.6	44.8
Titer (°C)	40.7–49.0	46.3	36.9–47.8	45.5	44.3–52.6	49.8
Unsaponifiable matter (wt %)	1.5–3.4	2.5	1.6–3.7	2.3	1.3–2.5	1.9
Fatty acid composition (wt %)						
C12:0	0.1–0.3	0.2	0.01–0.6	0.2	0.1–0.3	0.1
C12:0	0.9–1.5	1.2	0.3–1.5	1.2	1.2–1.6	1.3
C16:0	42.9–51.0	47.1	39.1–49.0	44.1	47.6–61.3	57.0
C18:0	4.1–4.9	4.5	3.8–5.1	4.5	4.2–5.4	5.1
C18:1	32.8–39.8	36.6	29.3–42.6	39.0	25.2–36.3	29.0
C18:2	8.6–11.3	9.6	7.1–12.8	10.2	6.0–7.4	6.8
C18:3	0.2–0.6	0.5	0.3–0.9	0.5	0.2–0.5	0.4

FRACTIONATION

The triglycerides of palm oil consist of a combination of fatty acids with different chain length as well as degrees of unsaturation. This results in the presence of substantial quantity of both low and high-melting triglycerides. Crystallisation of the oil under controlled cooling followed by separation will yield a low-melting liquid phase (olein) and a high-melting solids phase (stearine). Factors affecting the crystallisation process are oil composition, polymorphism and cooling condition.

Oil Composition

Palm oil contains about 4–8 per cent of diglycerides, which can form a eutectic mixture with the triglycerides resulting in lower solid content. This can slow down the rate of crystallisation. The monoglycerides, present at less than 1 per cent in palm oil, have no significant effect on the crystallisation.

Polymorphism

Palm oil triglycerides are polymorphic and thus can crystallise in several forms. The polymorphic forms are designated as α, β' and β in the order of increasing stability and melting points. Upon cooling, palm oil initially crystallises in the α-form, which gradually transforms in the order of $\alpha \rightarrow \beta' \rightarrow \beta$ form. To have a good separation, it is desirable to have β'-form crystals because β' crystals agglomerate into large, firm clusters resulting in good subsequent filtration.

Cooling Rate

Cooling rate affects the nucleation and crystal growth of the oil. As the oil is cooled, it becomes supersaturated. When the temperature is sufficiently low (about 32–36°C) saturated glycerides will crystallise and these crystals act as nuclei for further crystallisation of the lower melting glycerides, resulting in formation of larger clusters of crystals. Slow cooling rate and proper stirring speed is essential for the formation of the desired crystal form.

Process Description

There are three commercial methods for fractionating palm oil: dry, detergent and solvent process.

Dry fractionation

This is usually carried out semi-continuously using neutralised, neutralised and bleached, or fully refined palm oil. It does not require the use of any chemicals or additives. The oil is kept homogenised at about 70°C to destroy any presence of crystal in order to induce crystallisation in a controlled manner during subsequent cooling. Crystal formation and growth occur as the oil is agitated and cooled using chilled-water circulation. The cooling programme is controlled by setting the temperature differential between the oil and chilled water and also the time of cooling. When the temperature reaches the desired temperature, which is dependent on the quality of olein required (but usually about 20°C), the cooling is stopped and the thick partially crystallised mass is ready for filtration.

The different filtration systems now used in the industry are drum rotary filters, stainless steel belt florentine filters, and 'membrane filters'. Over the last decade, the membrane filter, which actually is a filter press equipped with a membrane plate, is increasingly used because it gives a higher yield of olein (about 70–75 per cent) and a harder stearine compared to that of about 65 per cent obtainable from florentine or rotary drum filters.

Detergent fractionation

Also known as the Lanza or Liprofrac process, detergent fractionation is normally carried out on crude palm oil. The oil is first cooled in the crystalliser with chilled water to allow the crystallisation of the higher melting glycerides. When the desired temperature is reached (usually about 20°C), the crystallised mass is mixed with an aqueous detergent solution containing about 0.5 per cent of sodium lauryl sulphate and magnesium sulphate as an electrolyte. The stearine crystals are wetted by the detergent solution and separate out into a suspension in the aqueous phase. On centrifuging, the olein is discharged as the lighter phase and the stearine forms part of the aqueous phase. The olein phase is then washed with hot water to remove excess detergent and vacuum dried before storage. The aqueous phase is heated to 95–100°C to break the emulsion for recovering the stearine, which is again washed with hot water and dried under vacuum before storage. Yield of olein is about 80 per cent.

Solvent fractionation

This process is the most expensive because of solvent loss, solvent recovery equipment, much lower temperature requirement and stringent safety features. The process involves the use of solvents such as hexane or acetone. The oil is first dissolved in the solvent followed by cooling to the desired temperatures to obtain the desired crystals. Cooling is effected by brine if very low temperature is required. The miscella containing the partially crystallised oil and solvent is then filtered under vacuum suction in an enclosed drum filter. The olein miscella and stearine miscella are then separately distilled to remove the solvent and recover the fractions. Yield of olein is about 80 per cent. The solvent process now-a-days is only viable in the production of high value products such as cocoa butter equivalent or other speciality fats.

Double fractionation

Double fractionation is carried out for the production of palm olein with higher iodine value of about 60 or for the production of palm-midfraction (PMF), which contains a high proportion of oleodipalmitin used for production of palm-based cocoa butter equivalent. Usually the first olein obtained is recycled back to the plant for further cooling, crystallisation and filtration. The second stearine obtained is termed palm-midfraction. Special and skillful control of the crystallisation of both stages is critical in achieving the desired quality of the products.

Fractionation of palm kernel oil

As in palm oil, palm kernel oil can also be fractionated via the dry, detergent and solvent processes. The principles applied are quite similar. The conditions of operation, however, are quite different because of the different triglyceride composition and crystallisation behaviour of palm kernel oil. In the dry fraction process, the separation of palm kernel olein from the palm kernel stearine is effected by hydraulic pressing under high pressure. In this case, the palm kernel stearine, which is an important material for production of lauric-basic cocoa butter substitute, is the premium product. Its yield ranges from 25 to 40 per cent depending on the process used.

QUALITY ASSURANCE

Palm oil is one of the most stable vegetable oils and this can be attributed to the presence of natural antioxidants and also to the balanced ratio of saturated to unsaturated fatty acids. Nevertheless palm oil, whether crude or refined, is still susceptible to quality deteriorations. Stringent preventive measures are necessary to ensure the production of refined palm oil products of superior quality and acceptability.

Quality Chain

The processing chain of palm oil begins with the harvesting of fresh fruit bunches (FFB) from the estates followed by processing of the FFB into crude palm oil. Thereafter, the crude palm oil is sent to the refinery for processing into various grades of refined products that are then transported to the bulking installations for export. Because any quality problem that may arise at any point in the chain will affect the other stages down the line, it is necessary that the right quality be attained right from the beginning of the process chain. The two main quality problems associated with palm oil are hydrolysis, leading to formation of fatty acids and oxidation, leading to rancidity.

Hydrolysis

The hydrolysis of palm oil is promoted by the presence of free moisture and heat and also by lipolytic enzymes endogenous to the plant tissue. This mode of deterioration occurs during the bruising of fruits in the harvesting and transportation of the FFB to the mill and also their extended storage under unfavourable conditions. In the case of oil during storage, the hydrolysis is attributed to a chemical reaction that is autocatalytic. The presence of high FFA in crude palm oil is undesirable as it (i) reduces the yield of RBD palm oil through higher loss of PFAD by-product; (ii) reduces the capacity of refining; and (iii) results in poor bleachability of crude oil and poor stability in the final production. Crude palm oil with high FFA content invariable also contains a high amount of partial glycerides, especially diglycerides. Interactions between the diglycerides and the triglycerides often lead to formation of eutectics resulting in poor crystal formation during fractionation, difficulty in separation of olein and stearine by filtration and also in production of olein with poor cold stability.

Oxidation

Oxidation of oils and fats is due to prolonged exposure to air. By virtue of the low polyunsaturated fatty acid content, palm oil is relatively more stable to oxidative deterioration than the polyunsaturated vegetable oils. However, in the presence of trace metals such as iron and copper, excessive oxidation at the olefin bonds of the oleic and linoleic acids can occur, resulting in rancidity. Highly oxidised crude palm oil is known to have poor bleachability and thus requires more bleaching earth and more severe refining conditions and final product will likely be of poor stability.

Quality Assurance Measures in Plantation and Milling

Availability of good-quality crude palm oil is prerequisite for the production of good-quality refined palm oil products. The criteria for good-quality crude palm oil are:

Low free fatty acid content
Low in oxidation characteristics
Good bleachability
Low in trace metals and insoluble impurities
Moisture content of about 0.15–0.20 per cent
High in deterioration of bleachability index (DOBI)

DOBI, which is defined as the ratio of the uncorrected absorbance values at 446 nm to that at 269 nm, was introduced as a quality parameter to differentiate the refinability of good and poor-quality crude palm oil. The relation to quality is DOBI > 3, good; 2.4–2.9, fair and < 2.3, poor.

In the palm oil industry, it is often said that 'good quality is made in the filed, not in the mill'. This statement clearly emphasises the importance of maintaining good harvesting practices of fruit bunches in the plantation. A good harvesting practice is one that gives the best compromise between oil yield, oil quality and harvesting cost. Field factors that determine quality of the palm oil include the degree of ripeness of the fruit bunches, the severity of bruising of the harvested crop, delays between harvesting and sterilisation, and contamination of FFB by sand, dirt, or stones. Data in Table 6.14 serve to indicate that the oil extracted from fresh unbruised fruits can have very low FFA and oxidative characteristics compared with the normally trade.

Table 6.14. Crude palm oil quality.

Component	*Ripe, fresh, unbruised fruit*	*Average traded PO*
Triglycerides (%)	98	< 98
Diglycerides (%)	2–3	4–8
Monoglycerides (%)	0.1	0.2
FFA (%, as C16:0)	0.1	3.5 (max. 5)
Phosphorus (ppm)	2–3	20–30
Tocopherols (ppm)	800	600–800
Carotene (ppm)	550	550
Totox	1	> 5
Iron (mg/kg)	0.1–0.3	5–10
Copper (mg/kg)	0.01	0.05

Precautions to minimise hydrolysis

Precautionary measures taken by mills to minimise hydrolysis, oxidation and contamination of the crude palm oil are summarised below:

FFB handling: Minimise bruising and sterilise as soon as possible (≤ 24 hr.)

Sterilisation: Optimise conditions, avoid overheating, do not mix boiler condensate with crude palm oil.

Clarification: Eliminate water and impurities; use hermetic system.

Drying: Reduce moisture to 0.17–0.2 per cent before storage.

Processing of Crude Palm Oil

The ultimate aim of the processing of crude palm oil is to obtain various products such as RBD palm oil, RBD palm olein, or RBD palm stearine that meet the requirement of the end users. An effective and efficient quality assurance programme in a processing plant is essential and should consist of the following monitoring activities.

Raw material

Each and every delivery must be carefully inspected to ensure that specifications are met and that the shipment is free from contamination. Good raw material is a prerequisite to good-quality product.

In-process materials: Regular analyses of important quality parameters will serve as a check that proper processing has been achieved. Good communication between laboratory and production personnel is of great importance to ensure success of the quality assurance programme.

Finished products: Regular checks on the finished products must be carried out to ensure compliance to quality specifications before the products are permitted for discharge to storage or shipment.

Process control: Proper processing conditions (dosage of processing aids, temperature, pressure/ vacuum, flow rate, etc.) must be closely adhered to and monitored to ensure the oil is processed correctly and to minimise undesirable side reactions. In order to asses process efficiency, the oil losses through spent clay or soapstock (as in the case of alkaline refining only) must also be monitored.

PALM OIL REFINERY EFFLUENT TREATMENT

The characteristics of palm oil refinery effluent vary according to the type of refinery operation (chemical or physical refining, fractionation process, etc.), process control and housekeeping programme. It is quite difficult to derive general characteristics for raw effluent. Therefore the choice of treatment system will depend very much on the complexity of the raw effluent, i.e. its flow and characteristics.

However, there has been a trend among refiners over the last two decades to reduce effluent and other forms of pollution by:

1. Changing from chemical refining to physical refining.
2. Automation and strict process control to prevent spillage and product loss.
3. Installation of new equipment that is based on low energy and water consumption.

There are two basic ways of providing effluent treatment facilities for the edible oil refining industry. One way is for the industry to construct a treatment system at the manufacturing plant site to treat its effluent to an acceptable level for discharge directly to rivers or other public water courses. The second is to discharge the untreated or partially treated effluent to sewers of a local government agency providing waste-water conveyance and treatment services or publicly owned treatment works. The latter practice is generally termed 'joint treatment'. The industry practicing joint treatment is required to provide control and pre-treatment to various degrees in order to use the publicly owned facilities.

The main reason for the joint treatment practice is that it costs less than the alternative. The other treatment costs are fairly shared among the users and beneficiaries of the system. Another important advantage is minimising the space necessary for the treatment facilities. This saving of space is particularly important for crowded industrial estates. Obviously, the joint treatment plant is expected to operate to achieve better efficiency because its operation is carried out by specialised, full-time and well-trained professionals. Understandably, in any commercial operation, an effluent treatment plant is way down on the priority list. The industry is quite reluctant to spend money on pollution control equipment. Therefore joint treatment is the most economical and practical choice unless such facility is not available.

Treatment Method

Oils and fats recovery

The oils and fats recovery system adopted obviously depends on the local circumstances. Typically, the first stage of pre-treatment is the use of a physical process to recover the free oils and fats. The most commonly used physical separation process for the removal of free oils and fats are fat traps, tilted-plate separators and dissolved air flotation units. In addition, centrifuge and electroflotation systems are occasionally used.

pH control and chemical treatment

pH adjustment is often necessary to prepare the effluent for subsequent treatment processes, especially the biological ones. Chemical treatment involves the use of coagulants and flocculants such as ferric chloride, aluminium sulphate, lime and polyelectrolytes to reduce the total fatty matters prior to the separation by floatation and sedimentation processes. pH adjustment is often required for optimum results. Chemical treatment is usually required for effluent from a chemical refining process.

Aerobic treatment process

Effluent from edible oil refineries has been shown to be amenable to biological treatment, both anaerobic and aerobic processes. The application of activated sludge process or aerated lagoon in this context is well established in the edible oil industry. Fig. 6.17 shows the process flow of a typical activated sludge process.

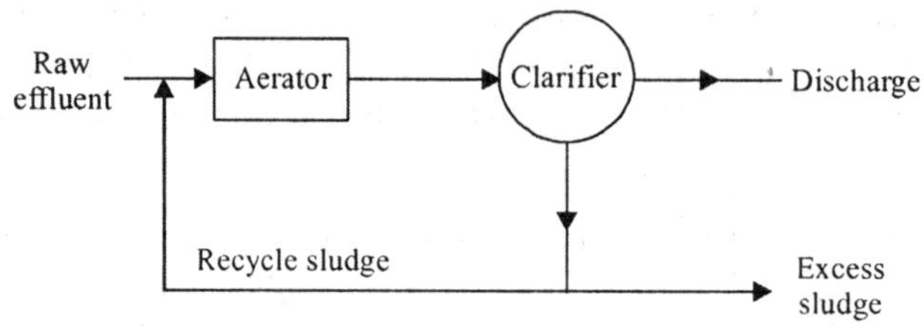

Fig. 6.17. Conventional completely mixed.

Aerated lagoon treatment was very popular in the United States. The main disadvantage of the aerated lagoon as compared to the activated sludge process was that it required large land areas and long hydraulic retention time (5–20 days). The long retention time was required because of the low mixed-liquor suspended solids (MLSS) concentration in the aerated lagoon. There was no recycle of the MLSS from the discharge to the aerated lagoon. On the other hand the MLSS in the activated sludge process is maintained at optimum level by recycling the MLSS, which is a standard feature of the process. The MLSS is normally maintained between 2000 and 5000 mg/l depending on the process requirement. Thus the retention time can be very short, a fraction of a day.

The effluent from edible oil refinery tends to be deficient in nitrogen for aerobic biological treatment. Nitrogen has to be added to fulfill the nutrient requirement. In general, ratio of BOD:N of 100:5 is required for biological treatment. Phosphorus is generally present in adequate amounts in the effluent.

This is because phosphoric acid is used in the refining pre-treatment process. A ratio of BOD:P of 100:1 is adequate for the same purpose.

Total fatty matter (TFM) concentrations in the raw effluent can lead to poor performance of the treatment process. Therefore, care should be taken to ensure efficient removal of TFM in the pre-treatment (physical/chemical) process. TFM is one of the main contributors to BOD.

Another important design criterion that affects the treatment process performance is the organic loading rate. Typically, good treatment efficiency can be achieved when the organic loading rate is less than 0.15 kg BOD per kilogram MLSS per day. Under good maintenance and operation conditions, the discharge BOD and suspended solids (SS) concentration of 20 and 30 mg/l, respectively, can readily be obtained. The treatment system for those refineries employing the chemical refining processes consists of a train of processes with balancing tanks with pH adjustment, chemical and physical treatment facilities (coagulation and flocculation as well as air flotation) and the activated sludge process. The main problems encountered in the operation of the activated sludge plant are the high fluctuations of the loading rates (both hydraulic and organic) and the requirement of close system monitoring and supervision by skilled operators. This was much lacking in the industry. Thus the process seldom achieved the expected treatment efficiency.

Sequencing batch reactor—A new treatment process for palm oil refinery effluent

An alternative aerobic process called the sequencing batch reactor (SBR) has gained much popularity in the treatment of various types of waste-waters. It is very similar to the old fill and draw (batch) system. There are essentially five modes of operation in the SBR process: fill, react, settle, draw and idle.

All these operations take place in a single reactor instead of two as in the conventional AS process (Fig. 6.18). These operations have been made simple by the advent of reliable and inexpensive microprocessor-based controllers. The process has been found to be very efficient in the treatment of a variety of waste-waters.

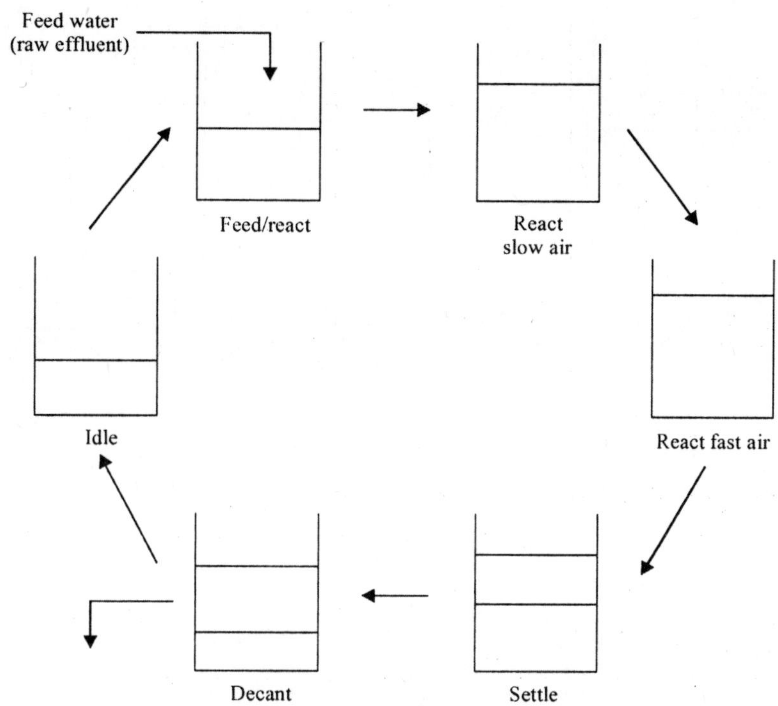

Fig. 6.18. Schematic representation of SBR cycle.

It has been reported that SBR has many advantages over the convention AS process. These advantages include equalisation, ideal settling, simple operation, compact labour and cost saving (capital and running costs). *Melcer* and co-workers have used SBR to treat waste-water from small communities. Excellent removal efficiency of BOD and nitrogen were recorded. Other co-workers have also successfully applied the laboratory-scale SBR treatment to milking parlour waste-water and piggery waste-water, respectively.

SBR treatment plant characteristics

The flow diagram of the SBR treatment plant used is shown in Fig. 6.19. It consists of one holding tank and two reactors. The effluent, consisting of circuit bleed water from the barometric condenser, cooling water, floor washing and cleaning water, is collected in a sump where any oil and fatty matter is recovered. It is then pumped to the holding tank where nutrient and pH adjustments are carried out, if necessary, before it is fed to one of the reactors by a centrifugal pump. Air is supplied by a compressor through cone-shaped diffusers installed at the bottom of the reactor. Discharge of the treated effluent is controlled by the operation of a valve. Time for each mode of operation is predetermined. All the operations including the switching on and off of the pumps and compressor, opening of discharging valve, etc. are controlled by a microprocessor-based sequencing controller as shown in Fig. 6.19. The programme can be easily changed on site to meet the operation requirements.

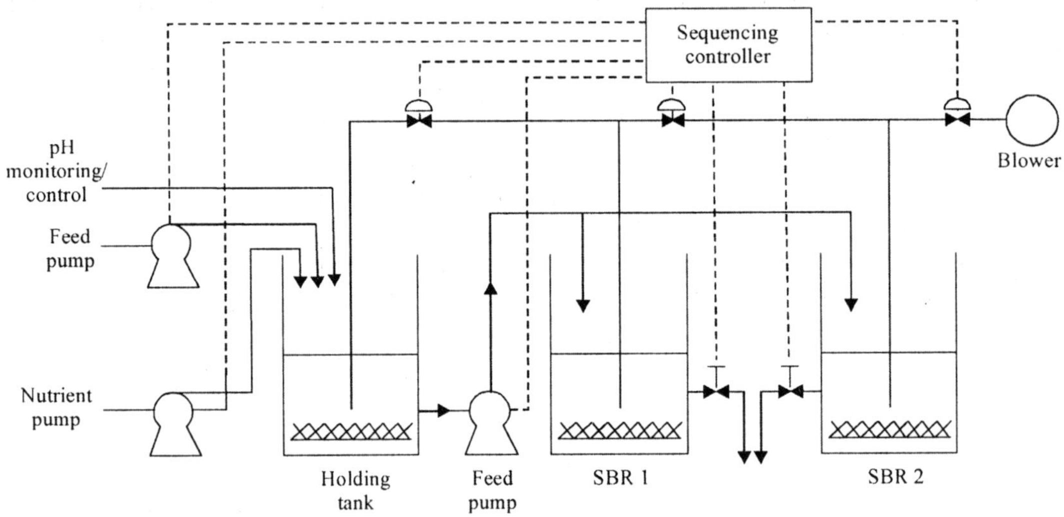

Fig. 6.19. Schematic diagram of full-scale SBR process.

Characteristics of palm oil refinery effluent

Table 6.15 shows the characteristics of the effluent from a typical palm oil refinery employing a physical refining process.

Table 6.15. Characteristics of palm oil refinery waste-water (physical refining and dry fractionation process).

Parameter[a]	*Range*	*Mean*
Temperature (°C)	30–45	35
pH	3.8–7.0	5.3
BOD	50–1500	530
COD	150–3000	900
TS	100–2000	580
SS	50–100	80
TN	0–20	10
P	1.0–10	4
O & G	25–600	200

[a] All parameters in milligrams per litre (mg/l) except pH and temperature. BOD, biochemical oxygen demand; COD, chemical oxygen demand; TS, total solids; SS, suspended solids; TN, total nitrogen; P, phosphorus; and O & G, oil and greases.

The effluent is slightly milky and is acidic in nature. It can be seen that the characteristics of the effluent vary quite widely. Nutrient (N and P) contents seem to be sufficient for biological process according to BOD:N:P of 100:5:1 ratio. It contains low suspended solids and fatty matter, which are mainly dirt and free oil from washwater and oil spillage.

END USES

Palm oil is used in both edible and non-edible applications (Fig. 6.20). Ninety per-cent of palm oil and its products are used for edible purposes. Currently palm oil is used in food preparation or food manufacture worldwide. The remaining 10 per cent of palm oil and its products are used for non-edible applications, mainly in the soap industry and in the manufacture of oleochemicals.

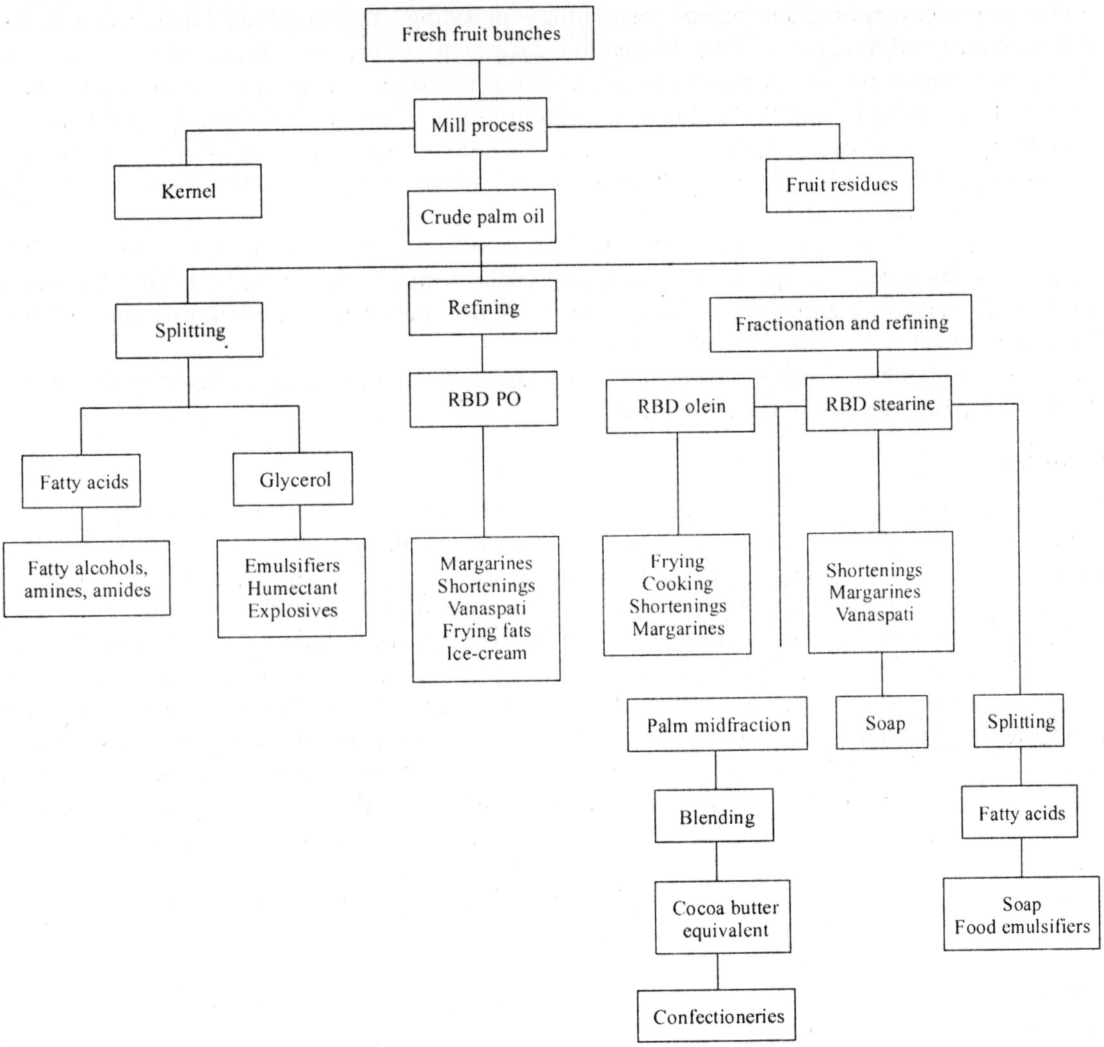

Fig. 6.20. Palm oil utilisation chart.

Food Applications

The range of palm oil applications in food are shortening, margarine, vanaspati, deep frying fat and speciality fats.

Shortening

At 20°C, palm oil has 22–25 per cent solid fat content and is a valuable ingredient for shortening formulations. Unlike margarine, which is an emulsion of about 80 per cent oil and 20 per cent water, shortening is pure (100 per cent) oil and fat.

There are many types of shortenings, each tailor-made for a particular application. There are also general-purpose shortenings that are used in the preparation of many foods: in cooking and frying and in the manufacture of bakery products such as cakes, cookies, rusks, wafer, pastries and bread.

Other related bakery products include cream fillings and icings. One important function of a shortening is to incorporate and hold air, whether beaten in a cake batter or creamed with sugar.

It has been found that a high-palmitic-acid content was food for aeration of fat/sugar mixtures. The results of the study indicated that palm stearine-cottonseed oil (3:2) shortening was best for aerated cream filling. A blend of palm stearine and low-erucic-acid rapeseed oil was very economical and performed excellently in cakes. For application both in cream fillings and baking, interesterified palm olein was the suitable material.

Palm oil can be blended with butterfat for use as a biscuit shortening or alternatively diacetyl flavouring can be added to palm oil shortening to give the desirable buttery taste. It has been found that biscuits made with shortenings containing palm oil and butterfat were not significantly different in flavour from those made with 100 per cent butterfat.

With higher solid fat content and firmer consistency, shortening based on hydrogenated palm oil produced firmer dough and harder as well as crispier cookies.

Margarine

Margarine is a type of emulsion consisting of fat and water. Although the original purpose in developing margarine was to imitate butter, there has since been a considerable diversification of margarine products, which now include: (i) table margarine in tubs; (ii) table margarine in block form; (iii) cream/cake margarine; (iv) margarine for tropical climates; and (v) puff pastry margarine.

There are also low-calorie spreads that are similar to margarine in their physical behaviour but have a much higher water content. The physical properties of margarines are largely determined by the fat component and these properties vary with the type of product. Thus tub margarines are soft and are spreadable straight from the refrigerator. Table margarines in packets are not as soft but are spreadable at room temperature, while cake or cream margarines are a little firmer than table margarines. At the extreme end, pastry margarines are much firmer, in order to give the flaky texture to the end product. Palm olein is suitable as the liquid component of margarine blends, while palm stearine or hardened palm oil can be used as the solid component. *Ward* recommended that at least 10 per cent palm oil be incorporated in canola-based margarines. Palm oil and palm oil products have also been found to be very good ingredients for puff pastry margarine.

Vanaspati

Vanaspati, or vegetable ghee, is a major commodity in countries such as India, Pakistan, Egypt, Saudi Arabia, Iraq and Iran. A higher percentage, more than 50 per cent of palm oil products, has been reported in Pakistan vanaspati formulations.

Deep frying fat

Palm oil is the most widely used industrial frying fat because it has no unpleasant room odour, has high resistance to oxidation, does not polymerise to gums and has a nutritionally good fatty acids composition (50 per cent unsaturated and no *trans* acids). The good frying properties of palm oil are due to its moderate degree of unsaturation, the absence of linolenic acid and the presence of tocopherol. The tocopherol (380–890 ppm) acts as an effective natural antioxidant. For industrial frying of instant noodles, palm oil is very suitable.

The liquid fraction of palm oil, palm olein, is also widely used for frying. In fact, in Malaysia, it is now the main cooking oil used in most households. During fractionation the tocopherols are somewhat concentrated in the palm olein, so that refined palm olein typically has 500–600 ppm total. The longer frying life and its reduced tendency to foam and polymerise make it a better frying oil than corn or soyabean oils. The ability of palm olein to produce fried foods of acceptable quality without the need for hydrogenation can be considered an advantage.

Speciality Fats

Palm oil and palm kernel oil are also ideal raw materials for the production of speciality fats. Speciality fats are particularly suitable for confectionery products, especially chocolates. Speciality fats can be classified according to their chemical composition into three types: (i) symmetrical; (ii) lauric; and (iii) high *trans*.

Symmetrical

Symmetrical-type speciality fats contain predominantly SOS-type triglycerides. The major triglycerides in cocoa butter are POSt, StOSt and POP. These triglycerides, comprising about 75 per cent of the total, are often summarised as SOS or symmetrical triglycerides. Solvent fractionation of palm oil (Fig. 6.21) produces a midfraction with a high content of the POP triglyceride but deficient in StOSt and POSt. This deficiency can be corrected by adding Illipe fat (or Borneo tallow), which contains these glycerides. Therefore the physico-chemical characteristics of cocoa butter can be closely matched by correctly blending palm midfraction with Illipe fat.

Because of the similarity in the chemical compositions of the symmetrical-type fats and cocoa butter, they are compatible with each other in almost any proportions and for this reason these speciality fats are usually called cocoa butter equivalents. In certain countries, legislation allows up to about 15 per cent of the cocoa butter in chocolate to be replaced by symmetrical-type speciality fat and the product may still be described as 'chocolate'. In terms of texture and flavour these products are very close to real (cocoa butter) chocolate.

Lauric

Lauric-type speciality fats are produced from oils containing mainly lauric and myristic acids. The simplest lauric-type fats can be made by hardening palm kernel oil to slip melting points between 32 and 41°C. Palm kernel oil can also be fractionated to give a stearine with much better melting properties than the hardened palm kernel fats. The palm kernel stearine with physical properties resembling that of cocoa butter, is called cocoa butter substitute, or hard butter. It is usually hydrogenated to further improve its meeting profile. Hydrogenated palm kernel oil or olein is used as a cheaper alternative toffee fat to replace the more expensive dairy butter, either completely or in combination with butter. Hydrogenated palm kernel oil is also a good general-purpose coating fat.

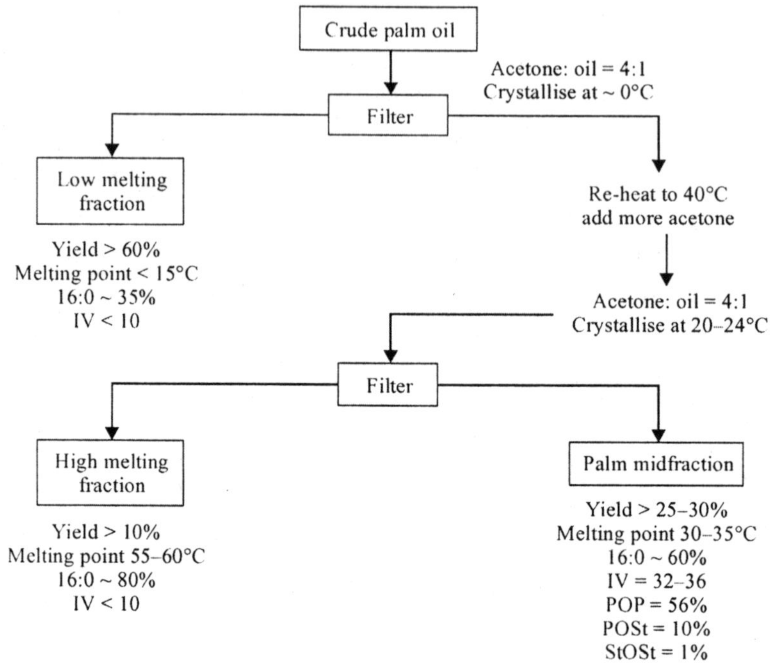

Fig. 6.21. Acetone fractionation of palm oil.

High trans type

High-*trans*-type speciality fats can be produced by a combination of selective hydrogenation and fractionation from liquid oils. These high-*trans*-type fats can be produced by selectively hydrogenating blends of soyabean oil and palm olein or palm olein alone. They are more compatible with cocoa butter than the lauric-type cocoa substitutes, thus they are sometimes called cocoa butter partial replacers.

Recent Food Applications

The newer applications of palm oil in foods include its use in emulsion-based powdered and convenience food products. Butterfat has been traditionally used in ice cream, but palm oil and palm kernel oil are now used commercially to replace it. Similarly, palm oil can also replace butterfat in the manufacture of milk, to give a product known as 'filled milk'.

Palm oil is used because it is more economical than other oils and is easily available. In addition, it is more stable to oxidation than butterfat. Filled-milk powder can be made from skimmed-milk powder re-combined with refined palm oil. Another use of a palm oil product is in infant food formulations. The low-melting olein has been found to be very suitable for use in infant food formulations when blended with other vegetable oils. Low-melting olein contains 10–15 per cent palmitic acid in the 2-position of the glycerol chain. This contributes to the high digestibility of the product.

Apart from the products mentioned, there are many other foods that contain palm oil and palm kernel oil products. These include soup mixes, cake and dessert mixes, 'rendang' or curry mixes, sardines, baked beans, breakfast cereals, shrimp-paste powder, bouillon, peanut butter and beverages. Palm oil products have also been used as a spray oil on biscuits.

New Potential Food Applications

An important future application of palm oil in food is the use of refined red palm oil in cooking. Refined red palm oil is a highly nutritious oil rich in vitamin E and β-carotene. The deep red colour of the oil blends well with ingredients such as chili and curry, making the dishes more attractive and appealing. The use of refined red palm oil is a possible alternative means of combatting vitamin A deficiency, which is prevalent in many countries. Another promising application of a palm oil product is the use of RBD palm olein of high IV as salad oil. The use of palm olein as salad oil can be made possible by blending it with other vegetable oils. Yet another potential use of RBD palm oil is as a barbecue oil. Its high stability and bland taste makes it a good choice for this application. The oil acts as a flavour carrier and it also prevents the barbecued meat from drying out, so that one gets a juicy and tasty end product. Palm oil is indeed a versatile oil. Its applications are varied and it can be used in almost any food.

Non-edible Applications

For simplicity the non-food uses of palm oil and palm kernel oil and their products will be divided into two categories, i.e. those where products are made directly from the oils (direct route) and those where they are obtained via the oleochemicals route (Fig. 6.22).

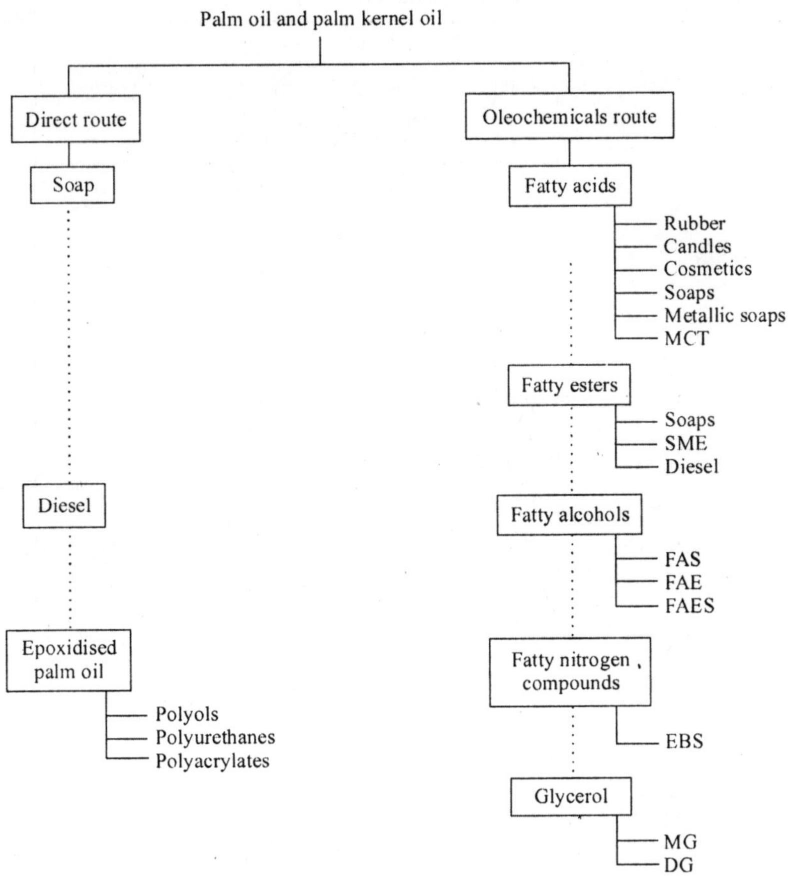

Fig. 6.22. Non-food applications of palm oil and palm kernel oil.

Direct route

Soaps

Soaps are derived from oils or fats by reacting them with caustic soda at 80–100°C in the process known as saponification. The use of soap as a laundering agent and for cleansing the skin is many centuries old. Although modern detergents have almost eliminated the use of soap for home laundry purposes, soap is still the main ingredient in toilet bars for personal use.

The incorporation of C16–C18 and C12–C14 fatty acids in soaps is important as they provide the cleaning, solubility and foaming properties required. Tallow and coconut oil have been the traditional sources of these fatty acids. A comparison between the fatty acids compositions of palm oil, palm stearine, tallow, palm kernel oil, palm kernel oleins and coconut oil (Table 6.16) indicates that the first three are rich in C16–C18 fatty acids while palm kernel and coconut oils are rich in C12–C14 fatty acids. However, for palm products to establish a niche in the market as raw materials, soap manufacturers have to be convinced that apart from price competitiveness, they will yield soaps with properties and performance comparable if not superior to those from tallow and coconut oil.

Table 6.16. Fatty acids compositions of selected oils/fats.

	Weight percentage							
Fatty acids	*Palm oil*	*Palm stearine*	*Tallow*	*Palm kernel oil*	*Palm kernel olein*	*Coconut*	*Palm olein*	*Soyabean oil*
C6	–	–	–	0.3	0.4	0.2	–	–
C8	–	–	–	4.4	5.4	8.0	–	–
C10	–	–	–	3.7	3.9	7.0	–	–
C12	0.2	0.3	–	48.3	41.5	48.2	0.2	–
C14	1.1	1.3	2.5	15.6	11.8	18.0	1.0	–
C16	44.0	55.0	26.6	7.8	8.4	8.5	39.8	6.5
C18	4.5	5.1	21.8	2.0	2.4	2.3	4.4	4.2
C18:1	39.2	29.5	42.8	15.1	22.8	5.7	42.5	28.0
C18:2	10.1	7.4	2.3	2.7	3.3	2.1	11.2	52.6
Other	0.8	0.7	4.0	0.1	0.1	–	0.9	8.0
IV	53.3	35.5	35–48	17.8	25.5	9.5	58.4	133
SAP. V	196	199	195	245	–	256	198	192

Palm stearine (POs) and palm kernel olein (PKOo) are produced, along with palm olein (POo) and palm kernel stearine (PKOs), when palm oil and palm kernel oil (PKO) are fractionated. While palm olein and palm kernel stearine have higher added value because of their specific food applications, POs and PKOo are normally sold at discount prices.

Several studies carried out revealed that POs and tallow can be formulated together with PKO to give soaps that are comparable with tallow-PKO blends (Tables 6.17, 6.18 and 6.19). Since POs is cheaper than tallow, the resulting soaps are expected to be cheaper. Perfume retention of palm-based soaps has also been found to be better than that of soaps made from tallow.

Table 6.17. Characteristics of soaps based on tallow-palm stearine-palm kernel oil blends.

Parameters	*Blends*				
	00:80:20	*60:20:20*	*40:40:20*	*20:60:20*	*80:00:20*
Acid value	223	222	220	219	216
Titer (°C)	49	44	41	39	39
IV	29	29	27	31	34
Free caustic (%)	0.3	0.3	0.2	0.2	0.1
Moisture	9	10	8	8	9
Hardness	23	22	22	23	21
Foamability	255	255	245	265	260
Whiteness (Hunter)	81	82	84	81	80

Table 6.18. Characteristics of palm oil-palm stearine-palm kernel fatty acids blends in soap.

Parameters	*Blends*				
	80:00:20	*60:20:20*	*40:40:20*	*20:60:20*	*00:80:20*
Acid value	214	215	217	219	219
Titer (°C)	42	44	46	47	48
IV	38	34	33	30	34
Free caustic (%)	2.1	0.1	0.8	6.7	2.3
Moisture	7	7.6	8.6	8.2	7.8
Hardness	13.5	12.0	12.3	10.5	16.2
Foamability	320	355	340	150	295

Table 6.19. Characteristics of soaps based on palm stearine-palm kernel olein blends.

Parameters	*Blends*				
	90:10	*80:20*	*70:30*	*60:40*	*50:50*
Acid value	179	185	188	191	196
Titer (°C)	47	44	42	41	40
IV	28	30	28	28	25
Free caustic (%)	0.12	0.05	0.2	0.09	0.1
Moisture	16.3	22.7	25.2	17.2	27.6
Hardness	10	12	11	16	7
Foamability	340	270	315	340	365
Whiteness (Hunter)	94	93	92	90	93

Soaps based on these were found to have comparable foaming power and better colour.

Diesel substitute

Vegetable oils were used as motor fuel by *Rudolf Diesel* in 1900 when demonstrating his compression engines. Since then many publications referring to similar usage of cracked products of oils and fats have been published.

Recent research has demonstrated that crude palm oil can be used directly as a fuel to run cars fitted with suitable diesel Elsbett engines. The exhaust fumes from crude palm oil engines were found to be cleaner than those from diesel engines, with essentially no sulphur or nitrogen oxides. It is also cheaper and safer to transport crude palm oil than diesel because of the higher flash point (crude palm oil at 240°C vs. diesel at 52°C).

Cost is always the main factor that determines large-scale utilisation. However, the initial results suggest that the use of crude palm oil as an engine fuel would be 30 per cent more costly compared with petroleum diesel under Malaysian conditions.

Epoxidised palm oil and products, polyols, polyurethanes, and polyacrylates

Epoxidised palm oil and palm oil products (EPOP) can be produced by reacting palm oil, palm stearine, or palm olein with peracids. Pre-formed peroxyacetic and peroxyformic acids, as well as peroxyacetic acid and peroxyformic acid generated *in situ*, to find suitable methods for the production of EPOP. The best procedures were found to be pre-formed peroxyacetic acid and peroxyformic acid generated *in situ* (Table 6.20).

Table 6.20. Reaction conditions for the production of epoxidised crude palm oil.

Reaction parameters	*Pre-formed peroxyacetic acid*	*Peroxyacetic generated in situ*	*Pre-formed peroxyflormic acid*	*Peroxyformic generated in situ*
Mole ratio acid: H_2O_2	2.06:1	1.1:1	4.6:1	0.35:1
Catalyst (%)	1.52	0.02	–	–
Temperature (°C)	40–45	–	RT	–
% Peracid formed	13.7	–	11.0	–
Epoxide temperature (°C)	60–65	60–70	50–60	50–60
Oxirane oxygen content (%)	2.62	1.87	Low	1.82
				2.08 (toluene)
				1.86 (CH_2Cl_2)

Epoxidised oils, especially epoxidised soyabean oil (ESBO), are used extensively as plasticisers and stabilisers for plastics, particularly polyvinyl chloride (PVC). A plasticiser increases the workability of a plastic while a stabiliser reduces the rate of degradation of a plastic by heat, light, or micro-organisms. Epoxidised oils can fulfill both functions and their compatibility with plastic increases with their epoxide content.

Because palm oil and its products have lower iodine values, the epoxide contents of EPOP are lower than that of ESBO. As plasticisers and stabilisers, EPOP are therefore not expected to perform better than ESBO, but their performance could be made comparable by slight modifications of the formulations. PVC jungle and rain boots plasticised and/or stabilised with EPOP have been produced that are comparable in performance to those plasticised and/or stabilised with ESBO.

The value of epoxidised oils lies in the versatility of the epoxide rings. Being labile, they can easily be converted to other useful functional groups, thus diversifying end uses. EPOP can be converted to various polyols by reacting them with short-chain polyhydric alcohols in the presence of catalysts. By changing the ratio of EPOP to polyhydric alcohols and the types of polyhydric alcohols, polyols with a range of hydroxyl values and viscosities can be produced (Table 6.21). Polyols when reacted with isocyanates produce polyurethane foams. The water formed in the reaction acts as an internal blowing agent, thus avoiding the need to use environmentally unfriendly blowing agents such as chlorofluorocarbons.

Table 6.21. Properties of palm-based polyols.

Parameters	*Ratio of EPOo-polyhydric alcohol* 1:1	2:1	4:1
OH value	350–450	200–300	150–200
Viscosity (mPa S)	980–1300	1300–2100	3500–4700

Polyols from EPOP react with isocyanates at a slower rate than do polyols based on petrochemicals. The resulting foams, however, have regular cell structures and exhibit good hydrophobicity. With suitable formulations these properties could be fully exploited to give rise to interesting products.

Polyacrylate resins can be produced from EPOP by reacting them with acrylic acids. These resins can be applied on solid surfaces and when they are cured by ultraviolet irradiation, clear and glossy finishes result. The hardness and the tackiness can be increased or reduced by varying the amount and types of cross-linkers and the strength of the irradiation used. Characteristics of some polyacrylates based on epoxidised palm olein are listed in Table 6.22.

Table 6.22. Characteristics of some polyacrylates coating based on epoxidised palm oil/products after UV curing[a].

Additives	*Formulations* 1	2	3	4	5
EPOLA	60	60	60	60	80
PUA	–	–	–	–	5
HPA	10	10	10	10	10
NPGDA	–	–	–	15	–
TMMTA	25	–	–	–	–
TMPTA	–	25	–	10	–
PETA	–	–	25	–	–
Viscosity (cps/25°C)	614	423	705	228	947
Pencil hardness	3H	4H	4H	4H	< 2B
	< 2B	2B	< 2B	< 2B	–
Gel fraction (%)	86	87	86	88	84

[a] In all the formulations 5% benzophenone was used as photoinitiator. EPOLA, epoxidised palm olein acrylates; PUA, polyurethanes acrylates; HPA, hydroxypropyl acrylate; NPGDA, neopentyl glycerol diacrylates; TMMTA, trimethalolmethane triacrylates; TMPTA, trimethalopropane triacrylates; and PETA, pentaerythritol triacrylates.

Oleochemical route

Oleochemicals

Oleochemicals are chemicals derived from oils or fats. They are analogous to petrochemicals, which are chemicals derived from petroleum. The hydrolysis or alcoholysis of oils or fats form the basis of the oleochemical industry. The hydrolysis of the triglycerides composing oils and fats produces fatty acids and glycerol. If oils or fats are made to react with an alcohol instead of with water, the process is alcoholysis and the products are fatty acid esters and glycerol.

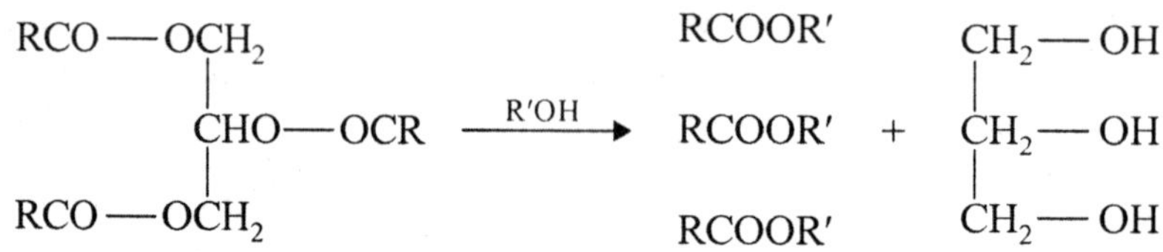

Oils or fats — Fatty acids or esters — Glycerol

In hydrolysis, R′ = H.

In alcoholysis, R′ = alkyl group.

Fatty acids or their esters can be used as the starting materials for making fatty alcohols and fatty nitrogen compounds. These products can be further modified to produce various derivatives. Hence oleochemicals are often divided into at least two categories: basic oleochemicals and derivatives. The five basic oleochemicals are fatty acids, esters, alcohols, nitrogen compounds and glycerol. There are various types of derivatives that can be produced from these through different chemical modifications. Fig. 6.22 shows some of them.

Raw materials for oleochemicals

Oleochemicals or derivatives based on C12–C14 and C16–C18 chain length have a variety of uses. Tallow and coconut oil have been the traditional raw materials used for the production of C16–C18 and C12–C14 chain lengths, respectively. While tallow is produced by the developed countries such as the United States, the world has to rely on the Pacific region for the supply of lauric oils (C12–C14 source). The Philippines has been the main supplier of lauric oils.

Palm-based oleochemicals

Palm kernel oil, like coconut oil, is a lauric oil. Its fatty acid composition is in fact very similar to that of coconut oil (Table 6.16).

Uses of oleochemical based on palm oil and palm kernel oil

Fatty acids

The most common method for the production of fatty acids adopted by the oleochemicals industry is high-temperature and high-pressure fat splitting. The fatty acid mixture produced is separated into broad cuts or pure fatty acids by simple or fractional distillations. Tables 6.23 and 6.24 lists examples of fatty acids derived from palm products. The exact specifications of the various fatty acids produced vary slightly depending on the exact raw materials and process used. The specifications could also change due to continuous upgrading of processes.

One of the traditional raw materials used for the production of stearic acid is tallow and very often, consumers or customers will ask for products equivalent to stearic acid from tallow. Single, double and triple pressed stearic acids from palm oil are in fact produced via distillation processes, but similar terminologies were used to indicate their similar characteristics. Besides their light colour, fatty acids derived from palm products have a low content of unsaponifiables, indicating excellent purity.

Table 6.23. Palm-based fatty acids.

Fatty acids	Fatty acid compositions								
	C8	*C10*	*C12*	*C14*	*C16*	*C18*	*C18:1*	*C18:2*	*C18:3*
Distilled PKO	3	3	50	15		9	16	2	–
Stripped PKO	–	–	52	17		10	16	3	–
Distilled PO	–	–	0.5	1.5		45.5	37.5	9.5	0.5
Distilled POs	–	–	0.5	1.5		61.6	25	6	0.5
Single pressed	–	–	0.5	1.5		50.0	10	–	–
Double pressed	–	–	0.5	1.5		53.0	3	–	–
Triple pressed	–	–	0.5	1.5		51.0	–	–	–
Lauric 70%	–	0.2	> 70	> 22		2	–	–	–
Lauric 92%	–	2–5	> 92	> 2		–	–	–	–
Lauric 98%	–	0–1	> 98	–		–	–	–	–

Table 6.24. Palm-based fatty acids.

Fatty acids	Quality parameters					Colour	
	Acid value	*Sap. value*	*Iodine value*	*Unsap. matter*	*Titer (°C)*	*R*	*Y*
Lauric 70%	270–275	270–275	0.1–1	0.5	33–35	0.5	2
Lauric 92%	277–281	277–281	0.1–0.5	0.5	41–42	0.2	2
Lauric 98%	278–282	278–282	0.1–0.2	0.2	43–44	0.2	1
Rubber g. stearic	> 195	> 196	< 1.0	–	> 52	5.0	50
Candle g. stearic	208–214	209–215	1.0	0.5	53–56	0.4	4
Cosmetic g. stearic	206–211	207–212	0.5	0.5	55–56	0.3	3

Without further chemical modification, fatty acids are used in rubber processing and in the manufacture of candles and cosmetic products.

Fatty acids for rubber processing

In rubber processing, fatty acids are added as processing aids with softening effect, as external lubricants and as vulcanisation accelerators. The chain lengths of the fatty acids have no effect on the performance of the fatty acids. However, a high degree of unsaturation will interfere with the process.

Fatty acids for candles

Fatty acids or mixtures of fatty acids and petroleum waxes can be used for the production of candles. For maximum shrinkage in order to ensure easy removal from the mould, about 7:2 ratio of C16:C18 fatty acids is required. This ratio favours fatty acids from PP since they have higher palmitic acids content.

Fatty acids for cosmetic products

Only good grades of fatty acids can be used for the production of cosmetic products. The normal types of fatty acids used are myristic, palmitic and stearic acids. They are used for various purposes such as lather improver, conditioners and to provide luster and sheen.

Fatty acids for soaps

The most important application of fatty acids is for the production of soaps via a neutralisation process. As discussed, white soaps cannot be prepared directly from PP due to their high SCVs. When palm fatty acids are distilled, part of the impurities are removed and the SCVs of palm fatty acids are usually lower than 3R. Good-quality soaps can therefore be derived from palm fatty acids. Besides the ease in production, the use of fatty acids allow soap formulators to blend their own ratio of fatty acids, thus allowing greater flexibility.

Lately, purely for aesthetic reasons, transparent and translucent soaps are gaining popularity, especially in South America. Stearic acid and triple pressed stearic acids can be used to produce soaps having good transparency. High palmitic acid content, however, appears to reduce transparency. The crystals of transparent/translucent soaps based on palm kernel oil and palm fatty acids are found to be in the β form.

Fatty acids for the production of metallic soaps

Another important applications of palm fatty acids is for the production of metallic soaps other than sodium soaps. The most common are the Ca and Zn palmitates or stearates. They can be prepared either via the fusion or the precipitation method. During the processing of rubber, the processability is improved regardless of the fatty acids used. However, Zn soaps were found to provide better internal lubrication.

Fatty acids for medium-chain triglycerides

When palm kernel oil is used as the starting raw material, the medium-chain fatty acids, i.e. C6–C10, present are normally stripped off since these acids are known to cause skin irritation. Originally considered as waste products, these medium-chain fatty acids can be re-synthesised into a new class of oil known as medium-chain triglycerides (MCTs). MCTs have many applications such as the flavour and fragrance industries, in surface treatment of confectionery products, as release agent in the baking industry and for the lubrication of machines.

Fatty esters

Esterification of fatty acids with alcohols and alcoholysis of triglycerides are two of the most common methods used for the production of fatty esters based on palm products. Fatty esters are found in several industries such as textiles, cosmetics, pharmaceuticals, plastics and lubricants. As synthetic lubricants, fatty esters are getting closer attention due to their good lubricity, minimum viscosity change with temperature, low-temperature fluidity and high thermal and oxidative stability.

Fatty esters for soaps production

Fatty esters are increasingly being used for the production of pure white soaps. In contrast to fatty acids, soaps produced from fatty esters are normally better in quality since the fatty esters can be better purified. In the process, alcohols (usually methanol) will be produced. Complete removal of alcohol is necessary before the soaps can be certified fit for use.

α-Sulphonated methyl esters

α-Sulphonated methyl ester (SME) is a new class of anionic surfactant. SME has received a lot of attention as an active ingredient for the production of washing and cleaning products due to several factors that include.

1. Easy production procedures.
2. Good detergency especially in hard water and in the absence of phosphates.

3. C16–C18 have been found to exhibit good detergency.
4. Good biodegradation characteristics.

Distilled fatty methyl esters with low iodine values are used as the starting raw materials for the production of SME. The fatty methyl ester is first reacted with sulphur trioxide at 80–90°C in a falling-film reactor. The dark product obtained from this process is bleached using hydrogen peroxide. After bleaching, the lighter colour product is neutralised with alkali to produce an α-sulphonated methyl ester.

Due to the good detergency properties of C16–C18 fatty methyl esters, palm stearine provides a suitable and cheap source of raw material for the production of SME. The detergency properties of SMEs derived from palm stearine have been found to be comparable to linear alkyl benzene sulphonates (LAS), the 'workhorse' of the detergent industry. It is interesting to note that SME based on palm fatty acids distillates (PFAD), a by-product of the physical refining industry (Fig. 6.23), performs as well as those derived from fractionated esters. These findings indicate that SME could very well be an important anionic surfactant for the future and palm fatty acids distillates can be a cheap source of raw materials.

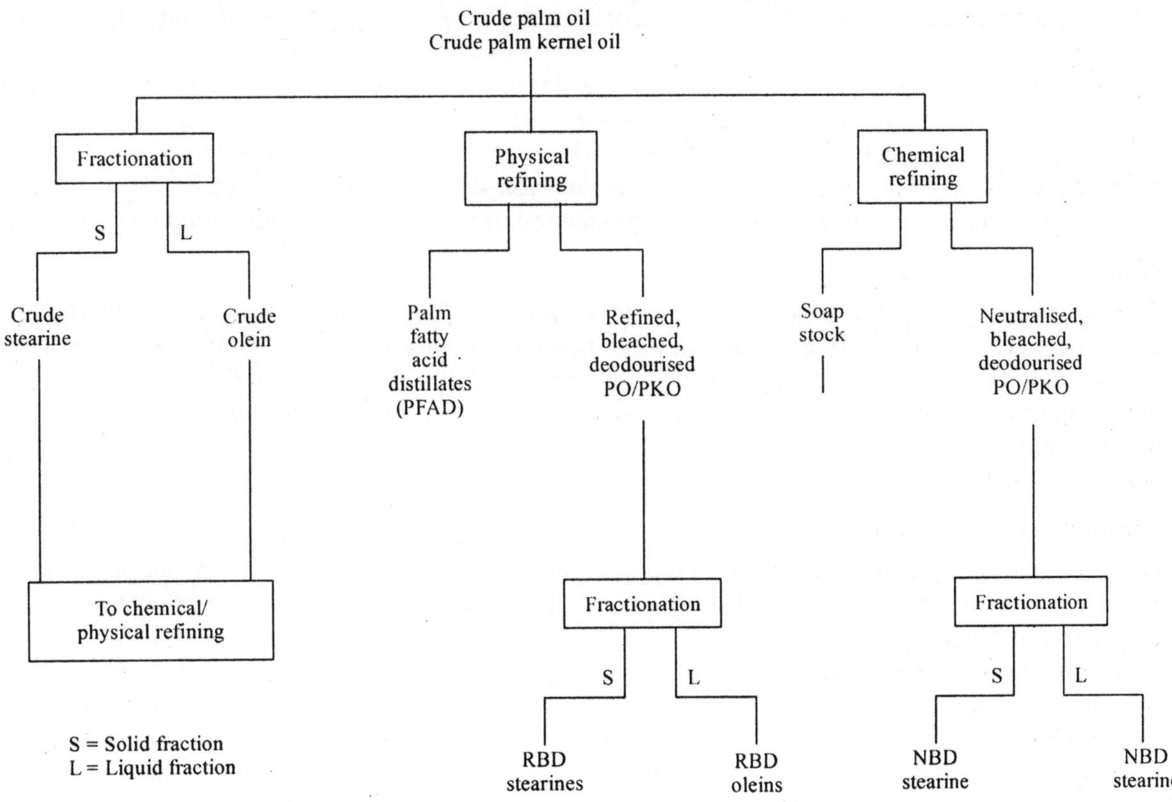

Fig. 6.23. Processing of palm oil/palm kernel oils.

Diesel substitute

Palm-based methyl esters have been extensively tested as diesel substitutes in taxis, buses, lorries, tractors and stationary engines. Methyl esters from crude palm oil and crude palm stearine have very similar fuel properties as petroleum diesel. Data available to date indicate that cold starting is easy, engines run smoothly with less unburnt hydrocarbon, CO, SO_2 and black smoke in the exhaust fumes. No diluation of the lubricating oil was observed and ignition lag was shortened. In contrast to crude palm oil, the use of palm methyl esters as diesel substitute does not require any modification of the engines. The economic viability of palm methyl esters as diesel substitute will depend on the cost of diesel, crude palm oil and glycerol.

Fatty alcohols

The most important application of fatty methyl esters is for the production of fatty alcohols. For economic reasons, three technologies for the production of fatty alcohols have gained worldwide acceptance:

1. High temperature and pressure hydrogenation of oils/fats.
2. High temperature and pressure hydrogenation of fatty methyl esters.
3. High temperature and pressure hydrogenation of fatty acids.

To date the most common method for the production of fatty alcohols is via high-temperature and high-pressure hydrogenation of fatty methyl esters using a copper-chromite catalyst in a fixed-bed reactor. If the unsaturation present in the molecule needs to be preserved, then a Zn-containing catalyst is used.

Fatty alcohols as such find limited uses. Cetyl and stearyl alcohols are used as suppressors of water evaporation in dry areas. Unsaturated alcohols are used as emulsifiers and textiles auxiliaries. More than 90 per cent of the fatty alcohols produced worldwide are used for the production of anionic (fatty alcohol sulphates and fatty alcohol ether sulphates) and non-ionic (fatty alcohol ethoxylates) surfactants. These derivatives are extensively used in the production of washing and cleaning products.

Fatty nitrogen compounds

The most common fatty nitrogen compounds are fatty amides, nitriles, amines and quarternary ammonium. The most important of these are the quarternary ammonium compounds, better known as quats. Lately manufacturers in the developed countries are voluntarily reducing or stopping the use of quats and imidazoline derivatives in softeners and conditioners in view of the findings that they may not be completely biodegradable and may involve the possible formation of nitrosamine in products containing them.

Glycerol

Glycerol is a valuable co-product of the oleochemical industry. Although glycerol can be produced synthetically, natural glycerol (i.e. glycerol derived from oils or fats) are preferred by the customers or consumers of today.

During the production of fatty acids via fat splitting or fatty ester via alcoholysis, mixtures of 10–30 per cent glycerol and water, known as sweet waters are produced. These sweet waters can be processed to pure glycerol via distillation or ion-exchange methods. To achieve the pharmaceutical grade further treatment with activated carbon is carried out. Table 6.25 gives some of the important characteristics of glycerols produced from palm oil.

Glycerol is a polyhydric alcohol that finds wide uses in several areas of applications. These include; as solvent or drugs carrier in pharmaceutical products; as humectants in cosmetics and tobacco; as ingredients for the production of explosives; as plasticiser/stabiliser for less polar polymers; as anti-freeze or heat transfer agent; as hydraulic fluid; for the production of polyesters that can be used in grease and/or lubricants; and for polyols and polyurethanes and mono and diglycerides, which are useful food emulsifiers.

Table 6.25. Types of glycerol derived from palm oil and products.

Quality parameters	*Crude*	*Glycerol* 99.5%	*Glycerol* 99.8%
Ref. density 20/20 C	–	1.2623	1.2631
Ref. index n_D^{20}	–	1.4731	1.4737
Glycerol %	88	99.5	99.8
α-Colour	–	5	5
Acid content	–	0.1	0.1
Saponification eqv.	–	1.0	1.0
Ash (%)	1.0	0.01	0.01
Cl (organic Cl) (ppm)	–	2(5)	2(5)
Heavy metals (ppm)	2	1	1

Ability to reduce the surface or interfacial tension of water and oil is one of the properties required of an emulsifier. Different chain length fatty acids in monoglycerides were found to have little effect on the interfacial tension between palm oil and water, in contrast C18:2 (linoleic acid) monoglyceride reduces the interfacial tension to greater than 50 per cent (Table 6.26).

Table 6.26. Interfacial tensions of RBD palm oil and water at 1% concentration of monoglyceride of various fatty acids.

Fatty acids	*C12*	*C14*	*C16*	*C18*	*C18:1*	*C18:2*
Interfacial tension (mN/m)	11.3	12.2	12.1	13.1	14.8	6.5

The amount of palm oil/palm oil products used in the production of soaps is expected to increase in the near future especially in the developing countries. Besides being price competitive and exhibiting good performances, soaps from palm oil/palm oil products are acceptable by all religions. There is a strong competition between products derived from oleochemicals and those derived from petrochemicals. With the current awareness on environmental issues and preference for environmentally friendly products, the utilisation of palm oil/palm oil products for non-food applications via the oleochemicals route is also expected to increase. Due to the ready availability of raw materials, technology, capital and market demand, the non-food applications of palm oil/palm oil products are expected to have a bright future.

CHEMICAL AND PHYSICAL PROPERTIES OF PALM OIL

Palm oil, like all oils and fats, is made up mostly of glyceride materials with some non-glyceridic materials in small or trace quantities. It is this chemical composition that defines the chemical and physical characteristics of palm oil, which in turn will determine the suitability of the oil in various processes and applications.

Chemical Properties of Palm Oil

Triglyceride and fatty acid composition

Triglycerides form the major component and bulk of the glyceridic material present in palm oil with small amounts of monoglycerides and diglycerides, which are artifacts of the extraction process. The fatty acid chains present in the palm oil triglycerides could vary in the number of carbons present in the chain (chain length) and in structure (presence of double bonds, i.e. unsaturation). It is the variations in the structure and number of carbons in the fatty acid chains that largely define the chemical and physical properties of palm oil.

The chain lengths of the fatty acids present in the triglycerides of palm oil fall within a very narrow range from 12 to 20 carbons as shown in Table 6.27.

Table 6.27. Fatty acid composition of Malaysian palm oil.

Fatty acids chain lengths	*% of total*	
	Mean	*Range*
12:0	0.23	0.1–1.0
14:0	1.09	0.9–1.5
16:0	44.02	41.8–46.8
16:1	0.12	0.1–0.3
18:0	4.54	4.2–5.1
18:1	39.15	37.3–40.8
18:2	10.12	9.1–11.0
18:3	0.37	0–0.6
20:0	0.38	0.2–0.7

It can be seen that about 50 per cent of the fatty acids present in palm oil are saturated and about 50 per cent are unsaturated. This even balance between saturation and unsaturation determines the iodine value of the oil and confers some stability against oxidation to the oils as compared to other vegetable oils. The three fatty acids in the triglycerides could be represented by the multitude of fatty acids listed in Table 6.27. The different placement of the fatty acids attached to the glycerol molecule can lead to a large number of different triglyceride.

Triglyceride analysis of *tenera* palm oil based on saturation-unsaturation criterion are shown in Table 6.28. From the computation data, it can be seen that the triglyceride molecules could be divided according to the number of saturated (S) and unsaturated (U) groups that they contain. The computational results were found to be very close to the analytical data obtained by Tan shown in Table 6.28.

Table 6.28. Triglyceride analysis of *tenera* palm oil (based on saturation-unsaturation criterion).

Triglyceride type	*Composition (%)*
Trisaturated (GS_3)	10.2
Disaturated (GS_2U)	48.0
Monosaturated (GSU_2)	34.6
Triunsaturated (GU_3)	6.8

Carbon number analysis by high-temperture gas-liquid chromatography provides a rapid but less detailed analysis of the triglyceride composition in terms of the total number of carbon atoms present in the three fatty acid chains of the triglyceride molecule. Carbon number analysis data of palm oil are shown in Table 6.29.

Table 6.29. Carbon number analysis of Malaysian palm oil.

Carbon number	*Mean*	*Range*
C46	0.8	0.4–1.2
C48	7.4	4.7–10.8
C50	42.6	40.0–45.2 (POP, PPO)
C52	40.5	38.2–43.8 (POO)
C54	8.8	6.4–11.4

Knowledge about the detailed structures of the triglyceridess present in palm oil is important because they defined some of the physical characteristics of the oil. The melting points of triglycerides are dependent on the structures and position of the component acids present. They also affect the crystallisation behaviour of the oil. The semi-solid nature of palm oil at room temperature has been attributed to the presence of the oleo-disaturated fraction.

As mentioned previously, partial glycerides are artifacts of the extraction process, especially the stages prior to sterilisation. Oil obtained from unbruised sterilised fruit shows trace levels of partial glycerides. Random analyses of samples of refined palm oil, palm olein and palm stearine have shown the presence of about 2 per cent of 1,2-diglycerides and about 4 per cent of 1,3-diglycerides with trace amount of monoglycerides. These partial glycerides are important as they are known to affect the crystallisation behaviour of the oil.

Minor components

The carotenoids, tocopherols, sterols, phosphatides, triterpenic and aliphatic alcohols form the minor constituents of palm oil. Though present in less than 1 per cent altogether in palm oil, nevertheless they play a significant role in the stability and refinability of the oil, in addition to increasing the nutritive value of the oil.

Crude palm oil contains between 500 and 700 ppm of carotenoids mainly in the forms of α and β-carotenes, the precursor of vitamin A. Unless extracted prior to refining, these carotenoids are thermally destroyed during the deodourisation stage in order to produce the desired colour for a refined oil. In crude palm oil, the presence of these carotenoids appears to offer some oxidative protection to the oil through a mechanism where they are oxidised prior to the triglycerides. Table 6.30 lists the carotenoid types that are present in crude palm oil.

Table 6.30. Carotenoid composition of palm oil.

Carotenoid	*Percentage*
Phytoene	1.27
cis-β-Carotene	0.68
Phytofluene	0.06
β-Carotene	56.02
α-Carotene	35.16
cis-α-Carotene	2.49
ζ-Carotene	0.69
γ-Carotene	0.33

(Contd...)

Table 6.30. Carotenoid composition of palm oil. *(Contd.)*

Carotenoid	*Percentage*
δ-Carotene	0.83
Neurosporene	0.29
β-Zeacarotene	0.74
α-Zeacarotene	0.23
Lycopene	1.30
Total carotene (ppm)	673

Crude palm oil contains tocopherols and tocotrienols in the range of 600–1000 ppm. Refined palm oil retains about 50 per cent of these products. Tocopherols and tocotrienols are antioxidants and provide some natural oxidative protection to the oil. Table 6.31 shows the types of tocopherols and tocotrienols present in palm oil.

Table 6.31. Tocopherols and tocotrienols in crude palm oil.

Type	*Percentage*
α-Tocopherols	21.5
β-Tocopherols	3.7
γ-Tocopherols	3.2
δ-Tocopherols	1.6
α-Tocotrienols	7.3
β-Tocotrienols	7.3
γ-Tocotrienols	43.7
δ-Tocotrienols	11.7

From Table 6.31, it can be seen that α-tocopherols and γ-tocotrienols account for the major portion of the total tocopherols and tocotrienols present in palm oil. *Gapor* confirms the presence of the above-listed tocopherols and tocotrienols by high-performance liquid chromatography (HPLC) and also indicated the probable presence of the esterified forms.

The combined effects of the properties of the carotenoids, tocopherols, tocotrienols and the 50 per cent unsaturation of the acids confer on palm oil a higher oxidative stability as compared to a lot of the other vegetable oils.

In terms of sterols, palm oil contains far less cholesterol than many other vegetable oils as shown in Table 6.32. Table 6.33 gives the sterol composition of crude and refined palm oil and their products. From Table 6.33, it can be seen that the low-cholesterol levels in crude palm oil and crude palm olein are further reduced to even lower levels upon refining. This low-cholesterol level, together with the antithrombotic and anticarcinogenic properties of some of the carotenoids, tocopherols and tocotrienols present add further to the nutritive value of palm oil and palm oil fractions.

Table 6.32. Cholesterol levels in crude oils and fats.

Oil type	*Average (ppm)*	*Range (ppm)*
Coconut oil	14	5–24
Cocoa butter	59	–

(Contd...)

Table 6.32. Cholesterol levels in crude oils and fats. *(Contd.)*

Oil type	*Average (ppm)*	*Range (ppm)*
Palm kernel oil	17	9–40
Palm oil	18	13–19
Sunflower oil	17	8–44
Soyabean oil	28	20–35
Rapeseed oil	49	25–80
Maize oil	50	18–95

Table 6.33. Sterol composition of crude and refined palm oil and their products (ppm).

Sample	*Cholesterol*	*Campesterol*	*Stigmasterol*	*Sitosterol*	*Unknown*
Crude palm oil	7–13	90–151	44–66	218–370	2–18
Degummed, bleached	5–10	49–116	22–51	113–286	Trace–8
RBD	1–5	15–16	8–30	45–167	Trace
Crude palm olein	6–8	57–104	30–51	149–253	24–28
Degummed, bleached	3–4	36–43	21–25	99–123	Trace–5
RBD	2	26–30	12–23	68–114	–

Inherent chemical properties of palm oil

Table 6.34 summarises some of the inherent chemical properties of Malaysian palm oil.

Table 6.34. Inherent chemical properties of Malaysian palm oil.

Chemical characteristics	*Mean*	*Range*
Saponification value (mg KOH/g oil)	195.7	190.1–201.7
Unsaponifiable matter (%)	0.51	0.15–0.99
Iodine value (Wijs)	52.9	50.6–55.1
Slip melting point (°C)	34.2	30.8–37.6

Physical Properties of Palm Oil

Table 6.35 shows some of the physical properties of palm oil. The apparent density is an important parameter from the commercial point of view since it is used for volume to weight conversions. It can also be used as purity indicator.

The solid fat content of an oil is a measure (in per cent) of the amount of solid fat present in the oil at any one temperture. It is measured by means of wide-line nuclear magnetic resonance (NMR) spectrometry after a standard tempering procedure for the samples.

The solid present in the oil at any one temperture is due to the process of crystallisation occurring in the oil as a consequence of its chemical properties. The different molecular triglyceride structures with their differing chemical characteristics manifest their physical states at different temperatures, thus imparting certain crystallisation and melting behaviour to the oil. These thermally associated processes can be followed by means of differential scanning calorimetry (DSC).

Table 6.35. Major physical properties of palm oil.

Property	*Mean (of 215 samples)*	*Range*
Apparent density at 50°C (g/ml)	0.889	0.888–0.889
Refractive index at 50°C	1.455	1.455–1.456
Solid fat content		
5°C	60.5	50.7–68.0
10°C	49.6	40.0–55.2
15°C	34.7	27.2–39.7
20°C	22.5	14.7–27.9
25°C	13.5	6.5–18.5
30°C	9.2	4.5–14.1
35°C	6.6	1.8–11.7
40°C	4.0	0.0–7.5
45°C	0.7	
Slip melting point (°C) (+)	34.2	31.1–37.6

From the thermal characteristics considered above, it can be seen that palm oil can be separated under controlled thermal conditions into two components, i.e. a solid (stearine) and a liquid (olein) fraction. This fractionation process can be affected either in the dry form in the presence of a detergent or solvent.

The method employed, to a certain extent, determines some of the chemical and physical properties of the oleins and stearines produced, especially the stearines. By varying the fractionation methods and conditions used, a range of stearines with different chemical and physical properties could be produced, yet keeping the chemical and physical properties of the oleins to within a very narrow range of values as shown in Tables 6.36, 6.37 and 6.38.

Table 6.36. Fatty acid compositions (%).

Fatty acids	*Oleins* *Range observed*	*Mean*	*Stearines* *Range observed*
12:0	0.1–0.5	0.2	0.1–0.6
14:0	0.9–1.4	1.0	1.1–1.9
16:0	37.9–41.7	39.8	47.2–73.8
16:1	0.1–0.4	0.2	0.05–0.2
18:0	4.0–4.8	4.4	4.4–5.6
18:1	40.7–43.9	42.5	15.6–37.0
18:2	10.4–13.4	11.2	3.2–9.8
18:3	0.1–0.6	0.4	0.1–0.6
20:0	0.2–0.5	0.4	0.1–0.6
Iodine value (Wijs)	56.1–60.6	58.0	21.6–49.4

Table 6.37. Triglyceride composition by carbon number.

Carbon number	*Oleins* *Range*	*Mean*	*Stearines* *Range*
C46	–	–	0.5–3.3
C48	1.3–4.0	2.3	12.2–55.8
C50	37.7–45.4	42.0	33.6–49.8
C52	43.3–51.3	45.7	5.1–37.3
C54	7.0–12.6	9.9	Trace–8.4

Table 6.38. Melting and solidification characteristics.

Test	*Oleins* *Range*	*Mean*	*Stearines* *Range*
Slip melting point (°C)	19.4–23.5	21.6	44.5–56.2
Cloud point (°C, crude)	6.6–14.3	10.4	–
Neutralised	5.4–11.9	8.1	–
Refined	6.0–11.5	8.8	–

Under normal fractionation conditions, soft stearines and oleins with cloud points in the range of 8–10°C are produced.

For more specialised usage such as in the confectionery industry, a more specific type of stearine is required or desired. This is catered to by using a double fractionation process as shown in Fig. 6.24.

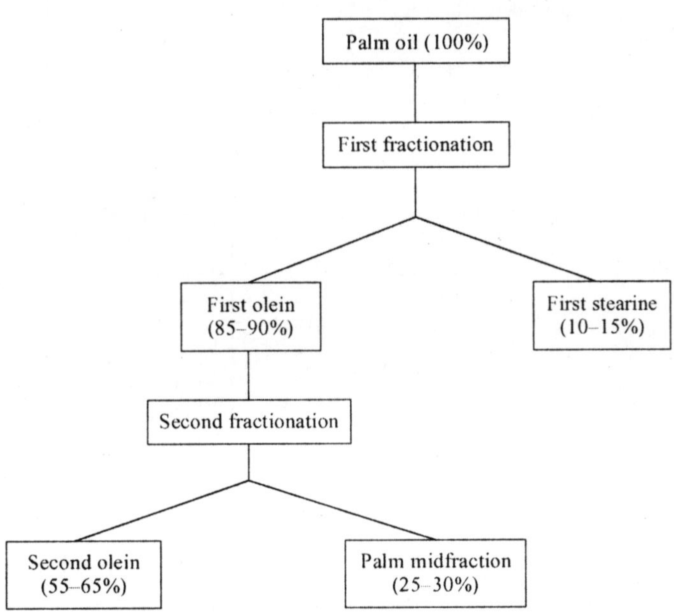

Fig. 6.24. Fractionation and palm midfraction.

ENVIRONMENTAL TRENDS: TOWARDS ZERO WASTE OPERATION

With the growing awareness for the protection of the environment, there is a greater need for producers to improve the environmental profile of their products. Consumers and regulatory bodies expect more information on the effects of products on the environment during their manufacture, use and disposal. The life cycle assessment (LCA) is used as a holistic approach to assess the impact of a product throughout its life cycle. An important aspect of LCA is recycling and waste management.

In oil palm plantations, the main residues that must be disposed of are the fronds that are pruned regularly and the biomass comprising the palm trunk and fronds at the end of the crop's economic cycle (about 25 years). In commercial practice, pruned fronds are re-distributed to the fields while the biomass, senescent palms, are recycled with the zero burning technique of re-planting. Without burning the palm residues, this approach does not pollute the air and it enhances soil fertility through recycling of organic matter and plant nutrients. The main by-products and wastes produced from the processing of palm oil are the empty fruit bunches (EFB), palm oil mill effluent (POME), palm fibre and palm kernel shell. EFB and POME has been used extensively as mulch and organic fertilisers in oil palm plantations while palm fibre and shell are used as fuel, making the palm oil mill self-sufficient in energy (Fig. 6.25). Excess shell has been used for road surfacing in estates.

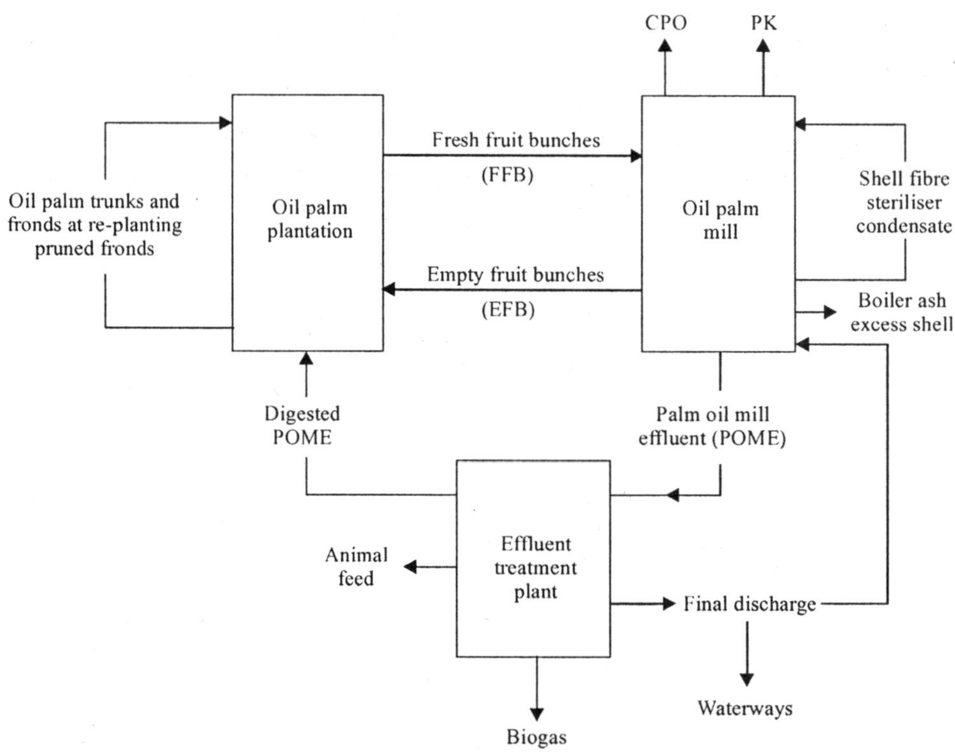

Fig. 6.25. Current use and recycling of by-products and wastes from oil palm plantings and conventional palm oil milling.

The underutilised wastes from the oil mill are biogas generated by anaerobic digestion of POME and clinker and boiler ash; however, the quantities of the latter two are insignificant to cause any environmental impact. In segments of the palm oil industry, biogas is being used to generate heat and electricity to supplement the energy demand by the subsidiary factories.

Although most of the by-products and 'wastes' from oil palm planting and palm oil processing are recycled or utilised, there is still room for improvement, the ultimate aim being toward zero waste. Concerted efforts by all sectors are being directed toward zero waste. Concerted efforts by all sectors are being directed toward improving the efficiency of oil palm cultivations and processing with a view to optimising the use of inputs and energy and reducing the production of wastes/effluents. Currently, the by-products from the palm oil industry are used mainly as organic fertiliser, soil ameliorants and fuel. Research and development effort has shown that these resources can be made into value-added products such as fibre boards, furniture and single-cell proteins. Table 6.39 shows the potential utilisation of oil palm by-products.

Table 6.39. Utilisation of oil palm/palm oil by-products and wastes.

Waste/by-products	*Quantity*[a]	*Where utilised*	*Present level of utilisation*	*Status*	*Potential new uses*
Oil palm plantations					
Pruned fronds	10.4 tonnes/ha	Recycled in plantation	Very high	By-product	Vitamin E extraction, fill board, etc.
Palm trunks and fronds at replanting	89.9 tonnes/ha	(i) Recycled in plantation	Very high	By-product	Wood product, pulp, paper, animal feed, palm hea glucose, cellulase, fuel
		(ii) Furniture	Very low	By-product	
Palm oil mill					
EFB	20–23%	Mulching in plantation	Very high	By-product	Fibre board; MDF
Fibre	12–13%	Fuel to boiler	Very high	By-product	Fibreboards
Shell	6–8%	Fuel to boiler	Very high	By-product	Activated carbon;
Decanter solid	2–3%	Land application as fertiliser	Moderate-high	By-product	Animal feed
Boiler ash	0.4–0.6%	Surface landfill	Low in plantation/fertiliser	By-product/	Fertiliser/soil waste
ameliorant					
Clinker	Small quantity	Landfill/disposal	Very low-low	Mainly waste	Surface landfill
Steriliser condensate	12–20%	(i) Feed to ETP	Very high	Waste/by-product[b]	Cellulase, single-cell protein
		(ii) Recycle for dilution	Low	By-product	Crude oil dilution
Centrifuge waste	40–50%	Feed to ETP	Very high	Waste/by-product[b]	Oil recovery for acid oil reduction
Decanter effluent	30–40%	Feed to ETP	Very high	Waste/by-products[b]	
Hydrocyclone/ claybath water	5–11%	Feed to ETP	Very high	Waste/by-products[b]	Recycling to reduce quantity

(Contd...)

Table 6.39. Utilisation of oil palm/palm oil by-products and wastes. *(Contd.)*

Waste/by-products	*Quantity*[a]	*Where utilised*	*Present level of utilisation*	*Status*	*Potential new uses*
Factory washing	4–8%	Feed to ETP	High	Waste/by-products[b]	De-oiling-oil recovery for acid oil production
Effluent treatment plant					
Sludge cake		(i) Land application as fertiliser	Moderate-high	By-product	
		(ii) Animal feed	Very low-low	By-product	
Anaerobic solid	5–10%	Land application as fertiliser	Very high	By-product	
Aerobic solid	< 5%	Land application as fertiliser	High	By-product	
Biogas	28 m^3/T FFB	Biogas engine	Very low	Mainly waste	Heat and power generation

[a] Figures in percentages refer to % to FFB.

[b] These are mainly waste products in the mill but as they are reused immediately after treatment in the effluent treatment plant (ETP), are considered by-products.

NUTRITIONAL EFFECTS OF PALM OIL

General Nutritional Properties

Archeological evidence shows that palm oil has been consumed by human for more than 5000 years. Its digestion and absorption rates in the human body are in excess of 97 per cent, which is very similar to other common edible oils and fats. In many communities palm oil is an important source of dietary energy and provides sufficient quantities of the essential fatty acid, linoleic acid (18:2, *n*-6) for normal healthy metabolic functions. Like all other edible oils of vegetable origin, palm oil is considered cholesterol free.

Effect of Palm Oil on Blood Lipids

Being arbitrarily classified as a saturated fat, palm oil has been postulated to increase serum cholesterol levels and hence enhance the risk of coronary heart disease (CHD). Recent findings, however, indicate that this hypothesis is not uniformly true. For instance, *Kris-Etherton* and co-workers demonstrated that feeding a palm oil diet to rats did not raise plasma cholesterol in comparison to a highly polyunsaturated corn oil diet. Similarly, *Sugano* and co-workers were unable to establish significant differences in plasma cholesterol in rats fed on a palm olein diet compared to other polyunsaturated oils. *Sundram* and co-workers compared the effect of palm oil and its fractions with two commonly used polyunsaturated oils, namely soyabean and corn oil. It was demonstrated that palm oil feeding did not elevate plasma cholesterol whereas high-density lipoprotein cholesterol (HDL-C) tended to be raised on the palm oil diet relative to the corn oil diet. Similar observations have also been recorded in other animal models including the rabbit, chicken and hamsters.

Human Studies Evaluating the Effect of Palm Oil on Blood Lipids

The effects of palm oil on serum lipids and lipoproteins recorded in animal studies have similarly been observed in several human studies. In some early human studies, it was reported that subjects on a palm oil diet had elevated plasma and low-density lipoprotein cholesterol (LDL-C) levels compared to a diet containing a polyunsaturated fat. However, on a critical re-assessment of these and other relevant studies, it was found that plasma cholesterol levels after the palm oil period were actually lower than at the point of entry of the experiments when the subjects were on their habitual diets.

A human study by *Marzuki* and co-workers using young volunteers evaluated the effect of consuming foods containing either palm olein or soyabean oil. In normal healthy volunteers the level of blood cholesterol was not changed by the palm olein or soyabean oil diets. Similarly both LDL-C and HDL-C levels were unaffected by these diets. When the same diets were fed to volunteers having high blood cholesterol levels (hypercholesterolaemia), the soyabean oil diet was found to induce higher cholesterol levels than the palm olein diet. Similarly LDL-C was also raised by the soyabean oil diet.

In a similar experiment conducted on a Malaysian population, diets containing palm olein, corn oil and coconut oil were evaluated for their potential to modulate serum lipids. A reduction in serum cholesterol was observed on administering a palm olein or corn oil diet relative to a coconut oil diet. A second study evaluated the effects of palm olein and olive oil on serum lipids and lipoproteins in comparison to a coconut oil diet.

Each test oil was served as the sole cooking fat and contributed 23 per cent of the total dietary energy or two thirds of the total fat intake. The coconut oil diet significantly raised all the serum lipids and lipoprotein parameters measured, i.e. TC, LDL-C and HDL-C. However, the one-to-one exchange between palm olein (rich in 16:0) and olive oil (rich in 18:1) resulted in identical TC (192, 193 mg/dL), LDL-C (130, 131 mg/dL) and HDL-C (41, 42 mg/dL). This indicates that in healthy normocholesterolemic humans, palm olein can be exchanged for olive oil without affecting the serum lipoprotein concentration or distribution.

Possible Mechanism for the Cholesterol-Lowering Potential of Palm Oil

It has long been recognised that the cholesterol-raising potential of the saturated fatty acids is variable. Thus, it has been shown that stearic acid (C18:0) does not raise serum cholesterol. The major saturated fatty acid in the human diet as well as in palm oil is, however, palmitic acid. This, together with lauric (C12:0) and myristic (C14:0) acids, is considered hypercholesterolaemic. *Hayes* and co-workers recently re-examined this hypothesis in non-human primates (monkeys), using dietary fats containing predominantly lauric and myristic acids (coconut oil) or palmitic acid (palm oil). They showed that compared to diets rich in lauric and myristic acids, diets containing palmitic acid were actually neutral in their effect on both total serum and LDL-C.

These human and animal studies provide strong evidence that the lipemic effects of the different saturated fatty acids are not equal. 16:0 is hypothesised to behave as a neutral fatty acid (does not raise cholesterol) in normocholester-olemic individuals (<5.2 mmol/l) and when dietary cholesterol intake is low (<300 mg/day). In such situations 14:0 appears to be the unique cholesterol-raising fatty acid. The lack of 14:0 in palm oil and the hypothesised neutrality of 16:0 gives credence for the use of palm oil as a dietary oil suitable for the majority of the world's populations.

Nutritional Properties of Minor Components in Palm Oil

Palm oil, both crude and refined, is a rich source of vitamin E, which consists of a mixture of tocotrienols and tocopherols. A technology for the preparation of a tocotrienol-rich fraction (TRF) from palm fatty acid distillate has been developed.

Palm Vitee (TRF in superolein and encapsulated) has been evaluated in a number of nutritional studies in both animals and humans. The nutritional properties of TRF are as follows:

Qureshi first isolated tocotrienols from barley and proved that they could suppress the hepatic production of cholesterol through their ability to suppress the activity of the enzyme HMG-CoA reductase, which regulates cholesterol synthesis in the liver.

Subsequently, *Qureshi* extended his investigations to TRF (Palm Vitee) from palm oil in both animal and human models. In a double-blind crossover study involving 20 hypercholesterolaemic human subjects (serum cholesterol > 294 mg/dl), Palm Vitee supplementation was found to cause a significant drop in serum TC and LDL-C. The LDL-associated apolipoprotein Apo B was also decreased by 9–11 per cent. Moreover, Palm Vitee supplementation resulted in a significant decrease (25 per cent) in serum thromboxane and platelet factor PF4 by 16 per cent. Similar cholesterol-lowering effects of Palm Vitee have also been indicated in genetically hypercholesterolaemic swine.

In a similar study *Tan* and co-workers fed volunteers one Palm Vitee capsule per day for 30 consecutive days. Each capsule contained 18 mg tocopherol and 42 mg tocotrienols. In these volunteers, Palm Vitee lowered both serum TC and LDL-C. The magnitude of reduction for serum cholesterol was up to 36 per cent while reduction in LDL-C ranged from 0.9 to 37 per cent when compared to their respective starting values.

These studies indicate that Palm Vitee is most effective in reducing cholesterol when subjects have elevated blood cholesterol levels. However, there observations have not been uniformly reproduced by different workers. For example, *Wahlqvist* and co-workers have reported that Palm Vitee has no effect on blood cholesterol levels in their hypercholesterolaemic subjects. As a result, it has been suggested that certain population groups behave as 'responders' and others as 'non-responders' when given Palm Vitee to manage their hypercholesterolaemia. Studies are presently in progress to evaluate the underlying mechanisms associated with these observations.

The structural differences between d-α-tocopherol and tocotrienol, viz. the unsaturated side chains in the latter, account for some differences in their physiological activities. *Serbinova* and co-workers have reported that in membranes palm oil tocotrienols had 40–60 times higher antioxidant potency than α-tocopherol largely due to a higher recycling efficiency and uniform distribution in membrane bilayers. Under oxidative stress, tocotrienols protected human LDL against oxidation and their protective potency was greater than that of α-tocopherol in the presence of ascorbate. This may be a key factor in protection against the onset of degenerative atherosclerotic disease. A tocopherol-tocotrienol mixture in a ratio similar to that present in palm oil has also been shown to depress the systolic blood pressure and increase the aortic production of prostacyclin in spontaneously hypertensive rats.

Effects of Palm Oil on Experimental Carcinogenesis

Sundram and co-workers, using a rate model treated with the chemical carcinogen DMBA, evaluated the effect of palm oil on the progression of mammary tumours. Both crude and refined palm oils were evaluated against the polyunsaturated corn oil and soyabean oils. Rats fed either 20 per cent soyabean oil or corn oil developed tumours first at 9 weeks following DMBA administration. The appearance of tumours was more rapid and enhanced in the polyunsaturated corn and soyabean oil fed rats compared to the palm oil groups. At the time of sacrifice tumour incidence was 90 per cent in the soyabean fed rats, 85 per cent in with corn oil and only 65 per cent with crude palm oil maintained rats. At the same level of fat intake, tumour incidence in animals fed polyunsaturated oils was significantly higher than in rats fed palm oil diets. Tumour yield in the palm oil groups was significantly lower than that in the corn or soyabean oil diet.

In an earlier study *Sylvester* and co-workers found that palm oil induced lower tumour numbers and tumour load per rat comparable to a low-fat control (5 per cent corn oil). In comparison to a palm oil diet (20 per cent by weight), diets containing 20 per cent corn oil, beef tallow, or lard all resulted in significantly higher tumour numbers and tumour load.

These studies indicate that palm oil exerts an inhibitory effect on the progression of chemically induced carcinogenesis.

The 10 per cent level of linoleic acid in palm oil seems ideal in meeting the nutritional requirements of this essential fatty acid, without eliciting growth responses in the tumour cells in comparison to more polyunsaturated oils. However, when compared to animal fats such as lard, which contain similar levels of the linoleic acid, it appears that the inhibitory effect of palm oil cannot be attributed to its fatty acid composition alone. The answer may lie in the various minor components present in both crude and refined palm oil. It is postulated that these minor components, more so the tocotrienols, which have antioxidant activity, may be involved in exerting the inhibitory effect on tumour development.

The effect of tocotrienols on cancer progression was evaluated by *Komiyama*. The antitumour activity of tocotrienols was evaluated in terms of the increase in the lifespan of mice inoculated with tumour cells. α-Tocotrienols and γ-tocotrienols were effective against the sarcoma cancer cell lines and Ehrlich carcinoma. When human lung carcinomas were challenged with these tocotrienols, a cytotoxic activity due to the tocotrienol was exhibited. Similarly, DMBA-treated rats responded with lower tumour numbers when their diets were supplemented with palm tocotrienols. Recently, α-carotene isolated from palm oil has been shown to have antitumour activity against mouse lung cancer and against skin cancer.

Palm Oil Delays Ageing

Palm oil exhibits anticancer properties and has greater efficiency in inhibiting the growth of human cells and thus delays ageing. Palm oil diet helped in raising the level of good high density lipoprotein (HDL) cholesterol and lowered the bad LDL cholesterol. There is only one per cent fatty acid in the palm oil. So palm oil based diet reduced blood clotting thereby minimising the risk of coronary diseases. In the area of cancer pathogenesis, the palm tocotrienols inhibited the growth of tumour cells in chemically induced breast cancer in rodents. It was observed that it is as effective as tamoxifen a standard drug for breast cancer. The diet derived and topically applied tocotrienols accumulate in the skin and protect the tissue from damage due to exposure to intense ultraviolet light. The protection of the integrity of skin tissues by tocotrienols meant that these powerful antioxidants delay effects of ageing. The palm oil is the richest natural source of tocotrienols, a member of the vitamin E family with positive health effects. No other common edible oil, except rice bran oil, contains this form in significant amount, according to various studies. The superior characteristics of palm oil make it an excellent raw material for the production of oleochemical products including fatty acids, fatty alcohols, glycerol and other derivatives. Oleochemical manufactured from palm oil and palm kernel oil were now popular ingredients in the manufacturing of environmental friendly detergents as they were biodegradable.

FUTURE PROSPECTS OF PALM OIL AND MARKET REQUIREMENTS

In the past decade, palm oil has become internationally well known as a vegetable oil suitable for various applications—both edible and non-edible. This is brought about by it being a versatile oil for the production of various products with technical and economic advantages over other oils and fats. Its price competitiveness and readily available supply is able to serve the needs of oils and fats consumers worldwide.

Versatility of Palm Products

Palm oil has the flexibility to be used as it is or in fractionated forms to produce a very wide range of products. Interesterification can further significantly modify its properties including crystallisation behaviour. It has good oxidative stability. It has long been known as a good heavy-duty frying medium because of its relatively low polyunsaturated and the slip melting point, which is low enough to avoid excessive waxiness in most applications.

Margarine blends can be developed containing higher levels of palm products and having solids content profiles close to those of popular commercial brands. Standard quality table margarine can contain palm products as high as 70 per cent (50 per cent palm and 20 per cent hardened palm oil). Table margarines (packet) can contain as high as 63 per cent palm oil, 30 per cent palm kernel oil and 7 per cent palm stearine. Blends for random interesterification could utilise as much as 60 per cent of hardened palm oil or palm stearine. Tub margarines could be formulated containing 50 per cent palm oil.

Another major area of use for palm oil due to it being a semi-solid fat is in the manufacture of shortening and vanaspati. Shortenings include a variety of products such as the fats used domestically for cooking, frying and flour confectionery and those used industrially in cake baking and in large-scale frying operations for products such as potato crisps and doughnuts.

In ambient temperature range of 20–35°C there is a close similarity between butterfat, palm oil and European shortenings. It has, so to speak, been hydrogenated by nature to be very close to many specifications for shortenings and vegetable ghee.

The world consumption of oils and fats can be classified into two equally important categories, i.e. 50 per cent of solid fats and 50 per cent of liquid oil markets. The supply pattern of vegetable oil and fats is not equally divided between the solid and liquid oils and fats. Only palm oil is semi-solid accounting for 20 per cent of the vegetable oils market. Under current pattern of consumer demand, there is a shortage in the supply of solid fats and oversupply in the liquid oil market. Most of the liquid oils have to be hydrogenated to turn them into solid fats as shortenings, margarines and ghee. Palm oil is naturally placed in an advantageous position with respect to the pattern of large consumer demand for solid fats. With skillful formulation, no hydrogenation is required for palm products. Interesterification methods may be used to improve the ability of different components of palm oil to be used to meet customer specifications in the fat products.

Palm oil can also yield liquid fractions for the liquid oil market, i.e. palm olein and stearine can be obtained through fractionation of palm oil. Palm olein, the liquid fraction of palm oil, behaves as a liquid oil in hot climates. It is very stable frying medium that is comparable to any other frying fat for resistance to breakdown. However, in temperate climates or during cold nights where temperatures are below 18–20°C, palm olein begins to solidify. This problem can be overcome by blending palm olein with more unsaturated oil. In Japan, a blend of 50 per cent palm olein and 50 per cent rapeseed oil is being marketed successfully.

Palm stearine, the solid fraction of palm oil, has the edge over tallow because of the assurance of expanding supply, while world production of tallow has stagnated. Applications where stearine could replace tallow are in shortenings, frying fats and soaps. Lard could also be substituted by palm stearine or RBD palm oil in most of its applications.

Palm stearine can be the cheapest source of C16–C18 fatty acids for soap. Palm stearine alone has a very high titer value (47–50°C) such that when a high proportion is incorporated into toilet soap formulation, the soap becomes hard and cracks easily. Experiments conducted indicated that

30–50 per cent of palm stearine could be incorporated with tallow and 20 per cent of lauric (palm kernel) fatty acid to obtain titer of the finished product of between 40.5 and 44°C.

Technical and Economic Advantages

Hence, the utilisation of palm oil in products requiring a proportion of solid fats in their formulation would offer technical and economic advantages. Substantial cost savings can be achieved when palm oil is used in place of other oils and fats in various applications. Savings are made through the lower costs of raw materials, through the reduction in use of chemicals to process the oil, reduction in the cost of hydrogenation, reduction in costs due to minimal process losses and savings resulting from the long life of palm oil during the frying applications.

The amount of chemicals required to process oils and fats depends on the level of FFA, colour and other impurities in the oils. Normally about 2 kg of phosphoric acid, 2.5 kg of caustic soda and 30 kg of bleaching earth are required to refine 1 tonne of crude palm oil. Since processed palm products are available from the markets, importers need not have to refine crude palm oil but can instead import and use processed palm products directly. If unrefined vegetable oils were to be imported, these oils need to be processed by the importers. In addition, liquid oils would require a catalyst, usually nickel, for hydrogenation. Hence, the cost of chemicals would be minimal when RBD palm oil is used.

Palm oil, being a semi-solid or consistent fat, results in a reduction in cost if utilised in products requiring a proportion of solid fats in their formulation because hydrogenation is not necessary. If liquid oils are used, they have to be hydrogenated to obtain the solid consistency, which in turn leads to the formation of both *cis* and *trans* isomer.

Chapter 7

Castor Oil and Its Derivatives

INTRODUCTION

Castor oil, which is also known as ricinus oil, oil of palma christi, neoloid, and tangantangan oil, is obtained from the seeds of *Ricinus communis L.* of euphorbiaceae family. This castor oil is characterised by the presence of large quantities of ricinoleic acid which is hydroxy fatty acid. Ricinoleic acid chain has 18 carbon atoms, one double bond at C9 and a hydroxyl group at C12 position. Castor oil is the only commercially important oil containing major amounts of the hydroxy acid. Because of this unique fatty acid, castor oil has many different applications in various industrial sectors, as well as in medicines. Castor oil is also converted into a large number of derivatives.

Castor is cultivated widely in the tropics and warm regions, and castor is one of the major oilseed crops of India. It is valuable since a substantial portion of the oil produced is exported. Andhra Pradesh, Gujarat, Karnataka, Tamil Nadu and Orissa are the major castor producing states in India. The present chapter discusses the characteristics of castor oil, the derivatives prepared from it, their applications and commercial aspects like production and usage pattern of castor oil and its derivatives in India and world.

PROPERTIES OF CASTOR OIL

The fixed oil of castor seeds is obtained by pressing the castor seeds in various types of expellers. The oil content of castor seeds is about 50 per cent, and gives commercial recovery of about 45 per cent. The raw oil is steamed under vacuum to eliminate odours and coagulate the toxic albumin. Fuller's earth or activated charcoal may be used for further purification. The raw oil is viscous and pale yellow in colour. The oil can vary in colour depending on the method of recovery and blending to meet specifications. It has somewhat characteristic odour. The crude castor oil tastes slightly acrid with a decidedly nauseating after-taste. Castor oil has excellent keeping qualities and does not turn rancid unless subjected to excessive heat. It is dextrorotatory.

Pure castor oil can be readily identified by its hydroxyl value, viscosity and specific gravity. The important properties of castor oil are listed in Table 7.1. Castor oil is miscible with absolute ethanol, methanol, ether, and glacial acetic acid. It dissolves in its own volume of petroleum ether or 95 per cent alcohol, but does not dissolve to any extent in mineral oil, unless mixed with another vegetable oil. When heated to 300°C for several hours, it polymerises and becomes miscible with mineral oil. Castor oil is soluble in boiling hexane.

Table 7.1. Properties of castor oil.

Property	*Value*
Density at 15.5°C	0.961–0.963
Weight of technical grades, lbs/gal	8.1–8.9
Refractive index at 25°C	1.473–1.477
Solidification, °C	– 10 to –18
Viscosity at 25°C, poise	6–8
Flash point, °C	230
Ignition temp., °C	449
Surface tension at 20°C, dynes/cm	39
Optical rotation	+7.5 to +9.7°
Acid value	Less than 4
Saponification value	176–187
Iodine value (Wijs)	81–91
Unsaponifiable matter, %	0.2–0.3
Reichert-Meissl value	Less than 0.5
Polenske value	Less than 0.5
Acetyl value	144–150
Hydroxyl value	161–169

Composition

Castor oil mainly contains the glycerides of ricinoleic acid, i.e. *cis*-12-hydroxyoctadeca-9-enoic acid, and 18-carbon hydroxylated fatty acid having one double bond. The average fatty acid composition of castor oil is given Table 7.2.

Table 7.2. Fatty acid composition of castor oil.

Fatty acid	*Percentage*
Ricinoleic acid	89.5
Dihydroxystearic acid	0.7
Palmitic acid	1.0
Stearic acid	1.0
Oleic acid	3.0
Linoleic acid	4.2
Linolenic acid	0.3
Eicosanoic acid	0.3

Castor oil is distinguished from other glycerides by its high specific gravity, viscosity, and hydroxyl value. Another unique feature is its alcohol solubility. Its slight solubility in petroleum ether is a characteristic distinguishing it from other fats.

Adulteration

Castor oil is sometimes adulterated with rosin oil, blown oils or other unheated oils, like groundnut, coconut, sesame, mahua, cottonseed, and poppyseed oils and lard. The addition of any of these lowers the acetyl value. Rosin oil is detected by the increase in the unsaponifiable matter, while addition of fatty oils which have not been blown, lowers the specific gravity and increases the product's solubility in petroleum ether.

Grades and Specifications

Bureau of Indian Standards (BIS) has recognised four grades of castor oil for various applications—medicinal, first special, first and commercial. Medicinal grade is used for medicinal purposes and for making hydraulic brake fluids; first special is for use in cosmetics and lubricants; first grade is for use in lubricants and the commercial grade finds industrial uses such as paints, sulphated oil, soaps and printing inks. Requirements of and methods of sampling and tests for castor oil are given in BIS 435 (1973) (Table 7.3).

Table 7.3. BIS specifications for castor oil.

Property	*Medicinal*	*First special*	*First*	*Commercial*
Lovibond colour	6.5	3.7	30	40
Y + 5R units, max.	(13 cm cell)	(←	2.5 cm cell	→)
Sp. gr. at 30°C	0.954–0.960	(←	0.954–0.960	→)
Refractive index at 40°C	← 1.470 – 1.474	→		–
Critical solution temperature °C, max.	0	0	–	–
Acid value, max.	1.5	2.0	4.0	6.0
Saponification value		← 177 – 185 →		
Iodine value		← 82 – 90 →		
Acetyl value		← below 143 →		
Unsaponifiable matter, % max.	0.8	0.8	1.10	1.10
Ash, % max.	0.05	0.10	0.10	–
Moisture and volatiles, % max.	0.10	0.25	0.75	1.00

Various standard specifications for castor seeds, castor oil grades and related materials are given in different booklets of BIS (Table 7.4).

Table 7.4. Indian standards related to castor.

Description	*BIS number and year*
Castorseed grading for oil milling	4618 (1981)
Castor cake for fertiliser	3029 (1987)
Castor oil, various grades	435 (1973)
Castor oil for cosmetics	11486 (1985)
Electrical grade for use as capacitor impregnant	9960 (1981)
Dehydrated castor oil for paints	81 (1971)
Turkey red oil	1044 (1970)
Castor oleine and oleostearine	10536 (1983)

The grades commonly recognised in USA are US No. 1 and US No. 3. The first is ordinarily the oil produced by cold pressing, and the No. 3 is the oil extracted from the press cake. The usual characteristics of these grades are given in Table 7.5. Specifications for the medicinal grade castor oil may be found in various pharmacopoeias.

Table 7.5. International specifications for castor oil.

Property	*AOCS*	*US No. 1*	*US No. 3*	*British first*
Acid value	–	3 max.	10 max.	4 max.
Saponification value	176–187	179–185	177–182	177–187
Iodine value	81–91	82–88	80–88	82–90
RM value	below 0.5	–	–	–
Polenske value	below 5	–	–	–
Acetyl value	144–150	–	–	140 min.
Hydroxyl value	161–169	–	–	156
Unsap. matter, %	below 1	0.5 max.	1 max.	1 max.
RI (°C)	1.466–1.473 (40) 1.473–1.477 (25)	–	–	1.477–1.481 (20)
Sp. gravity 15°C	0.958–0.968	0.961–0.963	0.957–0.963	0.958–0.969
Gardner-Holdt viscosity, 25°C	–	U+/–0.5	U+/–0.5	–
Colour	–	3 max. (Gardner)	7 max.	2.2Y–0.3R max. Lovibond 1 in. cell
Critical solution temperature in alcohol, °C	–	–	–	below 0

The specifications of commonly recognised grades of castor oil are listed in Table 7.6.

Table 7.6. Castor oil specifications.

Grade property	*Extra-pale special*	*Extra-pale*	*Ist processed*	*Medicinal (BP/IP)*	*B.S.S.*	*Commercial*
Colour	2Y + 0.2R	5Y + 0.5R	8Y + 0.8R	10Y + 1R	20Y + 2R	40Y + 4R
Acid value, max.	0.3	0.8	1.0	1.5	2–2.5	6
Iodine value	83–88	83–88	83–88	82–88	80–90	80–90
Hydroxyl value	158–178	158–167	155–170	178–187	176–187	174–184
Saponification value	180–187	180–187	180–187	178–187	176–187	174–184
Moisture max. %	0.1	0.2	0.25	0.25	0.3	0.8
Viscosity	U–V	U–V	U–V	U–V	U–V	U–V

Extra pale special is specially refined odourless oil with lightest colour and highest purity. It has minimum acidity for use in cosmetics and electric insulation oil. Medicinal grade castor oil should be cold pressed and has to conform to all standard values and pass chromatography test which is carried out to ensure the castor oil is indeed cold pressed under medicinal specification.

APPLICATIONS

The applications of castor oil may be classified into direct applications of the oil, applications of variously treated castor oil, and chemical derivatives of castor oil. The treated castor oils may be called as primary derivatives and the chemical derivatives like undecylenic acid may be called as secondary derivatives of castor oil. The major applications of castor oil are shown schematically in Fig. 7.1.

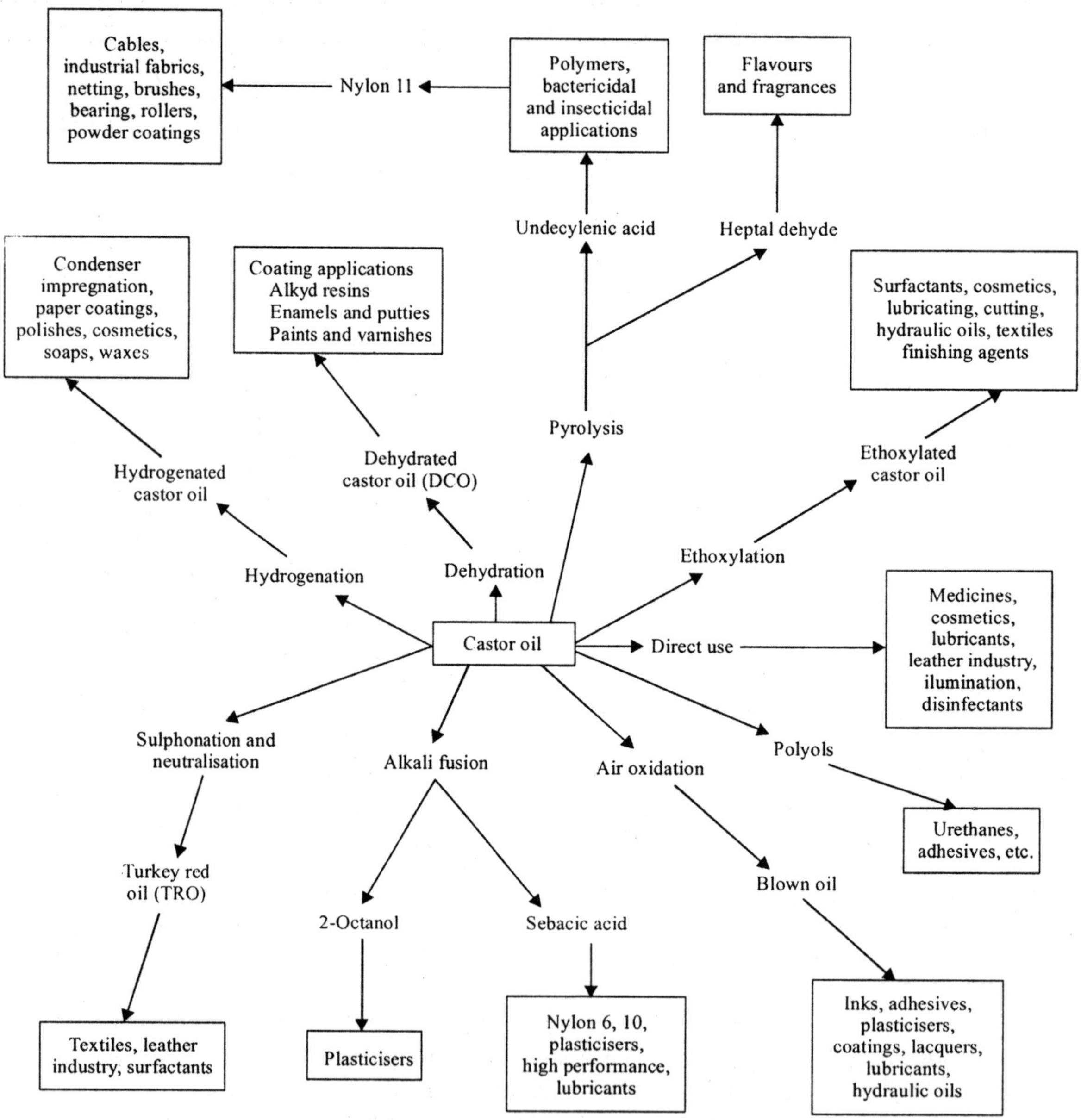

Fig. 7.1. Applications of castor oil.

Direct Uses

Castor oil is an excellent illuminant and has been used in India in lamps from early times. It is believed to give a cooler and brighter light than other vegetable and mineral oils, and to burn more steadily with the formation of very little soot. Till the 1960s, the Indian Railways used castor oil for signal lights and hand lamps. For this, the cold drawn castor oil was admixed with a little of kerosene (7:1) which reduces its high viscosity. However, today, castor oil is consumed for illumination in very small quantities due to its higher price as compared to that of kerosene and cooking gas which have become readily available. Castor oil is used in medicines as a cathartic agent and has earned an unenviable reputation in this regard. It is an official cathartic in IP and other pharmacopoeias. Castor oil has been preferred over other cathartics in the treatment of accute diarrhoea caused by any form of food poisoning, and also in the cleansing of bowel before roentgen examination.

Castor oil is usually given in doses of 4–16 ml. maximum safe dose being 60 ml. in 24 hours. Infants require a relatively large dose of the oil than adults. Castor oil is often given (60 ml) orally, alone or with quinine sulphate to induce labour in pregnancy at term. However, this is reported to be dangerous. The oil can be used as a vehicle for parenteral administration of steroidal hormones. It is also used in abortifacient pastes, and ricinoleic acid is used in contraceptive creams and jellies. The oil as such is a good emollient, and is used extensively in cosmetic formulations like lipsticks. Castor oil gives high gloss and high viscosity in stick cosmetics. Castor oil is sometimes applied externally as bland emollient in the form of a 5–10 per cent ointment, in seborrheic dermatitis and other cutaneous affections. It is also used in nail lacquer removers. Castor oil is employed as a solvent for removing irritating substances in the eyes. Medicinals such as atropine, cocaine, etc. are often suspended in the oil for opthalmic application. Castor oil has been used in making carbon black for kajal and also for massage, in villages.

In veterinary practice, castor oil is a safe and effective purgative for most animals, and may be given to pregnant animals, though it is considered unreliable for horses. Commonly administered doses of castor oil for different animals are: horses 250–500, cattle 500–1000, foals and calves 60–90, sheep and swine 60–120, cats 4–15 and dogs 8–30 ml. Castor oil is used in the preparation of liquid disinfectants like phenyles. An emulsion of the oil and soap is reported to be effective against some crop pests. Castor oil is applied as a preservative to food grains, and pulses in some parts of India. Castor oil is a hair tonic and is commonly used for the preparation of hair oils, hair fixers, and aromatic perfumes. It is used as an ingredient in hair lotions and tonics in concentrations of 0.5 to 20 per cent. Soaps/shampoos prepared from coconut oil, are rather irritating, and less irritating soaps are produced by using diluents like castor oil and olive oil in coconut oil before saponification. This practice was more prevalent in the 1930s and 1940s.

Vegetable oils find application in boundary lubrication mainly because of their outstanding property of 'oiliness'. They have greater adherence to metal surfaces than petroleum oils. Fatty oils have a tendency to undergo oxidation and decompose at higher temperatures. Their stability can be improved by the use of suitable additives. As straight lubricants fatty oils have limited application. They are used mainly for lubricating light machinery and delicate mechanisms like watches, clocks, scientific instruments, spindles, and looms, sewing machines, etc. Fatty oils are compounded with refined petroleum oil lubricants. They have considerable importance as ingredients of greases, cutting oils, wire drawing lubricants and many speciality lubricants. Castor oil (and rapeseed oil) is particularly important. It possesses excellent 'oiliness' and is used for bearings subjected to high loads. Castor oil is more viscous than ordinary oils and hence is suitable for lubricating fairly heavy machinery. Its viscosity changes only slightly with temperature, and it has a low-cold test or solidifying point. Hence, it was once considered more or less essential for the lubrication of piston-type airplane engines. In this and similar applications it has been replaced by the newer low-cold test mineral oils and especially by synthetic esters, which have become the lubricants of choice for jet aircraft engines and are now being used as long-life automotive lubricants. Castor oil is used as one of the wet lubricants in mechanical packaging. Historically, while castor oil has been an important automobile lubricant, its application in this use is now small. But, in principle, vegetable oil could be used to replace petroleum-based lubricants.

Castor oil is widely used in India as a lubricant for all types of heavy machinery especially for locomotive bearings, steam cylinders in railway engines, internal combustion engines in aeroplanes, and marine engines. In rural areas it is used for lubricating country carts, crane crushers, persian wheels etc. Its merits are extremely resistance to oxidation, retention of viscosity at high temperatures, and low solidification point (–10 to 18°C). Stable compound oils consisting of castor oil with 5 per cent of mineral oil or mineral oil containing 5 per cent of castor oil are also used as lubricants. Though castor oil has been almost replaced by the cheaper mineral oils and lubricant, it may gain renewed importance because of its eco-friendliness. A non-toxic lubricating cutting oil for metals at 700–1200°C can be prepared from aluminium powder, graphite, potassium soap, and castor oil. Castor oil is used as a base for fluids for hydraulic systems, particularly those exposed to low temperatures such as automobile brake systems and in airplanes. Castor oil has certain advantages over mineral oils and other fatty oils. It has relativley low-cold test, is compatible with polar liquids of low viscosity, and has little effect on rubber gaskets. Esters of dibasic acids, such as adipates and sebacates compete with castor oil in the hydraulic fluid market together with tall oil esters.

Castor oil is used as condenser oil due to its high dielectric constant (5.3) and low loss angle at 90°C (less than 0.01). Biphenyls can be replaced by castor oil or dicarboxylic acid esters. Castor oil and polychlorinated biphenyl (PCB) are principal impregnants used in energy storage capacitors. Castor oil is better than PCB for pulsed applications. Castor oil is reported to be used by tanneries in Uttar Pradesh, West Bengal, Tamil Nadu and Andhra Pradesh for preserving leather belting, buckets, shoes, etc. It has softening effect on leather, and it serves as a preservative. The sulphated product, viz., Turkey red oil is widely used for making fat-liquors for chrome leather, particularly shoe-upper leather. Fat-liquors made from castor oil and Turkey red oil are more commonly used than those prepared from a mixture of castor soap and neutral oil. Castor oil is used as lipid component in defoamers. It has been used as antifoam agent in low-froth dyeing. Castor oil being a non-drying oil finds only limited direct use in oleoresinous paints and in non-drying short-oil-alkyd resins, useful in stoving finishes. Castor oil (and coconut oil) is used for colour-retentive plasticising resins because of the non-oxidising character of the oil. It is used as a plasticiser, more particularly for spirit varnishes and for urea- and melamine-formaldehyde finishes. Castor oil (and coconut oil) is used for colour-retentive plasticising resins because of the non-oxidising character of the oil. It is used as a plasticiser, more particularly for spirit varnishes and for urea- and melamine-formaldehyde finishes. Castor oil is used as a vehicle in ball-point inks.

Vegetable oils in general are used as softeners in rubber processing and compounding. However, they are less effective softeners from the standpoint of solubility. They produce smooth, low shrinkage processing compounds. Insoluble or only slightly soluble softeners like castor oil reduce the tackiness and stickiness of the compounded mix. Many of the latter group are soluble in the rubber at processing or vulcanising temperatures but they bleed or bloom from the compound during storage at room temperature. Paraffin and castor oil are examples.

Castor oil has been used for rubber processing, especially for the preparation of technically specified rubber (TSR). The *Heveacrumb* process for manufacturing TSR was invented in 1964, and the first

Standard Malaysian Rubber (SMR) was produced by this process. Castor oil (0.7 wt. per cent) was added to the field latex prior to coagulation. During milling of the coagulum, the castor oil acted as a crumbing agent converting the blanket to a fine crumb. Some consumers were concerned about the possible use of excess castor oil which could bloom to the surface in certain vulcanisates. Because of this concern and with the development of improved mechanical granulating equipment, the amount of castor oil was drastically reduced. Castor oil is used mainly as a partitioning agent to prevent the crumbs from agglomerating after being mechanically converted.

Rubber that is produced with even small amounts of castor oil is commonly referred to as *Heveacrumb* rubber. If the rubber is produced by strictly mechanical processing without any castor oil or other incompatible oil, it is referred to as comminuted rubber. This comminuted rubber is preferred by some customers, especially those producing adhesives or using rubber in rubber-to-metal bonding. In practice, it is extremely difficult to show any difference between modern *Heveacrumb* rubber and comminuted rubber. The SMR scheme is the most complicated of the TSR schemes and is made up of the largest number grades.

A cigarette filter with improved nicotine and tar absorption characteristics is obtained by spraying glyceryl triacetate on crimped acetate fibre tow and subsequently applying castor oil and 75 per cent polyvinyl alcohol. Castor oil in combination with ethanol can be used successfully in replacement of diesel, and is claimed to be environment friendly. Another application of castor oil is the adulteration of various other vegetable oils, such as groundnut, sesame and linseed, because of the lower cost of castor oil. It is also used for adulterating essential oils like vetiver, clove, eucalyptus and lemon.

MODIFIED CASTOR OILS

Due to the presence of several functional groups, castor oil can be treated by many different chemical reagents like sulphuric acid, hydrogen peroxide, ethylene oxide, etc. to give various modified castor oils having extensive applications in many industrial sectors.

The important modified castor oils are as follows:

1. Sulphated castor oil (Turkey red oil).
2. Dehydrated castor oil.
3. Hydrogenated castor oil.
4. Epoxidised castor oil.
5. Ethoxylated castor oil.
6. Blown castor oil.
7. Gelled castor oil.

Products known in the trade as 'Floricin' or 'Dericin' prepared by treating castor oil at 300°C are soluble in mineral oils and are used for the preparation of compound mineral oils; however, they have fallen into disuse. Various treated castor oils find use in one form or the other in over 200 products. They include paints, enamels and varnishes, oil fabrics, linoleum, patent leather, flypaper, typewriting and printing inks, greases and special lubricants, polishes and waxes, cutting dielectric and condenser oils, softening agent for gelation in rayon sizing, nitrocellulose baking finishes, hydraulic brake fluids, urethane foams, and rubber substitutes, cosmetics, soaps, pharmaceuticals, and insecticidal formulations.

DEHYDRATED CASTOR OIL (DCO)

Dehydrated castor oil (DCO) is prepared by heating castor oil at the temperature of about 240–260°C and this DCO is useful as a drying oil in paints and surface coatings. The dehydration involves the removal of the hydroxy group from ricinoleic acid together with an adjacent hydrogen atom to yield both 9:10, 11:12-, and 9:10, 12:13-octadecadienoic acids. The presence of the conjugated double bonds in the former is responsible for the drying properties of the DCO. The dehydration is generally not complete. It proceeds only to the extent of 80–85 per cent and some hydroxy acids always remain in the product. Dehydration may be effected on the oil or its fatty acids or on a solution of either of them in a suitable solvent. Many processes, using a wide variety of catalysts for the dehydration of castor oil have been patented. The catalysts include sulphuric acid and alkali bisulphates, phthalic anhydride, alumina, and other oxides of metals and activated earths.

The acid catalysts are generally used at the level of 0.1–0.2 wt. per cent. A process has been developed by Indian Institute of chemical technology, Hyderabad, for dehydration of castor oil to give a light coloured DCO. The extent of conjugation introduced in the product during dehydration varies usually from 22 to 33 per cent. Tung oil, oiticica oil are natural conjugated oils and compete with dehydrated castor oil. Conjugated oils are considerably more reactive than non-conjugated oils. Non-conjugated oils can be isomerised at least partially to conjugated oils with a variety of catalysts, mostly alkali hydroxides. Dehydrated castor oil contains mixture of isomers of the 9,11-conjugated and 9,12-non-conjugated fatty acid esters. DCO dries relatively rapidly at room temperature, but upon further exposure the surface becomes tacky. The after-tack has been attributed to the presence of various geometric isomers formed during the dehydration. The conjugated double bonds in the fatty acids favour oxidation and polymerisation, and these oils dry more rapidly and are consequently in great demand for the manufacture of high quality finishes. In the manufacture of protective coating resins, conjugated oils contribute faster bodying, and the films dry more rapidly, have excellent, surface-dry, through-dry and hardness. Initial film colour and colour retention on ovenbake are outstanding. The films are flexible, do not tend to yellow, and show improved resistance to water, alkalies and weathering.

A commercial sample of DCO had the following characteristics (Table 7.7).

Table 7.7. Properties of dehydrated castor oil.

Property	*Value*
Specific gravity at 25°C	0.9420
Refractive index at 25°C	1.480
Viscosity, cp	300
Acid value	9.0
Saponification value	190.6
Iodine value (Wijs)	111
Hydroxyl value	18.9
Hanus value (400% excess)	130.7
Gardner colour	6
Gel time at 315°C, hours	2.25
Set-to-touch time (0.05% cobalt drier), hrs.	3.5
Conjugation, %	25.0
Fatty acid composition, %	
Saturated	0.5–2.5

(Contd...)

Table 7.7. Properties of dehydrated castor oil. *(Contd.)*

Property	*Value*
Hydroxy acids	3–8
Oleic	7.5–10.5
9:10, 12:13 - linoleic acid	59–64
9:10, 11:12 - linoleic acid	17–26

Some properties of dehydrated castor oil are compared with those of other drying oils in Table 7.8.

Table 7.8. Comparison of various drying oils.

Property	*DCO*	*Soyabean*	*Tung*	*Oiticica*
Sp. gr. at 25°C	0.931	0.920	0.935	0.970
Viscosity, cP	165–200	50	165–250	Solid
Iodine value	136	120–140	158–166	148
Acid value	4	0.5–1.6	0.3–4.0	4
Refractive index	1.482	1.473	1.515	1.510
Colour, gardner	4	–	6–10	8–10

The choice of a drying oil for a particular application depends on price and availability, reactivity towards resins, on drying characteristics and colour retention. The drying oils in order of decreasing hardness are listed in Table 7.9.

Table 7.9. Drying oils in order of decreasing hardness.

Tung
Oiticica
Fish (high IV fraction)
Perilla
Linseed
Dehydrated castor
Corn
Soyabean

The Indian Standards for polymerised DCO for use in paints are given by BIS-81 (1971). Tests for drying oils for paints are given by BIS 74 (1966) Table 7.10.

Table 7.10. BIS specifications for polymerised DCO.

Property	*Value*
Lovibond colour, 0.63 cm cell	14Y + 1.2R
Sp. gravity at 30°C	0.944–0.951
Acid value, max.	18.0
Saponification value	190–200
Iodine value	90–112
Aceto value	Not more than 30
Viscosity, St	12–25

Another set of specifications for dehydrated castor oil are given in Table 7.11.

Table 7.11. Specifications for dehydrated castor oil.

Property	*Value*
Description	Light yellow, thick, liquid oil
Specific gravity at 25°C	0.935–0.945
Refractive index at 25°C	1.4800–1.4824
Acid value, max.	6
Iodine value (Wijs), min.	130
Hydroxyl value	20–25
Saponification value	185–190
Colour gardner	6–8
Viscosity at 25°C, poise	1.86–2.25

The ASTM Specifications for dehydrated castor oil are given by D961–86. These are given in Table 7.12.

Table 7.12. ASTM specifications of dehydrated castor oil.

Property	*Unbodied*	*Bodied*
Viscosity at 25°C	F to I	Z2 to Z4
Sp. gravity at 25°C	0.926–0.937	0.944–0.966
Refractive index at 25°C	1.4805–1.4825	1.4860–1.4890
Acid value	6	6
Iodine value (Wijs)	125–145	100 min.
Colour no. (Gardner)	6	7
Gel time at 315°C., min.	145	63
Set-to-touch time, hr.	2.5	1.4

Bodied refers to polymerised castor oil of higher viscosity; unbodied refers to dehydrated castor oil that has not been additionally heat polymerised

The contents of conjugated and non-conjugated fatty acids in DCO and Tung oil are given in Table 7.13.

Table 7.13. Conjugated fatty acids in DCO and tung oils.

	Conjugated, %		*Non-conjugated, %*	
	Trienoic	*Dienoic*	*Trienoic*	*Dienoic*
Tung acids	71.7	4.5	0.0	11.6
DCO acids	0.1	29.6	0.8	52.1

Water can also be eliminated from free ricinoleic acid, yielding a mixture of 9,11- and 9,12-octadecadienoic acids, called dehydrated castor oil acid. However, the degree of conjugation in this reaction is higher, upto 65 per cent conjugated octadecadienoic acid. Resins can be produced from these dehydration products.

Dehydrated castor oil can be further chemically treated to give polymerised products like stand oil with viscosity of 100 to 150 poise. Dehydrated-dehydroxylated castor oil is a unique synthetic drying oil. Its non-yellowing properties make it highly suitable for use in super-white paints for fridge, fans, etc.

DCO is generally marketed after polymerisation or bodying. DCO is unique in its qualities and is considered often as an ideal drying oil. The drying property of fats and oils is related to their degree of unsaturation, hence iodine values. A generally accepted rough classification considers those oils with iodine values (Wijs) below 90 to be non-drying, between 90 and 130 to be semidrying, and above 130 to be drying. Linolenic acid is responsible for the high yellowing tendency of alkyds based on linseed oil fatty acids. Therefore, alkyds intended for making white or light coloured enamels should avoid high linolenic content fatty acids by choosing soyabean, safflower or DCO fatty acids.

DCO is roughly intermediate between linseed and tung oils in many of its properties such as drying time, rate of polymerisation when heated, and resistance to water and alkalies in varnish films. Its resistance to yellowing is outstanding. The films are softer than those of linseed, perilla or tung oils, but of superior elasticity. DCO films when applied to textiles provide a springiness not so apparent with other oil films. Dehydrated castor oil is used for colour-retentive baking alkyds, since the conjugated linoleic imparts rapid cross-linking and curing at elevated temperatures. Tung oil and oiticica oil are occasionally used in admixture with other oils to impart faster dry and early hardness. Tall oil fatty acids contain 43–46 per cent dienoic linoleic acid and their presence in alkyds imparts slower air drying than soyabean fatty acids. However, they give slightly better colour retention and are especially useful in alkyds for baking finishes.

DCO has established itself as a drying oil of considerable value in the alkyd type surface coatings and in stoving finishes. It is extensively used as a replacement for tung oil in the manufacture of quick-drying and water-resistant varnishes and enamels. In decorative varnishes, it can substitute linseed and tung oils. Tung and oiticica oils are derived from the seeds of subtropical tress. They are relatively high priced and with the increased use of synthetic surface coatings, these oils are becoming increasingly rare in commerce. DCO is of value in white surface finishes as in the case of refrigerators. Since it does not have any trienes, it can be used as the base for non-yellowing binders for highly flexible coatings with good weather resistance. In paints and varnishes, various viscosity grades of DCO are used. In combination with other drying oils and resin, high gloss vehicles and white enamles are obtained. Drying oil-based alkyds are used in primers over chalky surfaces where latex paints do not provide adequate adhesion. On a solids basis, about 15 per cent of the latex polymer is replaced by the drying oil in emulsion form. After application, the emulsion breaks, permitting the drying oil to penetrate between the loose pigment particles of the chalky surface where the latex cannot penetrate. The oil reaches the continuous substrate of the old paint, thus serving to cement the chalky particles together and provide adhesion.

DCO is also used in most other varieties of drying oil products including oiled fabrics, linoleum, patent leather, and inks and lithographic varnishes. Drying oils react with a variety of chemicals to yield resins that produce harder and more durable films. Phenolic resins, alkyd resins and urethanes are the most commonly used. Synthetic resins such as phenolic and cresylic resins are modified with drying oils such as DCO, elastomer, cardanol, an epoxy, phosphorus or boron-based compound and used as friction material binders in brake linings and clutch facings. Other resin systems are based on elastomers, drying oils, or combinations of the above and other polymers. Dehydrated castor oil was earlier used in varnishes used in electrical equipment. But, today it has been replaced by synthetic resins.

The drying and film properties of DCO can be improved by processes such as polymerisation, solvent fractionation, isomerisation, etc. Isomerised DCO (with about 50 per cent of conjugation) of a light colour and a low acid value (2.2) and viscosity, has been prepared. It produces varnishes possessing better qualities than DCO but inferior to tung oil in some respects.

Blown and stand oils have also been made from DCO. Blown oil is useful in wrinkle finishes, lightly blown oils giving fine wrinkles and heavily blown oils producing coarse wrinkles. DCO on styrenation yields a product with improved drying properties. Properly prepared styrenated oils give excellent films. A comparison of some styrenated products is given in Table 7.14.

Table 7.14. Comparison of styrenated products.

Base oil	*Viscosity St*	*Process*	*Properties of product*	
			Drying time, min.	*Hardness sward, 2wk*
Linseed	30	Bulk	30	30
Soyabean	30	Bulk	45	22
Dehydrated castor	14	Solution	10	26
Dehydrated castor	30	Bulk	30	32
Tung soyabean	2	Solution	12	26

Styrene in the presence of conjugated dienoic esters reacts to form a dibasic ester from one molecule of styrene and two of the ester. When it reacts with a polyol, this dibasic ester gives polyesters. These convert readily into urethane foams.

The conjugated double bond system of DCO undergoes Diels-Alder reaction with maleic anhydride to give a dianhydride. This maleated dehydrated castor oil or maleinised DCO reacts with polyols to give moderate molecular weight derivatives that dry faster than the unmodified drying oil. Water soluble resins and paints prepared with maleated dehydrated castor oil were used in electrodeposition of water-soluble primers containing 10–15 per cent solids for optimum paint thickness. Heat-bodied drying oils (e.g. DCO) are used as vehicle in steel-die inks.

Since 1920s, drying oils in general, have been increasingly replaced by other film-forming materials. In paints, they have, in large measure, been replaced by nitrocellulose, alkyds, acrylic co-polymers, polyesters, epoxy resin, vinyl acetate and acrylic co-polymer latices, among others. In artists' colours, acrylic latices have been the principal replacement.

In printing inks, hydrocarbon and phenolic resins, alkyds, nitrocellulose, and polyamides are among the replacements.

Oil cloth has been replaced by supported and unsupported vinyl films, and linoleum has been supplanted by a variety of polymer-based tile materials. Some natural drying oils continue to be used. Synthetic drying oils such as oxidising alkyds, epoxy esters, and urethane oils (uralkyds) are made from drying and semi-drying oils. Although use of such synthetic drying oils is substantially larger than that of natural drying oils, their consumption is also declining, primarily because oxidation leads to embrittlement and discolouration.

DEHYDRATED CASTOR OIL (DCO) FATTY ACIDS

Dehydrated castor oil is saponified and acidulated to give a mixture of fatty acids called as DCO fatty acids. The by-product glycerine-water is concentrated to pure glycerine. Light coloured distilled grade dehydrated castor oil fatty acids with high degree of conjugation (more than 55 per cent) are obtained by decomposition distillation of estolides of castor oil fatty acids. The specifications for DCO fatty acids are given in Table 7.15.

Table 7.15. Specifications for DCO fatty acids.

Property	*Value*
Description	Less viscous, oily liquid
Colour	Almost colourless with slight yellow tinge
Specific gravity at 25°C	0.902–0.905
Refractive index at 25°C	1.473–1.477
Acid value	195–200
Iodine value	150–156
Saponification value	195–200
Conjugation	30 to 35% or 55 to 65%

ASTM requirements for DCO fatty acids are listed in Table 7.16.

Table 7.16. ASTM requirements for DCO fatty acids.

Property	*Type-I*	*Type-II*
Acid value	195–200	187–195
Saponification value	195–200	193–199
Iodine value	150–156	138–143
Colour, gardner	1 max.	5–8
Spectrophotometric diene value	28–35	25–32

DCO fatty acids are used to make epoxy resins, special automotive paints with good weathering resistance and water resistance, high class enamels for copper and aluminium windings, electrical varnishes, epoxy special glues with insulating properties. DCO fatty acids are generally preferred for metal coating, durable printing inks, paint lacquers, etc. DCO fatty acids are also used as high trans-trans acid for speciality chemicals through Kjedahl reaction.

Dehydrated castor oil fatty acids have been co-polymerised with acrylic acid esters or allowed to react with hexamethoxy methylolated melamine resin and neutralised with ethyl amine to give an aqueous solution reportedly useful as a water soluble resin for an electrodeposition coating. Cathodically electro-depositable binders prepared using dehydrated and maleated dehydrated castor oil fatty acids resulted in binders that gave good adhesion of maximum film thickness. A co-polymer latex emulsion system utilising dehydrated castor oil fatty acids yields superior wall paints providing adhesion to varnishes and glossy surfaces. DCO fatty acids react with phthalic anhydride and other polyols to produce a water soluble, air-drying alkyd resin.

HYDROGENATED CASTOR OIL

Castor oil on controlled hydrogenation yields a tasteless, hard, brittle, waxy material with a characteristic fatty odour and a cream to milk-white colour. The product is known in the trade by such names as castor wax, opal wax, etc. It consists mainly of 12-hydroxystearin. The properties of hydrogenated castor oil are listed in Table 7.17.

Hydrogenated castor oil is extremely insoluble in water and in the more common organic solvents. Specifications for hydrogenated or hardened castor oil are given in Table 7.18.

Table 7.17. Properties of hydrogenated castor oil.

Property	*Value*
Melting point, °C	84–86
Colour	0.5 to 1.0R + 1.5 to 3.0Y
Iodine value	2–5
Hydroxyl value	150–58
Acid value	1–4
Saponification value	179

Table 7.18. Specifications for hydrogenated castor oil.

Property	*Value*
Description	White or cream coloured solid flakes
Colour, 5.25 in. cell	3Y + 1R
Melting point, °C	82–86
Acid value	2–5
Hydroxyl value, min.	155
Iodine value	2–5

Hydrogenation of castor oil at higher temperatures (200°C) and lower pressure (5 kg/sq.cm) for 38 hours, yields a product free from hydroxy acids and is useful as a source of stearic acid. Hydrogenated castor oil can be used as a substitute for carnauba wax. It is a valuable constituent in non-drying alkyds, plasticisers, antisagging and antistatic agents of pigments in paints. It is useful in polishes, candles, insulating materials, coating compositions for paper, textile and leather, moulding compositions (mould-release agent) in formed plastics and rubber goods, heat-sealing adhesives, water-repellant coatings, carbon paper, ointments and cosmetics, for impregnating paper, wood, cloth, for electrical condenser impregnation, as solid, special lubricant, and for the preservation of fruits and vegetables. Hydrogenated castor oil and 12-hydroxy stearate are important constituents of high grade lubricating greases and other speciality lubricants. A vinyl chloride-vinyl propionate co-polymer varnish at 40 per cent solids with TiO_2, talc, baryte, dioctyl phthalate (DOP) in ethyl acetate-xylene can be mixed with castor wax and gelton 50 (acid treated $CaCO_3$) to yield a stable coating for spray-coating glass plate, giving a 30 micrometer thick sag-free film. A metal working lubricant useful for drawing aluminium tubing can be prepared by mixing sulphur-free wool wax, aluminium stearate, castor oil wax, and 1,1,1-trichloroethane at 38–49°C. Hydrogenated castor oil can be employed for making 12-hydroxy stearic acid which is extensively used in the formulation of greases, in surface coatings and as a plasticiser.

SULPHATED CASTOR OIL

Sulphated castor oil is called as Turkey red oil, because of its use in Turkey-red dyeing. The sulphated vegetable oils are sometimes erroneously referred to as sulphonated oils. Sulphated castor oil, used since about 1875 in textile processing, was the first such commercial product. Apart from castor oil neat's foot oil, olive oil, and rapeseed oil are also sulphated and used in various applications. With its hydroxyl group and a double bond, ricinoleic acid is particularly suited for sulphation. Oleic acid is also suited for sulphation. Esters of these acids can be sulphated with a minimum of hydrolysis of the glyceride group. Treatment of castor oil with sulphuric acid under controlled low temperatures, yields a product known as Turkey red oil (TRO). It is used widely for dyeing, bleaching, and sizing in the textile industries, and also in leather tanning.

Turkey red oil is made commercially by treating castor oil with concentrated sulphuric acid (25–30 per cent on weight of oil) at 30–35°C for 5–6 hours, settling the product overnight, washing off the excess acid, and neutralising the product with caustic soda, caustic potash or ammonia. The commercial manufacturing process actually esterifies the 12-hydroxy group in ricinoleic acid, thus producing a true sulphate ester, and not a sulphonate. The quality of TRO depends on the conditions of production, such as proportion of acid used, temperature of treatment, extent of washing etc. A higher temperature of sulphation (50–60°C) yields a product with superior wetting out action and also reduces the time of reaction to about one hour. Various grades with different strength and acidity are marketed. Soda-finished grade having a neutral reaction and containing about 50 per cent of the original castor oil is the most popular due to the clarity of its aqueous solution. Commercially sulphated castor oil contains about 8.0–8.5 per cent combined SO_3, indicating that only one of the reactive sites in the unsaturated, hydroxy-bearing triglyceride has been sulphated. The sulphate group imparts excellent wetting, emulsification, and dispersion characteristics. The Indian Standards are available for TRO from BIS 1044 (1970) and are presented in Table 7.19.

Table 7.19. BIS for Turkey red oil.

Property	*Requirement* *Grade I*	*Grade II*
Degree of sulphation, % wt. on TFM, min.	6	6
Total alkali (as KOH), % wt, max.	3	3
Total fatty matter (TFM) % wt, min.	70	50
pH value	6.5–7.5	6.5–7.5
Total ash, % wt, max.	8	8
	Optional for both grades	
	Calcium soap dispersion value, min. 0.5	
	Unsaponifiable matter % wt. max. 2	
	Unsulphated oil, % wt, max. 7	

Despite many side reactions, most commercial products seem to be similar in properties. Sulphated oils ordinarily contain large amounts of free fatty acids and unreacted glycerides. However, high purity is not required in the formulations where these oils are used in fat liquoring of leather, in emulsification systems for fibre lubricants, as an aid in pigment dispersion and latex emulsification in paints, in paper coating systems, and in dyeing applications.

Turkey red oil is an anionic i.e. anion-active wetting agent and has a combination of wetting, emulsifying and mordant actions. It is widely used for the softening of cotton goods and as a dyeing aid in neutral or slightly alkaline baths. It enables the dyes to penetrate the cloth fibres and subsequently cling more tenaceously. It adds to the brightness and lustre of the dyed fabrics. It has been widely used with red alizarine dye and hence its name.

TRO finds use as an emulsifying agent in DDT preparations and in the preparation of scouring agents for wool, lake pigments, 'monopole' soaps, etching baths, soluble cutting oils, metal polishes, calico printing, in horticultural sprays, in the leather industry for fat-liquoring and as an antifoaming agent and peracid agent in the paper industry. Different polyamide fibres with different affinities for anionic dyes are dyed level shades by pre-treatment with aqueous acidic solutions containing sulphated castor oil. Other uses of sulphated castor oil include lubrication and imparting hydrophilicity to polyester tricot.

Substantial advancements in dispersion technology have been made since the initial introduction in 1923 of disperse dyes in paste form for cellulose acetate. Dyes were dissolved in sulphated fatty acids such as sulphoricinoleic acid (SRA), 12-hydroxy-9-octadecanoic acid sulphate, and then formed into a dispersed paste by mixing the solution with water giving rise to an entire line of SRA dyes for cellulose acetate. The Duranol dyes were prepared by grinding or dissolving aminoanthraquinone dyes in a solvent followed by drowning into water containing soap or Turkey red oil. Manufacturing procedures for producing dye dispersions are generally not disclosed. The principal dispersants in use today include long-chain alkyl sulphate, alkaryl sulphonates, fatty amine-ethylene oxide condensates, fatty alcohol-ethylene oxide condensates, naphthalene-formaldehyde-sulphuric acid condensates and the lignin sulphonic acids. The neutral sulphated castor oil is used in place of soap in cases of contact dermatitis resulting from air plane dope, machine cutting oil, plaster, etc. Sulphated castor oil and an emulsifiable polyethylene can be used to make a creamy paste pigment-dispersing agent. This paste can be blended with resins and pigments in high-flash naphtha to develop a white baking enamel with good non-settling readings after three months. A sulphated derivative of castor oil is used as a dispersant for Plaster of Paris. It reduces the amount of water needed to form a plastic slurry. Sulphated and sulphonated derivatives have also been used in electroplating zinc, as softeners in dishwashing compounds, and in glass cleaning.

Apart from castor oil, tallow, glycerol trioleate, soyabean oil, neat's-foot oil, butyl oleate, and tall oil are among the many oils that have been marketed in the sulphated form. Sulphation of castor oil or ricinoleic acid combined at some stage with heating results in the formation of inner esters (estolides), and yields the so-called honey-coloured, viscous, 'monopole' soaps which are used for surface active properties in desizing, bleaching, dyeing, printing, etc. in the textile industry. In patented processes, the oil or its fatty acids are either heated first and then sulphated, or vice versa. Commercial products like Avirol AHX, Surfax WO, and Parapon made by sulphation of butyl, propyl, and amyl ricinoleates possess greater foaming and wetting power than TRO. Sulphation of castor oil with anhydrous SO_3 at temperatures higher than the sulphuric acid treatment produces a product also containing 8.0–8.5 per cent combined SO_3 which has better hydrolytic stability and contains less inorganic salts and free fatty acids.

EPOXIDISED CASTOR OIL

Castor oil which contains ricinoleic acid is difficult to epoxidise to high yields due to the structural placement of the unsaturation and the presence of interfering groups. In order to predict the economic success of an epoxidised vegetable oil, one must look beyond the expense of the oil. Low iodine number-oils are less expensive but they are also less useful, very high iodine number-oils are costly to epoxidise since the peroxide required is one of the more costly ingredients in the epoxidation process. By reacting castor oil with hydrogen peroxide under suitable conditions, it is possible to get epoxidised castor oil having about 3 per cent oxirane oxygen. Epoxidised castor oil is used as plasticiser and stabiliser for PVC and in surface coatings.

ETHOXYLATED CASTOR OIL

Castor oil is a triglyceride high in ricinoleic esters. Castor oil reacts with varying amounts of ethylene oxide to form water-soluble surfactants useful as emulsifiers, dispersant and lubricants and in numerous other applications. Ethoxylation in the presence of an alkaline catalyst to a polyoxyethylene content of 60–

70 per cent yields water-soluble surfactants. Since alkaline catalysts also affect transesterification, ethoxylated castor oil surfactants are complex mixtures with components resulting from transesterification and subsequent ethoxylation at the available hydroxyl groups. The ethoxylates are pale amber liquids of specific gravity just above one at room temperature. They are hydrophilic emulsifiers, dispersants, lubricants and stabilisers that are used as textile additives and finishing agents, as well as in paper and leather manufacture. Castor oil ethoxylates are widely used in the textile industry in the preparation of microemulsions for spin finishes and as components of the emulsifier systems used as dye carriers. Some use is found in the preparation of latex paints. They are utilised by the cosmetic and toiletry manufacturers in the preparation of creams. The CFTA designation for these products includes the mole ratio of ethylene oxide as well as the name of the oil. For example, PEG 40 castor oil is the 40-mol ethylene oxide adduct on castor oil. Various ethoxylates of castor oil and their applications are listed in Table 7.20.

Table 7.20. Ethoxylated castor oils.

Trade name	*No. of EO groups*	*Conc. %*	*Functions*
Emulphor EL 719	40	96	Di E Em S Ta
Emulphor EL 980	200	100	Di E S Ta
Emulphor EL 719P	40	100	Di E S Ta
Emulphor EL 620P	30	100	Di E Em S Ta
Surfactol-318	–	100	E dF Ta
Surfactol-380	–	100	E St Ta
Trylox CO-5	5	100	E dF
Trylox CO-30	30	100	E Sp Ta
Trylox CO-40	40	97	E S Ta
Trylox CO-200	200	100	A Sp Ta
Trylox HCO-25*	25	100	E Sp

* Hydrogenated castor oil.

Functions: Di: Dispersion; E: Emulsification: Em: Emolliency; S: Solubilisation, Ta: Textile assistance/processing; dF: Defoaming; St: Stabilisation, Sp: Softening, A: Antistatic reduction or elimination.

Ethoxylated castor oil products containing from 5 to 200 moles of added ethylene oxide are marketed. These surfactants cover the HLB range from 5 to 18. Some examples of ethyoxylated castor oil samples and their HLB values are listed in Table 7.21.

Table 7.21. Relationship of No. of moles of ethylene oxide added to castor oil and the HLB of the ethoxylated oil.

Moles added	*HLB value*
5	4.9
10	7.2
15	9.7
25	10.8
30	11.7
40	13.1
80	15.8
200	18.1

The low level ethoxylated castor oil derivatives are water emulsifiable and are used as defoamers and de-emulsifiers for petroleum emulsions. The highly ethoxylated products are excellent solubilisers for water-insoluble oils in cosmetic compositions. Ethoxylated castor oils (Cremophor EL) and ethoxylated castor waxes (Cremophor RH, a 40/60 ethoxylated-hydrogenated castor oil) are used as solubilisers of hydrophobic substances such as perfumes in cosmetics.

Ethoxylated castor oil was the best solubiliser in a multicomponent system and the hydrogenated castor oil was more effective than the unsaturated oil.

Ethoxylated castor oil is also a good solubiliser for vitamin A palmitate. The highly ethoxylated castor oils are also used as components in detergents, lubricating and cutting oils, hydraulic fluids, textile finishing compositions and as antistatic agents for nylon carpets and apparel. Stable protease-containing tablets can be prepared with ethoxylated castor oil. The propoxylated derivatives of castor oil are mineral oil soluble and useful in lubricating oils and hydraulic fluid compositions. The quality of poly (phenylene sulphide) coating was improved by using polyoxyethylated castor oil as a surfactant (e.g. Emulphor EL-620 or Emulphor EL-719).

A hydrolysis-sensitive biocide was stabilised using an ethoxylated castor oil as the emulsion stabiliser for use in a pesticide. Alkylene oxide-modified castor oil has also been used in inks for pigment dispersion. Polyoxyethylated castor oil (42 moles ethylene oxide) has been approved by FDA in USA as an emulsifier in nitrocellulose coatings for paper and paperboard intended for use in contact with fatty foods at a maximum level of 8 per cent of the coating solids.

A partially saponified castor oil surfactant was prepared by reactions of an ethoxylated castor oil with alkali metal hydroxide. The product had excellent detergency, low foaming characteristics, stability in alkali and acid solutions, complete water solubility, and no cloud point in typical aqueous use solutions. A highly stable oil was prepared by the reaction of castor oil with an epoxy compound (e.g. ethylene oxide, butadiene oxide, styrene oxide).

The product was hydrogenated and dispersed in oils or water to yield a highly stable rolling oil. Oils suitable as lubricants for rolling of steels at high temperatures are prepared by mixing castor oil, alkylene oxide-containing castor oil, and C_4–C_{24} fatty acid esters.

Adducts prepared from hydrogenated castor oil show slightly lower HLB values when about 20 mol of ethylene oxide is reacted. At higher levels of addition, small or no differences in HLB values are noted. Readily dispersible antistatic agents are prepared by reaction of polyoxyalkylated hydrogenated castor oil and a dialkyl oxalate or malonate. A mixture of polyethylene terephthalate is spun with this product to give fibres having static discharge properties.

BLOWN CASTOR OIL AND OTHER PRODUCTS

Blown castor oils, also called as oxidised castor oils, are prepared by bringing air or oxygen into intimate contact with the oil at temperatures in the range of 80–130°C. Complex changes are produced, including an increase in the viscosity, specific gravity and saponification value of the oil. Air-blown castor oil is used as a plasticiser for varnishes, lacquers, artificial leather, polymers and adhesive mixtures. They are excellent plasticisers for cellulosic resins, polyvinyl butyral, polyamides, shellac, and natural and synthetic rubber. The high viscosity products are also used as tackifiers in gasket compounds and as adhesives because of their good oil and solvent resistance. They also serve as excellent pigment grinding media and a base for lubricating and hydraulic oils. The specifications for blown castor oil are given in the Table 7.22.

Table 7.22. Specifications for blown castor oil.

Property	*Value*
Description	Viscous, oily liquid
Colour	Pale yellow to reddish yellow
Viscosity, poise	15 to 75 poise
Acidity	5 to 25
Specific gravity	0.96 to 1.0

Another set of properties of oxidised castor oils produced in USA are given in Table 7.23.

Table 7.23. Properties of oxidised castor oils.

Property	*Value*
Colour, gardner	2–15 (pale to dark brown)
Viscosity at 25°C, stokes	11–2000 to gel
Sp. gr. at 25°C	0.970–1.028
Acid value	5–25
Saponification value	188–230
Iodine value	80–50
Hydroxyl value	160–125

Blown castor oil is used as brake fluid mixed with different additives and diacetone alcohol. It is used in lubricating oils. It is mixed with plasticisers to reduce their evaporation point. Aviation oils were prepared by the esterification of thermally treated or oxidised castor oil with alcohols in the presence of H_2SO_4 or HCl. Blown castor oil is further blown to form gelled castor oil. The specifications for gelled castor oil are presented in Table 7.24.

Table 7.24. Specifications for gelled castor oil.

Property	*Value*
Description	Solid, jelly-like material
Colour	Reddish yellow
Iodine value	65 to 75
Liquification point, °C	190 to 212
Saponification value	210 to 220

Gelled castor oil is used to lower the evaporation point of plasticisers like DOP in the manufacture of plastic products, leather cloth, etc. It is also used in boot polishes and leather finishes. The heavier blown or gelled castor oils are softening agents for rubber and are also used for making asphalt tile or as a plasticiser in nitrocellulose coatings. Blown castor oil serves as an excellent base for hydraulic fluids for automobile and aeroplane hydraulic systems. In India, both light duty and heavy duty hydraulic brake fluids for motor cars used to be made from castor oil during the 1960s.

Ozonation of castor oil also yields a product of superior qualities. Acrylic esters of castor oil are used in solution for the emulsion polymerisation coating process. 1-bromo-naphthalene is mixed with polymerised castor oil and used as a general immersion oil in microscopy. Aluminium lakes are mixed with castor oil, finely ground, and used as lipstick ingredients.

Castor oil is the only commercially available renewable vegetable oil resource having a hydroxyl group structure and functionality that leads to diverse oleochemicals. Castor oil is composed of triesters of ricinoleic acid and glycerol. On hydrolysis, i.e. splitting, castor oil gives mainly ricinoleic acid and glycerol. These two chemicals may be called as the basic derivatives of castor oil. Glycerol is a common chemical obtained from all the vegetable oils. Therefore, it is not discussed in detail here.

RICINOLEIC ACID

The major fatty acid present in castor oil, ricinoleic acid (141–22–0), has the following properties (Table 7.25).

Table 7.25. Properties of ricinoleic acid.

Property	*Value*
Acid value	180
Saponification value	186
Iodine value, Wijs	89
Melting point, °C	5.5
Boiling point, °C (at 10 mm Hg)	245
Optical rotation	+6.67
Gardner colour	5
Viscosity at 25°C., cSt	400
Pour point, °C	–23
Specific gravity at 25°C	0.940
Refractive index at 25°C	1.4699

Ricinoleic acid is soluble in alcohol, acetone, ether, chloroform. Ricinoleic acid cannot be distilled unless special precautions are taken via derivative formation to protect the hydroxyl group. Acid sulphate of ricinoleic acid, i.e. ricinosulphuric acid, is obtained by the action of chlorosulphonic acid. It is a viscous, brown liquid with weak blue fluorescence. It is soluble in water (about 10 per cent), alcohol, ether, and chloroform. Ricinoleic acid, a hydroxy acid, is used as pigment and dye disperser in printing inks, and to some extent in heat convertible alkyds. But its main use is as chemical intermediate to produce various oleochemicals. Ricinoleic acid is used in textile finishing. It is sometimes added to Turkey red oil and dry-cleaning soaps. It has been used in contraceptive jellies.

Ricinenic acid is obtained by dehydration of ricinoleic acid. Ricinenic acid is a mixture of about 30 per cent conjugated 9,11-linoleic acid and about 50 per cent 9,12-linoleic acid. The double bonds are partly in the trans configuration. By subsequent conjugation, a ricinene fatty acid with increased content of conjugated double bonds can be prepared. Alkaline cleavage of ricinoleic acid under pressure and at high temperature leads to the formation of sodium sebacate and 2-octanol. In industry, castor oil which contains about 87 per cent ricinoleic acid, is normally used instead of the pure acid. Suberic acid can be prepared from ricinoleic acid by HNO_3 oxidation. However, suberic acid appears not to be economically competitive with related diacids in most uses.

Commercially, several mono- and polyhydric alcohol esters of ricinoleic acid are produced and are further subjected to reactions like acetylation, epoxidation, sulphation, ethoxylation, urethane formation, hydrogenolysis and others. Methyl and butyl esters and their acetylated products find extensive use as plasticisers for many natural and synthetic resins and elastomers. These, in general, contribute to softening and flexibilising action at lower temperatures, lubricity, pigment wetting and better electrical properties. Epoxidised derivatives are used as plasticisers for PVC and nitrocellulose. Partial esters of polyols can be used in textile industry and can be used as surfactants after sulphation or alkylene oxide condensation. They are highly useful raw materials for the manufacture of polyurethanes and also in the manufacture of alkyds, surface coatings, and lubricants. Esters of ricinoleic acid are also used as chemical intermediates in the manufacture of other oleochemicals.

Ricinoleic acid can be ethoxylated not only at the carboxyl group but also at the hydroxyl group, depending on the conditions to obtain good emulsifiers. It can be reacted with ammonia or amines to get amides having hydroxy groups which find applications as wax and lubricant additives. Ozonolysis of ricinoleic acid (or mixed castor oil fatty acids) results in azelaic acid and hydroxynonanoic acid which can be used as a raw material for making nylons and lubricants.

Ricinoleic acid or mixed castor oil fatty acids react with gallic acid using a tetraisopropyl titanate catalyst to give a castor oil fatty acid adduct from which an amide hardner is made using diethylenetetramine (DETA). This hardner can be mixed with Epikote 1001 epoxide resin coating compound to give an anticorrosive film-forming coating composition. The adduct can also be used with commercial epoxy resin EPN 740 with triethylamine in mixed solvents. This will produce a thick anticorrosive film on rusted steel sheet with good salt-water resistance.

Several water-soluble and water-insoluble salts can be prepared from ricinoleic acid. Sodium, potassium and ammonium salts or soaps are water-soluble, and function as mild germicides and improve the action of disinfectants in formulations. Aluminium, lithium, and calcium salts are some of the water-insoluble salts used in industry. Barium, cadmium, and zinc salts are occasionally produced and used. Sodium salt of ricinoleic acid is white or slightly yellow, odourless or almost odourless powder. It is soluble in water or alcohol, and the aqueous solution is alkaline. Sodium salt of ricinoleic acid is used in soft brown soaps and transparent soaps. Solutions of castor oil soap, i.e. sodium ricinoleate, are used as injections in sclerosing treatment of varicose veins. Sodium ricinoleate and sulphoricinoleate have strong bactericidal action and are used in tooth formulations.

Soaps

On saponification with alkalies, castor oil forms soap, i.e. sodium (or potassium) ricinoleate (and glycerol). The castor oil soap has poor detergent qualities and is easily water soluble. Washing with castor soap is reported to be impart a shining silky appearance to jute fibres. The oil is used to a limited extent for making transparent soaps. Castor soap finds use as an emulsifying agent and sometimes as a dispersing and wetting agent in cotton finishing, preparation of insecticidal compositions, air deodourants, mining operations and for dispersion of pigments in paints. In these applications TRO can replace castor soap to a large extent. The potash soap of castor oil is used in shampoos. Emulsified non-creosotic pine oil disinfectant compositions were prepared using less than 7 per cent of an emulsifier mixture and soaps of castor oil and a mixture of soaps of tall oil. Metal soaps of castor oil fatty acids are used as thickeners in greases as gelling agents in concentrations of 8 to 25 wt. per cent. Lithium, calcium, sodium, aluminium and barium are the commonly used soaps.

Methyl Ricinoleate

Castor oil, on alcoholysis with methanol, gives methyl ester or ricinoleic acid (and glycerol). This methyl ricinoleate may be considered as a primary derivative of castor oil. The specifications for methyl ricinoleate are given in Table. 7.26.

Table 7.26. Specifications for methyl ricinoleate.

Property	*Value*
Description	Less viscous pale yellow oil
Colour	Yellow tinge
Specific gravity at 25°C	0.9248
Refractive index	1.4597–1.4610
Acid value, max.	4
Iodine value	85–90
Hydroxyl value	155–165

Methyl ricinoleate is mostly used as intermediate in processing of castor oil and other chemicals. It is used in paints, plasticisers and lubricants. Methyl ricinoleate is an excellent wetting and foam control agent. It is also used to make various perfumery chemicals and additives. Methyl ricinoleate is converted to acetyl and butyl esters which are used as plasticisers. Acetyl methyl ricinoleate has the following properties (Table 7.27).

Table 7.27. Properties of acetyl methyl ricinoleate.

Property	*Value*
Colour	Reddish yellow tinge
Acid value	3–6
Hydroxyl value	0.8
Saponification value	290–310

Methyl ricinoleate and *n*-butyl acetyl ricinoleate are used as plasticisers. Methyl acetyl ricinoleate is used as plasticiser in nitrile rubber. Catalysts and conditions have been established for making ricinoleyl alcohol from methyl ricinoleate.

Ricinoleyl Alcohol

Hydrogenation of ricinoleic acid with copper-cadmium catalyst at 220°C and 3800 psi yields 70 per cent ricinoleyl alcohol with wax esters and fission products. It can be prepared from methyl ricinoleate. Ricinoleyl alcohol made by sodium or hydrogen reduction, is an excellent wetting agent about 1.5 times as efficient as lauryl sulphate. Its foaming, suspending, and surface tension-lowering powers are, however, less. Ricinoleyl alcohol is a superior raw material for useful anionic and non-ionic surfactants, and is a unique diol. The disulphate salt of ricionleyl alcohol has good wetting, detersive, and emulsifying properties. The ethyoxylate alcohol is an excellent emulsifier.

Glycerol Monoricinoleate

When castor oil is reacted with excess quantity of glycerol, the triglyceride is converted to monoglyceride, i.e. glycerol monoricinoleate. This product has several applications. Monoglycerides of castor oil can be used in plasticisers, detergents and lubricants.

12-Hydroxy Stearic Acid

Flakes of hydrogenated castor oil are subjected to complete saponification and the soap is acidulated to give 12-hydroxy stearic acid and glycerine. Specifications for 12-hydroxy stearic acid are given in Table 7.28.

Table 7.28. Specifications for 12-hydroxy stearic acid.

Property	*Value*
Description	Yellowish or cream coloured flakes
Colour, 1 in. cell	20Y + 3.5R
Slip point, °C	70
Acid value, min.	175
Hydroxyl value, min.	150
Iodine value, max.	5

12-Hydroxy stearic acid is used in high quality greases, in surface coatings and as plasticiser. Its most important use is in lithium-based greases because of the improved stability during usage under shear and water resistance. It is also used in glycerol-free lithium and calcium multipurpose greases. It is also used in the pharmaceutical industry for ointment manufacture and in cosmetics. It is a good replacement for natural hard waxes. Esters of 12-hydroxy stearic acid after acetylation are used as lubricant additives, textile processing aids, and as plasticisers for vinyls. The glyceryl esters, methyl and stearyl esters and even partial esters of polyols are used as waxes. It is possible to get hydroxy waxes of desired melting point. These have limited compatibility with paraffins and are used as wax modifiers in cosmetics (lipsticks and hair setting preparations) and stiffening agents for other wax formulations and pharmaceuticals. Amides made from 12-hydroxy stearic acid are used in various wax formulations. Emulsifiers are obtained by ethylene oxide condensation of the acid or its esters. These ethoxylates have lower solubility than the corresponding surfactants obtained from ricinoleic acid. 12-Hydroxy stearic acid ethoxylates find use as processing aids for polymers particularly for natural and synthetic rubber.

Methyl 12-hydroxy Stearate

It is made by complete hydrogenation of methyl ricinoleate. The specifications for methyl 12-hydroxy stearate are given in Table 7.29.

Table 7.29. Specifications for methyl 12-hydroxy stearate.

Property	*Value*
Description	Yellowish solid
Colour, 1 in. cell	1Y + 1.5R
Slip point, °C	45–50
Acid value	2–5
Iodine value, max.	5
Hydroxyl value, min.	150

Most of the uses of methyl 12-hydroxy stearate are the same as that of 12-hydroxy stearic acid.

Azelaic Acid

Disruptive oxidation of ricinoleic acid produces azelaic acid. Azelaic acid is obtained in yield upto 35 per cent from the crude ricinoleic acid. The acid is dissolved in aqueous sodium hydroxide solution containing a 50 per cent theoretical excess of alkali which is then treated with an aqueous (4 molar excess) potassium permanganate solution. This reaction is exothermic, but is readily kept under control. The resulting potassium azelate solution is then acidified, filtered to remove manganese dioxide, concentrated and finally the acid product is recovered. Potentially a more efficient industrial synthesis of this acid may be the conversion of 1,7-heptanediol with carbon monoxide under heat and pressure to azelaic acid. Yields for similar acids average about 80 per cent by this process. Other alternate methods for the preparation of this acid include nitric acid oxidation of ricinoleic acid, ozonation of oleic acid and decomposition of the ozonide, dichromate or sulphuric acid oxidation of dihydroxystearic acid. However, today, it is produced from oleic acid by oxidation.

Azelaic acid has monoclinic prismatic needles. It distils above 360°C with partial anhydride formation. The properties of azelaic acid are listed in Table 7.30.

Table 7.30. Properties of azelaic acid.

Property	*Value*
Melting point, °C	106.5
Boiling point at 100 mm Hg, °C	286.5
Specific gravity at 110.6 °C	1.0291

One litre of water dissolves one gram of azelaic acid at 1°C, and 2.4 gram at 20°C. It is freely soluble in boiling water, and in alcohol. One kg of ether dissolves 26.8 gm at 15°C. Various esters of azelaic acid are used in many different applications. Azelaic acid is also known to have antiacne properties. The simple esters of azelaic acid like dioctyl, etc. find applications as plasticisers, especially for PVC resins. The branched chain esters (isoamyl, 2-ethylhexyl, isooctyl) are employed as lubricants in jet engines and other military applications. The diesters serve as bases for special greases, particularly when jelled with a lithium soap (lithium stearate or 12-hydroxy stearate) and as diffusion pump oils (dioctyl, isooctyl, etc.). Linear polymers of azelaic acid and glycerol are useful plasticisers for PVC resins, rubber, nitrocellulose, and other resins.

Azelaic acid is used in hydraulic fluids and lubricants as a complex ester condensate with glycols and branched-chain alcohols. Aluminium and lithium salts of azelaic acid and stearic acid are used to thicken siloxane lubricating greases. Azelate esters, mixed with diphenylsilanols are incorporated into high performance lubricants. *Bis* (3,7-dimethyloctyl) azelate has been used as a diffusion-pump oil. *Bis* (2-butoxyethyl) azelate is used to plasticise vinyl resins, similarly the *bis* (2-ethylbutyl) and *bis* (2-ethylhexyl) esters are used in lubricant blends as plasticisers in GR-S rubber. Azelaic acid derivatives are used widely in the plastics and synthetic fibre industries. With hexamethylenediamine, the acid gives filaments when extruded into solutions of isocyanates in silicone oils. In the presence of an acid catalyst (HCl), polymers of diethyl sebacate or diethyl azelate will undergo addition co-polymerisation with poly (methyl methacrylate). The dihydrazide of azelaic acid is used as a curing agent for epoxy resins.

Undecylenic Acid

Castor oil, when subjected to pyrolysis or destructive distillation at temperatures of over 450°C under vacuum, undergoes decomposition to yield mainly undecylenic acid and heptaldehyde, in roughly equal quantities (20 per cent). In this case, the scission is principally between the 11th and 12th carbon atoms of the fatty acid chain. The presence of water vapour has been found to enhance the yields of both acid and aldehyde. The spongy mass left behind, consisting principally of polymerised undecylenic acid can be used in varnishes, oil cloth and insulating tapes or for further regeneration of the acid. Continuous processes involving the passage of castor oil through variously filled heated tubes or impinging the oil against molten metallic surfaces have been developed for the preparation of undecylenic acid in increased yields (about 36 per cent). Castor oil is converted into methyl ricinoleate which is thermally cracked into methyl undecenate and heptanal. The specifications for pharmaceutical grade undecylenic acid are given in Table 7.31.

Table 7.31. Specifications for undecylenic acid.

Property	*Value*
Description	Oily liquid
Purity, min. %	98
Colour	Water-white to pale yellow
Freezing point, °C	21–23
Refractive index at 25°C	1.4880
Acid value	295–304

Undecylenic acid is also called as 10-undecylenic acid, undecenoic acid or 10-hendecenoic acid (112–38–9). It is a colourless crystalline mass or a colourless to yellowish liquid, insoluble in water, slightly soluble in chloroform, and miscible in all proportions with ethanol and diethyl ether. The acid has been used for many years as an intermediate in the manufacture of perfume ingredients, especially the alcohol obtained by hydrogenation of ethyl and butyl undecylenate and various derived ketones. Undecylenic acid is a useful starting material for the preparation of other acids and related compounds. On oxidative fission, undecylenic acid yields sebacic acid.

Undecylenic acid and its esters are used in perfumes and essences. Undecylenic acid is the parent substance for the synthesis of higher alcohols and aldehydes, especially those containing nine and eleven carbon atoms. Nonylic acid, nonylic alcohol and nonylic aldehyde can be obtained from undecylenic acid or heptaldehyde. Nonylic aldehyde is a natural constituent of rose and orange oils. Undecylenic acid can be used to make *n*-decylaldehyde which has a spicy orange-like odour. Reaction of undecylenic acid with sodium and alcohol converts it into undecylenic alcohol which has a floral odour with a persisting fatty note. Reduction of undecylenic acid by free hydrogen in the presence of colloidal palladium as catalyst saturates the double bond producing undecylic acid which can be converted to undecylic alcohol which is a valuable modifier and a top note compound. Undecylic aldehyde is also an equally important perfumery chemical. The aldehyde of undecylenic acid, undecylenic aldehyde, is used in modifying floral odours. Gamma-undecalactone which is known as C14 aldehyde, in combination with light fruity esters is a part of citrus note. Gamma-undecalactone is a component of peach flavour for candies. The lactone has a powerful and persistent odour of peaches. Esters of undecylenic acid such as ethyl ester are useful to a limited extent as flavouring agents. The more important ester of undecylenic acid to the perfumery chemist is the allyl ester which has a powerful odour of quinches and is used as a modifier. Undecylenic acid is widely used in France for making nylon-11 and is called as Rilsan.

The 11-carbon atom undecylenic acid is found to give optimum antimycotic effect. Undecylenic acid and its salts have unusually good fungistatic activity, particularly the sodium and zinc salts. The latter salt has been widely used in the treatment of skin dermatophytoses. Zinc and calcium salts of undecylenic acid are used in topical ointments. Zinc undecylenate is also used in combination with several drugs in various ointments. Zinc undecylenate and other salts have been in use for many years in topical treatment of smooth skin ringworm infections. One popular preparation, desenex containing 5 per cent of undecylenic acid and 20 per cent of the zinc salt, is available in several forms. Preparations of undecylenic acid are of value primarily in the treatment of hair or nail infections. Some foot powders are combination of zinc undecylenate, boric acid, hexachlorophane and salicylic acid. Claims have also been made that mercury, aluminium and bismuth salts also have valuable pharmaceutical applications. Undecylenic acid reduces the local pH on the skin to levels unfavourable to dermatophyte growth. It is the most widely used acid as dermatotherapeutic agent. However, it is effective only in 50 per cent of the patients. Undecylenic acid is one of the topical antimycotics with activity on dermatophytes and yeasts. Zinc undecylenate is produced by reacting undecylenic acid with zinc carbonate at 90°C. The specifications for zinc undecylenate are given in Table 7.32.

Table 7.32. Specifications for zinc undecylenate.

Property	*Value*
Description	Fine white powder
Solubility	Insoluble in water and alcohol
Alkalinity	Passes as per BP 81
Sulphate alkalinity, %, Max.	1.5
Loss on drying, %	98–101
Melting point, °C	117–121

Undecylenic acid (and salicylic acid) is used exclusively as preservatives for pharmaceutical products. Due to its easy conversion to esters, aldehydes, alcohols, containing 9 and 11 carbon atoms, undecylenic acid finds uses in pharmaceuticals, insecticides, perfumery chemicals. Undecylenic acid is used to make N-isobutyl undecylene amide, a valuable synergist which can be used with pyrethrum. Chlorine and sulphur derivatives of undecylenic acid are used as additives in the lubricant industry. Undecyl dodecyl phthalate (UDP), diundecyl phthalate (DUP), diisoundecyl phthalate (DIUP) exhibit superior high temperature performance as plasticisers, but they are required in more quantities.

Nylon 11

Nylon-11 (255–80–8) is a rarity in that its manufacture does not involve petroleum-based feedstocks. Nylon-11 is poly(w-undecanamide), i.e. a polymer of 11-aminoundecanoic acid. Nylon-11 was first manufactured commercially in France in 1955. Undecylenic acid is treated with hydrogen bromide and ammonia to form w-aminoundecanoic acid. The synthesis of nylon-11 is based on a five-step process from castor oil to 11-aminoundecanoic acid. A 30–40 per cent aqueous dispersion of the acid can be polycondensed in three steps, and the nylon-11 formed can be isolated and melt-spun. Addition of phosphoric or phosphorus acid facilitates spinning by raising the melt viscosity and limiting side-chain growth. Nylon-11 is made by a continuous process which is operated under inert gas at atmospheric pressure. The monomer, 11-aminoundecanoic acid, is suspended in water and mixed with the desired additives (antioxidants, stabilisers, etc.) before charging to a three-zone upright converter. The monomer is melted, water eliminated, and the polymerisation begun in the top section. The low polymer flows through a screw feeder to a tall middle section of the reactor. Near completion of the condensation occurs in the middle zone. Adjustment to the target molecular mass takes place in the bottom section which provides a melt pool for uniform discharge of the product. The product is formed into chips as it is withdrawn from the bottom of the reactor. The good thermal stability of the polymer and the low tendency of the monomer for cyclisation simplify the process design. Phosphoric acid catalyses the polymerisation as does hypophosphoric acid. The latter can, however, insolubilise the product, if used in excess with prolonged heating. The reaction goes essentially to completion, i.e. no residual monomer remains. Extrusion of the product into pellets occurs under inert gas to prevent oxidation and is similar to that of other pelletising operations. The polymer and the monomer have the same melting point, i.e. 189°C. High molecular weight nylon-11 is prepared in about 3 hours by melt condensation of w-aminoundecanoic acid, under nitrogen at 215°C. Extractable content is low (about 0.5 per cent). Typical properties of nylon-11 are shown in Table 7.33.

Table 7.33. Properties of nylon-11.

Property	*Value*
Sp. gravity of non-oriented polymer, gm/cc	1.10
Sp. gravity of dry as moulded	1.04
Water absorption at saturation, %	1.6
24 hours	0.3
Equilibrium at 50% RH	0.8
Softening temperature, °C	175
Melting temperature, °C	189
Brittleness point, °C	–70
Tensile strength, MPa*	54
Ultimate tensile strength, MPa	58.8–78.4
Yield stress, MPa*	34
Elasticity modulus, MPa	98–196
Elongation, %	90–120
Elongation at break of dry as moulded, %	200
Elongation at break, %*	330
Elongation at yield, %*	22
Modulus, MPa*	1730
Flexural modulus, kPa	1.2 million
Rockwell hardness, R scale	108
Surface hardness, Brinell (25 kg load), MPa	45.6
Izod impact strength, J/m of notch	40
Compression strength, MPa	1.07
Deflection temperature under load, °C	
At 455 KPa	150
At 1850 KPa	55
Volume resistivity, ohm-cm	10 exp. 14
Dissipation factor, MHz*	0.03
Dielectric constant, at MHz*	4.0
Dielectric constant, at MHz	3.1–3.2
Dielectric constant at 1000 Hz	3.7
Dielectric constant at 60 Hz	3.7

(Contd...)

Table 7.33. Properties of nylon-11. *(Contd.)*

Property	*Value*
Dielectric strength, kv/mm-short time	16.7
Tangent of the dielectric loss angle at MHz	0.02–0.03
Specific heat, J/(g.K)	1.26
Thermal conductivity, W/(m.K)	0.19
Coefficient of linear thermal expansion, per °C.	9.9# 10 exp. -5

Note: To convert MPa to psi, multiple by 45.
To convert kPa to psi, multiply by 0.145.
To convert J/m to ft. lbf/in. divide by 53.38.
To convert J to cal. divide by 4.184.
* Properties of nylon-11 conditioned at 23 °C and 50% rh.

The polyamide is hydrophobic and provides excellent insulating properties. As expected from the longer hydrocarbon chain, nylon-11 shows hardly any sensitivity or dimensional change in the presence of moisture. The fibres are too expensive for general use and are restricted to a few special applications in luggage, ropes, transmission belts, and bristles.

Principal uses include pressure mouldings and fibres. Properties of nylon-12 (24937–16–4) closely correspond to those of nylon-11 and it competes with nylon-11 in many applications. The recommended maximum moisture content for processing nylon-11 are: injection moulding 0.10 per cent, extrusion (tubing and film) 0.10 per cent. The water vapour transmission rates of barrier films made from selected polymers are compared in Table 7.34.

Table 7.34. Water vapour transmission rates of selected polymers.

Polymer	*WVTR,* nmol/(m.s)*
High density polyethylene (HDPE)	0.095
Polypropylene	0.16
Low density polyethylene (LDPE)	0.35
Polyethylene terephthalate (PET)	0.45
Polyvinyl chloride (PVC)	0.55
Nylon-6,6 and nylon-11	0.95
Polystyrene	1.8
Nylon-6	2.7
Nylon-12	15.9

Determined at 38°C and 90% rh.
To convert nmol/(m.s) to g.mil/100 sq. inch d multiply by 3.95 to g.cm/sq.m.d multiply by 0.155.

Nylon-11 and nylon-12 are used for making films with lower water vapour permeability. Nylon-11 and nylon-12 are available as film and sheeting material with thickness of 0.025–1.27 mm and 0.01–1.52 mm, respectively. Film heat sealing range is 166–199°C and 160–193°C, respectively. Continuous service may be carried out at 120 and 110°C, respectively. The gas permeability properties of nylon-11 are compared with nylon 6 and nylon 6.6 in Table 7.35.

Table 7.35. Permeability properties of nylons.

Polymer	*Permeability rates* O_2*	CO_2 *	H_2O#
Nylon-6, nylon-66	20	35	700
Nylon–11	102	252	180

*: At 23°C (100% rh), (cu.mm/sq.m.d.PPa).
To convert PPa to bar, multiply by 10 exp. 10.
#: kg.cm/sq.km.day, at 38°C(100% rh).

Nylon-11 is useful in making bristles. Monofilament, fibres and fabrics made out of nylon-11 are reported to have a 'hand' (feel) and shade similar to that of silk. Hence, it can directly replace raw silk. Nylon-11 is an industrial fabric and is used in filter cloth, bags netting, flexible fishing lines and varieties of brush applications where toughness and durability are required. It has applications in insulators (transformer, coil formers), gears (adding machines, computers), bearings and rollers in conveyor systems, stirrers, gaskets and pipes in the chemical industry. Flat and tubular film is used in packaging demanding superior quality nylon. Glass-reinforced nylon-11 provides an appreciable increase in rigidity and improvement in compressive strength and resistance to cold flow under load. The use of thermoplastic nylon-11 for powder coatings or dry coatings that require no solvents has been developed in response to a growing concern for the environment. These coatings find use where wear, abrasion, corrosion and impact resistance are required.

Nylon-11 is used as an engineering plastic in the automotive and transport industry in the form of fuel lines. Products are also used for extruded and moulded components of fuel systems such as filler necks, gas tanks, reservoir modules, filters, fuel rails and vapour recovery systems. Air brake hoses on large trailer truck transports are made of nylon-11. Nylon-11 tubing is used in the compressed air industry in pneumatic control systems.

Nylon-11 offers high dimensional stability, close moulding tolerance, high abrasion and impact resistance, excellent dielectric strength, low moisture absorption, electrical insulating properties, and overall chemical resistance and durability. Coating powders based on polyamide resins have been in fusion coating processes for a long time. Nylon-11 has been used almost exclusively. However, more recently, coating powders have also been sold based on nylon-12. The properties of these resins are quite similar. Nylon-6 and nylon-6,6 are not used because the melt viscosities are too high.

Polyamide powders are prepared by both the melt-mixed and dry-blend processes. In the latter, the resin is ground to a fine powder and the pigments are mixed in with a high intensity mixer. Melt–mixed powders have higher gloss, e.g. 70–90 per cent on the Gardner 60° gloss metre, whereas the dry blended powders have a gloss in the range of 40–70 per cent. Because the pigment is not dispersed in the resin in the dry blend process it must be used at very low concentrations, usually less than 5 per cent. Even in melt-mixed formulations, the concentration of pigment and fillers seldom exceeds about 30 per cent.

Nylon coating powders are available for electrostatic spray and fluidised bed application. Nylon coatings are very tough, resistant to scratching and marring, have a pleasing appearance, and are suitable for food contact applications when properly formulated. These coatings are used for chair bases, hospital furniture, office equipment, knobs, handles and other hardware. Because of expense, nylon is generally applied only to premium items. Nylon coatings have good solvent and chemical resistance and are used for dishwasher baskets, food trays, hot waterheaters, plating and chemical etching racks, and large diameter water pipes in power generating stations. For maximum performance of these power coatings, a primer is used. Most nylon primers are unpigmented resin solutions applied at only about 10 per cent solids. A dry film thickness of about 3–5 micrometer is obtained. If nylon is applied by electrostatic spray techniques, it is necessary to prime the parts. Although the exact reason is not known, it is speculated that either the primer partially insulates the coating from the substrate and prevents the charge in the electrostatically applied powder layer from leaking off too quickly, or the primer is tacky at the fusion temperature and causes the coating to adhere to the substrate. The physical and coating properties of some thermoplastic powders are compared in Table 7.36.

Table 7.36. Physical and coating properties of thermoplastic powders.

Property	*Vinyls*	*Polyamides*	*Polyethylene*
Melting point, °C	130–150	186	120–130
Pre-heat/post-heat temps. °C	290–230	310–250	230–200
Sp. gravity	1.2–1.35	1.01–1.15	0.91–1.00
Adhesion	G–E	E	G
Surface appearance	Smooth	Smooth	OP
Gloss gardner 60°	40–90	20–95	60–80
Hardness, Shore D	30–55	70–80	30–50
Resistance			
Impact	E	E	G–E
Salt spray	G	E	F–G
Weathering	G	G	P
Humidity	E	E	G
Inorganic dilute acid	E	F	E
Inorganic dilute alkali	E	E	E
Inorganic dilute solvent	F	E	G

Note: All powders require a primer and pass the flexibility test, which means no cracking under a 3mm dia mandrel bend. E = Excellent; G = Good; F = Fair; P = Poor; and OP = Orange peel effect.

Undecylene Aldehyde

Undecylene aldehyde (112–45–8) or 10-undecenal or undecylenic aldehyde is the most important unsaturated C11 aldehyde. It is not known to occur naturally. Undecylene aldehyde is a colourless liquid with a stiffling, farry green, slightly metallic heavy-flowery odour. Undecylene aldehyde is prepared by Rosenmund reduction (hydrogenation) of undecylenoyl chloride which in turn is obtained by cleavage of castor oil. A further possibility is the reaction of undecylenic acid with formic acid in the vapour phase over a titanium dioxide catalyst. The aldehyde is an important modulating agent in the perfume industry. In perfumery, 10-undecenal is one of the aldehydes essential for creating the 'aldehydes top note'.

Undecyl Alcohol

The correct technical name of undecyl alcohol is 10-undecen-1-ol (112–43–6). This straight-chain primary alcohol has not not yet been found in nature. It is a colourless liquid with a fatty green, slightly citrus-like odour. It can be synthesised from 10-undecylenic acid by hydrogenation. It is used to give flower perfumes a fresh note.

Undecanedioic Acid

It is a C11 dicarboxylic acid without any unsaturation. Undecanedioic acid is formed by catalytic oxidation of castor oil or by auto-oxidation of methyl oleate. It can also be formed by the action of sodium hydroxide and calcium oxide on 12-hydroxyoctadecanoic acid, i.e. 12-hydroxy stearic acid. Undecanedioic acid has not proved important in large scale industrial products but some unusual applications have aroused interest. Undecanedioic acid (as well as other dicarboxylic acids having three to ten carbon atoms) has shown complete suppression of ascites forms of lymphosarcoma in animals, and is under investigation for treatment of various cancers. Its esters with aromatic alcohols have potential in the plasticiser and lubricant fields. 6-amino-2-octylundecanedioic acid has been used for polyester investigation work, and polyester amides have been prepared from the amino alcohol salts of this acid.

Heptaldehyde

Heptaldehyde is a seven-carbon atom straight chain aldehyde. It is also called as heptanal or enanthal. This aldehyde occurs naturally in ginger oil. It is formed as a by-product during the pyrolysis of castor oil to get undecylenic acid. It is also obtained by the hydroformylation of 1-hexene in the presence of modified rhodium catalysts. In this case, heptanal is formed almost exclusively. A minor by-product is 2-methylhexanal, which can be prepared specifically by reaction of hexanal with formaldehyde followed by hydrogenation. Heptaldehyde, like all the lower C1-18 aldehydes has a pungent, penetrating and unpleasant odour, some of which may be attributed to the presence of the heptanoic acid that is formed by oxidation. The specifications for heptaldehyde are given in Table 7.37.

Table 7.37. Specifications for heptaldehyde.

Property	*Value*
Description	Pungent smelling oily liquid
Colour	Colourless to pale yellow
Acid content, max. %	5
Aldehyde content. min.	90–95
Refractive index at 25°C	1.415
Specific gravity at 25°C	0.815
Boiling point, °C	152–154

Heptaldehyde is a starting material for many perfumery products. It is used principally for the preparation of alpha-amyl cinnamic aldehyde, a popular fragrance chemical especially in the field of soap perfumery. This aldehyde is also called as jasmine aldehyde. This is useful in artificial jasmine compounds where it gives the basic note previously attainable only by the use of jasmine absolute.

Heptaldehyde is also used to make heptyl alcohol and its esters. Heptaldehyde is a starting material for heptanoic acid. In perfumery synthetics, heptaldehyde is used for the synthesis of nonylenic acid series, the enanthic (heptanoic) acid series, and the 'Verts de Violette'. Nonylenic acid is converted to 8-*n*-amyl butyrolactone, also called as coconut aldehyde due to its persistent and remarkable coconut smell. In the enanthic acid series, methyle-*n*-heptyl ketone and the ester of 3-noninic acid are the two compounds obtained from heptaldehyde. Methyle-*n*-heptyl ketone (iso-decylic ketone) is a natural constituent of oil of rue and is used in violet perfumes. It is also used to impart a fresh green note to aromatic blends. It is made from heptaldehyde by a series of reactions wherein the aldehyde is converted successively into enanthic acid, heptyl chloride, heptylacetoacetic ester and finally to methyl-*n*-heptyl ketone. The acetoacetic ester can also be converted to 3-noninic acid. The two of the most important groups of Verts de Violette are the esters of heptine and octine carboxylic acids. They belong to the group of acetylenic esters used in the manufacture of violet perfumes due to the smell of fresh violet leaves they impart on dilution. Knoevenagel condensation of malonic acid with heptaldehyde followed by ring closure, gives gamma-nonanoic lactone, which is a fragance material. Heptaldehyde is also useful for making vulcanisation accelerators by condensation with ammonia, aniline or amines, and in plastics. Heptaldehyde is mostly used in rubber and perfumery chemicals.

Heptanoic Acid

Heptanoic acid is the seven-carbon atom straight-chain fatty acid. It is prepared from heptaldehyde by oxidation. Heptanoic acid and pelargonic acid are made commercially be carboxylation/oxidation of straight-chain alpha-olefins. Heptanoic acid and nonanoic acid (pelargonic acid) are currently derived from natural fats and oils through non-conventional technology. Heptanoic acid, also called as enanthic or heptylic acid, is a by-product of nylon-11manufacture. It is produced in France by the oxidation of heptaldehyde which results from the destructive distillation of the sodium or calcium salts of ricinoleic acid or castor oil. A potential source of heptanoic acid is by oxidation of isooctanol (2-octanol)produced in the manufacture of sebacic acid by the pyrolysis of castor oil with excess alkali. The specifications of heptanoic acid are given in Table. 7.38.

Table 7.38. Specifications for heptanoic acid.

Property	*Value*
Description	Oily liquid
Odour	Rancid, disagreeable
Colour	Colourless to pale yellow
Density	0.9090–0.9099
Refractive index	1.4216
Acid value, min.	410
Purity, %	95–97

Esters of heptanoic acid have applications in lubricants.

Sebacic Acid

Sebacic acid is the common name for decanedioic acid, HOOC (CH_2) 8COOH, having CAS registry No. (110–20–6). It is also referred to as 1,8-octane-dicarboxylic acid. Castor oil when subjected to alkali fusion produces sebacic acid and secondary octyl alcohol (2-octanol). The cleavage takes place between the 10th and 11th carbon atoms of the fatty acid chain. The caustic alkali may be used in solid form, in aqueous solution, or suspended in a high-boiling organic liquid such as white mineral oil. With aqueous caustic, batch operations in an autoclave are often used in pressurised operations. In a related process, castor oil is slowly added to a caustic solution at 245°C in a kettle connected to a distilling column. The chief by-product, 2-octanol, passes off with some water vapour into a condensing system. The hydrogen gas evolved is vented and a separator returns the water to the reaction mixture, thereby maintaining a constant boiling point. After the run, the residue containing the alkali salt of sebacic acid is removed and dispersed in water. On acidification of the medium, sebacic acid is obtained as a fine white precipitate. Yields upto 40 per cent of the castor oil input are obtained.

The use of lead is claimed to increase yields and the presence of barium is said to provide a process which enables the recovery of sebacic acid without neutralisation of the fusion bath. Yields as high as 74 per cent of sebacic acid and 72 per cent of the alcohol on the weight of the castor oil can be obtained by improved processes.

Commercially, sebacic acid is produced by caustic oxidation of castor oil or ricinoleic acid in either batch or continuous process. The castor oil or ricinoleic acid and caustic are fed to a reactor (usually monel or nickel) at a temperature of 180–270°C where the ricinoleic acid undergoes a series of reactions with evolution of hydrogen to give di-sodium sebacate and capryl alcohol (2-octanol). When the reaction is complete, the soaps are dissolved in water and acidified to a pH of about 6. At this pH, the soaps of the monobasic acids (C16 to C20 plus dimer acids) are converted to free acids that are insoluble in water. The disodium sebacate is only partially neutralised to the half acid salt which is water soluble. The oil and aqueous layers are separated. The aqueous layer containing the half salt is acidulated to a pH of about 2, causing the resulting sebacic acid to precipitate from the solution. It is then filtered, water-washed and finally dried.

A number of process improvements have been described, and include the use of white mineral oil having boiling range of 300–400°C or the use of a mixture of cresols. These materials act to reduce the reaction mixture's viscosity, thus improving mixing. Higher sebacic acid yields are claimed by use of catalysts such as barium salts, cadmium salts, lead oxide and salts. An electro-oxidation process was developed by Asahi Chemical Co. and also piloted by BASF.

It produces high purity sebacic acid from readily available adipic acid in three steps. Due to fluctuations in the prices of castor oil, there has been considerable interest in the electrochemical route to sebacic acid based on adipic acid. The electrochemical step involves the Kolbe-type or Brown-Walker reaction where anodic coupling of the monomethyl ester of adipic acid forms dimethyl sebacate. This diester is then hydrolysed to give sebacic acid.

BASF, Asahi Chemical Industry and a CIS group have carried out pilot plant studies of the electrochemical route to sebacic acid. Yields of dimethyl sebacate reported for the BASF and CIS processes are in the range of 80–85 per cent. Asahi claims product yields as high as 92 per cent at current efficiencies in the range of 85–90 per cent.

The use of high electrolyte velocities (3 m/s) through the cell is evidently the key to achieving these superior results. A 20 wt per cent aqueous solution of monomethyl adipate having 30 mole per 100 mole of the organic acidity neutralised with sodium hydroxide, comprises the electrolyte and cell temperature is 55°C. A bipolar plate and frame cell with a platinum-plated titanium anode and a steel cathode is used in the Asahi process.

Typically, the cells are operated at a current density of 20 A/dm^2 and 14 V. The electrode spacing is 2 mm. The economic study by Chem Systems showed a slightly better than 10 per cent cost advantage of the electrochemical route over the conventional process from castor oil in an 11000 tpa plant. In this electrochemical process for sebacic acid, the product separation and refining dominate the cost model. Here the demands of the cell require the half ester which has to be recycled.

This technology was demonstrated by Asahi at a 100 tpa scale. As of 1990, Asahi had not commercialised the process. It is reported that sebacic acid has been produced by the electrochemical route on a commercial scale in the former USSR. Other methods used to manufacture sebacic acid are oxidation of stearic acid by N_2O_4, oxycarbonylation of butadiene, carbonation of disodiooctodiene, oxidation of 1,10-decanediol and various fermentation processes. In some cases, butadiene is also used as feedstock. Sebacic acid can also be obtained by distillation of oleic acid.

A mixture of sebacic acid and other dicarboxylic acids are formed during ozonolysis of palmitic acid. Bicyclo-decane, i.e. decalin, is obtained by catalytic hydrogenation of naphthalene. Liquid phase oxidation of decalin with oxygen gives the corresponding tertiary hydroperoxide, which is eventually converted into 1,10-decanedicarboxylic acid i.e. sebacic acid, by a four-step sequence. The specifications for sebacic acid are given in Table 7.39.

Table 7.39. Specifications for sebacic acid.

Property	*Value*
Description	White powder
Purity, %. min.	98
Colour, APHA(45 in SDA, 3-A-190) max.	30
Melting point, °C, min.	12
Moisture, % by wt.	0.3
Ash, ppm by wt. max.	1000
Acid value	544
Iodine value (Wijs), max.	0.2

Commercial sebacic acid is sold in two grades. Their specifications and typical analysis is shown in Table 7.40.

Table 7.40. Sebacic acid : specifications and typical analysis.

Property	*Purified grade*		*High purity grade*	
	Specifin.	*Analysis*	*Specifn.*	*Analysis*
Appearance	Small white beads		Small white beads	
Acidity, % min. as sebacic	98.0	98.8	99.0	99.3
Acid value, min.	544	548	550	551
Ash, ppm max	1000	200	40	10
Colour, APHA (4% in SDA 3-A-190) max.	50	15	10	5
Moisture, % max.	0.5	0.05	0.5	0.05
Iron, ppm	–	<1.0	–	<1.0
Iodine value	–	<0.5	–	<0.5
GLC Assay (approximately)				
C10 dibasic acid	–	95.0%	–	93–96%
C11 dibasic acid	–	3.5	–	2.5–4.5
C12 dibasic acid	–	1.5	–	1.5–3.0

Now-a-days, three grades viz. purified, CP and nylon, are marketed by Union Camp and are packed in multiwall bags or super sacks. The nylon grade is >95 per cent sebacic acid with low ash content and low colour. In USA, sebacic acid is used for the production of moulded nylon-6, 10 and plasticisers. The nylon is used in moulding powder, gears, textile machinery, oilfield pipes, brush bristles, fibres, monofilaments, extruded and moulded articles and protective coatings for electric wiring. Sebacic acid is also used for making dioctyl sebacate, an important lubricant for jet engines. For low pressure applications, sebacate esters are mixed with metal soaps in many formulations. Sebacic acid is an important intermediate in the manufacture of polyamides. Azelaic and sebacic acids are used in dimer-based polyamides to give higher melting resins along with higher tensile strengths when these properties are desired. These dibasic acids are also used as modifiers in polyethylene terephthalate polymers to obtain specific properties such as melting point, flexibility and improved dye receptivity. Hydroxyl terminated polyesters prepared from C9, C10 and C12 dibasic acids and glycols can be used to prepare polyurethane polymers. These polyurethanes find application in fabric coatings, Spandex fibres, and speciality films. Recently, C9, C10, C12 and C13 dibasic acids have been used in preparing co-polyether-esteramides, which are useful in heat sealing of textiles.

The sebacates of various oxo and straight chain alcohols are important plasticisers for PVC and other film forming polymers. Their main characteristics are high migration resistance and good low-temperature resistance. With polyglycols, sebacic acid forms polyesters such as polypropylene glycol sebacate and extremely stable plasticiser which shows little tendency to migrate from the polymer in which it is incorporated. An example of sebacic acid polyester is paraplex G-25. The esters are also used as diluents. Because of their low toxicity, sebacates are important components of packaging films.

Nitrile rubber compounds contain softeners which aid processing and enhance particular properties e.g. low temperature flexibility. Monomeric esters are superior to other plasticisers for imparting low temperature properties to nitrile rubber. However, certain monomeric plasticisers are better than others in this respect. Apart from sebacates, adipates, tributoxyethyl phosphate monomeric fatty acid ester, methyl acetyl ricinoleate are recommended for low temperature properties. The use of a monomeric plasticiser results in less swelling because the plasticiser is extracted and the rubber is diluted. Sebacic acid and its derivatives have a variety of industrial uses as plasticisers, lubricants, diffusion pump oils, and in the manufacture of alkyd resins, polyamide esters, polyurethanes, urea resin, cosmetics, fruit essences, candles, etc. Dibutyl sebacate is a cosmetically useful lipid and is used as emollient. Ethyl sebacate is used as a flavouring agent. Sebacic acid is used in the synthesis of mascons, i.e. 3-methyl cyclopentadecanone, which is an essential constituent of musk. In alkyd resin applications, sebacic acid possesses inherent flexibility characteristics which render them suitable for many applications that exclude more rigid alkyds, for example, textile coatings and paint and lacquer systems. Sebacic acid and azelaic acid are primarily used in alkyds designed for application as plasticisers.

Azelaic acid, and more particularly sebacic acid, is used as partial or complete replacement of phthalic anhydride in the formulation of oil-modified glycerol or pentaerythritol alkyd resins. Sebacic acid also forms polyamide systems with ethanolamine and terephthalic acid. A polyamide of piperazine and sebacic acid forms a tough water-proof resin which can be used to impregnate textile braid for covering electrical conductors. A patented preparation covers polyamides of piperazine formed from esters of dicarboxylic anhydrides and phosphoric acid. Most of the applications of sebacic acid are same or similar to those of azelaic acid.

Esters of the linear sebacic and azelaic acids are used in PVC applications for the same reason as the corresponding adipates, i.e. low viscosity and good low-temperature performance. Azelates and sebacates are useful primarily due to their low volatility. Azelates and sebacates are considerably less sensitive to water than adipates. Di-2-ethylhexyl sebacate (DOS) and di-2-ethylhexyl azelate (DOZ) are the most widely used. In comparison to adipates, however, these esters tend to have a significantly higher cost and consequently only find use in extreme low-temperature applications. Because of performance-price considerations, the production of adipates with various alkyl moities, both linear and branched, has grown substantially. Di-n-hexyl azelate and dibutyl sebacate are approved for use in plastics directly in contact with food, such as polyvinyl chloride food wraps. Dibenzyl sebacate has found application in buna-N type printing-roll rubbers because of its resistance to printing ink solvents.

Monomeric plasticisers from C9, C10 and C12 are used with other plasticisers, such as dioctyl phthalate, because of cost and compatibility limits. The polymeric plasticisers of C9, C10, and C12 dibasic acids have molecular weights of about 500 to 3000 and are viscous liquids. They are made from difunctional alcohols such as 1,2-propylene or 1,3-butylene glycols. Generally they are less efficient and have inferior low-temperature properties than those of the monomeric plasticisers. Because of their higher molecular weights, the polymerics have better resistance to oils and solvents, lower volatility, and generally less migration than do the monomerics. In some formulations adipic acid is used to reduce cost. Many desired properties can be achieved in the polymerics by proper termination with either a monobasic fatty acid or a monohydric alcohol. Azelaic acid and sebacic acids are used for making thermoplastic polyesters.

Sebacic acid and its derivatives are used in the plastics industry for plasticisers, alkyd resins, polyester resins, polyurethanes, and polyamide moulding resins. The diesters are used as lubricants, and as stable plasticisers for PVC and other film-forming polymers. Well-known esters are methyl, iso-propyl, octyl, iso-octyl, 2-ethylhexyl, nonyl and benzyl.

Polyesters from polyglycols and sebacic acid yield non-migrating plasticisers. Typical glycols that react with the diacids to give polyesters include 1,2-propylene glycol, 1,3-butylene glycol, and 1,4-butanediol. Aliphatic polyesters, unsaturated polyesters and co-polyesters find applications in fibres, films, casting and potting resins, lubricants, resins, adhesives and laminates. The plasticisers offer greater resistance to solvents, lower volatility, and less migration, but are generally less effective and have poor low temperature flex characteristics than those of the corresponding aliphatic monomerics. Azelaic and sebacic acid based polyester fibres and films exhibit excellent low temperature flexibility and high tensile strength.

In lubricants, the most important dicarboxylic acids are sebacic, adipic, and azelaic acids. Of particular importance are the complex esters from dicarboxylic acids and glycols (or polyglycols) and monocarboxylic acids and monoalcohols, respectively. Synthetic lubricants based on azelaic, sebacic and dodecanedioic acids exhibit a number of advantages over conventional petroleum lubricants. The dioctyl, didecyl and ditridecyl esters are some of the more commonly used synthetic lubricants and they find application in aircraft turbine engines. They are used as additives in lubricants and in the preparation of low temperature greases. They have extremely good viscosity/temperature relationships, have high flash points, low viscosity rise during oxidation, excellent lubricity, low coking, low peroxide values, and good additive response.

The lubricant needs of jet engines have been met primarily with diesters derived from adipic, azelaic or sebacic acids that have been esterified with C6-C9 branched chain alcohols. A widely used diester fluid for synthetic lubricants is di(2-ethylhexyl) sebacate which has been employed in formulations of the MIL-L-7808 jet engine oils and other military lubricants. The recent demands and price competition have, however, led to the use of a variety of diesters and their mixtures. Organic esters came into importance as lubricants during World War II. They were first used in Germany in mineral oil blends to improve low temperature properties and to supplement scarce supplies of petroleum. They have been used as jet engine lubricants since the early 1950s and they are now the base for essentially all aircraft jet engine oils. An ester is an organic, oxygen-containing material formed by the reaction of an alcohol and an organic acid. Esters used in lubricants have excellent physical properties compared with an SAE 10 weight grade mineral oil. The ester and the mineral oil have similar viscosities at 100°C. The ester however, has a very low pour point because it contains no wax, so it retains its fluidity at much lower temperatures. It has a much higher viscosity index, which is indicative of its excellent viscosity-temperature characteristics. The ester also has a very low volatility, apparent from the high flash point and low percentage distilled at 400°C.

It readily dissolves most additives and helps retain deposit precursors in solution. This solvency property renders ester-based lubricants unsuitable for use with various compounds commonly found in paints and elastomers. Here, the term ester includes dibasic acid esters, polyol esters, and polyesters. A comparison of physical properties of an organic ester and mineral oil is given in Table 7.41.

Table 7.41. Physical properties of organic ester and mineral oil.

Property	*Ester*	*Mineral oil*
Viscosity, cSt,		
at 100°C	4.6	5.2
at 40°C	21.0	29.5
at –17.8°C	474	Too viscous to measure
at –40°C	5465	Too viscous to measure
Viscosity index	140	102
Pour point, °C	– 57	+20
Flash point, °C	243	218
Distillation, per cent overhead at 400°C	3	20

The performance of mineral oil and a dibasic acid ester as lubricant is compared in the Table 7.42.

Table 7.42. Lubricant performance: mineral oil and dibasic acid ester.

Property	*Mineral oil paraffinic*	*Dibasic acid ester*
Viscosity index	Fair	Good
Low temperature, °F (depends on starting torque)	0 to 25	–40 to –60
High temperature, °F (intermittent service)	250 to 350	375 to 550
Oxidation stability	Fair	Good
Volatility	Fair	Excellent
Hydrolytic stability	Excellent	Fair
Additive response	Excellent	Very good
Compatibility with mineral oil	Excellent	Good
Compatibility with most paints and finishes	Excellent	Poor

Di-2-ethylhexyl sebacates are well-known derivatives used in high performance lubricants including aviation lubricants.

Their advantages are:

1. At very low temperatures, lower viscosity and lower pour point, enable easier engine start.
2. High viscosity index provides good lubrication at high temperatures.
3. Low volatility reduces oil consumption.
4. Longer life due to good thermal stability.

5. Excellent chemical and thermal stability.
6. It is non-toxic.
7. It is non-dermatitic.

Recently, the diesters used in aircraft turbines have been replaced by other compounds. The first fluids, with sebacic acid esters, slowly gave way to esters of adipic acid (synthetically derived from petroleum). This change took place because adipates were less expensive than sebacates. However, there was controversy over the effect on performance. Subsequently, both types of diesters have largely been replaced with esters made from monobasic fatty acids and hindered (no beta-hydrogen) polyhydric alcohols. Dibasic acid ester (diesters) are used as hydraulic fluids. Typical acids and alcohols that are used are adipic, azelaic and sebacic acids, and 2-ethylhexyl, 3,5,5-trimethylhexyl, isodecyl, and tridecyl alcohols. Dibasic acid esters have excellent low temperature fluidity and very low pour points. Viscosity indexes usually are high, some above 140, and the products are shear stable. Generally, their hydrolytic stability is not as good as mineral oils. Diesters have good lubricating properties, good thermal and oxidation stability, and lower volatility than comparably viscous mineral oils. They also have the ability to suspend potential deposit-forming materials so that hot metal surfaces in contact with them remain clean. However, diesters cause more seal-swelling than mineral oils. Also, they may not have good solubility for additives that have been developed for use in mineral oils, although good results are obtained with additives that are developed especially for these fluids. They may affect paints and finishes more than mineral oils.

Sebacic acid diesters are also used in metal cutting fluids. Typically, 95 per cent water and 5 per cent cutting fluid are used. Fatty acids play a major role when reacted with caustic or amines to form soaps that act as anionic emulsifiers. Lower chain length fatty acids such as capric, caprylic and pelargonic, when neutralised with alkanolamines tend to improve ferrous corrosion inhibition properties. Typical synthetics may be comprised of greater than 50 per cent water, alkanolamines, short chain fatty acid soaps such as caprylic or pelargonic, dicarboxylic acids such as sebacic, fatty alkanol amides and proprietary lubricity additives.

Many other applications for the diesters of dibasic acids are known. Du Pont's dibasic ester (DBE) is an effective industrial cleaning solvent and paint stripper. DBE is marketed as an environmentally superior substitute to the chlorinated solvents. Sebacic acid can be used as an alternative raw material to replace adipic acid in the synthesis of nylon-12. Sebacic acid is used in the manufacture of detergents, wetting agents, insecticides and waxes. Polyvinyl butyral resin is used as the interlayer in safety glass laminates. For many years, the plasticiser used almost universally was triethylene glycol *bis* (2-ethyl butyrate). This has been supplanted by adipates, tetraethylene glycol derivatives, dibutyl sebacate, ricinoleates and others. Polymeric diacyl peroxides are prepared from the reaction of dibasic acid chlorides, e.g. succinoyl, fumaryl, sebacoyl and terephthaloyl chlorides, with sodium or hydrogen peroxide. Cyclic monomeric diacyl peroxides can form with suitable dibasic acid chlorides, especially in dilute solutions.

A fatty acid amide called sebacamide can be prepared from sebacic acid. Random co-polymers can be synthesised from sebacamide by step-growth co-polymerisation. Perfluoroalkanedicarboxylic acids have been prepared and are stable compounds, e.g. perfluoroadipic and perfluorosebacic acid. These fluorinated diacids offer a convenient method for introducing a perfluoro moiety into organic molecules. They are of potential interest in the preparation of polyamides and other fluorinated polymers. Many carboxylates and other organic acids cause the precipitation of thorium from solution, but the solids are generally basic, hydrated and of indefinite composition. They can serve for analytical ignition to ThO_2. The oxalate has been particularly useful for this purpose.

Stoichiometric salts of sebacic acid and other acids have been prepared. A commercial product called isosebacic acid is also available. It is produced synthetically from butadiene, and consists of mixture of normal and branched-chain sebacic acids confirming approximately to the composition: 6–10 per cent normal sebacic acids, 12–18 per cent, 2–5-diethyladipic acid and 72–80 per cent,2-ethylsuberic acid.

The product is claimed to have application for the production of vinyl resin plasticisers, polyamide moulding and extrusion plastics and adhesives, alkyd resins, ester lubricants for jet propelled aircraft, polyester used in the manufacture of glass reinforced plastics, polyurethane resins for foamed plastics and synthetic rubber.

Nylon-6,10

Nylon-6,10(9008–66–6) is prepared under conditions similar to those for nylon-6,6. The polyamide nylon-6,10 is obtained by reaction of sebacic acid with hexamethylenediamine. Because of low water absorption and high elasticity, nylon-6,10 is especially suitable for production of bristles. The cost of sebacic acid, produced from castor oil, is fairly high and limits the use of this polymer. However, this polyamide no longer has great industrial significance. The properties of nylon-6,10 are given in Table 7.43.

Table 7.43. Properties of nylon-6,10.

Property	*Value*
Specific gravity	1.09
Water absorption, % 24 hours	0.40
Saturation	3.50
Melting point,°C	222
Tensile strength, kPa	59000
Elongation at break, %	50–200
Drop impact (toughness)	Good
Creep (cold flow)	Moderate
Heat distortion temperature, °C	More than 90

Nylon-6,10 has high toughness and lower moisture absorption than nylon-6 and nylon-6,6 as well as better dimensional stability and electrical properties. Nylon-6,10 is one of the high-barrier polymer. Nylon-6,10 is translucent in nature and has fair processability. Nylon-6,10 has better moulding properties and resistance to moisture than adipic acid-based nylon-6,6 and is used in mouldings, extrusion, filaments, bristles, and some textile fibre applications.

2-Octanol

2-Octanol is an eight-carbon atom secondary alcohol, obtained as a by-product during the alkali fusion reaction of castor oil. It is also called as secondary caprylic alcohol, methyl hexyl carbinol, or hexyl methyl cabinol. It can be prepared by heating sodium ricinoleate with caustic soda in a copper vessel and distilling. 2-Octanol is an oily, refractive liquid.

It has an aromatic and yet somewhat unpleasant odour, particularly on heating. Properties of 2-octanol are listed in Table 7.44. 2-Octanol is miscible with alcohol and ether. The specifications for 2-octanol are given in the Table 7.45.

Table 7.44. Properties of 2-octanol.

Property	*Value*
Specific gravity at 20°C	0.8193
Melting point, °C	–38.6
Boiling point, °C	178.5
Refractive index at 20°C	1.42025
Flash point, °C	60
Solubility in water, ml/100 ml water	0.096

Table 7.45. Specifications for 2-octanol.

Property	*Value*
Description	Clear liquid, without suspended particles
Colour, Hazen units, max.	50
Acid value, max.	0.5
Hydroxyl value, max.	6
Saponification value, min.	260
Ester content, (as DEHS) % of DOS, min.	98.9

This material can be re-distilled to get purer grades. 2-Octanol is used in plasticisers in the form of diesters of various dibasic acids, lacquers, disinfectant soaps, and perfumery. It is also used as solvent for fats and waxes, and as antifoaming agent i.e. to prevent foaming.

Urethane Foams

Urethane is a chemical polymer obtained by the reaction of polyisocyanate with polyhydroxy compound. Urethane foam obtained from urethanes is a cellular polymer which is blown to voluminous material by incorporating gases in the polymer during the reaction. Castor oil, is a useful natural polyol. The trifunctionality of castor oil, i.e. double bond, hydroxyl groups and the carboxyl groups, contributes toughness to the polymer structure and the long fatty acid chain imparts flexibility and water resistance. Different types of urethanes such as flexible, semi-rigid and rigid can be prepared from various derivatives of castor oil.

Castor oil-based urethanes are characterised by their excellent hydrolytic stability and shock absorbing and electrical insulating properties. Castor-based polyols such as propylene glycol monoricinoleate, ethylene glycol monoricinoleate, epoxidised castor oil, castor oil monoglyceride, are prepared and used for urethane making. These polyols increase the reactivity of castor oil during the reaction. Urethane foams have also been prepared from blown and elaidinised castor oils, and ricinoleic acid polyols.

Castor oil on reaction with isocyanates and subsequent oxidation, yields cellular urethane foams having good elastic and other properties superior to rubber foams. These foams containing sometimes as high as 80 per cent of castor oil have been made in USA by patented processes. They are useful for furniture, automobiles, mattresses, etc. Tough plastics were derived based on castor oil-elastomer-reinforced vinyl polymers and simultaneous interpenetrating networks prepared by cross-linking castor oil with sebacoyl chloride and polystyrene. Water-resistant polyurethane-based encapsulants were prepared from castor oil polyols and plasticisers. The cured elastomer is used to encapsulate plastic insulated telephone cables and seal the interior spaces of the electrical connections from the incursion of water.

Urethane casting resins prepared from castor polyols exhibit a broad range of properties from hard plastics to soft elastomers. These two component liquid urethanes are reported to have good electrical and mechanical properties, good shock absorption, low exotherm, low shrinkage, and excellent hydrolytic stability and heat ageing properties. These castor oil-based casting resins are used principally for potting and encapsulation of electrical components. Low-viscosity urethane polymers have been prepared from castor oil and polymeric isocyanates. These low mix viscosity systems are extremely useful for potting electrical components where fast penetration without air voids, and fast dispensing cycles are desirable. Very low viscosity urethane systems containing castor polyols have been prepared for use in reclaiming water-logged buried telephone cable and for encapsulating telephone cable splices.

By varying molecular weight and functionality of the castor polyols and the type of isocyanate, a variety of clear and pigmented urethane coatings can be prepared. Co-polymer of vinyl and castor-based urethanes have also been reported for use as exterior coatings for polywood and flexible substrates and for application over steel, concrete and wood substrates. Polymerisation of castor oil, chemical or oxidative, results in higher viscosity or bodies oils that are more useful in urethane coatings than the untreated castor oil. Other castor oil derivatives used to prepare urethanes are amides prepared by reaction of castor oil and alkanolamines, amides of ricinoleic acid with long-chain di- and triamines, and butanediol diricinoleate.

Major applications of polyurethanes is in flexible foam for cushioning material for various end-uses such as furniture, upholstery, carpet and garment underlays, footwear lining, transportation and packaging. Polyurethane coatings are used in leather goods, rubber, wood, sports goods, glass and metal substrates, coatings for nylon fabrics, umbrella cloth, tarpaulins, etc. The products have excellent adhesion, hardness and flexibility. Polyurehane rollers for printing presses manufactured by National Chemical Laboratory process distribute ink uniformly with proper tack, they are not affected by climatic changes, have wear resistance, and are more durable than the commonly used gelatine rollers. Polyurethane soles can be used in the shoe industry. Major research is going on to replace sheet metal in car and autobody, metal rollers in rice mills, etc. by polyurethanes.

Aliphatic polyurethanes coating compositions are prepared from trimethylolpropane, dehydrated castor oil fatty acids, 4,4-dicyclohexylmethane diisocyanate, and other components to prepare a weather resistant finish suitable for use on motor vehicles. Coatings from castor oil-toluene diisocyanate co-polymers exhibited high abrasion resistance on steel plates and flexibility on aluminium plates and PVC cloth. Castor oil's resistance to hydrolysis, pigment dispersion ability, and compatibility with polyether polyols has also made it useful as a modifier for polyether based foams. Recently, considerable emphasis has been placed on developing fire retardant urethane foams. Brominated castor oil has been investigated as a modifier for preparing fire-resistant urethane foams.

Dimer Acids

The term dimer acids (61788–89–4) is applied to the dicarboxylic acids formed by the reaction of two or more C18 unsaturated fatty acids such as oleic and linoleic acids. The only higher dicarboxylic acids of commercial importance are unsaturated cyclic C36 dicarboxylic acids known as dimer acids. Commercial dimer acids are mixtures of C36 dibasic acids containing some trimer (C54), higher oligomers, and also some amounts of monomer acids (C18 saturated, unsaturated, and structurally modified fatty acids). The main difference among various standard porprietary products is the content of trimer compounds. The ratio of dimer to trimer acids can vary from 36:1 to 0.7:1. The composition of dimer acids depends to a large degree on the feed materials and the manufacturing process.

Commercial processes yield mixtures of dimer acids, higher polycarboxylic acids, and various isomeric monomer acids, which are separated by distillation using film evaporators. After removal of the monomer acid fraction and the polycarboxylic acids, the remaining acids can be separated by further distillation, into dimer and trimer acids. Dimer acids are produced by intermolecular condensation of unsaturated C18 carboxylic acids or their esters. Dehydrated castor oil fatty acids are one of the raw materials used. In USA, tall oil fatty acids are the main feed materials. Oleic and linoleic acids can also be used. The reaction is conducted preferably in the presence of special aluminium silicates (montmorillonites) at 190–240°C, thermal dimerisation at 270–290°C is also possible. After the reaction is completed, the mixture is cooled, a metal complexing agent, usually phosphoric acid, is added, and the crude dimer acids are filtered. They contain about 40 per cent monomeric acids, which are removed by distillation with a simple still or a wiped-film evaporator. If higher purity is required, the trimer and higher homologues are removed by a second distillation with a molecular still or a wiped-film evaporator.

A process for the preparation of dilinoleic acid from ricinoleic acid has been patented by the Sri Ram Institute of Industrial Research, Delhi (India). The product is useful for synthesising polyamide resins and polyesters, in surface coatings, and in making lubricating oils and mould-releasing compounds. Typical physical properties of dimer fatty acids derived from tall oil fatty acids are shown in Table 7.46.

Table 7.46. Properties of dimer acids.

Property	*High dimer*	*Intermediate dimer*	*High trimer*
Composition, %			
Monomer	0–1	0–1	0–1
Dimer	95	83	79
Trimer	4	16	20
Acid number	194–198	189–196	189–196
Neutralisation equivalent	283–289	285–299	285–299
Saponification number	196–200	193–199	192–199
Unsaponifiables, %	0.3–0.5	0.3–0.6	0.3–0.6
Colour, gardner	4–5	7–8	7–8
Sp. gr. 25–20°C	0.947	0.941	0.947
Viscosity, sq.m/s at 25°C	0.72	0.84	0.94
Flash point, °C	275	295	295
Fire point, °C	320	320	320

The properties of distilled dimer acids are given in Table 7.47.

Dimer and trimer acids, as well as monomer acids derived from dimer acid processing, are neither flammable nor combustible. The physical properties are affected by the raw materials used in the manufacture, reaction conditions, and content of monomer, dimer, and trimer. All dimer acids are liquid at 25°C, even though their molecular weight is about 560. This phenomenon can be attributed to the presence of a large number of isomers which tend to lower the melting point. The structure of the dimer acids depends on the feedstocks used. Oleic or elaidic acid gave large amounts of acyclic dimer, while linoleic acid gave significant amounts of cyclic dimer and only small amounts of acyclic trimer. The detailed analysis of dimer acid structure obtained from different feedstocks is presented in Table 7.48.

The dimer acids produced on an industrial scale are used in the manufacture of polyamides, polyesters, epoxy resins, lubricants, plasticisers, and pesticides. Dimer acids are used to impart special properties to alkyd resins. The most important use of dimer acids is the manufacture of polyamides. Polyamide resins based on dimer acids fall into two general categories, non-reactive or neutral polyamides and reactive polyamides. The major applications of neutral polyamides are in printing inks, hot-melt adhesives, and coatings. Reactive polyamides find their applications as curing agents for epoxy resins which are used as thermosetting surface coatings, adhesives and potting and casting resins.

The non-nylon dimer based polyamides are characterised by lack of crystallinity, relatively low softening points, adhesiveness, hydrophobicity, and generally, relatively low transition temperature ranges. These properties contrast sharply with the crystallinity and high melting temperatures of the nylon polyamides based on C6 to C12 dibasic acids. Non-reactive polyamide resins are prepared from dimer and trimer acids by condensation with a large variety of diamines, including aliphatic and aromatic diamines, heterocyclic diamines, and polyamines. Dimer acids impart flexibility, corrosion resistance, chemical resistance, moisture resistance, and adhesion to non-reactive polyamides.

Dimer-based non-reactive polyamide resins are especially suited for printing on plastic films such as polyethylene bread wrappers and metallic foil laminates because the resins adhere well to these substrates.

Table 7.47. Properties of distilled dimer acids.

Property	*Hydrogenated*	*Unhydrogenated*
Composition		
Dimer	95	95
Trimer	4	4
Monomer	1	1
Monobasic acids, max.	1.5	1.5
Neutralisation equivalent	289–298	283–289
Acid number	188–194	194–198
Saponification number	190–197	198–202
Unsaponifiables, %	0.1	0.5
Colour, gardner, max.	1	5
Iodine value	20 max.	–
Viscosity at 25°C, cSt	7000–8000	7000–8000
Sp. gr. at 25°C	0.98	0.95
Refractive index at 25°C	1.478	1.483
Pour point, °C	– 11	– 8

* Hystrene series of dimer acids, Humko Sheffield Chemical.

Table 7.48. Dimer structure from different feedstocks(%).

Feedstock	*Acyclic*	*Monocyclic*	*Polycyclic*
Oleic or elaidic acid	40	44	5
Tall oil fatty acid	15	70	15
Linoleic acid	5	55	40

These resins can be formulated to give an ink with high gloss, excellent adhesion, good alcohol dilution, low gelation, good drying and low odour properties. Another large use for netural polyamide resins is in hot-melt adhesives. Because of modern high-speed assembly requirements, polyamide resins based on dimer acids are used widely in shoe construction, packaging and bookbinding. Frozen orange, apple and other juice containers are of a new construction and also utilise polyamide adhesives. Two relatively new applications of hot-melt polyamide adhesives are in fabric bonding and in bonding electrical cable sleeves. Polyamide-based hot melts are also used in laminating metal films and various plastics. Some other uses for neutral polyamides include the preparation of thixotropic coatings, water-based glues for household use, antistatic agents for textiles, transparent candles and in making detergents. The most important coating application for the non-reactive polyamide resins is in producing thixotropy. Typical coating resins such as alkyds, modified alkyds, natural and synthetic esters oils, varnishes and natural vegetable oils can be made thixotropic by the addition of dimer acid-based polyamide resin. Speciality high performance coating applications often require the properties imparted by dimer acid components.

A separate class of polyamide resins is based on condensation of diamines or triamines, e.g. ethylenediamine or diethylenetriamine, with relatively high molecular weight dibasic acids or esters, e.g. the dimer acids. The polyamides vary from tacky to high-melting (175°C) resins and are used as heat-sealing adhesives and as moisture barrier coatings for paper, wood, cellophane, etc. The polymer coatings are applied as solutions, dispersion, or powders.

Reactive polyamide resins are made from dimer acids and polyamides such as diethylenetriamine (DETA), triethylenetetramine (TETA), and pentaethylenehexamine (PEHA). The resins thus formed are capable of undergoing further reaction and are used as curing agents for epoxy resins. Extensive cross-linking occurs, giving a thermosetting resin. The use dimer-based polyamide curing agents imparts excellent adhesion, flexibility, toughness, chemical resistance, moisture resistance, and surface gloss to epoxy resins. The reactive polyamide resins are used extensively to react with epoxy or phenolic resins, yielding adhesives that are useful in casting and laminating, in structural work, for patching and sealing compounds, and for protective coatings. The largest uses for dimer-based polyamide-epoxy resins are in adhesives, surface coatings, potting and casting resins.

Dimers have the ability to absorb on the surface of metals, apparently because of their unique properties and form a hydrophobic barrier that protects the metal from corrosion. Salts and other derivatives, in addition to dimer acids *per se*, are used in many areas of the oil production and transportation industries for corrosion protection. Compositions based on dimer acids are useful in protecting iron or steel tubing and metal equipment in oil and gas wells which come in contact with such corrosive agents as carbon dioxide, hydrogen sulphide, brines, weak inorganic acids, and organic acids. Dimer acids are components of downwell corrosion inhibitors for oil drilling equipment. A combination of the dimer derived from 9,11-octadecadienoic acid and dodecylbenzyltrimethyl ammonium chloride was claimed to give excellent protection against general corrosion. Dimer acids have provided lubrication in casting operations and serve as corrosion inhibitors and lubricants in cold rolling of metal. They have been known to act as antiwear agents providing EP in metal cutting operations. Dimer acids have achieved some use in speciality lubricant applications such as gear oils and compressor lubricants. The dimer esters, compared to dibasic acid esters, polyol esters, and poly-alpha-olefins, are higher in cost and of higher viscosity. The higher viscosity, however, is an advantage in some specialities and the dimer esters are very stable thermally, and can be made quite oxidatively stable by choice of proper additives.

Recent patents describe the use of dimer acids in a fuel composition containing ethanol to prevent corrosion. It has also been found that dimer acids added to gasoline fuel containing cyclopentadienyl manganese antiknock agents result in reduced hydrocarbon emissions. Dimer acids have also been found to act as a friction and wear-reducing additive in diesel fuels.

Dimer acid esters of trimethylolpropane are used in a synthetic lubricant formulation and dimer derivatives of amines have found application as aqueous metal-working fluids. Dimer-based water-soluble polyesters useful as textile and yarn lubricants, metal-working lubricants, and in aqueous hydraulic systems has been described. There are several drawbacks in the use of dimer acids in metal working. Excessive foam has been noted, particularly where high molecular weight acids have been used, although excellent lubricity and rust protection is found. Also, tackiness and some further polymerisation at the tool will cause deposits that may interfere with the metal-cutting process. Current awareness on environmental impact has prompted the metal-working industry to head toward the use of synthetics, which will run clean while rejecting hydraulic tramp oil. Dimer acids tend to emulsify tramp oil just as many other long-chain aliphatic fatty acids do. This tendency may preclude their use in metal-working applications in today's markets.

Dimer acids are used to make polyesters used in coatings. Dimer acids have found extensive use in alkyd paints, urethane coatings, and epoxy ester coatings. Dimer acids are also used as intermediates for nitriles, amines, and diisocyanates, and in fibre glass manufacture. Other areas where dimer acids are useful are in cleaning formulations and shampoo products, and in water-resistant sunscreen compositions. Many other uses for dimer acids have been found and are too numerous to mention here. Dimer fatty acids pose no toxic effects either by injection or as primary skin irritants.

Miscellaneous

The fatty acids, alcohols and glycerides from castor oil have been tried in ester and alkyd type surface coatings. The glycerides give good products. A process has been patented for the preparation of a syndet, similar to dodecylbenzene sulphonate, from castor oil. Castor oil condenses with triethylene tetramine to give diamides which, when neutralised with H_3PO_4, act as emulsifiers for nitrocellulose lacquers.

HYDROLYSIS OF CASTOR OIL

Castor oil (8001–79–4) is derived from the bean of the castor plant *Ricinus communis* belonging to the family Euphorbiaceae. Castor oil is also known as Ricinus oil, oil of Palm Christi, Tangantangan oil and Neoloid. Typical of vegetable oils and most fats, castor oil is a triglyceride of various fatty acids (Table 7.49). Its uniqueness stems from the very high (87–90 wt. per cent) content of ricinoleic acid $C_{18}H_{34}O_3$, structurally *cis*-12-hydroxyoctadeca-9-enoic acid, an 18-carbon hydroxylated fatty acid having one double bond.

Table 7.49. Fatty acid composition of castor oil.

Fatty acid	*CAS registry No.*	*Molecular formula*	*Weight %*
Ricinoleic acid	[141–22–0]	$C_{18}H_{34}O_3$	89.5
Dihydroxystearic acid	[26248–43–1]	$C_{18}H_{36}O_4$	0.7
Palmitic acid	[57–10–3]	$C_{16}H_{32}O_2$	1.0
Stearic acid	[57–11–4]	$C_{18}H_{36}O_2$	1.0
Oleic acid	[112–80–1]	$C_{18}H_{34}O_2$	3.0
Linoleic acid	[60–33–3]	$C_{18}H_{32}O_2$	4.2
Linolenic acid	[436–40–1]	$C_{18}H_{30}O_2$	0.3
Eicosanoic acid	[506–30–9]	$C_{20}H_{40}O_2$	0.3

Ricinoleic acid

$$CH_3—(CH_2)_5—\underset{\displaystyle |}{\overset{\displaystyle OH}{CH}}—CH_2—CH=CH—(CH_2)_7—COOH$$

Castor oil, sometimes described as a triglyceride of ricinoleic acid is the only commercially available renewable vegetable oil resource that contains hydroxyl functionality, in such a high percentage of one fatty acid.

Castor Fatty Acids

The major fatty acid product of castor oil is ricinoleic acid, although it is not produced on industrial scale. Industrially ricinoleic acid is substituted by castor oil with around 87 per cent acidity. This is because it cannot be distilled unless special precautions are taken via derivative formation to protect the hydroxyl group. It is soluble in alcohol, acetone, ether and chloroform.

Ricinoleic acid has found applications as pigment and dye disperse in printing ink and to some extent in heat convertible alkyds. Ricinoleic acid is used in textile finishing and sometimes added to turkey red oil and dry-cleaning soaps. It has also been used in contraceptive jellies. However, it's main use is as chemical intermediate to produce various oleochemicals.

Various ricinoleic acid derivatives viz. ricinenic acid (dehydrated ricinoleic acid), suberic acid (oxidation of ricinoleic acid with HNO_3), azelaic acid (ozonolysis of ricinoleic acid), methyl ricinoleate, ricinoleyl alcohol etc. have been used in the preparation of plasticisers, emulsifiers, soaps etc.

The other important fatty acid products of castor oil are 12-hydroxy stearic acid, sebacic acid, *n*-heptaldehyde/*n*-hepanoic acid and undecylenic acid. 12-hydroxy stearic acid is used in speciality greases, sebacic acid for nylon-6,10 and speciality plasticisers and undecylenic acid as a precursor for nylon-11. Dehydrated castor fatty acids are produced in minor amounts from castor oil.

General Considerations for Fat-splitting

Splitting of fats to produce fatty acids and glycerol has been practiced since, before the 1890s. Processes for fat splitting have been patented as early as 1854 by *Tilghman, Twitchell* etc. with a large diversity in the methodologies.

The techniques range from enzymatic, room temperature methods to very high temperature (300°C) catalysed superheated steam splitting and from atmospheric pressure to those carried out at the pressures of around 3000 psig. The various direct hydrolytic splitting methods are those customarily employed for the production of fatty acids from common fats and oils. One other route available is the saponfication (Reaction-I) of a fat or an oil.

Reaction–I

$$\begin{array}{l} CH_2\,OCOR' \\ | \\ CHOCOR'' \\ | \\ CH_2\,OCOR''' \end{array} \xrightarrow{3\ NaOH} \begin{array}{l} CH_2\,OH \\ | \\ CHOH \\ | \\ CH_2\,OH \end{array} + R'COONa + R''COONa + R'''COONa$$

$$\begin{array}{l} \ R'COONa \\ +\ R''COONa \\ +\ R'''COONa \end{array} \xrightarrow{3\ HCl} \begin{array}{l} \ R'COONa \\ +\ R''COONa \\ +\ R'''COONa \end{array} + 3\ NaCl$$

This method is more suitable for certain heat sensitive fatty acids and it is also considered to be the method of choice for laboratory preparations of fatty acids. The overall chemical equation for fat splitting is shown in Reaction-II.

Reaction–II

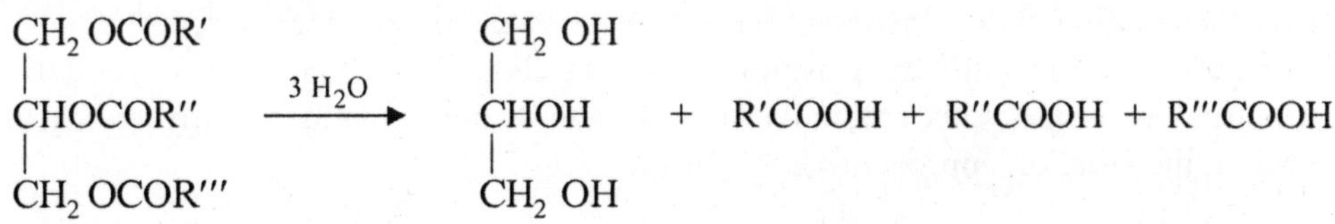

The above equation indicates very little of the high degree of complexity involved in the hydrolysis reaction and hence from a theoretical point of view, it is better to consider fat splitting as occuring in stages (Reaction-III).

Reaction–III

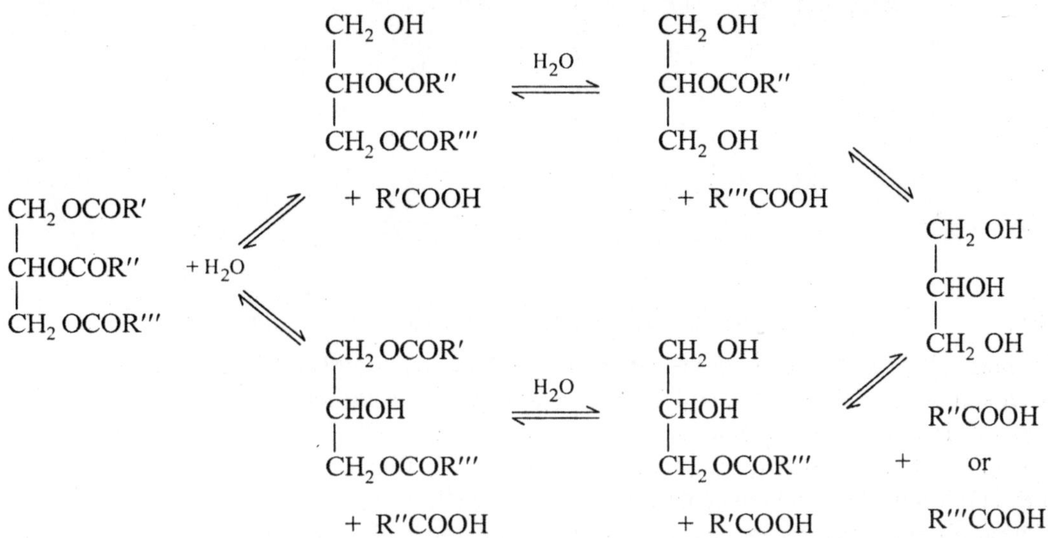

Fat splitting has been realised as a homolytic reaction occuring in the fat phase as a consequence of small amounts of water solubilised in the fat phase and not as a heterogeneous reaction at the interface of fat and water globules. It has been shown that the solubility of water in fat phase at temperatures below the boiling point of water is very less and it is for this reason that it is necessary to adjust the temperature and pressure values to obtain optimum solubility of water in the fat phase. The use of presssure is primarily to keep water in the liquid phase at temperatures above its boiling point.

In addition to achieving fat-water homogenicity, it is important for complete fat hydrolysis, to remove glycerol from the reaction. This is in accordance with the law of mass action, which states that unless this is accomplished, fat hydrolysis cannot be carried out to 100 per cent completion. In aqueous thermal fat splitting methods, diffusion of glycerol into or extraction of glycerol by excess water is an effective means of eliminating glycerol from the fat phase in which hydrolysis is carried out.

However, in some processes, equilibrium is developed between glycerol in the fat phase and glycerol in the aqueous phase and the hydrolysis eventually slows down and stops at a certain point short of 100 per cent completion.

All the fat-splittng methods employ maximum water concentration in the fat phases through solubility at optimum temperatures consistent with the reaction conditions, and all the methods attempt to remove glycerol by-product from the fat phase as completely as possible for optimum degree of reaction. Pressure is employed in some methods in order to maintain water in the liquid phase. Availability of excess, hot, liquid water in the reaction also helps to sufficiently remove the by-product, glycerol.

It has been observed that while hydrolysing fats containing a low proportion of free fatty acids, the rate of reaction is slow to start with, gradually increasing to its normal value, finally ending in a reduced rate period, when the reaction approaches equilibrium (Fig. 7.2).

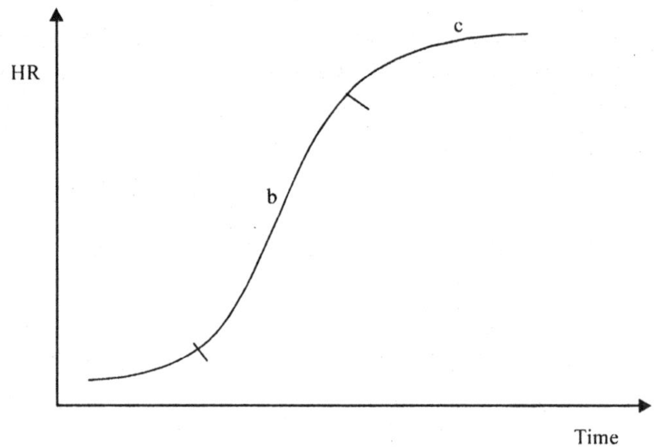

Fig. 7.2. Evolution of the rate of hydrolysis with respect to time.

The reason behind this behaviour is believed to be the low solubility of water in natural fats compared to higher solubility in fatty acids, and as described earlier for splitting to proceed at a satisfactory rate, water must dissolve in the oil phase. Thus, the induction period (a) may be reduced or eliminated by addition of free fatty acid to the initial reaction mixture. The final equilibrium of the reaction, however, depends only on the glycerol concentration in water, which is in contact with fats. *Lascaray* has drawn from these observations, an empirical equation, valid for batch hydrolysis which shows that in the hydrolysis reaction, formation of sweet water rich in glycerol, prevents attaining a high rate of hydrolysis:

$$HR = 100 - 0.8G$$

where,

G – Percentage of glycerol in water

HR – Hydrolysis rate

On the contrary, for an enzyme catalysed reaction, the rate of hydrolysis has been found to be low in the presence of free fatty acids and is less affected by the presence of glycerol.

Temperature does not affect the equilibrium composition of the reactants, but does affect the reaction rate. Increasing the temperature serves to speed up the reaction, besides markedly increasing the solubility of water in the fatty phase. It is due to this reason that the length of the induction period is mainly dependent on the splitting temperature.

Apart from temperature, the reaction rate is increased by the use of certain catalysts. In the autoclaves and some high-pressure processes, catalytic bases like metallic oxides of zinc, calcium, magnesium or hydroxides of sodium, potassium and ammonium are used. These combine with fats/fatty acids to form metal soaps which acts as emulsifiers and increase the solubility of water in the oil-phase. In some processes, such as the *Twitchell* process, acid catalysis is employed. These are different sulphonated compounds of benzene or naphthalene with fatty acids. The acid in Twitchelling is believed to lower the solubility of the splitting agent in the aqueous phase, thereby increasing its solubility in the oil phase. This serves to increase the hydrogen ion concentration in the fat phase where hydrolysis occurs.

To summarise, fat hydrolysis is a stepwise, homogeneous reaction, which takes place in the oil phase. The rate of the reaction is dependent on the amout of water present in the solubilised form in the fat phase while the final equilibrium split depends on the concentration of glycerine. The rate of the reaction is affected chiefly by the operating temperatures and the catalyst.

Methods of Fat Splitting

As the hydrolysis reaction is reversibe, as mentioned earlier, it is necessary to remove either the glycerol or the fatty acids to shift the equilibrium. Efficient contact between the fat and water would help obtain highest yields of fatty acids and glycerine through hydrolysis. The various methods of fat splitting differ only in the means of obtaining an intimate contact between water and the oil or fat during hydrolysis and in the physical method of removing the glycerol and fatty acids. However, fatty acids are essential or the presence of fatty acids in the sphere of the reaction is desirable due to the higher solubility of water in fatty acids i.e. for autocatalysis.

The various methods of hydrolysis of natural fats involve splitting with water and with or without the use of catalyst. These have been categorised under two broad heads:

1. Pressure splitting; that includes autoclave process, which may be further, classified as batch, semi-continuous or continuous process.
2. Atmospheric splitting; that includes enzymatic and *Twitchell* process.

Each of these processes has advantages and disadvantages. Selection of a particular process in commercial practice depends upon the type and quality of raw material, size and scale of operation. Almost all the splitting techniques have been extensively reviewed by several authors and a wide range of literature is available on this subject.

Pressure splitting

The use of steam under pressure raises the reaction temperature, thereby effecting a marked increase in the reaction rate during hydrolysis. Generally, for every 10°C rise in temperature, the reaction rate is increased by approximately 33 per cent. This gave rise to the development of autoclave process for splitting of fats under pressure involving the use of pressure vessels. Pressure splitting may be futher classified as:

1. Low pressure (50–150 psig)—high temperature spliting with catalyst.
2. Medium pressure (150–500 psig) autoclave splitting with or without catalyst.
3. Very high pressure (2000–3500 psig) concurrent fat spliting.
4. Continuous, uncatalysed, high-pressure (700–750 psig) countercurrent splitting (*Colgate-Emery* process).

The high pressure, high temperature techniques are unsuitable for the hydrolysis of castor oil due to the susceptibility of the castor fatty acid, ricinoleic acid to thermal degradation. These techniques are only suitable for the hydrolysis of dehydrated castor oil, where the hydroxyl function imparting degradation susceptibility to ricinoleic acid is removed. Under high temperature, high-pressure conditions, ricinoleic acid undergoes a series of different chemcial transformations, viz.:

Splitting

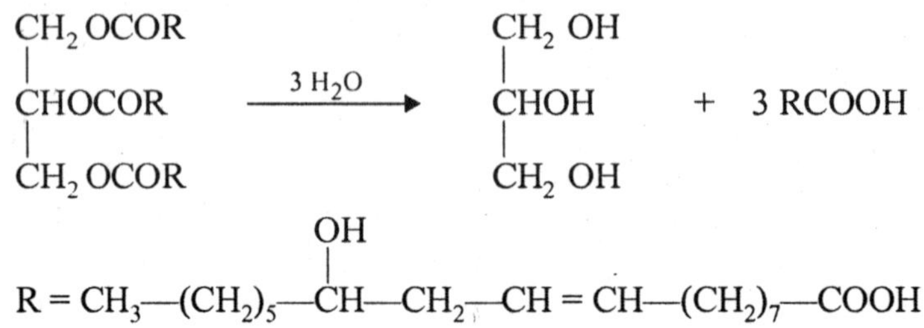

Dehydration

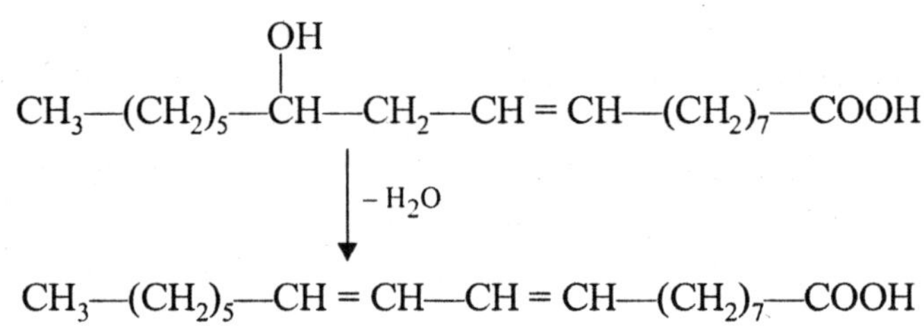

Estolide formation

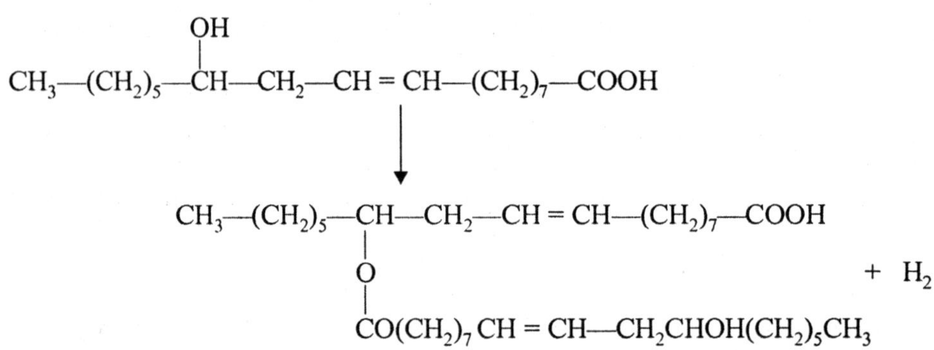

Diels–Adler polymerisation

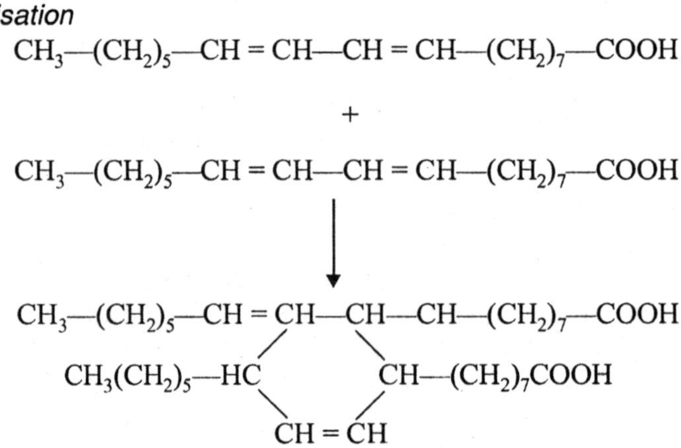

Hydrogenated castor oil undergoes corresponding reactions of types I, II and III. It is due to these reaction susceptibilities, to undergo undue dehydration, conjugation and polymerisation, that castor oil cannot be hydrolysed by the high temperature methods like the Colgate-Emery process. *Kallyanpur* reported a high pressure splitting method for the splitting of hydrogenated castor oil. He performed the reaction at three different temperatures (225, 250 and 300°C) in an autoclave. At 225°C, hydrolysis and estolide formation was found to occur concurrently and reach equilibrium with their respective backward reactions after 4 hours. At higher temperatures the equilibrium was reached in still lesser time.

Lakhminarayana carried out the high pressure hydrolysis of dehydrated castor oil and studied the effect of variations in pressure (20–40 kg/cm^2), oil to water ratio (1:0.4 to 1:1 w/w) and time (0–8 hours) on the extent of hydrolysis. It was found that, higher pressure; lower oil to water ratio and longer duration gave higher percentage splits along with concurrent estolide formation. However, when the oil was hydrolysed at lower pressure (20 kg/cm^2) in two stages, with the first split yielding 77 per cent in 6 hours and the second split yielding 96 per cent in four hours proved acceptabe in affording fatty acids with less than 2 per cent conjugated unsaturation. Thus it has been noted that although the high pressure, high temperature methods yield higher and faster rates of reaction, the concurrent formation of estolide results in a drop in the net yield of the process. Alternatively various other methods employing atmospheric pressure conditions and lower temperature conditions have been proposed for hydrolysis of castor oil.

Atmospheric splitting

Saponification and acidulation

Saponfication constitutes one of the oldest methods for fat hydrolysis. It is the most suitable method for the hydrolysis of heat sensitive fats. It is one of the ideal methods for castor oil hydrolysis due to the milder reaction conditions used. Typically, the splitting is performed in two steps (Reaction-I, Fig. 7.3); the first being the saponification of the fat (triglyceride) with caustic soda by refluxing at about 80–90°C, to generate glycerol and sodium salts of the fatty acids (in the case of castor oil mainly ricinoleic acid).

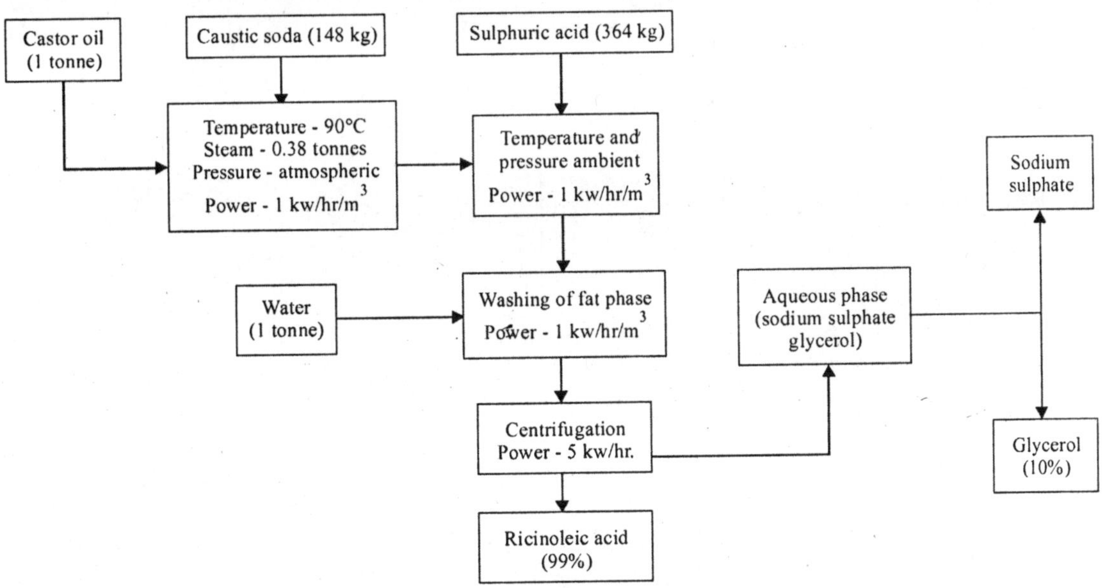

Fig. 7.3. Saponification and acidulation.

The second step comprises of acidulation of the reaction mixture with sulphuric acid to produce the fatty acids and sodium sulphate. The fat phase is washed with water and the centrifuged to separate the two phases. The non-aqueous phase is used as the final product and the aqueous phase consisting of glycerol and sodium sulphate is further processed to recover the salt and glycerol as the by-products. The risk of estolide formation is less due to the acidification. The method suffers several drawbacks, such as, generation of large quantities of salts and problems associated with the disposal of these salts. Also the process yields higher coloured products, due to the inherent charring property of sulphuric acid. However, to date saponification and acidulation has been found to be the most suitable and economical method for the hydrolysis of castor oil.

Twitchell process

The Twitchell process of fat spltting, was descibed in a patent (US 601, 603) granted to *Twitchell* in 1898. The patent described a sulpho-fatty acid, produced from oleic or saturated fatty acids and sulphuric acid, as a fat splitting agent. The method is a batch process of hydrolysis conducted at atmospheric pressure in open or vented tanks of lead lined or Monel construction at a temperature of 100°C, in presence of sulphuric acid and an acid catalyst, namely, alkyl arly sulphonic acid, which is a surface active agent. It is customary to pre-treat the fat stock before Twitchellising by boiling with sulphuric acid to remove or decompose foreign materials, such as metal soaps and coagulate and remove proteinaceous and albuminous material which may poison the catalyst.

Typically, in operation (Fig. 7.4), the charge consists of fat with 25–50 per cent of its own weight of water, 0.5–1.25 per cent of the splitting reagent and about 1 per cent of sulphuric acid. Sparge steam is directly introduced to agitate and boil the mixture for 15–20 hours to attain a hydrolysis of 80–85 per cent.

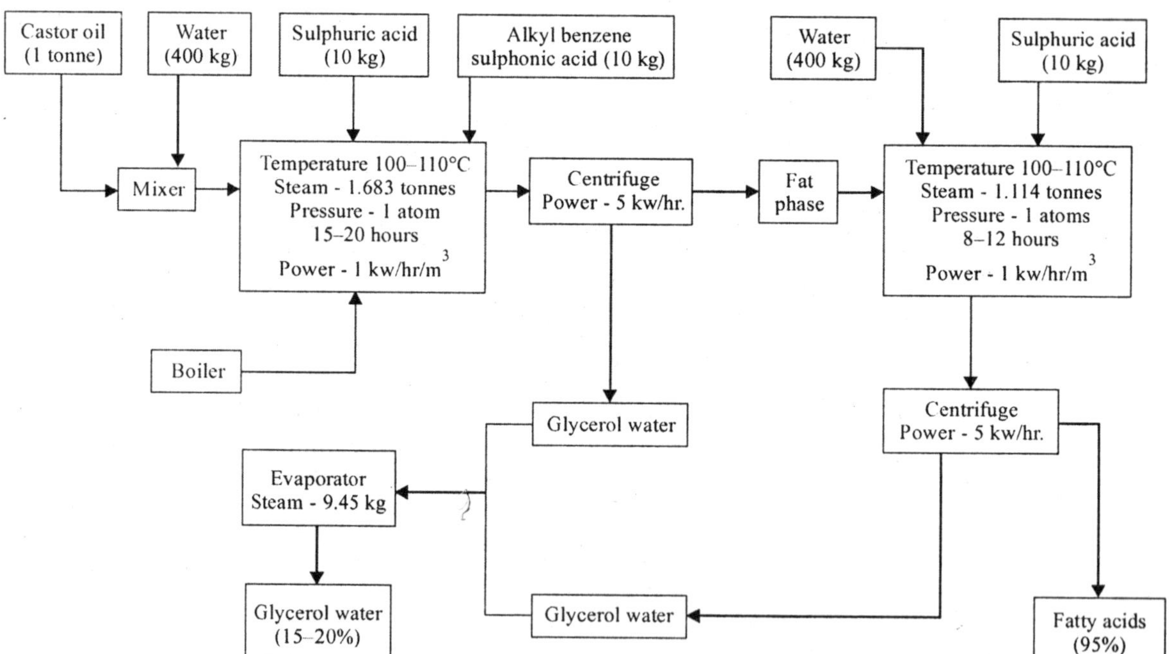

Fig. 7.4. Twitchell process.

A second boil with fresh water containing sulphuric acid after operation of another 8–12 hours enhances the degree of split to as high as 95 per cent. To reduce the overall operating period and to obtain a more concentrated glycerine solution, spltting in more than one stage is adopted along with counter-current operation. In this manner, glycerol water with a concentration of 15–20 per cent can be obtained. The rate of the splitting has been found that to be dependent on the temperature of splitting and, the type of raw material and catalyst used. *Hartman* found that the rate of the reaction almost doubled with a 10°C rise in the splitting temperature in the vicinity of 100°C. It was found by *Gueniot* that by using phosphoric instead of sulphuric acid the degree of hydrolysis was high along with lesser discolouration of the product. A study by *Ueno* revealed that unsaturated fats get hdyrolysed faster as compared to saturated fats. It has also been shown in a study that the rate of hydrolysis was dependent on the distribution pattern of the catalyst. They reported that the rate of fat splitting was faster when the catalyst was dissolved in the fat phase than in the aqueous phase.

However, the same considerations do not apply for hydrolysis of castor oil by Twitchelling. *Kallyanpur* carried out the hydrolysis of castor oil using alkyl bezene sulphonic acid catalyst and found that 93 per cent split could be achieved in the first 6 hours, however, the acid content was found to reduce with time due to the formation of estolides in the reaction mixture. The formation of estolides was confirmed by the increase in the optical rotation with the increase in the reaction period. They concluded that the estolide formation takes place right from the start of the reaction, however, the initial increase in the acid value is observed because of the higher rate of the hydrolysis reaction as compared to the estolide formation reaction in the initial period. The reaction reaches equilibrium with respect to estolide formation after 6 hours, after which the rate of estolide formation is more and the acid value drops. It was noted that, the Twitchell reagent which was dispersed in the fat phase initially to the extent of 98 per cent, whether added to oil or water phase, with the formation of estolides slowly diffuses into the aqueous phase from the fat phase because of which the rate of hydrolysis is reduced. At equilibrium 45 per cent of the reagent was observed in the aqueous phase. Also the solubility of water in fat phase was found to be reduced by estolides, thereby reducing the rate of hydrolysis.

Enzymatic splitting

Historically castor oil has been hydrolysed by the enzymatic splitting method at room temperature using castor bean lipase. *Connstein* reported this application of castor bean lipase as early as 1902. However, this method could not get industrial acceptance due to several limitations, such as, slowness of the hydrolysis, difficulties in recovery of optimum glycerol liberated, allergenic protein toxicity associated with castor proteins and most importantly economic inviability due to loss of the catalyst.

In later years, with the advent of new microbial sources for lipase and with the concurrent development of various enxyme immobilisation techniques, this old idea was revived to replace the costly and heavy technology of modern thermal splitting. Fatty acids obtained by this technique are very pure and Japanese soap manufacturer (Mig oshi oila.fat) uses enzymatic hydrolysis for manufacturing a special soap. The method is especially; highly suitable for hydrolysing castor oil, due to its mild operating conditions (atmospheric pressure and temperature ranging from 15–40°C).

A typical enzymatic batch process for castor oil hydrolysis, using castor beans as the source of enzyme is described by *Narula*. The oil is agitated along with 30–40 per cent of its weight of water and the enzyme (fermented castor beans), with air to form an emulsion in open lead lined tanks. The pH is adjusted to 5 with acetic acid and about 0.2 per cent of an activator (usually calcium or manganese sulphate) is added. After holding the batch at about 35°C for 36–48 hours with occasional stirring, a split of 85–90 per cent is obtained following which the emulsion is broken down by agitation with steam and addition of a little sulphuric acid to separate and collect the split fatty acids from the emulsified layer. The enzyme process gives optimum results at 35°C and produces fatty acids of good quality with fairly good colour without dehydroxylation or estolide formation of ricinoleic acid.

In an enzymatic process developed in one of laboratory (given in Fig. 7.5), the reactor is charged with 1:1 mixture of castor oil and isooctane. To this, 8 parts of water along with the enzyme (Lipolase 0.2kg/100 kg oil, is added). The temperature and pH are maintained at 55°C and 7.0, respectively. The reaction is carried out for a period of 6 hours after which a second dose of enzyme (Lipolase 1.3 kg/100 kg of oil) is given and the reaction further continued for 18 hours. A total of 24 hours of the reaction yields 90 per cent conversion. The fatty phase is separated from the aqueous phase as the product i.e. ricinoleic acid. The aqueous phase consists of glycerol and the enzyme with 31 per cent residual activity and can be re-used as the enzyme source for another cycle. The process yields colourless product with around 94 per cent acidity. The major factors affecting the conversion rates were found to be the amount of enzyme, pH, temperature, product inhibition and the presence of isooctane.

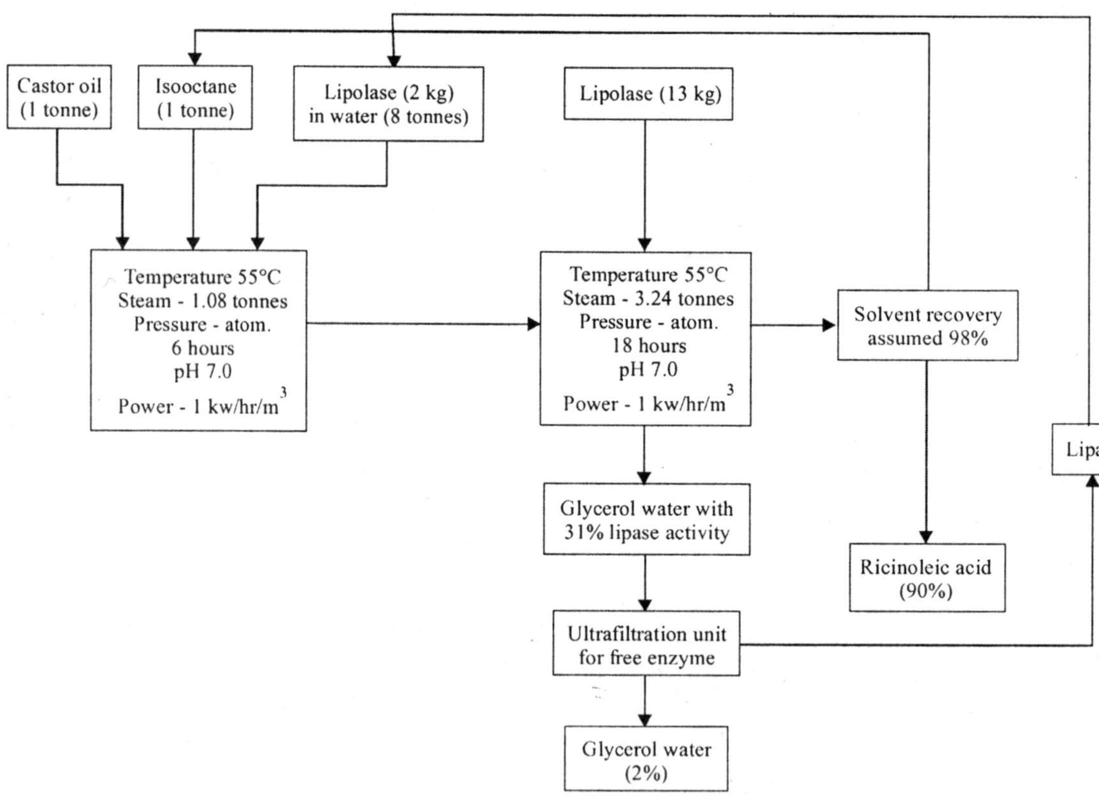

Fig. 7.5. Enzymatic process (Developed in laboratory).

Bhirud, achieved 95 per cent split of castor oil using a commercial lipase. He found that around 56 per cent hydrolysis occurred in the first 4 hours of the reaction and further-on 20 more hours were required for the split to reach 97 per cent. They reported total absence of estolide formation and production of better quality fatty acids with no discolouration of the final product.

Kinetics of enzyme catalysed hydrolysis

Kinetics of enzymatic hydrolysis does not exactly conform to that of thermal splitting. Even in batch equipment, the hydrolysis rate per cent (HR per cent) vs time graph does not have the same shape as in Fig. 7.2. With a mixing reactor, the induction period is very brief or even not-existent and the lipolysis typically follows a logarithmic behaviour (Fig. 7.6). As compared to high pressure splitting, in enzymatic spliting, for the same HR per cent, the concentration of glycerol in the sweet water is higher. However, the final HR depends on the quality of water and the speed of hydrolysis depends on several factors such as, effective concentration of lipase in the reaction, accumulation of fatty acids (imparts product inhibition) and interfacial area. Lipase shows the unique property of interfacial activation and it is for this reason that enzyme catalysed hydrolysis necessarily occurs at the oil-water interface (Fig. 7.7). This is in distinct contrast to the mechanism of water hydrolysis by the Twitchell, autoclave and Colgate-Emery process of conventional fat splitting, where the overwhelming tendency to hydrolyse is homogeneous reaction of dissolved water in the fat phase.

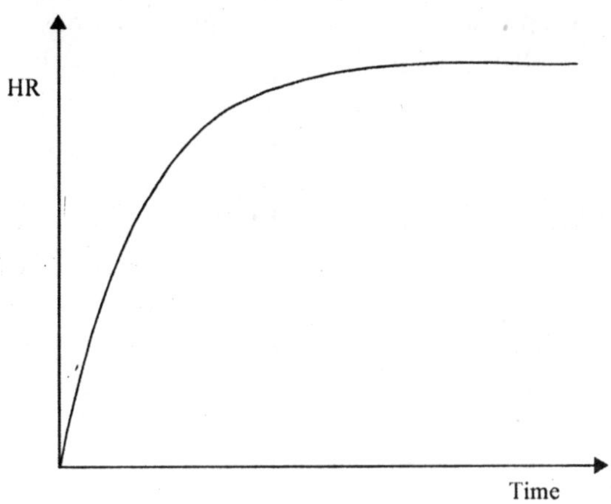

Fig. 7.6. Development of hydrolysis rate during enzymatic hydrolysis.

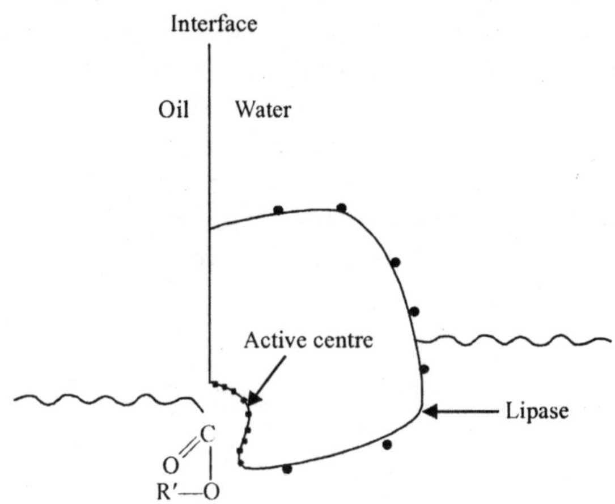

Fig. 7.7. Hypothetical model of orientation of lipase oil-water interface.

Enzyme catalysed hydrolysis as a technology

In general, the enzymatic hydrolysis encounters a number of limitations as far as technology is concerned. Firstly, there is little information originating from the companies using this procedure on a large scale. Further, extrapolation from the laboratory scale to the industrial scale necessitates many precautions. The water for the reaction should be completely free of minerals. Reliable machinery for this exists, and the cycle is automated with the help of a fixed ohmmeter. A final pH of about 5.0–7.0 is not a problem, but this water is very corrosive even towards certain steel alloys. Cleaning of fats is more difficult and calls for very efficient degumming and filtration techniques. The fatty phase should be as less polluted as possible. The stock of lipases should be managed carefully, keeping in mind cleanliness and favourable conditions (temperature, conditioned air).

Regarding machinery, there is first of all batch reactor that is less complicated. It adapts itself to frequent product changes in order to regularly manufacture special fatty acids, which often have a high added value, and hence withstand loss of a portion of lipases which is inherent to this technique. This reactor resembles an esterification reactor. A vacuum installation degases the raw material and a nitrogen insufflation is given to carry out the reaction. Reaction water is added and a calibrated pump introduces a suspension of lipase. A double wall with thermostatic control maintains the reaction temperature and a strong agitator brews the mixture. The agitator plays an important role in determining the rate of the reaction by governing the interfacial area between the water and oil phases. Once the desired HR is attained, decantation is done. Fatty acids and sweet water are removed from lipase by ultrafiltration, although its re-utilisation is not always evident (unless the enzyme is rendered re-usable by immobilisation, in which case ultrafiltration is replaced by a simpler filtration technique).

Continuous methods follow the example of splitting tower(Fig. 7.8). The column with double-wall heating contains water and lipase suspension continuously fed through a diffuser. At the bottom of the column, a diffuser of the same principle, pushes up, against the current, finer droplets of oil whereas, at the bottom of the column, glycerine-water and lipase which is recovered by ultrafiltration. In practice, this part of the installation is more complicated, because a portion of fats is carried forward to the aqueous phase. A suitable centrifugation technology and utilisation of selective membranes for ultrafiltration may be employed to improve results. This does not mean that glycerine-water, if they are not immediately concentrated, should be sterilised. The small fraction of lipases carried forward by fatty acids is destroyed during distillation. Performance of such an installation is such that the only element, which can increase the cost price, is lipase. Therefore, rendering the enzyme re-usable by immobilisation can help minimise the cost. However, the reaction system then resembles a heterogenous catalysis system and requires a reactor suitable for such a system e.g. packed bed reactor, fluidised-bed reactor, multi-tubular reactor, etc.

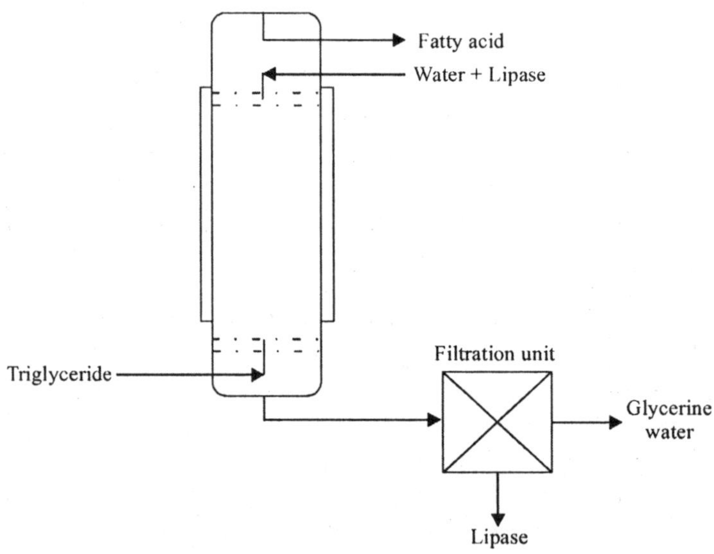

Fig. 7.8. Schematic representation of tower for enzymatic hydrolysis.

Commerical potential of enzymatic fat splitting

Adoption of enzymatic fat splittting over conventional Colgate-Emery splitting in case of thermostable oils, is motivated by the need to cut on the energy costs, to conserve energy and to produce better quality products. In the present context (hydrolysis of castor oil), it has to compete with the cost of the establised process of saponification and acidulation and Twitchelling with respect to the rate of the reaction, concentration of glycerol produced, quality of product etc. Comparison in the process conditions of saponification, Twitchell and enzymatic processes and the yields of all the three processes are given in Table 7.50.

Table 7.50. Comparison between saponification, Twitchell and enzymatic processes of castor oil hydrolysis.

Process	*Chemical used*	*Qty. per tonne of castor oil*	*Time (hours)*	*Conditions*		*% Split*	*Glycerol*	*Quality of product*
				Temperature	*Pressure*			
Saponification	Caustic soda Sulphuric acid	148 kg 364 kg	6	80–90°C	Atmospheric	99	10%	Highly coloured product Lesser estolide formation Low cost as compared to the enzymatic process
Twitchell	Water Alkyl benzene sulphonic acid Sulphuric acid	400 kg 10 kg 10 kg	32	100–110°C	Atmospheric	93	15–20 %	High estolide formation Low cost as compared to the enzymatic process
Enzymatic	Water Isooctane Enzyme	8 tonnes 1 tonne 15 kg	24	55°C	Atmospheric	90	2%	High quality product Less estolide formation Higher cost with non-coverable enzyme

The conventional process operates as a batch, at 80–90°C and involves the use of corrosive chemicals such as caustic soda and sulphuric acid. The product formed is of poor quality due to the developement of excessive colouration. The enzymatic process offers a promising alternative as it operates at moderate temperatures (around 35–40°C) and to produce better quality products. However, the major discouraging factor is the rate of the reaction and the cost of the enzyme. A rough cost structure for the three processes for the purpose of comparison is given in Table 3.51 (the capital cost, the cost of maintenance and of labour have not been considered). The enzymatic reaction is known to be very slow and may take upto 1–2 days for achieving 90–95 per cent hydrolysis whereas, about 99 per cent hydrolysis is achieved by the conventional process in much lesser time (actual process time –5 hours, reaction time –2 hours). Such a lengthy reaction process, although requires less magnitude of energy, it requires this energy over an extended length of time, because of which the net cost of the process due to the utilisation of large amount of utilities is raised. As can be observed from Table 7.51, to process 1 tonne of castor oil, the cost incurred by the amout of enzyme constitute the major bulk of the cost of the process. Hence, the cost of the enzyme is the most important factor responsible to large extent for the difference in the cost of the enzymatic and the other proceeses. Therefore, by using the enzyme in a re-usable immobilised form it is possible to considerably reduce the cost of the process. By using an immobilised enzyme re-usable up to 10 cycles (assuming there is no loss of activity), it is possible to reduce the cost of the process by almost 37 per cent. The other factor responsible for higher cost of enzymatic process is the amount utilities required which is dependent on the time of the reaction for which the utilities have to be supplied.

Table 7.51. Comprative cost structure for the saponification, Twitchell and enzymatic processes of castor oil hydrolysis.

	Saponification process		*Twitchell process*		*Enzymatic process*	
Description	*Quantity required*	*Cost Rs.*	*Quantity required*	*Cost Rs.*	*Quantity required*	*Cost Rs.*
Raw materials						
Castor oil	1 tonne	37,000	1 tonne	37,000	1 tonne	37,000
Caustic soda	0.15 tonne	1894	–	–	–	–
Sulphuric acid	0.36 tonne	1019	0.02 tonne	56	–	–
Alkyl benzene sulphonic acid	–	–	0.01 tonne	400	–	–
Lipase	–	–	–	–	0.01 tonne	20,700
Isooctane	–	–	–	–	0.02 tonne	3000
Water	1 tonne	16	0.8 tonne	12	8 tonne	120
Utilities						
Steam	0.38 tonne	306	2.8 tonne	2245	4.32 tonne	3457
Electricity	17.5 kw	9	56.7 kw	28	28.4 kw	12
Total		40,244		39,741		64,288
Cost of ricinoleic acid per kg		42		43		73

A comparative cost analysis study conducted by the research team with pressed castor seeds as the source of lipase, revealed that the enzymatic process does not provide an economically viable option. Better, immobilised lipases are available today and use of these immobilised enzymes can considerably influence the cost structure. Immobilised enzymes themselves are very costly in nature and in order to estimate the cost advantages of immobilised lipases, it is necessary to look at the individual manufacturing steps and their contribution to the overall costs. These include the cost of biomass (from different microbial sources) obtained by fermentation. This cost is determined mainly by the fermentation scale and the expression rate of the enzyme. Downstream processing is needed to achieve the required purity but is accomanied by loss in activity. The use of larger fermentation scales might be necessary in order to compensate for loss of activity and may lead to some increase in the cost. The costs of the immobilisation procedure further increase the manufacturing costs. Thus, leaving aside potential advantage of easier removal of the enzyme from the product, immobilised enzyme may not provide any cost benefit. However, cost benefit will be achieved by the repeated re-use of the immobilised enzyme. The costs of the enzyme must not be higher than a few per cent of the production costs for the desired product. In conclusion, only repeated use will lead to dramatic cost reductions and this can be easily tracked by monitoring the amount of enzyme required per kilogram of product formed.

In general, for enzymatic process to be economic, it has to satisfy several requirements, such as:

1. The rate of the reaction should be as high as possible to save on the energy cost and to increase the yield of the process per unit time.

2. The enzyme should be re-usable for sufficient cycles to enable cost-effective operation.
3. The quantity of water used should be as low as possible to produce higher concentration of glycerol.
4. The enzyme should not pose any environmental problems.

This castor oil hydrolysis is of major significance due to the wide potential applications of ricinoleic acid and the other fatty acids of castor oil. The conventional hydrolysis protocols are unsuitable in the case of castor oil hydrolysis due to the thermal instability of ricinoleic acid. Alternatively, the hydrolysis of castor oil is carried out by saponification followed by acidulation or by using the Twitchell reagent. This process produces low quality products due to the generation of coloured products and high degree of estolide formation. An enzymatic process therefore promises an effective alternative to the saponification and Twitchell processes but is marred by the drawbacks of slow reaction rates and high enzyme cost. With the advent of several enzyme immobilisation techniques and the concurrent development in the production of enzymes with higher reaction rates and better stability profiles to enable extensive re-use, it may be possible to reduce the cost of the enzymatic process and substitute the conventional hydrolysis protocols by enzymatic processes.

CASTOR CAKE

The cake remaining after crushing or extracting castor seeds for oil is known as 'poonac' or 'pomace'. The oil content of the cake can usually be reduced quite low, 6 to 6.5 per cent by efficient crushing, whereas solvent extraction brings it down to below 2.0 per cent. Typical composition of decorticated and undecorticated cake and pressed solvent-extracted cake are shown in Table 7.52.

Table 7.52. Percentage composition by weight of castor cake (pomace).

Description	Indian			USA	
	Pressed cake	*Solvent extracted*	*From whole seed*	*From decorticated seed*	*Solvent extracted*
Moisture	9.9	11.7	9.8	10.4	11.8
Ash	15.0	6.1	15.0	10.5	6.0
Oil	6.8	1.1	5.2	8.8	1.1
Protein	21.6	36.6	20.4	46.4	30.6
Crude fibre and carbohydrate	46.8	44.5	49.4	24.0	50.5

Castor seed contains three harmful components that severely limit the usefuless of the cake left after extracting the oil. The first is a mildly toxic alkaloid, ricinine, which is present in minor amounts. The second compound which is present in large amount is ricin, which is virulently poisonous and the third is a protein polysaccharide with low molecular weight is very dangerous to those allergic to it. The allergen, named CB-A is non-toxic but exhibits an extraordinary capacity to sensitive individuals exposed to small concentration of dust from castor-seed or castor cake.

Castor seed and cake are highly poisonous to man and to most animals. If castor cake is to be used as a feed ingredient, it must be first rendered non-toxic and non-allergenic. At present the cake is used entirely for manurial purposes.

It is excellent as a fertiliser because of its high content of nitrogen (6.4 per cent), phosphoric acid (2.55 per cent) and potash (1 per cent) and its moisture retention, slow decomposition and humus forming quality. Indian Standard Specification for castor-seed cake for fertiliser purpose, BIS: 3029–1960 is given in Table 7.53.

Table 7.53. Requirements for castor seed cake for fertiliser purposes (BIS: 3029–1960).

The material shall be uniform in texture, clear and free from adulterants, such as sand, dust and metallic pieces.

The material shall pass wholly through 2.36 mm IS sieve of which not less than 70 per cent shall pass through 850 micron IS sieve.

Characteristics	*Requirements for*	
	Decorticated	*Undecorticated*
Moisture, % by wt. maximum	8.0	8.0
Water insoluble org. nitrogen, % by wt. minimum	4.5	3.5
Total ash, % by wt. maximum	9.5	10.0
Acid insoluble ash, % by wt. maximum	2.4	2.5

The proteins from castor cake have been isolated and have the following amino acid composition (Table 7.54).

Table 7.54. Composition of protein in castor cake (G of amino acid per 16_G of N).

Lysine	2.85	Alanine	3.92
Histidine	1.89	Cystine	2.14
Arginine	7.17	Valine	5.47
Aspartic acid	8.50	Methionone	1.67
Threomine	3.18	Isoleucine	4.66
Serine	5.30	Leucine	5.80
Glutamic acid	17.82	Tyrosine	1.82
Probine	3.39	Phenyl alanine	3.41
Glycin	4.13	Tryptophan	0.80

Castor cake has been used to produce distempers, oil-bound water paints, adhesives for wood and products similar to casein plastics. Castor cake has also been used in drilling fluids, lubricating plastics and fire-fighting foams.

Castor seed contains a lipase capable of hydrolysing glycerides and esters; this is the most active of several enzymes present (an amylase, invertase, phytase, maltase and endotrypsin). High purification of the enzyme lowers its stability; a mixed enzyme preparation obtained as a fluffy powder by decorticating castor seeds, thoroughly extracting the fat and then milling keeps its activity even for years. The enzyme method was actually used for commercial hydrolysis of fat especially in Germany, in spite of troublesome emulsion formation. The active enzyme is rapidly inactivated in an acid medium in the absence of fats, but in their presence it is more stable. It is active only within a very narrow pH range, 4.7–4.8. On the basis of the number of molecules hydrolysed, the enzyme shows no difference in rates of specificity for different oils. Various Indian castor seed varieties have been used to obtain active enzyme preparations, the best of which hydrolysed 89 per cent of an oil within 24 hours. Hydrolysis of unsaturated oils showed selective liberation of the more unsaturated acids, especially in the early stages.

Castor Hulls

Castor hulls constitute 25.6 to 33.2 per cent of the seed and contain 1.64 per cent N, 0.82 per cent P, 3.81 per cent K and 10.38 per cent ash. They also contain 7–7.6 per cent of a viscous dark green oil the fatty acid composition of which is—Palmitic acid 1.8–4.8 per cent, stearic acid 0.8–1.2 per cent, oleic acid 5.3–5.8 per cent, linoleic acid 6.7–7.2 per cent, ricinoleic acid 81.4–83.9 per cent and dihydroxy stearic acid 0.6 per cent. Castor hulls have also been utilised for preparation of good quality active carbons.

Castor Stalks

Castor stem-pulp is believed to be used for the manufacture of paper; in India, laboratory and pilot plant trials have shown that it can be used for production of writing, printing and wrapping papers. The use of 4 per cent alkali for digestion at 80 lb per sq. inch gave a 40 per cent yield of bleached pulp leaving 30 per cent resistant fibre while in other work a 57 per cent yield of pulp was obtained from unbarked castor stalks by the sulphate process. The beaten pulp is short-fibred and mixed with about one-third of bamboo-pulp has been used to obtain writing, printing and wrapping papers of good formation and satisfactory strength and like mechanical wood-pulp which is also short-fibred, may be used alone to produce newsprint where strength is not of primary importance.

APPLICATION OF CASTOR OIL AND ITS DERIVATIVES IN LUBRICANT INDUSTRY

Castor seed is produced in India, Brazil and China. India is the major castor producing country in the world. As only one third of the quantity of this oil goes for domestic consumption, the balance has to be exported.

Therefore there is a need to look into the value addition of castor oil and its products so that value added products can be exported which is likely to earn more price. Also there is a need to look at alternate uses of this oil in lubricant industry. In the present section attempts have been made to examine the other possible applications of castor oil/its derivatives in lubricant industry.

Chemistry of Castor Oil

Castor oil is a triglyceride of Recinoleic acid and consists approximately 90 per cent of this acid. The other fatty acids present are linoleic (4.2 per cent), Oleic (3.0), Palmitic (1.0), stearic acid (1.0), dilhdroxy stearic acid (0.7) linolenic acid (0.3) and eicosanoic acid (0.3). The molecule ricinoleic acid has three functionalities viz., ester, hydroxyl group and a double bond. Therefore the molecule shows three types of the reactions based on these functionalities. The various types of reactions and their products are given as under:

1. Derivatives from ester linkage:

Hydrolysis	Fatty acid, glycerol
Esterification	Esters
Saponification	Soluble/insoluble soaps
Alcoholysis	Mono and diglycerides
Reduction	Alcohols
Amidation	Amine salts, amides
Halogenation	Fatty acid halogens

2. Derivatives from double bond:

Oxidation/polymerisation	Polymerised oils
Hydrogenation	Hydroxystearates
Epoxidation	Expoxidised oils
Halogenation	Halogenated oils
Addition reactions	Polymerised oils
Sulphonation	Sulphonated oils

3. Hydroxyl group.

Caustic fusion	Sebacic acid, capryl alcohol
Pyrolysis	Undecylenic acid, heptaldehyde
Halogenation	Halogenated castor oil
Alkoxylation	Alkoxylated castor oil
Sulphation	Sulphated castor oil
Esterification	Esters, phosphate esters, etc.
Urethane reactions	Urethane polymers

The above reaction described the versatile reactivity of the molecule which indicates that this oil can be used for a number of applications either as such or its derivatives.

Applications of Castor Oil/Derivatives in Lubricant Industry

Castor oil is the only renewable vegetable oil resource having a hydroxyl group and other functionality that leads to diverse oleochemicals. This oil is unique among all the fats due to its following features:

1. It is the only source of an 18-carbon hydroxylated fatty acid with one double bond.
2. Ricinoleic acid (12-hydroxyoleic acid) comprises approximately 90 per cent of the fatty acid composition.
3. Product uniformity and consistency are relatively high for a naturally occurring material. .
4. It is a non-toxic, biodegradable and renewable source.

The hydroxyl groups in castor oil account for a unique combination of physical properties, i.e. relatively high viscosity and specific gravity, solubility in alcohols in any proportion, limited solubility in aliphatic petroleum solvents. Because of its higher polar hydroxyl groups, it is not only compatible with but will plasticise a wide variety of natural and synthetic resins, polymers and waxes. Castor oil has excellent emollient and lubricating properties. In the form of chemical derivatives, castor oil's application versatility is further enhanced.

Castor oil/derivatives find applications in lubricating grease production, synthetic esters, textile auxilliaries, 2T/4T oil pre-mixes, metal forming industries, aviation lubricants and mould and casting lubricants.

Synthetic Esters from Castor Oil

1. Sebacic acid, produced by alkali fusion of castor oil is the raw material for a series of esters used as non-volatile, high temperature lubricants for jet engines. The synthetic esters produced from sebacic acid and 2-ethylhexanol (di-2-ethylhexyl sebacate) provides excellent wide temperature applications as the ester which has very low pour point (<–60°C), high viscosity index (150) and low volatility. These properties make it an excellent lubricant for aviation applications as the requirements placed on the jet engine lubricant in terms of stresses cannot be met by hydrocarbon oils.
2. Diesters of sebacic acid are used to make low temperature plasticisers for PVC. These esters provide improved flexibility and resistance to cracking at low temperature. These are usually added as a partial replacement for primary plasticisers. Dioctyl sebacate is also used as corrosion inhibitor in long life automobile radiator coolants.

Castor Oil Derivatives in Textile Industry Lubricants

1. Castor oil is used as starting material for many textile finishing agents. Fatty acids obtained by pyrolysis of castor oil are used for production of polyamides (nylon 11) which is used as an engineering plastic in automotive and textile industry in the form of fuel lines.

2. Sulphated castor oil (Turkey Red Oil) is used in compositions for treatment of fabrics, animal skins and furs to clean and revive colours.
3. Sulphated castor oil is incorporated with mineral oil and isopropylamine salts of C12–C18 fatty acids as a lubricant for polyester fibers. The lubricant minimised undesirable abrasion and facilitates subsequent weaving operations.
4. Castor oil can be transformed from an oil to a water soluble surfactant depending on the moles of ethylene oxide added to its hydroxyl group. Different types of castor oil ethoxylates are used as textile finishing agent (Table 7.55).

Table 7.55. Castor oil ethoxylates as textile finishing agents.

Castor oil exthoxylate	*Applications*
POE (5) castor oil	Emulsifier for mineral oil, dispersant for pigments and clay
POE (16) castor oil	Water dispersible emulsifier finish additive for emulsifying esters
POE (30) castor oil	Emulsifier, dispersant stabiliser for latex emulsions
POE (80) castor oil	Dyeing assistant and fabric lubricant
POE(200) castor oil	Fabric softener, good antistatic agent
Ethoxylated hydrogenated castor oil	Softener and finishing agent

Use of Castor Oil in Metal Working Operations

1. Due to the presence of ricinoleic acid and hydroxyl group, castor oil as such has excellent lubrication properties. In emulsifiable rolling oil, castor oil is used in combination with mineral oils or other fatty oils to get the desired lubricity.
2. Castor oil has been used as a component for the formulation of a non-toxic lubricating cutting oil for metals at 700–1200°C.

Though castor oil has good lubrication properties, it is not used as neat rolling oil formulations due to its poor oxidation stability and tendency for rancidification.

Applications of Castor Oil in Biodegradable and Environmental Friendly Lubricants

Castor oil is a renewable resource made up of long chain fatty acids and glycerine. The oil is biodegradable which makes it a highly desirable raw material for many applications subject to environmental regulations. Esters made from natural fatty acids in castor oil are used for formulations of biodegradable hydraulic fluids, compressor oils, Gear oil, chain saw oil and wire rope lubricants. These biodegradable lubricants are used in environmentally sensitive areas like forests, mines, agriculture, railway tracks, shop floors, etc.

Vegetable oil based hydraulic fluids have a high flash point, high viscosity index and good wear characteristics. In order to achieve significant market penetration in this area, technical challenges of generally poor oxidative stability must be overcome. To deal with this issue, three avenues are being pursued:

1. Genetic modification of vegetable oil to higher mono-unsaturated compounds.
2. Chemical modification (partial hydrogenation) of the oil.
3. The use of various additives and property enhancers.

Castor Oil Derivatives in Grease Industry

Hydrogenated castor oil/hydroxystearic acids are used in the lubricating grease formulation to make lithium/lithium complex, calcium hydroxy stearate and Lithium-calcium greases. The main advantage of castor oil in grease preparation is its advantage of making twisted soap like structure compared to straight chain structure of stearates, which results into higher yield and better oil retention capability. Approximately more than 70 per cent of greases world wide are made with hydrogenated castor oil/12-hydroxy stearic acid. The characteristics of hydrogenated castor oil (HCO) and 12-hydroxy stearic acid are given in Table 7.56.

Table 7.56. Quality of HCO and 12-hydroxy stearic acid for grease formulations.

Property	*Hydrogenated castor oil*	*12-Hydroxy stearic acid*
Iodine value gl/100g, maximum	2.0	2.0
Acid value mg KOH/g, maximum	2.0	180
Hydroxyl value	155	155
Melting point (°C) minimum	85	> 72
Saponification number (mg KOH/G)	176–186	176–186

The importance of 12-hydroxy stearic acid has been recognised by the grease industry for the last 40 years. The long chain fatty acid structure, the ability to form hydrogen bonds and its excellent compatibility characteristics make it the premier choice for heavy duty grease applications. The presence of hydroxyl groups allows hydrogen bonding with other soap molecules, which causes the formation of crystalline aggregates and long textured fibres upon cooling of the grease. Both of these properties impart strength and workability to a grease formulation. Table 7.57 shows the improved properties obtained in grease using 12-hydroxystearic acid based soaps.

Table 7.57. Properties of metal soap thickened greases.

Type	*Dropping point (°F)*	*Shear stability*	*Water resistance*	*Texture*
Hydrated calcium stearate	205–220	Fair	Excellent	Buttery, small fibres
Calcium 12-hydroxy stearate	275–290	Fair/good	Excellent	Smooth, long fibres
Sodium stearate	325–350	Fair/good	Very poor	Long fibres
Lithium stearate	350	Good	Very good	Short fibres
Lithium 12-hydroxy stearate	375–400	Excellent	Very good	Short fibres
Calcium complex 12-hydroxy stearate	500+	Fair/good	Fair/excellent	Smooth
Lithium complex 12-hydroxy stearate	500+	Good/excellent	Good/excellent	Smooth

Sebacic acid is also used in making lithium complex grease which has the dropping point of more than 260°C and enhances the operating temperature range of lithium grease by about more than 50°C. It was recognised early in the development of lithium based greases that under conditions of high shearing forces, these lubricating greases begin to breakdown and loose their lubricating properties. But if esters of sebacic acid were added to the lubricating greases as substitutes for mineral oils, then the greases offered a comparatively uniform viscosity over a wide temperature range and were not subjected to evaporation loss. The addition of diesters of sebacic acid imparted shear resistance at very low temperatures such as in refrigeration equipment, arctic service equipment and air craft operation and for food industry.

Future Projections: Castor Oil/Synthetic Blends as Fuel

Castor oil/synthetic blends have been projected as fuels after incorporating extreme pressure, friction modifier and antifoam additives. The castor oil used for this application is modified castor oil which is first pressed and filtered to give the unique lubricating properties of castor without excessive build up and carbon. This fuel improves power and performance through a dramatic decrease in friction.

Thus castor oil is a versatile molecule which opens a number of ways for its modifications into different desired molecule. Attempts should be made to export value added products instead of the oil. Also studies should be conducted to modify the molecule to get its applications in fuel area. Future projections for castor oil uses are: (i) synthetic base fluids for biodegradable oils; (ii) conversion to make it suitable for biofuels; (iii) production of sebacic acid to cater for growing automobile industry and other applications; and (iv) looking for use in 2T/4T oils along with synthetic fluids.

CASTOR OIL EMERGING AS VITAL CHEMICAL INDUSTRY INPUT

Castor oil has emerged as a very promising raw material for chemical industry. Due to its significance in industry emphasis is now being given to cultivation of castor trees. Castor is one of the major oil seed crops of India. Crude castor oil is yellow to brown in colour, while the refined castor oil is nearly colourless or very pale yellow viscose liquid having mild taste and odour which soon becomes unpleasant. The oil is distinguishable from most other oils by its high viscosity, specific gravity and acetyl value, its solubility in absolute alcohol and glacial acetic acid and poor solubility in petroleum ether, gasoline, kerosene etc. Castor oil is dextro-rotatory due to the asymmetric carbon atom of ricinobic acid present in it.

The major fatty acid component of castor oil is ricinobic acid as high as 90 per cent and is responsible for high viscosity and other unique characteristics of oil. Castor oil processed in various forms has been found to have applications in few hundred products. Significant research and development work has been carried out in various laboratories of CSIR. Regional Research Laboratory (RRL) has developed a modified process for the refining of castor oil. It is well known that castor oil emulsifies partly or completely to a gel in the presence of strong alkalies usually used in the refining of other vegetable oils and fats. The refining losses of castor oil are, therefore, extremely high. Hence, the term refining, when applied to castor oil, usually implies merely precipitation of mucilage after hydration, bleaching and filtration. The laboratory has successfully refined this oil using aqueous sodium carbonate, whereby a reduction in free fatty acidity of the oil to a very low level is accomplished with minimum losses in the form of solvent. The oil refined this way responds more effectively to the subsequent bleaching procedure as compared with the direct bleaching of the oil conventionally employed. Refined castor oil is used in cosmetics, pharmaceutical and insecticidal applications. Castor oil is often administered orally alone or with quinine sulphate to induce labour pains in pregnancy. It has variety of industrial applications, for example, castor oil is used as a lubricant in engineering industry.

Hydrated castor oil (HCO) is a valuable constituent in organic surface coatings such as non-drying alkyds, plasticisers, antisagging and antisettling agents of pigments in paints. It is also used in the production of insulating materials, coating composition for paper, textile and leather, plastic moulding compositions, heat sealing devices, etc. HCO is the starting material for the production of 12-hydroxy stearic acid which is extensively used in the formulation of multipurpose and other high quality greases, pharmaceutical preparations, ointments, cosmetics and as replacement for natural hard waxes. The product has tremendous market requirement by these industries.

Sulphated castor oil or turkey red oil (TRO) as it is commercially called, is an important textile auxiliary and it is used on a large scale by almost all cotton and silk manufacturing units as a wetting out agent. In leather industry, it finds use in fat-liquoring. The product is manufactured by the reaction of 98 per cent sulphuric acid with castor oil (25–30 per cent by weight) for five to six hours.

National Chemical Laboratory (NCL) has developed two types of polyurethane coatings based on castor oil as raw material. It is a versatile coating for leather goods, rubber, wood, sport goods, glass and metal substrates. Second type, finds application for nylon fabrics which are used in the manufacture of umbrella, cloth, tarpaulins, high weight high altitude tents and book binding. The products have excellent adhesion, hardness and flexibility. The process of manufacture consists of reacting toluene di-isocyanate with polyol such as castor oil under agitation for the former applications. The product is further diluted before use with suitable solvents. The second formulation is prepared by reacting polyester resin with isocyanate. The resin is prepared from chellac, glycol, castor oil and adipic acid.

Most of the printing presses use both gelatine (glue) and rubber type rollers which have several inherent disadvantages. The other type of rollers used in modern presses are based on synthetic rubbers, PVC and more recently polyurethane. The later type of rollers were not available indigenously till NCL came forward with the development of polyurethane printing rollers based on castor oil. The product is now in production. The rollers manufactured according to the NCL process have been tested on small, big and faster printing machines by leading printers in the country and found to be satisfactory.

These rollers distribute ink uniformly with proper tack on low and high speed machine. These are not affected by climatic changes, have wear resistance and more durable than the gelatine rollers.

In the field of cellular plastics, the technology of rigid urethane foam has developed rapidly during the last few years. Manufacture of polyols has been standardised by NCL.

The preparation of microcellular flexible polyurethane on laboratory scale has been carried out. Several mouldings of soles and sheets were made and the physical properties, of the rubber such as shore 'A'-hardness, density, tensile, strength, tear strength and abrasion properties—were evaluated. Work is in progress for the standardisation of conditions for the preparation of urethane elastomer.

It has been observed that polyurethane will be used for varied applications including replacement of sheet metal in car and auto body making. Due to high demand in coming years, cultivation of castor seeds should be encouraged to achieve greater yield in semi-arid climatic conditions.

QUALITY ASPECT

Oil Content

More than 2000 genotypes have been assessed for oil content. The oil content among genotypes ranged from 25–54 per cent. The majority of genotypes contain oil content from 45 to 50 per cent. Since wide genetic variability exist, oil content can further be improved following appropriate breeding method.

Physico-Chemical Property of Castor Oil

The castor oil is unique among vegetable oils with regard to its physical property. The distinct properties of castor oil are high acetyl or hydroxyl value, high specific gravity, high refractive index, high viscosity, low saponification value, low iodine value and strong extra rotatory nature (Table 7.58).

Table 7.58. Physico-chemical property of castor oil.

Parameters	*Range*
Iodine value	82–90
Saponification value	179–187

(Contd...)

Table 7.58. Physicochemical property of castor oil. *(Contd.)*

Parameters	*Range*
Acid value	3–4
Specific gravity (at 25°C)	0.958–0.969
Refractive index (at 25°C)	1.477–1.487
Optical rotation (in a 20 mm tube)	+ 7.6–9.7

Fatty Acid Profile of Castor Oil

Fatty acid composition of over 200 germplasm lines including some of accessions collected revealed that the most important fatty acid ranged from 79.0–94.6 per cent. This shows the existence of wide genetic variability for this important fatty acid and thus, the ricinoleic acid content can be improved in castor genotypes following hybridisation and selection (Table 7.59).

Table 7.59. Fatty acid profile of castor oil.

Fatty acid	*Range (%)*	*Mean (%)*
Ricinoleic acid	79.0–94.6	87.0
Linoleic acid	2.4–9.4	6.1
Oleic acid	1.7–7.3	4.0
Arachidic acid	<3.0	1.2
Palmitic acid	<1.9	1.1
Stearic acid	<1.4	0.6

Castor Seed Meal

The protein content of castor seed meal varies from 21–48 per cent depending upon the extent of decortication. It has an ideal amino acid profile with moderately high cystine, mithionine, and isoleucine. But its antinutritional substances, ricin, ricine and allergen restrict its use in poultry feed even at low level of inclusion. Processing of castor meal by heating at 80°C temperature in the presence of 6 N ammonia or an acid or alkali reduces the toxins substantially. Detoxified meal need to be effectively corrected for lysine. Its net protein utilisation (49.39 per cent) is higher than other oilseed meals used in poultry.

Castor cake is an excellent fertiliser because of high content of N (6.4 per cent), phosphoric acid (2.55 per cent) and potash (1 per cent) and moisture retention.

Anti-nutritional Factor

The ricinine and ricin are two antinutritional factors found in castor seed. The ricinine is the alkaloid of castor oil, a product of specific nitrogen metabolism. It is monocyclic derivative of pyridine with the cyanide group. The toxicity of ricinine is selective. The caterpillars of Torticidae die due to it but for human beings it is less toxic. The endosperm contains only a small quantity of ricinine (30.8–70.7 mg per cent).

The quantity is much greater in seed hull (257.6 to 431.2 mg per cent) and the capsule shell (738.9 to 1664.4 mg per cent). There is negative correlation between the contents of ricinine in castor seeds and oil content (r = –0.76). Great deal of genotypic differences exist for this alkaloid in castor seed. Agro climatic conditions also affect the content of ricinine in castor seeds. Toxic protein in castor seeds is called ricin. Ricin is a phyto-toxin and is one of the most poisonous substances known.

The biogenic relationship exist between the synthesis of ricinine and ricin. Synthesis of toxic protein (ricin) is accompanied with the decrease in the concentration of ricinine of the seeds. Ricin is a solid solution of several proteins: of allergen and toxin. Protein allergen acts selectively causing asthma, dermatitis and eczema. The toxin is a protein of the globular type capable of causing agglutination of suspension of erythrocytes. After extraction of oil, ricin remains in the oil cake. The biological role of ricin remains inadequately studied. There are data that high ricin containing genotypes are less contaminated by fusarium and are less affected by leaf damaging pest. There is negative correlation between ricin and oil content in castor seed. Ricin is concentrated in the endosperm of the seed and is totally absent in other parts, as well as in the vegetative organs of plants. Varietal differences are reported to exist with respect to toxicity of the protein.

Since varietal differences exit with respect to the alkaloid, ricinine and toxic protein, ricin, there is possibility to reduce these chemical constituents through selection following hybridisation with suitable donors. The toxic protein and ricinine can also be reduced/eliminated following biotechnological approaches. The work in the direction has already been initiated in US with great success.

Industrial Uses

Castor oil is among the most versatile natural products. There are several hundreds of industrial applications of castor oil and its derivatives. The major derivatives of castor oil used in the industry of hydrogenated castor oil (HCO), dehydrated castor oil (DCO), sebacic acid, ethoxylated castor oil, polyols, 2 octanol, undecylenic acid, heptaldehyde, turkey red oil etc. and major uses of these derivatives are in the production of lubricants, polymers, plasticisers, plastics, adhesives, waxes, polishes, coating, inks, paints, varnishes, cosmetics, perfumes, flavours, fragrances, textile dyes etc. There is urgent need to evaluate the castor oil for its derivatives and industrial uses. If genotypes differences exist, suitable/ appropriate varieties can be developed to fulfill specific industrial requirement.

ORIGIN ANALYSIS AND TECHNOLOGY DEVELOPMENTS OF CASTOR OIL AND ITS DERIVATIVES

India is the large grower and exporter of castor oil in the world. The span of these exports range from South east Asia, Japan, USA, and many European countries. The product range includes various grades of castor oil, hydrogenated castor oil, hydroxy stearic acid, ricinoleic acid etc. The end applications of these oils and derivatives cover a range of products such as pharma-grade oils, resins, paints, plasticisers, lubricants and food emulsifiers. The quality requirements for these applications vary significantly. The international quality standards are available, however, in some cases the quality specifications can also be decided mutually between the consumer and supplier.

Stringent Quality Requirements for Exports

Lately there is increasing demand from the consumers of castor oil and derivatives on making specifications for them more and more stringent e.g. consumers from Europe and Japan are requesting COAs for pesticide residues, PAH, trace metals and for presence of poisonous substances like ricin. Each of these groups contain various components and requires specific and specialised technique for analysis. Eventually this may amount to meeting EEC/EFET/GSCL specifications and legislation.

In this section we plan to review some of the aspects related to origin, analysis and technology development for measurements and control of these substances from castor oil and derivatives.

Polycyclic Aromatic Hydrocarbon (PAC)

Origin of PAH

Polycyclic aromatic hydrocarbons originate from partial burning of organic matter. They are most commonly derived from burning of hydrocarbon fuels especially diesel. There are two categories of these hydrocarbons: light and heavy. Heavy PAHs are known to be carcinogenic to laboratory animals and consequences of PAHs in food, in terms of consumer safety are unclear at this moment. The European Scientific Committee on Foodstuffs is working to asses these risks.

PAHs were first brought to notice of oil industry when high levels of these were detected in crude coconut oil. More recently they have been found in rapeseed oil, sunflower oil and olive oil. It is now recognised that the main source of contamination is diesel fired seed dryers and through soot intake from environment. They were also found in the oil derived from seeds/pomace dried using direct heat from solvent extracted cake/extracted pomace. Temperatures above 350°C are known to produce much greater concentration of PAH. Type of process used, temperature of combustion and time of the exposure to the fumes determines the level of PAHs accumulated in the oil.

Analysis of PAHs

There are about 40 known compounds in this category and based on the molecular weight they are divided in two categories; light PAHs and heavy PAHs. 13 members have been detected in various vegetable oils.

Light PAHs

Light PAHs (i) fluorene; (ii) phenanthrene; (iii) anthracene; (iv) fluoranthene; (v) pyrene; (vi) benzo; and (vii) chrysene.

Heavy PAHs

Heavy PAHs (i) benzo (b) fluoranthese; (ii) benzo (k) fluoranthene; (iii) benzo (a) pyrene; (iv) indeno (1,2,3 cd) pyrene; (v) dibenzo (ah) anthracene; and (vi) dibenzo (ghi) perylene.

Light PAHs are found at higher levels in oil and are considered to be non-carcinogenic, heavy ones are at lower level but are difficult to remove and believed to be carcinogenic. The accepted maximum levels of PAHs by European trade association are 25 ppb as total PAH, 5 ppb for heavy and 1 ppb for benzo (a) pyrene. Statutory limits have not yet been set.

Isolation procedures

Isolation of PAH involves extraction of oil using acetonitrile-acetone mixture and then separation of PAHs using HPLC equipped with fluorescence detectors.

Methods/technologies for removal of PAHs

Steam de-odourisation of oils

This technique helps to remove light PAHs. Heavy PAH are not steam volatile. Bleaching with at least 1 per cent activated special grade charcoal is also known to adsorb some of the PAHs but again the method is not foolproof.

Indian scenario

Castor oil is produced in India using three processes: (i) cold pressing; (ii) standard expelling after cooking and steaming of seeds; and (iii) solvent extraction. In none of the above operations, seeds or oil gets exposed to direct heat from fuel burning or exposed to very high temperatures. Therefore chances of PAHs occurring due to these reasons are minimal. However, exposure to boiler soot through air might lead to presence of PAH in castor oil. Similar situation exists in manufacture of derivatives.

Regardless of the cause, once consumer demands certification of the oil, analysis must be performed and certificate must be issued based on the analysis results.

Trace Metal Impurities

There are customers who seek the certification of the export consignment for the analysis of heavy metals in it. These metals are lead, arsenic, mercury, nickel and copper. They are known to be toxic in nature.

During manufacture of castor oil/derivatives the heavy metal contamination can occur from the soil, or from processing chemicals used for bleaching, hydrogenation etc. Contamination can also originate from various grades of water/acids used in the process.

Methods for isolation and estimation

A sample of oil/derivative is ashed completely and the residue is dissolved in nitric acid. Diluted nitric acid solution can then be analysed by atomic absorption spectroscopy or by internal combustion plasma analysers. As these metals are present in very small amounts (ppm-ppb), use of sophisticated instruments is necessary to provide accurate and consistent measurements.

Indian scenario

Qualitative tests for analysis of total heavy metals up to 10 ppm are available and many standard laboratories have capability to perform the tests. However, quantitative analysis of individual trace metals need sophisticated instruments as mentioned earlier.

To begin with, there is a need to develop qualitative and easy to do tests for individual elements, which can be implemented at the production site.

Pesticide Residue Analysis

Some customers are demanding certification of the products for analysis of pesticide residues. There are several pesticides used from organo chlorine, organo phosphorus and organo sulphur compounds. Government legislation in European countries requires analysis of various pesticides in food grains, oil seeds, vegetables and vegetable oils. Castor beans or oil are not listed in any of the American or European list of the products requiring residue analysis.

In future these demands are projected to become more challenging and Indian industry must be prepared to take up this issue.

Indian scenario

In India, not many farmers are known to use pesticides for castor crop. Those who use are also known to use them in a very limited fashion. However, seeds from these crops might get mixed with other seeds in the market. Therefore castor oil from Indian industry might have a risk of having pesticide residues. It is very difficult to trace back the origin of oil to seeds and then to conclude on presence or absence of pesticides in the resultant oil.

Currently there are sporadic queries to meet the EEC standards by some customers and this may become a general practise. There are 32 pesticides listed in EEC. Standard Indian laboratories have capability to analyse 16-pesticide residues from this list. Methods for others need to be developed and standardised.

Toxic Substances

Castor beans contain one toxic substance ricin and two allergens. All these materials are proteinaceous in matter. Ricin, a cyctotoxic glycoprotein is one of the most potent toxic substances known to man. The poisoning symptoms involve gastrointestinal, hepatic, renal toxicity and malfunctioning of central nervous system.

There has been concern in the world about misuse of ricin. Customers world-wide have been raising queries about the presence/absence of these toxic substances in the castor oil and its derivatives and seeking certification for its absence.

Isolation and detection methods

Ricin being a protein, is hydrophilic in nature and is water-soluble. Due to this property it is very unlikely that it will be present in castor oil. In addition, normal degumming and bleaching processes in castor refining further ensure its denaturisation, if at all it finds its way in oil. However, on customer request one must analyse the sample and then issue the report based on the analysis results. Unfortunately as on today there is no conclusive evidence/method to indicate the level of ricin (if at all there) in the oil and the derivatives.

There is need to develop technique to detect ricin at this level in the oil and derivatives. Further there are other two potent allergens in castor. It is therefore necessary to develop methods for estimation of these allergens as well.

Techno-Commercial Issues

As it is clear from the above discussion, certification of the castor products after analysis of the trace impurities requires sophisticated instruments, trained manpower and well equipped laboratories and factories. Moreover at the moment, there are no standard methods commonly accepted by Indian manufacturers for estimation of these impurities. An estimated analysis for one PAH would cost Rs. 2000/ analysis and this could prove to be commercially unjustifiable.

It is therefore suggested that:

1. A common agreed method for analysis of these trace impurities should be developed which can be done without involving expensive chemicals and equipments.
2. These methods should be implemented in the accredited laboratories.
3. Technologies need to be developed to eliminate or at least minimise these impurities from the castor oil and derivatives.

Chapter 8

Rice Bran Oil

INTRODUCTION

Fat or oil is the important ingredient of human diet. Human body daily requires requisite quantity and quality of this essential nutrient for smooth running of its all activities and overall good health. Judicious choice of oil with its controlled usage is a must for sound physical and mental health in today's action packed, stressed life-style. At present, no single available oil can be labelled as 'ideal oil' for cooking. However, rice bran oil (RBO) with balanced fatty acid profile and other nutritional compounds can respond to an ever increasing consumer interest in the health attributes of food.

Value added products like oryzanol, squalene, wax, gums, lecithins, tocopherols and tocotrienols from RBO refining should be encouraged in order to rejuvenate RBO industry with good economic returns and more jobs for the people in this situation of unemployment.

Rice bran oil is a 'heart friendly health oil' and it is available in every city and village in consumer packs with the help of extensive marketing network all over the country.

Blending of other edible oils like sunflower, safflower, soyabean and rapeseed/mustard to be encouraged with Rice Bran Oil. This will enhance the nutritional properties, balanced fatty acid composition, oxidative stability and frying performance of blended oils. This will be more beneficial to the consumers as regards their overall health than their previous established oil in use by them. Micro-nutrient profile of rice bran oil are given in Table 8.1.

Table 8.1. Micro-nutrient profile of rice bran oil.

Micro-nutrient	*Amount %*	*Advantage*
Tocopherol	0.02–0.08	Antioxidant, free radical scavenger, reduce risk of cardio vascular diseases, arthritis, cancer, cataract, antitumour activities.
Tocotrienol	0.025–0.17	Cholesterol reduction, reversing atheroscelerosis, anticancer (breast, liver) tumour supression, antioxidant.
Oryzanol	1.2–1.7 (physically refined RBO)	Elevate HDL cholesterol, decrease plasma cholesterol, treat nerve imbalance, menopause disorder, retard ageing effects, antidandruff and anti itching agent
Squalene	0.3–0.4	Antioxidant

India is the second largest producer of rice in the world next to China and has a potential to produce about 10 lakh tonnes of rice bran oil (RBO) from about 68 lakh tonnes of rice bran. Saturated/unsaturated fatty acid profile of rice bran oil are given in Table 8.2.

Table 8.2. Saturated/Unsaturated fatty acids profile of rice bran oil.

Oil	*Fatty acid % wt.*		
	Saturated (SFA)	*Mono-unsaturated (MUFA)*	*Poly-unsaturated (PUFA)*
Mustard/Rapeseed	6	67	27
Cottonseed	28	22	50
Sunflower	12	21	67
Safflower	10	15	75
Soyabean	16	24	60
Palm	51	39	10
Olive	14	77	9
Canola	6	58	36
Corn	13	20	62
Coconut	92	6	2
Palm kernel	86	12	2
Groundnut	20	50	30
Rice bran	**18**	**45**	**37**
Recommanded by WHO	**28.6**	**42.8**	**28.6**

Note: Rice bran oil with 'ideal' SFA/MUFA/PUFA ratio which is the closest to WHO recommendation as compared to other edible oils.

Rice bran oil is a unique gift of nature and is a miracle product with unique cholesterol lowering property of bad cholesterol, (LDH) and increasing the good cholesterol (HDL) which is beneficial to heart. It has other outstanding nutritional properties and contains micro-nutrients like tocopherols, tocotrienols, oryzanol, squalene and phytosterols. In addition it has a nicely balanced profile of saturated and unsaturated fatty acids which helps to maintain good health. Rice bran oil also gives excellent value added products like oryzanol, IP-6, squalene, ferulic acid esters, which are very useful for cosmetic and pharmaceutical industries. Hypocholesterolemic activity of various edible oils are given in Table 8.3.

Table 8.3. Hypocholesterolemic activity of various edible oils.

Edible oil	*Linoleic acid %*	*Cholesterol level*
Safflower	77.1	-16
Sunflower	61.4	-12
Cottonseed	58.0	0
Soyabean	50.2	+3
Sesame	45.9	+2
Corn	43.0	-15
Rice bran	**36.0**	**-17**
Groundnut	35.0	+5

Note: Rice bran oil is the 'heart friendly' cooking oil due to its remarkable cholesterol lowering property in comparison with other edible oils. (+) increase and (–) decrease in cholesterol level.

Chemical composition and specifications of rice bran oil are given in Tables 8.4 and 8.5.

Table 8.4. Chemical composition of rice bran oil.

Composition	%
Triglycerides	80.5
Free fatty acids	6.8
Diglycerides	4.8
Monoglycerides	1.7
Oryzanol	2.0
Phosphatides	1.3
Wax	2.9

Table 8.5. Rice bran oil—specifications.

Characteristics of refined rice bran oil	
Iodine value	99–108
Saponification value	180–190
Smoke point	213°C
Fire point	352°C
Cloud index	17°C
Refractive index 25°C	1.470–1.473
Specific gravity 25/25°C	0.916–0.921
Unsaponifiable matter	3–5%
Total tocopherols	200 mg/kg (0.02%)

NUTRITIONAL SIGNIFICANCE OF RICE BRAN OIL

Basics

Fat is an important ingredient of human diet. The functional role of fat in the diet is manifold. Fat is a most concentrated source of energy. One gram of fat gives 9 kcal of energy. Carbohydrates and proteins provide only 4 kcal of energy per gram. Hence, fat provides calorie density to the diet. Secondly, fats are essential in the diet for the absorption and mobilisation of fat-soluble vitamins such as vitamin A, vitamin E and fat-soluble antioxidants. These vitamins are not utilised by the body if fat is not available in the diet. Thus, fat works as a vehicle to carry the fat-soluble vitamins, nutrients and antioxidants in the body. Thirdly, vegetable oils are the only sources of essential fatty acids to the body.

Essential fatty acids (EFA) are those fatty acids, which the body cannot synthesise and need to be supplied through diet. EFA's are the precursors for a group of chemically related compounds, known as prostaglandins that are synthesised in the body from EFA. If EFA is not supplied through diet to the body, the body cannot synthesise prostaglandins. Prostaglandins play a key role in regulating many physiological processes in the body, such as controlling blood pressure, vascular damage in the brain and the heart, preventing blood clot in the arteries, lowering cholesterol, uterine contractions during child birth and menstrual cycles, inflammation and other conditions.

Fourthly, fat also helps in raising HDL the so called good cholesterol. Low-fat diets can result in reduction of HDL cholesterol. Fifthly, fat in the diet imparts certain textural qualities, taste and palatability to the food. Fats and oils are integral lubricants of foods in two ways: (i) through use as release agents as a part of the cooking process; and (ii) as a lubricant during mastication (chewing). Fats and oils modify flavour release and improve palatability.

Thus fat is an essential nutrient in our balanced diet. Since several studies have directly implicated the amount and type of fat intake to specific diseases such as cardiovascular disease, cancer, high blood pressure and obesity, so one has to be very particular as to quantity as well as quality of fat in the diet to maintain good health.

The quantity and quality of fats determine the intakes of various fatty acids in the total diet. Several studies have indicated that the hypocholesterolemic action of a dietary fat depends on the fatty acid composition and also to some extent on the micro-nutrients like tocopherols and tocotrienols present in them. The optimal balance of saturated, polyunsaturated and monounsaturated fatty acids make an ideal oil.

Quantity of Fat

According to the American Heart Association, the optimum intake of fat for an adult is 30 per cent of its total caloric intake. Therefore for an adult man consuming 2000 calories, the upper limit of total fat (visible and non-visible) is 65 grams/day.

Apart from visible fat some amount of fat is present in food items like cereals, pulses, milk, eggs, meat etc. as invisible fat. From the data base generated on fatty acid composition in Indian foods and National Nutrition Monitoring Bureau diet survey, the daily invisible fat intake is estimated to be 15 grams in rural poor and 25 grams in the urban middle and high-income groups. Thus the daily requirement of visible fat intake works out to be some where between 20 to 50 grams depending upon physical activity and physiological status.

Quality of Fat

Choice of a healthy cooking oil is a controversial subject. The ideas keep on changing as new evidence accumulates. Desi ghee (butter oil) used to be the conventional cooking oil of the Indians. There was nothing wrong with this most favourite cooking medium of Indian households. But the problems started with the change in our eating and living habits. Desi ghee being rich in saturated fats has the tendency to increase total cholesterols including HDL (so called 'good cholesterols') and LDL and VLDL (so called 'bad cholesterols'). As such consumption of desi ghee coupled with fibre rich diet which reduces LDL bad cholesterol and physical activity which reduces VLDL bad cholesterol brings about a favourable lipid profile. In the good olden days, our forefathers used to take fibre rich diet such as chapatties made out of flour of wholegrains and lots of leafy vegetables along with desi ghee and the nature of work involved lots of physical labour. All this used to neutralise the adverse effects of desi ghee on LDL and VLDL while retaining its positive effects on HDL and thereby favourite cooking medium used to bring about a favourable lipid profile.

But with the lack of physical activity in the recent years coupled with intake of diet devoid of fibre such as chapatties and other snack foods made from flour of decorticated foodgrains, consumption of desi ghee started raising total cholesterol levels. Therefore, the search started for an alternative cooking oil which could reduce the cholesterol levels. Short-term scientific studies conducted in this regard found that vegetable oil rich in polyunsaturated fats had the ability to reduce total cholesterols as such the recommendations tilted in favour of PUFA rich oils such as sunflower oil and safflower oil.

In due course of time evidence began to accumulate establishing that although polyunsaturated fats were effective in reducing LDL cholesterol, very high content of poly-unsaturated fats had the undesirable affect of reducing HDL cholesterol as well. It further came to light that since oils with high levels of polyunsaturated fats oxidise faster than those with MUFA or SFA profiles, therefore, the excessive intake of these oil could impair the ability of the antioxidants in the human system to control free radicals and thereby enhance the risk of certain cancers, cataracts, rheumatoid arthritis, Parkinson's disease and contribute to the ageing problem. All these findings have led to a change with respect to the earlier recommendations and a delicate balance between cholesterol balancing (instead of cholesterol lowering) properties and oxidative stability has become an important concern now-a-days.

The American Heart Association now recommends use of oils having an equal proportion of saturated, monounsaturated and polyunsaturated fats. The World Health Organisation (WHO) recommends poly-unsaturated fatty acid (PUFA)/saturated fatty acid (SFA) ratio of 0.8 to 1.0 and linoleic acid (omega-6) alpha linolenic acid (omega-3) ratio of 5–10 in the diet. The most recent recommendations according to the National Research Council, Washington DC are 10 per cent saturates, 12 to 13 per cent for monounsaturates, not more than 7 to 8 per cent for polyunsaturates. Japan's Ministry of Health and Welfare suggests fatty acid ratio of saturated/monounsaturated/polyunsaturated as 1:1.5:1 and a 4:1 ratio of omega-6 to omega-3 fatty acids. A further noteworthy point in this recommendation is that the intake of polyunsaturated fat should be accompanied by vitamin E, vitamin C or carotene consumption to help prevent lipid peroxidation.

A perusal of these recommendations would reveal that there are basically three parameters to adjudge any oil as the healthiest cooking oil., i.e. ratio of saturated/monounsaturated/polyunsaturated fatty acid, ratio of essential fatty acids (omega-6/omega-3) and presence of natural antioxidants .

If we look to the composition of the various vegetable oils commonly used in India, we shall find that no single oil could meet the requirement of these recommendations but some are nearer to the recommendations. Composition of common vegetable oils is given in Table 8.6.

Table 8.6. Composition of common vegetable oils.

Oil	*Fatty acids % weight*			*Essential fatty acids (EFA)*	
	SFA	*MUFA*	*PUFA*	*Omega-6/omega-3*	*Antioxidants*
Mustard oil/Rapeseed oil	6	67	27	2	Tocopherols
Sunflower oil	12	21	67	57	Tocopherols
Safflower oil	10	15	75	69	Tocopherols
Soyabean oil	16	24	60	10	Tocopherols
Groudnut oil	20	50	30	32	Tocopherols
Rice bran oil	20	45	35	15	Tocopherols Tocotrienols and oryzanol
Recommended	**Below 33%**	**Above 33%**	**About 33%**	**5–10**	**Maximum possible**

The use of any one of the oils like safflower, sunflower and soyabean furnish very high PUFA/SFA ratio. The use of safflower and sunflower also provide high ratio of omega-6/omega-3. Though the levels of omega-3 will be in the desirable range when mustard and rapeseed oils are used, these oils furnish very high levels of erucic acid. The use of groundnut or rice bran oils ensure optimal intake of MUFA and ratio of PUFA/SFA. But rice bran oil a unique cooking oil produced from rice, is the only oil which besides having balanced fatty acid composition and an ideal omega-6/omega-3 ratio, contains three categories of natural antioxidants i.e. tocopherols, tocotrienols and oryzanol as against only one category i.e. tocopherols found in all other conventional oils. These natural antioxidants help in controlling production of free radicals in the body which can cause injury to the inner lining of blood channels and initiate coronary diseases, certain cancers, cataracts, rheumatoid arthritis, Parkinson's disease and contribute to the ageing process.

High content of natural antioxidants present in the rice bran oil, impart higher oxidative stability and a longer shelf-life as compared to other edible oils. In a recent study conducted by the Central Food Technological Research Institute, Mysore, oxidative stability of natural antioxidants enriched refined rice bran oil has been found to be five times more than the groundnut oil.

Although the PUFA content of rice bran oil is much lower than the other conventional oils, but it has higher cholesterol reduction power than even the PUFA rich oils.

Human trials conducted by the National Institute of Nutrition, Hyderabad, have confirmed significant reduction in total cholesterols particularly the triglyceride levels by using rice bran oil as compared to conventional cooking oils. In this study 21 subjects with high total cholesterols (>225 mg/dl) or high triglycerides (>190 mg/dl) were randomly divided into experimental and control groups. Twelve subjects, who constituted experimental group, were instructed to use rice bran oil in place of their usual cooking oil which they were using earlier. The remaining nine who served as controls, were allowed to continue the use of their habitual edible oil. There was a significant reduction in total cholestrol and triglyceride levels 15 and 30 days after the use of rice bran oil. However, in control subjects there were no changes in serum lipids. (Table 8.7).

Table 8.7. Serum lipids in control and experimental subjects.

Subjects	*Cholesterol (mg/dl)*	*Triglycerides (md/dl)*
Control group Basal	244.0 ± 14.0	295.5 ± 33.0
After 30 days	249.6 ± 16.1	286.3 ± 34.9
% Change from Basal	2.3	–3.1
Experimental group Basal	247.3 ± 10.5	349.8 ± 42.4
15 days after RBO	204.0 ± 6.6	236.5 ± 31.9
30 days after RBO	182.7 ± 8.4	212.9 ± 20.0
% Change from basal after 15 days	–17.5	–32.4

Similar findings have been reported in the animal studies coupled with the human clinical trials recently completed and published by the University of Lowell, USA. These studies have confirmed that rice bran oil has the ability to reduce blood cholesterol levels up to 40 per cent and has significantly better cholesterol lowering than the popular olive oil. Experiments conducted by the Human Nutrition Research Centre Washington St., showed a greater reduction in the total and LDL cholesterol in animals fed with the RBO diet than those consuming canola oil diet.

All these studies conducted in India and abroad have attributed the cholesterol reducing properties or rice bran oil to the presence of a unique component in this oil known as 'oryzanol', which is not found in any other edible oil. This fact has been re-confirmed in a most recent study conducted by the Central Food Technological Research Institute, Mysore. In this study it was observed that although the fatty acid composition of groundnut oil and rice bran oil is almost similar but use of rice bran oil significantly raises good cholesterols and reduces bad cholesterols due to 'oryzanol' present in the rice bran oil. Serum lipid levels of rats fed with diet containing 10 per cent groundnut oil and refined rice bran oil containing about 10,000 ppm, of oryzanol produced through the latest techniques of refining, were reported as under after 11 weeks of feeding (Table 8.8).

Table 8.8. Serum lipid levels of groundnut oil and refined rice bran oil.

Parameters (mg/100/ml)	*Groundnut oil (control)*	*Refind rice bran oil*
HDL cholesterol (good cholesterol)	33.8 + 6.69	50.57 + 3.57
LDL + VLDL (bad cholesterol)	23.48 + 8.51	15.75 + 6.68
Triglycerides	272.8 + 59.7	118.7 + 34.55

Besides beneficially influencing the lipid profile, oryzanol is also known to have anti-itching, antidandruff and anti-ageing properties. It is effective in treating a broad range of gastrointestinal disorders including stress-induced gastric and duodenal ulcers. Oryzanol is also sold abroad as an agent for body building in humans and for treatment of nerve imbalance and disorders of menopause.

Tocotrienols which are found only in rice bran oil and in no other conventional Indian oil, besides being a cholesterol reducing agent and a powerful antioxidant are also known to have anti-thombotic, anti-cancer and anti-ageing properties. Compared to other edible oils, rice bran oil has high content of squalene which is reported to be a quencher of singlet oxygen and a free radical scavenger and it has been shown to maintain the texture of skin. It is well known for its anti-wrinkle properties.

Thus rice bran oil is the healthiest cooking oil having desirable fatty acid composition with higher oxidative stability along with better cholesterol reducing power than all other edible oils. It contains certain unique micro-nutrients which are important for promotion and maintenance of good health.

'The world's healthiest oil' is also the best tasting. A properly refined rice bran oil has a pale yellow colour and is completely odourless and tasteless. In a blind taste test at the university of Oregon USA, comparing refined rice bran oil, palm oil, safflower oil, soyabean oil, corn oil and sunflower oil, rice bran oil won hands down. Rice bran oil is extensively used in Japan, Korea, China, Taiwan and Thailand as a premium edible oil. It is the conventional and the most favourite cooking medium of Japanese, the longest living human beings on this earth and is popularly known as a 'heart oil' in Japan. In recent years US scientists have also shown a tremendous interest in the cholesterol lowering properties of rice bran oil and this oil has acquired the status of 'health food' with Americans.

Only recently some of the Indian markets too have started witnessing some brands of refined rice bran oil. As India is the second largest producer of crude rice bran oil in the world, some of the leading National and multi-national players in this field are having an eye over this largely untapped potential and are planing to enter in a big way in the branded refined rice bran oil segment. The time is not very far when the leading brands of refined rice bran oil will be available in all the markets of India and consumers in India too could enjoy the benefits of this unique gift from the mother nature.

India has now unique rice bran oil brand now as 'Ricela' which is soon being available in western India. It will be the cooking medium of first choice and preferance to bypass the dietary way.

ROLE OF FATTY ACIDS AND MICRO-NUTRIENTS ON CARDIOVASCULAR DISEASE WITH SPECIAL REFERENCE TO RICE BRAN OIL

Cardiovascular Disease in Indians

The epidemic of cardiovascular disease emerging in the developing countries during the past two decades has attracted little public health response. People from the Indian sub-continent have been found to have the unique distinction of having the highest prevalence of coronary artery disease among all ethnic groups of the world. Of particular concern is their early age of affliction with coronary disease compared to that in the west, and its associated rise in mortality rates.

One of the important task is to identify the modifiable risk factors in the Indian pretext and take necessary steps to launch a fight against this dreaded disease. Amongst the modifiable risk factors, hyperlipidemia is one of the important ones which can be easily controlled by adequate and timely intervention and this can only be achieved by a multipronged approach increasing the awareness amongst physicians, scientists, public at large and policy makers.

Several protective (antioxidants, dietary fibre, small amounts of alcohol) and deleterious (excessive intake of saturated fats, simple sugars) dietary components have been proposed, though the strength of the evidence supporting the relationship between each of these factors and CHD is variable. The effect of nutrition transition has also influenced India with globalisation of food production and marketing leading to increased consumption of energy dense foods poor in dietary fibre and several micro-nutrients. Fatty acid composition of common vegetable oil is given in Table 8.9.

Table 8.9. Fatty acid composition of common vegetable oils.

Oil	*Fatty acid % wt*		
	Saturated	*Monounsaturated*	*Polyunsaturated*
Groundnut oil	20	50	30
Sunflower oil	12	21	67
Mustard oil/Rapeseed oil	6	67	27
Safflower oil	10	15	75
Soyabean oil	16	24	60
Cottonseed oil	28	22	50
Ricebran	**18**	**45**	**37**
Recommended	**28.6**	**42.8**	**28.6**

Hence, the first and the foremost step towards control of hyperlipidemia is dietary regulation of the quality and quantity of fat intake. Despite the fact that fat intake is deletenous, it also serves several important functions of the body and hence its intake has to be controlled rather than stopped.

Several studies have indicated that the hypocholesterolemic action of a dietary fat depends on the fatty acid contents and also to some extent on the micronutrients like tocopherols and tocotrienols. The optimal balance of saturated, polyunsaturated and monounsaturated fatty acids make an ideal oil. No single available oil is ideal in this regard, however, some are near ideal and in this section we will discuss rice bran oil in this context.

Cholesterol and Other Blood Fats

Cholesterol is a fat like substance of soft and waxy consistency. It is an integral component of animal cell membranes. It is also a precursor of bile acids, vitamin D and several hormones.

Though a normal constituent of all animals including humans, excess of blood cholesterol increases the risk of CHD. This supports the age old dictum that 'excess of everything is bad'.

Triglycerides: Another Important Blood Fat

Triglyceride is a specific form of fat. It is so named, because it consists of one molecule of glycerol and three molecules of fatty acids. Cooking oil, fat spreads and invisible fat in meat and dairy products largely consist of triglycerides. Triglycerides are also synthesised in liver. Like cholesterol, high blood triglyceride levels also increases the risk of CHD. The term 'fat' is used interchangeably for triglycerides.

Cholesterol and Triglycerides are Lipids

Both cholesterol and triglycerides are lipids. Lipids are a group of chemicals which are insoluble or poorly soluble in water. In biological systems, lipids provide energy and also serve as components of cell membranes or precursors of hormones.

Cholesterol and Triglycerides Circulate in Blood as Lipoproteins

Just as oil and water do not mix, fat soluble cholesterol and triglycerides also don't mix well with blood. So, lipid transportation in blood relies on carriers called lipoproteins which are complexes of cholesterol and triglycerides with blood proteins. Lipoproteins escort cholesterol and triglycerides in the blood from one part of the body to another where they serve various functions.

Types of Lipoprotein Particles

Lipoprotein particles vary in size, density, and significance. Based on their density, these are classified as:

1. Low density lipoprotein (LDL): Circulating LDL tends to dump its cholesterol in arterial walls. Accumulation of cholesterol in major arteries gradually results in their narrowing and eventual blockage. Therefore, LDL cholesterol is nicknamed as 'bad cholesterol'. About 2/3 of the cholesterol in blood is transported as LDL.
2. High density lipoprotein (HDL): HDL scoops cholesterol from arteries, and takes it back to liver for disposal. This property has earned HDL cholesterol the tag of 'good cholesterol'. A low HDL cholesterol level increases the risk of CHD.
3. Very low density lipoprotein (VLDL): VLDL carries fat and cholesterol from liver into circulation. Like LDL cholesterol, high VLDL cholesterol level also enhances the tendency of arterial narrowing. Therefore, it is also a form of 'bad cholesterol'.
4. Chylomicrons: Chylomicrons are triglyceride rich particles, which carry dietary fat from intestines to the liver and sites of fat deposits in the body after food digestion and absorption.

Sources of Cholesterol

As noted above, cholesterol is a normal constituent of animal cells. It is derived from:

1. Cholesterol synthesis in the liver: Liver forms about 1500 mg of cholesterol everyday. This amount is much more than, what is already present is a day's diet.
2. Diet cholesterol content of diet varies from person to person and from day to day, depending on the type and amount of food.

Sources of Triglycerides

Like cholesterol, blood triglycerides are also derived from: (i) dietary fats; and (ii) endogenous production in liver from fats and sugars.

Fate of Dietary Cholesterol and Trigycerides

Dietary fat goes to liver

After ingestion of food, dietary cholesterol and triglycerides pass through the stomach and intestine. Triglycerides are digested and absorbed in the small intestine.

They are taken up by liver and fat deposits in the body. Dietary cholesterol is also absorbed and gets assimilated in the body rapidly.

Liver loads triglycerides and cholesterol into VLDL

Triglycerides and cholesterol manufactured in the liver are incorporated in VLDL, and released into the blood. Circulating VLDL unloads triglycerides in body tissues, where they serve either as fuel for energy, or stored in fat deposits. Triglyceride unloading converts VLDL into cholesterol rich LDL.

LDL cholesterol is dumped

LDL dumps excessive cholesterol in arterial walls, which blocks the passage of blood over a variable period of time.

HDL cleans out stuck cholesterol

HDL brings cholesterol released by constant wear and tear of body tissues and excessive deposits of cholesterol in arterial walls, back to the liver. Here, it is either recycled into VLDL or converted into bile acids and secreted with bile juice into the gut.

Only two oils comes close to recommended levels—groundnut oil and rice bran oil. Groundnut oil, due to its high content of long chain fatty acids have been found to be atherogenic in rats.

On the contrary rice bran oil has a highly beneficial minor component viz., Y-oryzanol which along with high contents of tocopherols and tocotrienols imparts excellent oxidation stability to the oil. Y-oryzanols (ferulic acid esters of triterpene alcohols and sterols) are uniquely found in rice bran oil. Ingestion of RBO causes a cholesterol lowering effect and reduction of the aortic fatty streaks. It was demonstrated that in monkeys fed rice bran oil as the sole dietary fat, a 40 per cent reduction occurred in LDL cholesterol without affecting HDL cholesterol. A 5 week regimen with RBO led to a 20 per cent reduction in blood LDL cholesterol in humans.

Robert Nicolosi at the University of Massachusetts recently provided a remarkable demonstration of potential benefits of a nutrient-modified diet when hypercholesterolemic subjects on a special nutrient bas containing guar gum, soya protein and unsaponifiable-rich rice bran oil had their LDL cholesterol levels reduced by 25 per cent. It was also demonstrated that the cholesterol lowering property of the oil is greater than other oil blends of similar fatty acid composition. Oryzanols and tocotrienols found in RBO have been investigated for possible health benefits.

Physiological effects associated with Y-oryzanol intake are decreasing plasma cholesterol, decreasing platelet aggregation, decreasing hepatic cholesterol biosynthesis, increasing faecal bile acid secretion and decreasing cholesterol absorption. Oryzanol has also been used to treat nerve imbalance and disorders of menopause. Effect of dietary factors on serum lipids are given in Table 8.10.

Table 8.10. Effect of dietary factors on serum lipids.

Dietary component	*Total cholesterol*	*LDL*	*HDL*	*TG*
Saturated fats	↑↑	↑↑	↑	↑↑
Omega-3 polyunsaturates	↓	↓	↑→	↓↓
Omega-6 polyunsaturates	↓↓	↓↓	↓	–
Monounsaturates	↓	↓	↑→	↓

(Contd ...)

Table 8.10. Effect of dietary factors on serum lipids. *(Contd.)*

Dietary component	*Total cholesterol*	*LDL*	*HDL*	*TG*
Dietary control	↑	↑	↑	
Soluble fibre	↓	↓	–	–
Alcohol	–	–	↑	↑↑
Excess calories	↑↑	↑↑	↓	↑↑

Thus fatty acid composition of RBO comes close to the recommended levels with high oxidative stability and high content of antioxidants like oryzanol, tocotrienols and tocopherols. Oryzanol exerts a high additional hypolipidemic effect. RBO enjoys the status of a health food from a cardiovascular view point and routine use of RBO can be recommended as a sole cooking medium.

Dietary Fat Recommendations

WHO and the NIH, USA have specified guidelines for the nature and quality of fats in diets. The American Heart Association step I diet recommend that not more than 30 per cent of the total energy intake should come from fats and the fatty acid composition to be saturated/PUFA/MUFA ratio of 1:1:1. In the AHA step 2, the ratio of PUFA is further reduced to 7–8 per cent. Japanese ministry of health recommends 20–25 per cent energy intake from fats and saturated/PUFA/MUFA ratios 1: 1: 1.5 and 4: 1 ratio of omega-6 to omega-3 fatty acids.

VALUE ADDED PRODUCTS FROM RICE BRAN OIL INDUSTRY

Rice bran oil is gaining commercial importance in the world as it has many beneficial nutritive and biological effects. Rice bran oil is similar to groundnut oil in fatty acid composition, but the oil has some extra components which are commercially and nutritionally important and are carried away into the sludges and are underutilised. They are oryzanol, lecithin, wax and tocopherols and sterols. The composition of a typical commercial rice bran oil industry is given in the Tables 8.11 and 8.12 and Figs. 8.1 and 8.2.

Table 8.11. Chemical composition of rice bran oil.

Chemical composition of oil	*%*
Triglycerides	80.5
Free fatty acids	6.8
Diglycerides	4.8
Monoglycerides	1.7
Oryzanol	2.0
Phosphatides	1.3
Wax	2.9

As more number of industries are processing edible grade vegetable oil from rice bran, entrepreneurs are looking forward to recover the value-added products from the refinery wastes of the oil to promote products such as oryzanol, lecithin and wax for varied end uses. Many industries in India, do not recover lecithin for example, from the gum-sludges because of the absence of technology with the entrepreneurs. The recovery of these products of the rice bran oil industry viz., oryzanol, lecithin and wax fetch good revenue to the industry and somewhat more than or equal to the total value of crude and refined rice bran oils produced in the country.

Table 8.12. Waste products of rice bran oil industry requiring further processing.

Waste products	*% of oil*	*Composition*
Wax sludge	4	Wax and entrained oil
Gum sludge	4	Lecithin and entrained oil
Soapstock	2.5 × FFA	Fatty acids as soap, oryzanol and entrained oil
Deodouriser distillate (DOD)	< 0.0001	Tocopherols, sterols, free fatty acids and odouriferous substances

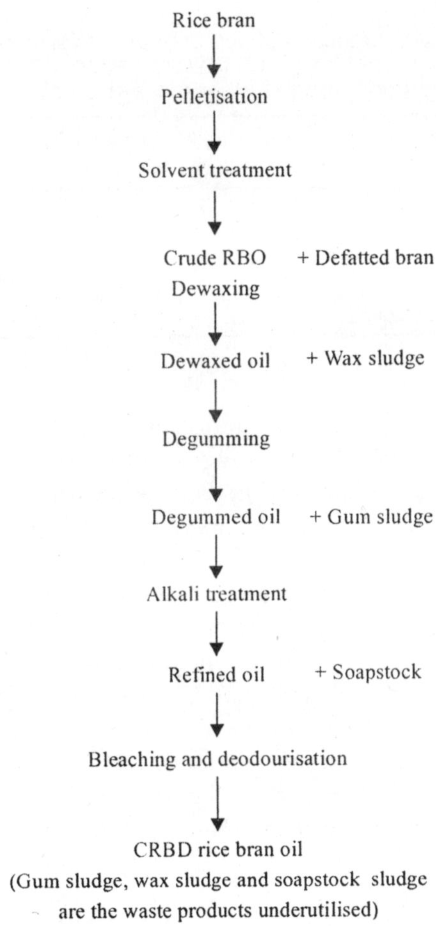

Fig. 8.1. Rice bran oil production (chemical refining).

The soapstock, the gum-sludge and wax-sludge are the starting materials for the recovery of oryzanol, lecithin and wax from the rice bran oil industry. At present, these are partially processed to recover some intermediate products which need to be further processed into value added products, viz. oryzanol, lecithin and wax and also recover the entrained oil in case of gum and wax sludges (Table 8.13).

Degummed oil

↓

Bleaching

↓

Distillation and deodourisation

↓

PRBD RBO + fatty acids
(Gum sludge and wax-sludge are the
waste products underutilised)

Fig. 8.2. Production of rice bran oil (physical refining).

Table 8.13. Value added through further processing of waste products.

Waste products	*Value addition through recovery and purification*
Wax sludge	Pure wax and oil recovery
Gum sludge	Lecithin and oil recovery
Soapstock	Oryzanol
Deodouriser distillate (DOD)	10–50

Oryzanol

Oryzanol is the name given to the class of phenolic compounds naturally occurring in rice bran oil. Principally, these are ferulic acid esters (Ferulic acid esterified to sterols, viz., cycloartenol, 24-methylene cycloartanol, *beta*-sitosterol and campesterol). Oryzanol is known to have a number of biological effects such as hypocholesterolemic, anti-itching, anti-dandruff, anti-ageing and anti-oxidative effects as demonstrated in experimental animals. Oryzanol has also found application as a food additive for imparting cholesterol lowering, antioxidant, preservative and antibiotic effects. Therefore, it is important to recover oryzanol and use it for food/pharmaceutical/cosmetic applications in the country. The cost of oryzanol varies depending on the grade and the manufacturing method in the international market. Oryzanol is mainly used in pharmaceutical preparations in the USA and Europe and in Japan it is used in foods.

Rice bran oil contains 1.5–2.9 per cent of oryzanol and during refining, about 90 per cent of the oryzanol goes into the soapstock as a refinery waste. Therefore, oryzanol needs to be recovered from the soapstock of rice bran oil. With the refined rice bran oil, there is a possibility of recovering as much as 4000 tonnes of oryzanol. Oryzanol has been produced commercially in a rice growing country like Japan and China.

The commercial products of rice bran oil industry had a varied oryzanol content of 1.8–2.0 per cent for crude oil, 0.19–0.22 per cent in chemically refined oil, 2.21–6.71 per cent in the soapstock. Physically refined RBO has retained about 85–90 per cent of oryzanol in the oil itself as no soapstock is produced during removal of fatty acids. Processing of soapstock to recover oryzanol has been made difficult due to use of varied conditions of refining of the RBO. Oryzanol has been isolated from the commercial soapstock of rice bran oil by using solvent extraction methods and is given in Fig. 8.3.

Further purification of the concentrate to get oryzanol of >90 per cent purity for commercial production is in progress at CFTRI, Mysore.

Soapstock

↓

Solution in solvent mixture (recovered and re-used)

↓

Extraction of oryzanol

↓

Oryzanol concentrate (20–30% purity)
Fatty acids + potassium chloride (by-products)

(Recovery of oryzanol ranges from 20 to 100%
for different types of soapstock)

Fig. 8.3. Preparation of oryzanol concentrate from rice bran oil soapstock.

Lecithin

Commercially produced lecithin is a by-product of soyabean oil industry. Lecithin usage in foods can bring about not only a reduction in the fat which is desirable in the present context but also revolutionise the Indian food technology in terms of new and better products. The reduction of fat usage brought about by lecithin is expected to improve the availability of the oil to the vegetable oil pool of the country. As such, the technology for the recovery of lecithin even from soyabean oil is not well-developed in the country. Most of the lecithin plants are underutilised and for rice bran gum-sludge processing technology needs to be developed. The processing of rice bran gum sludge into lecithin is made difficult due to the carry over of wax into it and occludes lot of neutral oil causing entrapment of oil into the sludge. The advantages of rice bran lecithin over soya lecithin is that the former has a lower amount of linolenic acid (< 2 per cent) (fishy odour causing fatty acid) than the latter (>5 per cent) (soya lecithin, >5 per cent of linolenic acid and rice bran lecithin, < 2 per cent), thereby the stability is also expected to be much better. In order to recover this valuable by-product from rice bran oil industry, a method using 10–20 kg of wet gum sludge has been developed at CFTRI, Mysore for the preparation of food grade rice bran lecithin (commercial lecithin equivalent) and also for pure lecithin (80–85 per cent purity).

The process of Lecithin preparation from the gum-sludges is shown in Fig. 8.4.

Crude RBO

↓ Degumming

Wet gum-sludge

↓

Drying

↓

Lecithin (40–60% purity)

Fig. 8.4. Preparation of crude lecithin from wet-gum sludges of rice bran oil industry.

Upgradation to 65 per cent purity (which is equivalent to commercial soya lecithin)

The crude lecithin so obtained could also be purified further to get a good grade, wax free rice bran lecithin (80–85 per cent purity).

Lecithin produced in western countries is a by-product of soyabean and rapeseed oil industries. In India, a few industries processing soyabean oil are recovering crude lecithin and some brands of the product are available in the market which have dark colour as compared to imported lecithins. However, not much quantity of lecithin is recovered from the sludge although it has a good potential (about 10,000 tonnes from soyabean and 5000 tonnes from rice bran oil industries). Lecithin has multifunctional uses and has also been shown to improve health through nourishment to the brain of living beings. It is also used in the synthesis of drugs apart from its usage in cosmetics and foods.

In order to recover this valuable by-product from soyabean and rice bran oil industries and its utilisation in foods, existing technology needs to be developed/improved further thereby demand for lecithin is created in the domestic market.

The role of lecithin in food industry particularly in traditional food product development has been established as a fat reducer, texture softener and emulsifier. Therefore, there is a need for examining the possibilities/processes where lecithin can be incorporated to get better products which might revolutionise the Indian food industry. Technology development for the large scale preparation of lecithin and lecithin added processed foods is an emerging area in the Indian vegetable oil and processed traditional foods industries, although large number of products based on soya/rapeseed lecithin are available in the western and Japanese food markets.

Wax

Rice bran oil has about 2–5 per cent of wax which has a number of industrial applications, such as for polishes, ingredient in cosmetics, and food applications. The wax is more or less similar to bee's wax and carnauba wax in properties and therefore could be used as a substitute for these waxes. Rice bran wax has a potential of 20,000 tonnes and could capture the export and import market of carnauba, bee's and sugarcane waxes.

The chemical composition of Japanese rice bran wax contains four classes of wax lipids viz., steryl ester, longer alkyl ester, shorter alkyl ester and hydrocarbons. The main component of fatty acids were linoleic, oleic and palmitic acid for steryl ester (same as for RBO); behenic, lignoceric and palmitic acids for longer alkyl ester and oleic and palmitic acid for shorter alkyl ester. The representative molecular species for longer alkyl esters were dotriacontanyl behenate, octacosanyl palmitate. Squalene also has been found under hydrocarbons. Rice bran wax has been chosen as one of the waxes of the plant source for application as a natural additive in foods, cosmetics and other industries (polishes, carbon paper, lubricants).

Rice bran wax consists almost exclusively of saturated monoesters with a typical carbon chain length distribution of C46 to C60. This monoester distribution makes its intermediate between the ranges of the esters found in bee's wax and carnauba wax. However, the low level of unsaturation makes rice bran wax nearly as hard as carnauba, with almost the same melting point.

The existing processes on rice bran wax production involves use of low temperature crystallisation of wax from the oil followed by filtration. The filtered wax is treated with different solvents to prepare a wax suitable for the industry.

These methods are practiced only in advanced countries such as Japan and the wax so produced is exported to international markets. India has a great potential to compete with Japan in trading good grade wax in the international market but due to absence of technology, the wax sludge is purified partially and is sold which is soft (60°C–75°C) compared to the required grade rice bran wax with a melting point of 80°C–82°C.

A purified rice bran wax having natural composition without alteration of its chemical nature has been prepared at CFTRI, Mysore at 100 gram batch levels of purified wax (>97 per cent purity, 5–6 X harder than starting soft wax currently produced in the industry, white powder or compact brown mass and is similar to carnauba wax with respect to melting profile with a melting point of 80°C–82°C) (Fig. 8.5).

Rice bran wax sludge

↓

Conditioning and solvent extraction

↓

Wax residue + oil in solution

↓ Desolventisation

Entrained oil + wax (powder or compact mass)

Fig. 8.5. Preparation of purified rice bran wax from wax sludge.

The advantage of the process is that the entrained oil is also recoverable and can be re-used for cooking purposes fetching additional revenue. The process may be scaled up for individual entrepreneurs through sponsored projects at CFTRI, Mysore for a sponsorship fee. Also processes for manufacture of pure wax has not picked up due to absence of awareness of commercial value in the international markets for high quality wax and therefore entrepreneurs are not recovering wax, and the recovery of entrained oil fetches more money than the wax itself.

The tocopherols and sterols extraction from DOD has not been attempted due to the presence of very low amounts of these components in the DOD. But these are recovered from the DOD of soyabean and sunflower oil refining industries. From the waste products/intermediate products produced in the RBO industry, it is possible to get additional revenue, if proper technology is used to recover the value added products viz., oryzanol, lecithin and wax from the sludges of the industry.

INTEGRATED TECHNOLOGY FOR REFINING RICE BRAN OIL AND BY-PRODUCTS RECOVERY

Rice bran oil (RBO) is considered the most difficult oil to refine meeting with edible oil specifications in terms of its physical and chemical characteristics that govern the marketability of the product. The fact that edible grade rice bran oil has not been in the market currently with consistent quality and consumer acceptability is primarily attributed to techno-commercial problems starting from rice milling, bran quality, solvent extraction, refining techniques employed etc. that are unique to rice bran oil. Attempts have been made to solve refining problems of crude rice bran oil (CRBO) in the past by many workers with partial success by chemical refining. The CRBO commercially available vary significantly in terms of physico-chemical characteristics such as colour, free fatty acids (5–20 per cent), gums, wax etc. Normally high free fatty acids is associated with dark colour due to degraded pigments indicating the longer storage of bran before solvent extraction. High content of gums and wax are associated with rice variety and solvent extraction conditions. The degraded pigments and residual gums (phospholipids) are responsible for final colour and residual wax due to incomplete dewaxing, results in turbidity and consequent higher cloud point. Chemical refining yields better products in terms of colour and cloud point. However, high refining loss makes this process uneconomical. For instance CRBO with 10 per cent FFA, the loss is between 25–30 per cent considering refining factor of 2.5 to 3.0. Yet another loss is due to degumming and dewaxing. With 5 per cent gums and wax, the loss on this count will be 10 to 15 per cent. Therefore CRBO with 5 per cent gums and wax and 10 per cent FFA would end up with total loss of 35 to 45 per cent by chemical refining that makes this process economically unviable. Besides oil, loss of phytoceuticals are other disadvantages of chemical refining. RBO is unique among edible oils due to its oryzanols (1.5–2.0 per cent) a strong hypocholesterolemic agent and a potent antioxidant. Tocopherols and tocotrienol contents (1500–2000 ppm) also makes RBO, the most nutritious and health promoting edible oil. Almost the entire oryzanols and significant amount of tocopherols are lost by chemical refining. The unique selling point of RBO, based on the current knowledge, is its exceptionally high content of these micro-nutrients. Any process technology for refining CRBO should be designed to retain them to their original level in refined product as well.

Physical refining only can address these techno-commercial and quality issues of RBO and therefore is the only alternative available for the commercial exploitation of rice bran as a healthy edible oil.

A Novel Approach to Physical Refining of Rice Bran Oil

Extremely high levels of phosphatides and wax present in the CRBO and appropriate process to remove them to meet the physical refining requirements are the major problems associated with refining. The phosphatides in terms of inorganic phosphorus need to be brought down from 500–700 ppm to less than 5 ppm and wax content from 2–3 per cent to less than 50 ppm. Residual phosphatides and wax in excess undergoes charing at high temperature of deacidification (240–250°C) resulting in brownish end product. Excess wax also imparts cloudiness to the refined oil on storage. The conventional degumming and dewaxing employed for other edible oils such as sunflower, soyabean, mustard etc. has not been found feasible with CRBO because of high levels of gums and wax. Separate process therefore for degumming and dewaxing was required for CRBO. Of the several methods tried simultaneous degumming and dewaxing method using calcium chloride followed by two stage controlled crystallisation (winterisation) in addition to the other standard process steps such as bleaching, deacidification and deodourisation was found most suitable for CRBO of new process was developed based on this. Flow sheet of the new process is presented in Fig. 8.6.

The advantages of this process is that it is amendable to CRBO containing, FFA, 5–15 per cent, phosphatides 1–2.5 per cent, wax 1–3 per cent. The simultaneous degumming and dewaxing steps demand certain additional facilities such as crystallisers and chilling unit, (two stage), separation of gum and wax by centrifugation at the first stage followed by second winterisation (after bleaching) and separation of the wax by filter pressing. After the second stage winterisation, the phosphorus content is brought down to less than 5 ppm and wax content to less than 50 ppm. RBO with these specifications is subjected to deacidification at less than 250°C at 1 mm Hg vacuum. The final oil after polishing has been found to have colour value less than 12 Lovibond unit in 1″ cell and cloud point less than 6°C as obtained in the laboratory trials.

The other advantages of this process is that there is no lye wash or water wash involved and hence zero-discharge and therefore the process is eco-friendly. So far no physical refining process for rice bran oil to produce RBO with high quality specifications has been developed with proven economic viability. The process developed by RRL, Trivandrum is first in the case of RBO and is technically sound and commercially feasible as demonstrated here at commercial scale operations.

Recovery of By-products from RBO Refining

Process streams

Crude rice bran oil composition is given in Table 8.14. A typical CRBO containing 4 per cent gums and wax (1.5 and 2.5 per cent respectively) will carry 75 per cent by weight of neutral oil. In a 50 TPD rice bran refinery, per day loss of oil is about 1.5 tonnes. When oil is mixed with gum and wax, on simultaneous degumming and dewaxing a low value by-product is generated. A laboratory process now has been developed by RRL, Trivandrum to recover oil, high purity wax and lecithin. By this process the gum-wax-oil mixture containing 50 per cent water by weight, is first dehydrated under vacuum.

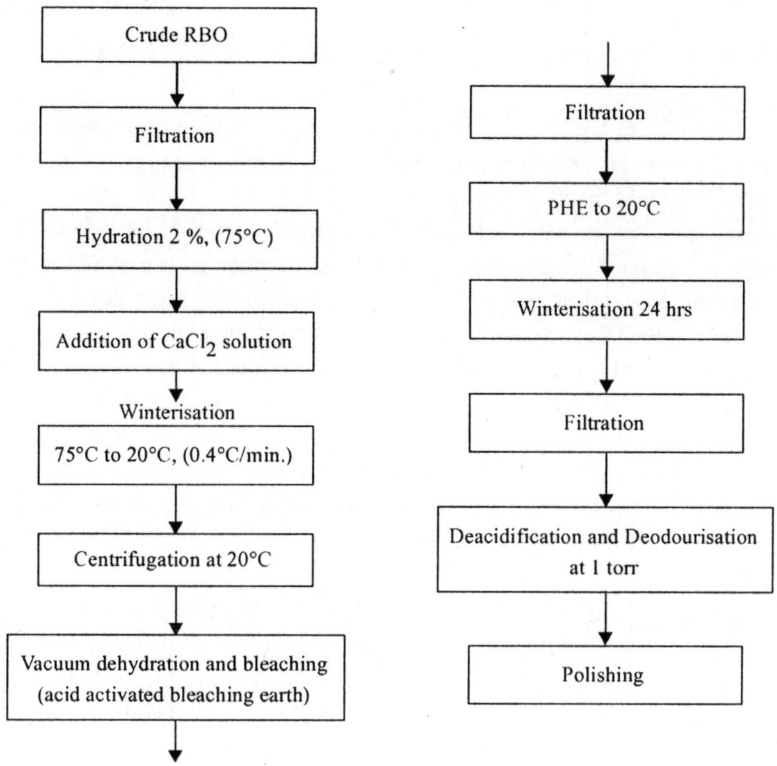

Fig. 8.6. Physical refining of rice bran oil (Regional Research Laboratory, Trivandrum).

Table 8.14. Physico-chemical characteristics of CRBO.

Appearance	*Brownish to dark brown*	*Fatty acid composition (average, % by wt)*	
Colour value (¼″)	30–70	$C_{16:0}$	25
Free fatty acid (%)	5–25	$C_{18:0}$	2
Gums (%)	1–3	$C_{18:1}$	41
Wax (%)	2–5	$C_{18:2}$	32
Unsaponifiable matter (%)	4–6		
Sterols (%)	0.2–0.5		
Tocopherols (ppm)/Tocotrienols	300–2000		
Oryzanol (%)	1–2		

The oil is extracted selectively using solvent A. The residue containing wax and gum further separated using solvent B. A typical 50 TPD refinery would yield about 3.5 tonnes of gum-wax oil mixture per day.

From this about 1.5 tonnes of good quality rice bran oil can be recovered per day. 1 tonne of wax (95 per cent) and 0.5 tonne of lecithin can also be recovered. The oil is of good quality and therefore recycled into the mainstream refining process. Considering 200 days of rice bran refining per year at 50 TPD, the annual recovery of rice bran oil otherwise going waste is about 300 tonnes, in addition to 200 tonnes of wax with 95 per cent purity and 100 tonnes of lecithin.

APPLYING INNOVATIONS AND IMPROVEMENT IN RICE BRAN OIL PROCESSING TECHNOLOGY

During the last few years there has been considerable progress in the development and understanding of the pre-treatment processes for rice bran oils and fats before the final high temperature deodourisation stage. This has enabled physical refining processes to be developed for processing rice bran oil. However, during these few years, the criteria used for establishing final oil quality have undergone some changes.

Today more attention is paid to the overall nutritional quality of the product. Oils should contain low *trans* fatty acid levels, low polymeric triglycerides and oxidation products and still contain adequate levels of natural antioxidants. These demands are in direct conflict with the high temperatures generally required in the physical refining process to remove free fatty acids. Certain changes have been made to the deodourisation process and equipment design to achieve the demands for better oil nutritional quality. This involves a comprehensive study of the temperature and steam profiling and optimisation of the distillate recovery systems to provide a more valuable by-product.

Deodourising is in most cases the final stage in an oil refining process. Its main purpose is the removal of those substances that are responsible for off-flavours and odour. In physical refining, it also removes the free fatty acids.

Rice bran oil is processed through chemical or physical refining processes. Each type of refining process has its importance in the industry from the standpoint of its viability and performance in specific types of oil. Rice bran oil refined chemically or physically must exhibit good stability in the applications to be made. Stability of oil can be defined in terms of its autooxidative stability, hydrolytic stability and thermal stability.

Often in the oil processing industry, an oil with good oxidative stability is assumed to have good flavour stability. In reality flavour stability is not necessarily connected to its oxidative stability. Oil loses its stability due to reaction with oxygen in the presence or absence of light. In some applications hydrolysis becomes a limiting factor for choosing an oil. There are many different types of oil application and not every oil is suited for all types of applications.

However, by taking certain precautions during the processing of an oil, a satisfactory application may be anticipated. A comprehensive understanding of the various reaction phenomena involved is essential to prevent problems arising during both the processing of oilseeds and refining of the oil. With a better understanding of these factors any type of oil may be processed effectively to enhance its performance.

Innovation and research in the edible oil sector is a very vast subject. Only foremost areas of research are being highlighted in this section. Licensed technologies developed on commercial scale in association with research institutions are also discussed. The recent technological innovation has been able to pare the processing cost significantly. Although the discussion is limited to refining technologies of rice bran oil. However, to ensure that a better quality oil is produced one needs to maintain quality right from rice milling and extraction stages and hence the same are also covered in this section, briefly.

There are several areas where we have been able to improve over conventional system and design. Few of them are highlighted below:

1. Hexane recovery by vent air purification system.
2. Miscella separation at low temperature.
3. Low temperature separation of hexane from defatted meal.
4. Special filter media to reduce impurities.
5. Recirculation of waste-water wherever vacuum pump is used to create vacuum in the process.

Further innovations in design of vegetable oil refining being shared on enzymatic degumming, simultaneous degumming and dewaxing for high wax content oils, twin bleaching and deacidification and deodourisation.

Let us explain about our first innovation and research in solvent extraction plant, the vent air purification system. It allows to recover hexane from the vent air discharged in a solvent extraction plant, instead of letting it escape into the atmosphere. The innovation resulted in reduction in hexane consumption, elimination of refrigeration process in the stream and improve the efficiency of all processing equipment substantially. By adoption of magnetic pump, air intake is reduced in the process, thereby, augmenting reduction in hexane consumption. Moreover, it helps to recover hexane at start- up and desolventising stages besides offsetting the opertor's mistakes.

Presently vent air which is discharged into the atmosphere, is taken back to a column where high pressure mixing will take place, which dissolves air in water and hexane is arrested and recovered through a water separator. The bubble cap column provided in the system will maintain the pressure consistently. A formula has been devised to calculate profit in your own way no matter what amount of hexane your plant uses currently. Adopting this technology will bring the consumption of hexane down by atleast 50 per cent.

Let us elaborate upon our second innovation, miscella separation at low temperature. This is achieved by means of azeotropic distillation using a reflux column. It refers to a process in which a component known as entrainer is added to the top of the column, to form a homogenous complex known as an azeotrop. In this case the process is quite different from the normal stripping distillation because the solvent that appears as a distillate can be separated faster at a lower temperature.

The reflux column consists of number of trays fitted with bubble caps, curb plates and down comers. The area and number depends upon the mass and heat balance calculated on the basis of isothermal distillation, where no physical or chemical changes take place in any component present. The complex formed in the vapour filter is fed to the reflux column, where the hexane separation takes place isothermally at less than 80°C temperature.

As explained in azeotropic distillation, since there are no physical or chemical changes, it gets deteriorated or converted into undesired component. Hence, saving of 0.1 to 0.2 per cent compared to conventional system, which is converted as non-hydratable phosphatides, free fatty acid, peroxide value among many others. Another benefit is of steam saving, since higher temperature is not applied in the process, condensate line from the desolventising toaster is used for distillation.

The competitive advantage of such process includes prevention of deterioration of oil at the higher temperature due to sparged steam and sediments. Moreover, higher bleachability is achieved, as there is no damage below oxidation temperature since the temperature is lower than 80°C. Among other advantages, oil processed in such conditions will benefit reduction in chemical consumption and neutral oil loss in the refining unit.

Innovation in design adopted in the desolventisation process results in reduction of steam consumption and superior quality of deoiled meal. It is a counter-current process with acceleration of hexane separation through a blower. This helps to reduce the temperature and steam utilisation. As you know, 75 per cent of steam consumption in desolventisation is applied through open steam, blower acceleration results in significant reduction of steam consumption. By using blower acceleration a reduction of at least 30 per cent takes place in steam consumption.

The counter current desolventisation will create an azeotrope as well as a contact of sparged steam from bottom compartment. This will effect all compartments as a carry over, which will result in reduction of steam consumption.

The advantage of filter design reduces impurity level below 0.1 per cent which eliminates power driving and choking of pre-heater tubes. Besides this, no water storage is required in the miscella tank to arrest and drain out sediments. It is also an eco-friendly process since there is no water drainage from the miscella.

The sediments and oil coming along with the miscella are separated by cyclonic separation and the trace sediments escaped from the cyclonic separator are arrested in the highly compressed special filter media to reduce the impurity level in the distillation column. Arrested sediments in the cyclonic separator and filter will be drained out to the extractor.

The above research and development can be implemented in a new project or be incorporated in an existing project based on environment-friendly technology, for reduced utility consumption and improved product quality. These advantages are accompanied by substantially reduced cost of processing and increased efficiency.

Let us now continue with innovation in refining. An effective process to reduce the phosphorus level below 5 ppm by using commercially available enzymes. The oil loss during the enzymatic degumming process technology developed on pilot scale by Indian Institute of Chemical Technology, Hyderabad, a government of India laboratory will be lower than any other method adopted by the industry. Other methods carry along 50–60 per cent neutral oil along with the gums quantity, whereas the enzymatic process carried barely 30–35 per cent. Moreover, it does not alter the fatty acids composition or micro-nutrients presents in the oil. It is also a process which is eco-friendly since it does not generate waste- water, unlike other processes. Even the wash water centrifuge can be eliminated from the stream.

Simultaneous degumming and dewaxing technology developed on a lab scale by Regional Research Laboratory, Trivandrum, a national laboratory of the government of India, is suitable for high quantity wax present in the oil. The process has been commercialised for removal of the phosphorus level to below 5 ppm using calcium chloride. The advantages are not only the reduction of phosphorus, but also include reduced neutral oil loss in the process.

The twin bleaching adopted in the process will help to reduce a significant level of bleaching earth consumption depending on the oil processed. 95 per cent of the spent earth will be re-used in the process to remove the volatiles with the bleaching earth resulting in reduction of bleaching earth consumption. There is less expenditure on the filter cleaning and requirement for the filter cake disposal. This also results in lower oil losses.

The new technology adoption in the twin bleaching to inactivate the volatiles with the bleaching earth results in increase of its consumption. Specially designed multi-cyclonic separator recover 80 per cent of the spent bleaching earth to the first bleacher. A modified plate and frame filter in close circuit is incorporated to accommodate the higher dosage of activated bleaching earth and carbon.

Degummed or neutralised oil is pumped to a plate heat exchanger where final oil will be passed for an economisation and pumped to the first bleacher where spent earth is used from a cyclonic separator and taken to a close circuit plate and frame filter process. The oil purified will be taken to a second bleacher where fresh earth will be added and discharged to a bleached oil tank through a leaf filter, after separating 80 per cent through a cyclonic separator.

The new process technology developed in dewaxing and winterisation by gravitational force without using any pump results in better quality of dewaxed and deodourised oil than any other method used by the industry. The cloud point of oil processed can be brought to desired levels using appropriate cooling of oil and winterisation.

The principle of deacidification and deodourisation is that the pressure of each component is proportional to its motor concentration. The newly designed bubble cap columns with recirculation through sealless pumps, allows to make a thin film which results in elimination of shallow bed process. Consistent increase of temperature in each tray helps to avoid local heat effect and more surface area for better deacidification and deodourisation in the new design.

The efficient economisation system provided in deacidication and deodourisation results in 70 per cent to 80 per cent of reduction in energy consumption. The new concept of design helps to reduce the downtime and maintenance. The process results in less *trans* fatty acid development since oil is not exposed to its evaporation temperature for a longer period.

New design and development incorporating bubble cap column create thin film for faster separation of volatile and fatty acid at lower temperature. Through this design sparge steam consumption is reduced by 25 to 30 per cent. By reducing the sparge steam, it subsequently helps to reduce the steam consumption in booster ejector.

The other advantages in deodourisation are that temperature is raised gradually not more than 50°C in one tray. Sensors interpret to control the rise in temperature during the process. The retention time and temperature in the process can be reduced drastically. Satisfactory results have already been achieved in rice bran oil. Further innovation has facilitated significant colour reduction in difficult oils like rice bran by over 20–24 units. A reduced level of neutral oil loss was found in deacidification process.

The deodouriser with 16 trays and circulation systems structural packing is more effective in the process of vegetable oil for quality and reduced utility consumption. A substantial amount of energy saving helps the process to make it more competitive.

There are numerous opportunities waiting for us on the horizon. It is up to us to team up to innovate for the collective betterment of the industry by bringing new technology into focus and utilisation.

RICE BRAN OIL—A BETTER CHOICE

Choice of a healthy cooking oil is a controversial subject. The ideas keep on changing as new evidence accumulates. It was generally believed earlier that polyunsaturated fats (PUFA) prevented coronary heart disease by lowering blood cholesterol and the dietary advice was to use vegetable oil rich in PUFA such as sunflower oil and kardi oil. In due course of time evidence began to accumulate establishing that prolonged usage of PUFA rich oils had the undesirable affect of reducing the 'good cholesterol'. Composition of common vegetable oils is shown in Table 8.15.

It further came to light that since oils with high levels of polyunsaturated fats oxidise faster than those with monounsaturated fats (MUFA) or saturated fats (SFA) profiles, therefore the excessive intake of these oils could lead to production of free radicals in the human system and thereby enhance the risk of certain cancers, cataracts, rheumatoid arthritis, Parkinson's disease and contribute to the ageing process. All these finding have led to a change with respect to the earlier recommendations and a delicate balance between cholesterol balancing instead of cholestrol lowering properties and oxidation stability has become an important concern now-a-days.

Table 8.15. Composition of common vegetable oils.

Oil	*Fatty acids % weight*			*Essential fatty acids*	*Antioxidants*
	SFA	*MUFA*	*PUFA*	*Omega-6/Omega-3*	
Kardi oil	10	15	75	69	Tocopherols
Sunflower oil	12	21	67	57	Tocopherols
Cottonseed oil	29	20	51	100	Tocopherols
Soyabean oil	16	24	60	10	Tocopherols
Palm oil	50	40	10	–	Tocopherols
Groundnut oil	20	50	30	32	Tocopherols
Corn oil	16	37	47	46	Tocopherols
Mustard oil/Rapeseed oil	6	67	27	2	Tocopherols
Olive oil	16	75	9	–	Tocopherols
Canola oil	8	59	33	2	Tocopherols
Rice bran oil	20	43	37	15	Tocopherols and oryzanol
Recommended	**Below 33%**	**Above 33%**	**About 33%**	**5–10**	**Maximum possible**

The American Heart Association now recommends use of oils or their blends having an equal proportion of saturated, monounsaturated and polyunsaturated fatty acids. Some other recommendations allow even higher content of monounsaturated fatty acid (MUFA) than that suggested above as mono-unsaturated fatty acids have higher oxidation stability and unlike polyunsaturated fatty acids, these fatty acids do not adversely affect the good cholesterols. Certain amount of saturated fatty acids is also considered desirable as it helps to increase good cholesterol and improves oxidation stability of the oil. But saturated fatty acids (SFA) content beyond 33 per cent is considered undesirable due to its tendency to raise total cholesterols.

Although higher intake of polyunsaturated fatty acid (PUFA) content is no longer recommended but even the present day recommendations do suggest that about 33 per cent of the total intake of fat should be derived from polyunsaturated fatty acids as poly-unsaturated fatty acids are the only source of essential fatty acids commonly known as omega-6 and omega-3.

These essential fatty acids play a key role in regulating many physiological processes in the body, such as controlling blood pressure, vascular damage in the brain and the heart, preventing blood clot in the arteries, lowering cholesterol, inflammation and other conditions. The World Health Organisation (WHO) suggests that the ratio of omega-6 and omega-3 in the diet should be 5–10. Further a higher content of antioxidants present in the vegetable oil is also considered desirable to help prevent lipid peroxidation.

Precisely there are three parameters to adjudge any oil as the healthiest cooking oil i.e. ratio of saturated/monounsaturated/polyunsaturated fatty acids, ratio of essential fatty acids (omega-6/omega 3) and presence of natural antioxidants. Now let us see where the various cooking mediums stand vis-à-vis these recommendations.

Butter/Desi Ghee

Butter/Desi ghee has very high content of saturated fatty acids than recommended levels. Unless coupled with intake of fibre rich food and physical activity it is likely to increase total cholesterols. In view of the present day life style, use of butter/desi ghee as a sole cooking medium is no more recommended.

Vanaspati

Vanaspati produced by the process of hydrogenation of vegetable oils, also has very high level of saturated fatty acid content than the recommended levels. Moreover, *trans*-fatty acids produced in vanaspati as a result of hydrogenation of soft oils, have been found to raise bad cholesterol and reduce good cholesterols in recent scientific studies. FDA of US is in the process of formalising labelling guidelines for *trans*-fatty acids. In view of these developments, use of vanaspati produced through the process of hydrogenation of vegetable oils, in not being suggested as a sole cooking medium.

Kardi Oil/Sunflower Oil

Both kardi oil as well as sunflower oil have very high content of polyunsaturated fatty acids than the recommended levels. Both the oils have high ratio of omega-6/omega-3. Sole and prolonged usage of these oils is no longer recommended because of their adverse impact on the good cholesterols in the long run and poor oxidation stability which may lead to number of health complication such as arthritis.

Cottonseed Oil

Cotton seed oil also has a high level of polyunsaturated fatty acids than the recommended levels. The EFA ratio found in cottonseed oil is also far away from the recommended levels. So like sunflower oil and kardi oil, sole use of cottonseed oil can not be considered ideal.

Soyabean Oil

Although the EFA ratio of soyabean oil is quite ideal due to higher content of omega-3 in this oil, but polyunsaturated fatty acid content in this oil is much higher than the recommended levels. So use of soyabean oil needs to be coupled with use of some other vegetable oil having higher content of saturated/monounsaturated fatty acid to arrive at the desired ratio of SFA/MUFA/PUFA.

Palm Oil

Palm oil has high content of saturated fatty acids than the recommended levels. It needs to be used along with some other oil of high polyunsaturated fatty acid content and higher level of omega-3 essential fatty acids to ensure optimum intake of all fatty acids and their ratios. Tocotrienols found in palm oil help to reduce 'bad' cholesterol (LDL) and maintain 'good' cholesterols (HDL) levels in the blood. Beta-carotene found in red palm oil is a powerful natural antioxidant.

Groundnut Oil

Although the SFA/MUFA/PUFA ratio of groundnut oil is quite ideal, but the EFA ratio of this oil is higher than the recommended levels. So the use of groundnut oil needs to be supplemented with some other source of omega-3 essential fatty acids to ensure ideal balance of omega-6/omega-3 intake.

Corn Oil

Corn oil has almost ideal fatty acid profile except that its polyunsaturated fatty acid content is slightly on the higher side. It will be more appropriate to use corn oil along with some other vegetable oil of high monounsaturated fatty acid content and higher level of omega-3 essential fatty acid with a view to ensure optimum intake of SFA/MUFA/PUFA ratio.

Mustard Oil/Rapeseed Oil

Mustard oil has the most ideal SFA/MUFA/PUFA ratio as well as EFA ratio. But unfortunately most of the varieties of mustard oil found in India have very high content of erucic acid ranging between 35 to 48 per cent. High content of erucic acid present in the mustard oil has been held responsible for hampering the conduction of electrical impulses in the heart. Recent studies conducted by the Indian Council of Medical Research have held high content of erucic acid in mustard oil as single factor responsible for the high incidence of heart diseases in eastern India. So it is advisable to use only those varieties of mustard oil which have very low erucic acid content.

Olive Oil

Olive oil also has an ideal SFA/MUFA/PUFA ratio with highest content of monounsaturated fatty acids than any other cooking oil. As such it is also an ideal cooking oil and is a preferred cooking medium of the Europeans. But olive oil needs to be supplemented with any other vegetable oil with high polyunsaturated fatty acid content and ideal ratio of omega-6/omega-3 to achieve the desired intake of essential fatty acids.

Canola Oil

Canola oil belongs to mustard oil family and is mainly found in Canada. Like mustard oil, it has an ideal fatty acid profile and EFA ratio closer to the recommended levels. The plus point of canola oil over mustard oil found in India is that it has very negligible level of erucic acid i.e. below 2 per cent. So canola oil is considered an ideal cooking oil world over. But canola oil like olive oil lacks some anti-oxidants and micronutrients which are unique to rice bran oil because of which rice bran oil have shown to provide better protection against coronary heart disease than canola oil and olive oil.

Rice Bran Oil

Rice bran oil is the only cooking medium which besides having an ideal SFA/MUFA/PUFA ratio and EFA ratio quite closer to the recommended levels, contains three types of natural antioxidants as against only one category found in most of the other vegetable oils, leading to comparatively better oxidation stability desirable for promotion of overall health. A number of scientific studies conducted in India have confirmed that using rice bran oil as a cooking medium could significantly reduce bad cholesterol without adversely affecting the good cholesterol due to the presence of a unique component in this oil known as 'oryzanol' which is not found in any other vegetable oil. Studies conducted by the University of Lowell, USA, have revealed that rice bran oil has a significantly better cholesterol lowering properties than popular olive oil. Similar studies conducted by Human Nutrition Research Centre, Washington ST, have confirmed that rice bran oil provides better protection against coronary heart disease than canola oil.

In view of the most ideally balanced fatty acid profile or rice bran oil and findings of a number of scientific studies as to its better oxidation stability along with better cholesterol balancing properties than other vegetable oils, rice bran oil can definitely said to be a better choice. Use of rice bran oil as a sole cooking medium can ensure optimum intake of all fatty acids and their ratios which in turn will ensure better cardio as well as overall health.

RICE WAX—UTILISATION

Rice wax being a costly item will not only improve the economics of rice oil processing but will also improve the quality of edible cooking oil and lower its cloud point.

However, some of the units working on the recovery of crude wax by filtration are facing a bottleneck on the utilisation of crude dark rice wax. This was mostly dumped in soaps, vanaspati and other commercial products. The recovered wax was even hydrogenated by enterprising industrialists but the cost realisation is still not upto the mark.

This section highlighted the versatile nature of waxes and the branches of industries in which waxes have been used. Branches of industry in which waxes are used have been summarised in Table 8.16.

Table 8.16. Industries in which waxes are used.

Name of industry	*Examples of applications*
Adhesives, hot melts	Viscosity regulation, lubricants, surface hardening
Building	Modification of bitumen, anti-graffiti treatment
Candles	Fuel, drop point regulation
Ceramics and metal	Binders and consistency regulation for ointments, pastes, creams, lipsticks
Electrical and electronics industries	Release agents, insulating materials, etching bases
Explosives	Stabilisation
Foods	Citrus fruit and cheese coating, chewing gum base, confectionery
Matches, pyrotechnics	Impregnating, fuel
Medicine and pharmaceuticals	Moulding and release agents in dental laboratories, retardants, surface hardening of pills
Office equipment	Dispersing agents and binders for carbon paper and self duplicating paper, anti-offset for toners for photocopiers
Paints and coatings	Matting, surface protection
Paper and cardboard	Surface hardening
Plastics	Lubricants (PVC), release agents (PA), pigment carriers (master batch)
Polishes	Surface protection of leather, floors, cars
Printing inks	Improvement of rub resistance, slip
Recycling	Compatibilising
Rubber industry	Release agents enhancing rigidity, surface hardening.

It had been mentioned that waxes are highly versatile material. Particular features for many industrial applications, are their good absorption capacity and their ability to bind solvents, hot wax solutions from stable homogenous paste on cooling and can be applied to surface and polished after evaporation of the solvent. The resultant wax films are shiny or glossy, and resistant to mechanical stress. When used in release, lubrication, binding, imparting, compatibility or flexibility, they also regulate viscosity, adjust consistency, and drop point. Waxes are also used as combustibles and in illuminating materials as in candles. They have good dielectric properties. Some of them are approved for edible application. Rice wax, is not only suitable for food grade applications but also raises the melting point of blend, and prevents crystallisation of behaviour of paraffin and similar waxes.

Due to their widespread use and uncertainty of availability from natural sources, waxes have been obtained from mineral sources such as paraffins, fossil and earth waxes, such as Monton wax etc. Various grades and properties of Monton wax are shown in Table 8.17.

Table 8.17. Grades and properties of Monton wax grades .

Hoechst wax grade	*Melting point °C*	*Point °C as per DIN 51556*	*Acid value*	*Saponification value*	*Unsaponifiable. matter*	*Specific gravity at 20°C*
S	80–83	76–79	140–155	160–180	7–10	1.01–1.02
L	80–83	76–79	125–145	150–170	7–10	0.99–1.00
O, OP	102–106	80–85	10–15	105–120	7–10	1.03–1.04
OM	93–98	70–75	17–22	105–120	7–10	1.02–1.03
Special	96–100	72–77	13–18	90–105	20–22	1.00–1.01
F	74–77	70–73	6–10	95–105	42–47	0.98–0.99
E	78–82	73–75	15–20	145–165	7–10	1.01–1.02
CR	82–85	75–78	30–35	110–125	12–16	1.00–1.01
DPS	80-83	73–76	20–30	135–150	12–14	1.02–1.03
SPE	78–81	73–76	18–25	120–130	20–25	1.02–1.03
SP	80–83	72–75	20–30	135–150	12–14	1.02–1.03
BJ unbleached	70–73	66–69	17–25	135–150	22–25	0.95–0.96
BJ bleached	70–73	66–69	65–85	155–170	21–23	0.96–0.97
N new	73–76	66–69	40–50	120–135	22–26	0.96–0.97

Grade	*Melting point °C*	*Acid value*	*Ester value*	*Saponific value*	*Matter insoluble in benzene*	*Ash*
Crude	81–84	28–31	34–38	62–69	0.1–0.2	0.3–0.4
Highly bleached grade 1	72–76	82–89	2–6	84–95	0.0	0.0
Riebeck Germany grade 2	72–75	68–72	2–6	70–78	0.0	0.2–0.3
Grade 3	72–76	72–76	48–58	15–24	0.0	0.0
Grade 4	60–63	20–24	2–5	22–29	0.0	0.0
Czech crude	80–81	22–35			0.03	.18
Montan black						
Wax						
Bohemia						

Synthetic waxes have also been tailor-made for various industrial applications. Synthetic waxes have been obtained from the different sources with varying composition and properties to meet the requirements of end-user. Low molecular weight, polythene has wax like structure, with low viscosity and is made both in emulsifying and non-emulsifying grades. Many of us will be aware of their use in wax coating of bread and biscuit wrappers. Specification of different grade of polyethylene wax are show in Table 8.18.

Emulsifying grades are important for use in polishes, cosmetics, food preservation, pharmaceutical and other purposes. Rice wax will be particularly suitable for such uses.

Table 8.18. Specifications of different grade of polyethylene wax.

Specification	*PA 119°C (non-oxidised)*	*PAD 521 (emulsifiable)*	*PAD 522 (emulsifiable)*
Drop point	114–119°C	98–102°C	95–100°C
Setting point	101–106°C (×)	86–90°C (×)	83–88°C (×)
Acid number	0	14–18	23–28
Saponification number	0	20–35	40–55
Specific gravity (200°C)	0.94–0.96	0.93–0.95	0.93–0.95

Some of the possible approaches for raising the melting point and modifying other properties will be as follows:

1. Fractionation in soft and hard grades by dry and solvent fractionation. Crystallisation from solvents like isopropyl alcohol at different temperature could be commercially feasible.
2. Esterification of free fatty acids present in waxes by alcohols with or without the presence of catalyst.
3. Reaction with metallic oxides like zinc or calcium and other additions of metal soaps and compounds like ethyl cellulose castor wax etc.
4. Hydrogenation of rice wax raises the melting point. This is being commercially followed in India and other countries.

Rice wax for proper utilisation has to be processed in different grades for specific application. Some work in this directions has been done in other countries, like Japan, China etc.

A sample of rice wax prepared on lab scale from filter press mud was purified by crystallisation from solvent Isopropyl alcohol, chemically bleaching was duly approved by a Japanese company with following analytical parameters:

Acid value	37.3
Saponification value	119.8
Iodine value	18.3
Melting point °C	78.0

Crude rice wax produced by different manufacturers in this country has to be refined and processed by wax refinery units either indigenously or abroad, so as to get different grades for specific applications.

Rice wax so far has not been able to play a major role in the economics of rice oil processing.

RICE BRAN OIL—HEART-FRIENDLY COOKING OIL

Rice is the staple diet for majority of the Indians. India is the second largest producer of rice in the world next only to China. Rice (*Oryzae sativa* L.) like other cereals contains only a small amount of oil (2–3 per cent). However, when paddy is milled, the germ and bran layers separate from the endosperm and form the milling residue which is commonly known as 'bran'. The rice bran oil (RBO) is obtained from this brown layer of rice, which constitutes about 5 per cent of the paddy and is enriched in oil to the extent of 10–25 per cent. Bran from parboiled paddy yields more oil than the bran from fresh paddy. The oil extracted from the bran has a high free fatty acid content.

Role of Fat in Human Body

Fat is an important ingredient of human diet. The functional role of fat in the diet is manifold. Fat is a most concentrated source of energy. One gram of fat gives 9 kcal of energy. Carbohydrates and proteins provide only 4 kcal of energy per gram. Fat carries out following functions in our body.

1. Fat provides calorie density to the diet.
2. Fats are essential in the diet for the absorption and mobifitation of fat-soluble vitamins such as vitamin A, vitamin E and fat-soluble antioxidants. These vitamins are not utilised by the body if fat is not available in the diet. Thus fat works as a vehicle to carry the fat-soluble vitamins, nutrients and antioxidants in the body.
3. Vegetable oils are the only sources of essential fatty acids (EFA) to the body, which the body cannot synthesise and need to be supplied through diet. EFA are the precursors for a group of chemically related compounds, known as prostaglandins that are synthesised in the body from EFA. Prostaglandins play a key role in regulating many physiological processes in the body.
4. Fat also helps in raising HDL (high density lipoprotein), the so-called good cholesterol. Low-fat diets can result in reduction of HDL.
5. Fat in the diet imparts certain textural qualities, taste and palatability to the food. Fats and oils are integral lubricants of foods in two ways: through use as release agents as a part of the cooking process and as a lubricant during mastication (chewing).
6. Fats and oils modify flavour release and improve palatability.

Daily Fat Intake

According to the American Heart Association, the optimum intake of fat for an adult is 30 per cent of its total calorie intake. Therefore for an adult man consuming 2000 calories in diet should be getting about 600 calories (30 per cent) or 60–65 grams of total fat (visible and non-visible) per day.

Apart from visible fat some amount of fat is present in food items like cereals, pulses, milk, eggs, meat etc. as invisible fat. From the data-base generated on fatty acid composition in Indian foods and National Nutrition Monitoring Bureau diet surveys, the daily invisible fat intake is estimated to be 15 grams in rural poor and 25 grams in the urban middle and high-income groups. Thus the daily requirement of visible fat intake works out to be somewhere between 20 to 50 grams depending upon physical activity and physiological status.

Quality of Fat

Choice of a healthy cooking oil is a controversial subject. The ideas keep on changing as new evidence accumulates. Desi ghee (butter oil) used to be the conventional cooking oil of the Indians. There was nothing wrong with this most favourite cooking medium of Indian households. But the problems started with the change in our eating and living habits. Desi ghee being rich in saturated fats has the tendency to increase total cholesterols including HDL (so called 'good cholesterols') and LDL and VLDL (so called 'bad cholesterols'). As such consumption of desi ghee coupled with fibre rich diet which reduces LDL bad cholesterol and physical activity which reduces VLDL bad cholesterol brings about a favourable lipid profile.

But with the lack of physical activity in the recent years coupled with intake of diet devoid of fibre such as chappaties and other snack foods made from flour of decorticated foodgrains, consumption of desi ghee started raising total cholesterol levels. Therefore, the search started for an alternative cooking oil which could reduce the cholesterol levels. Short-term scientific studies conducted in this regard found that vegetable oils rich in polyunsaturated fats had the ability to reduce total cholesterols as such the recommendations tilted in favour of PUFA rich oils such as sunflower oil and safflower oil.

In due course of time evidence began to accumulate establishing that although polyunsaturated fats were effective in reducing LDL cholesterol, very high content of polyunsaturated fats had the undesirable affect of reducing HDL cholesterol as well. It further came to light that since oils with high levels of poly-unsaturated fats oxidise faster than those with MUFA or SFA profiles, therefore, the excessive intake of these oil could impair the ability of the antioxidants in the human system to control free radicals and thereby enhance the risk of certain cancers, cataracts, rheumatoid arthritis, Parkinson's disease and contribute to the ageing problem.

The American Heart Association now recommends use of oils having an equal proportion of saturated, monounsaturated and polyunsaturated fats. The World Health Organisation (WHO) recommends polyunsaturated fatty acid (PUFA)/saturated fatty acids (SFA) ratio of 0.8 to 1.0 and linoleic acid (omega-6) alpha linolenic acid (omega-3) ratio of 5–10 in the diet. The most recent recommendations according to the National Research Council, Washington DC are 10 per cent saturates, 12 to 13 per cent for monounsaturates, not more than 7 to 8 per cent for polyunsaturated fat should be accompanied by vitamin E, vitamin C or carotene consumption to help prevent lipid peroxidation.

Rice Bran Oil (RBO)

The rice bran oil is almost similar in fatty acid composition to that of groundnut oil and ranks fourth in production next to groundnut, rape/mustard, soyabean oils which are among the major edible oils produced in the country. Rice bran oil contains a high proportion of monounsaturated fatty acids (40–50 per cent) and polyunsaturated fatty acids (29–42 per cent). The rice bran oil contains about 36 per cent of linoleic acid and a very low content of linolenic acid (2.2 per cent) which ensures good shelf life. The fatty acid composition of the common vegetable oils is compared with that of rice bran oil as shown in Table 8.19.

Table 8.19. Fatty acid composition of common vegetable oils.

Oil	*Fatty acids % weight*			*Antioxidants*
	SFA	*MUFA*	*PUFA*	
Mustard oil/Rapeseed oil	6	67	27	Tocopherols
Sunflower oil	12	21	67	Tocopherols
Safflower oil	10	15	75	Tocopherols
Soyabean oil	16	24	60	Tocopherols
Groundnut oil	20	50	30	Tocopherols
Rice bran oil	20	45	35	Tocopherols Tocotrienols and oryzanol
Recommended	**Above 33%**	**Above 33%**	**Above 33%**	**Maximum possible**

If we look to the composition of the various vegetable oils commonly used in India, we shall find that no single oil could meet the requirement of these recommendations but rice bran oil and groundnut oil are nearer to the recommendations.

The use of any 'one of the oils like safflower, sunflower and soyabean furnish very high PUFA/SFA ratio. The use of safflower and sunflower also provide high ratio of omega-6/omega-3. Though the levels of omega-3 will be in the desirable range when mustard and rapeseed oils are used, these oils furnish very high levels of erucic acid. The use of groundnut or rice bran oils ensure optimal intake of MUFA and ratio of PUFA/SFA.

HEALTH BENEFITS OF RICE BRAN OIL (RBO)

1. The research institutes in India and abroad have found rice bran oil 'heart friendly health oil' with unique properties beneficial for maintaining good health.
2. Rice bran oil contains unique component oryzanol which is present only in rice bran oil. It increases HDL (good) cholesterol and lowers LDL (bad) cholesterol and triglycerides. Prolonged use of this oil reduces total cholesterol and triglycerides within 15 to 30 days. This oil is beneficial for heart patients and for people suffering from high level of cholesterol.
3. It has the ideal ratio of saturated, monounsaturated and polyunsaturated fatty acids and is the closest to World Health Organisation recommendation.
4. It has a natural antioxidant tocopherol (vitamin E), which helps in protecting the human body against diseases and maintaining balance of nervous system.
5. It has tocotrienol which has anti-thrombotic and anti-cancer properties.
6. It is good for skin. It contains squalene which improves skin tone and delays wrinkle formation. It has 4 hydroxy 3 methoxy cinnamic acid which stimulates hormonal secretion and rejuvenates health.
7. Food fried in refined rice bran oil absorbs 15 per cent less oil, ensuring lower calorie intake.

Rice bran oil has remarkable cholesterol lowering property in comparison with other cooking oils which can be seen from Table 8.20. Some value added product from rice bran oil are shown in Fig. 8.7.

Table 8.20. Cholesterol lowering activity of various edible oils.

Edible oil	*Linoleic acid %*	*Cholesterol level*
Safflower	77.1	– 16
Sunflower	61.4	– 12
Cottonseed	58.0	O
Soyabean	50.2	+ 3
Sesame	45.2	+ 2
Corn	43.0	– 15
Rice bran	36.0	– 17
Groundnut	35.0	+ 5

(+) increase; (–) decrease in cholesterol level.

Human trials conducted by the National Institute of Nutrition, Hyderabad, have confirmed significant reduction in total cholesterol particularly the triglyceride levels by using rice bran oil as compared to conventional cooking oils. In this study 21 subjects with high total cholesterols (>225 mg/dl) or high triglycerides (>190 mg/dl) were randomly divided into experimental and control groups. Twelve subjects, who constituted experimental group, were instructed to use rice bran oil in place of their usual cooking oil which they were using earlier. The remaining nine who served as controls, were allowed to continue the use of their habitual edible oil. There was a significant reduction in total cholestrol and triglyceride levels 15 to 30 days after the use of rice bran oil. However, in control subjects there were no changes in serum lipids.

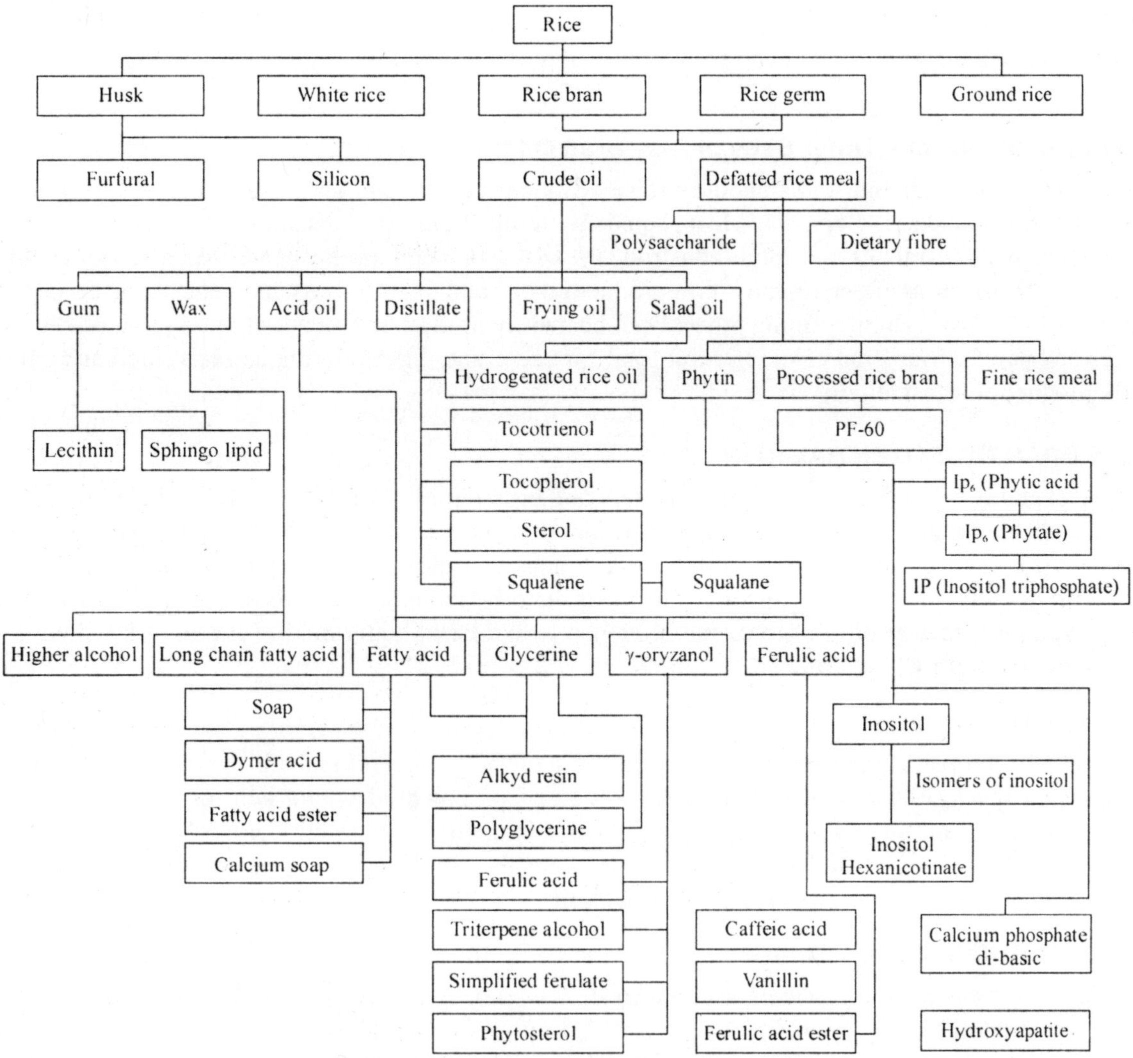

Fig. 8.7. Value added products from rice bran oil.

Cooking Advantages of Rice Bran Oil

1. More stable at higher temperature than other oils and food items prepared in rice bran oil are safe for human consumption.
2. Longer shelf life. Research has shown that nutritional value of oil is intact for 6 months of storage at room temperature.
3. Oil is less sticky, saves soap while cleaning the utensils.
4. It is economical in daily use as cooking oil as it is 15 per cent less absorbed during frying of food items.
5. Gives better taste and flavour to food items.
6. Frying takes less time, saves cooking gas.

Edible Uses of Rice Bran Oil

(i) Good cooking medium; (ii) deep frying fat; (iii) salad oil; (iv) bakery fat; and (v) manufacture of vanaspati.

Pharmaceuticals and Other Uses of Rice Bran Oil

As in Japan and Thailand, India should utilise advanced technology and research inputs to produce value added products like oryzanol, IF-6, squalene, ferulic acid esters which are very widely used in pharmaceutical, cosmetics and food industries. Product like calcium phosphate—dibasic can be used as organic fertiliser in effective, long term production of good quality fruits and vegetables. Soapstock obtained from rice bran oil refining process can be mainly used as raw material for the manufacture of soaps and paints. Rice bran cake is useful food for cattle and other milking animals, increasing their milk production substantially.

RICE BRAN OIL POTENTIAL IN INDIA

100 kg of paddy rice will generate approximately 5–8 kgs of bran. Currently India has potential of more than 10.0 lakh tonnes of rice bran oil per annum based on 5 per cent bran availability from paddy or 7.5 per cent from rice on production of 136 million tonnes of paddy or 91 million tonnes of brown rice. However, if entire quantity of brown rice milling is done by modern rice mills or sheller or modernised huller rice mills then quality of rice bran produced would be of better quality and in turn would give excellent grade of rice bran oil.

Status of RBO in India

In India, rice bran oil is the most important oil among the non-conventional oils in terms of its potential to augment the availability of oils. Full realisation of its potential would help in reducing the gap between demand and supply of indegenous edible oils in the Indian market. There is lot of awareness in government and industry circles about the goodness of this oil. But there is less awareness among the consumers about this oil as 'heart friendly health oil'.

Status of RBO in Other Countries

1. Rice bran oil is extensively used in Japan, Korea, China, Taiwan and Thailand as a 'premium edible oil'.
2. In Japan, rice bran oil is more popularly known as a 'heart oil'.
3. In recent years, US scientists have also shown a tremendous interest in the cholesterol lowering properties of rice bran oil (RBO).
4. In western countries rice bran oil has acquired the status of a 'health food'.

Finally, in todays 'Health is Wealth' scenario, this wonderful 'heart friendly' cooking oil is the new way to effective heart care. Its prolonged use reduces the risk of coronary heart disese (CHD) in humans by reducing total cholesterol and triglycerides. It has also anti-cancer properties which helps in allround health. The need of the hour is to promote this oil as 'Health Oil' in India with the help of the government, industry and other associated agencies. Television, electronic media and press can play a major role in its popularity as 'Health Oil' for the masses. Family physicians, dieticians and health experts will be useful as important members of our society in their guidance to the common man about the choice of cooking oil to be used in future. Easy availability of RBO in every village level in consumer packs will help for better health of our people in the national interest, thereby turning every house in India into 'Happy Home'

Chapter 9

Rapeseed Oil

INTRODUCTION

Rapeseed oil is a vegetable oil derived from rapeseed by expression or solvent extraction. It is a viscous, brownish liquid, though when refined it is yellow; specific gravity 0.913–0.916; solidifies at about 0°C; flash point 325°F (162°C); autoignition temperature 836°F (446°C); subject to spontaneous heating. It is high in unsaturated acids, especially oleic, linoleic and erucic and non-toxic.

It is edible oil for salad dressings, margarine, etc.; lubricant additive; substitute for soyabean oil; soft soaps; blown oils.

Rapeseed meal is the ground press-cake left from expression of rapeseed oil. Its major use is as an animal feed ingredient, but the presence of harmful glucosides, which may be goiter-inducing, limits its application for this purpose. Research efforts are being directed to removal of this constituent from meal produced in Canada. The meal also has some use as a fertiliser. If fed to animals, it should be blended with other feeds.

The technological aspects of processing rapeseed for obtaining crude oil and meal will be dealt in this section. The process involves the separation of oil from the oil-bearing material, which may be achieved by mechanical means (pressing), chemical means (solvent extraction), or a combination of both. The process also comprises the steps of seed cleaning, crushing, cooking and oil degumming and in the case of solvent extraction also desolventising and solvent recovery. The storage of crude oil and meal will also be dealt.

There is no principal difference between the extraction of rapeseed and that of other oil seeds. All extraction processes have certain objectives in common, viz., to obtain the oil undamaged and pure, to obtain the oil in as high a yield and as economically as possible and to produce an oil residue of the greatest possible value. In processing rapeseed, equipment originally developed for other oil seed is generally used. However, different oil seeds may demand different processing parameters. Thus, with rapeseed, for example, the small seed size, the high oil content and the presence of glucosinolates influence the process, especially the operation of the preparatory equipment.

The oldest method of rapeseed oil extraction to obtain edible oil is the cold pressing operation, which gives an oil of mild flavour and is probably still used in some parts of the world.

The process of expeller pressing was then developed. With the use of high pressures, a good yield of oil was obtained that was mainly used as industrial oil. However, to produce high quality edible oil and meal, the pressures had to be lowered, but this also lowered the yield. In order to economise the process, especially for plants of high capacity, solvent extraction was adopted. Direct extraction as well as pre-pressing followed by solvent extraction of the press residue, have been tried. All three oil extraction methods are still used for rapeseed, but the pre-pressing process is the most common today.

In the following section, the process, as it is used for rapeseed, will be dealt with step by step. In the last section the influence of the process on the products are discussed.

STEPS INVOLVED IN PROCESSING RAPESEED OIL

The steps involved in processing rapeseed oil (Fig. 9.1) are given below.

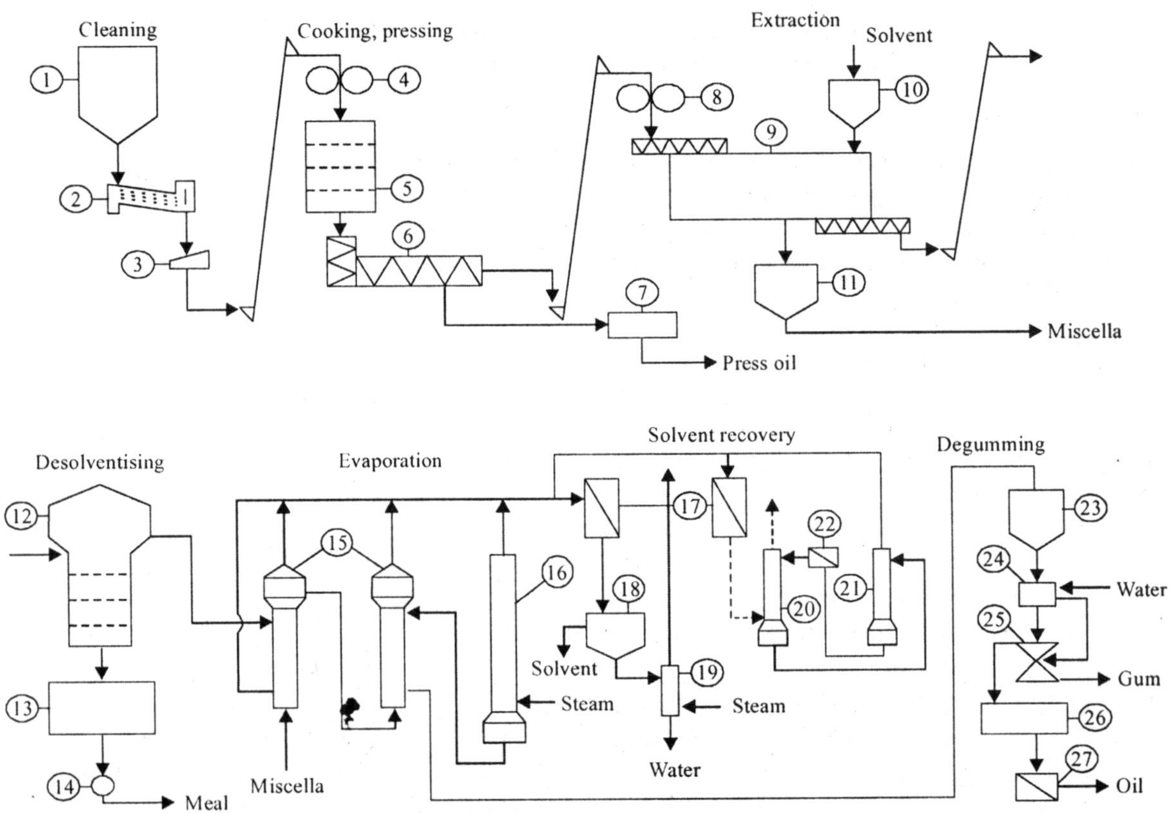

1. Silo
2. Cleaner
3. Stoner
4. Crushing mill
5. Cooker
6. Press
7. Filter
8. Flaking mill
9. Extractor
10. Solvent tank
11. Miscella tank
12. Desolventiser
13. Meal cooler
14. Mill
15. Long tubular evaporators
16. Stripping column
17. Condensers
18. Separator tank
19. Stripper
20. Absorbator column
21. Stripping column
22. Heat exchanger
23. Tank
24. Mixer
25. Degumming separator
26. Oil dryer
27. Oil cooler

Fig. 9.1. Flow diagram of a pre-press extraction plant.

Seed Cleaning

The first step in the processing of rapeseed is cleaning to separate foreign material, such as sticks, stems, leaves, other seeds, sand and dirt, which is removed by screening and aspiration. Permanent or electromagnets are used for the removal of iron objects.

Seed cleaning is an important operation in the preparation of rapessed for extraction. Proper seed cleaning will increase capacity, reduce maintenance and improve oil and meal quality. Final cleaning of the seed is sometimes carried out before storing but usually it is done at the extraction plant just before processing.

Dehulling

Rapeseed is presently not dehulled in commercial plants. The small size and the high oil content of the seeds make it rather difficult to dehull rapeseed efficiently, since the oil losses with the hulls would be too great. The hull content is about 15–20 per cent and its oil content is about 16 per cent (dry basis) giving losses of about 6–7 per cent of the oil. The hull content mainly depends on seed size, the smaller seeds having the higher hull content. Furthermore, the hulls may act as supporting material during the extraction providing better conditions and higher oil yields.

There is, however, a growing interest in dehulling rapeseed. This is primarily due to the high fibre content and dark colour of meal from undehulled seed, which lowers the feeding value of the meal. The glucosinolate content of the meal has, however, a greater influence on the meal quality. Extraction of dehulled seeds may also positively affect oil quality. Furthermore, there is a gain in the capacity of the extraction equipment. The whole problem of dehulling is of course dependent on the economic balance of yield and quality.

The normal way of dehulling oil seeds is first to crush the seeds in a mill. Then the hulls and the meats are separated by screening, air classification and gravity separation in several steps. This method may also be used for rapeseed. The problems are most likely to be found in the separation of the smaller fragments of hulls and meats. The high oil content of the seeds also causes the meats to adhere to the hulls. A usual technique for reducing this adherence is to dry the seeds slightly before dehulling, i.e. the moisture content is lowered 1–2 per cent-units.

Hull separation can be carried out before or after solvent extraction. *Eapen* and others have tried both alternatives in connection with detoxification of rapeseed in laboratory and semi-pilot plant experiments. Grinding wet-heat treated seeds in a stream of water resulted in squeezing out the meats intact from the hulls. After detoxification, drying, flaking and solvent extraction, the meat-hull mixture was air classified into a meat fraction and a hull fraction. *Eapen* thus obtained 44 per cent of the crude material (moisture and oil-free basis) as a white powder containing 60 per cent protein and a hull fraction containing 33 per cent protein.

Crushing

After cleaning, the rapeseed is crushed by roller mills to facilitate the following cooking and extraction. Usually one or a few pairs of rolls are used, which may be smooth or corrugated. The rolls in one pair can be run at equal or different speeds. Sometimes the crushing is done in a roll assembly consisting of five rolls placed one above the other and also in this case both corrugated and smooth rolls are used. Too intense crushing will produce large amounts of fines, leading to difficulties in the extraction process and with the clarification of the oil.

Essential with rapeseed is that the seed must have been dried to a moisture content of 6–9 per cent before crushing. With too high a moisture content the myrosinase in the seeds will start hydrolysing the glucosinolates when they are brought together during the crushing.

Cooking

Before pressing and extraction the crushed seed is subjected to a heat-treatment which is called conditioning or cooking. There are many reasons for the positive effect of cooking. The following objectives of cooking are generally considered completion of the breakdown of oil cells, coagulation of the proteins to facilitate the separation of the oil, reduction of the affinity of the oil for the solid surfaces, insolubilisation of phosphatides, increased fluidity of the oil by an increase in temperature, destruction of moulds and bacteria, inactivation of enzymes and drying of the seed to a suitable moisture content. The inactivation of myrosinase is the most important object of rapeseed cooking. This is essential for obtaining oil and meal of good quality and free of hydrolysis products from the glucosinolates.

The cooking is generally carried out in 'stack cookers', which consist of a series of closed cylindrical steel kettles stacked one on the top of the other. Each kettle is equipped with a sweep-type stirrer for agitating the crushed seeds during the cooking. The kettles are also jacketed for steam heating. Automatically operated gates provide a continuous flow of the seed down through the kettles. Moisture may be added through spray jets in the top kettle and removed through exhaust pipes in the lower kettles. Thus, it is possible to control the moisture content of the cooking seed at each stage of operation.

The cooking conditions used are largely dependent on the subsequent extraction procedure and the meal quality desired. The primary operation parameters of rapeseed cooking are residence time, temperature and moisture. The residence time in the cooker is usually 30–60 minutes. The temperature may be different in the different kettles, usually rising down the cooker as the seed becomes dryer. Temperatures of 75–120°C have been used, but usually the temperature is about 85–90°C in the top kettle. 80°C is the minimum temperature for myrosinase inactivation. The highest temperatures are used when the seed is to be straight expeller pressed; for example 105–110°C has been used followed by 5 minutes conditioning at 130–135°C.

Since the moisture content of the seed affects the affinity between the meat and the oil, control of the moisture in the cooker is very important. Water or steam is sometimes added to the top kettle, but often there is no such addition in cooking rapeseed. The moisture content in the top kettle is 8–12 per cent and is reduced in the lower ones to 4–8 per cent. The lesser moisture the more powder is produced. For straight expeller pressing the seed is dried to a lower moisture content (4–5 per cent) than when the seed is to be solvent extracted.

The method of cooking markedly determines the quality of both the oil and the oil-cake. Cooking is particularly important for the maintenance of a low level of oil refining losses. The meal quality depends on the content and nutritional value of the protein and the content of goitrogenic substances. The protein of the meal is often extensively denatured in the cooker, which usually gives a better feeding value. Too much heating, however, causes losses in amino acids, especially the basic ones. Lysine, which is an important essential amino acid, appears to be the most heat sensitive.

Screw Pressing

Extraction of rapeseed oil by means of screw presses (expeller presses) can be done as straight pressing or pre-pressing for subsequent solvent extraction. In the former case the oil content of the cake will be 4–7 per cent and in the latter 12–20 per cent.

The mechanical screw press has five essential elements: the main worm shaft and worms, the drainage barrel, the choke mechanism, the motor transmission and thrust bearings and the cooling system. The main worms shaft and worms are designed to exert a pressure of 1000–1400 kg/cm^2 on the seed being processed and at the same time to convey the seed through and out of the pressure chamber. The choke mechanism permits a final adjustment of the pressure. The drainage barrel is made up of rectangular bars, which fit into a heavy barrel bar frame. The individual bars are separated by spacing clips which provide spaces of about 0.25 mm. This permits the drainage of oil from the seed being pressed and keeps the cake within the barrel which also acts as a coarse filter medium. The screw press can be built up of a single horizontal shaft or a combination of a vertical and a horizontal shaft. A great deal of heat is produced during the pressing operation, especially when straight pressing is employed. In order to keep the temperature down various cooling devices are required. The screw press may be equipped with water-cooled shafts and ribs in the bar cage or some of the expressed oil may be cooled and circulated over the exterior of the barrel.

Principally there is no difference between presses used for straight pressing and those used for pre-pressing. Ordinary high-pressure screw presses may be operated at low pressure and at an increased capacity, but especially designed machines are now normally used for pre-pressing.

The process of straight pressing has the advantage over solvent extraction in being simple and of lower investment cost. Disadvantages are, however, the high power consumption and the wear and tear of the machinery, giving high working costs. Another major drawback is that it is only possible to press down to a residual oil content of 4–7 per cent, giving high losses in yield. There is also the effect of the high temperature generated during straight pressing causing an oil and particularly a meal of lower quality. The cooking conditions employed are also more severe when straight pressing is used.

The process of straight expeller pressing was the first method used for obtaining rapeseed oil and is even today applied in some cases. Now, however, the solvent extraction process is dominant, but it is most often combined with a pre-pressing step using screw presses. This utilises the advantages of both techniques and is especially useful in the case of seed with high oil content such as rapeseed.

After pressing, the oil contains fine meal particles. Before further treatment the oil is thus settled in a tank and filtered in filter presses with added filter aids. The amount of fines and thus the filtration rate, is highly dependent upon the cooking process. The fines may be returned to the cooker or, preferably, fed directly to the solvent extractor.

Flaking

After straight expeller pressing the cake is cooled, moistured and milled to a meal. If a pre-pressing operation is used the crushed cakes are instead flaked for solvent extraction. This step is very important because the seed particles must be thin enough to be extracted readily and yet large and coherent enough to form a mass through which the solvent will freely flow.

The flaking mill consists of one pair of smooth rolls by which a flake thickness of 0.2–0.3 mm is obtained. The moisture content of the seed material must not be too low, if thin coherent flakes are to be formed. Sometimes when the oil content in the press cake is low (12–14 per cent) the flaking is not so important and the cake is just milled. This may also be the case if the flaking is done before the cooking.

Solvent Extraction

Extraction of rapeseed with solvents constitutes the most efficient method of obtaining the oil from the oil bearing material. The advantage over mechanical pressing is greater for oil seeds with low oil content. Oil seeds high in oil content such as rapeseed are, however, also most economically solvent extracted, especially in high capacity plants. Generally pre-pressing is used to give a press cake containing 12–20 per cent oil, which has the most advantageous properties for solvent extraction. This process also gives the lowest residual oil content, about 0.5 per cent. However, direct solvent extraction of rapeseed is also used and will be dealt with below.

In spite of the extensive application of solvent extraction of oilseeds, there is no universally accepted theory of the manner in which the basic process occurs. The rate of extraction is profoundly affected by the flake thickness and the flake area, the temperature, the solvent and moisture. *Karnofsky* concluded from extraction studies that the bulk of the oil is extracted rapidly but the remainder of the oil only slowly and with increasing difficulty. Based on the porous nature of the flakes, two entirely different mechanisms have been advanced to explain solvent extraction of oils. These are the theory of molecular diffusion and the theory that the rate of extraction is determined by the rate of solution of undissolved oil. It is generally accepted that extraction takes place by both mechanisms simultaneously, but that dissolution of the difficult soluble materials eventually determines the extraction rate.

The object of solvent extraction is to remove as much of the oil as possible from the meal with a minimum of solvent. This is most effectively achieved by continuous counter-current extraction. The principle is that the solvent should be passed in several stages through the travelling mass of the oil-bearing material. The characteristics of a good extractor are that it produces a miscella (solvent plus oil) of high oil content and with low content of fines, has relatively small dimensions, high capacity, reliability and safety of operation and has a low labour requirement.

Continuous extractors fall into two classes: percolation extractors in which the solvent flows by gravity through a bed of solids and immersion extractors in which the solids are immersed in the solvent or miscella. There are also extractors that comprise a combination of these two types. Modern extractors are of the percolation type, which gives higher miscella concentration and allows use of the solid bed as a filter giving very small amounts of fines in the miscella. This is achieved by recirculation of miscella within the percolation extractor. In addition, the percolation type usually incorporates much better drainage of liquid from the spent solids. This fact and the increased miscella concentration decreases steam requirements for solvent recovery.

There are many modern percolation extractor designs, but the general principle is mainly the same. This may become clear from the following brief descriptions of some extractors.

The oldest type is the vertical basket extractor (Bollman extractor) still built in America and Europe, which consists of a vertical chamber in which a number of baskets with perforated bottoms are carried on a chain. The baskets are moved vertically. The flakes are charged at the top and are concurrently extracted by miscella (half concentrated) on the descending side. On the ascending side the flakes are countercurrently extracted by fresh solvent, which is added just below the last drainage part. The meal is discharged at the top and transported away by a screw conveyor.

This type of extractor has the advantage of producing miscella free of fines because of the effective filtration in the concurrent extraction step. The large volume and height of the extractor is, however, an important drawback. Modifications have thus been developed resulting in rectangular and horizontal basket extractors.

Lurgi has developed a frame belt extractor, composed of two screen belts supporting the frame basket (Fig. 9.2.). These have no bottoms and move at the same speed as the screen. The prepared seed material is delivered at one end of the upper part of the extractor and at the other end the material falls down into the lower returning frame baskets. From the other end of this screen belt the meal is discharged. The solvent passes the seed material in several steps countercurrently. Under each step there is a hopper collecting the miscella. This is then partly pumped and sprayed over the material in the same step and partly pumped to the next step. The apparatus is divided into three zones—the washing zone, the extraction zone with circulating miscella and the filtering zone in which the concentrated miscella is percolated through the bed until free from flakes. There are also drip-off zones between the steps, preventing the miscella of low concentration from diluting the more concentrated one. The shifting of the material when falling to the lower belt prevents any formation of channels and secures uniform de-oiling.

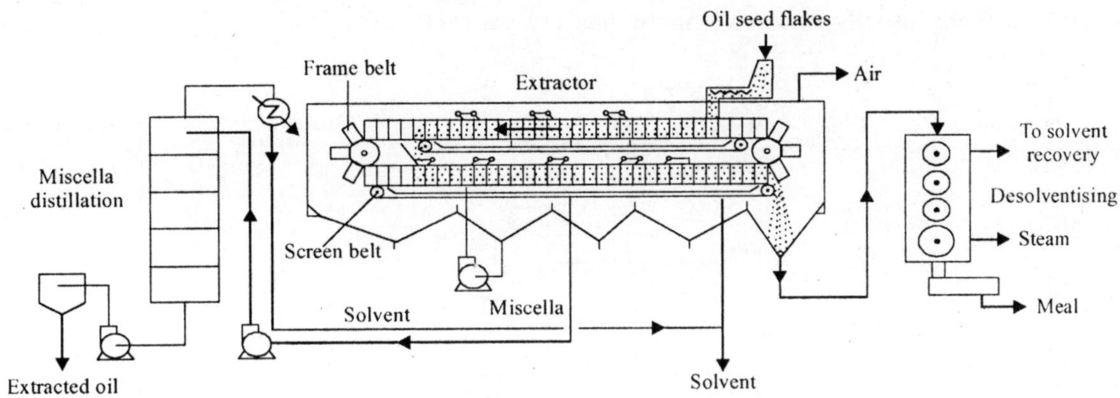

Fig. 9.2. Lurgi frame belt solvent extractor.

Another horizontal extractor is the De Smet endless belt type extractor. The main differences as compared with the Lurgi extractor are that there is only one belt and no baskets.

Many constructions of extractors are based on the principle of turning sector-shaped compartments (baskets) in a horizontal plane around a vertical axis. The best known example is the Rotocel extractor. This construction principle provides a simple, compact and flexible extractor. As each compartment passes the feeder pipe, the seed material is deposited therein. As the filled compartment continues to revolve, it is sprayed at intervals mainly countercurrently with solvent from overhead nozzles. The resulting miscella drains continuously through the mesh type floor of the compartment and collects in troughs. At the completion of the circuit the floor of the compartment opens, thus dumping the residual meal into a hopper. As soon as the floor is restored to its original position, the compartment receives another load of unextracted flakes and the cycle is repeated. The whole extractor is entirely enclosed in a vapour tight cylindrical tank and makes one complete revolution in about 90 minutes.

French has developed a similar extractor called the 'stationary basket extractor' where the baskets are stationary and the filling spout, miscella distributors, miscella collection pans and the flake discharge hopper are rotating.

With the percolation type of extractor it is difficult to extract seeds of high oil content such as rapeseed directly. The volume reduction of the seed material and the tendency to produce fines causes improper packing of the bed and channelling, giving a lower yield of oil. Only if the extraction is performed in steps with re-packing of the bed, as in the *Lurgi* extractor built in several floors from Extraktionstechnik, can direct extraction be used.

The filtration extraction process was developed for direct extraction of various oilseeds and is also used for rapeseed (Fig. 9.3). The extraction is carried out in two steps, slurrying and filtration. Another distinctive feature of this process is that the cooking step is essential to the process. The cooked flakes are rapidly cooled in an evaporative step called 'crisping' and are then re-rolled, producing a material with good extraction and filtration properties. The 'crisped' seed material is continuously fed together with one of the miscella filtrates to the slurry mixer, which is a mildly agitated horizontal tank. The slurry is kept there for 30 to 45 minutes at a temperature of 50–60°C and is then filtered on a rotating horizontal screen vacuum filter. The concentrated miscella drains through the filter leaving a 5 cm thick bed. The cake is countercurrently washed with solvent, but the most concentrated miscella is re-filtered through the bed to reduce the content of fines. The meal, which only contains 25–30 per cent solvent, is removed continuously. The residual oil content, which can be obtained economically by using direct extraction methods, usually amounts to more than 1–2 per cent.

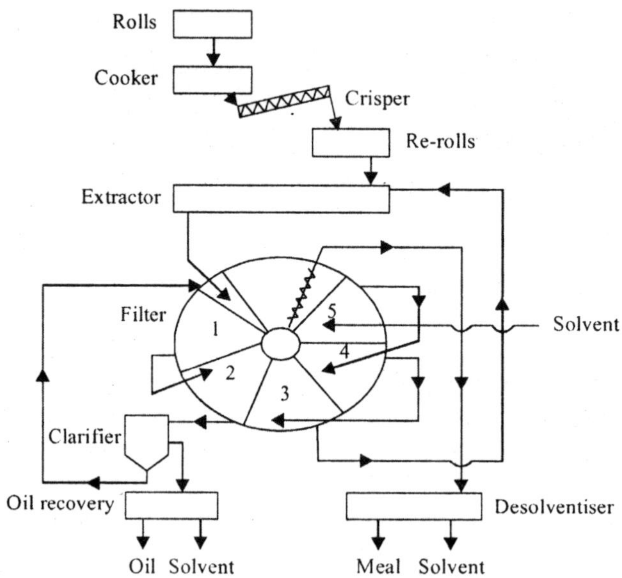

Fig. 9.3. Schematic diagram of filtration extraction unit.

The most widely used solvent for extraction is commercial *n*-hexane, a petroleum fraction with a boiling range of 63–66°C, especially refined for use in the vegetable oil extraction industry. Although other solvents are sometimes used for other oilseeds, only hexane is used for rapeseed. The high erucic acid content of rapeseed oil makes its solubility in polar solvents such as absolute ethanol or isopropanol less than that for other oils. *Loury* and *Feng* reported that acetone, ethanol and isopropanol gave poor yields and that dichloroethane and trichloroethylene gave red-coloured oils. Cyclohexane gave the same results as *n*-hexane. *Franzke* found that the higher the aromatic content in the hexane, the darker was the extracted oil.

The solvent to meal solids ratio in modern extractors is about 1.1–1.3. The concentration of oil in the miscella leaving the extractor is 20–35 per cent, somewhat dependent on the oil content of the extracted seed material. By extracting high oil content seeds in the Lurgi extractor the figures 45–50 per cent have been obtained. The amount of solvent left in the meal after extraction in a percolation type extractor is about 30–35 per cent.

Meal Desolventising

After completed solvent extraction, the extracted material is discharged from the extractor and conveyed to the desolventiser, where the hexane is removed by indirect heating and by direct steam injection. Besides a good desolventising effect leaving a meal with not more than 0.01 per cent hexane, this step should give the meal an appropriate heat treatment, so that the feeding value of the meal will remain high.

The oldest equipment for desolventising consists of a series of horizontal steam jacketed conveyer tubes. The wet meal is introduced at one end at the top and is moved downwards in zig-zag fashion. In the bottom tube sparging steam is sometimes introduced to evaporate the last traces of solvent. This desolventising system is, however, rather inefficient and suffers from dust control problems and other problems of operation, but it usually gives a light coloured meal.

The most common desolventiser equipment today is the desolventiser-toaster, originally developed to give high quality soyabean meal. The equipment is similar to a cooker but is totally enclosed and with a dust collector at the top. Live steam is the main source of heat for the vapourisation of the solvent from the meal and it is introduced in the top stages. Solvent vapours replace the steam, which condenses and replaces the liquid solvent. The moisture content of the meal is then about 20 per cent. The toasting process taking place in the lower stages denatures the proteins and reduces the amount of glucosinolates in the rapeseed meal, by the action of moisture and heat, in order to give the meal a high feeding value. In the toasting section the meal is also dried to a moisture content of 6–11 per cent. The temperature of the meal in the desolventiser-toaster is about 100–110°C and the residence time about 30 minutes. The meal will usually be rather dark coloured after this process.

Another modern system is flash desolventising, in which superheated solvent vapours are recycled over the meal to evaporate the solvent (Fig. 9.4). The desolventising time can be reduced to a few minutes or even lower. A final steam stripping is required to take away the last traces of solvent from the meal. This is carried out in a special 'deodouriser' or in an ordinary desolventiser-toaster. The meal will have a light colour and the protein solubility can be easily regulated by this process.

1. Feed conveyor
2. Desolventising tube
3. Flake separator
4. Circulating blower
5. System pressure control valve
6. Vapour scrubber
7. Scrubber discharge lock
8. Solvent condenser
9. Recirculating control valve
10. Vapour superheater
11. Separator discharge lock
12. Flake stripper

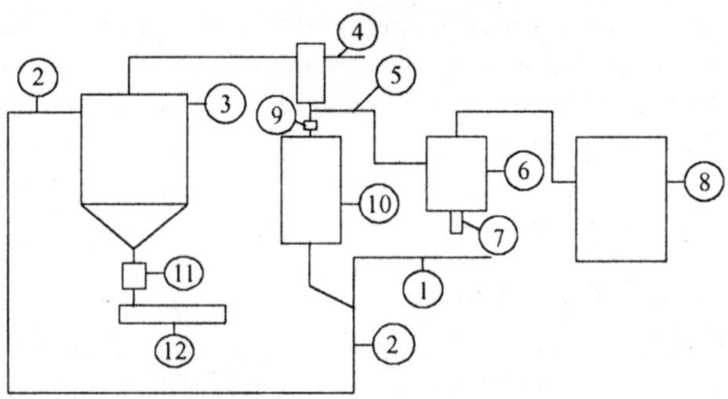

Fig. 9.4. Flash-desolventiser.

After desolventising the meal is immediately cooled in a meal cooler, where the final drying also takes place. The moisture content should be 9–10 per cent for good storage and transport quality. Finally the meal is milled in a hammer mill. Fines in the miscella are first removed in a filter press or in modern plants by recycling the miscella through the bed in the extractor. To keep the evaporation costs low the miscella should be as concentrated as possible.

The evaporation is ordinarily carried out in a long tubular evaporator where 90 per cent of the solvent is recovered. The oil is then passed through a stripping column which can be either a packed column or a bubble cap column. There the last traces of solvent are removed by applying sparging steam at the bottom of the column. To obtain good heat economy of the process the evaporation is divided into two stages and may be done under vacuum. The miscella in the first (second) stage is heated by the hot vapours from the desolventiser and the hot oil from the stripper is used for heating the second (first) stage evaporator. This system will lower the steam consumption to about 25 per cent. The residual amount of solvent in the oil is less than 0.1 per cent. A separate problem is the recovery of solvent from the final discharge of vent gases. Refrigerated vent condensers have been used extensively. A new system of solvent recovery has been developed based on the principle of solvent absorption by a vegetable oil. This equipment is almost always installed as a part of new plants. The main advantage is the simplicity of the equipment, especially since a complicated compressor for the cooling medium is unnecessary. Sometimes a water scrubber is used for solvent recovery. A solvent loss as low as 0.06 per cent based on the raw material feed can be achieved in large plants.

Degumming

The crude solvent extracted oil contains about 2 per cent 'lecithin' gums, which are usually removed by the degumming (desliming) process. If the gums are not removed, they will cause trouble by settling out in storage tanks and they will tend to produce large refining losses.

The oil is treated with a small amount (4 per cent) of water or a dilute water solution of acids or salts in order to precipitate the gums (mainly phosphatides). Often degumming with water at the extraction plant is followed by a degumming with phosphoric acid at the refining plant. The precipitated gums and the water are separated from the oil in continuous centrifuges, which are of the same type as those used in continuous refining. The advantage is especially that these centrifuges are hermetical and all contact with air can be avoided. This is important for preventing oxidation since the oil temperature is high (80°C). If the removed gums are to be used for commercial lecithin production, they should be dried as soon as possible. This is carried out in vacuum with agitation or in a thin film evaporator at 70–110°C and the drying can be combined with hydrogen peroxide bleaching. Usually, however, the separated gums are transported back to the desolventiser-toaster, since rapeseed lecithin is of low value and soyabean lecithin dominates the market. In the toaster the gums make the meal powder agglomerate more easily. The degummed oil, which still contains about 0.5 per cent phosphatides, is mixed with the press oil and dried in vacuum at 80–90°C. After cooling, the oil can be stored for a long time if kept cold.

Storage of Crude Oil and Meal

The crude degummed oil may be stored for a longer or shorter time before refining. The longest time the oil can be stored depends mainly on its properties and stability but also on the way in which it is stored.

Rapeseed oil is usually stored in tanks. If these are large the temperature must be kept high enough to avoid crystallisation of the oil. Rapeseed oil stored in this way at 25–30°C will already after 3–4 months, be highly oxidised. It is then better to store the oil in small tanks, in which it may solidify and then heat the oil to melt it, when it is to be used. The heating must be controlled closely in order to avoid local overheating and this is best done by indirect heating.

The amount of free fatty acids (FFA) and the degree of oxidation of the oil as measured by the peroxide value (PV) and benzidine value (BV), are the main properties which change during the storage. The amount of free fatty acids in crude oil of good quality should not be greater than 1.5 per cent in order to avoid high refining losses. The amount is usually lower and the increase during the storage is also rather low and thus there is usually no problem with free fatty acids.

Oxidative stability has the greatest significance for the storage of rapeseed oil. The linolenate content of the oil makes it prone to oxidation. The tocopherol content in rapeseed oil, however, makes the stability acceptable. It is important that the rapeseed is fully ripened and of good quality to give an oil of high stability. The tocopherol content increases and chlorophyll (which is prooxidative) decreases in amount during ripening. Germinated and mouldy seeds also possess higher lipase activity and the oil produced is less stable.

Rapeseed meal is stored in bulk at a maximum of 10 per cent moisture in silos or metal tanks with conical bottoms. There is no problem in storing meal for a short time if properly done and with an appropriate percentage of moisture. Moisture adsorption due to high relative humidity causes heating, loss of etherextractable fat, increases in free fatty acids and slight decreases in protein and non-reducing sugars.

Influence of the Extraction Process on the Quality of Rapeseed Oil and Meal

To obtain a rapeseed oil of high quality, it is always necessary to start with well ripened high quality seeds. Furthermore it is essential to use optimal processing parameters throughout the production. The earlier the oil and meal get any damage the more serious it is for the end products. In other words, the basic quality of the oil is not improved, only more or less damaged.

One of the most important considerations with regard to quality is to have a continuous and rapid flow of material from one piece of equipment to another. In crushed and flaked seeds deterioration takes place rapidly, particularly if the material has a high moisture content or is of poor quality. It is important to avoid oxygen contact with the oil, to keep the temperatures low, to make the hold-up times short, to eliminate sources of catalysis and to make the treatments in the dark.

There are two enzyme systems in rapeseed that may influence the oil and meal quality, viz., lipase and myrosinase. They must be inactivated at an early stage in the process and this is done in the cooker.

The lipase activity of rapeseed may result in high free fatty acid values of the oil. The optimum temperature for lipase activity is 37–40°C. Lipase inactivation is highly dependent on moisture and temperature. Inactivation starts at 60–70°C and is rapid at 90–100°C in an aqueous medium and even more rapid in an acidic medium. In the dry state activity decreases only very slowly and temperatures of 140–160°C are needed for rapid inactivation.

The myrosinase in rapeseed catalyses the hydrolysis of the glucosinolates to give glucose, sulphate, isothiocyanates and (*l*)-5-vinyl-2-oxazolidinethione. The latter product is goitrogenic, thus yielding meal of low feeding value. They also act as a poison for hydrogenation catalysts if they are dissolved in the oil. It is thus important to keep the glucosinolates intact by effectively inactivating myrosinase by proper cooking. Myrosinase and glucosinolates are brought together by the crushing of the seed, but the hydrolysis can only take place if the moisture content is high enough (13 per cent). The enzyme has the greatest activity at 40–70°C.

Inactivation of myrosinase starts at 70–80°C if the moisture content is 6–10 per cent and is usually performed at 80–90°C in the cooker. *Eapen* and others have performed various laboratory heat treatments in order to evaluate their effect on myrosinase inactivation. Dry-heat treatment (30 minutes, 105°C) proved unsatisfactory, but steam blanching (5 minutes), microwave heating (3 minutes) or immersion of the seed in boiling water (1.5 minutes) were effective. If the moisture content of the seed material is high and the temperature rise is not quick enough, glucosinolates may be hydrolysed before the myrosinase is inactivated.

Heat treatments during the process may also have a negative effect on oil and meal quality. If cooking temperatures above 110°C are employed the extracted oil may be difficult to hydrogenate, which has been attributed to chemical breakdown of the glucosinolates to give oil-soluble sulphur-containing compounds. Microwave heating (1.5 minutes) and steam blanching (5 minutes) have resulted in dark coloured oils, while dry cooking (30 minutes, 105°C) and immersion in boiling water (1.5 minutes) gave lighter-coloured oils. Rapeseed stocks with green seeds often yield dark coloured oil, but an immersion of the seed in boiling 0.5 per cent sodium hydroxide has been found to give lower colour values.

The colour of the meal is also affected by the heat treatments, especially in the desolventiser-toaster. High temperature and moisture favour non-enzymatic browning reactions (Maillard reactions) between reducing sugars and amino acids. At the same time the feeding value may be lowered by a decrease in the amount of lysine, which is the most sensitive amino acid in this respect. This decrease in lysine is also the reason that the straight expeller pressed meal is of lower nutritional value than solvent extracted meal. A small lowering in oxidation stability of the oil may take place during the extraction process, together with a slight increase in the peroxide value and the benzidine value. Heat treatments of the seeds and exposure of the oil to light and heat result in higher peroxide values.

The fatty acid composition of the oil can also be affected by the extraction technique used. The fatty acid composition in the neutral oil is different from that in the phospholipids. These are most difficult to extract and are thus obtained only in the solvent extracted oil. By using repeated extraction of white mustard with different solvents *Iverson* found that the last extracted oil was different from the first oil. *Zeman* also found that the oil remaining in meal extracted with hexane contained a higher amount of polyunsaturated fatty acids and a lower content of erucic acid.

PROCESSING OF RAPESEED OIL

Rapeseed oil is mainly processed in the same way as other vegetable oils such as soyabean oil, although some differences exist due to its rather high content of erucic acid. This may be changed in the near future since varieties with low or negligible erucic acid content have been developed by plant breeders. One aim of the various processing steps is to remove substances which can give unpleasant flavour to the oil. The natural impurities consist of free fatty acids, mono and diglycerides, phosphatides, steroids, vitamins, hydrocarbon pigments, proteins, sulphur compounds and oxidation products from the triglycerides. The crude oil must be of good quality, since an improperly pre-treated and stored crude oil needs more elaborate refining and in spite of more intense and expensive treatment, such an oil is also less stable than is normal for an oil originally of good quality. First of all the extraction must be done carefully, since incorrect heat treatment can give dark oils, especially if the seed stock contains green seeds. Furthermore, if the enzyme myrosinase has reacted with the glucosinolates, some oil-soluble sulphur compounds may be obtained. Even a good quality crude rapeseed oil contains about 10 ppm sulphur.

Besides being responsible for flavour disturbances, the sulphur compounds are severe catalyst poisons in the hydrogenation step. In the hydrogenation the melting point of the oil is increased and products with varying consistency can be obtained for various purposes, such as the production of margarine and shortening. The processing of rapeseed oil normally consists of degumming and refining, bleaching, hydrogenation and post-bleaching (if desired) and deodourisation.

Degumming and Alkali Refining

Lecithins and other gums are often removed at the end of the extraction process to avoid tank residues. Nevertheless, there are still some mucilaginous components in the oil, mainly phosphatides, which can be measured by the amount of phosphorus present. The presence of phosphatides will increase the risk for losses due to the formation of emulsions during the alkali treatment and decomposition to coloured products may occur during the deodourisation. Various methods for degumming the oil have been suggested, for example treatment with sulphuric acid, hydrochloric acid and also nitric acid. Although an effective degumming agent nitric acid is found to give a dark-coloured deodourised oil with bad flavour. For edible fats it is more advisable to treat the oil with 0.02–0.1 per cent of phosphoric acid (85 per cent technical grade) directly followed by alkali treatment according to the Alfa Laval Short-Mix process, or to treat the oil at a temperature of 60–90°C with 0.25–0.3 per cent by volume of 85 per cent phosphoric acid for 5–30 minutes under vacuum and then separate the precipitates by centrifugation according to the Zenith process. One advantage with these two methods is the removal of calcium and magnesium ions by the phosphoric acid, which decreases the soap content in the refined oil.

Neutralisation

Normally the process of neutralisation is carried out with sodium hydroxide of various concentrations, but also with other alkaline solutions, e.g. sodium carbonate (Clayton-De Laval process). However, the latter process has the disadvantage that carbon dioxide is formed. The sodium hydroxide combines with the free fatty acids to form sodium soaps, but if used in excess it will also react with the triglycerides. A stronger alkaline solution (4 M) will increase this risk, while the use of more dilute alkali increases emulsion formation. Together with the reaction time these factors must be borne in mind in order to secure low neutral fat losses. The efficiency of the refining can be expressed in various ways as shown in Table 9.1.

Table 9.1. Definition of different efficiencies.

$$\text{Refining factor} = \text{RF} = \frac{\text{Per cent total refining loss}}{\text{Per cent fatty acid content of crude oil}}$$

$$\text{Fatty acid factor} = \frac{\text{Per cent total fatty matter in soapstock of crude oil}}{\text{Per cent fatty acid content of crude oil}}$$

$$\text{Refining efficiency} = \frac{\text{Yield of neutral oil} \times 100}{\text{Neutral oil content of crude oil}}$$

When different factories or processes are to be compared, it is important to know whether or not the total refining loss or the yield of neutral oil includes the bleaching step. To obtain a rapeseed oil of high quality it is often necessary to perform another treatment with alkali, using either a weak solution of sodium hydroxide or a mixture of sodium hydroxide and calcium carbonate. After the alkaline treatment the rapeseed oil has to be washed thoroughly with soft water, since the calcium and magnesium ions in the hard water can be dissolved in the oil phase. It is practical to use either water from the condensers or desalted water from ion exchangers for this purpose. Immediately after washing, the oil must be dried under vacuum to avoid an increase in free fatty acids due to hydrolysis. A refined and dried rapeseed oil may be stored for a few days before bleaching, although it is of course preferable to make this storage as short as possible.

Closed tanks in which the oil is protected by nitrogen are recommended for storage in order to avoid contact with air during the processing, especially when the oil is warm. These precautions become more necessary after the oil has been refined since some substances with antioxidative effect are removed during the processing. The addition of antioxidants to rapeseed oil has no effect when the oil is properly stored and is only of limited value when stored under improper conditions. The method of refining greatly influences the oxidative stability.

Batch process

In batch refining, good results have been obtained with Swedish rapeseed oil under the following conditions: the degummed oil is heated to 70°C and 1 volume per cent of a solution containing 10 per cent acid (e.g. oxalic, citric or phosphoric acid) is added under stirring. The agitation is continued for one hour, whereupon sodium carbonate (25 per cent solution) is added in an amount of 0.3 volume per cent and the agitation is continued for another hour while the temperature is increased to 80°C. The oil is then neutralised with 3.5 M sodium hydroxide (20°Be). After one hour of separation the oil is washed with 10 volume per cent of water and the mixture is left for two hour's sedimentation. The separated soap phase is drained and the oil is heated to 95°C. The oil is refined with 3 per cent 1.5 M sodium hydroxide (6°Be) under agitation for three hours. In order to break the emulsion, 0.25 per cent sodium chloride is added and stirring is continued for another hour. The oil is washed three times with 10 per cent of water. Finally, 0.05 per cent of a citric acid solution in ethanol (1 : 1) is added in order to split the remaining soaps. This also increases the bleaching effect, probably due to the fact that iron and other complexing metals form chelates with citric acid.

In order to avoid the emulsion losses, suggestions have been made to use electrolytes or alcoholic alkaline solution. In the Marchon process the use of sodium xylene sulphonate in the alkali has proved to decrease the refining losses. From one to three per cent of the hydrotrope is used for each unit of acid value in the crude oil. A 40 weight per cent hydrotrope solution is mixed with the normally required amount of lye and is added to the heated oil (70°C–90°C) with vigorous agitation. Compared to normal batch refining the settling of the soap stock is considerably promoted by the high specific gravity and the low viscosity of the hydrotrope solution with the dissolved soap. After separation the soap stock is split with sulphuric acid and the fatty acids separate. The hydrotrope solution has to be concentrated by evaporation to 40 per cent before recycling to the process. The sodium sulphate obtained by splitting the soaps accumulates and the excess will crystallise, leaving a deposit which must be removed by filtration. The advantage with the Marchon process is more evident for batch refining. The improvement of continuous refining may be too small to pay the extra capital and processing costs.

Continuous process

Emulsion in the soapstock causes high refining losses, even if the emulsion layer is separated as such and afterwards split by treatment with sodium chloride. Better results have been achieved with continuous processing using separators, e.g. Alfa Laval, Sharples and Westfalia. With the continuous system a uniform quality is obtained compared to that from batchwise neutralising.

Alfa Laval Short-Mix equipment is especially well designed for the refining of rapeseed oil and gives an oil of high quality with low losses.

In the most recent design called the Alfa Laval. All Hermetic Self Cleaning Short Mix, a new type of separator is used. At regular intervals the sludge is discharged from the bowl, which reduces the frequency with which the bowl must be washed.

The crude oil is treated with 0.1–0.3 volume per cent of phosphoric acid in a mixer and then neutralised in the next mixer with 3.5 M sodium hydroxide (20°Be), somewhat in excess of the stoichiometrically calculated amount. The soap phase is removed in the first separator and the oil is treated once more in a mixer with about 1–3 per cent of 2 M sodium hydroxide (12°Be) before entering the second separator.

Although phosphoric acid pre-treatment is preferred , it is also possible to use a mixture of 10 per cent sodium carbonate together with the sodium hydroxide in the re-refining step. Due to a decrease in the content of lecithin this will increase the decolourising and oxidative stability. Finally, in a third step the oil is washed with 10 per cent of water, separated and dried in a vacuum drying tower.

Rapeseed oil refined in batch or continuously according to Alfa Laval was found by *Rutkowski* to give oil with the same quality, while *Moller* reported a greater increase in the peroxide value of the oil directly after batch refining compared to hermetic continuous refining. The short-mix process counteracts the saponification which occurs at higher lye concentrations by a short contact time (30–40 seconds). Another method to decrease the refining loss is to neutralise with a weak lye, which is used in the Zenith process. Fig. 9.5 shows a refining installation designed for rapeseed oil and other oils containing lecithin.

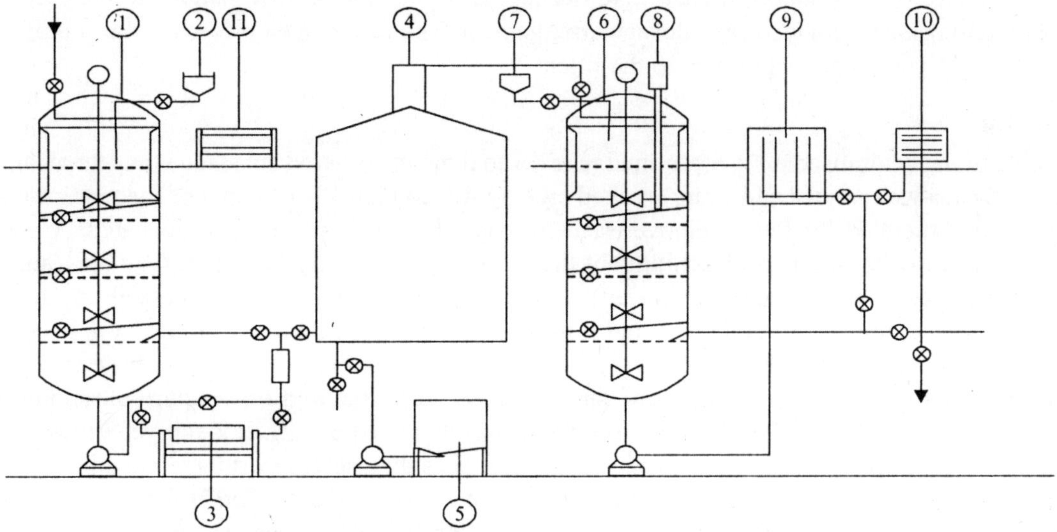

1. Phosphoric acid treatment vessel
2. Phosphoric acid dosage
3. Centrifuge for sludge
4. Neutralising vessel
5. Soapstock draining tank
6. Bleaching vessel
7. Citric acid dosage
8. Bleaching earth dosage
9. Bleaching filter
10. Polishing filter
11. Panel board

Fig. 9.5. Zenith process.

By pre-treatment with 0.15–0.3 per cent of phosphoric acid (85 per cent) the impurities will give a precipitate, partly due to denaturation of phosphatides. The acid also reacts with magnesium in the chlorophylls, thus decreasing the green colour of the oil. The precipitate formed is removed by a separator. The pre-treated oil is continuously introduced at the lower end of a vertical column filled with 0.3 M sodium hydroxide. In the bottom of the neutralising column, the oil stream is distributed into small droplets which rise to the top of the column. Above the caustic lye an oil layer is built up and continuously discharged to the next step. The lye can be used until the concentration has decreased to about 0.1 M and then has to be renewed. The neutralised oil contains about 10–100 ppm of soap, the last traces of which would not be eliminated by a normal washing with hot water. Instead a solution of citric acid in water is added in an amount corresponding to 0.05 weight per cent of citric acid. The oil is then dried in the so called C-unit, where free fatty acids and sodium citrate are formed. The amount of free fatty acids formed is negligible and in any case is removed in the deodourising step, while the citrate is adsorbed on the bleaching earth. For rapeseed oil a refining factor of 1.8, a fatty acid factor of 1.5 and a refining efficiency of 99.0 per cent are reported.

In the Zenith process the crude oil is semi-continuously supplied to the plant while the bleached oil is continuously discharged. The lye can easily be renewed by using two neutralising vessels. Other systems for weak lye treatment are described in which both the oil and the lye are continuously supplied.

A system for bleaching is also included in the Zenith process for refining. The system is designed with a vacuum vessel holding three or more trays and the whole system is kept under vacuum. The oil is sucked into the upper tray through a spraying device for de-aerating and drying the oil. In the first tray the oil is mixed with a solution of citric acid during the drying and is then pumped to the second tray where bleaching earth is added by a dosage arrangement. The lower-most tray serves as a buffer from which the fat is continuously pumped to the filter.

Bleaching

Bleaching of the neutralised oil is preformed in order to remove coloured substances and small amounts of soaps. Even a well refined and water-washed oil may still contain 10–100 ppm of soap. It is especially important that the bleaching step remove traces of soap if the oil is to be hydrogenated. A low soap content is favoured by vacuum bleaching, but the adsorbed soap partly inactivates the bleaching earth and will thus decrease the decolourising effect. The coloured compounds in rapeseed oil are among others the green pigments chlorophyll A and B as well as their products of decomposition, pheophytins A and B, which have been determined by *Niewiadomski*. Seed stocks containing as little as 3 per cent unripened seeds normally give noticeably dark oil. Such dark coloured oils require bleaching with a larger amount of bleaching earth in order to give an acceptable edible oil. It is not yet known whether this is due to decomposition products of the colour pigments themselves or if there are other colourless products which increase as the disagreeable flavour increases. However, *Taufel* and co-workers have shown that chlorophyll has a prooxidative effect on unsaturated fats exposed to light. An acid-activated earth is more advantageous because the pigments are unstable under acidic conditions.

There is some risk that antioxidants are removed during the bleaching. *Rutkowski* has found decreasing amounts of tocopherols and phosphatides (Table 9.2), together with a decreased oxidative stability for rapeseed oil according to the AOM test after each processing step and also after bleaching at 70–80°C.

Table 9.2. Phosphatide, tocopherol and iron contents of rapeseed oil.

Description	*Phosphatide (ppm)*	*Tocopherol (ppm)*	*Iron (μg/100 g)*
Pressed oil	101–166	442–589	36–65
Oil in miscella	166–261	452–656	217–519
Mixed crude oil	97–140	417–585	126–384
Neutralised oil	17–34	326–521	101–217
Bleached oil	10–23	249–423	30–87

On the contrary *Taufel* and co-workers report an increasing stability of bleached oil compared to neutralised oil, but they also found the best stability for crude oil.

The choice of bleaching earth is a question of economy where decolourising ability, filterability and oil retention have to be compared to the price. A non-activated earth may be economically feasible for a rapeseed oil of good quality with light colour but as soon as the colour of the crude oil increases, an activated earth will become more economical. The addition of activated carbon did not increase the bleaching power. The advantage of using only one quality of bleaching earth in a factory is, however, so evident that normally an activated earth is used for all kinds of oils. The amount of earth must be adjusted to the quality of the oil to minimise the cost of earth and oil retenion. The formation of free fatty acids increases with the amount of bleaching earth.

Vacuum bleaching has proved to be advantageous, especially for the removal of soaps, but also to protect the oil against oxidation. Nonetheless, it has been found that the bleaching earth should not be too well dried, since an earth completely dried at high temperature is inactive. According to *Kogler*, there is an optimal water content for bleaching earth. For bleaching at 95–98°C under atmospheric pressure, the optimal water contents for soyabean and cottonseed oil were 8–12 per cent and 9–16 per cent, respectively. Further increase in water content decreased the bleaching power. In spite of this, an earth with 24 per cent water was reported to give better decolourising than a fully dried earth.

A good idea of the amount of bleaching earth required for a certain oil can be obtained by testing on a bench scale, but it is always advisable to follow the results in the plant.

Equipment

Open bleaching tanks are still used, but most factories endeavour to bleach under vacuum. Such equipment should be air-tight, especially below the oil level and furnished with agitation in order to provide intimate contact between the oil and the bleaching earth. There are also various systems for continuous bleaching, e.g. the Lurgi system (Fig. 9.6).

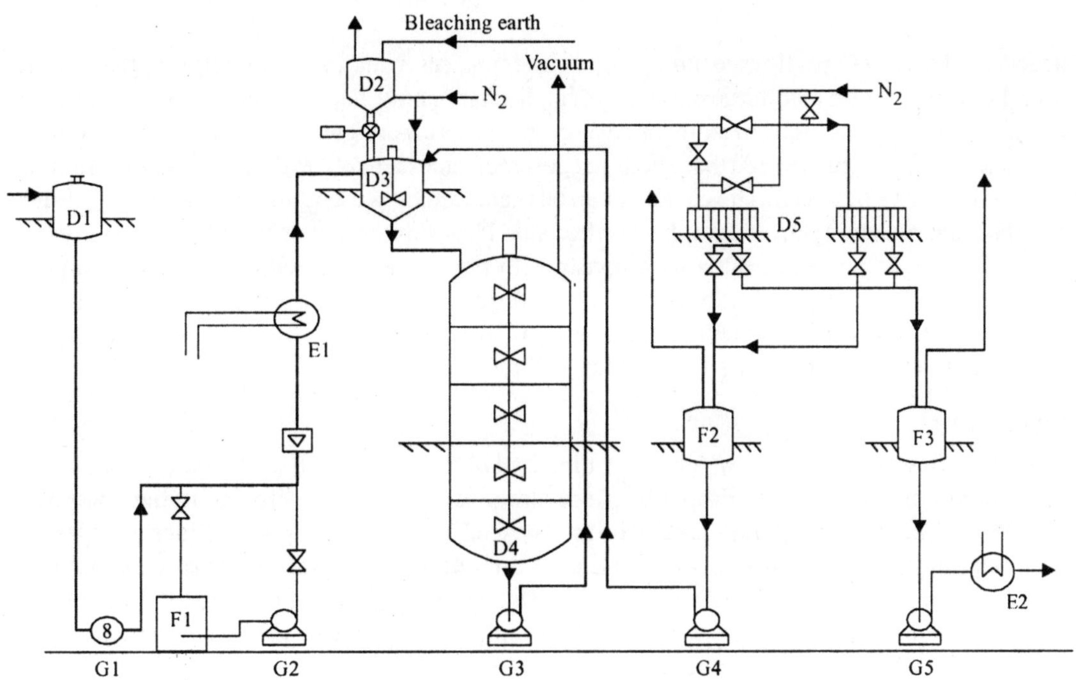

D1 Drying tank
D2 Clay hopper
D3 Mixer
D4 Bleaching vessel
D5 Filter press

E1 Heater
E2 Cooler
F1 Holding tank
F3 Holding tank
G1-G5 Pumps

Fig. 9.6. Lurgi continuous bleaching system.

The dried oil and the deaerated bleaching earth are dosaged with agitation in a mixing tank. The mixture runs through a bleaching reactor with a certain residence time and is then filtered. Together with the Zenith process for refining, a system for bleaching has already been described. An entirely automatised bleaching system is the Alfa Laval auto bleach process. The bleaching plant comprises three main sections: bleaching reactor with automatic control device, dosing equipment for bleaching earth and a filtering unit. The oil is heated to a suitable temperature in the heater and goes to the upper compartment of bleacher where deaerated bleaching earth is added continuously. The amount of bleaching earth and the oil flow are ratio controlled, which ensures a constant percentage of bleaching earth in the oil. A programme controls the residence time independent of the capacity and the only change in the bleacher at different throughputs is some variation in the level in the compartments.

From the bottom of the bleacher the oil is pumped to one of the filters by means of pump and the filtered oil is then cooled in the heat exchanger.

The filter unit consists of two alternatively working Funda filters, each one containing a number of horizontal filter plates on a common shaft. When one filter has to be emptied, the oil in this filter is pumped by means of pump to the second filter. The bleaching earth on the plates of the first filter is deoiled and discarded. When changing from one oil to another, all oil, even the residual volume, can be completely filtered in one filter, thus avoiding mixing between two different oils.

Filtration

The bleached oil is sometimes filtered through open filter presses, which are manually emptied and cleaned, but several new types with automatic or mechanical discharge have been presented. At the same time it is still desirable to protect the oil from exposure to the air. A large amount of soap in the refined oil can disturb the filtration and it has been reported that a badly degummed rapeseed oil can decrease the filtration capacity to 1/5 of that normally obtained. The capacity can be increased by pre-coating with a slurry of filter aid in a clean oil or just by having the right type of filter aid. The oil retention in the filter cake may be rather high, about 25 weight per cent, even after blowing with steam. Various methods have been suggested to counteract retention losses, for example extraction of the cake in the filter itself or in a special plant. More attractive is extraction with hot water (95°C) directly in the filter, the so-called Thomson process. For the Funda type filter about 10–12 per cent oil is then left in the cake.

Hydrogenation

The previously described processes, alkali refining and bleaching as well as deodourisation, attempt to clean the oil without changing its properties. In hydrogenation there is a change in the composition of the fatty acids. The amount of linolenic acid in rapeseed oil is reduced, which increases its stability against oxidation. Hydrogenation also increases the melting point more or less depending on how many double bonds are reduced. Fat hardening is still an empirical process, but a great deal of systematic work has been done which is of great help in the estimation of suitable process parameters, although most of the work has been done for other oils than rapeseed oil.

Selectivity and isomerisation

In the triglycerides, linolenic, linoleic and oleic acids are hydrogenated at different rates. A simplified reaction model makes it possible to define selectivity.

$$\text{Linolenic} \xrightarrow{k_1} \text{Linoleic} \xrightarrow{k_2} \text{Oleic} \xrightarrow{k_3} \text{Stearic}$$

The linolenic selectivity (S_{LI}) is the ratio between k_1 and k_2, $S_{LI} = k_1/k_2$, and the linoleic selectivity is the ratio between k_2 and k_3, $S_{L2} = k_2/k_3$.

In this model the concentration of hydrogen at the surface of the catalyst is not considered. The rate constants will thus be influenced not only by the temperature and catalysts but also by those process parameters which influence the concentration of hydrogen at the catalyst surface. The importance of mass transport processes has been demonstrated by *Coenen* and others, who have presented a comprehensive discussion of various selectivity concepts , in which they also consider the triglycerides themselves and the effect of pore diffusion. Attempts have recently been made to mathematically formulate a rate equation including the external mass transport steps.

A low concentration of hydrogen on the surface of the catalyst favours selective hydrogenation and can be obtained by low hydrogen pressure, a high concentration of catalyst and high temperatures (above 180°C). At the same time a low concentration of hydrogen promotes conjugation, which causes migration of double bonds and the formation of *trans* isomers. Under these conditions the conjugation of trienoic acid can cause polymerisation and cyclisation. The amount of aromatic fatty acids (AFA), about 0.1 per cent, found by *Coenen* and for rapeseed oil hydrogenated at 200°C and with shortage of hydrogen, is not found in commercial hydrogenated rapeseed oil processed in Sweden (AFA below 0.01 per cent) and probably appears only under conditions such that extremely high amounts of *trans* acids are produced.

The conjugation and *trans* isomerisation of especially linolenic acid will after oxidation give a special objectionable flavour to the hydrogenated rapeseed oil as it does for all other oils containing linolenic acid. This flavour disappears when the oil is deodourised but after long storage the hydrogenation flavour will appear again. *Chang* characterised some of these flavours in hydrogenated soyabean oil as saturated aldehydes and methyl ketones.

Keppler separated the volatile substances from hydrogenated linseed oil by gas-liquid chromatography and could show the presence of 6-*trans*-nonenal and identify this as the principal carrier of hydrogenation flavour.

However, the selectivity concept, as defined above, relates to the changes in the amounts of trienoic, dienoic and monoeoic acids and not to the formation of *trans* isomers or changes in physical properties such as melting point, dilatation curves etc. but these properties are of course influenced by the selectivity. A low linoleic selectivity means a larger amount of stearic acid at a certain iodine value, which increases the melting point. On the other hand the content of *trans* isomers influences the dilatation to a greater extent than the selectivity ratios (Fig 9.7). In rapeseed oil it is also possible that the erucic acid and oleic acid in the triglycerides hydrogenate at different rates. *Hilditch* reported that erucic acid and oleic acid in the triglycerides hydrogenate at about the same rate, but as methyl esters there was a preference for methyl oleate. *Hilditch* conclusions were confirmed by *Coenen*. *Jakubowski* stated that in rapeseed oil erucic acid hydrogenates more slowly than oleic, while *Koman* found the reverse rate order. In any case the difference in hydrogenation rates which is often found between rapeseed oil and soyabean oil is probably due to the suphur compounds in the former.

In the hydrogenation process there are three phases involved, each of which has to be taken into account.

Hydrogen

There are several ways of producing hydrogen but today the choice normally is between the hydrocarbon reforming process and electrolysis of water. Electrolytic hydrogen is of high purity, normally more than 99.8 per cent and no purification is required. In the reforming process desulphurised hydrocarbons are used and the hydrogen produced contains about 0.05 per cent methane, 0.3 per cent carbon monoxide and 0.1 per cent carbon dioxide after scrubbing with ethanolamine. Passage through a methanation catalyst increases the purity to about 99.5 per cent hydrogen with less than 10 ppm carbon oxide.

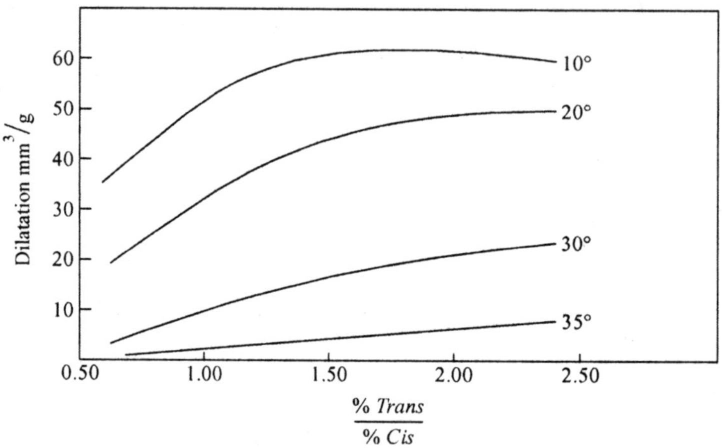

Fig. 9.7. Dilatation *vs. trans: cis* content for rapeseed oil hydrogenated under different conditions to an iodine value of 75.

Oil

The isothiocyanates and oxazolidinethiones in rapeseed oil are not only objectionable for refined oil but also function as poisons for the catalysts. *Rutkowski* found that a catalyst is deactivated in proportion to the amount of sulphur present in the oil. An amount of 0.016 ppm sulphur had an almost negligible influence on the hydrogenation, while 5 ppm reduced the catalytic activity by about 50 per cent. *Rutkowski* also found an addition of 6 ppm phosphorus as rapeseed lecithin harmful for the hydrogenation and that this effect dominated sulphur poisoning in rapeseed oil. The decomposition products of the sulphur compounds extracted together with the oil are not fully removed by alkali refining or bleaching. One alternative is to treat the oil with a small amount of catalyst at 150–200°C together with a slow flow of hydrogen in order not to reduce the double bonds. This method has been recommended for fish oil. The best solution is to pre-treat the seeds so that no sulphur compounds are extracted with the oil. *Reynolds* and *Youngs* describe a technique for rapeseed which gives an oil that needs about 0.03–0.08 per cent nickel for hydrogenation. A comparable figure is found for Swedish rapeseed oil. Deodourisation will only be helpful if the sulphur compounds are volatile, but in that case they are also removed if the bleaching is done under vacuum.

Catalyst

In most cases a nickel catalyst, such as nickel formate reduced in oil at about 250°C or dry-reduced nickel carbonate on a support, may be used. One advantage with the supported, dry -reduced catalyst is its better filterability. A normal analysis of such a catalyst gives 20–25 per cent nickel, 10–15 per cent silica and 60–65 per cent hydrogenated fat, while the wet-reduced catalyst normally ranges from 22 to 30 per cent nickel in hydrogenated fat. One of the most important demands upon a catalyst is uniformity between different lots. The catalysts can be compared to a standard catalyst by hydrogenation under standardised conditions. *Cocks* and *Van Rede* describe a method in which they add 0.07 per cent nickel to the oil and compare the decrease in refractive index for hydrogenation of a standard sesame oil for half an hour. The activity is expressed as the relative refractive index decrease:

$$\text{Relative refractive index decrease} = \frac{\text{Refractive index decrease with sample}}{\text{Refractive index decrease with standard}} \times 100$$

Still more important than good activity is that the catalyst gives a hardened product of constant quality. The simplest way to check the quality of the product is to compare the dilatataion curve for the same type of oil hydrogenated to the same iodine value with the sample and the standard catalyst or to compare the fatty acid composition and *trans* acid content. Catalysts based on other metals have hitherto been of little practical interest.

Copper catalysts, however, give a better linolenic acid selectivity, but due to their low activity the hydrogenation has to be performed under a higher pressure, 5–10 atmosphere. Traces of copper in the hydrogenated oil are objectionable and would demand a special production line for copper hydrogenated products, equipped with an effective post-bleaching process.

Equipment and systems

For batch hydrogenation there are reactors with and without recirculation of hydrogen, both of which types are used for the hydrogenation of rapeseed oil. The so called 'dead-end' system (i.e. without recirculation) sometimes requires purging during the reaction. A normal batch in Europe is 10,000–15,000 kg, while in the USA, a normal charge for soyabean oil is about 30,000 kg.

As continuous processes for refining and bleaching have become more common, interest in continuous hydrogenation has been aroused. From Russia, for example, work concerning stationary catalysts has been reported. However, the only tests on rapeseed oil seem to be have been carried out only on a bench scale. Recently two commercial systems have been introduced, one (Lurgi) working with constant hydrogen consumption and another (Buss) working at constant pressure with normal powder catalyst.

Lurgi system for continuous hydrogenation

The installation consists of one or more pressure reactors, divided by a series of plates into a number of reaction chambers, which can be cold or heated for adjustment to the desired temperature.

The refined and bleached oil is mixed with new or re-used catalyst and fed to the bottom chamber of the first tower by a volumetric pump. The hydrogen is introduced at a constant flow to the feed line, from which it bubbles through the oil and catalyst mixture in the reactor. From the drop chamber in the last reactor the oil phase and the unused hydrogen flow through a cooler into a separating vessel, where the liquid fat and catalyst separate from the hydrogen gas. By the gas circulation pump the unused hydrogen is returned to the first reaction chamber (Fig. 9.8). Due to the gas circulation no agitation is necessary. In relation to the volume of the liquid phase there is a small gaseous volume, so small changes in reaction rate will immediately increase or decrease the gaseous pressure and thus neutralise the change. The decrease in the iodine value is directly affected by the flow rates of hydrogen and oil.

Buss system for continuous hydrogenation

The equipment is composed of four or more reactors in cascade. In each reactor the contact between oil, catalyst and hydrogen is carried out in mixing nozzle, by rapid circulation of the oil and catalyst mixture. The refined and bleached oil is dried under vacuum and fed to the reactor. The four to six chambers have the same gaseous space but the oil is separated by walls between the chambers. By an overflow the oil is transported through the reactors. The hydrogen is directly fed to the reactor through a flowmeter, or the hydrogen can also be circulated from the last to the first reactor by a ring pump. A suspension of catalyst in oil is fed in a fixed amount by a pump, to the first chamber and to the third chamber by another pump, which is controlled by the hydrogen flowmeter, thus making the addition of hydrogen constant.

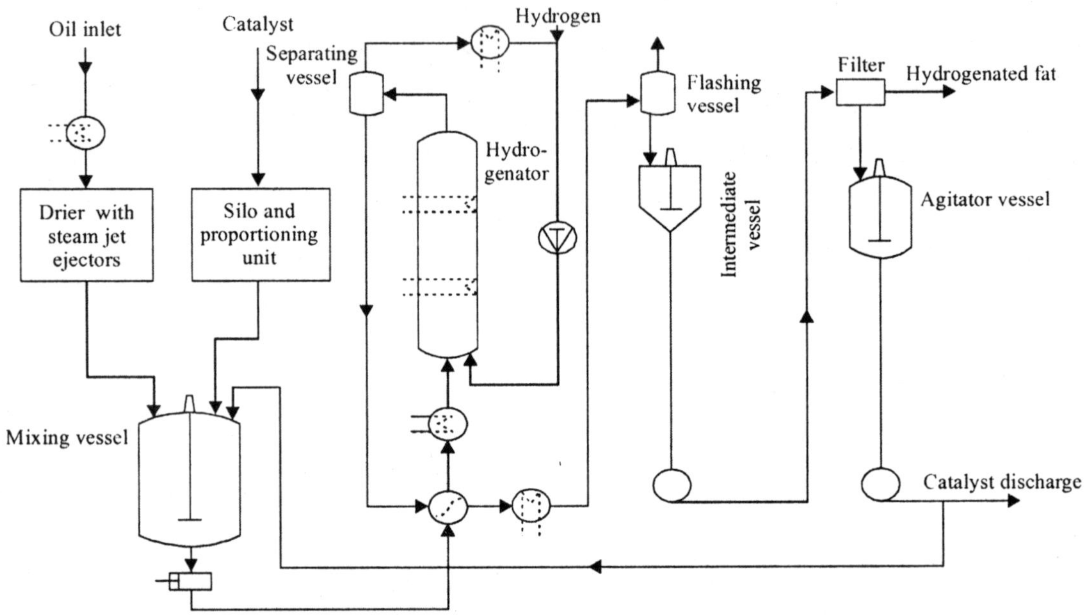

Fig. 9.8. Lurgi system for continuous hydrogenation.

For cascade reactors such as Lurgi and Buss it is necessary to have a sufficient number of steps, so the residence time distribution will not be too unfavourable. Changes in reaction time by alterations in temperature, pressure and amount of catalyst will probably influence the end product in the continuous processes in a manner similar to that observed in batch hydrogenation. The advantage of such reactors would then be a more uniform product, a better heat economy and perhaps less operation maintenance.

Practical aspects of the hydrogenation of rapeseed oil

In order to reduce the iodine value of 1000 kg of oil by one unit, 39.4 moles of hydrogen are required, which corresponds to 79 g or 0.883 m^3 at 0°C and 760 torr absolute pressure. There are no published data on the heat of reaction in the case of rapeseed oil, but the corresponding values for peanut oil and soyabean oil are 24.5 and 26.1 kcal/mole and double bond respectively.

The alkali refined and bleached oils heated under vacuum in the reactor to about 140°C and then a pre-determined amount of catalyst is added. At about 150°C the hydrogen is supplied and the temperature rises to about 180°–200°C. At this temperature the reaction is continued until the desired iodine value and melting point have been reached. By altering the amount of catalyst, temperature, pressure and agitation, a fat with the same melting point but with different plasticity can be produced. Hydrogenated fat for margarine or shortening is normally described by the amount of solids, dilatation or solid fat index (SFI) at various temperatures. Normally there are certain differences in the products of different companies with regard to dilatation due to variations in hydrogenation techniques and crude oils. In common is, however, the desire to increase the stability of the oil by eliminating linolenic acid. *Jakubowski* has reported that industrially hydrogenated Polish rapeseed oil contains less than 1 per cent linolenic acid at an iodine value of 85. The amount of octadecenoic acid found consists of iso-linoleic acid to a great extent. According to *Jakubowski* a hydrogenated rapeseed oil of iodine value 70 (m.p. 30°C) contains only about 2 per cent *cis,cis*-9,12-octadecadienoic acid.

Laboratory hydrogenation of a normal blend of Swedish rapeseed oil showed that no conjugated trienes, but almost 1 per cent conjugated dienes could be detected at the beginning of the hydrogenation (Table 9.3). In the same table the gas chromatographic analyses and the amounts of polyunsaturated fatty acids after conjugation with alkali and lipoxidase are given. In Fig. 9.9 the fatty acid composition as determined by gas chromatography is given for another blend of Swedish rapeseed oil.

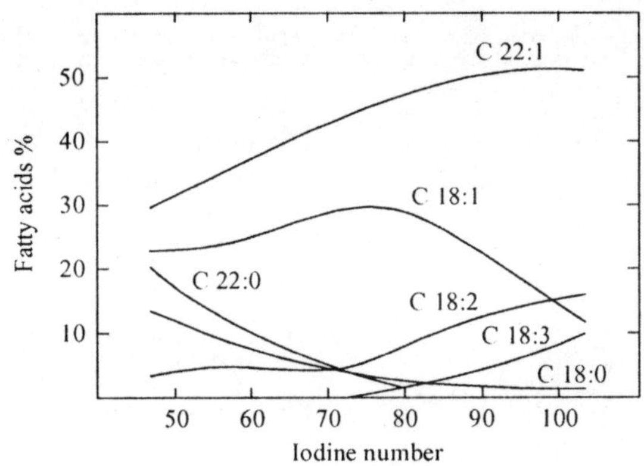

Fig. 9.9. Fatty acid content of rapeseed oil hydrogenated under standard conditions.

The total *trans* content increases during the hydrogenation up to an iodine value of 60 and will then decrease with decreasing iodine value (Fig. 9.10).

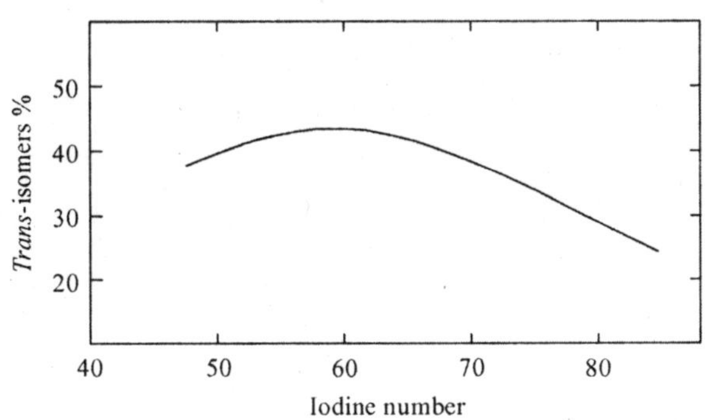

Fig. 9.10. *Trans* content *vs.* iodine number for hydrogenated rapeseed oil.

An oil with the indicated composition and *trans* content gives the dilatation and melting point curves shown in Figs. 9.11 and 9.12. Rapeseed oil contain oils from various cultivated rapeseed species. A comparison between different Swedish varieties has shown that there is only a small change in the dilatation for a given variety compared to an average (Table 9.4).

Table 9.3. Analysis of rapeseed oil hydrogenated with 0.07 per cent unsupported nickel at 185°C.

		*Conjugated**			*Non-conjugated***		*Fatty Acid Weight Percentage Composition****													
Time min	*Iodine value (Hanus)*	*Tans isomers wt. %*	*Dienes wt. %*	*Trienes wt. %*	*Dienes wt. %*	*Trienes wt. %*	*16:0*	*16:1*	*18:0*	*18:1*	*18:2*	*18:3*	*20:0*	*20:1*	*20:2*	*22:0*	*22:1*	*22:2*	*24:0*	*24:1*	%
0	105.7	2.7	0.26	0.00	13.8	9.7	3.0	0.3	0.8	12.2	13.9	10.3	0.0	8.6	0.5	0.0	47.7	1.1	0.4	1.2	26.0
3	103.4	3.0	0.30	0.00	12.0	8.6	3.0	0.4	0.9	13.4	13.8	9.8	0.0	9.5	0.6	0.0	47.5	0.3	0.4	1.3	23.4
6	99.9	6.2	0.51	0.00	13.0	7.3	2.9	0.3	0.8	14.6	12.9	7.3	0.0	8.7	0.4	0.0	49.3	1.1	0.3	1.2	19.8
9	98.3	8.4	0.65	0.00	9.7	5.4	3.1	0.3	1.0	16.5	13.0	7.0	0.0	8.6	0.5	0.0	48.0	0.6	0.3	1.1	17.3
12	95.6	10.2	0.66	0.00	8.3	3.5	2.9	0.3	0.9	17.5	12.3	5.4	0.0	8.7	0.4	0.3	48.9	0.8	0.3	1.3	13.7
15	92.7	13.5	0.65	0.00	6.7	3.2	3.4	0.4	1.2	19.2	11.7	5.1	0.0	8.5	0.6	0.5	46.9	0.9	0.2	1.3	11.0
25	85.5	24.1	0.28	0.00	3.6	1.3	3.1	0.4	1.3	24.4	9.4	1.7	1.0	8.1	0.0	0.8	48.4	0.0	0.2	1.2	4.2
35	77.7	35.6	0.00	0.00	1.2	0.0	3.1	0.5	2.6	27.6	6.3	0.3	1.0	8.1	0.0	2.0	47.0	0.0	0.2	1.2	0.7
45	71.3	41.5	0.00	0.00	0.0	0.0	3.0	0.4	4.5	27.1	4.7	1.0	1.8	7.3	0.0	3.7	45.2	0.0	0.4	1.1	0.2
60	62.1	43.8	0.00	0.00	0.0	0.0	3.2	0.4	7.8	25.1	2.9	0.0	2.3	6.8	0.0	9.3	40.8	0.0	0.7	0.9	0.0
75	53.9	42.0	0.00	0.00	0.0	0.0	3.1	0.3	11.7	22.4	1.9	0.5	3.5	5.6	0.0	15.4	33.9	0.0	0.7	1.0	0.0
90	46.4	38.0	0.00	0.00	0.0	0.0	3.1	0.5	15.8	18.0	1.2	0.5	4.6	4.5	0.0	21.2	29.1	0.0	0.9	0.8	0.0
105	40.3	32.9	0.00	0.00	0.0	0.0	3.1	0.6	17.2	14.1	1.7	0.0	5.0	4.1	0.0	23.5	29.0	0.0	0.9	0.8	0.0
120	30.0	26.5	0.00	0.00	0.0	0.0	3.2	0.4	20.0	14.1	1.1	0.4	5.8	3.7	0.0	28.8	20.8	0.0	0.9	1.2	0.0

* Determined by ultraviolet spectroscopy according to *Mehlenbacher*.

** Reported values are measured according to *Mehlenbacher* multipliced with a factor 1.2.

*** Determined by gas-liquid chromatography.

**** Determined by ultraviolet spectroscopy after oxidation with lipoxidase.

Estimation of *cis*-methylene-interrupted polyenoic acids according to *MacGee*.

% Polyunsaturated fatty acid, %

Table 9.4. Dilatation for different varieties of rapeseed oil and mustard seed oil hydrogenated in the laboratory under the same conditions with a commercial catalyst.

Variety	*Rapeseed oil m.p. 34°C dilatation*						*Rapeseed oil m.p. 38°C dilatation*						*Rapeseed oil m.p. 40°C dilatation*					
	J.V.	*10°C*	*20°C*	*30°C*	*35°C*	*40°C*	*J.V.*	*10°C*	*20°C*	*30°C*	*35°C*	*40°C*	*J.V.*	*10°C*	*20°C*	*30°C*	*35°C*	*40°C*
Brassica napus, winter type	71.5	49.5	35.0	10.0	3.0	0.0	67.5	53.0	49.0	21.5	10.0	2.0	63.8	63.0	58.5	34.5	20.0	8.0
Brassica campestris, winter	71.3	48.0	33.0	10.0	4.0	0.0	67.5	57.5	47.0	20.5	10.5	2.0	63.5	62.0	58.0	34.0	20.5	8.0
Brassica napus, summer type	72.5	49.0	34.0	10.0	2.5	0.0	68.0	59.0	50.0	23.0	11.0	2.0	64.0	65.0	61.0	36.0	21.0	8.0
Sinapis alba	72.5	46.5	31.0	10.0	3.5	0.0	68.5	54.5	43.5	19.5	9.0	2.0	64.4	62.5	56.5	33.0	19.5	8.0
Brassica campestris, summer	74.5	49.5	32.0	10.0	3.0	0.0	70.5	58.0	45.5	21.0	9.5	2.0	66.0	64.0	57.5	34.0	20.0	8.0
Brassica napus, winter type	72.5	42.5	31.0	10.0	3.0	0.0	69.5	56.5	44.5	20.0	9.0	2.0	65.3	63.5	57.0	33.0	19.0	8.0
Brassica napus, winter type	74.0	48.5	30.5	10.0	4.5	0.1	70.5	56.0	41.0	17.5	8.5	2.0	66.0	63.5	53.0	29.0	16.5	8.0

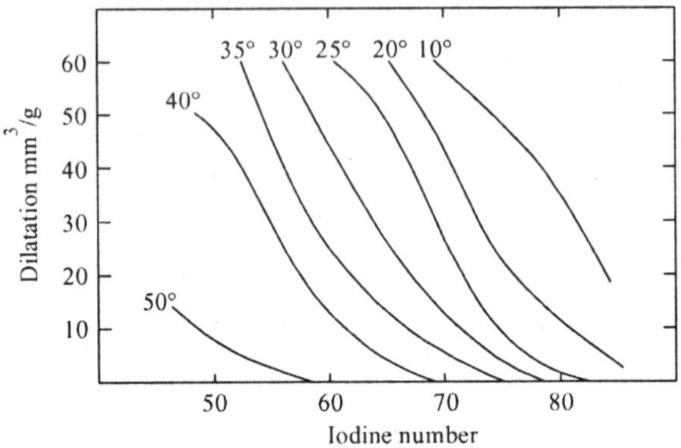

Fig. 9.11. Dilatation *vs.* iodine number at different temperatures for rapeseed oil hydrogenated under standard conditions.

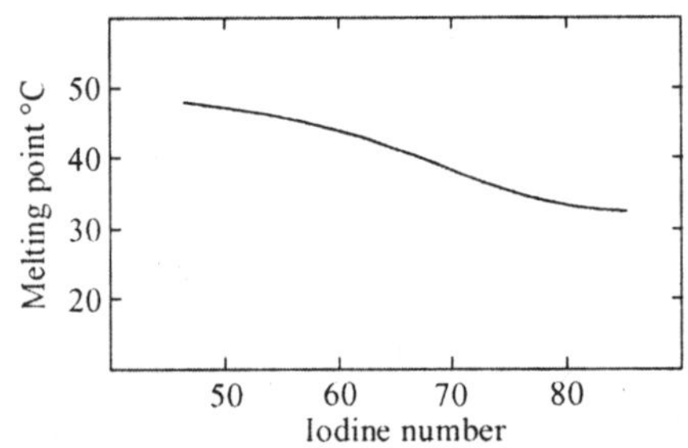

Fig. 9.12. Melting point (AOCS) *vs.* iodine number for hydrogenated rapeseed oil.

Controlling the hydrogenation process

A common way of following and controlling the degree of hydrogenation is to determine the refractive index and melting point. There are different methods for the determination of melting points, but normally a rapid method such as slip-point is used. Some factories combine slip-point with congeal-point, which provides a notion of the dilatation. A good control of catalyst quality, the amount charged and the hydrogenation temperature together with a proper pre-treatment of the oil will give a product which meets the demand of uniform dilatation, assuming that the hydrogenation is stopped at the right iodine value. Modified more rapid methods for iodine value, dilatation or solid fat index can be used. By continuously determining the amount of hydrogen the degree of hydrogenation can be followed, but the leakage in the seal for the agitator shaft has to be considered and this method is often combined with iodine value or more easily refractive index. There is a good correlation between refractive index and iodine value (Fig. 9.13).

Post-bleaching

When the hydrogenation has reached the desired end point the oil is cooled and the catalyst removed by filtration. In many places this is still done in open filter presses, but as in the case of the filtration of bleaching earth, there is successive change to closed automatic filters. The catalyst has a tendency to clog the filter clothes and must therefore be mixed with filter aids. Supported catalysts are often found to be more easily filtered than formate catalyst. A great deal of the producer's efforts are devoted to increasing filterability.

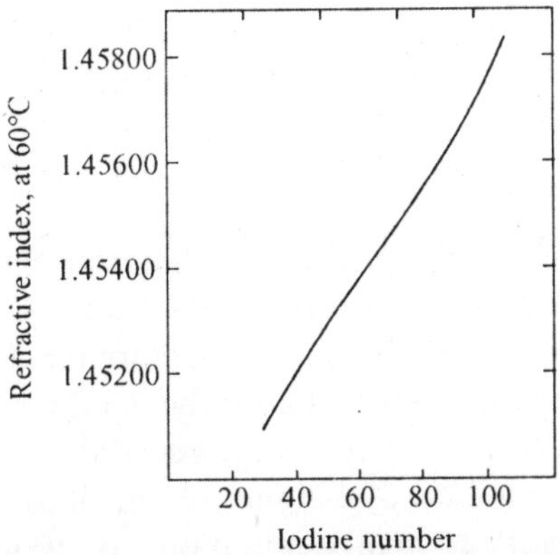

Fig. 9.13. Refractive index at 60°C vs., iodine number for hydrogenated rapeseed oil.

After the catalyst is filtered off the oil contains mainly nickel as soaps. To remove 5–10 mg nickel per kg of oil, one per cent Fullers' earth was needed and the addition of citric or lactic acid considerably improved the efficiency. 100 g acid per 1000 kg of oil decreased the need of bleaching earth to 0.5 per cent for an oil containing 5 mg nickel per kg oil. The best efficiency is obtained with one kg of a 25 per cent solution of citric acid in water to 10,000 kg hydrogenated fat added to the fat after the bleaching earth. Alkaline refining after hydrogenation seems to be unnecessary if the post-bleaching is carried out as described with citric acid.

Deodourisation

Whether the oil has been neutralised and bleached or hydrogenated the fat contains minor amounts of volatile compounds with a certain flavour. Especially unhydrogenated rapeseed oil may have a strong seedness flavour, but also other undefined flavours may exist, e.g. a pungent flavour arising from glucosinolates and their decomposition products. In the deodourisation, which is a steam distillation, these volatile compounds are removed together with the free fatty acids (about 0.05 per cent) obtained in the bleaching and hydrogenation steps. At the same time, the colour is improved and peroxides are decomposed partly to volatile products.

It is difficult to give general rules for rapeseed oil in addition to those for other oils due to variations in equipment from factory to factory. Normally rapeseed oil needs a more efficient deodourisation than other vegetable oils. On the other hand, if the temperature is too high the natural antioxidants, such as tocopherols, may be distilled or decomposed. From laboratory tests it was concluded that for rapeseed as well as for soyabean oil there is no formation of conjugated dienes at 210°C, but that they appear in amounts up to one per cent when deodourisation is carried out at 280°C. The content of *trans* isomers increases from 2–3 per cent at 210°C to about 10 per cent at 280°C. The experiments were carried out at 5 torr with a decreasing residence time, t, as the temperature was increased following the expression $t = \text{const} \times 1/P_T$, where p_T is the vapour pressure of stearic acid at temperature T. A shorter deodourisation time seemed to be more favourable for rapeseed oil than for soyabean oil.

The stability of the deodourised oil with regard to oxidation and flavour as well as colour depends on the storage temperature and on whether it is protected by nitrogen or not. However, earlier steps in the process influence the stability. A minor increase in the peroxide value of the neutralised oil will decrease the flavour stability of the deodourised oil.

According to *Harada* rapeseed oil from pressure milling was stable, while that from solvent extraction showed some colour reversion. For soyabean oil it is stated that the colour reversion is dependent on the amount of 'tocored' in the crude oil. During the processing the 'tocored' seems to transform to a colourless compound which remains in the deodourised oil and will transform back to the red 'tocored' in the reverted oil.

Rapeseed oil for margarine may be deodourised together with the other oil components or may be separately deodourised. A batch deodourisation normally takes about eight hours with heating and cooling and reaches a temperature of about 195–200°C. Before the deodourisation, 0.01 per cent citric acid may be added to the oil and the same amount after the deodourisation in the cooling step.

A great step forward was achieved when *Bailey* introduced the Votator principle (Fig. 9.14), which made it possible to deodourise rapeseed oil at a higher temperature (230°C) for a shorter time (i.e. a cycle time of 27 minutes, giving an effective processing time of about two hours).

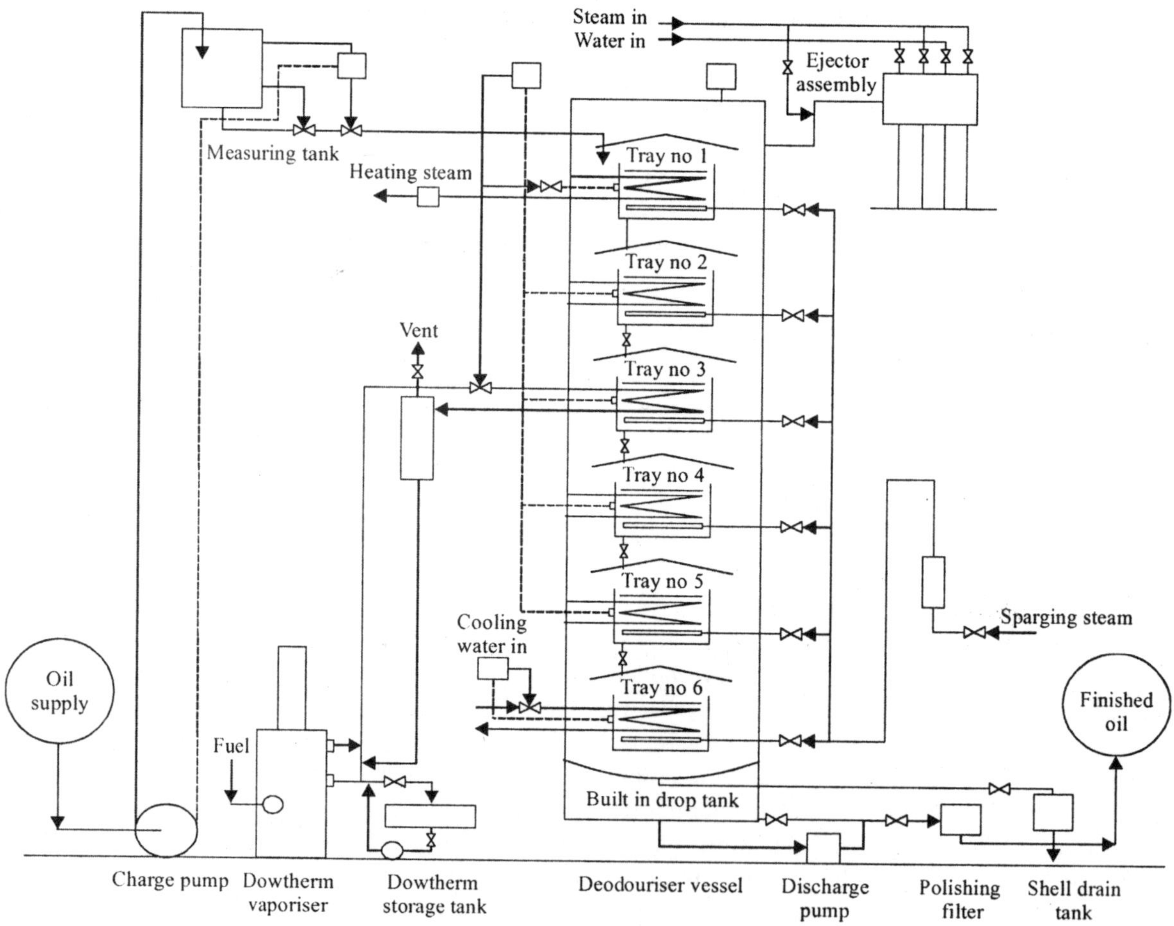

Fig. 9.14. Six tray semi-continuous deodouriser.

A great number of equivalent designs of the same principle have appeared, among them the Lurgi plant (Fig. 9.15) which can operate either continuously or semi-continuously.

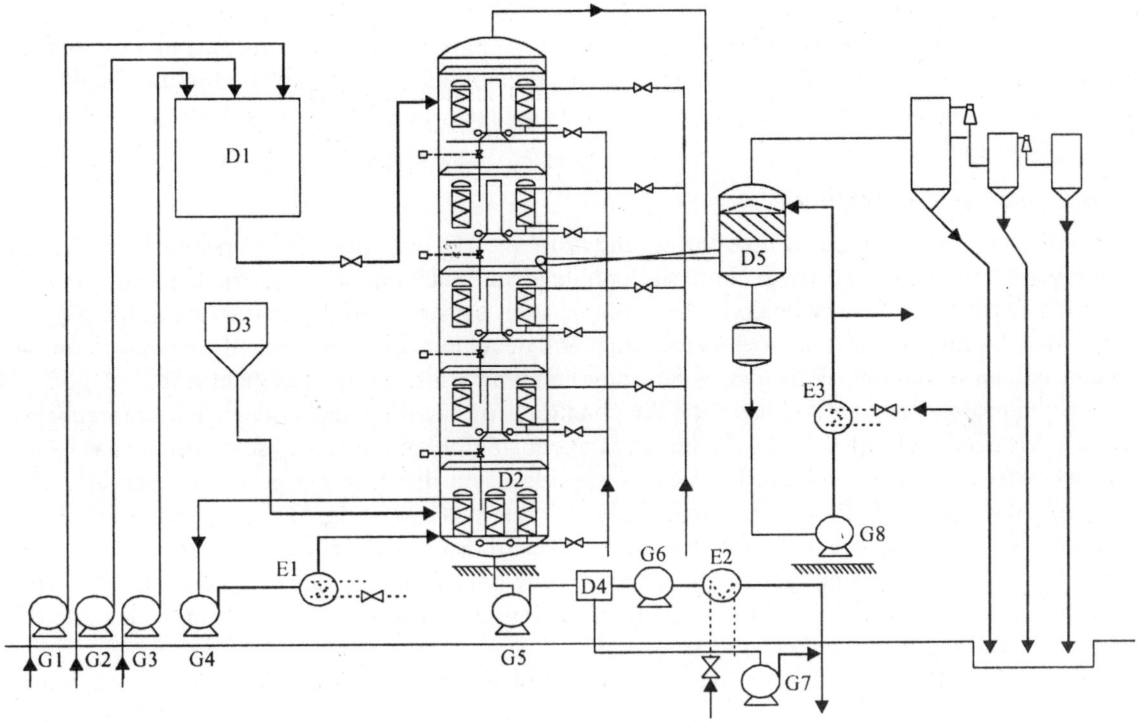

D1 Scale
D2 Deodouriser
D3 Tank for citric acid
D4 Filter

E1, E2, E3 Cooler
G1, G2, G3, Pump for bleached oil
G5 Pump for deodourised oil

Fig. 9.15. Lurgi semi-continuous or continuous deodouriser.

EDIBLE PRODUCTS FROM RAPESEED OIL

It is still a rather common misconception that rapeseed oil is mainly an inedible oil. Rapeseed is used to some extent for edible purposes and for soap making. It is also a source of oil, which is used chiefly in mixtures with mineral oils for lubrication, or alone for tempering steel plates. The refined oil is known as colza oil.

The purpose of this section is to review the use of rapeseed oil as food with emphasis on quality factors and food technological aspects that are specific for the *Brassica* oils.

Quality Factors

Rapeseed oil and related Cruciferae seed oils are classified in Bailey's Industrial Oil and Fat Products as erucic acid oils but could as well have been included under the title linolenic acid oils together with e.g. soyabean oil. These two compositional factors, i.e. the presence of glycerides containing erucic and linolenic acids, thus constitute the major difference between rapeseed oil and other fats and oils of importance in food uses. There are other specific minor components in rapeseed oil that distinguish this oil from others. One such example is the presence of the sulphurous compounds, glucosinolates, in the seeds. These are, however, eliminated to a great extent by proper processing and can therefore perhaps be disregarded when end uses of the oil are discussed. The influence of other seed components such as moisture, chlorophyll, enzymes, free fatty acids, phospholipids etc. that can affect the oil quality and/or processing yields markedly. The lack of appropriate technological know-how and proper facilities for cultivation, seed handling and processing have at times certainly prevented the production of high quality edible rapeseed oil.

Flavour and flavour stability

A properly produced rapeseed oil is odourless and tasteless. The presence of 7–11 per cent linolenic acid in the glycerides makes it, however, more susceptible to autoxidation than e.g. groundnut oil and places it in the same category as soyabean oil. The linolenic acid as a source of the off-flavour of these oils has been studied by *Moser*. The rate of auto-oxidation and development of off-flavours depend as for other fats and oils on a number of factors which may be summarised as the resultant effect of glyceride composition prooxidants, antioxidants and the conditions of handling and storage. The comprehensive work that has been performed on the flavour and flavour stability of the related soyabean oil has recently been reviewed by *Cowan* and can be used as a guide when discussing rapeseed oil stability. *Holm* reported on the type of off-flavour in rapeseed oil and other oils by stating: 'Thus rapeseed and soya oils with aldehyde contents over 2 had a noticeable oily, green off taste after a couple of days while cottonseed oil and peanut oil formed a nutty, metallic, or bitter taste only at higher aldehyde contents. The type of off-flavour that develops can thus be characterised as similar to that found in soyabean oil and therefore comparisons can be fruitful between these two oils.

In the work of *Moser* the flavour characteristics and the oxidative stability of crambe seed, mustard seed, rapeseed and soyabean oils are subjected to further study with the purpose of establishing similarities or dissimilarities among these linolenate-containing oil. The oils were produced in the laboratory and the flavour evaluation were made by a test panel of 20 trained judges and analyses of variance and F-tests were used to test the means on a 10-point scale. The description of the flavour of freshly deodourised oil was dominated by the term 'buttery', with nutty 'or 'beany' flavours present in minor amounts. After 4 days storage at 60°C the terms 'rancid', 'beany', 'painty', 'grassy', were given to oils when they did not contain citric acid. The presence of citric acid markedly protected the oils and there was no general difference between soyabean oil and the samples of Cruciferae seed oils. However, when samples with added citric acid were exposed to light for two hours the soyabean oil was given the same descriptive terms as the oil without citric acid, while the Cruciferae oils exhibited a significant drop in the flavour scores and developed a definite 'rubbery' flavour that was often accompanied by a 'garlic' or 'onion-like' flavour (Table 9.5).

Table 9.5. Effect of citric acid on Cruciferae oils exposed to light for 2 hours.

Oil	*Score*	*Without citric acid description*	*Score*	*With citric acid description*
Crambe seed	7.2	Buttery, beany	5.7	Rubbery, grassy, buttery
Mustard seed	4.6	Rancid, buttery, grassy	2.1	Rubbery, rancid, grassy
Rapeseed	6.9	Buttery, grassy	4.8	Rubbery, buttery
Soyabean	6.1	Grassy, buttery, rancid	6.5	Grassy, buttery, rancid

It was found that this effect may be due to isothiocyanate compounds introduced into the oils from the meal during extraction and not completely removed by deodourisation. Whatever the cause, the compounds in question must exhibit specific photochemical reactions in the presence of citric acid. In any case it is a remarkable and specific effect caused by minor components in the Cruciferae seed oil which, 'project a new problem for study.' *Moser* also reports results on industrially produced Swedish and Canadian rapeseed oils obtained in alkali refined and bleached form and deodourised in the laboratory. These results are reproduced in Table 9.6 in order to indicate the extrapolation of the results with the laboratory produced samples to products obtained from large scale operation.

Table 9.6. Effect of citric acid on the keeping quality of Swedish and Canadian rapeseed oil.

Treatment	*Continuous screw-pressed*		*Swedish extracted*		*Canadian extracted*	
	Without citric acid	*With citric acid*	*Without citric acid*	*With citric acid*	*Without citric acid*	*With citric acid*
Initial flavour	7.8	8.4	8.0	8.1	7.7	7.4
4 Days at 60°C	4.2	7.0	4.9	6.9	4.5	6.1
2 Hours' light exposure	6.8	4.8	5.8	2.9	4.0	5.4
AOM-value*	20.0	9.0	15.1	4.6	22.4	21.4

* Peroxide values after 8 hours under AOM conditions.

The positive effect of citric acid on the flavour stability when the oils were kept in darkness is demonstrated in all three samples as well as the deleterious effect on the rapeseed oil produced in Sweden when exposed to the standardised illumination for two hours. The Canadian oil, however, does exhibit a higher flavour score with citric acid than without, being the only exception to the off-flavour inducing effect of citric acid in the presence of light. The characteristic 'rubbery' flavour was found in all samples but evidently less pronounced in the Canadian oil.

In Fig. 9.16 the peroxide values after 8 hours' treatment under the AOM test conditions are reported for the laboratory samples of soyabean oil and the Cruciferae oils.

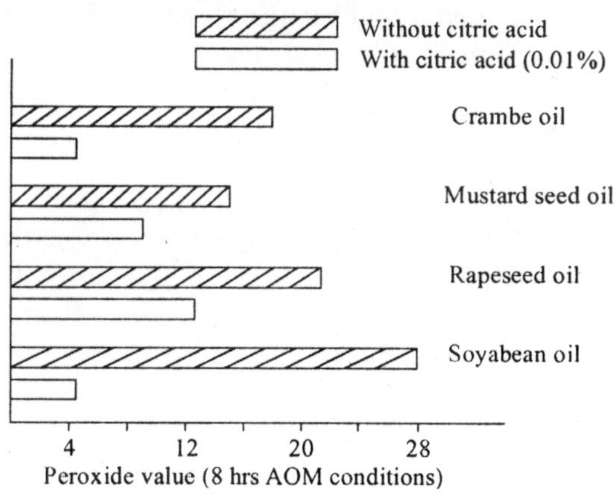

Fig. 9.16. Effect of citric acid on the AOM-values of soyabean oil and Cruciferae oil.

It may be noted that the effect of citric acid treatment is most marked for the soyabean oil investigated. The oxidative stability is indicated to be less for the soyabean oil than for the rapeseed oil before treatment with citric acid but superior after treatment with citric acid. These stability tests, performed at ca. 100°C, are in good agreement with the flavour evaluations except for the stability to light.

Comparison of these data with other reports on the flavour and flavour stability of rapeseed oil in the literature is not very easy to make since different methods of evaluation have been used. A recent monograph reports among the analyses of typical commercial samples of salad and cooking oils the keeping quality for soyabean oil as 'AOM stability, 8 hours', which is the time needed for the sample to reach the peroxide value 125 under the AOM test conditions. Taking into consideration the fact that the peroxide value in Fig. 9.16 after 8 hours for a citric acid-treated soyabean oil is found to be *ca.* 4 and that *Cowan* has reported values as low as 3.3 mekv/kg under the same experimental conditions for so-called protected soyabean oils, this value may be suspected to be representative of a processing technology that is obsolete. As a comparison, 19 lots of rapeseed oil produced in Sweden were followed with AOM tests. The 8-hour peroxide value for citric acid -treated oils were between 9.2 and 11.5, in good agreement with Fig. 9.16 and the AOM stability was 22–23 hours. The addition of 0.01 per cent ascorbyl palmitate reduced the 8-hour peroxide value to 0.8 and increased the AOM stability to 33 hours. These results indicate, as can be expected, that the kinetics of the auto-oxidation of soyabean oil and rapeseed oil are different. Three reasons for this deserve to be pointed out here.

The fatty acid composition

Frequently rapeseed oils contain 8–10 per cent linolenic acid and 14–15 per cent linoleic acid while soyabean oils contain 6–9 per cent linolenic acid and *ca.* 50 per cent linoleic acid. If other conditions are assumed to be equal the soyabean oil reacts faster with oxygen than the rapeseed oil due to the larger amount of dienoic compounds.

The glyceride composition

It has been found that the position of fatty acids in the triacylglycerol molecules influences the rate of auto-oxidation in the liquid state. When more than one reactive group is present in the molecule the rate is faster than when the groups are located in different molecules. Randomisation of mixtures of trilinoleoylglycerol and tridecanoylglycerol decreased the rate of auto-oxidation while randomisation of cocoa butter, Borneo tallow, corn oil and soyabean oil caused an increase in the rate of auto-oxidation. As shown by *Mattson* and others the dienes and trienes of rapeseed oil are preferentally located in the 2-position of the triacylglycerols which should be advantageous from the above point of view. On the other hand the position of linolenic acid in soyabean oil is nearly random, whereas the dominant linoleic acid has a slight preference for the 2-position.

The composition of tocopherols

The content of different tocopherols in rapeseed oil is reported and it appears that both the content and composition are rather favourable from the point of view of auto-oxidation. In contrast the most advantageous concentration of tocopherols in soyabean oil is lower than the natural concentration and therefore processing is directed towards a reduction of the tocopherols present.

In several studies *Holm* and others have established the importance of the degree of pre-oxidation, i.e. the content of primary and secondary oxidation products in crude oil on the flavour stability of deodourised rapeseed oil.

The non-volatile carbonyl compounds that are present in the oils as a result of the decomposition of primarily formed hydroperoxides and are not eliminated during the deodourisation, were measured by the benzidine test and the results correlated with the flavour stability of the oil. These results emphasise the importance of a protected storage and handling of rapeseed oil, quite in accordance with the principles established during years of research and development work on soyabean oil. It is therefore evident that comparisons between flavour and flavour stabilities of different oils must be made with great consideration of the pre-treatment of the samples. A deficiency in the carefully performed work of *Moser* is that the degree of pre-oxidation of the crude oils is not documented.

The degree of pre-oxidation is of course not the only factor that influences shelf life. Previously it was noted that the addition of ascorbyl palmitate had a markedly retarding effect on the rate of auto-oxidation of rapeseed oil. Antioxidant effects have further been reported on the addition of propyl gallate, butylated hydroxyanisole (BHA), butylated hydroxytoluene (BHT), nordihydroguaiaretic acid and mixtures of BHA and BHT. *Sedlacek* compared the antioxidative efficiency of 0.2 per cent ascorbyl palmitate, 0.02 per cent BHA and 0.01 per cent propyl gallate, respectively, on rapeseed oil and sunflower oil by storing samples at 28°C for 35 weeks and at 2–4°C for 52 weeks. The samples containing 0.2 per cent ascorbyl palmitate were most effectively protected and were given the flavour score 6 on a 10 to 0 point scale after 30 weeks at 28°C compared to 2 for the blank. At the earlier stages of storage the off-flavour was characterised as harsh and in addition a somewhat bitter off-flavour was noticed in samples containing ascorbyl palmitate. In these experiments the rapeseed oil reacted slower with oxygen than the sunflower oil. This work was also extended to shortenings manufactured from rapeseed oil and sunflower oil, giving essentially the same results. The effect of BHA, propyl, lauryl, dodecyl and cetyl gallates and some mixtures with these compounds and citric acid on the auto-oxidation of hydrogenated rapeseed oil has also been evaluated. As a generalisation of the reported work with antioxidants in rapeseed oil and hydrogenated rapeseed oil it can be stated that BHA and BHT have a rather small retarding effect on the rate of autoxidation while the addition of 0.2 per cent ascorbyl palmitate markedly reduces it. When choosing antioxidants, specific off-flavours and heat resistance of the compounds as well as local regulations have to be considered. These factors are not specific to rapeseed oil and will not be commented on further. For information on the heat resistance of antioxidants reference is made to the literature.

In conclusion the flavour and flavour stability of properly processed, protected and stabilised rapeseed oil is completely satisfactory for an oil to be used in today's food industry. There are several reports showing that rapeseed oil is superior to oils such as soyabean and sunflower. There are also reports in which rapeseed oil was found to be inferior to soyabean oil, e.g. in the presence of citric acid and light. This only stresses the well-known point that many factors influence the flavour of vegetable oils and fats, some of which are general for all and some of which are specific for a certain oil.

Physical properties

Among the physical properties the viscosity and the behaviour on solidification and melting can be of specific interest for the use of rapeseed oil for edible products.

Viscosity

The viscosity of an oil is of significance not only for the chemical engineer in the design of pumps, piping and other pieces of equipment, but also for the food technologist as it affects the consistency of foods. Due to its composition the viscosity of rapeseed oils is higher than that of other vegetable oils used commercially at the present time. At 20°C the dynamic viscosity of rapeseed oil of usual composition is around 90 centipoises compared to about 60 centipoises for soyabean oil and to 77 centipoises for olive oil. The dependence of the viscosity of rapeseed oil on temperature was shown to follow the Walther formula for mineral oils.

$$\log \log (\Gamma + 0.8) = m(\log T_1 - \log T) + \log \log (\Gamma_1 + 0.8)$$

where Γ is the kinematic viscosity in centistokes and T the absolute temperature in °K. With $m = 3.075$ good agreement between calculated and observed values was found in the temperature range 20–150°C.

The calculation of the viscosities of rapeseed oil mixtures with other oils is tested by mixing rapeseed oil and palm kernel oil and applying the formula used by *Kendall* and *Monroe*. Perhaps a satisfying degree of accuracy for food technological purposes may be reached by using the same methods that are applied to mineral oils, i.e. linear interpolation of the function log log (Γ + 0.8) when this is plotted against the volume fraction of one of the two oils.

Kinematic viscosities at 65–95°C have been determined by *Tremazi* for a hydrogenated rapeseed oil, iodine value 1.5, containing *ca.* 60 per cent behenic acid and it is stated that hydrogenated rapeseed oil is probably the highest melting and most viscous fat which may be incorporated in food products.

Melting and solidification

The high content of erucic acid in rapeseed oil would imply a high melting point of this oil. The melting point of the most stable polymorphic form of trierucylglycerol is *ca.* 30°C[50]. It has, however, been shown that erucic and eicosenoic acids are preferentially situated in the 1-and 3-positions of the glycerol residue, thus leading to mixtures of molecules containing oleic, linoleic and linolenic acids in the 2-position. These compounds have not yet been characterised, but melting points between 4°C and 5.8°C have been reported for rapeseed oils from Pakistan, Sweden and Canada. *Riiner* investigated a number of Cruciferae seed oils with X-ray and thermal methods and found generally the same polymorphic behaviour for oils containing between 8 per cent and 63 per cent of erucic acid. It was concluded that the requirement for a Cruciferae oil to remain clear at 0°C is that the erucic acid content is below *ca.* 39 per cent for oils from *Brassica napus* and below *ca.* 43 per cent for oils from *Sinapis Alba*. The behaviour on melting and crystallisation was shown to vary regularly with the erucic acid content (Fig. 9.17) and thus the phase behaviour of a certain oil can be predicted from its composition. This can perhaps be of value when oils for foods that are stored and used at low temperatures are under consideration.

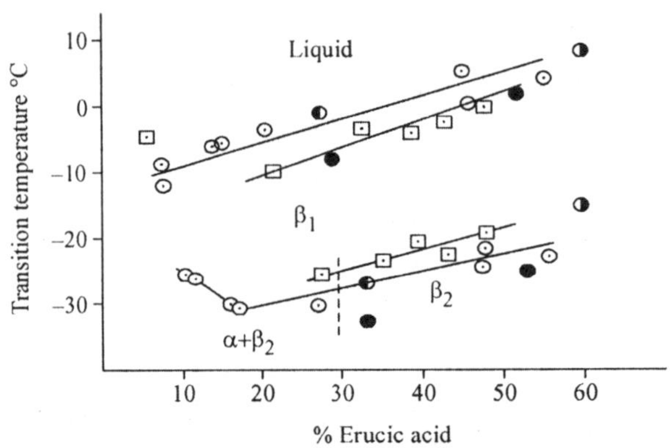

○, *Brassica napus;* □, *Sinapis alba;* ●, *Brassica campestris;* ◑, *Crambe abyssinica; and* ◐, *Conringium orientalis.*

Fig. 9.17. Phase transitions in Cruciferae oils. (Erucic acid c ntent > 8%).

Riiner and *Melin* have reported on the solidification characteristics of hydrogenated rapeseed oil. The initially formed α-phases were shown to have a longer lifetime in rapeseed oil than in other hydrogenated vegetable oils, which can influence the consistency and processing conditions of margarine and shortenings.

Specific Edible Products

In the following section specific uses of rapeseed will be reviewed. Once again it is stressed that the quality of an oil, as well as the prepared food containing the oil, varies with the processing of the raw materials, hygienic factors and the quality criteria applied. Therefore the selection of material cannot be made on a completely objective basis nor can the evaluation of results be expressed in internationally generalisable terms. It is, however, still believed that reviewing the literature on rapeseed oil can give guidance for the selection of oils and fats for use as food.

Salad and cooking oils

It is natural to begin with the use of rapeseed oil as such for cooking, frying, food dressings, mayonnaises and as a table oil. Taking into consideration the fact that the major part of rapeseed oil at present is produced and used in Asia, where the food industry is rather undeveloped, it is perhaps true that the larger part of the world production is used directly in cooking.

Without doubt the rapid growth of rapeseed oil utilisation in Canada has brought about or been promoted by the careful evaluation of the use of rapeseed oil in foods. From other sources such evaluation are hardly available. The work done by *Lips* who compared sample of rapeseed oil and mustard seed oil with corn oil in mayonnaise, pastry and for frying doughnuts. Disregarding the processing parameters included in this work, the flavour of the Cruciferae oil was found comparable with that of corn oil in cold use (oil, mayonnaise) but inferior in hot uses (frying). The auto-oxidation susceptibility was higher for Cruciferae oils than for corn oil. It can be of interest to compare the work of *Lips* with two recent evaluations. In the first samples of rapeseed, soyabean, sunflower, groundnut and corn oils, obtained from commercial manufacturers, were compared in cold and hot uses. A panel of 14 persons did not find any significant difference between salad dressings prepared from these oils. French fries fried in corn oil were preferred to those fried in rapeseed oil or soyabean oil, but no significant differences were established between French fries prepared in rapeseed, soyabean, sunflower or groundnut oil. The number of panelists was 8 in these evaluations. It was noticed that the corn oil and the groundnut oil had the strongest flavours before frying. In the more extensive evaluation by *Vaisey* and *Shaykewich* two brands of rapeseed oil were found comparable in frying performance to corn, sunflower and soyabean oils. In this case doughnuts were fried at 175°C twelve successive times, each time the total frying time being approximately 10 minutes. Neither the odour scoring of the oils used nor the flavour scoring of doughnuts distinguished the oils from each other. The smoke point determinations indicate that all the oils were of comparable quality from the beginning and that the breakdown of the soyabean oil after 12 fryings had perhaps advanced more than that of the other oils. These two contradictory results indicate the relativity of the measurements but also seem to permit the conclusion that a rapeseed oil which compares favourably with other vegetable oils in hot uses can be manufactured. In an 8 day's accelerated storage test at 60°C (Schaal oven test) one brand of rapeseed oil remained unchanged in odour and was in this respect equal to corn oil and soyabean oil. The other brand was superior to sunflower oil though less stable than corn oil. One zero-erucic acid oil that also was evaluated was the least stable of the oils examined. *Hrdlicka* and *Pokorny* identified some of the volatile compounds formed in rapeseed oil on cooking and frying.

The expression zero-erucic rapeseed oil refers to seed oils from *Brassica Strains* obtained by recent plant breeding work in Canada. *Downey* gives a summary of this work and reports that the oil is essentially free from erucic acid and is named Canbra oil, a contraction of the words Canada and Brassica. The single flavour evaluation reported above cannot be taken as evidence for a statement that Canbra oil is inferior to ordinary rapeseed oil in flavour stability. As the melting point of Canbra oil is below –10°C it is, however, possible to hydrogenate it in order to increase the oxidative stability without too large an increase in the content of solids at low temperatures. *Teasdale* claims that the reduction of the linolenic acid content to 3 per cent improves the flavour stability only slightly whereas a reduction to levels below 1 per cent gives a substantial improvement. Winterisation of hydrogenated Canbra oils with 0.4–0.9 per cent linolenic acid at 7–8°C was reported to give salad oil yields of approximately 95 per cent. This high yield is compared to yields from rapeseed oil with 25–40 per cent erucic acid and to yields from soyabean oil. At 1 per cent residual linolenic acid these yields are of the order of 75 per cent. Description of the manufacture of hydrogenated cooking and frying oils from Canbra oil without winterisation is also found in the patent literature. The stabilities obtained are illustrated in Table 9.6.

Table 9.6. The relationship between linolenic acid content and stability of hydrogenated canbra and soyabean oils.

Oil	*Linolenic acid per cent*	*AOM stability hours*	*Schaal oven test days*
Canbra oil	3.2	35	18
Canbra oil	2.9	50	21
Soyabean oil	2.1	25	14

Salad and food dressings often contain liquid oil (40–50 per cent) dispersed in an aqueous phase containing salt, acetic acid, egg yolk and other ingredients. In mayonnaise the oil content is often 70–80 per cent. As the products are preferentially stored under refrigeration the oil used should not crystallise, otherwise the emulsion will break. From the results of the study by *Riiner* it can be concluded that Cruciferae oil with erucic acid contents above 40–45 per cent can be hazardous to use in the industrial production of food dressings and mayonnaise distributed under refrigeration conditions. There is, however, in the patent literature a specific application of hydrogenated rapeseed oil in the manufacture of food dressings with an oil phase in fact containing crystalline phases. Solid fats (2–8 per cent) are melted with liquid oil (92–98 per cent), mixed with the aqueous phase and chilled. The solid fat portion, with iodine value $\leqq 12$, consists of a blend of a so-called β-tending fat and hydrogenated rapeseed oil in proportions ranging from 1: 4 to 4 : 1. The expression 'β-tending' refers to the polymorphic form that is attained after tempering.

Margarine

An important use of rapeseed oil in Europe and Canada is in margarine formulations. A standard formula in Germany reported as under.

10 per cent animal fats, 10 per cent hydrogenated rapeseed oil, melting point 40–42°C, 38 per cent hydrogenated rapeseed oil, melting point 30–32°C, 42 per cent rapeseed oil.

Under the prevailing extreme conditions the product quality must have been of secondary importance, after product quality. At the present time, it is reported that rapeseed oil or hydrogenated rapeseed oil is seldom used in margarine in West Germany. The reason for this is said to be that the products in question give the margarine a harsh mouth-feel and that this is caused by specific phosphatides in the oil. This should apply particularly to oil from rapeseeds grown in West Germany. In other localities, in any case, this characterisation of margarine containing rapeseed oil has not been given the same emphasis. It must be borne in mind, however, that consumer acceptance as well as the factors that influence oil quality vary considerably from country to country. One such important quality factor is the requirement on the shelf-life of the product with respect to time, temperature and packaging. A consequence of this is e.g. that in some countries liquid rapeseed oil and soyabean oil are accepted for margarine manufacturing whereas only hydrogenated forms of these oils are considered appropriate in others.

In a recent monograph on margarine it is pointed out that hydrogenated rapeseed oil is preferentially used in two forms, one with a melting point 34°C (iodine value 72) and the other with a melting point of *ca.* 42°C (iodine value 60). A great range of hydrogenated fats can be manufactured from rapeseed oils, as from other oils, ranging from practically liquid to very hard. *Teasdale* reported that margarines with a butter-like consistency are produced by blending two types of hydrogenated rapeseed oil, one soft and one hard. Examples of soft and hard components are given in Table 9.7. Soft margarines, spreadable from the refrigerator, were reported to be composed of three to six fat components, of which two can be chosen from either group in Table 9.7.

Table 9.7. Hydrogenated rapeseed oils for margarine in Canada.

Type	*I.V.*	*Fatty acid composition %*											*Trans acids%*	*S. F. I.**			
		16:0	*16:1*	*18:0*	*18:1*	*18:2*	*18:3*	*20:0*	*20:1*	*22:0*	*22:1*			*°C* *10*	*21.1*	*26.7*	*33.3*
Soft																	
A	93.3	2.5	0.3	2.1	42.7	14.1	2.9	1.0	10.8	0.4	23.2		17.6	3.3	0.8	–	–
B	87.5	2.5	0.3	2.9	47.0	10.9	1.0	1.0	10.8	0.8	22.2		29.9	6.4	2.0	–	–
Hard																	
C	70.7	3.6	0.3	8.9	38.0	3.8	–	1.2	8.6	1.7	33.9		–	58	40	33	15
D	64.2	2.7	0.2	14.0	44.5	2.9	–	2.7	8.1	4.8	20.1		–	59	45	37	19

*Official and tentative methods of the AOCS, Chicago.

Transesterification of hydrogenated rapeseed oil with fats containing short chain fatty acids can, however, be justified under certain circumstances. *Rudischer* reported that a substantial improvement in organoleptic properties could be attained by the transesterification of hydrogenated rapeseed oil blends with coconut oil whereas no significant improvement could be established by treating the hydrogenated rapeseed oil alone. *Seiden* further emphasises this point by showing that transesterified blends of hydrogenated rapeseed oil and coconut or palm kernel oils preferentially in combination with liquid oils, can be favourably used for the production of margarine with improved oiling-off properties and good eating quality. Another application of transesterification is the reduction of the melting points of high melting fats by the rearrangement of blends of the fat in question and a liquid oil. Thus high melting animal tallows could be made useful for margarine production. Although this application of liquid oils is by no means specific for Cruciferae oils, the properties of the resulting fat depend on the composition of the oil used.

A specific use of highly hydrogenated rapeseed oil, i.e. with iodine values below 30 and preferentially below 10, in retarding the oiling-off of margarines is also claimed by *Seiden*. Small quantities of hydrogenated rapeseed oil (0.2–2.5 per cent) in the formulae are reported to reduce the oiling-off tendency considerably, without too much adverse influence on the eating quality.

Shortenings or compound cooking fats

The definition of different types of edible fat products is not completely clear. The word shortening was originally used in American literature for fats employed in baking, making the bread shorter in texture. It has since then been transferred to fat products in general constituted by 100 per cent or close to 100 per cent fats and oils, used not only in baking but also in domestic and institutional cooking, as well as in the rapidly increasing different uses in the food industry.

As pointed out previously systematic evaluations of rapeseed oil and hydrogenated rapeseed oil in edible products were performed in Canada when rapeseed growing was started. The work on hydrogenated rapeseed and mustard seed oils in baking and frying applications, made with samples produced on a pilot-plant scale, showed that these were as satisfactory as the commercial reference material for the preparation of pastry and doughnuts. The acceptability tests were made by a 24 member panel and supported by chemical and physical measurements of several properties of the samples. For the doughnut preparations the shortenings were used both in the dough and as frying fats. No significant differences were established between commercial and pilot-plant samples. Odour and flavour of pastry made with the experimental fats were rated as slightly preferable to those of pastry made with the commercial shortening. Baking value tests, performed with a standard biscuit batter, showed that plasticising i.e. controlled crystallisation and agitation of the fats, was required to achieve volumes comparable to commercial shortening when the fats were used in non-melted form. A more recent evaluation of shortenings containing rapeseed oil made with commercial products in Canada revealed no differences between these and other types when used for the preparation of cakes but gave a less tender and flaky pastry than lard. Compared to other vegetable shortenings, no difference was found.

In shortenings, as well as in margarines, a variety of hydrogenated products manufactured from rapeseed oil can be used. This situation is not specific for rapeseed oil alone. There are, however, some specific properties of hydrogenated Cruciferae oils that are desirable in some of the food preparations where shortenings have traditionally been used. It should be mentioned here that the tradition in some countries is to almost exclusively use margarines in baking. The composition and properties of the fat phase can of course be the same whether a dispersed water phase is present or not. Therefore the following discussion can also be valid for the fat phases of e.g. bakery margarines in some countries.

Tremazi compared the temperature ranges within which hydrogenated rapeseed oils and cottonseed oils could be expected to be plastic. It was stated that at comparable iodine values hydrogenated rapeseed oil shows a broader range of plasticity than hydrogenated cottonseed oil. This can be a consequence of the presence of the triacylglycerols containing erucic and eicosenoic acids, which upon hydrogenation, in part, are transformed to triacylglycerols containing behenic and arachidic acids. The action of these is then the same as that of high melting triacylglycerols intentionally added as plasticisers to the fat blends, a procedure sometimes used in the manufacturing of shortenings. This effect is further elaborated by adding highly hydrogenated rapeseed oil as such as a plasticiser to fat blends intended to be used for shortening.

Besides being useful to provide an increased range of temperature at which the products remain plastic, hydrogenated rapeseed oil is also reported to have favourable aerating properties in baking applications where this is desirable, e.g. in cake batter mixing operations where fat and sugar are first beaten. It may be seen that even without the addition of emulsifier, which is common practice in aerating shortenings and margarines, the Cruciferae oils hydrogenated to low iodine values gave fair batter and cake volumes. When emulsifier was added, these 5 per cent suspensions were completely satisfactory, in contrast to the other fats investigated. It was also found that fractions made from rapeseed and mustard seed oils and certain peanut oil fractions possess the properties desired for the production of a fluid shortening having aerating properties. The triacylglycerols indicated to be especially effective in this respect were of the behenoyldistearoylglycerol and behenoylstearoylpalmitoylglycerol types. The hydrogenated rapeseed oil, iodine value 1, was, however, almost as effective as the triacylglycerol fractions used. It is interesting to note that no correlation was found between the polymorphic forms of the solid phases and their aeration properties in the systems. It has been a rather common opinion that fats in the β' form in general have a better creaming ability than fats in the β form. In this work the opposite was clearly demonstrated: when hydrogenated rapeseed oil was precipitated from liquid cottonseed oil at 21°C, the solid phases were in the β′ form and when precipitated at 29°C they were in the β form. However, no difference in aeration properties was found.

An additional requirement of a fluid shortening is that the viscosity (fluidity) remains unchanged on storage at different temperatures and that the product remains homogeneous. Dispersions of hydrogenated rapeseed oil, iodine value 1, were stable between 16°C and 37°C for shorter periods but thickened upon cyclic tempering and prolonged storage. This solidification tendency was assumed to be caused by a leaching out of components initially present in solid solution and their re-deposition as long interlocked crystals. This effect could be minimised by using blends of hydrogenated rapeseed oil (IV= 1) and hydrogenated mustard seed oil (IV = 30), perhaps by the formation of more stable solid solutions. Settling of solid phases and the formation of oil layers on the surface was inhibited by the addition of finely divided (particle size 1–10 micron) hydrogenated soyabean stearine.

Another route to fluid shortening containing, among other fats, hydrogenated rapeseed oil in minor quantities (0.5–4.0 per cent) is described by *Matsui*. In this case the triacylglycerols do not seem to play a decisive role for the determination of the physical properties. Instead combinations of triacylglycerols with surface active lipids are used to achieve functionality.

Other food uses

Confectionary fats and substitutes for cocoa butter, have received considerable attention both from fundamental and applied points of view. The fats can be divided into two groups, i.e. fats miscible with cocoa butter and fats immiscible with cocoa butter. The former group is preferentially prepared from fat sources containing triacylglycerols similar in composition to cocoa butter and the other group traditionally from lauric acid fats. It is evident that the average chain length of rapeseed oil fatty acids is too long to permit the isolation of compounds structurally similar to 2-oleoylpalmitoylstearoylglycerol, the main component of cocoa butter. The other group of fats is used together with defatted cocoa powder, thus imitating the properties of chocolate without the requirement that the fat must be miscible with cocoa butter.

Using combinations of hydrogenation and fractional crystallisation, it has been reported that the physical properties such as hardness and rate of melting of cocoa butter can be imitated with rapeseed oil. The oil is first hydrogenated under condition of high *trans* fatty acid formation and then subjected to fractionation in solvents. It is claimed that chocolate with good bloom resistance, hardness and mouth-feel can be produced. In another procedure for the production of hard butters chiefly from lauric acid fats, e.g. coconut and palm kernel oils, by the vacuum distillation of transesterified blends, it is reported that the use of up to 15 per cent hydrogenated rapeseed oil, iodine value less than 5, is suitable.

In a new hydrogenation procedure which is not specifically concerned with Cruciferae oils, it is pointed out that blends suitable as ice-cream coating fats can be prepared containing 50–80 per cent hydrogenated rapeseed oil.

A specific use of monoacylglycerols derived from hydrogenated rapeseed oil iodine value less than 8, is claimed by *Bedenk*.

Future uses

It can be concluded from this review, on edible use of Cruciferae seed oils, that they compete favourably with other vegetable oils and animal fats for use as food. The balance between different oils depends in part on differences in quality, but also on other technological and economical factors, that have nothing to do with the quality of the oils itself. There are, however, also some specific physical properties of hydrogenated rapeseed oil, not found in other vegetable oils, that are or will be utilised in fluid or plastic fatty products. These properties are due to the presence of longer fatty acids in the glyceride molecules. On the other hand the high content of erucic acid-containing triacylglycerols is in some applications a limitation. It seems, however, that rapeseed oil in its present form can uphold a place among the raw material for the food industry of tomorrow with its increasing number of diversified products or product lines.

NON-NUTRITIONAL USES OF RAPESEED OIL AND RAPESEED FATTY ACIDS

Rapeseed oil is primarily used as food, but it has many applications for other than nutritional purposes. Rapeseed oil, rapeseed fatty acids and their derivatives are employed in these connections, as will be described in this section.

Rapeseed Oil

The specific uses of rapeseed oil are based on its content of long-chain molecules or molecules with double bonds. In the past rapeseed oil was used for illumination purposes and mustard seed oil has been used as a fuel oil.

The rapeseed oil used for industrial products is usually refined. Certain modifications in properties can also be achieved, e.g. by blowing. This process has long been practised to increase viscosity and consists of passing a stream of air through the heated oil. A blown rapeseed oil is appreciably soluble in paraffins and the solubility varies inversely with the initial iodine value as well as with the extent of blowing. The alcohol solubility of a blown oil tends to be low and decreases with increased blowing time.

Rubber additives (factice)

This is probably the principal industrial use of rapeseed oil in the USA, at present and rapeseed oil is the classical oil for the production of factice in e.g. England, Germany and Sweden.

Factice, originally introduced as a rubber substitute, is now frequently compounded with rubber. It is used in the rubber industry to increase stability towards ageing and changes in shape. Rapeseed oil and sulphur are reacted together to form a polymer in the same way as in the vulcanisation process of rubber.

There are two varieties of factice: white and brown. White factice is prepared by reacting a relatively saturated oil, such as rapeseed oil, with liquid sulphur monochloride . The reaction takes place readily in the cold and is quite complex.

There is no particular relation between the unsaturation of the oil and the amount of reagent absorbed. Substitution also appears to take place. White factice is a light-coloured, compressible, but more or less crumbly material, which has been used as an extender or modifier for rubber. An important use of this material is in the manufacture of erasers, where it confers the degree of friability essential in this product.

Brown factice is made by first blowing and then bodying a drying oil until thickened, followed by reaction with about 5–30 weight per cent sulphur in a closed vessel, with stirring at a temperature of about 120–175°C for 1–2 hours.

Depending upon the percentage of suphur, the temperature and the reaction time, the product varies in consistency from a dark, viscous and sticky semi-solid to a hard and relatively brittle solid. The characteristics desired depend upon the specific use to which the factice is to be put. The chemical reactions that take place in the manufacture of brown factice are undoubtedly even more complex than those involved in making white factice, but again they may be compared to those which occur in the vulcanisation of rubber. Brown factice has been used not only as a rubber extender but also to modify the properties of drying oil products, such as varnishes and linoleum.

Factice-like products, with excellent flame-proofing properties, were prepared by polymerising rapeseed oil and treating the product with phosphorous halides. The products could be used as additives for rubber, lacquers, varnishes, plastics, adhesives and lubricants.

Lubricants for metal surfaces

This is a traditional use of rapeseed oil, but although it has very favourable lubricating properties, current usage of rapeseed oil solely for this purpose seems to be rather limited. Changes in industrial techniques and the availability of synthetic derivatives have caused this decline. For instance, rapeseed oil was employed to lubricate steam locomotives, but other mixtures are now used.

An exclusive use of fatty oils is exceptional today, but compounded oils of fatty and mineral oils are commonly used. Actually the fatty oil orients itself on the metal surfaces and the mineral oil is sandwiched between the layers. The coefficient of friction of fatty oils is generally lower than that of mineral oils. Mixing of mineral oil, vegetable oil and fatty acids reduces also undesirable features, such as the tendency of vegetable oil to become gummy and to carbonise. In blends with mineral oils rapeseed oil is used as a cutting oil, because such blends have a higher dipole moment than either of the single components. There is probably a relation between this phenomenon and the prolonged tool life obtained with the use of rapeseed oil blends. In lubricants for things that are shock loaded, e.g. locomotive axles, rapeseed oil also has special advantages as the viscosity of the oil can be increased to take care of the maximum load. Marine reciprocating engine bearings are usually lubricated with oil containing up to 20 per cent blown rapeseed oil.

In the metal working industry rapeseed oil-factice is used for slow processes and cutting oils of emulsion type with sulphated oil or soap for rapid processes.

Quenching of metals is the process of withdrawing heat from metal rapidly to obtain a required structure. Rapeseed oil is used as an additive in quenching oil, because it increases the cooling rate.

Rapeseed oil is used in mixtures with beef tallow to give an excellent cold rolling lubricant. The coefficient of friction varies with pass reduction.

The high coefficient of friction at low reductions is attributed to a predominance of boundary lubrication and the increasing efficiency of hydrodynamic lubrication with increasing pass reduction presumably accounts for the decrease in friction coefficient. As the load increases beyond the load carrying capacity of the hydrodynamic oil wedge, the efficiency of lubrication decreases and finally breaks down and, as a result, friction increases.

Rapeseed oil is as useful as ordinary commercial core oils for making castings. Core oils are used as binding agents for the sand cores of hollow metal castings. The cores are prepared by mixing approximately 50 parts by volume of sand with 1 part of oil, moulding the mixture in a wooden form and baking at a temperature of about 200° to 230°C until a hard, coherent mass is formed. The sand is usually dampened to make it more easily moulded and a water-soluble binder, such as casein or dextrin, is sometimes added to assist in maintaining the core in the proper form until it becomes hardened through heat polymerisation of the oil.

The oils used in core making are usually bodied to different viscosities and are used without dryers. The process of continuous casting is in a unique position in steel-making from the standpoint of lubricants. Without dependable and continuous lubrication, the process either slows down or comes to a complete halt. The prevention of ruptures or cracks in the solidified skin which forms in the mould is extremely important for the process and proper lubrication is an important factor in the formation of the skin. The most widely used lubricant for this purpose is rapeseed oil. About 5 oz., of oil are needed per tonne of steel for adequate mould lubrication.

In the manufacture of additives for lubricants, rapeseed oil is preferably used in the form of sodium, lithium and calcium soaps and patents are reported for non-foaming lubrication grease, heat resistant bearing lubricants, antifriction bearing grease, heavy-duty gear lubricants and for the improvement of the properties of these lubricants.

Fat-liquoring of leather in tanneries

Leather, the product of cattle hide, is processed either by vegetable tanning or by chrome tanning. Rapeseed oil finds use in the fat-liquoring step that always follows chrome tanning, primarily in the eastern parts of Europe.

The purpose of the fat-liquoring of leather is to lubricate the fibres of the material to give it strength and flexibility. Actually this step is performed to restore oils that were removed during the tanning process. The fat-liquoring solution is composed of a mixture of the crude oil and a soap or sulphated oil. The latter acts as the carrying agent for the crude oil and the mixture is absorbed into the leather. The oil may also be oxidised before sulphonation and mixed with non-oxidised oils before or after. The sulphonation is accomplished with 25 per cent sulphuric acid at 15–25°C and the resulting sulphoesters are washed with a 20 per cent solution of sodium sulphate.

Varnishes and lacquers

Rapeseed oil has been classed as semi-drying but it has too low an iodine value to be used directly. However, derivatives of rapeseed oil, or its fatty acids may be very suitable for this purpose.

Rapeseed oil has been used in specialised inks. An embossed effect with a transfer ink is obtained by the incorporation of 13 per cent blown rapeseed oil, which acts as a drying oil and forms a film on the surface thereby preventing excessive penetration of the ink.

Blown rapeseed oil has also been recommended in processes for printing with a glossy finish on leather. The blown oil is recommended as a wetting agent and as a vehicle for emulsion inks. It is often used in the formulation of gold and silver sizes to impart the desired tack and adhesion. In the preparation of metallic pigments, rapeseed oil behaves as a lubricant and prevents welding of the powdered metal during preparation. Crude rapeseed oil has the peculiar property of bringing out the 'reflex' or bronzy cast of the iron blues in inks.

Textile chemicals

Chemicals based on rapeseed oil are employed for three general purposes in the textile industry: scouring, washing and dyeing.

Scouring

In the scouring of wool or worsteds, a solution of a rapeseed fatty acid amine condensate, mixed with various glycols and water, is employed to remove the oil and grease present in wool. A similar solution is used to remove the warp size gums and oils present in raw 'greige' goods, leaving a clean fabric for subsequent dyeing.

Washing

A solution containing a rapeseed oil fatty acid amine condensate is used to wash a fabric before and after printing. Prior to this operation, the fabric is washed to reduce cracking and bleeding of the print to be applied . After printing, the treated fabric is washed again to eliminate the gums and thickeners employed in connection with the printing, in order to obtain a soft, pliable fabric.

Dyeing

The amine condensate is also used as a dyeing assistant. It reduces and most often completely eliminates, the formation of surface scum during the dyeing operation. The goods are left soft and conditioned for additional processing. Sulphonated erucic acid oils have been widely used in the dyeing of textiles.

Other uses in the textile industry for chemicals based on rapeseed oil are as finishing agents (soaps) and for greasing wool (felting).

Detergent additives

Sulphated rapeseed oil alcohols (18–22 carbon atoms) have shown very good foam and detergent properties and are used for the preparation of washing compounds. Epoxidised fatty acid triglycerides are used as plasticisers and stabilisers in e.g. PVC and co-polymers of PVC and vinyl acetate, polyvinylidene chloride and chlorinated rubbers. They are also used for extension and modification of epoxy resins. An epoxidised oil is used as plasticiser-stabiliser together with a primary plasticiser, *e.g.* dioctyl phthalate (DOP) and a metal stabiliser. Normally the amount of epoxy plasticiser used is 5–10 parts per hundred parts of resin for stabilising and 15–60 parts as softener. The epoxy plasticisers have a stabilising effect against heat and light and they also increase the effect of the primary stabiliser. The epoxy compound has a function in reacting with the hydrochloric acid from the decomposing PVC to give a chlorohydrin. The metal stabiliser then reacts with the chlorohydrin to give a metal chloride and a new epoxy ring.

A larger number of epoxy groups increases the solubility in PVC but cold resistance is decreased. A compound with an epoxy group in the centre of the molecule has better properties than one having a terminal epoxy group.

A rapeseed oil with high content of polyunsaturated fatty acids can easily be epoxidised and used as mentioned above.

Other reported applications

Rapeseed oil has been employed to protect neoprene against heat-ageing and as slip agent for paper tubes. The addition of rapeseed oil factice instead of sulphur during calendering of elastic packing material produces a material with decreased tendency to stickiness.

Production of surface-active agents from rapeseed oil which have special applications in floatation have been described.

A plastic gasket compound with improved tensile strength and elasticity at low temperatures is prepared from polybutadiene rapeseed oil, chalk and asbestos or mineral or textile fibres. Opaque iodinated oils have found use in röntgenographic exploration of the internal organs. Iodinated rapeseed oil has proved to be a highly satisfactory preparation which was tolerated by all types of tissues.

Rapeseed oil has been used to stabilise fungicidal and insecticidal suspensions in paraffin oils. A fairly new use for rapeseed oil is as a spray for weed control instead of commonly used petroleum oils. One major advantage of using vegetable oils in post emergence weed sprays is that by themselves they offer little hazard of crop injury or harmful residues.

Rapeseed Fatty Acids

In some of the above mentioned technical uses it is possible to use the rapeseed oil fatty acids as raw materials instead of the original oil. No specific uses of the combined rapeseed oil fatty acids are reported, but industry requires large amounts of fatty acids derived from agricultural fats and oils in the preparation of the following products: (i) soaps; (ii) surface active compounds; (iii) lubricating agents; (iv) polishes; (v) alkyd resins; (vi) buffing compounds; (vii) candles; and (viii) rubber goods. Sodium and aluminium salts of rapeseed oil fatty acids have been used as thickening agents for liquid fuels.

A coating for vinyl polymer films consisting of a sodium or potassium salt of a fatty acid with 12–22 carbon atoms per molecule improves the heat seal strength of the films.

The fatty acid chlorides are useful as intermediates in e.g. preparation of esters. Erucyl chloride can be condensed with aromatic hydrocarbons, cyclic terpenes, aniline, phenols and nitrated aromatics. The products are used as pour point improvers for machine oils, gear oils and other lubricants.

Erucic Acid and its Cleavage Products

On account of its special properties (long chain-high percentage of carbon), erucic acid (*cis*-13-docosenoic acid) is used e.g. as sodium salt in hot water detergents and in water-repelling agents and the corresponding amine as corrosion preventative. Erucic acid can easily react with *N*-bromosuccinimid to yield bromoacetoxy behenic acid. The epoxide and other derivatives have also been prepared.

The production of hard waxes, suitable for the manufacture of shoe creams and floor polishes, by the esterification of 'behenic acid' with ethylene glycol, has been reported.

Erucyl amide

Certain long-chain fatty amides of erucic acid and rapeseed oil fatty acids have proved to be good plasticisers for vinyl chloride resins. Physical characteristics of vinyl chloride-vinyl acetate co-polymer plasticised with *N,N*-disubstituted amides of erucic acid are reported. Erucyl amide is also used in polyethylenes and polypropylenes to solve intricate surface characteristic problems, such as antiblock slip, anti-static susceptibility to heat scaling and the ability to be formed or moulded.

The plasticiser evaluation data show that erucic acid derivatives in general provide more desirable characteristics in the plastic stock than an equal concentration of DOP control plasticiser. This superiority is most apparent in low-temperature performance and in characteristics such as brittle point and volatility losses. Extractability of all the erucic acid derivatives in soapy water is rather high but decidedly inferior to that of DOP.

Products of ozonolysis

The use of ozone to effect oxidative cleavage of an unsaturated fatty acid is common. Ozonolysis of erucic (*cis*-13-docosenoic) acid yields pelargonic and brassylic acids:

$$\underset{\text{Erucic acid}}{CH_3(CH_2)_7CH=CH(CH_2)_{11}COOH} + \underset{\text{Ozone}}{O_3} \longrightarrow \underset{\text{Pelargonic acid}}{CH_3(CH_2)_7COOH = COOH} +$$

$$+ \underset{\text{Brassylic acid}}{HOOC(CH_2)_{11}COOH}$$

There is a great demand from industry for both of these products.

Pelargonic acid

This acid is already industrially available as a cleavage product from the ozonolysis of oleic acid. Widespread use of this 9-carbon monobasic acid is reported in the following field plasticisers, alkyd resins, vinyl stabilisers, hydrotropic salts, pharmaceuticals, synthetic flavour and odours, floatation agents, insect repellants and jet-engine lubricants. Pelargonic acid is applied primarily in the form of esters, amine condensates and metallic soaps.

Esters

The compatibility of certain pelargonic esters with a wide range of synthetic resins is much greater than that of esters of the higher molecular weight acids. Reaction with polyfunctional alcohols, branched chain 'oxo' and other alcohols give esters that possess properties required in superior vinyl and synthetic rubber plasiticisers and for fluids that are suitable for the rapidly expanding field of synthetic lubricants.

Amine condensates

Amine condensates can be made by reacting pelargonic acid with low molecular weight amines such as monoethanolamine or isopropanolamine. Such condensates are characterised by good water solubility, which can be extended to blends made with other amine condensates. Thus, gelation tendencies of other low molecular weight amine condensates can be reduced by partial replacement with the pelargonic counterpart.

Metallic soaps

The metallic soaps of pelargonic acid may be made by direct fusion or by decomposition. The latter method is generally preferred, giving lighter coloured end products.

The mixed barium-cadmium soaps of pelargonic acid have been found to be very effective as stabilisers for polyvinyl chloride resins, particularly when used in conjunction with a chelating type stabiliser. The use of this mixed soap has been found to be more effective than the customary barium-cadmium laurate, giving plasticised PVC resins with better resistance to ultra-violet and heat degradation.

Alkali metal salts of pelargonic acid are all very soluble in water, but are insoluble in most organic solvents, even with an excess of the fatty acid. Detergency of the short chain length salts is quite limited. The calcium salts and those of the heavy metals are insoluble in water.

Brassylic acid

Brassylic acid (1,13-tridecanedioic acid) is a longer straight chain aliphatic dicarboxylic acid than those hitherto readily obtainable in quantity. Esters of brassylic acid are also reported to be suitable as lubricants over a wide range of temperature. The macrocyclic ester ethylene brassylate has been prepared from dimethyl brassylate for use as a synthetic musk in the fixing of perfumes. Another intriguing possibility is the production of new nylons from brassylic acid.

In making nylon 13, erucic acid can be cleaved at its double bond by ozonolysis, followed by reductive amination to the *omega* acid, which is then polymerised by amide condensation. To make nylon 1313, erucic acid is cleaved by ozonolysis, followed by hydrolysis, to brassylic acid. Then part of the brassylic acid is converted to a diamine.

RAPESEED LECITHIN

Rapeseed 'lecithin' is a current expression for a chemically complex mixture of substances, which originates from the polar lipid matrix of membraneous structures. These substances occur in various sub-cellular organelles in the seeds, in conjunction with e.g. triacylglycerols, carbohydrates, proteins, glucosinolates etc. Rapeseed oil manufactured by processing of rapeseed, i.e. by pressing and solvent extraction, contains, along with neutral oil (triacylglycerols), 'lecithin' components and the amounts present will to a certain extent depend both qualitatively and quantitatively on the processing parameters. Thus, what is commonly known as 'lecithin' includes besides phospholipids in the proper sense also other phosphorus as well as non-phosphorus-containing compounds.

Besides in the general sense of 'gums' obtained during the processing of vegetable oils 'lecithin' is also used in the following ways:

1. To designate the acetone precipitable lipids from crude lecithin, which are soluble in ethanol.
2. As a synonym for phosphatidyl choline.

Industrial Production of Rapeseed Lecithin

Technical rapeseed lecithin is obtained as a by-product in the production and processing of rapeseed oil. Crude oil from solvent extraction or from pressing and solvent extraction contains a certain amount of mucilaginous materials or so-called gums, which are removed in the degumming procedure, generally consisting of a treatment of the crude oil with 2–4 per cent water at moderate temperatures. This procedure leads to precipitation of the gum, from which crude lecithin is subsequently obtained after separation from the oil.

Crude rapeseed lecithin is a brown or dark brown highly viscous paste (Table 9.8).

Table 9.8. Characteristics of technical rapeseed lecithin.

Water	1 (% wt.)
Acid value	20–25 (mg KOH/g)
Acetone insoluble material	65–70 (% wt.)
Colour (Gardener)**	10–11
Viscosity (η)*** 25°C	200–300 P
50°C	30–40 P

* Approximate figures; ** 5 per cent solution in mineral oil. *** 'falling ball' method.

The amount of crude lecithin produced by extraction plants at least partly depends on the processing parameters and probably on differences in the source. Normally the yield is likely to be 0.2–0.5 per cent. *Goss* has reported a 0.2–0.4 per cent yield of crude rapeseed lecithin as calculated on seeds. A considerably lower figure is about 0.1 per cent phosphatides in 'crude' rapeseed oil.

The precipitation of phosphatides from the crude oil with water is not complete, as judged from the phosphorus content of the degummed oil. The remaining phosphorus-containing compounds in the oil are commonly designated non-hydratable phosphatides and they are structurally different from those which constitute the main part of the precipitated lecithin sludge. These non-hydratable phosphorus compounds are still less known than the hydratable as for as their chemical structures are concerned and have not yet been systematically investigated in the case of rapeseed oil. In soyabean oil for instance these non-hydratable phosphatides have been found to consist mainly of phosphatidic acids, glycero- phosphates and inorganic phosphates, primarily as Ca-Mg salts, as reported by *Nielsen*. Most likely the same types of compounds are present in the non-hydratable portion of the phosphorus compounds in rapeseed oil. The phosphorus content of degummed rapeseed oil manufactured in Sweden is of the order of 0.015 per cent (equivalent to 0.4 per cent lecithin). On the basis of the phosphorus content before degumming, approximately 0.03 per cent, the over-all precipitation efficiency is about 50 per cent.

The figures given represent an average calculated on the basis of the total oil yield, i.e. pre-pressed and solvent extracted oil. The phosphorus content in pressed oil is about 0.02 per cent (0.5 per cent 'lecithin') and about 0.05 per cent (1.2 per cent lecithin) in the solvent extracted oil with the former representing 2/3 and the latter 1/3 of the total oil yield. Taking into account the fact that only the extracted part is normally degummed, efficiency increases to about 70 per cent, i.e. from about 0.05 per cent P to 0.015 per cent P.

Industrially produced crude lecithin contains about 30–40 per cent triacylglycerols. The reason for this inclusion of oil in the precipitated 'lecithin' sludge has been explained by *Desnuelle* by the assumption that so-called mixed double-layers are formed on the surface of the water droplets in contact with the crude oil. These double layers consist of neutral acylglycerols and phosphatides, with a maximal stability of the mixed double-layers at a 1:2 molar ratio between the neutral acylglycerols and phosphatides, which may thus explain the inclusion of about one-third neutral oil into the lecithin sludge.

Laboratory Preparation

Extraction of lipid material from the seeds

The total amount of phosphorus-containing compounds in rapeseed oils obtained from seeds extracted on a large scale, as well as the yield of crude lecithin, has to some extent been referred to previously. The figures given concerned hydrocarbon solvent extraction.

If other extraction media are taken into consideration, which can be used primarily in laboratory scale preparations, completely different results can be obtained. *Shinozaki* thus report a content of 2 per cent phosphatides insoluble in acetone and soluble in chloroform/methanol 7:1 from rapeseed *(Brassica napus)* extracted with $CHCl_3/CH_3OH$ 2:1.

The influence of the properties of the solvent on the extractability of phosphorus compounds can be very strikingly demonstrated: rapeseed without any pre-treatment extracted (on a laboratory scale) with *n*-hexane yielded about 9 parts of oil and a subsequent extraction of the residue with chloroform/ methanol (2:1) yielded about one part of oil. The phosphorus contents of the two extracts, however, showed that the chloroform/methanol extract contained ten times the phosphorus of *n*-hexane extracted oil, calculated on the total yield of oil. The same observations have also been reported for soyabeans. Compared to an oil from industrial pressing and extraction, the phosphorus content of the *n*-hexane extract (laboratory scale) was considerably lower (0.05 per cent and 0.015 per cent respectively). Since the properties of the solvents should be approximately identical, other parameters than solvent must influence the extractability of phosphorus. These must primarily arise from the processing pre-treatments of rapeseed in extraction plants, humidity conditioning, heat treatments etc. *Nielsen* for instance has shown that the water content in soyabeans influences the phosphorus content in the extracted oil in such a way that a higher humidity in the seeds produces a higher phosphorus ratio in the oil. Since technical extraction procedures are more efficient than laboratory extraction with *n*-hexane with respect to the yield of oil, this seems to indicate that the less available fatty material contains larger amounts of phosphorus than the readily extractable oil.

Isolation of the polar lipids from crude extracted oil

The further isolation of phosphatides is performed on a large scale by precipitation with water, as mentioned above. The conditions of smaller scale preparations are, however, much more variable. Besides water treatment, which of course is generally applicable, numerous other modifications of precipitation and isolation procedures have been adopted. Precipitation with acetone is commonly used with or without the addition of water. Other precipitation agents are for example acids and certain salts and occasionally dialysis has been used. Numerous extraction procedures with suitable solvents or solvent mixtures are also frequently applicable with or without a precipitation treatment. Modern chromatographic techniques such as column and thin layer chromatography are now-a-days frequently used for the isolation of the polar lipids.

Analysis of Rapeseed Lecithin

Before proceeding with more detailed discussion of the analysis and composition of rapeseed lecithin, it is advisable to start with some general remarks about the limitations of the data presented in the following sections.

1. Polar lipids, which constitute the major part of the 'lecithin', are on the whole a rather unknown and undefined material, as far as their origin in the seed is concerned.
2. The constitutional characteristics of these components depend as pointed out above on several processing parameters and methods during the manufacturing of oil from the seeds.
3. Many different varieties and strains often not specified in the literature are usually included under the name 'rape'. Constitutional differences between species must also be considered, together with variations arising from environment and cultivation.

Separation and analysis of rapeseed oil phospholipids

The isolated crude lecithin is eventually further purified. Commercial (or technical) lecithin contains, as mentioned above, about one-third neutral oil which can be removed by e.g. solvent fractionation.

The classical procedure still in use for the separation of vegetable phosphatides is based upon a solvent fractionation in ethanol. The two fractions thus obtained, i.e. the soluble and the insoluble parts, were designated lecithin and cephalin respectively, which previously were considered to be two different, uniform substances. Rather soon, however, it became clear that the fractionation did not give such well defined separation, but at the most two groups of substances which in addition were rather ill-defined. Furthermore the two fractions were sometimes more or less contaminated with each other. *Folch*, for instance, isolated pure phosphatidyl ethanolamine (cephalin), which proved to be readily soluble in ethanol, while *Bear* showed that synthetic phosphatidyl ethanolamine containing saturated fatty acids was on the contrary only slightly soluble in ethanol. There are probably many factors governing the solubilities, which are more or less marked for each sample and in natural mixtures certain mutual interactions are of importance.

Thus, *Rewald* investigated rapeseed phosphatides by fractionation in ethanol and found that the main portion of the phosphatides was insoluble in ethanol. He further characterised a third intermediate fraction, insoluble in cold but soluble in warm ethanol and established that the phosphatides were associated with polysaccharides. *Heiduschka* and *Neumann* using the same fractionation method, likewise reported that the insoluble phosphatides were the dominant substances in rapeseed 'lecithin'. They did not however, contrary to *Rewald*, observe any association with polysaccharides.

Sulser analysed a commercial rapeseed lecithin sample. After hydrolysis of the phosphatides, the water soluble compounds which primarily characterise the various phosphatides were analysed by paper chromatography.

The major phospholipids of Canadian rapeseed gum have been investigated by *Weenink*. The crude gum was first purified by solvent fractionation and acetone precipitation and subsequently separated on a DEAE-cellulose column followed by a silicic acid column separation. The fractions thus obtained were further analysed by e.g. thin layer, paper and gas chromatography and by IR spectroscopy. A qualitative test also indicated sterol glycosides in the phosphatidyl choline fraction. The presence of sterol glycosides in rapeseed phosphatides has also been reported by *Aylward* and *Nichols.*

Commercial Swedish rapeseed lecithin (mainly from *B. napus*) was investigated by *Persmark.* After preliminary fractionation and precipitation with acetone, the components were repeatedly separated on silicic acid columns by elution with a gradient mixture of $CHCl_3/CH_3OH$ (Fig. 9.18).

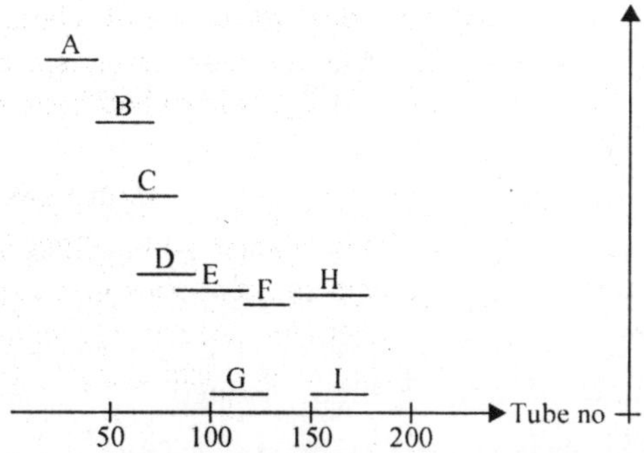

Fig. 9.18. Elution pattern of rapeseed lecithin components: A, neutral oil; B, sterol glycosides; C, phosphatidyl ethanolamine; D, phosphatidyl inositol; E and H, phosphatidyl choline; F, lysophosphatidyl ethanolamine; and G and I unidentified.

Further identification of the fractions was carried out by TLC, paper chromatography after hydrolytic splitting, IR, NMR, GLC and elemental analysis. Besides neutral oil ($CHCl_3$ fraction on the silicic acid column) four major components were found. Among the minor constituents isolated, all of them not fully identified, was lysophosphatidyl ethanolamine, accounting for about 2 per cent.

Linow and *Mieth* investigated and compared various analytical data for rapeseed and soyabean phosphatides. The technical phosphatides were fractionated in acetone, ethanol and ethanol/chloroform, and the fraction obtained compared with regard to phosphatides, fatty acids, carbohydrates, tocopherols, copper and oxidative substances. They concluded that only small differences existed in the composition of the phosphatides of soyabean and rapeseed. Differences in the fatty acid pattern, taking into account the normal differences between the oils, were found in particular with respect to linolenic acid (lower in soyabean phosphatides). Further differences were reported for minor substances such as carbohydrates, copper and oxidative material.

Fatty acid composition

The fatty acid patterns of rapeseed phosphatides are in most respects different from those of the triglycerides. The major fatty acids reported were palmitic, oleic and linoleic acids.

Utilisation of Rapeseed Lecithin

The gross composition of rapeseed lecithin does not to any substantial degree differ from that of soyabean lecithin and it thus seems probable that even the physical properties are approximately equivalent. *Jakubowski* and *Bialecka* have reported that the surface activity of rapeseed lecithin was less than that of soyabean lecithin only at relatively low levels (0.01 per cent). The difference could be explained, by a lower activity of 'cephalin' in rapeseed lecithin.

In spite of their close resemblance, however, rapeseed lecithin has not been given the same widespread use in the food industry as soyabean lecithin. The main reason for this is that rapeseed lecithin has been found inferior to soyabean lecithin in colour as well as in flavour, taste and general appearance. Nevertheless it has been reported by *Stanley* that rapeseed lecithin has been successfully utilised by margarine and chocolate manufacturers in England and Germany. Presumably it is technically possible to produce acceptable preparations of rapeseed lecithin which are comparable to soyabean lecithin by the use of suitable refining treatments. Efforts in this direction will be briefly described below.

A method for purification of vegetable lecithins based on acetone washing has not been quite successful. Thus, *Pardun* reports that although somewhat better flavour was obtained compared to the untreated rapeseed lecithin no essential improvement in colour was achieved. As an alternative *Pardun* proposes a purification procedure of lecithin in binary solvent mixture. The method was originally worked out for soyabean lecithin but later patented for other vegetable lecithins, e.g. that from rapeseed. Good recoveries and products were reported. Preparations of modified vegetable phosphatides, including rapeseed phosphatides, by treatment of the hydratable lecithins with dilute acids at pH 2–5, yielded a product which was reported to have improved emulsifying properties. In Poland a new degumming process for producing high-grade lecithin has been developed. No subsequent refining is required, the moisture content is less than 2 per cent, the acid number 4–8 and the flavour is described as 'perfect'. A process by *Bolleman* including steam deodourisation considerably improved the flavour quality, but colouring matter was not eliminated and thus an additional bleaching was necessary in order to improve colour characteristics. For the latter treatment hydrogen peroxide has been frequently used, but this method of bleaching has been reported not to have any substantial effect on rapeseed lecithin. Furthermore, a considerable increase in the peroxide value is obtained. In order to obtain a lighter lecithin, bleaching of the micella by active earth has been proposed. Large scale applications of this method have, however, not been reported. Hydrogenation may also be a possible method for improving qualitative characteristics and properties. Although used to some extent, no specific application to rapeseed lecithin has been found.

In spite of the fact that several possible methods for the improvement of rapeseed lecithin quality have been described and investigated, at the present time the main question concerns economy. Since there is a good supply of soyabean lecithin on the market the process must be cheap and preferably also provide a lecithin with better properties than products which are now commercially available.

NUTRITIONAL VALUE OF RAPESEED OIL

A discussion of the biological value of a nutrient for human beings often culminates in a number of general statements because the final conclusions in most cases are based on a limited selection of all the information necessary to give a specific answer.

Rapeseed Oil/Erucic Acid in Experimental Diets

The suitability of rapeseed oil as a dietary fat has been discussed for many years. One of the main subjects in most of these discussions has been the fact that rapeseed oil has a remarkably high content of erucic acid, i.e. Δ^{13}-*cis*-docosenoic acid.

Effect of rapeseed oil on growth of experimental rats

The nutritive value of rapeseed oil has primarily been studied in experiments with rats. Thus, it appears reasonable to review first the studies carried out with this species.

Discussion on the dietary value of rapeseed oil reached one of its peaks in connection with a debate on the superiortity of butter over margarine in nutritive value and vice versa. Much work has been concentrated on a possible relationship between the presence of erucic acid and the nutritive effects of rapeseed oil. However, these discussions were hampered by the fact that available analytical techniques were rather primitive, especially regarding the unsaturated fatty acids. *Boer* found young rats growing at a slower rate on a diet containing rapeseed oil than on a similar diet with butter. *Beznak* and *Hajdu* suggested that the weight inhibiting effect of dietary rapeseed oil was not related to its content of erucic acid but to the presence of a toxic compound. However, this postulate was invalidated in a subsequent experiment by the same authors. In studies on the growth promoting effect of summer butter, *Boer* used rapeseed oil for comparison in some of the experiments. Again rapeseed oil promoted a poor growth rate in young rats. These experiments were primarily concerned with a suggested specific growth effect of vaccenic acid, which *Boer's* group claimed to be responsible for the special effect of summer butter.

These experiments prompted *Deuel* to study the dietary effects of vaccenic acid and of rapeseed oil. Vaccenic acid was shown to be without specific effects on the growth of young rats. The gain in weight of animals on diets containing 10 per cent rapeseed oil was in most cases significantly lower than the increase in weight of the animals receiving butterfat or cottonseed oil over a 6-week period. The efficiency of utilisation of the rapeseed oil was poorer than that of the other two dietary fats. It was suggested that this may be attributable to a poor digestibility of rapeseed oil, which again may be due to its high content of erucic acid. *Deuel* also mentioned that crude rapeseed oil has an especially unappetising taste. However, it is very difficult to judge whether such an argument has any importance in experiments with rats. *Von Euler* found the growth rate of young rats fed on rapeseed oil inferior to that of animals reared on butterfat. The growth of young animals may be influenced not only by the type of dietary fat but also by the amount of fat in the diet. *Thomasson* considered this fact and carried out a large experiment with 20 different types of fats varying in amounts from 10 to 73 calories per cent (i.e. 8–48 weight per cent fat in the diet) over a period of 6 weeks. Compared to butterfat, a significantly poorer growth was obtained on diets containing 20 to 73 calorie per cent of rapeseed oil. Of the rats receiving the diet with 73 calorie per cent of rapeseed oil all were dead after 37 days of experiment (mean value: 17 days.). *Von Euler* suggests that the rapeseed oil effect on growth may be ascribed to the presence of fatty acids with 20 or more carbon atoms, i.e. erucic acid in particular. It is further suggested that these fatty acids do not differ from other fatty acids in caloric value, but that their metabolic rate is different. This latter statement is based on the fact that the calculated food efficiency constants were about the same for all dietary lipids tested. Finally, it is pointed out that the biological value of oils and fats cannot be judged only on such criteria as growth and food intake, because the animals on 50 calorie per cent rapeseed oil lived significantly longer than those on 50 calorie per cent butterfat.

The suggestion that the growth-retarding effect of rapeseed oil may be related to its high content of erucic acid was further studied by *Thomasson* and *Boldingh*. Diets were prepared with 20 to 73 calorie per cent rapeseed oil (about 50 per cent erucic acid), 30 calorie per cent Nasturtium seed oil (86 per cent erucic acid), 30 per cent of the fatty acids of the two oils and a mixture of groundnut oil and trierucin (50:50), or of groundnut oil and erucic acid (50:50). With 73 calorie per cent of rapeseed oil in the diet, all experimental animals died within a mean life span of 16 days. No death occurred within any of the other groups. By interpolation it appeared that 30 calorie per cent of Nasturtium seed oil had the same effect on growth as 57 calorie per cent of rapeseed oil. All of the experiments supported the conclusion that the poor growth obtained with rapeseed oil is actually and exclusively due to the presence of erucic acid. The unsaponifiable fractions of rapeseed and groundnut oils did not affect the growth of the animals.

Carroll and *Noble* verified the growth and appetite-inhibiting properties of erucic acid and its methyl and ethyl esters. They also found similar effects of eicosenoic acid, whereas oleic acid permitted relatively good growth. In another experiment, *Carrol* and *Noble* used a stock diet (masters meal), the composition of which is not given. To this diet was added 25 per cent rapeseed oil, 10, 15 or 30 per cent crude erucic acid, or 15 per cent oleic acid. Normal growth curves were obtained in all cases, whether the diets were given to young or adult male rats. However, the animals suffered a progressive reduction in spermatogenesis. Female rats grew similarly well and their estrous cycle remained undisturbed, but there was some interference with parturition. *Beare, Murray* and *Campbell* studied the effect of a Canadian rapeseed oil on young Wistar rats, which were fed for 9 weeks on diets with 10 and 20 per cent fat. The dietary fat consisted of corn oil, rapeseed oil or mixtures thereof containing 10, 20, 40 or 80 per cent rapeseed oil. When the ordinary oils were fed as 10 per cent by weight of the diet, there were no significant differences in weight gain. At the 20 per cent level there was a significant reduction in weight in both sexes with the 80 per cent rapeseed oil mixture, equivalent to 16 per cent of the total diet and still further reduction for rapeseed oil as 20 per cent of the diet.

Murray and others studied the growth effect of various long-chain fatty acids in young male and female rats. They used diets with 5 per cent of pure methyl esters of oleic, eicosenoic and erucic acids compared to diets with 5 per cent corn oil or no fat. After 10 weeks on the respective diets, no significant growth depression was found amongst the males, whereas the growth of the females was depressed significantly by each of the pure esters. However, *Murray* mentions that a higher dietary level of erucate results in a significant growth depression in males. Symptoms of EFA-deficiency occurred only among rats fed the fatty acid esters, indicating an acceleration of the depletion of EFA in these animals.

Roine and *Uksila* found that young rats, given a free choice, eat a little more of a diet with rapeseed oil than with soyabean oil at a 10 per cent dietary fat level. However, at a 20 per cent level the animals showed a distinct preference for the soyabean oil diet. Erucic acid at a 20 per cent level was disliked as much as was rapeseed oil. Diets with 10 and 20 per cent butter were preferred to similar diets with rapeseed oil. The growth rate for the first 30 days on diets with 10 per cent was about the same with rapeseed oil, soyabean oil and butter. During the following 2 weeks growth was slower on rapeseed oil than on the other two diets. At the 20 per cent level, growth rate was distinctly slower on rapeseed oil than on the other two types of fat. Food intake was also lowest in the rapeseed oil group. With 20 per cent erucic acid in the diet, growth was as poor as with rapeseed oil in spite of a definitely higher food intake. Experiments with rapeseed and soyabean oils emulsified in skim milk indicated that the poor growth on rapeseed oil was not due to low emulsifiability of this oil in the digestive tract. *Murray* conclude that rapeseed oil in relatively small amounts in the diet of young rats has no deleterious effect on growth rate. When the amount of rapeseed oil or erucic acid becomes higher, the food is less palatable and growth rate is retarded. At very high rapeseed oil levels the animals lose weight and even die. Growth retardation also occurred on a feed restriction regimen, which implies that reduced food intake is not the only reason for poor growth on rapeseed oil.

Beare and others compared the utilisation of rapeseed oil (Golden strain) and corn oil in young and adult Wistar rats. The oils were given as 20 per cent of the diet over a period of 5 weeks. Further, paired feeding of the two diets was carried out with young rats. In all three experiments, animals feed rapeseed oil showed smaller weight gains and less food consumption than those on corn oil, but when body weight gains were adjusted for food consumption by co-variance analyses, the differences largely disappeared. They concluded that there was no practical difference in the utilisation of the two oils in rats. The chief effect of rapeseed oil was an apparently depressed appetite early in the feeding period followed by adaption to the oil during the experiment.

The fatty acid spectrum of rapeseed oil is characterised by a higher content of erucic acid and a lower content of saturated fatty acids (especially palmitic acid) than most other vegetable oils. *Beare* studied the effects of saturated fat in rats fed on rapeseed oil diets. Young male Wistar rats were fed each test oil as 20 per cent by weight in a semi-synthetic diet for 4 weeks. The dietary fats comprised two Canadian-grown varieties of rapeseed oil (Polish and Swedish) supplemented with palm oil to reduce the concentration of erucic acid and increase that of palmitic acid, or a mixture of lard and olive oil (3:2). The animals fed Polish rapeseed oil ate less and grew significantly poorer than those on the other two diets. Analyses of co-variance indicated that the variation in weight gain could be accounted for by food consumption. It was evident that the diet containing Swedish rapeseed oil and palm oil was more acceptable to the rats than that containing Polish rapeseed oil.

The saturated fatty acids comprised approximately 30 per cent of the total fatty acids in both the mixture of Swedish rapeseed oil and palm oil and the mixture of lard and olive oil. This proportion (1:2) of saturated to unsaturated fatty acids in diets containing 10 per cent of the fat as linoleic acid was shown by *Hopkins, Murray* and *Campbell* to give the greatest weight gain in rats. The proportion 1:2 of saturated to unsaturated fatty acids is approximately equal to that found in fat depots of rats.

In the same series of experiments *Beare* added methyl laurate, mayristate, palmitate or oleate to Swedish rapeseed oil to dilute its content of erucic acid to the same level as that of Polish rapeseed oil. These dietary fat mixtures were compared with Polish rapeseed oil. None of the added methyl esters increased the body weight gain beyond that obtained with Polish rapeseed oil. A mixture of Swedish rapeseed oil and palm oil promoted greater weight gains than those obtained with Polish rapeseed oil of a similar erucic acid content (20 per cent). The relative effects of Polish and Swedish rapeseed oils alone in depressing appetite and weight gain were directly related to the content of erucic acid. Animals fed on a diet containing an oil low in palmitate (3 per cent palmitic acid) with no erucic acid grew approximately as well as those fed on Polish rapeseed oil (2.7 per cent palmitic and 19.5 per cent erucic acid). Feeding of rapeseed oil (Swedish), corn oil or the methyl or ethyl esters of the fatty acids of these two oils showed that the methyl esters produced a significantly lower food intake and weight gain than the respective glycerides or ethyl esters. The ethyl esters of Swedish rapeseed oil were more acceptable to the animals than the original oils but were not so efficiently utilised.

Acceptability of the diet was improved by increasing the saturation of rapeseed oil by the addition of palm oil and lowered by decreasing the saturation of an oil containing fatty acids of no more than 18 carbon atoms. Nutritional properties of such vegetable oils are therefore influenced by the degree of saturation as well as by the chain length of their fatty acids. It is concluded that the effect of rapeseed oil in experimental diets can be ascribed both to its low content of saturated fatty acids and its high content of erucic acid.

All the experiments described so far have shown that rapeseed oil has a poorer nutritive value than most other oils and fats. Because of its high content of erucic acid, it has been natural to suggest that this particular acid may be responsible, at least to some extent, for the low dietary value of rapeseed oil. For this reason it was very interesting when *Stefanson*, *Downey* and *Craig* succeeded in breeding *Brassica napus L.*, with no erucic acid in the seed oil.

Rocquelin and *Cluzan* compared the effect of two types of rapeseed oil, one with 45 per cent erucic acid and another with no erucic acid (1.9 per cent docosaenoic acid), to that of arachis oil. The oils were given as 15 per cent of the diet to young male and female rats for 6 months. Arachis oil and the rapeseed oil free from erucic acid gave the same growth in both sexes. The ordinary rapeseed oil caused growth retardation in both sexes compared to the two other oils.

Experiments with partially hydrogenated rapeseed oil

Partially hydrogenated oils are widely used, especially in human diets. It is well known that many such treated oils are very low in essential fatty acids and they have therefore become of special interest in relation to studies on essential fatty acid deficiency. However, rapeseed oil has rather reasonable amounts of linoleic and linolenic acids. Thus, a partial hydrogenation should leave fairly high amounts of essential fatty acids in the oil. Partially hydrogenated rapeseed oil has been studied to some extent in relation to the general effect of rapeseed oil in experimental animals. Gas chromatography of the methyl esters of the partially hydrogenated rapeseed oils revealed that the partially hydrogenated rapeseed oil, which increased weight gain, was still relatively rich in linoleic acid.

The other two samples of partially hydrogenated rapeseed oil contained lesser amounts of linoleic acid. However, all three samples contained appreciable amounts of $C_{18:2}$ positional isomers. Partial hydrogenation also decreased the content of $C_{22:1}$ acid and increased the amount of stearic acid. The alterations in the C_{18} fatty acids resulting from hydrogenation of rapeseed oil are responsible for the differing responses in weight gain.

Determination of the gross calorific value of rapeseed oil as heat combustion in a bomb calorimeter was found to be practically the same as that of lard. Determination of the available caloric contents of rapeseed oil, partially hydrogenated rapeseed oil and corn oil by rat growth were carried out by *Middleton* and *Campbell*. Growth responses to the three oils and to lard were similar and the calculated caloric content of the oils showed no significant difference from that of lard under the experimental conditions employed.

Cheniti studied the nutritive value of sunflower oil, rapeseed oil, partially hydrogenated rapeseed oil with and without a supplement of sunflower oil as a source of essential fatty acids (7 g sunflower oil/kg food). The oils were given at a level of 20 per cent of the diet to young rats for 7 months and 4.5 months. The growth rate on partially hydrogenated rapeseed oil was significantly poorer than on sunflower oil, but was brought to about the same level as obtained with rapeseed oil by supplementation of the diet with essential acids, i.e. 7 gram of sunflower oil per kilogram of diet. Both of these latter two groups grew somewhat less than the controls. Calculations of the dietary efficiency (grams of weight gain/grams of diet ingested × 100) as a function of time showed a levelling off on all four diets. However, the efficiency was highest for the sunflower diet throughout the experiments. *Cheniti* conclude that the decreased efficiency of hydrogenated rapeseed oil may not only be a matter of an increased requirement for essential fatty acids, but also related to the presence of *trans* isomers and of behenic acid.

In a third experiment, lasting only 39 days, young rats were fed on a fat-free diet supplemented with about 8 per cent of ethyl oleate, erucate or behenate and 0.4 per cent ethyl linoleate. Erucate gave the lowest weight gain and efficiency. The biological value of rapeseed oil was inferior to that of other vegetable oil. *Cheniti* suggest that this inferiority is located at the metabolic level, in part as a direct effect of specific components in the rapeseed oil, e.g. erucic acid and in part indirectly as an interference with other dietary elements.

Experiments with heat-treated rapeseed oil

Diets containing 10 per cent heated rapeseed oil or hydrogenated vegetable fat had a depressive effect on the growth of rats during a 6 months' test period (*Nikonorow* and *Homrowski*). These studies were followed up recently by *Homrowski* in studies on heat-treated rapeseed oil, margarine and hydrogenated shortening. Both of the latter two products contained rapeseed oil. Rapeseed oil and hydrogenated shortening were heated to 170°C and 200°C respectively, for 30 minutes, or to 170°C for 12 hours. Margarine fat was heated to 130°C or 170°C for 30 minutes, or to 130°C for 3 hours. The fresh and heated fats were fed as 10 per cent of the diet to rats for 12 months. All rats gained weight, but statistically significant differences were found in the growth of animals fed on heated fats compared to those on the non-heated products or normal laboratory diet. Fresh rapeseed oil alone also retarded growth. *Homrowski* found no significant weight deficit in rats receiving 18.5 per cent erucic acid from unheated margarine or 40.5 per cent of this acid from hydrogenated shortening, the latter amounting to approximately the same content as in fresh rapeseed oil. *Homrowski* suggests that the effect of rapeseed oil on growth may be connected with a component other than erucic acid, or the variations in chemical composition of margarine and hydrogenated vegetable shortening may influence the reduction of the growth depressive effect of erucic acid when contained in these products.

Effect of dietary rapeseed oil on other species than the rat

In general, the rat has been the experimental animal par excellence in studies on the nutritional effects of rapeseed oil. In the following a brief account of the results obtained in various species of experimental animals is given.

Pigs have been used as experimental animals in some studies involving dietary rapeseed oil. *Paloheimo* carried out two experiments with young pigs. They added 100 g/kg diet of rapeseed or soyabean oil and divided the experiments into a period of restricted feeding (about 4–5 weeks) and a period of *ad libitum* feeding (about 3–5 weeks). During the period of restricted feeding, no marked differences were seen between the two groups, but during the *ad libitum* feeding in the second experiment the animals on soyabean oil showed a better appetite and grew more than those fed on rapeseed oil. Further, the latter animals had a larger water consumption than those on soyabean oil. *Crampton* found that swine were not able to utilise dietary rapeseed oil efficiently. Thus, a significant growth depression occurred in swines fed on a diet containing 20 per cent rapeseed oil when compared to those on a similar diet containing butterfat.

Dogs were used as experimental animals by *Lindlar* in a study on the presence of erucic acid in organs and depots after rapeseed oil feeding. *Lindlar* found large quantities of erucic acid in the depot fat, lesser amounts in various organs and none in brain, bone marrow and erythrocytes. He did not find erucic acid in the glycerophospholipids. *Crampton* found puppies to thrive well on a diet with 20 per cent rapeseed oil. *Bernhard* gave 3 dogs a load of rapeseed oil (60–100 g/day) during 50 to 76 days. He found erucic acid accumulated in the depot fat, whereas it could for the most part not be detected in the phospholipids. *Carroll* found a normal growth rate in dogs on a 25 per cent rapeseed oil diet for 24 days.

Chickens reared on a diet containing rapeseed oil had depot fat which differed sensibly from ordinary chicken fat. Normal growth of chickens on a diet containing 25 per cent rapeseed oil for 4 weeks was observed by *Carroll*. *Sell* and *Hodgson* kept chickens on rations with 4 or 8 per cent rapeseed oil, tallow, soyabean oil or sunflower seed oil for 8 weeks. Rapeseed oil supported weight gain to the same extent as the other vegetable oils. Equivalent levels of tallow, rapeseed oil and sunflower oil resulted in ration metabolisable energy values which were approximately equal but significantly lower than with soyabean oil in the diet.

Ducklings grew well on diets with 20 cal. per cent rapeseed oil, but at 40 cal. per cent their growth rate was significantly poorer than that of duckling on 40 cal. per cent butterfat. With 60 cal. per cent rapeseed oil a very pronounced growth retardation occurred, indicating a progression of this effect with increasing content of rapeseed oil in the diet.

Guinea pigs as test animals for dietary rapeseed oil were used by *Crampton*. A diet with 20 per cent rapeseed oil reduced growth in guinea pigs compared to those on a similar diet containing 20 per cent butterfat. Guinea pigs given a diet containing 25 per cent rapeseed oil did not gain weight during an experimental period of 4 weeks (*Carroll*). *Thomasson* found a significant growth retardation in guinea pigs fed on a 50 cal. per cent rapeseed oil diet compared to those on a similar diet with butterfat.

Mice fed on a diet containing 25 per cent rapeseed oil showed no gain in weight after 4 weeks on the diet *Carroll*. *Thomasson* found that mice grew significantly more poorly on diets with 40 cal. per cent, or more, of rapeseed oil than on butterfat.

Rabbits have generally not been used in rapeseed oil experiments. However, *Wigand* gave diets containing 8 per cent of rapeseed oil, butterfat or corn oil to rabbits. He found no significant differences in growth rate among the three groups of animals during experimental periods of 7.5 and 15 weeks. *Carroll* found a normal growth rate in rabbits fed on a diet with 25 per cent rapeseed oil for 4 weeks.

Turkeys showed a poor growth on a 10 per cent rapeseed oil diet.

Geese were given barley for 12 days, then an addition of 40 ml rapeseed oil daily for another 16 days and finally barley only for another 12 days. The subcutaneous and internal fat assumed the characteristics (iodine number, melting point, smell) of rapeseed oil after only 8 days on the diet. These characteristics disappeared again after a week on barley alone.

Digestibility and Absorption

In order to evaluate the nutritive value of dietary fats it is necessary to consider their digestibility and absorbability. The overall utilisation of dietary fat is often defined as the coefficient of digestibility calculated from the ingested amount of fat and the fecal excretion of fat: [Fat ingested–Fat excreted corrected for metabolic fat)] × 100/Fat ingested. Fat absorption commonly describes the removal of fat from the small intestine as a function of time. Thus, the absorption coefficient usually describes the degree of absorption as measured by the amount of the ingested fat present in the gut several hours after ingestion of a single dose. Therefore, the absorption rate can be determined only in experiments of short duration.

Rapeseed oil can be digested and absorbed, at least to some extent. By feeding rapeseed oil to rats, erucic acid was found in the depot fat. The average coefficient of digestibility in four persons was found to be 98.8. More rational and elaborate experiments on the utilisation of dietary rapeseed oil in experimental animals were started in 1940 by *Deuel. Deuel* compared the rate of absorption of various natural fats, including rapeseed oil, in rats. In order to obtain the most uniform and comparable results, the absorption rates were calculated on the basis of body surface area. Of the oils tested, i.e. cotton seed oil, butterfat, coconut oil and rapeseed oil, the latter was found to have the slowest absorption of all.

During the following years a number of reports indicated poor growth of rats on rapeseed oil. This prompted *Deuel* to study the digestibility of crude and refined rapeseed oil in rats. The coefficient of digestibility was found to be 77 and 82 for crude and refined rapeseed oil, respectively. The calculation of the coefficient included correction for metabolic fat, i.e. excreted fat in feces from animals on a fat-free diet. This endogenous fat, 'metabolic fat', comprises not only triglycerides and fatty acids, but also soaps which are not extractable with ether. This had been ignored in previous studies, but was shown by *Augur* and *Crockett* and *Deuel* to be a factor of importance.

Further, the amounts of fecal soaps are related to the composition of the salt mixture. Thus, calcium-magnesium rich salt mixtures increase the amount of excreted soaps. A similar effect is seen with poorly digested fats. The presence of fecal soaps implies that acidification before extraction is necessary. It has been later shown by *Hopkins* that acidification with 5 per cent acetic acid in petroleum ether is preferable to acidification by means of hydrochloric acid which result in the extraction of considerable amounts of non-lipid material also. It should be pointed out that in order to calculate the actual food intake in experiments with *ad libitum* feeding, it is very important that possible feed spillings are carefully controlled.

The fatty acid composition of dietary and chyle fat was studied in a child with a chylothorax condition. The assumption was made that similar absorption coefficients of fatty acids indicated the absence of splitting of the glyceride molecules. On this background it was suggested that certain fats, e.g. rapeseed oil, olive oil and sunflower seed oil, are absorbed almost without hydrolysis.

However, fatty acids are normally present in food as triglycerides. Thus, *Carroll* and *Richard* studied possible differences in digestibility between triglycerides and non-esterified fatty acids. Triglycerides of unsaturated fatty acids were more completely digested than the corresponding non-esterified fatty acids. Thus, the coefficients for trierucin and erucic acid were 63 and 48, respectively. Palmitic and stearic acids were poorly digested either as triglycerides or as free acids. Further, the calcium content of the diet was found to play an important role in determining the digestibility of fatty acids. The poor digestibility of the free fatty acids may be caused by the formation of calcium soaps and calcium-phosphate containing complexes which are not readily absorbed. This suggestion may be of special importance for erucic acid, which is slowly absorbed. On the other hand, the digestibility of erucic acid appeared to improve as the protein level in the diet was increased. Finally, the digestibility of erucic acid seemed to be lower in old than in young rats.

Carroll compared the digestibility of rapeseed oil in rats, guinea pigs and rabbits. He used a fox chow diet supplemented with 25 per cent rapeseed oil by weight and found coefficients of digestibility to be 58, 61 and 58, respectively. *Beare* fed Wistar rats on a diet with 20 per cent rapeseed oil and found an apparent digestibility (no correction for endogenous fat) of 93. *Beare* investigated the response of two strains of rats, inbred Wistar and Sprague-Dawley, to dietary rapeseed oil and corn oil. The digestibility of corn oil was not significantly different in the two strains of rats, but the digestibility of rapeseed oil was significantly lower in the Sprague-Dawley than in the Wistar (coefficients 65 and 83, respectively).

Effect of dietary components other than fat on the digestibility of rapeseed oil

Carroll and *Richards* found that the calcium content of the diet plays an important role in determining the digestibility of fatty acids. On a calcium-free diet the coefficient of digestibility of erucic acid was 92, whereas with a calcium-phosphate ratio of 1.4:1 in the salt mixture (5 per cent of the diet) it was 21. With only 2 per cent salt mixture high in calcium the coefficient was 48. An increase of the dietary content of calcium primarily resulted in a large increase in the fecal soaps and of the calcium-phosphate containing fatty acid complexes. Further, high dietary protein (65 per cent) improved the digestibility of erucic acid compared to a low protein level (5.5 per cent). Young animals appeared to digest erucic acid better than older animals. These studies were followed up by *Richards* and *Carroll,* who found that the ether-soluble calcium-phosphate complex is formed with a number of saturated fatty acids, but not with e.g. palmitic and stearic acids.

The poor digestibility of rapeseed oil may also be related to its emulsifiability in the digestive tract. *Augur* found that 3 per cent lecithin lowered the susceptibility to diarrhea in rats fed on diets with cottonseed oil or hydrogenated cottonseed oil. Simultaneously, lecithin increased the digestibility of the hydrogenated oil, probably because of an increased rate of absorption. *Roine* and *Uksila* mystified rapeseed oil and soyabean oil with skim milk powder in skimmed milk. The milk preparation contained 3.1 per cent fat, equivalent to 38 per cent of the dietary calories. This preparation fed to rats caused diarrhea, but the milk preparation given together with a low-fat basal ration was tolerated. After 30 days of experiments the rats on the rapeseed oil milk preparation grew significantly more poorly than those on the soyabean oil preparation. *Roine* and *Uksila* conclude that retarded growth on the rapeseed oil diet is not due to poor emulsifiability in the digestive tract. *Thomasson* and *Gottenbos* found poor growth of rats on diets high in rapeseed oil (72 per cent) to be due to the inhomogeneity of the diet, which was a result of the oil being separated off. This could be prevented by the addition of 1 per cent sawdust (saturated with water) which resulted in improvement of growth and decreased mortality. Several years later *Alexander* and *Mattson* found that gelatinisation of the liquid oil with ethyl cellulose overcame this difficulty.

Effect of rapeseed oil/erucic acid on 'metabolic fat'

In the discussion on digestibility and absorption of dietary fat the fact that not only is dietary fat digested in and absorbed from the digestive tract, but a secretion of fat out into the gut is also taking place has been referred to. Some of this fat is re-absorbed and some is found in the feces. Thus, this endogenous or 'metabolic fat' poses a problem in itself.

Norcia and *Lundberg* studied the effect of dietary fat (tripalmitin, triolein or a tripalmitin-triolein mixture) on the composition of fecal fat. They concluded that endogenous fat was unaffected both in quantity and composition by the dietary fats, although the depot fat was markedly different depending on whether the rats had been pre-conditioned on tripalmitin or olive oil. The composition of the depot fat and changes in composition induced by dietary changes exerted no influence on the composition of the endogenous fecal fat. From these data they concluded that endogenous fat does not result from secretion of fat, from desquamation of epithelial cells of the intestinal mucosa but is synthesised by intestinal bacteria. A similar view was advanced by *Holasek.* He found that rats on a low-fat diet with cellulose and terramycin produced faces in which longer chain fatty acids were essentially absent. Similar results were obtained in rats with a biliary fistula and with humans treated with antibiotics. He concluded that fecal fat was not secreted into the gut but was of bacterial origin.

However, this conclusion is invalid since fat secretion does take place. Thus, *Bernhard* fed rats for prolonged periods with daily doses of 2–3 ml of rapeseed oil followed by a similar period of feeding a normal fat-free diet. He found in these rats, as well as in rats with a biliary fistula and the same dietary regimen, that erucic acid occurred in the feces long after the rapeseed oil feeding was stopped. He concluded that erucic acid is absorbed, some is deposited and can be subsequently secreted into the gut together with other endogenous fatty acids. A similar secretion was observed in rats with biliary fistual. The collected bile contained no erucic acid, indicating that the secreted fat is not transported out into the gut by means of the bile. The fat secretion is more marked in the absence of bile as the fat absorption is then hindered. In this case the fatty acids from the fecal lipids contained much more stearic acid and erucic acid than the fat of the wall of the intestine or of the depots. In the case of a fat-deficient diet the fecal lipids consist of modified body fat together with dietary fat which has not been absorbed. When the supply of bile to the intestine is inadequate body fat appears in increased amounts in the feces.

It appears reasonable to suggest that the excretion of endogenous lipids into the gut is influenced to some extent by the dietary fat but it is difficult to estimate this effect. In studies on the digestibility of dietary fats, corrections should be made for 'metabolic fat'. Further, the correction factor should be kept as low as possible. Thus, short-chain fatty acids cause less endogenous fat excretion than a fat-free diet, whereas long-chain fatty acids increase the amount of fecal lipids and certain monoenoic acids increase the content of fecal sterols. It should be pointed out that co-prophagy apparently has not been considered in studies concerning erucic acid and rapeseed oil.

Appetite and palatability related to rapeseed oil diets

Beare concluded from studies of rapeseed oils and their ethyl esters that the effect of dietary rapeseed oil on appetite is not solely related to its content of erucic acid. Acceptability of the diet was improved by increasing the saturation of rapeseed oil by the addition of palm oil and lowered by decreasing the saturation of an oil containing fatty acids of not more than eighteen carbon atoms.

Beare compared diets with 20 per cent rapeseed oil or corn oil during 5 weeks by *ad libitum* and paired feeding. The reduced rates of weight gain and food consumption on the rapeseed oil diet were ascribed to an apparent depressed appetite early in the feeding period followed by an adaption to the oil during the experiment.

In another experiment, eating habits were controlled. With rapeseed oil and butterfat diets the rats of both groups had an average of 10 meals per day, each lasting for about 5 minutes. However, the animals consumed 13 g and 9 g of the butter and rapeseed oil, respectively. Thus, the difference in food intake arises during the meals. The physiological mechanism underlying the food-intake-reducing action of rapeseed oil is unknown, but is apparently not localised in the taste organs. It is suggested that the reduced consumption of rapeseed oil diets is related to a low rate of transport and/or metabolism of rapeseed oil fatty acids, especially erucic acid. In order to follow up this idea it is necessary to study the intestinal absorption of rapeseed oil and butterfat. On investigation the absorption rate was about twice as high for the former of the two fats as that of the latter. The decreased rate of absorption for rapeseed oil was suggested to be the result of a feed-back mechanism originating from other phases of the metabolic pathway. However, this view was not supported by studies of *chyle* from cannulated thoracic duct rats given rapeseed oil or corn oil, which indicated that rapeseed oil enters the blood stream more slowly than corn oil. Thus, these results indicate that the rate of intestinal absorption is partly or entirely a local process. Thus, this does not rule out the possibility of rapeseed oil also reducing appetite via other mechanisms.

Physiological Changes Induced by Rapeseed Oil

Effect of rapeseed oil on depot fat

It is well-known that dietary fat has some influence on the composition of depot fat. Studies on the relationship between rapeseed oil and depot fat have of course been mainly concerned with the presence of erucic acid, which normally occurs in very low amounts, if at all, in depot fat.

Wagner kept adult male rats on a rapeseed oil diet for 6 months. The contents of erucic acid and several other fatty acids were determined in neutral fats and phospholipids from numerous organs and depot fat. A true accumulation of erucic acid could not be determined. A study of the biological half-life of erucic acid in lipids from various organs and depots showed it to be about 18–31 days. *Wagner* concluded that erucic acid behaves like a fatty acid normally utilised in metabolism.

Sell and *Hodgson* found substantial portions of eicosenoic and erucic acids in the adipose tissue of chicks receiving 4 or 8 per cent dietary rapeseed oil for 8 weeks.

Craig and *Beare* found erucic acid in lipids from skin, head, carcass and 'organs' (liver, heart, lungs, spleen, intestine) when rapeseed oil was fed to rats. The concentration of erucic acid in the fatty acids from the head and organs was lower than that from the skin and carcass, which is in agreement with previous data showing that erucic acid is less readily incorporated into phospholipids than into glycerides. Eicosenoic acid occurred in each body fat in somewhat higher concentrations than did erucic acid, although the dietary fat contained only half as much eicosenoic as erucic acid. In this experiment, 11-docosenoic acid was also studied and the incorporation of this acid was similar to that of erucic acid. Thus, apparently the rats did not distinguish between the two isomers.

Thoron fed boars on a diet with 15 per cent rapeseed oil for 7 months. Although the dietary level of erucic acid was high and the experiment lasted for 7 months, erucic acid occurred in the depot fat, liver and testes only in very small amounts: 2, 1 and 0 per cent, respectively.

Effect of rapeseed oil on the suprarenal glands

In has long been known that rapeseed and rapeseed oil in the diet of experimental animals result in changes in their adrenal glands. *Kennedy* and *Purvis* noticed hypertrophy of the adrenal cortex with a large increase in the lipid content of the adrenal cortical cell in rats fed on rapeseed. *Abelin* found an increase in the concentration of adrenal cholesterol to 6 per cent in rats on a diet containing rapeseed oil, whereas other dietary fats failed to produce a similar effect. *Carroll* fed young Sprague-Dawley rats rapeseed as 45 per cent by weight of the diet and found a slight increase in adrenal size and very marked increase (3–4 fold) in both the percentage and the absolute amount of adrenal cholesterol. Nearly the same effect was obtained by feeding rapeseed oil, turnip seed oil, or Nasturtium seed oil as 25 per cent of the diet. Various other fats and oils fed at the same level resulted in a very small increase of the cholesterol content. The increase in cholesterol is mostly in the esterified form. Rapeseed oil had very little effect on the concentration of ascorbic acid in the adrenals. Feeding rapeseed oil to hypophysectomised rats did not cause a significant increase in their adrenal cholesterol and the adrenals atrophied as a result of the hypophysectomy.

Carroll suggested in view of the high content of erucic acid in the above seed oils, that this acid may be responsible for the effect on the adrenals. Thus, rapeseed oil and the fatty acids from rapeseed oil were fractionated and tested during three weeks in male rats fed on diets containing 10–25 per cent of the various fractions. The results indicated that methyl erucate and to a lesser degree also nervonic acid (a higher homologue of erucic acid), cause a considerable increase in the weight of the adrenals and in their content of cholesterol. A number of individual fatty acids were also tested, e.g. oleic, eicosenoic, ximenic, lumequic and *trans*-nervonic acids, none of which showed an effect on the adrenals comparable to that of erucic and nervonic acids.

Carroll also studied the relation between dietary rapeseed oil and adrenal cholesterol in different species. He found the cholesterol content of the adrenals significantly increased in rats and mice, whereas only a small change or no such change was observed in rabbits, guinea pigs, chickens and dogs. The differences in response to rapeseed oil do not appear to be due to species differences in digestibility of rapeseed oil. *Carroll* points out that dietary cholesterol tends to be deposited more readily in the latter four species than in the rats and mice.

Effect of rapeseed oil on ovaries, testes, reproduction and milk secretion

The rapeseed oil in the diet may be of some importance in relation to reproduction was indicated by studies of *kennedy* and *Purves*, who found that female rats fed on rapeseed oil matured sexually much later than normal females, as they would not become pregnant until 5–7 months of age. Further, the ovary and the adrenals are the only organs that contain most of their cholesterol in the esterified form. Preliminary studies by *Carroll* and *Noble* indicated that the ovaries of rats fed on rapeseed oil contained more cholesterol than those of controls. Therefore, immature female rats were fed on a diet containing 25 per cent rapeseed oil for 60 days, during which period the rats reached maturity. Compared to the controls the rats on rapeseed oil grew less well, their vaginas opened one to two weeks later, but at the same approximate body weight. After opening of the vagina, daily vaginal smears were taken and no irregularities in the cycles were observed. At the end of the feeding period the cholesterol content of the ovaries from the rats on rapeseed oil was significantly higher than that of the controls. A similar feeding experiment with 25 per cent rapeseed oil for 28 days was carried out on adult female rats. Again the sex cycles were normal, but the ovarian cholesterol concentration was significantly increased by the rapeseed oil diet. In both of these experiments the ovaries were smaller in the treated than in the control animals.

Beare noticed that the testes in young Wistar rats fed on 20 per cent rapeseed oil for five weeks were significantly smaller than the testes of rats on a similar diet with corn oil. Histological studies of the testes revealed no alteration in cellular characteristics, but a reduction in tubular size. Similar results were obtained in experiments with older rats. In a subsequent study, *Beare* compared the effect of diets with rapeseed oil, corn oil, or a stock diet (fox chow) on reproduction in weanling Wistar rats. The females showed no irregularities in their estrous cycles that could be attributed to diet. During the production of three litters it appeared that the rats fed a commercial diet had a superior reproductive performance, particularly in the first litter. Statistical analyses revealed insignificant differences between the effects of the diets containing corn oil and rapeseed oil, both with respect to the number of young born and the number weaned. With rapeseed oil in the parental diet, the weight of the weanling rats was considerably lower. *Beare, Carroll* and *Noble* suggest that this is related to decreased milk production resulting from the reduced food intake of animals receiving rapeseed oil and to the presence of some erucic acid in the milk. In the male rats of these latter experiments, no disturbances were observed in spermatogenesis as they sired three litters. However, it should be recalled that rapeseed oil contains a substantial fraction of essential fatty acids.

Recently, *Bourdel* and *Jacquot* studied the effect of rapeseed oil on the testes of rats. They concluded that rapeseed oil does not have any harmful effect on the genital tract of male rats. *Thoron* found that the relative weight of testes and epididymes was much smaller in pigs fed on rapeseed oil than in those fed on a mixture of olive and palm oil.

Hilditch and *Thompson* studied the influence of different dietary oils on the fatty acid composition of cow's milk. Feeding of 4 oz., of rapeseed oil twice daily to cows for a week or more resulted in an increase in the proportion of oleic acid and the appearance of erucic acid to the extent of 3.7 weight per cent of the total milk fatty acids.

Effect of rapeseed oil on the thyroid gland

Anti-thyroid compounds are known to occur in a variety of seeds from plants of the Cruciferae family. The active compound was originally thought to be one of the mustard oils. Amongst the compounds with goitrogenic effect is *l*-5 vinyl-2 thiooxazolidone. In studies of the toxic reactions produced in rats by phenylthiourea and α-naphthylthiourea, *Carroll* and *Noble* found that rats which had been pre-treated with anti-thyroid compounds or fed certain goitrogenic diets developed a marked resistance to these toxic compounds. The goitrogenous substances are apparently found in much higher concentration in rapeseed meal than in rapeseed oil. Crambe meal given to chicks results in enlarged thyroids, poor growth and high mortality. Ammoniation of the meal caused improved growth and all body organs including thyroids appeared normal on autopsy.

Roine fed pigs on a diet with 28 cal. per cent rapeseed oil or soyabean oil. The pigs appeared in general more sensitive than rats to a high fat diet, as histological examinations of the thyroid gland revealed definite changes indicative of hyperfunction in both series of experiments. In the same experiment, rats were fed on diets providing 0, 15, 30, or 70 per cent of the calories as rapeseed oil. The control rats received 30 or 70 cal. per cent of soyabean oil. No changes were observed in the thyroids of any of the rats.

Sell and *Hodgson* found a small, if any, effect on the weight of the thyroid gland of feeding cockerels on a diet containing 4 per cent or 8 per cent rapeseed oil for 8 weeks. The rapeseed oil supplied about 2.5 per cent of erucic acid to the total ration when the oil comprised 8 per cent of the diet. They concluded that the goitrogenic material normally found in whole rapeseed and rapeseed oil meals probably is not present in the rapeseed oil tested.

Effect of rapeseed oil on the liver

The weight of the liver, the fatty acid composition of the liver fat and in some cases the content of liver cholesterol have been studied in relation to rapeseed oil diets.

Beare found that the liver weight of male rats was increased by feeding of rapeseed or corn oil heated to 200°C for 120 hours. No such effect was observed in female rats. *Wagner* studied the content of erucic acid in the liver of rats fed on choline-free and choline-containing diets containing 10 per cent rapeseed oil for 18 days. Choline deficiency results in fatty liver. *Wagner* found 7.8 and 3.8 per cent of the liver fatty acids to be erucic acid in the choline-deficient and control animals, respectively. *Wagner* examined the content of erucic acid in neutral and phospholipids of livers from rats fed 10 per cent rapeseed oil containing 42 per cent erucic acid for 6 months. Erucic acid occurred in the neutral lipids of all organs tested, but no special accumulation of this fatty acid was seen, although the dietary content was high. The percentage of erucic acid appeared lower in the liver than in any of the other organs tested. Erucic acid apparently was not present in the phospholipids. Similar results were obtained in dogs (*Bernhard*).

Partially hydrogenated rapeseed oil as 20 per cent of the diet of Wistar rats caused a slightly lower liver weight, but a significantly lower concentration of liver lipids than rapeseed or corn oil diets (*Beare*). In a subsequent study, *Beare* again found that dietary rapeseed oil does not influence the weight of the liver compared to other vegetable oils. Further, dietary fatty acids had more influence on the carcass fat than on the liver fat, where a large proportion of the fatty acids are constituents of the phospholipids which have a relatively constant composition. Erucic acid amounted to 5.6 per cent of the liver fatty acids compared to 11.3 per cent of the carcass fatty acids in rats fed rapeseed oil. *Craig* made a detailed examination of the tissue fatty acids of Wistar rats fed a diet containing 20 per cent rapeseed oil or corn oil for 21 weeks. The content of oleic acid was 48 per cent higher in the dietary corn oil than in the rapeseed oil, but in the liver the oleic acid concentration was highest on the rapeseed oil. Linoleic and linolenic acids showed a more direct relationship to the proportions of these acids in the dietary oils. Of the monounsaturated C_{20} and C_{22} acids present in rapeseed oil and essentially absent in corn oil, eicosenoic acid more closely reflected the content in the dietary fat than did erucic acid. Eicosenoic acid occurred in the greatest concentrations in the phospholipids, whereas erucic acid did not appear in this lipid fraction. From the fact that erucic acid was not found in tissue fat to the extent that it existed in the dietary fat and that an unusually high proportion of oleic acid was found in rats fed rapeseed oil, *Beare* and *Craig* suggested a conversion of erucic to oleic acid.

Beare added palm oil or methyl esters of fatty acids to rapeseed oil to study the effect of altering the intake of saturated fat on the fatty acid composition of rat organs. Liver weights did not differ significantly irrespective of the type of dietary fat. When rapeseed oil was present in the diet, the level of arachidonic acid in the liver was less than when the diet contained no erucic acid. Thus, it appeared that in the presence of erucic acid the conversion of linoleic acid to arachidonic acid was decreased. *Beare* studied the relative effects of rapeseed oil and corn oil in rats subjected to adrenalectomy, cold, or pyridoxine deprivation. The weights of the livers of rats maintained in the cold (4°C) were lower than those of unoperated or sham-operated animals at room temperature, irrespective of the dietary fat. Adrenalectomy did not influence the liver weight, whereas vitamin B_6-deficiency resulted in very small livers, the weight of which was not further affected by the type of dietary oil.

Effect of rapeseed oil on blood cholesterol

A possible relationship between dietary rapeseed oil or erucic acid and blood cholesterol has been studied showing very little effect of erucic acid on blood cholesterol and thus, only a few of the more pertinent data on this subject are mentioned below.

Carroll found no effect on the contents of free and esterified blood cholesterol of 25 per cent rapeseed oil in the diet of young rats compared to controls on a commercial diet. Similar results were obtained by

Carroll and *Noble* by adding 0.15 per cent oleic or erucic acid to a fat-free diet fed to male Sprague-Dawley rats for three weeks. Comparison of the effect of rapeseed oil on cholesterol metabolism in different species (*Carroll*) indicated no change in the blood cholesterol of rats, mice, rabbits and dogs, whereas a tendency to hypercholesterolemia was observed in guinea pigs and chickens.

Linko found a hypocholesterolemic action of rapeseed oil in man on a diet containing butter as the only lipid.

Beare compared the utilisation of rapeseed oil and corn oil in young and adult rats. They found no specific influence of either of the two oils on plasma cholesterol in young or adult Wistar rats. Similar results were obtained by comparing Wistar rats with Sprague-Dawley rats (*Beare*). On the other hand, partially hydrogenated rapeseed oil fed to Wistar rats resulted in significantly lower serum cholesterol levels than those obtained with rats fed on either rapeseed oil or corn oil. The highest serum cholesterol value were observed on the corn oil diet (*Beare*).

Horst found no influence on plasma lipids and cholesterol of diets containing cholesterol alone or with rapeseed oil although the liver cholesterol was increased. Similar results were obtained by *Rozynkowa*, but an almost fat-free diet or one containing hydrogenated rapeseed oil both caused a marked increase in serum cholesterol.

Pathological Changes Induced by Rapeseed Oil

In his studies on the longevity of rats fed on diets containing rapeseed oil or butterfat, *Thomasson* noticed slight indications of fat deposition and degeneration in the central parts of the liver lobules of some of the animals fed on rapeseed oil. The livers from rats on butterfat were normal. The increase of liver fat did not prevent the rapeseed oil group from attaining a longer life span. No important abnormalities were noticed in the other organs.

Murray found no histopathological symptoms in the lungs, heart, arteries, liver, kidneys and bladder of rats fed on a diet containing 5 per cent corn oil or the methyl esters of oleic, eicosenoic, or erucic acids for 12 weeks.

Young Wistar rats fed on 20 per cent rapeseed oil for 5 weeks had testes which were significantly smaller than the testes from rats on a similar diet with corn oil *(Beare)*. Histological examinations of the testes revealed no changes in cellular characteristics but a reduction in tubular size. Similar results were obtained in older rats.

Alexander and *Mattson* found no abnormalities on gross autopsy of the heart, lungs, liver, spleen, kidneys, thymus and thyroid from Holtzman rats fed on a diet containing 20 or 73 cal per cent gelatinised rapeseed oil. However, a slight hyperplasia was noticed in the adrenal of the animals that received the high level of dietary fat.

Homrowski found no histopathological changes in the liver, spleen, kidneys and adrenals of rats fed for 12 months on diets containing 10 per cent rapeseed oil or various types of heated fats containing rapeseed oil.

Effect of Rapeseed Oil on Vitamin Requirement

The liver storage of vitamin A was determined in young Wistar rats fed on a diet containing either 20 per cent rapeseed oil or corn oil. Vitamin A in doses of approximately 200 IU in 0.1 ml of rapeseed oil or corn oil was administered by mouth twice weekly (*Beare*). Analyses of the livers after 5 weeks showed that the storage of administered vitamin A was similar with both oils. Thus, there appeared to be no adverse effect of rapeseed oil on the absorption of vitamin A and consequently no indirect effect of the vitamin on weight gain.

Murray studied the effect of corn oil, olive oil and rapeseed oil on the rate of vitamin A depletion of liver and kidneys. Initial stores of vitamin A in the liver and kidneys were 92 ± 4 and 1.5 ± 0.1 IU, repectively. During the course of vitamin A depletion during two months following a single oral dose of 1000 IU, vitamin A in 0.1 ml of cottonseed oil, the total content of vitamin A in the liver was similar for each dietary oil. As the liver stores of vitamin A declined, the relatively small amount in the kidneys increased throughout the depletion period but less markedly in the group fed rapeseed oil. *Beare* and *Murray* conclude that rapeseed oil has only a slight influence on vitamin A metabolism.

Beare found that pyridoxine deprivation of rats had a more adverse effect on weight gain and food intake when the animals were fed on a diet containing 20 per cent rapeseed oil than when corn oil was used as dietary fat, although the total vitamin B_6 in the liver was independent of the type of dietary oil. Further, rats deprived of vitamin B_6 and fed rapeseed oil deposited a smaller proportion of fat than those fed corn oil. Erucic acid was somewhat more concentrated in the carcass fat of rats not supplied with pyridoxine than in those that were.

Carroll found the concentration of ascorbic acid in the adrenals to be somewhat lower in rats fed rapeseed oil than in the controls (fox chow), but the absolute amount was nearly the same in the two groups of animals.

Removal of one adrenal from normal rats produced a marked depletion in the concentration of ascorbic acid and cholesterol of the second adrenal. *Carroll* and *Noble* compared this response in rats fed on rapeseed oil to that of normal rats. They found that the lowering of ascorbic acid and cholesterol is much the same in both groups in spite of a marked difference in cholesterol content between the treated and control groups.

Metabolism of Erucic Acid

The preceding sections of this chapter have illustrated that feeding of rapeseed oil to experimental animals produces a variety of effects, growth retardation, occurrence of small amounts of erucic acid in body lipids, changes in cholesterol metabolism and reproduction disturbancies. Erucic acid is a very prominent fatty acid in rapeseed oil but is not a common acid in animal lipids except when it has been included in the diet. For these and several other reasons a number of experiments have centered on the metabolism of eruic acid.

Bernhard and *Vischer* isolated erucic acid from rapeseed oil and saturated it with deuterium, thus forming deuteriated behenic acid. The ethyl ester of this saturated fatty acid was fed to rats either as such or mixed with olive oil (1:1) for 7 and 8 days, respectively. Approximately 40 per cent of the dietary behenic acid was absorbed. Isolation of the fatty acids from liver, gut and body fat showed that stearic, palmitic, myristic and oleic acids were labeled with deuterium. The highest concentration of deuterium in the fatty acids isolated occurred in stearic acid. Behenic acid was found only in trace amounts in the body fat. *Bernhard* and *Vischer* interpreted their results as an indication that behenic acid is metabolised by β-oxidation.

Thomasson suggested that the long-chain fatty acids in rapeseed oil are metabolised to the same extent and in the same way as the C_{18}-acids, but that the rate at which metabolism takes place is different for the various fatty acids. *Hopkins* studied the deposition of eicosenoic and erucic acids in comparison to oleic acid when feed as the methyl esters to male and female Wistar rats. Eicosenoic and erucic acids were found to be deposited on the body fat. However, erucic acid was deposited inconsiderably in smaller amounts (circa 12 per cent of the fatty acids) than was eicosenoic acid (circa 28 per cent). Neither of the two acids appeared to be as readily deposited in body fat as the linoleic acid of corn oil. There was no evidence of conversion of eicosenoic acid to arachidic or stearic acids, or of erucate to behenic, arachidic or stearic acids in any appreciable amounts. Similarly, there appeared to be little or no conversion of eicosenoic acid to oleic acid or of erucic acid to eicosenoic acid. *Wagner* kept adult rats on a rapeseed oil diet for 6 months. A true accumulation of erucic acid could not be determined. The decrease of the biological half-life or erucic acid after termination of the rapeseed oil diet, was calculated to be about 18–31 days in various organs and depots. *Wagner* concluded that erucic acid behaves like fatty acids normally utilised in metabolism.

Effect of Rapeseed Oil on Longevity

A classical general parameter of a diet being sufficient and well balanced is its effect on growth rate. It should be pointed out that parameters such as growth and weight gain must be used with forethought in nutritional experiments. In principle weight gain can take place throughout the whole life span. Such a situation is not desirable though and especially not if good health and longevity are considered simultaneously.

Longevity, or expected longevity, has been studied in relation to various food components, proteins of animal or vegetable origin, dietary fat, the dietary ratio of protein and carbohydrates and *ad libitum* versus restricted access to food. A number of valuable references to some of the early work on longevity in relation to various food constituents is given by *French*. The present discussion is confined to dietary fat and longevity.

French fed male and female Wistar rats on a stock diet alone, or 80 per cent stock diet plus 20 per cent sucrose, or 20 per cent corn oil. The male and female rats grew well on the diet high in fat, but the life span of the males decreased markedly and there was also a less pronounced but significant decrease in the longevity of the females. Increased efficiency of utilisation of the diet was correlated with decrease in life span, but increased caloric intake per cent was not so correlated. The histopathological data and the frequency of disease and abnormalities, however, revealed nothing that could be interpreted as a fat-induced cause of premature death.

Roine and *Uksila* have shown that rapeseed oil at the 30 cal per cent level did not cause any pathological changes in long-term experiments. It may also be mentioned that *Murray* found that weight gain in vitamin A deficient rats was reduced somewhat earlier when rapeseed oil was fed, but the survival time of rats fed corn oil or rapeseed oil was similar.

Thomasson studied longevity in mice which were fed on diets containing 10, 20, 40, 50, 60 and 72 cal per cent rapeseed oil or butterfat over a period of 89 weeks. For animals in all groups receiving 40 cal per cent or more fat in the diet, the 50 per cent mortality was reached significantly later with rapeseed oil than with butterfat. At the end of the experiment less than 50 per cent of the mice fed on a diet with 10 cal per cent rapeseed oil had died.

In general it may be stated that the factors determining life span of man and experimental animals are multifarious and the type and amount of dietary fat may be two of them.

NUTRITIONAL VALUE AND USE OF RAPESEED MEAL

Seed of rape *(Brassica napus)*, turnip rape *(Brassica campestris)* and white mustard *(Sinapis alba)* contain ca. 47, 44 and 35 per cent oil, respectively. After removal of the oil from the seed by a pre-press and solvent extraction procedure a protein-rich meal is obtained.

The nutritional value of a food or feed is dependent on several factors: protein content, amino acid composition, crude fibre content, content of minerals, vitamins and the possible presence of toxic substances. In the following sections each of these factors will be considered.

Protein Content

Some data on the chemical composition of rapeseed meals are given in Tables 9.10 and 9.11.

Table 9.10. Compostion of rapeseed meal and soya meal.

Feedstuff		*Chemical composition, % of dry matter*						
	Dry matter %	*Protein (N × 6.25)*	*Fat*	*N-free extract*	*Crude fibre*	*Ash*	*Ca*	*P*
Rapeseed meal:								
B. napus	89	43.1	2.3	36.9	10.7	7.0	0.8	1.2
B. campestris	89	41.1	2.0	37.8	11.6	7.5	0.6	1.3
Sinapis alba	89	45.5	1.2	36.2	9.8	7.3	0.8	1.2
Soya meal	87	50.4	0.5	35.4	6.9	6.8	0.3	0.7

Table 9.11. Vitamin content of rapeseed meal and soya meal (mg per kg dry matter).

Feedstuff	*Niacin*	*Pantothenic acid*	*Riboflavin*	*Thiamine*
Rapeseed meal				
Expeller	168	17	3	9
Soya meal				
Expeller	23	15	4	4
Solvent	24	20	4	9

Although the protein content of rapeseed meal is less than that in soya meal, it represents the most concentrated vegetable protein feed produced in the cool temperate zones. Turnip rape appears to be a little lower in protein content than rape.

It should be noted that protein content can be greatly influenced by environmental factors. A large amount of nitrogen fertilisers increases the protein content.

Amino acid composition

Amino acid composition of the commercial meal is largely influenced by the way in which the meal is produced from the seed. Previously the oil was pressed from the seed in an expeller pressing mill and in order to obtain a maximum oil yield the crushed seed was exposed to high pressures and temperatures. This resulted in extensive destrucition of lysine, often too low in the diet. In modern processing the meal is usually produced by a combination of pre-pressing and solvent extraction or by sovent extraction alone. Since lysine is often the first limiting amino acid in feed of vegetable origin, this represents an important improvement in the nutritional value of rapeseed meal.

In Table 9.12 the amino acid compositions of rape, turnip rape, white mustard and soya meals are shown. The contents of most of the amino acids in the first three meals are similar to those in soya meal.

Besides lysine, methionine is the amino acid most often lacking in vegetable feed and therefore the contents of these two amino acids are of great importance. The lysine content of rapeseed meal is slightly lower than that in soya meal, whereas the methionine content is higher in rapeseed meal. Since cystine can in part replace methionine, the higher content of this amino acid in rapeseed meal is also of importance. According to the provisional pattern of FAO, isoleucine and methionine are the limiting amino acids of rapeseed.

Table 9.12. Amino acid composition of meals of rape, turnip rape, white mustard and soyabean (grams amino acid per 16 GN).

Description	*Rape*	*Turnip rape*	*White mustard*	*Soyabean*
Reference:	8	8	9	3
Number of samples:	4	4	1	27
Amino acid:				
Arginine	5.6	5.9	6.1	6.9
Histidine	2.6	2.7	2.9	2.7
Lysine	5.8	6.1	5.9	6.1
Tyrosine	2.6	2.7	3.3	2.9
Tryptophan	–	–	–	1.0
Phenylalanine	3.5	3.8	4.1	5.0
Cystine	2.4	2.4	–	1.7
Methionine	1.8	1.9	1.7	1.5
Serine	3.7	3.8	4.3	5.2
Threonine	3.8	4.0	4.6	4.0
Leucine	6.3	6.5	7.3	7.6
Isoleucine	3.7	3.8	4.1	4.8
Valine	4.8	4.9	5.6	5.0
Glutamic acid	16.6	16.8	18.0	20.6
Aspartic acid	6.2	6.6	7.9	11.9
Glycine	4.3	4.6	6.1	4.4
Alanine	3.9	4.0	4.5	4.5
Proline	6.4	6.7	6.4	–

Carbohydrates

Only a few studies have been made concerning the carbohydrates of rapeseed meals. When the content of protein, fat, crude fibre and ash in the meal have been determined the rest is designated as 'N-free extracts' in the conventional feed analysis. This is a poorly defined fraction mostly containing carbohydrates and 'crude fibre' is also an incompletely defined fraction. The crude fibre content is higher in rapeseed meal than in meals of most other oil crops, which may cause a lowering of the metabolisable energy.

Hrdlicka and others studied the saccharides in rapeseed meals as affected by different processing methods: laboratory extraction according to *Soxhlet*, or extraction under industrial conditions according to a battery method or according to the methods of *Bollmann, Lurgi* or *DeSmet*. They found no appreciable differences in crude fibre, *n*-free extract or pentosan content caused by extraction methods, but some differences in starch content were observed. Four samples extracted by the Soxhlet method gave the mean values 10.6 per cent crude fibre, 9.9 pentosans and 6.4 starch. The corresponding values for commercial methods were on the average, 10.7, 10.0 and 5.8 per cent respectively. The lower starch contents found in meals produced by the methods of Lurgi and Bollmann were assumed to be caused by conversion of starch to dextrins. The mean value for reducing sugars was 4.1 per cent in the meals produced according to Soxhlet, while it was only 3.2 per cent in the meals processed by the commercial procedures. When the meal was processed according to the method of DeSmet, the content of reducing sugars was not appreciably different from that of the Soxhlet-extracted meal. On the other hand, when the meal was extracted by the battery method the content was only 2.5 per cent. This was explained as a result of the Maillard reaction, in which condensation products are formed by reaction between basic amino acids, especially lysine and reducing sugars.

Besides a lower content of reducing sugars it results in lower contents of basic amino acids and a browning of the meal. Since the content of available lysine will be considerably lowered by this reaction it is detrimental to the nutritional value of the meal and thus it is important to process the meal under conditions (time and temperature) that do not favour the Maillard reaction. When the rapeseed meal had been hydrolysed before the determination of reducing sugars, the meals produced according to Soxhlet were found to contain 11.8 per cent of these compounds, whereas the meals extracted by commercial methods contained 9.5 per cent. A great part of this sugar was derived from the glucosinolates and since a high content of reducing sugars favours the Maillard reaction it is important to inactivate the glucosinolate-splitting enzyme, myrosinase, at an early stage of the extraction process.

Content of Minerals and Vitamins

As shown in Table 9.10 the ash content of rapeseed meal is similar to that of soya meal, while the contents of calcium and phosphorus are higher in rapeseed meal.

The figures in Table 9.11 indicate that rapeseed meal contains more niacin than soya meal, while the contents of pantothenic acid, riboflavin and thiamine are similar. According to *Klain*, rapeseed meal contains about three times as much choline as soya meal (6730 compared with 2170 mg/kg).

Glucosinolates

Chemical properties

The most limiting factor for the use of rapeseed meal as feed seems to be its content of glucosinolates. Earlier these compounds were most often named mustard oil glucosides, but later they were called thioglucosides, an expression which is sometimes still in use.

According to *Guignard* the glucosinolates are distributed diffusely in the parenchymal tissues of the seed. An enzyme system, myrosinase, with the ability to split glucosinolates, seems to be separated from the glucosinolates in the intact tissue by being located in special cells. If the cells are disintegrated and the water content is sufficiently high the glucosinolates will be spilt by the myrosinase, giving rise to glucose, sulphate and isothiocyanates. Figure 9.19 shows this reaction for 3–butenyl glucosinolate, one of the most predominant glucosinolates in rapeseed.

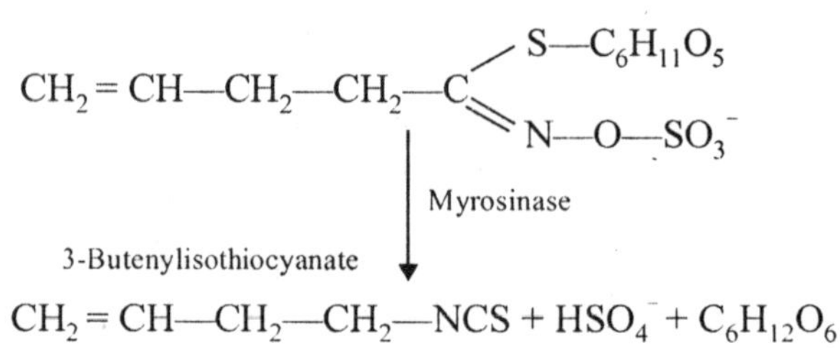

Fig. 9.19. Enzymic hydrolysis of 3-butenylglucosinolate.

The glucosinolates which have been reported to occur in meals of rape and turnip rape are enumerated in Table 9.13.

Table 9.13. Glucosinolates present in seed of rape (*Brassica napus*) or turnip rape (*B. campestris*).

Glucosinolate
3-Butenylglucosinolate
4-Pentenylglucosinolate
4-Methylthiobutylglucosinolate
5-Methylthiopentylglucosinolate
4-Methylsulphinylbutylglucosinolate
5-Methylsulphinylpentylglucosinolate
2-Phenylethylglucosinolate
2-Hydroxy-3-butenylglucosinolate
2-Hydroxy-4-pentenylglucosinolate

Two of these glucosinolates contain an OH-group connected to the second carbon atom of the aglucone. After the enzymic splitting of the glucosinolate molecule these aglucones cyclise spontaneously and rapidly, giving rise to 5-vinyloxazolidine-2-thione (Fig 9.20) and 5-allyloxazolidine-2-thione respectively.

In *Brassica napus* 2-hydroxy-3-butenylglucosinolate (progoitrin) is the predominant glucosinolate, with 3-butenylglucosinolate (gluconapin) in second place. In *B. campestris* 3-butenylglucosinolate predominates, while glucosinolates yielding oxazolidinethiones are most often low in amount, or lacking, in this species. Figures for the range in glucosinolate content of the seed meals of some varieties cultivated in Europe are given in Table 9.14.

Table 9.14. Range in glucosinolate content in seed meals of some Cruciferae oil crops cultivated in north or middle Europe.

The formula weight of the potassium salts of the glucosinolates has been used in the calculations with the exception of *Sinapis alba* for which the formula weight of the *p*-hydroxybenzylglucosinolate ion was used.

Species and type	*Number of varieties*	*Number of samples*	*Range in glucosinolate content (% in dry matter)*
Brassica napus			
Winter type	16	43	3.39–7.77
Summer type	5	15	3.72–5.75
Brassica campestris			
Winter type	4	8	3.09–4.30
Summer type	2	7	3.01–3.59
Sinapis alba	5	23	7.91–10.12

Only *p*-hydroxybenzylglucosinolate has been shown to occur in white mustard seed. The hydrolysis product *p*-hydroxybenzyl isothiocyanate is relatively labile in neutral and still more in alkaline solutions, giving rise to *p*-hydroxybenzyl alcohol and thiocyanate ion (Fig. 9.21).

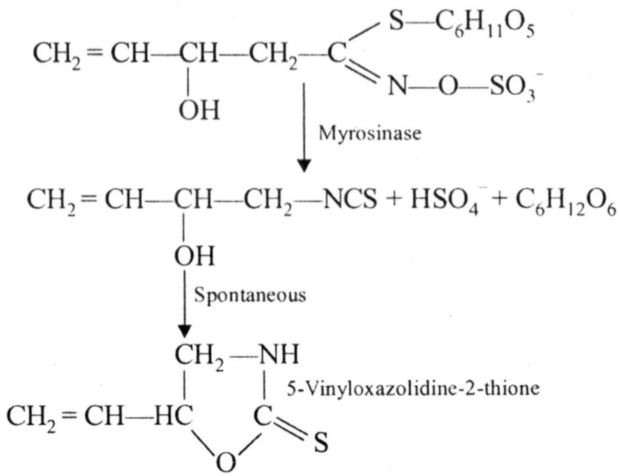

Fig. 9.20. Enzymic hydrolysis of 2-hydroxy-3-butenylglucosinolate and the conversion of the aglucone to oxazolidinethione.

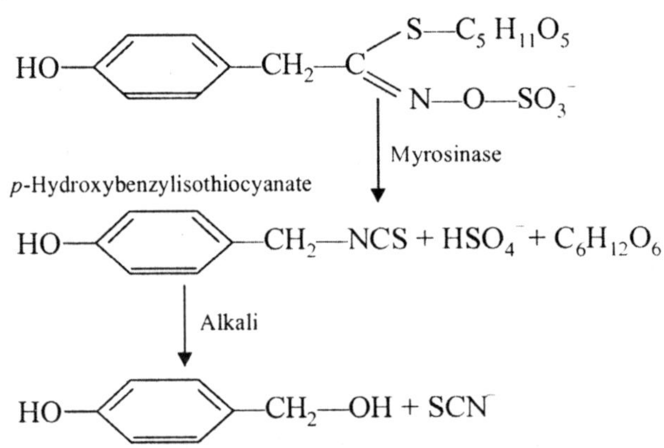

Fig. 9.21. Enzymic hydrolysis of *p*-hydroxybenzylglucosinolate, showing products formed and further conversion in alkaline solution.

3-Butenyl-, 4-pentenyl- and phenylethyl-isothiocyanate are volatile with steam. This may be of practical importance for meals containing mainly these glucosinolates, since the cleavage products after hydrolysis may be removed by steaming.

The glucosinolates are easily soluble in water and in ethanol, methanol and acetone. They are more soluble in 10–20 per cent water mixtures of these solvents than in the pure solvents. The isothiocyanates and oxazolidinethiones are soluble in ethyl ether and in chloroform. Steam volatile isothiocyanates are lipophilic and soluble, for instance in hexane, while the non-volatile isothiocyanates are less lipophilic and the oxazolidinethiones relatively hydrophilic.

Toxicity of various cleavage products

According to *Belzile* presence of myrosinase is important for the undesirable properties of rapeseed meal. The products formed by the enzymic reaction, isothiocyanates, oxazolidinethiones, nitriles and thiocyanate ions show harmful effects. The effect of the oxazolidinethiones seems to be that most studied.

Kennedy and *Purves* reported that *Brassica* seed fed to rats produced enlarged thyroids in these animals and they also showed that this effect could not be wholly counteracted by feeding iodine. When the rats were hypophysectomised their thyroid glands weighed only ca. 6 mg as compared to ca. 44 mg for the intact controls. These results indicated that rapeseed causes thyroid enlargement by interfering with the synthesis of thyroxin, whereupon the anterior pituitary is stimulated to produce TSH (thyroid-stimulating hormone) which acts on the thyroid glands causing hypertrophy and hyperplasia. Recent results by *Matsumoto* suggest that oxazolidinethione also interferes with the secretion of the thyroid hormone in the thyroid gland into the blood. *Frolich* reported that more than 5 per cent rapeseed meal fed to growing chicks caused thyroid hypertrophy and depressed growth. *Nordfeldt* found similar results when rapeseed meal was used at levels of 10–20 per cent in swine diets. No evidence has been reported for goitrogenic effects in cattle.

Subsequent to the work of *Kennedy* and *Purves*, several authors have shown by experiments on various animal species that the goitrogenic effects of rapeseed meal can only be partly corrected by feeding iodine. Use of iodinated casein or thyroxine counteracted the goitrogenicity of rapeseed when added to the diet of rats, poultry, or pigs, but the growth-depressing effect of the rapeseed meal was not always prevented. However, we now have reason to believe that when growth depression was noted, it was probably due to heat damage to the protein of the meal during processing or to failure to adjust the metabolisable energy value of the experimental feeds.

The thiocyanate ion which might be split off from the *p*-hydroxybenzyl isothiocyanate produced in white mustard seed meal is also goitrogenic. It exercises its effect by competition with the iodine uptake in the thyroid gland and can be counteracted by feeding iodine.

Several times palatability problems have been reported from feeding with rapeseed meals, which may be caused by several of the cleavage products of the meal as well as by other substances. Since the isothiocyanates have pungent tastes and odours they may be greatly responsible for the palatability problems. In high concentrations they damage skin and mucous membranes and by injuring the surface of the alimentary canal they may depress the weight gain. *Bell* has found that the reduction on weight gain of mice fed with rapeseed meal is approximately correlated with the sum of contents of glucosinolates yielding oxazolidinethiones and isothiocyanates. In a recent study *VanEtten* fed rats on a diet containing 10 per cent crambe seed meals that had been treated in various ways. A meal containing *epi*-progoitrin, however, with the enzymes destroyed by heat treatment, gave a weight gain that was 77 per cent of that of the control group. When the enzymes had not been destroyed the growth was poorer or the rats died. Meals containing active enzymes and autolysed under such conditions that mainly oxazolidinethione was formed yielded 85 per cent growth, while meals autolysed to form mainly nitriles caused death of the rats. Rats fed meals containing active enzymes or meals autolysed to contain nitriles showed enlargement and microscopic lesions in the liver and kidneys. Since it seems probable that nitriles formed from the glucosinolates of rapeseed behave similarly, more attention should be paid in the future to the production of nitriles in rapeseed meal.

Counteracting the Toxicity

Enzyme inactivation

Since the glucosinolates do not seem to be toxic a meal with the enzymes completely destroyed should not be toxic if it is not exposed to enzymes from other sources. In modern procedures of extracting rapeseed oil a heat treatment is included, which causes inactivation of glucosinolate splitting enzymes. After the seed has been crushed it is heated at approximately 80°C. This enzyme inactivation step is important not only for the quality of the meal but also for the quality of the oil. If the myrosinase is not destroyed in an initial stage of the extraction process extensive enzymic splitting of the glucosinolates would occur. Hydrolysis products, for instance isothiocyanates, would dissolve in the oil and poison the catalyst used for its hydrogenation.

The effect of the heat treatment on the myrosinase inactivation depends on temperature, humidity and time of treatment. *Josefsson* found in laboratory studies that the myrosinase of seed with 8 per cent moisture content was completely inactivated by heating at 90°C for 15 minutes. It is, however, possible that the animals may get myrosinase from other sources than the meal, for instance by eating fresh parts of Cruciferae plants.

Although myrosinase inactivation does not completely inhibit the hydrolysis of glucosinolates, it greatly improves the meal. This was demonstrated by *Bell* in a swine feeding experiment, in which he added a myrosinase source to an inactivated rapeseed meal and showed that the growth inhibition was three times larger than without addition of myrosinase.

Technical methods of removal or destruction of glucosinolates or cleavage products

Autolysis and distillation

Goering added water to meal of *Brassica juncea* (brown or yellow mustard) together with crushed seed as a source of myrosinase. The glucosinolates were split during this treatment and steam volatile isothiocyanates were removed by distillation. When the product obtained replaced 50 per cent of the soya meal in a diet for rats, the nutritional value of the diet was found to be at least as good as with soya meal alone. *B. juncea*, however, contains only glucosinolates that yield steam-volatile isothiocyanates, while only *ca*. 30 per cent of the glucosinolates of rapeseed (*B. napus*) meal yield volatile products. In the winter turnip rape variety most cultivated in Sweden approximately 90 per cent of the products are volatile. If it is possible to distinguish between *B. napus* and *B. compestris* seed in the trade, an improvement of the *B. campestris* meal might be obtained by using the distillation technique.

Autoclaving and steam stripping

Bell and *Belzile* reported that most of the glucosinolates were destroyed in a meal that had been autoclaved for 60 minutes at a pressure of 1.2 kg/cm^2. When the meal was fed to rats, the animals showed normal weight gains, demonstrating that the main part of the toxicity had been removed.

In another study *Bell* and *Belzile* passed steam through rapeseed meal which at the same time was maintained at a temperature of 110°C. Meal that had been treated in this way for at least 1 hour caused good growth responses when tested in diets for mice.

Bell and *Belzile* also pointed out, however, that there was a gradual deterioration in protein quality as the time of autoclaving or steam stripping was extended. Meal that was autoclaved for 2 hours at a pressure of 1.2 kg/cm^2 did not support growth when used as the only protein source in an otherwise adequate diet. It seems probable that this was an effect of lysine destruction, but this was not proved.

Treatment with ammonia and heat

Kirk reported that treatment of meal of *Crambe abyssinica* with ammonia and heat destroyed the glucosinolates effectively. Preliminary feeding test indicated a great improvement of the nutritional value. In recent experiments, however, there have been palatability problems on feeding ammonia-treated meal to cattle.

Soda cooking

Mustakas treated crushed seed, or defatted meal of crambe, by cooking with sodium carbonate under pressure. The glucosinolates were destroyed completely by this treatment, but the reaction products formed remain to be elucidated. Feeding tests with cows demonstrated that the palatability of the soda-treated meal was better than that of untreated or ammonia treated meal.

In feeding experiments with poultry the soda-treated as well as the ammonia treated meal was more palatable than meal which was only toasted. The palatability was, however, not quite satisfactory when the ratio of rapeseed meal in the feed was as high as 20 per cent. Since the soda treatment is inexpensive it may be of practical value.

Treatment with ferrous salts

Youngs and *Perlin* found that when certain metallic salts, especially ferrous sulphate, were added to Brassica seed meals, the glucosinolates decomposed to yield nitriles. When this treatment was followed by steam treatment the nitriles from the glucosinolates of *B. juncea* were removed, while nitriles formed from the glucosinolates of *B. napus* and *B. campestris* were not completely removed. Since the toxicity of the nitriles is still higher than that of isothiocyanates or oxazolidinethiones the feeding tests did not show any improvement in the nutritional value as a result of the ferrous sulphate treatment.

Extraction of the toxic factors

Extractions aimed at removing glucosinolates have been made partly with water or with buffers in water solution and partly with various organic solvents.

Rapeseed meal has been extracted with cold or warm water or with buffers. By extraction of myrosinase-free meal with buffer, *Belzile* and *Bell* could remove more than 80 per cent of the glucosinolates. When the meal contained active myrosinase the water extraction was less effective since the cleavage products are less water-soluble than the intact glucosinolates.

In bio-assays which involved feeding the myrosinase-active extracted meals to mice the weight gains of the animals were less than expected according to the assays of isothiocyanates and oxazolidinethiones. This may indicate that these meals contained other split products from glucosinolates which were harmful to the animals.

In a recent study myrosinase-free meals of rape and white mustard were extracted with 80 per cent ethanol. The results of these extractions on the bio-assays with mice are shown in Table 9.15.

Table 9.15. Effect of meals of rape (*Brassica napus*) and white mustard (*Sinapis alba*), extracted with 80 per cent ethanol, on mice growth and protein efficiency.

Protein	*Gain 20 days (g)*	*Consumed feed (g)*	*Gain per g protein*
Egg albumin	10.0	53	1.90
Casein	9.5	57	1.65
Brassica napus seed meal	9.1	49	1.87
Brassica napus seed meal, supplemented with methionine and isoleucine	9.5	50	1.89
Sinapis alba seed meal	7.6	50	1.58
Sinapis alba seed meal, supplemented with methionine and isoleucine	9.0	50	1.81

The weight gain per gram protein of the mice fed on rapeseed meal extracted with 80 per cent ethanol exceeded that of the mice fed on casein.

Attempts at removing glucosinolates, combined with chemical analyses and bio-assays of the products with rats, have also been performed on seed meals of *Crambe abyssinica.* The glucosinolate content of dehulled seed of crambe is on a level similar to that of winter rape or somewhat higher. More than 90 per cent of the crambe seed glucosinolates are composed of epi-progoitrin that yields oxazolidinethione on enzymic hydrolysis under suitable conditions. When acetone was used as solvent the water content had to be at least 20 or 25 per cent, if the glucosinolates were to be removed effectively. An increase in the water content of the acetone, however, resulted in a greater extraction of dry matter from the meal. When crambe meal extracted with 80 per cent acetone was fed to rats at a 23 per cent level the gains in weight were close to normal. On the other hand when untreated crambe meal was fed to rats at a 25 per cent level all the rats died within 21 days.

Extraction procedures of crambe meal or rapeseed meal with water or water solutions of ethanol or acetone have generally resulted in a loss of approximately 20 per cent of the dry matter of the meal. Approximately one-fourth of this represents glucosinolates. As indicated above the losses of dry matter in crambe seed meal were less when the meal was extracted after hydrolysis. This procedure has, however, been shown to be not very successful when reproduced on rapeseed meal. The losses of dry matter from meal in combination with the costs for evaporation of the solvent used in extraction methods seem to cause these methods to be too expensive for preparing meals for animal feeding.

If the detoxified meal product is intended to be used for human consumption it may be sold at a price which can bear higher production costs. *Eapen* have studied the possibilities of making a meal that could be used for this purpose. In his procedure he started with an enzyme inactivation by cooking the seed in water for two minutes. Glucosinolates are removed by aqueous extraction prior to or after the defatting process. The crude fibre content of the meal is lowered by removing the seed coat by air classification, a procedure that is facilitated by the previous hot water treatment of the seed. A white glucosinolate-free meal, with an increased content of protein and a decreased content of crude fibre is obtained according to this method.

Isolation of the protein

It should be possible to produce protein isolates of rapeseed meal according to procedures similar to those used for soya meal today, in which the proteins are extracted into a buffered water solution. The extracted proteins are then precipitated by adjustment of the pH to the point at which most of the proteins in question have their lowest solubility. After filtration or centrifugation the precipitate is dried. The losses of protein in these procedures amount to 30–40 per cent and the protein content of the protein isolate obtained is ca. 90 per cent. Although the price per kg protein obtained as isolate is considerably higher than the price per kg protein in meal because of losses and manufacturing costs, it is still low compared to the price of animal proteins.

Removal of glucosinolates by plant breeding methods

There are two requirements that must be fulfilled if the chemical composition of a plant material is to be changed by plant breeding methods. First, analytical methods must be available which are sufficiently accurate, precise, sensitive and suitable for serial analysis so that a large number of individual plants can be analysed. Secondly, there must exist a plant material that deviates genetically in the direction of the desired property.

An accurate analytical method used for glucosinolates in *Crambe abyssinica* was published by *VanEtten*. The method was based on photometric titration of enzymatically released suphate and consequently was a determination of total glucosinolate content. It was combined with the determination of steam-volatile isothiocyanates. *Appelqvist* and *Josefsson* published another method including a separation of isothiocyanates and oxazolidinethiones by selective extraction and the determination of these two components by spectrophotometric measurements.

A method that included gas chromatographic determinations of individual isothiocyanates and spectrophotometric determination of oxazolidinethiones was published by *Youngs* and *Wetter*. By the use of these methods it was possible to study a large number of samples including different species, varieties, lines and individual plants. It was found independently at the Prairie Regional Laboratory of the National Research Council at Saskatoon, Canada and at the Swedish Seed Association, Svalöf, Sweden that the Polish summer rape variety *Bronowski* had a genetically determined glucosinolate content that was much lower than the average.

Although this variety is the one which contains the lowest amounts of glucosinolates known hitherto in rapeseed, other varieties have also been found that are considerably lower in glucosinolate content than the average. Some results on material low in glucosinolates found at the Swedish Seed Association are shown in Table 9.16.

Table 9.16. Examples of low content of glucosinolates in seed meal of rape (*Brassica napus*) and *turnip rape* (*B. campestris*).

Species and variety	*Glucosinolate content % in dry matter*
Brassica napus, summer rape, cv., Ukraine	3.03
Brassica napus, summer rape, cv., Bronowski	0.64
Brassica napus, summer rape, selected plant from Bronowski	0.11
Brassica campestris, summer turnip rape	1.76

Kondra has found that there exist qualitative as well as quantitative differences in glucosinolate pattern between varieties. This may open possibilities for obtaining varieties without glucosinolates or low in such compounds by a plant breeding programme.

The seed meal of Bronowski has been compared with meal from normal varieties in mice tests. Preliminary results of such tests are shown in Table 9.17.

Table 9.17. Effect of different rapeseed meals on mice growth.

Protein	*Gain 20 days (g)*
Casein	11.2
Meal from summer rape, cv., Rigo, no heat treatment	–1.4
Meal from summer rape, cv., Bronowski, no heat treatment	5.6
Egg albumin	10.4
Meal from summer rape, cv., Rigo	
The seed was heat treated at 90°C for 15 minutes before defatting	8.9
Meal from summer rape, cv., Bronowski	
The seed was heat treated at 90°C for 15 minutes before defatting	10.6

Similar results have been obtained by *Bell* who only used heat-treated meals. The reason for the relatively large growth inhibition due to the meal of Bronowski that was not derived from heat-treated seed still remains to be elucidated. It does not seem probable that isothiocyanates and oxazolidinethiones released from the glucosinolates of the Bronowski meal would have resulted in such a growth inhibition since *Bell* did not obtain any significant difference in weight gains when he fed mice on heat-treated meal with and without added myrosinase. Since it has been found that glucosinolates in meals that have not been heat-treated often give rise to large amounts of nitriles these highly toxic products might have been released in the digestive tract of the animals.

The seed yield of Bronowski is considerably lower than that of the modern Swedish summer rape varieties. It is also important to obtain winter varieties with a low glucosinolate content. For these reasons Bronowski is now used in plant breeding programmes at several European institutes, being crossed with high yielding summer rape varieties as well as with winter varieties which are high yielding and winter resistent.

Digestibility

Rutkowski and *Kozlowska* reported that the digestibility coefficients of rapeseed meal protein obtained in experiments were similar to those of other oil seed meals. They also found that processing methods favouring formation of melanoids caused a lower digestibility coefficient.

Jarl reported a digestibility coefficient *in vivo* (ruminants) for the organic matter of rapeseed meal as 76 per cent and for crude protein 83 per cent. *Nordfeldt* found corresponding values of 76 and 85 per cent.

When *Bell* replaced equal parts of linseed and soya meal with *Brassica campestris* meal in a diet for swine he found that the energy and protein components of this meal were at least as digestible as those in the soya-linseed meal mixture. In studies with chicks, *Lodhi* obtained nitrogen absorbability values of 80 and 85 per cent for rapeseed meal and soya meal, respectively.

Rutkowski found a coefficient of digestibility of the rapeseed meal protein that was ca. 10 per cent lower in experiments *in vivo* (poultry) than *in vitro*. This might have been due to the glucosinolate content.

Feeding Value

Ruminants

Fewer studies have been made concerning the feeding value of rapeseed meal for ruminants than the number of studies made on swine and poultry. The reasons for this may be that studies with cattle are relatively expensive to perform and that sheep are not a large animal group in the rapeseed growing areas. The fact that the problems of feeding rapeseed are less for ruminants than for non-ruminant animals may be another reason. In contrast to swine, poultry and other non-ruminants, the ruminants do not develop goiter when fed on rapeseed meal, nor do there seem to be any toxic effects produced on these animals. The reason for this was clarified in an experiment by *Virtanen* in which 500 mg vinyl oxazolidinethione was goiter when fed to a cow and it was found that only 0.05 per cent of this quantity was transferred to the milk. Approximately two-thirds of this amount was transferred to the milk. Approximately two-thirds of this amount was transferred within two hours. Therefore, the substance had apparently passed through the rumen wall into the blood. When the cow was fed crystalline progoitrin only traces of oxazolidinethione were found in the milk, thus demonstrating that the progoitrin was not hydrolysed in the rumen or the oxazolidinethione formed was destroyed there. It has recently been reported that feeding rapeseed meal to sheep affected the thyroid gland.

Since the cleavage products of glucosinolates do not exercise toxic effects on ruminants, rapeseed meal is mainly used for these animals.

Young ruminants (below 6 months)

Few studies have been made on feeding rapeseed meal to young calves and lambs. In the studies made, rapeseed meal was compared to linseed meal. Generally rapeseed meal was less palatable initially than linseed meal, but after the initial stage there were no differences in feed consumption between the rapeseed groups and the linseed groups and the weight gains were equal.

Growing and fattening ruminant animals

The results of the studies made on growing and fattening ruminant animals are similar to those for the youngest ruminants: there seem to be some palatability problems in the beginning of the feeding period, but the animals consume their rations and gain in weight at approximately the same rate as on other oilseed meals.

Milk cows

In experiments with milk cows *Jarl* daily fed the animals ca. 2.5 kg of an oilcake mixture; containing 0, 25, 50 and 60 per cent rapeseed meal. There were some palatability problems in the beginning of the feeding period but not when the cows had become accustomed to the rapeseed. The milk production was less when 50 or 60 per cent rapeseed was used compared to 0 or 25 per cent. There was a small decrease in the fat content of the milk in all the combinations with rapeseed meal. The greater part of this decrease was explained, however, by the fact that the feed without rapeseed meal contained coconut oil. *Jarl* concluded that 2 kg of rapeseed meal could be given per cow per day and suggested that the meal should be given in a dry state.

Nordfeldt compared meals of rapeseed and white mustard seed with two different fat levels (ca 1.5 and 7.0 per cent) with soya meal. The quantities of rapeseed and white mustard meals amounted to ca 1.5 kg per cow per day. 1.2 kg soya meal was given to the cows in the control group. The meals of rapeseed and white mustard seed were less palatable than soya meal, especially in the beginning of the experimental period. Rapeseed meal gave a higher milk production than soya meal, but also a lightly lower fat content of the milk. Therefore, there were no differences between groups in fat production, nor were there any differences in the taste of the milk produced.

Whiting concluded that rapeseed meal can be considered to be equivalent in nutritional value (on an equivalent protein basis) to linseed and soya meals when it makes up to 10 per cent of the total dry matter of the ration.

Swine

For non-ruminant animals such as swine the amino acid composition of the protein fed is more important than for ruminants. When the amino acid composition of rapeseed meal is compared to the amino acid requirements of the young pig it is found to be quite favourable. According to results of feeding experiments, swine and especially young ones, are relatively sensitive to the cleavage products of glucosinolates, which cause goiter and growth inhibition. Therefore, in Sweden the ratio of rapeseed meal in concentrated feed for swine has most often been below one per cent. It seems possible that this figure is less than necessary. *Bowland* concludes that up to 4 per cent of the total ration may be composed of rapeseed meal for young pigs up to 25 kg in weight. Market pigs from 25 to 90 kg liveweight could efficiently use up to 10 per cent rapeseed meal in the ration. For mother-sows during pre-gestation and lactation rapeseed meal should not be fed at a level above 3 per cent of the total ration. Recently *Schuld* and *Bowland* reported that when 8 per cent rapeseed meal, substituted on an isonitrogenous basis for soya meal and wheat, was fed to young sows, there was a decreased number of pigs weaned per litter. In the second reproductive cycle, however, no differences were found that were caused by differences in diet. The same authors studied the progenies from these sows and found that the feed given to the sows did not influence the progeny. Eight per cent rapeseed meal given to the progeny depressed the rate of gain and efficiency of feed utilisation in one experiment, but not significantly in another one.

Results of experiments with market pigs indicate that the levels of rapeseed meal recommended will have no adverse effects on carcass characteristics.

Poultry

It is known from several reports that feeding of rapeseed or rapeseed meal to poultry causes thyroid enlargement. *Clandinin* concluded that the initial effects of this feed on the thyroid gland of poultry includes a decrease in the uptake of iodine by the gland and an increase in the release of iodine from the gland. After the poultry have become accustomed to the rapeseed meal (3–4 weeks), the uptake of iodine by the enlarged thyroid gland is increased while the secretion rate from the glands becomes normal. Although the thyroid gland will still be enlarged, a physiological equilibrium seems to have been reached.

Extensive studies of the effect of rapeseed feeding on poultry have been performed by *Clandinin* and others. When comparing expeller-processed rapeseed meals with soya meal, they found that the former resulted in decreased growth rate and depressed feed efficiency, even when only small amounts of the supplementary protein in the ration was composed of rapeseed meal. When the meal was processed at lower temperatures the nutritional value increased. Amino acid analyses revealed that the poor results, of feeding expeller-processed rapeseed meal largely reflected lysine destruction. In a series of experiments *Clandinin* established that the growth rate and feed conversion of chicken were the same when 15 per cent prepress-solvent rapeseed meal was used in the diet or when a diet isonitrogenous and isocaloric with soya meal was used. Therefore he concluded that 10–15 per cent rapeseed meal could be used in chick starter rations.

With regard to laying and breeding chicks and turkeys *Clandinin* concluded that 10 per cent rapeseed meal yields just as satisfactory production, feed conversion, fertility and hatchability as corresponding amounts of protein from soya meals.

Characteristics for rapeseed oil are flown in Table 9.18.

Table 9.18. Characteristics of Rapeseed oil.

Description	*Typical values*
Compressibility, $\times 10^{-6}$	60.9
Dielectric constant, cgs units	3.06
Flash point, open cup, °C	282
Iodine number	97–108
Pour point, °C	–12
Refractive index at 25°C	1.470–1.474
Refractive index at 40°C	1.465–1.469

(Contd...)

Table 9.18. Characteristics of rapeseed oil. *(Contd.)*

Description	*Typical values*
Saponification number	170–180
Specific gravity at 25°/25°C	0.906–0.914
Specific heat at 20°C	0.488
Surface tension, dynes per cm at 12°C	36.6
Thermal conductivity, cals/sq. cm. sec for	
1°C and 1 cm thick	41
Titer, °C	11.5–15.0
Viscosity (Seybolt) 38°C	260
100°C	54

Physical properties of erucic acid and methyl erucate are shown in Tables 9.19 and 9.20.

Table 9.19. Physical properties of erucic acid.

Description	*Values*
Specific gravity, d_4^{20}	0.85321
Specific gravity, d_4^{55}	0.860
Melting point, °C	34.7
Boiling point, °C	281 (30 mm)
Refractive index n_D^{20}	1.44438
Iodine value	74.98

Table 9.20. Physical properties of methyl erucate.

Description	*Values*
Density, d_4^{20}	0.8706
Density, d_4^{40}	0.8565
Molar volume, V_m^{20}	404.96
	411.63

Chapter 10

Sesame Oil

INTRODUCTION

Sesame (*Sesamum indicum L.*, synonymous with *S. orientale L.*, also known as sesamum, gingelly, sim sim, benniseed and til) is probably the most ancient oilseed known and used by human as a food source. It also appears to be the earliest condiment used. Its domestication is lost in the mists of antiquity. Oil from cultivated sesame seeds is markedly different from all other vegetable oils in many chemical, biological and physiological properties. Most of these unusual characteristics are due to its content of the unique unsaponifiable constituents, viz. sesamol, sesamin and sesamolin. These endogenous antioxidants provide sesame oil with unusual stability to oxidative deterioration. Sesame oil is a natural salad oil, requiring little or no winterisation and is one of the few vegetable oils that can be used directly without refining.

Only a minor portion of low-grade sesame oil is used industrially in the manufacturing of soaps, pharmaceuticals, insecticides, paints and varnishes. In scented oils used in perfumery. Because of the excellent quality of the edible oil it produces, sesame is often called the 'queen of the oilseed crops'.

The meal remaining after oil extraction has unique nutritional properties. It is rich in methionine, cystine and tryptophan. Sesame protein complements very well most oilseed and vegetable proteins. The seed cake is also an excellent protein supplement in the animal feed industry.

India is by far the largest producer with approximately 35 per cent of the world's sesame area and 27 per cent of its production. Different varieties of sesame seed (black, white and brown) are cultivated in India both as a rainfed and as an irrigated crop.

Clean and dry sesame seeds are stored in bulk. Seed that becomes heated or is contaminated with extraneous material produces discoloured or rancid oil. Because of its small size, sesame seed can be stored more economically than many other oilseeds. It can also be moved easily and efficiently by modern conveyors without causing damage to seeds.

Sesame seeds need to be stored under conditions of low moisture to prevent attack from storage fungi and insect pests. In India, the most frequently used kerosene can or grease drum with tightfitting lids are quite convenient to handle and for storing the seed in small quantities. According to *Prieto* and *Leon* sesame seed retains its viability well under controlled conditions. Storage of sesame seeds at 18°C and 50 per cent RH retained the germination capacity for over one year. *Van Rheenen* recommended the use of silica gel in sealed containers to preserve viability of sesame germplasm collections for long periods.

Sesame oil is extracted from the seeds of *Sesamum indicum*, an herbaceous plant of the Pedaliaceae family. The seeds are enclosed in the fruit (capsule).

Seed oil content fluctuates between 35 and 57 per cent but generally lies between 50 and 57 per cent. Seeds are of white, brown or black colour. Clear-colour seeds have higher oil content whereas black seeds have thicker hulls. Refined oil, remarkably stable due to the presence of natural specific antioxidants (sesamol and sesamoline), is used for frying and seasoning and also after hydrogenation in margarines and shortenings. It has also found applications in the cosmetics and pharmaceutical industries.

Sesame meal has great nutritional value, thanks to their high protein (43 per cent) and significant methionine (2.5–4 per cent) content, thus constituting a good supplement to soyabean cakes.

Sesame oil is obtained by traditional processes of the oil industry: heat treatment of seeds, pressing and solvent extraction. Oil can also be obtained by 'cold' pressing.

One portion of the production of seeds (about 20 per cent) is reserved for direct consumption after roasting or for the preparation of traditional dishes (gomasio, tahin or 'mashed sesame'). For these uses the seeds are first dehulled by chemical or physical methods (soaking, flotation, drying).

CHEMICAL COMPOSITION

Sesame is a rich source of oil and protein and has a high food value. Proximate composition of different types of sesame seed is summarised in Table 10.1. Table 10.2 gives the principal chemical and physical characteristics of sesame oil.

Table 10.1. Proximate composition of sesame seeds (%, moisture-free basis).

Sesame type	*Fat*	*Protein*	*Carbohydrates*	*Crude fibre*	*Ash*	*Oxalic acid*
Sudanese white sesame						
Whole	53.3	25.0	9.5	4.1	5.4	2.7
Dehulled	57.5	29.9	5.7	3.0	3.5	0.4
Indian black sesame						
Whole	54.3	20.2	12.4	4.5	6.2	2.5
Dehulled	63.4	23.4	8.2	2.5	2.4	0.1
Husk	10.7	8.4	23.0	19.3	23.8	14.9
Dehulled (expeller-pressed)	10.6	57.8	19.4	5.4	6.6	0.3
Dehulled (hexane extracted)	0.4	60.2	27.3	5.3	6.5	0.3
Indian white sesame						
Whole	53.1	20.9	15.7	3.4	5.1	1.8
Dehulled	57.7	22.5	15.7	1.5	2.5	0.2

Table 10.2. Main physical and chemical characteristics of sesame oil.

D_{20}	0.915–0.923
V_{20} (cp)	64–67
n_D^{20}	1.474–1.477
Solidification point (°C)	–3 to –6
Iodine value	104–120
Saponification value	187–195

The composition is markedly influenced by genetic and environmental factors. The seeds contain about 45–63 per cent oil, averaging about 50 per cent; 19–31 per cent protein, averaging about 25 per cent; 20–25 per cent carbohydrates including the crude fibre; and 4–6 per cent ash. In general, the Indian varieties tend to be lower in protein and higher in oil than Sudanese varieties, such as those generally appearing in the export market and are commonly used in the United States.

The hull content averages about 17 per cent of the sesame seed and contains larger quantities of oxalic acid, calcium, other minerals and crude fibre. Thus, when using sesame for human food, it is advisable to remove the hull. When the seed is properly dehulled, the oxalic acid content is reduced from about 3 per cent to less than 0.25 per cent of the seed weight. Screw-pressed, dehulled sesame contains about 56 per cent protein, while the solvent extracted meal contains more than 60 per cent protein. This is mostly used in feed except in India where it used as a food.

LIPIDS

Compared with many other oilseeds, sesame seeds contain more oil with a greater yield of oil on a per hectare basis. The oil content varies with genetic and environmental factors. A wide range of oil contents, from 37–63 per cent, has been reported for sesame seed. It also varies considerably among different varieties and growing seasons. The oil content is also related to the colour and size of the seed. White or light-coloured seeds usually contain less oil than the dark seeds and smaller seeds more oil than larger seeds. Rough seeded cultivars generally tend to have lower oil contents than smooth seeded types. Agronomic factors also influence the seed oil content. It increases with increasing length of photoperiod and early planting dates. Similarly, the seeds from plants with a short growing cycle tend to have higher oil contents than those from plants with a medium-to-long growing cycle. High rates of nitrogen fertiliser application reduce the oil content of sesame seeds.

Classification

The lipids of sesame seeds are mostly comprised of neutral triglycerides with small quantities of phosphatides (0.03 to 0.13 per cent with lecithin: caphalin ratio of 52:46). The phosphatides also contain about 7 per cent of a fraction soluble in hot alcohol but insoluble when cold. Sesame oil, however, has a relatively high percentage (1.2 per cent) of unsaponifiable matter.

The glycerides are mixed type, principally oleo-dilinoleo, linoleo-dioleo triglycerides and triglycerides with one radical of a saturated fatty acid combined with one radical each of oleic and linoleic acids. The glycerides of sesame oil, therefore, are mostly triunsaturated (58 mol per cent) and diunsaturated (36 mol per cent) with small quantities (6 mol per cent) of monounsaturated glycerides. Trisaturated glycerides are practically absent in sesame oil.

The unsaponifiable matter in sesame oil includes sterols (principally comprised of β-sitosterol, campesterol and stigmasterol), triterpenes and triterpene alcohols which include at least six compounds of which three were identified (viz., cycloartanol, 24-methylenecycloartanol and α-amyrin), tocopherols, and sesamin and sesamolin (described below) which are not found in any other edible vegetable oils. Among the pigments spectrophotometrically identified, pheophytin A (λ_{max} 665–670 nm) was found to markedly predominate over pheophytin B (λ_{max} 655 nm). The pleasant aroma and taste principles contain C_5–C_9 straight-chain aldehydes and acetylpyrazine.

Composition

Sesame oil is classified as a polyunsaturated, semi-drying oil; it contains about 80 per cent unsaturated fatty acids. Oleic and linoleic are the major fatty acids and are present in approximately equal amounts

(Table 10.3). The oil belongs to oleic-linoleic group which are present in equivalent quantities. The linolenic acid content is low. The saturated fatty acids, principally comprised of palmitic and stearic acids, account for less than 20 per cent of the total fatty acids. Arachidic and linolenic acids are present in very small quantities. Heptadecanoic acid (0.2–0.3 per cent) and hexadecenoic acid (0.0–0.5 per cent) have also been reported from some sesame seed oils Composition of unsaponifiable fraction of sesame oil is given in Table 10.4.

Table 10.3. Fatty acid composition of sesame oil.

Nature	*% Total fatty acids*
C12:0	< 0.1
C14:0	< 0.1
C16:0	8–11
C16:1	< 0.3
C17:0	< 0.3
C18:0	4–6
C18:1	37–42
C18:2	39–47
C18:3	< 0.6
C20:0	< 1
C21:1	< 0.4
C22:0	< 0.5

Table 10.4. Composition of unsaponifiable fraction of sesame oil.

Unsaponifiable fraction: 1.0 to 1.5%			
Sterol content (mg/100 g fats)	539–636	Tocopherol content (mg/100 g fats)	20–50
Sterol composition (% total sterols)		Tocopherol composition (% total tocopherols)	
Cholesterol	< 0.3	α-Tocopherol	4
Brassicasterol	< 0.1	β-Tocopherol	2
Campesterol	18–19	γ-Tocopherol	83
Stigmasterol	6–7	δ-Tocopherol	11
β-Sitosterol	59–62		
Δ5-Avenasterol	10–11		
Δ7-Stigmasterol	0.6–0.8	Sesamoline (mg/100 g oil)	20–480
Δ7-Avenasterol	0.8–1.1	Sesamine (mg/100 g oil)	70–610
Fucosterol	< 0.5		
Hydrocarbon content (mg/100 g fats) (crude oil)	200	Triterpenic alcohol content (mg/100 g fatty substances) (crude oil)	180

The ranges of fatty acid composition tentatively adopted by the Food and Agriculture Organisation (FAO) and the World Health Organisation (WHO) Codex Alimentarius Committee on Fats and Oils for sesame oil are summarised in Table 10.5. Codex standards also advise values for some of the characteristics of sesame oil.

Table 10.5. Codex standards (FAO/WHO) for fatty acid composition and characteristics of sesame oil.

Parameter	*Range*
Fatty acids (%)	
C < 14	< 0.1
C14:0	< 0.5
C16:0	7.0–12.0
C16:1	< 0.5
C18:0	3.5–6.0
C18:1	35.0–50.0
C18:2	35.0–50.0
C18:3	< 1.0
C20:0	< 1.0
C20:1	< 0.5
C22:0	< 0.5
Characteristics	
Iodine value	104–120
Saponification value	187–195
Unsaponifiables (%)	2.0, maximum
Acid value (%)	
Virgin oil	4.0, maximum
Non-virgin oil	0.6, maximum
Peroxide value (meq/kg)	10.0, maximum

Endogenous Antioxidants

Among the commonly used vegetable oils, sesame oil is know to be most resistant to oxidative rancidity. It also exhibits noticeably greater resistance to auto-oxidation that would be expected from its content of tocopherols (vitamin E). This unusual stability to oxidation is often attributed to the presence of a large proportion of unsaponifiable matter (approximately 1.0–1.2 per cent) in sesame oil compared with those in other vegetable oils. Moreover, the unsaponifiable matter itself includes substances, such as sesamol and phytosterol, that are normally not found in other oils.

The remarkable stability of unrefined sesame oil is now widely attributed to the presence of endogenous phenolic antioxidants, viz., sesamin, sesamolin, and sesamol (or sesaminol). Their concentrations in sesame oils and their absorption characteristics are summarised in Table 10.6.

Table 10.6. Concentrations of sesamin, sesamolin and sesamol in sesame oil and their absorption characteristics.

Parameter	*Sesamin*	*Sesamolin*	*Sesamol*
Concentration (mg/100 g oil)	293–885	123–459	Trace–5.6
λ_{max}	287	288.5	296
ε_{max}	23.0	21.8	29.7
λ_{max}	236	235	233
ε_{max}	26.0	24.9	21.2

Sesamolin differs from sesamin in having an oxygen atom connecting one of its methylene dioxyphenyl groups to the central tetrahydrofurfurofuran nucleus. Sesame oil has a characteristics UV absorption with two maxima, at 287.5 and 235 nm. This absorption of sesame oils is, most likely primarily if not exclusively, due to its content of sesamin, sesamolin and sesamol.

Presumably, the superior oxidation stability of sesame oil is due to sesamol. It is a very minor component of the oil and only small quantities of the free substance are found in the natural oil. Sesamol, however, is capable of generating from sesamolin by intermolecular transformation during the industrial bleaching process of unroasted sesame oil with acid-type bleaching earths, by dilute mineral acids, by hydrogenation, or during the frying process. Alkali neutralisation, washing and deodourisation diminish released sesamol. After the action of dilute acids has released sesamol from sesamolin, the remaining compound is sesamin. In contrast, sesamin gives episesamin during the bleaching process. Refined deodourised sesame oil usually contains only traces of free sesamol and is, thus, no more stable than other similar unsaturated oils.

The sesamin content of the oils ranged from 0.07 to 0.61 per cent, averaging 0.36 per cent, whereas the sesamolin content ranged from 0.02 to 0.48 per cent, averaging 0.27 per cent. The highest sesamin value found in this study conducted was for a sample of white-seeded strain having a high oil content of 55 per cent, whereas the lowest value was for a black-seeded strain having a low oil content of 44.6 per cent. In contrast, the highest and the lowest values for sesamolin were found in samples from the white-seeded strains.

Properties

A number of parameters are used to characterise vegetable oils. Sesame oil is dextrorotatory, which is unusual for an oil devoid of optically active fatty acid glycerides. The unsaponifiable fraction of the oil, however, does contain optically active minor constituents, which are responsible for the optical rotation of the oil.

Nutritional Quality

Sesame oil is practically free of toxic components. The oil contains more unsaturated fatty acids than many other vegetable oils. The high proportion of unsaturated fatty acids renders it an important source of essential fatty acids in the diet. Linoleic acid is required for cell membrane structure, cholesterol transportation in the blood and for prolonged blood clotting properties. Sesame oil is rich in vitamin E but is deficient in vitamin A. The crude oil contains a relatively low amount of free fatty acids. The minor constituents, sesamin and sesamolin, present in sesame oil protect the oil from oxidative rancidity.

PROTEINS

Composition

Sesame seed contains 19–31 per cent protein, with an average of 25 per cent. The proteins in the seed are mostly located in the outer layers of the seed. Based on their solubility, sesame proteins have been classified as albumin (8.6 per cent), globulins (67.3 per cent), prolamin (1.3 per cent) and glutelin (6.9 per cent) fractions.

Globulin is the predominant protein fraction in the sesame seed. It is composed of two components. α-Globulin is the major fraction and accounts for about 60–70 per cent of the total seed globulin, while β-globulin is a minor component contributing 25 per cent to the globulin fraction. α-Globulin is a high molecular weight protein (2,50,000–3,60,000 MW) and has a sedimentation coefficient of 11–13 S. It is an oligomeric protein composed of six dimeric units of molecular weight of about 50,000–60,000. The dimeric unit is of the A-B type linked by a disulphide bond. The quaternary structure of α-globulin has been well established. β-Globulin is the minor component of sesame seed globulins. It has a molecular weight of 15,000 and is rich in acidic and hydrophobic amino acids.

Solubility

An understanding of the solubility characteristics of sesame protein is essential for developing methods of preparing protein concentrates and isolates as well as for assessing their applications as functional ingredients in various food systems.

The major proteins of sesame flour, being globulins, are salt soluble. About 84 per cent of the total meal protein can be solubilised by extraction with successive water and salt solutions. The ionic strength of the solution affects the solubility of sesame protein by causing dissociation of electrostatically associated sub-units. Protein extractability is generally much lower from screw pressed meal cake than from whole meal. This is primarily due to denaturation of protein by the high temperatures during processing.

Maximum extraction of sesame protein is obtained above pH 9 at solvent (water)/meal ratios of 15:1. Very little protein is extracted between pH 2 and 8. However, sesame protein can be maximally extracted at pH 6 if the ionic strength of the solution is increased to 1 M with sodium or calcium chloride.

Isolated sesame protein is insoluble in water between pH 6 and 7. It becomes progessively more soluble as the pH is made acidic or basic. The minimum solubility in water around pH 6–7 reflects the lowest net charge or the isoelectric point. The solubility of sesame α-globulin is markedly affected by ionic strength. Thus, increasing ionic strength from 0 to 1 M at pH 7 progressively increases solubility to almost 90 per cent. Apparently the ions favour salting-in of the protein by reducing the electrostatic-based protein-protein interactions. The pH of minimum solubility decreases as the ionic strength is increased, reflecting increased cation binding and neutralisation of the anionic protein.

Nutritional Quality

The amino acid compositions of different preparations of sesame seeds and isolates are summarised in Table 10.7. Sesame proteins are rich in sulphur-containing amino acids and low in lysine, which is unusual for oilseed proteins.

Table 10.7. Essential amino acid composition of sesame products.

Amino acid	*Flour*	*Meal*	*Protein isolates*			*FAO/WHO reference protein*
			Alkali extraction	*Salt extraction*	*Water extraction*	
Tryptophan	–	2.0	–	–	1.8	1.0
Threonine	3.4	3.9	3.6	3.7	3.3	4.0
Valine	4.7	4.6	4.9	5.2	4.6	5.0
Methionine + cystine	5.8	5.6	3.2	2.1	3.7	3.5
Isoleucine	3.9	4.7	4.0	4.1	3.6	4.7
Leucine	6.7	7.4	6.7	6.6	6.6	7.0
Phenylalanine + tyrosine	8.2	10.6	8.7	8.2	7.9	6.0
Lysine	2.6	3.5	2.4	2.2	2.1	5.5

Among other essential amino acids, sesame protein is borderline deficient for threonine, isoleucine and valine contents compared with FAO reference values. During the preparation of a protein isolate (>90 per cent protein), there is some loss of methionine, cystine and tryptophan. This may reflect the selective recovery or elimination of certain proteins by the isolation methods employed.

The protein efficiency ratios (PER) of sesame seed, meal and isolated protein are 1.86, 1.35 and 1.2, respectively. Commercially prepared flour and presscake showed PER values of 0.9 and 1.03. Supplementation of sesame seed protein with lysine can increase its PER to 2.9. The biological value of sesame seed protein is 62, which is lower than that of soyabean.

The amino acid composition of sesame complements that of most other oilseed proteins. Tryptophan, which is limiting in many oilseed proteins, is adequate in sesame. The availability of amino acids from sesame seed protein is affected by the method of processing. Digestibility is enhanced by heat treatment under moist conditions, while screw pressing for oil recovery apparently has little adverse effect on available lysine. However, *in vitro* digestibility was reportedly the same for the isolated sesame protein before and after autoclaving, indicating a lack of trypsin inhibitors.

The high level of sulphur-containing amino acids in sesame seed protein is unique. It suggests that sesame protein should be more widely used as a supplement for methionine and tryptophan and should be an excellent protein source for baby and weaning foods. The use of sesame seed protein would eliminate the problems encountered when foods are supplemented with free methionine, which is unstable.

CARBOHYDRATES

Sesame seed contains 14–18 per cent carbohydrates comprised of glucose (3.2 per cent), fructose (2.6 per cent), sucrose (0.2 per cent), raffinose (0.2 per cent), stachyose (0.2 per cent), planteose (0.6 per cent) and several other minor oligosaccharides. In addition, it also contains 3–6 per cent crude fibre present mostly in the hull or seed coat.

Wankhede and *Tharanathan* reported 0.58–2.34 per cent and 0.71–2.59 per cent hemicellulose A and B, respectively, in defatted flour. Hemicellulose A was found to contain galacturonic acid and glucose in the ratio of 1:12.9, while hemicellulose B contained galacturonic acid, glucose, arabinose and xylose in the ratio of 1:3.8:3.8:3.1.

MINERALS AND VITAMINS

Sesame seed is a good source of minerals, particularly calcium, phosphorus, potassium and iron (Table 10.8). The seeds contain about 4–6 per cent minerals. *Deosthale* reported 1 per cent calcium and 0.7 per cent phosphorus in the seeds. Calcium is mostly present in the seed coat, which is lost during dehulling. Furthermore, the bioavailability of calcium from sesame is also less because of the high concentration of oxalate and phytate in the seed.

Erdman reported the bioavailability of calcium from some food products, relative to $CaCO_3$ as: non-fat dry milk 100 per cent, whole wheat bread 95 per cent, sesame seed 65 per cent, almond powder 60 per cent and spinach 47 per cent. Sesame grown on selenium-rich soils also contain high selenium, although most of it is present in the hulls. Sesame seeds are an important source of certain vitamins, especially niacin, folic acid and vitamin E (Table 10.8). The vitamin A content of the seed is, however, very low. Sesame oil is rich in tocopherols. However, the proportion of γ and δ tocopherols is more than that of α-tocopherol. The latter has the highest vitamin E activity. Hence, the vitamin E activity of sesame oil is less than that of other vegetable oils.

Table 10.8. Mineral and vitamin contents of sesame seeds[a].

Component	*Content*
Minerals (mg/100 g)	
Calcium	1000–1483
Phosphorus	570–732
Iron	10–56
Sodium	60–80
Potassium	725–831
Vitamins	
Vitamin A (IU)	Trace–60
Thiamin (mg/100 g)	0.14–1.0
Riboflavin (mg/100 g)	0.02–0.34
Niacin (mg/100 g)	4.40–8.70
Pantothenic acid (mg/100 g)	0.6
Ascorbic acid (mg/100 g)	0.5
Folic acid (μg/100 g)	51–134
Vitamin E (mg/100 g)	
Total (α-tocopherol equv.)	29.4–52.8
α-Tocopherol	1.0–1.2
β-Tocopherol	0.005–0.6
γ-Tocopherol	24.4–51.7
δ-Tocopherol	0.05–3.2

ANTINUTRITIONAL FACTORS

Sesame seed is almost free of antinutritional factors and is suitable for human consumption as such or after processing. However, it contains high amounts of oxalate and phytate, which adversely affect mineral bioavailability in human nutrition. Oxalic acid is mostly present in the hulls, which impart a slightly bitter taste to the whole seed or meal because of chelation of calcium. Dehulling reduces the oxalic acid content of sesame seed. Oxalic acid may also be removed from sesame meal by treating it with hydrogen peroxide at pH 9.5. Sesame seeds contain a substantial amount of phosphorus, most of which is tied up in phytic acid or as phytin (a calcium and magnesium salt of inositol hexaphosphate). Compared to soyabean which may contain 1.5 per cent phytic acid, sesame seed contains up to 5 per cent. The phytate in sesame meal is insoluble in water. *Boland* extracted phytate from the meal by dilute HCl (0.3 M) and precipitated it with NaOH. The insoluble phytate had a composition of NaMg-phytate, suggesting that it exists in sesame meal as magnesium phytate and not as phytin (CaMg-phytate).

The sesame plant seems to have a unusual capacity for lead accumulation in the seed. *Yannai* and *Haas* reported that whole sesame seeds and kernels contained lead at the level of 0.13–0.22 mg/100 gram. Excessive consumption of sesame (>200 gram/day) is, therefore, considered to be harmful to human.

PROCESSING

Sesame seeds are generally processed or used without removing the cuticle or the seed coat. This is especially true in areas where sesame is processed primarily for its oil content. The presence of cuticle, however, contributes to the dark colour, bitterness and high fibre and oxalate contents of the resultant screw pressed meal. Such meal is not useful as a protein source for human and other monogastric animals and is, largely used as a cattle feed or manure.

Dehulling sesame seed is essential to improve the quality of utilisation of the meal as a source of human food. In India, where sesame meal is an important food, dehulling is an integral part of modern oil extraction plants. Various routes used for commercial processing of sesame seed are shown in Fig. 10.1. These operation and technologies are briefly described below.

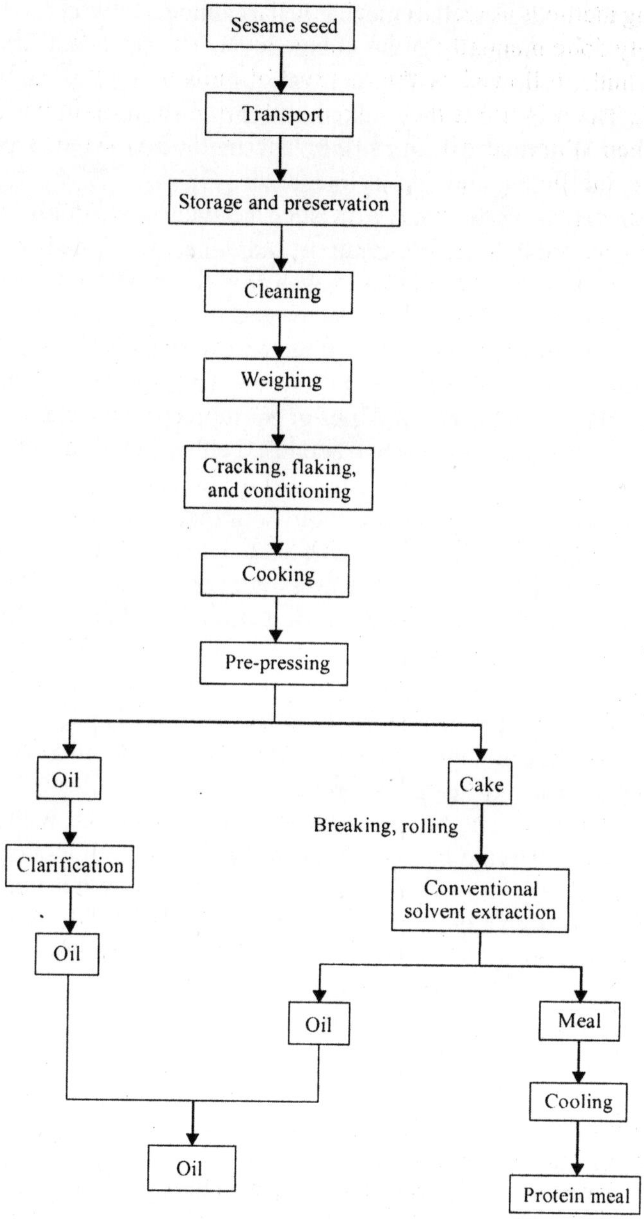

Fig. 10.1. Various methods of processing sesame seed.

Dehulling

Dehulling sesame is essential because the hull contains larger amounts of undesirable oxalic acid and indigestible fibre. Oxalic acid lowers calcium bioavailability in human nutrition, while the fibre impairs the digestibility of protein in the meal. The presence of hulls also imparts a dark colour and bitterness to the meal. However, dehulling still remains the single most important problem worldwide in the processing of sesame seed.

Several wet processing methods as well as mechanical treatments have been tried. In India, dehulling sesame seed is still largely done manually at the village level. The seeds are first soaked in water until they swell and burst the hulls, followed by the removal of hulls by light pounding or by rubbing on a stone or a wooden block. The mixture is then mixed with brine solution in which the hulls sink to the bottom, the seeds are then skimmed off and dried. The method is a batch process and is tedious, cumbersome, unhygienic, inefficient and labour intensive.

To overcome the drawbacks associated with such conventional dehulling processes, the Oil Technological Research Institute (OTRI) in Anantpur, India, has developed two mechanical methods. In one such process, cleaned seeds are soaked in water, with or without wetting agents, for a pre-determined length of time. The soaked seeds are then passed through two vertically mounted disks with hard surfaces. One disk is stationary, while the other one is rotated. This subjects the wet seed to rubbing action by which the hulls are freed from the kernels. The mixture is then spread on a wire mesh and the hulls separated, either by a current of water or by subjecting the mixture to a shower of water with scrubbing. The dehulled seed is then washed and dried either in the sunlight or by artificial means.

In a second method developed by OTRI, sesame seed is first soaked in water with or without added chemicals. The soaked seed is then passed continuously through a mechanical helical screw-press of special design without the application of pressure or heat. The moist mass of the mixture of dehulled seed and hulls that comes out of the cage of the press is spread over perforated sheets and the hulls washed down by means of a jet of water. The washed, dehulled seed is then dried.

As compared to the traditional manual methods, these two decorticating methods are mechanical, simple, continuous and employ only such equipment as would be available with a conventional oil mill in India. Only the seed washing equipment and the dryer are additionally required for both these methods.

Alkali treatments or lye peeling methods have also been used for dehulling sesame seed. *Ramachandra* and co-workers have reported a lye treatment process in which the cleaned seed is given a hot lye (0.6 per cent NaOH) treatment for one minute. The seed is then washed with excess cold water. The ruptured seed coats are then separated by scrubbing in a suitable equipment. The dehulled seeds are then dried. The quality of the oil was not affected by the lye treatment during dehulling.

Moharram and co-workers used a lye solution (6 per cent NaOH at 60°C) to dehull sesame seeds of 5 cultivars (Giza 23, Giza 24, Giza 25, Strain 22/4 and Sudany). A seed-to-lye ratio of 1:3 (w/v) was sufficient to decorticate all cultivars in 10 sec. However, the Giza 24 variety required a 12:1 lye-to-seed ratio at 60°C to obtain satisfactory dehulling in 10 sec. In this study, there was no appreciable loss in protein and oil contents and the lye treatment actually improved the kernel colour.

Since, most dehulling method involve the use of water and its removal from the wet, dehulled seed, the drying costs are considerable. It is estimated that the cost of the dehulled seed is generally 30–40 per cent higher than that of the commercial seed.

The removal of hulls results in a significant change in the chemical composition of seed. The dehulled seed contains significantly more fat and less crude fibre, calcium, iron, thiamin and riboflavin and slightly less phosphorus than the whole seeds. The oxalic acid being present mostly in the seed coat, is significantly decreased after dehulling. *Sastry* and co-workers observed significant improvement in the digestibility of proteins as a result of dehulling sesame seed. Heat treatment during dehulling as well as subsequent processing of the flour did not lower the available lysine.

Oil Extraction

Traditional methods

The most popular method of oil extraction from sesame seed in India is by ghani, which is basically a large pestle and mortar. The mortar is an inverted cone and the pestle a heavy bulk of timber seated at the bottom of the cone. The pestle is inclined at an angle so that it crushes the meal against the sides of the mortar during rotation. The timber bulk is pulled against the side of the mortar by weight and is rotated by a pole inserted through it at right angles. The ghani is driven by one or two bullocks. In earlier days, it was made from wood, but now-a-days may also be of wood-lined concrete or cast iron. The Burmese hsi-zon is similar to the Indian ghani.

Oil extraction by ghani consists of cleaning the seeds to remove dust and dirt particles and other contaminants and dehulling to obtain seed kernels which are then directly used in the ghani. In many parts of India, water or jaggery (brown sugar) is added to sesame seed to facilitate oil extraction. Following extraction, the oil is removed from the ghani, allowed to settle, skimmed and sometimes strained through a cloth before sale. Sometimes the residual meal is double-passed to obtain more oil.

The most serious drawback of the village ghanis is their comparative inefficiency of production of oil, production which is considerably lower than that from power mills, particularly screw presses. The yield of sesame oil from traditional Indian ghanis is only about 40–45 per cent, while that from Burmese hsi-zon is about 33 per cent, thereby leaving behind a higher percentage of oil in the cake. Other major drawbacks include inefficient operations, thereby resulting in low outturn; the resulting oil contains more moisture and suspended and mucilaginous impurities; and the unhealthy sanitary conditions under which the ghanis work.

The traditional bullock-driven Indian ghanis and the Burmese hsi-zon are now being replaced by power-driven mills. The power drive is simple in design and could be fitted to any type of ghani, whether traditional or improved, single-bullock or double-bullock.

In other developing countries, traditional methods for extracting oil from sesame are still being followed. For example, in Central Africa, sesame seeds are boiled to make them soft and then squeezed in a tube made from the fibres to extract the oil. In parts of Asia, sesame seed is parched until very dark and crushed to make a dark sesame oil that is highly flavoured and aromatic. Dark sesame oil is valued as a flavouring agent for meat, fish and vegetable dishes.

Modern methods

Most modern commercial methods for extraction of oil from oilseeds employ one of three basic designs: batch hydraulic pressing, in which the oil is expressed by hydraulic pressure from a mass of oil-bearing material; continuous mechanical pressing in which the oil-bearing material is squeezed through a tapering outlet, the oil being expressed by the increasing pressure; and solvent extraction in which the oil-bearing material is taken into solution with a solvent, which is then separated from the insoluble residue and the oil recovered from the solvent solution. These methods are also applicable for extracting oil from sesame seed. Sometimes, these methods are modified or combined to suit specific conditions. For example, combinations of these processes in commercial operations include continuous mechanical pressing and continuous solvent extraction and batch hydraulic pressing and solvent extraction.

Commercially, sesame seed is extracted using a continuous screw press, hydraulic press and pre-press solvent extraction. Direct solvent extraction is not suited for sesame seed because of its high oil content.

The sesame seed produced by farmers is not of uniform size, colour, or maturity, often being a mixture of different varieties. Seed may also be contaminated with soil particles. Moreover, because of the small size of seed, it becomes difficult to clean the seed. The oil quality is greatly affected if the seed is not properly cleaned. Similarly, prolonged storage under unsuitable conditions results in a loss of oil quality.

In Europe and Asia, sesame oil is usually extracted in three stages. The first pressing is cold. The oil obtained after filtration is of very good quality, high grade and ready for use. It has a light colour and agreeable taste and odour. The second pressing is made after the residue is subjected to high pressure. It yields a highly coloured oil which needs refining before using for edible purposes. The residue left behind is then used for the third extraction under similar conditions. It gives an oil of inferior quality, which is not suitable for human consumption. Hence, oil from the third extraction is generally used for manufacturing of soaps.

In parts of Europe, the 'Direx' extractor has been used for the extraction of oil from sesame. However, its commercial applications are not known. The processing of sesame involves certain difficulties due to the fact that it is a small sized material covered by a very hard skin and the kernel contained therein breaks up very easily when subjected to grinding, thereby creating a lot of fines. These tend to form with the miscella, a suspension extremely difficult to purify even when large filtration equipment is provided. For this reason, the rolling and cooking (flaking) steps need to be carried out with utmost care using suitable equipment, primarily because the successful outcome of the process depends on the efficiency of the latter. In case of pre-pressing, sesame seed from the final cleaning machine is cracked or flaked and then steam-cooked to about 110°C for 30 minutes. The incoming moisture content is normally 5–7 per cent, while the cooked seed leaving the last pan has 9–10 per cent moisture. The seed thus prepared is fed to the pre-press which reduces its oil content down to 18–19 per cent. The cake from the screw press is not subjected to granulating as it is extremely fragile and tends to form fines very easily. It is, instead, directly sent to the solvent extraction plant, which will bring its residual oil content down to 0.6–0.7 per cent.

During the preparation stage of direct extraction of sesame oil, it is important that just a coarse cracking be carried out and light flaking pressure be exerted. The broken and pre-flaked material is then passed into a stack cooker, where it is cooked to about 110°C for 25–30 minutes. The cooked material is finally conveyed to the direct extraction plant. It should preferably be of the dual-stage type with intermediate wet flaking of the partially deoiled material so that the oil cells remaining still intact are ruptured and the residual oil can be completely removed in the final extraction stage. The latter is of the immersion type and hence designed to handle the fines which may be produced during the light flaking steps.

Refining

Crude, cold-pressed sesame oil is used directly in cooking wherever it is produced and is often a favoured oil. The crude oil varies from yellow to a dark amber colour. Refined sesame oil, which is usually pale yellow in colour, is used commercially as a bland salad oil and for purposes where an edible oil of exceptionally good keeping qualities is required.

Sesame oil does not require extensive purification and refining. In fact, it is one of the few vegetable oils that can be used without refining. These are factors of increasing importance as energy costs escalate. The crude oil usually contains suspended meal particles ('foots') which are removed by settling, screening and filtering. The filtered crude oil from the extraction plant contains impurities like phosphatides, resins, free fatty acids and colouring matters. To obtain a fully refined oil from a crude oil, it is taken through degumming, neutralising, bleaching, winterising and deodourising steps.

Alkali-refining of sesame oil removes gums, free fatty acids and some of the colouring matters. An easy method of refining sesame oil with alkali-enriched dry sodium metasilicate (SMS) precludes emulsion formation and therefore, soapstock can be easily separated. This dry refining method was found to be superior to the conventional sodium carbonate process where liberated CO_2 creates operating problems. The extent of free fatty acids (FFA) reduction by this method was dependent on the degree of the alkalinity of SMS-NaOH mixture. The colour of the oil, however, was not markedly improved by the process. The process did appear to be useful for readily preparing low-FFA, medium-colour triglyceride oils that may find applications where the colours of final products do not pose a problem.

After neutralising, the alkali-refined oil is bleached with a relatively lower quantity bleaching earth than for most other vegetable oils. Bleaching produces a light-coloured oil. Deodourisation is necessary to produce bland oil. It is usually done by treating refined oil in vacuum with steam at 200–250°C. For use as a base for salad dressing, the oil must be stable under refrigeration. It is, therefore, winterised by cooling to remove components with higher melting points that may settle out at low temperatures. Sesame oil, however, requires little or no winterisation. The hydrogenation process considerably increases the stability of the oil.

Sesame Cake and Meal

Sesame cake is obtained as a by-product after extraction of oil. When powdered, the cake is converted into a meal or flour. There is no change in the chemical composition of cake when it is converted into meal.

Sesame cake is rich in protein (35 to 45 per cent, depending upon the oil extraction method and whether the seed is hulled or unhulled) and is a valuable livestock feed. The proteins are rich in methionine and cystine but low in lysine. Sesame cake varies in colour from light yellow to greyish-black, depending on the seed colour type. Dark-coloured cakes are frequently more bitter than lighter-coloured cakes and hence the latter are preferred. Sesame cake can be distinguished microscopically by the distinctive presence of calcium oxalate crystals in the epidermal cells.

Sesame cake is often used as an animal feed in India when the oil is extracted at the village level. The FFA content of Indian ghani cake is high and its keeping quality poor. Therefore, it must be fed to livestock as soon as possible or its rapidly becomes rancid and unpalatable. Where the oil is extracted in large mills, the cake is often considered too expensive for animal feed, and is processed to obtain a protein-rich flour that can be mixed with other ingredients (e.g. soyabean, corn, or chickpea flours) to produce very nutritious human foods. Sesame meal is also sometimes added to bread to improve nutritive value and give a distinctive flavour.

Four types of meals can be obtained from sesame seeds: whole seed meal, dehulled seed meal, defatted whole seed meal and dehulled, defatted meal. Of these, the dehulled, defatted meal is the most common. A flow diagram for the production of sesame flour is shown in Fig. 10.2.

Protein Concentrates and Isolates

Sesame is widely processed into several high-protein products such as flakes, flour, protein concentrates and isolates. The protein concentrate and isolate contain, respectively, 70 and 90 per cent protein. Unlike many oilseeds, protein concentrates and isolates prepared from dehulled, defatted sesame flour do not contain any undesirable pigments, off flavours and toxins.

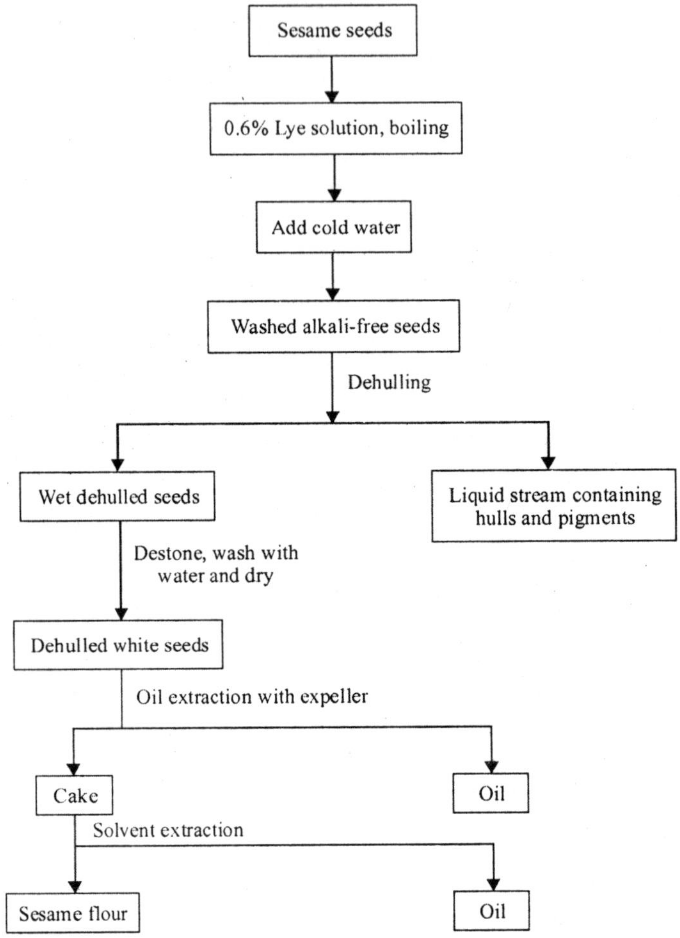

Fig. 10.2. A flow diagram for the preparation of defatted sesame flour.

Sesame proteins are extracted with various salts and alkaline solutions. The extractability of proteins varies with the extraction medium, pH and time. Sodium hydroxide (0.04 M) appears to be the most suitable solvent, extracting about 90 per cent of the meal nitrogen.

Since, protein exhibit minimum solubilities at their isoelectric point, most protein isolates are prepared by extracting the protein in a suitable solvent and precipitating them at or near their isoelectric point. The isoelectric point of sesame protein is 4.5 to 4.9. Sesame protein extracted with salt and alkali show minimum solubility in the region of pH 5.7. A low-phytate protein isolate can be prepared by dissolving the protein in an alkaline solution in a countercurrent procedure and precipitating it at pH 5.4. At this pH, 50 per cent of the phytate was removed while only 17.5 per cent of the protein was dissolved. The resulting protein isolate contain 91.4 per cent protein and is almost free of phytate.

Roasting

Sesame seeds are often roasted prior to their use in confections. Roasting reduces the moisture content, develops a pleasant flavour and makes the seed or meal more acceptable for consumption. Sesamol, an antioxidant, was detected only in roasted sesame oil. 2-Furfuryl alcohol is considered to be one of the most characteristic components giving a pleasant aroma to sesame seeds. It is present in higher concentrations in red and white sesame.

UTILISATION

Almost the entire world production of sesame seed is utilised for culinary purposes. However, in India, about 78 per cent of sesame produced is used for oil extraction; 20 per cent for domestic use including preparation of sweets, as condiments and confectionery; while 2–3 per cent are kept by farmers for seed purpose. Major uses of sesame oil, seed and other by-products are described below.

Oil

In India, of the total production of sesame, about 75 per cent oil is used for edible purposes as vegetable oil for culinary purposes, 5–10 per cent goes to the vanaspati industry for vegetable ghee (a type of shortening) manufacture and 4 per cent for industrial use as paints, soaps, perfumes, etc.

Sesame oil is straw-yellow-coloured with a pleasant, mild taste and remarkable stability. Its nutty flavour is well appreciated. It is a common constituent of Burmese dishes and is used in frying, roasting and stewing of meat, fish and vegetables. Foods fried in sesame oil, e.g. potato chips keep much longer than if fried in other oil. Sesame oil also makes an excellent frying medium for chicken and meat and is a good replacement for peanut oil. It is also used as a substitute for olive oil, primarily as salad and cooking oils. In Asia, dark sesame oil prepared from parched, dark sesame seed is valued for its flavour and aroma. It is used as a flavouring for meat, fish and vegetable dishes. Sesame oil, flavoured with hot chilli peppers and rice vinegar, is the spicy accompaniment to Chinese steamed or fried dumplings.

In addition to endogenous antioxidants, such as sesamol and sesamolin, sesame oil is popular because of its high content of polyunsaturated fatty acids. It is a natural salad oil requiring little or no winterisation.

Because of its quality and high price, sesame oil is frequently adulterated with groundnut, rape, or cottonseed oils. In India, particularly in the Mumbai area, groundnut, safflower and sesame seeds are extracted together to produce the so-called ‘sweet oil’. This sweet oil is cheaper than sesame oil and has better stability than groundnut or safflower oil. Sesame oil can be readily hydrogenated to medium melting fats of different textures for use in margarine, shortenings and vanaspati.

The industrial uses of sesame oil are minor. Only a small proportion of low-grade oil is used for the manufacture of soaps, pyrethrum-based insecticides, paints and perfumes and for various other purposes for which the non-drying oils are generally adapted. Its relative scarcity and higher price normally render it uncompetitive for large-scale industrial utilisation.

Sesame oil forms the basis of most of the fragrant or scented oils in India as it is not liable to turn rancid or solidify and it does not possess objectionable odour. In the perfumery industry, it is used as fixative. Scenting oil can be extracted from wetted sesame seeds that have been covered with layers of scenting flowers for 12 to 18 hr. One kg of strongly scented flowers is enough to perfume 6 litre of sesame oil.

Sesame oil is used in pharmaceuticals of many countries, primarily because of its high stability. Its high stability makes it a desirable vehicle for medicines which are administered subcutaneously or intravenously. Several medical products employ it as a solvent and vehicle for estrogenic substances. It is also employed in the preparation of liniments, plasters, ointments, and soaps. Like other fixed oils, sesame oil acts as a laxative when taken internally in sufficient dose. In India, it is also used for conditioning the skin.

Sesame oil has synergistic activity with insecticides such as pyrethrums and rotenone. The presence of sesame oil reduces the concentration of the insect toxin required to produce 100 per cent mortality. The synergistic activity of sesame oil has been attributed to the presence of sesamol and sesamolin.

Sesame Seed and Other By-products

Sesame is offered in several forms to the food industry for human consumption. These include: (i) dry, cleaned, unhulled sesame seed; (ii) white, hulled sesame seed; (iii) toasted sesame seed; (iv) toasted, flaked sesame seed; (v) partially defatted sesame flour; (vi) solvent extracted sesame flour; and (vii) sesame butter made from toasted sesame seed.

Dehulled sesame seeds are sweet and oleaginous and are used directly in different types of foods in various parts of the world. In the United States, sesame is used primarily in the bakery industry, where, whole dehulled sesame seed is used as a decoration and flavouring agent on speciality breads, rolls, buns, cakes, cookies, candies and other delicacies. Sesame is also used in high-protein snack foods and granola. Most sesame seeds on bakery products are pearly white, oval in shape, shiny and quite small, about 3 mm in length. The flavour after baking is described as rich and nutlike. Toasted sesame seed of both black and white varieties are sprinkled on hot rice, chicken, fish, or hot and cold vegetables.

In the Orient, sesame is used in various ways. In India, it is used in the manufacture of traditional confections such as *halva, laddu* and *chikki.* They are also eaten whole after roasting. A confection called *laddu* is prepared from roasted groundnuts and sesame seeds by pounding with jaggery (brown sugar) in the proportion of 2:1:2. Small balls are prepared by hand. *Laddu* is also prepared from sesame seeds by mixing them with hot jaggery syrup. The confection prepared by mixing sesame seeds with jaggery or sugar has an auspicious connotation in Maharashtra and other parts of India. The confection is distributed or exchanged one person to another to signify a great deal of sharing of goodwill. *'Chikki'* is another confection popular in western India. It is prepared by pouring sesame seeds in boiling jaggery solution to obtain a thick slurry. The slurry is spread uniformly on metallic sheet or table, cooled and then cut into small rectangular pieces.

Ready-to-eat instant foods using sesame seeds have been developed by the Indian Council of Medical Research for use in rural areas. *Bajra* instant food is prepared by mixing roasted Bajra (pearl millet) flour (60 gram) with roasted green gram dhal (15 gram), roasted groundnut (10 gram) and sesame seeds (5 gram). The mixture is pounded to obtain a flour. When required, the powder is mixed with boiling water or milk to the desired thickness. Sugar or salt are added to taste. *Ragi* instant food is prepared in the same way by replacing *bajra* flour with *ragi* flour. The Bajra instant food gives 18.6 gram protein, 389 cal and 8 per cent net dietary protein (NDP) cal per 100 gram flour. The *ragi* instant food gives 16 gram protein, 369 cal and 8 per cent NDP cal per 100 gram.

In the Middle East, dehulled sesame seeds are mainly utilised in the production of *tahini* (*tehineh* or *tachini*) and *halvah. Tahini* (sesame butter) is made from a paste of dehulled, roasted sesame seeds. *Tahini* is used in a variety of food dishes and added to bread and bakery products. A popular dip called *hummus* is made from *tahini* mixed with chickpeas, garlic and lemon juice. Another popular dish is *tahini* mixed with eggplant to make *baba ghonnooj*. *Tahini* also makes an excellent dressing for salads and other foods. The Mexicans mix it with chocolate and chilies and make mole that is poured over chicken.

Halvah is a traditional confectionery food of Greek, Turkish and other Eastern people and is made from *tahini*, sugar, honey, citric acid, vegetable oil, egg albumin and *Saponaria officinalis* root extract. It is believed to be one of the world's oldest confections.

Sesame meal has become increasingly important for human food due to the following unique properties: the presence of a high level of sulphur-containing amino acids, methionine and cystine; its lack of trypsin inhibiting factors; and its pleasant flavour.

Sesame meal is very palatable to human. In India, it is used extensively in human foods, especially as a protein supplement for cereal flours and as methionine supplement for soya and legume proteins. A number of ready-to-use infant foods (e.g. cholam and samai) have been developed by the Indian Council of Medical Research. There is also increasing interest in fortifying bread and cookies by replacing a portion of wheat flour with sesame meal.

Blends of peanut/chickpea, wheat/chickpea, rice/chickpea, peanut/soyabean, sunflower/maize and cowpea/rice have all shown improved nutritional qualities with supplementation of sesame meal. Even more significant, however, is the finding that a simple blend of one part each of sesame and soya protein has about the same protein nutritive value as casein, the main protein of milk. The high lysine and low methionine content of soya protein is complementary to sesame protein. Sesame meal is sometimes fermented for food in India and Java. In some European countries, it is also used as an ingredient in comminuted meat products.

In Indian villages, defatted sesame meal from ghanis is traditionally used as animal feed. The cake is available as stock feed and is considered equal to cottonseed cake or soyabean meal as a protein supplement for livestock and poultry. In the United States, most of the meal is used for livestock feed. The inferior quality sesame cake or meal is used as a manure in China and Korea.

Future Research

Sesame has a long history of cultivation and use as human food. It also has several desirable agronomic characters. It does relatively well on poor lands requiring very little attention during growing season and is resistant to drought. Compared with many other oilseeds, it also contains more oil with greater yield of oil on a per hectare basis. Moreover, sesame oil has excellent stability and its protein is rich in sulphur-containing amino acids and tryptophan.

In spite of a long history of usage and desirable nutritional qualities, sesame cultivation has been limited to countries where labour is plentiful and inexpensive. The major drawback associated with sesame cultivation in the developed western countries appears to be the unavailability of non-shattering varieties that are amenable to mechanical cultivation and harvesting. The lack of a large number of wild species of cultivated sesame and the hybridisation barriers that exist in the available gene pool for effective gene transfer may have contributed to this fact. Recent advances in gene transfer and the successful production of transgenic plants will greatly help in developing indehiscent sesame types in the near future.

Compared with most edible vegetable oils, sesame oil has several unique properties. It is the most stable naturally occurring liquid vegetable oil that will keep for several years without turning rancid. Crude sesame oil also requires very little, if any, refining. These are important considerations in today's world as energy costs continue to escalate.

More research should be focused on increasing the levels of sesamolin and sesamol in the oils from cultivated sesame and understanding their relations to oil and seed yield. Several studies have indicated considerable quantitative variability in both these traits. Thus far, strains having very high oil contents or markedly high sesamolin contents, which would increase the quality of both the seed and oil, have not been identified.

In the developing countries, opportunities still exist for improving the traditional oil extraction methods that will yield both a better quality of oil and defatted oilseed meal. Improved and easier methods of dehulling sesame seed are also required. More research concerning optimisation of oil extraction and protein preparation is required, with particular emphasis on developing techniques for minimising the oxalic acid content of the flour. Greater emphasis also needs to be placed on the utilisation of defatted sesame flour and meal in human nutrition. At present, India appears to be the only country where the oilseed cake after further processing, is utilised as human food. This is unfortunate given the fact that most of the sesame production in the world comes from developing countries where chronic malnutrition is still a major problem. Development of milder processing methods for processing sesame oilseed cake that will eliminate problems associated with dark colour will also help in preparing acceptable sesame products for human use.

The high level of sulphur-containing amino acids and tryptophan as well as the lack of trypsin inhibitor activity in sesame seed protein is unique and suggests that it should be used more widely as a supplement for these essential amino acids. It is also a valuable protein source for use in baby and weaning foods. Its use would eliminate problems encountered when foods are substituted with free methionine, which is unstable and imparts a bitter taste to the food. More research to develop nutritious foods from sesame protein, therefore, seems justified.

Chapter 11

Soyabean Oil

INTRODUCTION

For centuries soyabeans have been used by the Chinese, Japanese, Korean and Southeast Asian people in various forms as one of the most important sources of dietary protein and oil. The soyabean [botanical name: *Glycine max(L) Merrill*] is believed to have originated in China and one of the oldest crop of Far East. After flowering, the first stage in the formation of fruit is followed by the development of a pod containing from 1 to 4 seeds.

Like other leguminous crops, the soyabean seed is essentially composed of a smooth envelope (hull) and an embryo. Generally, its diameter varies between 5 and 10 mm and its weight (depending on the variety) between 50 and 400 mg. The shape of the seed varies with the cultivars. Front view of soyabean seed in shown in Fig. 11.1.

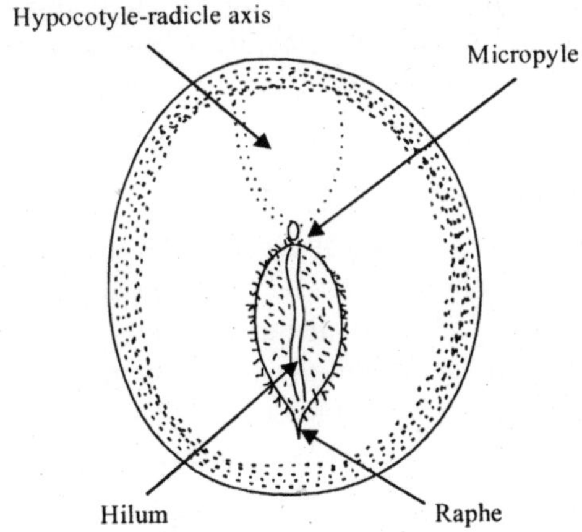

Fig. 11.1. Front view of soyabean seed.

The amount of protein produced by soyabeans per unit land area is higher than that of any other crop. This little bean has been called 'meat of the field', 'nature's miracle protein' and 'yellow jewel'. In recent decades soyabean has generated great interest as a weapon against hunger, a protein of the future and recently as a possible weapon against chronic diseases.

SOYA CHEMISTRY

Despite its possible usefulness this wonder bean suffers an image problem for its unfamiliar flavour and taste. Only a small portion of annual soyabean production is processed into traditional soyafoods. Most of the remaining portion is crushed into oil and defatted meal. Although the oil is used mainly for human consumption, only a small portion of protein-rich meal is further processed into protein products for food ingredients and the remaining large portion goes as animal feed.

However, the use as value-added food protein ingredients, as industrial paints to table oils and spreads, is increasing. This has resulted the development of a new type of soyabean crop with wide regions of adoptability, herbicide tolerance, pest resistance and altered chemical composition. Soyabean has emerged from limited regional cultivation to expanded worldwide production.

Food Beans and Oil Beans

In the Far East, traditionally, soyabeans are made into various food for human consumption, including tofu, soyamilk, soya sprouts, miso, natto and tempeh. In the west, most soyabeans are extracted for oil and defatted meal. Although soya oil is wholly for human consumption, the meal is mainly used as animal feed. Only a small portion, as mentioned earlier, is processed into soya protein ingredients like soya flour, concentrates, isolates and textured soya protein. These ingredients in their turn, have functional and nutritional applications in various types of bakery, dairy and meat products, infant formulas and so-called new generation of soyafoods.

Because of this difference in soyabean use, two different types of soyabeans have emerged: food beans and oil beans. Oil beans include all the commonly produced soyabeans. Food beans, on the other hand, have been selected and bred over the past several decades for direct food consumption. They do not differ fundamentally from field varieties (oil beans), except that they generally have a lighter seed coat, a clear hilum, higher protein content, lower oil and yield. Although most beans are yellow, some variation exists in seed coat colour, including green, dark brown, purplish black and black, some are even specked, such as brown with spots. A general outline of soyabean food utilisation based on oil beans and food beans may be seen in Fig. 11.2.

Composition

Among cereals and other legume species, soyabean has the highest protein content (around 40 per cent); other legumes have a protein content between 20–30 per cent, whereas cereals have a protein content in the range of 8–15 per cent. The soyabean also contains about 20 per cent oil, the second highest content among all food legumes. (The highest oil content is found in peanut, which is about 48 per cent on the dry matter basis). Other valuable components found in soyabeans include minerals, phospholipids and vitamins. Moreover, soyabeans contains many minor substances, some of which, such as trypsin inhibitors, phytates and oligosaccharides, are known to the biologically active. Other, such as, isoflavones, are beginning to be recognised for their ability to prevent human cancer and other diseases. On an average, oil and protein together constitute about 60 per cent of dry soyabeans. The remaining dry matter is composed of mainly carbohydrates (about 35 per cent) and ash (about 5 per cent). Since the water content of stored mature beans is usually about 13 per cent to ensure stability, on wet basis, soyabeans contain about 35 per cent protein, 17 per cent oil, 31 per cent carbohydrate and 4.4 per cent ash. In general, cultivated soyabeans comprise approximately 8 per cent hull, 90 per cent cotyledons and 2 per cent hypocotyl axis. Cotyledons contain the highest percentage of both protein and oil whereas the hull has the lowest values of these components.

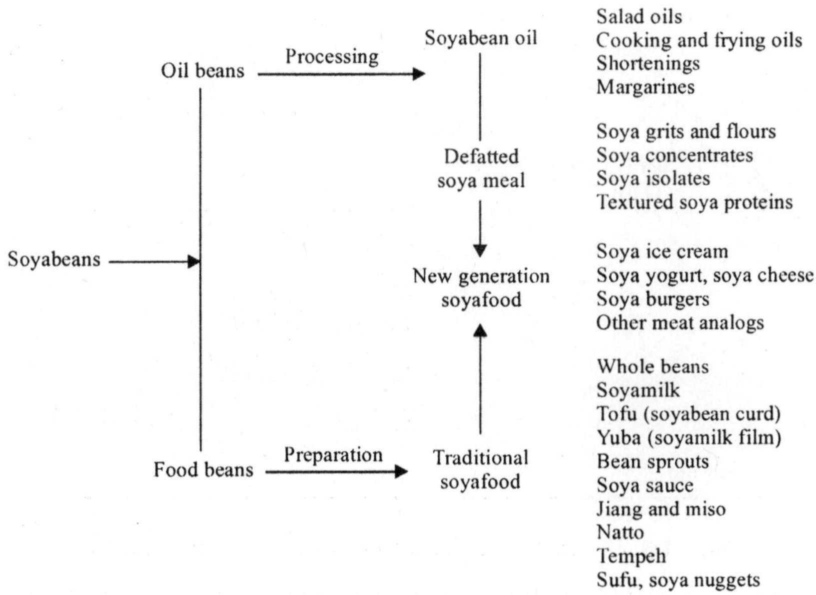

Fig. 11.2. A general outline of soyabean food use based on classification of oil and food beans.

During processing, components extracted from soyabeans by organic solvents such as hexane are called crude oil. Major components of crude oil are triglycerides. Minor components include phospholipids, unsaponifiable material consists of tocopherol, phytosterols and hydrocarbons. The concentrations of these minor compounds are reduced after processing the oil in the usual way. Refined soyabean oil contains more than 99 per cent triglycerides, which are natural lipids.

Carbohydrates

Being the second largest components in soyabean, on an average, as mentioned earlier, moisture-free soyabean contain about 35 per cent carbohydrates. However, the economic value of soya carbohydrates is considered much less important than soya protein and oil. The principle use of soyabean carbohydrate has been in animal feeds where it contributes calories to the diet. Soya carbohydrate contain trace amounts of monosaccharides, such as glucose and arabinose and measurable amounts of di- and oligosaccharides, with sucrose in the range of 2.5–8.2 per cent; raffinose 0.1–0.9 per cent; and stachyose 1.4–4.1 per cent. Basically, the oligosaccharides in soyabean are nonreducing sugars, containing fructose, glucose and galactose as two or more units, linked by beta-fructosidic and alpha-galactosidic linkages (Fig. 11.3).

Humans are not endowed with the enzyme called alpha-galactosidase which can hydrolyse the alpha-galactosidic linkage present in these oligosaccharides. So when consumed, the oligosaccharides cannot be digested in human duodenal and small intestine. The intact sugars go directly into lower intestine where they are metabolised by micro-organisms that contain the enzyme. This results in production of carbon dioxide, hydrogen, nitrogen, methane etc. Consequently, the host experiences flatulence. In fact, flatulence is one of the major factors limiting soyabean utilisation as food. Apart from flatulence in human reports have also shown that oligosaccharides may have a detrimental effect on the nutritive value of soya meal to animals. Removal of oligosaccharides from the soya meal increases the polysaccharide digestibility. A very effective method for removal of them has been aqueous ethanol extraction. This method is mainly used in the production of high quality protein concentrates. It produces a by-product known as soya molasses. This syrupy heavy brown product is used as cattle feed additive.

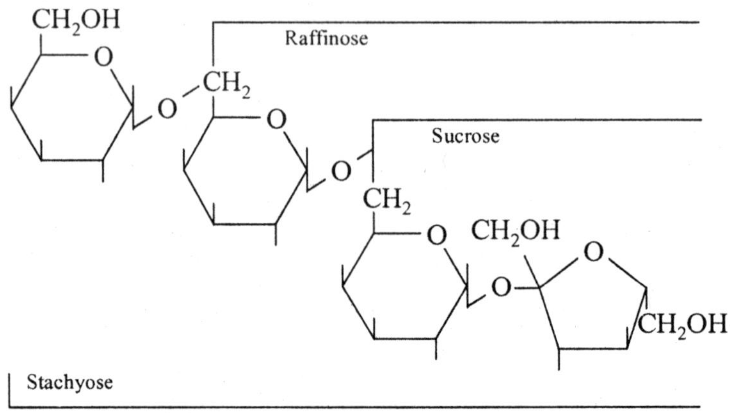

Fig. 11.3. Structure of oligosaccharides found in soyabeans.

Promising results have been obtained to produce lactic acid by fermentation of soya molasses. It has been reported that germination results in almost complete disappearance of oligosaccharides from soyabeans and tempeh fermentation produces a similar flatus-removing effect. As soluble carbohydrates are degraded during fermentation, soya protein isolates, tofu, tempeh and soya sprouts are virtually devoid of flatus activities. Although the presence of oligosaccharides in soyabeans and soya products is generally considered undesirable in terms of flatus activity, recent studies have established beneficial effects of dietary oligosaccharides in humans. These are mainly:

1. Reducing toxic metabolites and detrimental enzymes.
2. Preventing constipation due to production of high levels of short-chain fatty acids.
3. Protecting liver function due to reduction of toxic metabolites.
4. Reducing blood pressure.
5. Having anticancer effects.
6. Producing nutrients such as vitamins.

Consequently, oligosaccharides have been developed into one of the most popular functional food components, particularly in Japan.

Insoluble carbohydrates

In soyabeans these include cellulose, hemicellulose, pectin and trace amount of starch. Most soyabean carbohydrates (oligosaccharides and complex polysaccharides) fall into a general category known as dietary fibre, which are actually the plant components in food that are resistant to digestion by humans. The effect of dietary fibre in human diets on nutritional response has received increasing attention by scientists and public, during the last two decades.

Health and Soyafoods

The processing of soyabeans produces a remarkable array of foods, as mentioned earlier. Traditional soyafoods have been consumed for centuries by South East Asians and more recently by Western vegetarians.

Traditional soya foods

Soyamilk is the aqueous extract of cooked whole soyabeans properly filtered to removed refuse. This is used as a base in a wide variety of products including tofu, soya yogurt, soya cheese and also directly used like dairy milk. Tofu (soyabean curd) is made by adding a coagulating agent such as calcium sulphate, magnesium chloride etc. to the liquid soyamilk. Tempeh (fermented soyabean cake) is made by fermenting cooked soyabeans with 'Rhizopus oligosporus'; the mycelium of the bacteria causes the soyabeans to a firm cake.

Tempeh has a very distinctive flavour, taste and texture. Miso (fermented soyabean paste) is made by mixing cooked soyabeans with a grain, such as rice or barley that have been earlier fermented with 'Aspergillus oryae'. This is used as a broth for soups and tea or as a condiment.

Western soya foods

Soya protein isolate is prepared from defatted and dehulled soyabeans and contains not less than 90 per cent protein on dry basis. This is used in infant formulas and as a functional ingredient in a number of commonly consumed products. Soya protein concentrate is prepared from defatted and dehulled soyabeans and contains not less than 65 per cent protein on dry basis. This is also used primarily as a functional ingredient in a number of food products. Soya flour/grits are prepared from defatted soyabean flakes in a variety of particle size; have protein content of 40–54 per cent on dry basis. Used as a ingredient in food product. Textured soya protein is produced from isolates, concentrates and flour/grits by imparting a structure of fibre or chunk through extrusion or steam texturisation; frequently referred as a TVP for textured vegetable protein. These are made to resemble meat, seafood or poultry and is used as a meat extender or replacement.

Though soyabeans and soya foods have been primarily identified with their protein content, during the past decade there has been much interest among researchers in the potential role of soyafoods in preventing and treating chronic diseases. This recent upsurge in interest in soyabeans is due to the presence of isoflavones in soyabeans (apart from oligosaccharides describe earlier). Basic structure of isoflavones is shown below:

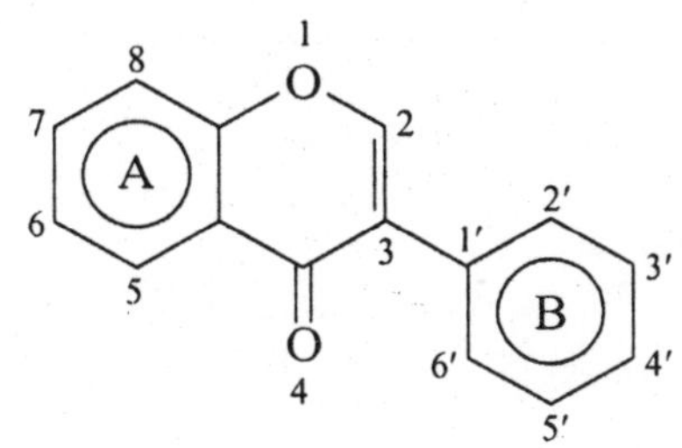

Genistein has -OH groups at positions 4′, 5 and 7: daidzein at 4′ and 7: and glycitein at 4′ and 7 and -OCH_3 groups at position 6. Isoflavone glycones have a sugar attached at position 7.

Basic structure of isoflavones.

Isoflavones have been shown to inhibit the growth of cancer cells and bone resorption and to lower cholesterol. Recent understanding of biological activities of isoflavones in soyafoods has established the relationship between them and the risk and treatment of chronic diseases like cancer, heart and kidney disease and osteoporosis.

Soyabean oil is the most important vegetable oil produced in the world because of its high quality and low cost. Soyabean oil has many advantages but also some disadvantages compared to other vegetable oils.

The advantages include: (i) a high level of unsaturation is present; (ii) the oil remains liquid over a relatively wide temperature range; (iii) it can be hydrogenated selectively for blending with semisolid or liquid oils; (iv) when partially hydrogenated, it can be used as a pourable, semisolid oil; (v) phosphatides, trace metals and soaps in soyabean oil can be removed without much difficulty to obtain a high-quality product; and (vi) the presence of naturally occurring antioxidants (tocopherols)—which are not completely removed during processing–also contribute to its stability.

The disadvantages include: (i) phosphatides are present in relatively larger amount (about 2 per cent) and must be removed by processing; and (ii) soyabean oil contain relatively high level (7–8 per cent) of linolenic acid, which is responsible for its flavour and odour reversion. Some of these disadvantages are offset, however, in that the recovered 'gums' are the source for commercial lecithin and by partial and selective hydrogenation of the oil the linolenic acid concentration can be lowered to below 3 per cent with the result of greatly improve stability.

COMPOSITION OF SOYABEAN OIL

The soyabean has the following approximate composition:

Protein, 40 per cent
Lipid, 20 per cent
Cellulose and hemicellulose, 17 per cent
Sugars, 7 per cent
Crude fibre, 5 per cent
Ash (dry weight basis), 6 per cent

Average compositions of crude and refined soyabean oils are given in Table 11.1. The refining processes does not affect fatty acid composition of the glyceride; it only removes most of the free fatty acids and colour bodies and lowers the levels of some of the lesser constituents, such as the tocopherols (by 31–47 per cent), the sterols (by 25–32 per cent) and squalene (by 15–37 per cent).

Table 11.1. Average composition for crude and refined soyabean oil.

Component	*Crude oil*	*Refined oil*
Triglyceride (%)	95–97	> 99
Phosphatides (%)	1.5–2.5	0.003–0.045[a]
Unsaponifiable matter (%)	1.6	0.3
Plant sterols (%)	0.33	0.13
Tocopherols (%)	0.15–0.21	0.11–0.18
Hydrocarbons (squalene) (%)	0.014	0.01
Free fatty acids (%)	0.3–0.7	< 0.05
Trace metals		
Iron (ppm)	1.3	0.1–0.3
Copper (ppm)	0.03–0.05	0.02–0.06

[a] Corresponding to 1–15 ppm of phosphorus.

Triglycerides constitute more than 95 per cent of the lipidic compounds of the soyabean seed, its other compounds being phospholipids and diacylglycerols.

Triglycerides composition is shown in Table 11.2, fatty acid composition in position 2 is shown in Table 11.3 and the total fatty acid composition is shown in Table 11.4.

Table 11.2. Composition of soyabean oil triglycerides (HPLC).

Nature	*% of total TG*	*Nature*	*% of total TG*
LLnLn	0.8	OOL	7.6
LLLn	7.5	SLL	4.2
OLnLn	0.3	POL	7.8
LLL	19.3	PPL	1.5
OLLn	5.0	OOO	1.9
PLLn	2.8	SOL	3.0
OLL	20.4	POO	1.3
OOLn	0.8	PPO	1.0
PLL	13.1	PSO	0.7
POLn	0.7	–	–

Table 11.3. Distribution of fatty acids (FA) on the 1, 2, 3 positions of soyabean oil triglycerides.

Nature of FA	*% of total FA in 2*	*% of total FA in 1*	*% of total FA in 3*
C16:0	1	14	13
C18:0	< 1	6	6
C18:1	21	23	28
C18:2	70	48	45
C18:3	7	9	8

Table 11.4. Composition of soyabean oil fatty acids.

Nature	*% of total FA*
C14:0	< 0.2
C16:0	8–13
C16:1	< 0.2
C17:tot.	–
C18:0	2–5
C18:1	17–26
C18:2	50–62
C18:3	4–10
C20:0	< 1.2
C20:1	< 0.4
C22:0	< 0.5

The unsaponifiable fraction of oil represents 1.6 per cent crude oil and 0.6 to 0.7 per cent **refined** oil. It is composed mainly of sterols and tocopherols (Table 11.5).

Table 11.5. Composition of unsaponifiable fraction of soyabean oil.

Unsaponifiable fraction: 0.5–1.6%			
Sterols (mg/100 g)	250–418	Hydrocarbon (mg/100 g)	
		Tocopherols (mg/100 g)	80–167
Sterol composition (% total sterols)		Tocopherol composition (% total tocopherols)	
Cholesterol	< 1	α-Tocopherol	5–10
Brassicasterol	–	β-Tocopherol	2–3
Campesterol	19–23	γ-Tocopherol	44–60
Stigmasterol	17–19	δ-Tocopherol	30–43
β-Sitosterol	47–59	Tocotrienols	–
Δ5-Avenasterol	2–4		
Δ7-Stigmasterol	1–3		
Δ7-Avenasterol	1–2	Triterpenic alcohols (mg/100 g)	
Ergosterol	< 3		

Triglyceride Structure

Because of the high unsaturated fatty acid content of soyabean oil, nearly all of the glyceride molecules contain at least two unsaturated fatty acids and di- and trisaturated glycerides are essentially absent. The fatty acids are not randomly distributed in a single glyceride molecule and the glyceride is biosynthesised with a definite configuration. Each hydroxyl group of glycerol is unique and has been assigned a stereospecifically numbered (*sn*) position'. The following is illustrative of a general structure:

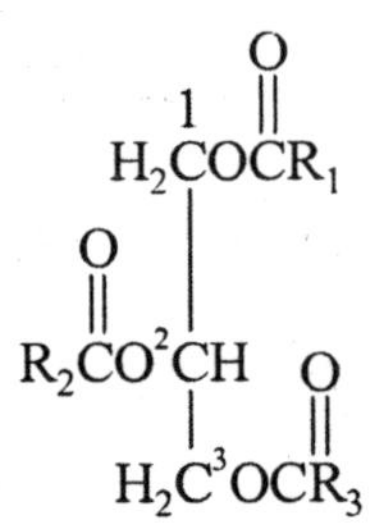

A typical distribution for the soyabean oil triglyceride structures has been given as follows: (S = saturated, U = unsaturated fatty acid):

SSS	0.07
SUS	5.2
USS	0.4
USU	0.7
UUS	35.0
UUU	58.4
Total	99.8

Unsaponifiable Matter

The unsaponifiable matter in soyabean oil has several minor constituents that have commercial value, such as the phytosterols and tocopherols.

Following is average composition for crude soyabean oil:

Unsaponifiable matter, 1.6 per cent.

Hydrocarbon (squalene), 0.0135 per cent.

Sterols, 0.365 per cent.

Tocopherols, 0.124 per cent.

The majority of vegetable oils contain 100–500 mg of sterols per 100 gram of oil, with soyabean oil containing an average of 327 mg per 100 gram. The plant sterols in the oil include β-sitosterol, compesterol, stigmasterol and others in lesser amounts. Also, in addition to the phytosterols, there are small amounts of 4-methyl-sterol and tri-terpene alcohol compounds present.

Composition for the hydrocarbon fraction of unsaponifiables in soyabean oil is given in Table 11.6.

Table 11.6. Composition for the hydrocarbon fraction of unsaponifiables in soyabean oil.

n-Alkane		4%
Nonacosane	27%	
Triacostane	8%	
Untriacontane	65%	
Alkenes		> 50%
Squalene	50%	
Branched chain		45%
Alkanes C_{30}–C_{35})		

Squalene is the biosynthetic precursor of sterols. There are at least seven types of tocopherols in soyabean oil; they have antioxidant properties of different degrees. The α, γ and δ-tocopherols are present in both crude and in refined soyabean oil. Although γ-tocopherol is present in the largest amounts, α-tocopherol has the highest vitamin E activity and δ-tocopherol functions better as an antioxidant. β-Tocopherol represents less than 3 per cent of the total tocopherols.

PHYSICAL PROPERTIES OF SOYABEAN OIL

Physical properties of soyabean oil depend on climate and variety. Processing the oil also has an influence on the presence and level of the many minor constituents.

Important physical properties of soyabean oil are summarised in Table 11.7. The values reported are not absolute because the composition and many of the properties depend on degree of unsaturation and other factors.

The *densities* of soyabean oil at several temperatures are given in Table 11.8. The densities of oils depend on temperature; they have an inverse relationship with molecular weight and a direct one with the degree of unsaturation.

Lund has developed an equation based on saponification and iodine values of vegetable oils to predict their specific gravity as follows:

Specific gravity = 0.8475 + 0.00030 (saponification value) + 0.00014 (iodine value)

Saponification values and iodine values can be determined by AOCS Methods Cd 3–25 and Cd 1–25.

Pryde states that the refractive indices of liquid fats have a direct relationship to molecular weight (and fatty acid chain length), to the degree of unsaturation and particularly to the degree of conjugation but have an inverse relation to temperature. Glycerides have considerably higher refractive indices than those of individual fatty acids.

Table 11.7. Selected physical properties of soyabean oil.

Property	*Value*
Specific gravity, 25°C	0.9175[a]
Refractive index, n_D^{25}	1.4728[b]
Specific refraction, r^{20}	0.3054
Viscosity, centipoises at 25°C	50.9[a]
Solidification point (°C)	−10 to −16°C
Specific heat (cal/g at 19.7°C)	0.458
Heat of combustion (cal/g)	9478[c]
Smoke point [°C (°F)]	234 (453)
Flash point [°C (°F)]	328 (623)
Fire point [°C (°F)]	363 (685)

[a] IV = 132.6; [b] IV = 130.2; and [c] IV = 131.6.

Table 11.8. Density of soyabean oil at several temperatures.

	Density, g/ml	
Temperature	*Magne and Skau*[a]	*Johnstone*[b]
−10.0	0.9410	–
0.0	0.9347	–
10.0	0.9276	–
25.0	0.9175	0.9171
37.8	–	0.9087
40.0	0.9075	–
50.0	–	0.9004

[a] Commercial, edible soyabean oil having an iodine value of 132.6 and containing 0.10 per cent free fatty acid.

[b] Soyabean oil that had been refined, bleached and deodourised had the following characteristics: iodine value, 130.1, peroxide number, 5.1: acid number, 0.11; thiocyanogen number, 80.0; and phosphorus content, 0.00095%.

An equation relating to refractive index to iodine value has been proposed by *Zeleny* and *Coleman*.

$$n_D^{25} = 1.45765 + 0.0001164 \text{ (iodine value)}$$

Since many fats and oils may be partially solid at 25°C (77°F), refractive indices should normally be determined at 40°C (104°F). In order to estimate the refractive index of an oil at other temperatures, the following equation from AOCS Method Cc 7–25 may be used:

$n''_D = n'_D + 0.000385\,(t - t')$, where t is the temperature in degrees Celsius at which the experimental determination was made and t′ is the temperature for which an estimate of the refractive index is needed.

Viscosities for soyabean oil and for its solution in hexane for six different temperature [–20 and –10°C (–4 and 14°F)] have been reported by *Magne* and *Skau*. *Interesse* and D' *Avella* have determined viscosities of soyabean oil over a 20–160°C (68–320°F) temperature range at 10°C intervals. *Detwiler* and *co-workers* have shown an inverse relationship between viscosity and unsaturation.

The melting point of triglycerides is usually determined by the component of fatty acids. Melting points increase with increasing chain length and decrease with increasing concentrations of *cis* double bonds but rise to an intermediate value with increasing concentrations of *trans* unsaturation. In some instances, melting points are also dependent on the position of the double bond; they rise as the double bond approaches either end of the molecule. In case of products with highly saturated fatty acids, the triglyceride melts above the fatty acid; with unsaturated-type products, the reverse is true (Table 11.9). Mixtures of triglycerides have depressed melting points compared to that for the homogeneous triglyceride.

Table 11.9. Melting points of fatty acids and their triglycerides present in soyabean oil and partially hydrogenated soyabean oils.

Fatty acid		*Triglyceride*		
Name	*Melting point (°C)*	*Name or composition*	*Melting point (°C)*	
			I (β)	*II (β')*
Palmitic	62.9	Tripalmitin	65.5	56.0
Stearic	69.6	Tristearin	73.0	65.0
		18:0–16:0–16:0	62.5	59.5
		16:0–18:0–16:0	68	65
		18:0–16:0–18:0	68	64
Oleic	16.3	Triolein	5.5	-12
		16:0–18:1–16:0	35.2	30.4
		18:0–18:1–18:0	41.6	37.6
		16:0–18:1–18:1	19.0	–
		18:0–18:1–18:1	23.5	–
Elaidic	43.7	Trielaidin	42	37
Linoleic	– 6.5	Trilinolein	– 13.1	–
Linolenic	– 12.8	Trilinolenin	– 24.2	–

Soyabean oil, like other fats, requires a lower temperature for solidification than for melting, the difference being only a few degrees when a barely melted oil is cooled again. The solidification point of soyabean oil is –16 to –10°C (3–14°F).

With an increase in the degree of hydrogenation, the melting as well as the solidification point of soyabean oil also increase. A fully hydrogenated soyabean oil can be expected to have a melting point no higher than about 68°C (154°F), i.e. at a melting point temperature that is about intermediate between tripalmitin [65.5°C (150°F)] and tristearin [73.0°C (163°F)].

The specific heat of soyabean oil varies directly with the chain length and temperature but inversely with degree of unsaturation (Table 11.10). Smoke, flash and fire points to a major extent are functions of the free fatty acid content in soyabean oil because the fatty acids are much more volatile than the glycerides. The temperatures for these properties are generally similar for all vegetable oils having comparable fatty acid composition and low free fatty acid content. The temperatures for smoke, flash and fire points of soyabean oil are listed in Table 11.11.

Table 11.10. Specific heats for soyabean oil with an iodine value of 128.3.

Temperature		
°C	*°F*	*Specific heat (cal/g/°C)*
1.2	34.2	0.448
19.7	67.5	0.458
38.6	101.5	0.469
60.9	141.6	0.479
70.5	158.9	0.490
80.4	176.7	0.493
90.4	194.7	0.504
100.4	212.7	0.508
120.8	249.4	0.527
141.3	286.4	0.531
161.9	323.4	0.550
182.7	360.9	0.567
200.1	393.6	0.594
250.5	483.8	0.621

Table 11.11. Smoke, flash and fire points of soyabean oil.

Soyabean oil				
Description	*Free fatty acid content (%)*	*Smoke point [°C (°F)]*	*Flash point [°C (°F)]*	*Fire point [°C (°F)]*
Refined and bleached	0.010	234 (453)	328 (623)	363 (685)
Refined and bleached	0.010	228 (453)	329 (625)	363 (685)
Crude, expeller pressed	0.51	185 (365)	296 (565)	349 (660)

Soyabean oil is miscible with many organic solvents, particularly those that are apolar and aprotic, i.e. those that do not form hydrogen bonds. Accordingly, hydrocarbons, esters, ethers, ketones and chlorinated solvents are miscible with soyabean oil. Ethanol, isopropyl and isobutyl alcohols, ethylene dichloride, trichloroethylene, carbon tetrachloride and hexane have been examined in the laboratory as potential extraction solvents. The solubility characteristics of vegetable oils can be roughly estimated from their dielectric constants or solubility parameters as a measure of polarity (Table 11.12). Hexane is the preferred solvent for the commercial extraction of oil from soyabean flakes.

Furfural and soyabean oil are not miscible at normal temperatures and their mixtures separate into two phases: the solvent-predominant phase containing more the unsaturated glycerides and the oil-predominant fraction containing the more saturated glycerides.

Physical properties for solutions of soyabean oil in chlorinated solvents and in hexane have been reported by *Johnstone*. Trichloroethylene (TCE) boiling at 86.7°C (188°F) was applied many years ago as a safe nonflammable solvent for soyabeans; however, serious toxicity problems arose from the extracted oil meal when it was fed to cattle.

While acetone and methyl acetate are miscible with soyabean oil, at low temperatures the more saturated glycerides crystallise, leaving the more unsaturated glycerides in solution—this will make their use in a commercial operation impractical.

Anhydrous or aqueous ethanol is not a good solvent for soyabean oil at normal temperatures. As the temperature increases, however, the solubility also increases until the critical solution temperatures is reached, at which point the components become miscible. Methanol is a poor solvent because of its polarity (a dielectric constant of 32.6 compared to 24.3 for ethanol).

Table 11.12. Dielectric constants and solubility parameters of solvents vs. vegetable oils.

Solvent	*Dielectric constant*	*Solubility parameter*[a]
Water	78.5	23.53
Furfural	41.9	10.09
Methanol	32.6	14.50
Ethanol	24.3	12.78
1-Propanol	20.1	12.18
2-Propanol	18.3	11.44
Acetone	21.4	9.62
Sulphur dioxide	14.1	–
Trichloroethylene	3.4	9.16
Vegetable oils	3.0–3.2	–
Cyclohexane	2.05	8.19
Hexane	1.89	7.27
2-Methylpentane	–	7.03
Propane	1.61	5.77
Carbon dioxide	1.60	–

[a] Solubility parameter, δ, is defined as $\delta = (\Delta E/V)^{1/2} = [(\Delta H - P\,\Delta V)/V]^{1/2}$.

RECOVERY OF OIL FROM SOYABEANS

After pretreatment, the oil is extracted from the soyabean. The three most common procedures are: hydraulic pressing, expeller pressing and solvent extraction. The most prevalent extraction method today in modern processing facilities is solvent extraction using *n*-hexane.

There are four parts to the solvent extraction process: (i) storage and preparation of beans; (ii) extraction of oil from the beans; (iii) reclamation of solvent from the oil and meal; and (iv) oil storage. A flow sheet of a solvent extraction plant is shown in Fig. 11.4.

Storage and Preparation of Beans

Before they are ready for processing, the beans are cleaned, dried to about 10 per cent moisture and preferably conditioned by storage for about 10 days to facilitate dehulling. For optimum operation of the solvent extraction process, 9.5–10 per cent moisture is desirable and preparatory flaking is accomplished most satisfactorily at 74–79°C (165–175°F).

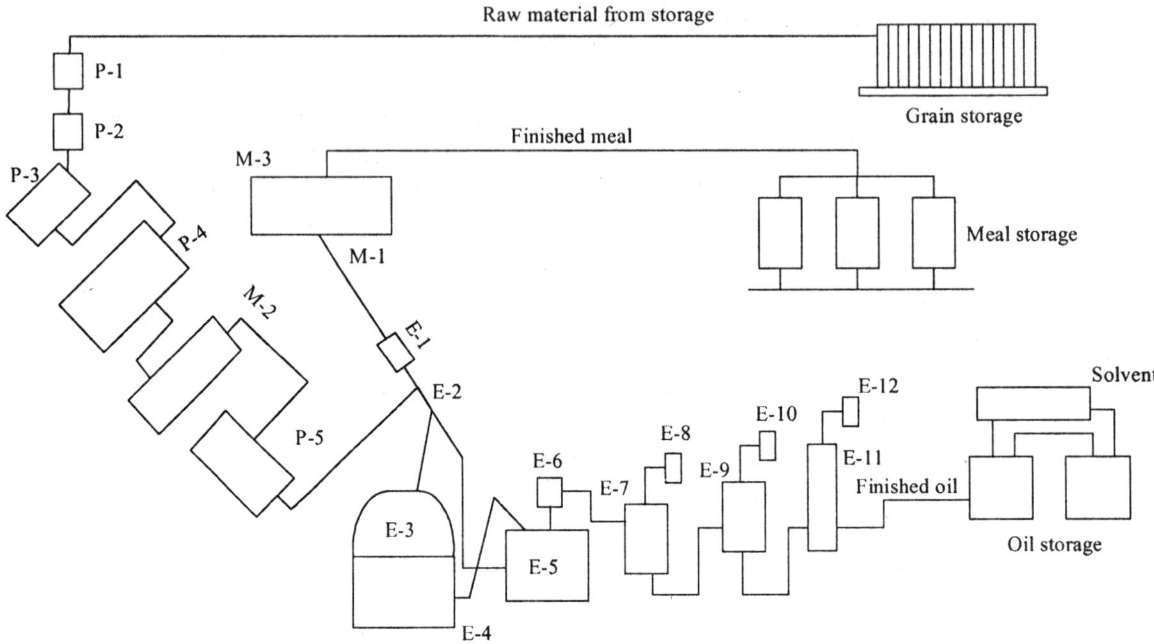

Direct extraction preparation

P-1 Surge bin
P-2 Scale
P-3 Cracking mill
P-4 Meats conditioner
P-5 Flaking mill

Meal handling

M-1 Louvered meal cooler
M-2 Meal grinder
M-3 Meal screen

Solvent extraction

E-1 Raw flake elevator
E-2 Extractor feed conveyor
E-3 Stationary basket extractor
E-4 Spent flake elevator
E-5 Desolventiser toaster
E-6 Vapour scrubber
E-7 1st stage evaporator
E-8 1st stage condenser
E-9 2nd stage evaporator
E-10 2nd stage condenser
E-11 Final oil stripper
E-12 Vacuum condenser

Fig. 11.4. Flow sheet of a soyabean extraction plant; preparation, extraction and meal handling.

The beans are conveyed to flaking rolls (Fig. 11.4). The roll stand consists of a pair of smooth-surface rolls regulated to produce flakes about 0.010 inch (0.254 mm) thick in one pass. Large-volume extractors can utilise somewhat thicker flakes; the thicker flakes pack better and require an increase in extraction time. After leaving the rolls, the flakes are conveyed by mass flow-type enclosed conveyors designed to minimise flake breakage (Fig. 11.4).

Solvent Extraction

In the solvent extraction process, the oil-bearing material is soaked in hexane in an extractor so that the oil is dissolved in the solvent to form a miscella, which drains from the meal. The basic types of commercial extractors in current use today fall into the categories of either percolation type or immersion types, the percolation process being the most efficient.

Percolation extractors

Percolation extractors are used worldwide and differ only in mechanical design. They are well suited to larger-capacity operations where limited space is available. For this reason, they are considered to be more efficient than the immersion type. In percolation extractors the liquid solvent or miscella is pumped over a bed of soya flakes, percolates down through the bed and leaves at the bottom through a perforated plate or mesh screen (Fig. 11.5).

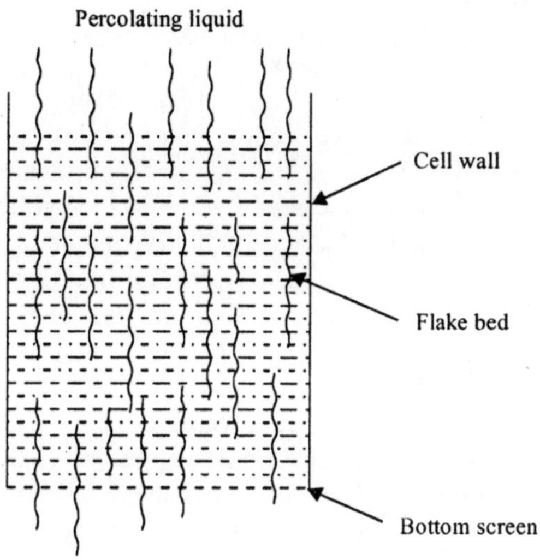

Fig. 11.5. Percolation extraction cell.

Rotary and stationary basket types consists of a vertical cylindrical shell within which are baskets or cells, either rotating or stationary, around a central shaft. Their small size is a definite advantage., but they require a more elaborate technical layout.

Chain and basket types are similar to percolation extractors. Separate baskets having fixed perforated bottoms are used to transport and extract the flakes. Both countercurrent and cocurrent principles are employed in their operation. The horizontal type is now used more commonly than the vertical unit. The miscella flow and design in the horizontal version are similar to those found in a rotary-type percolation extractor. The Crown extractor combines both percolation and immersion principles. Employing a vacuum filter system wherein the natural drainage force of gravity is assisted by vacuum under a filter cloth separation medium represents another form of perforation extraction.

Immersion extractors

Immersion extractors are not used commercially for soyabeans. The wet spent flakes leaving the extractor contain approximately 35 per cent hexane, 7–8 per cent water and 0.5–1.0 per cent oil and they are carried into the meal desolventising system (Fig. 11.4) commonly referred to as the D–T (desolventiser–toaster) unit. Miscella coming from the extractor is a mixture of crude oil, solvent moisture and fines; the latter must be removed by filtration before the oil freed of solvent is pumped to storage. Removal of hexane from the miscella exposes the oil to elevated temperatures that can be detrimental to crude oil quality. Prime concerns are: (i) avoiding oxidation of the oil; (ii) preventing removal of tocopherols that serve as natural antioxidants; and (iii) preventing formation of colour bodies and/or colour fixation.

Most US processors employ a dual-storage evaporator followed by a low-pressure steam stripping system for miscella handling. Full miscella from the extractor (20–30 per cent oil) is concentrated to 65–70 per cent oil (Fig. 11.4) at below atmospheric pressure in the first-stage evaporator, after which steam at 10–20 psig is used in the second-stage evaporator (Fig. 11.4) to further concentrate the miscella under atmospheric pressure to about 90–95 per cent oil, which then enters the final oil stripper (Fig. 11.4)–a steel cylindrical vacuum column—where it is flashed to about 98 per cent oil. Finally, low-pressure stripping steam is used to remove final traces of hexane. To prevent autoxidation, oxygen-free steam should be used in this step. The finished oil product leaves the column at about 99.8 per cent or higher oil content and is pumped to a storage tank.

Some soyabean extraction plants send part of their stream through a flash desolventiser to preserve the high-nitrogen solubility of the flakes for the production of refined protein products while the balance will go through a desolventiser–toaster to serve the animal feed industry.

Modern solvent distillation and recovery systems are designed so that moisture and other volatile materials in crude oil are within commercially acceptable levels, i.e. 0.15 per cent or less.

Since miscella handling has a definite effects on crude oil quality, miscella stripping should be carried out at the lowest possible temperatures and shortest residence time. Higher temperatures may also partially remove natural antioxidants, which in turn lower oxidative stability. Temperatures should never exceed 110°C (230°F) during stripping.

Extraction system either are equipped with oil-cooling devices or they can be designed to include them. For example, crude oil coming from the final stripper at 93.3–104.4°C (200–220°F) can be cooled to 65.6°C (150°F) maximum after passage through an oil dryer and cooler. More recent advances in extraction technology include exhaustive extraction of full-fat soyabean flakes with supercritical carbon dioxide (SC-CO_2), which yields an oil that is comparable to hexane-extracted oil; the oil, however, has a significantly lower chromatographic refining loss and phosphorous content.

Reverchon and *Osseo* have reported that there are three different process alternatives for the production of soyabean oil by supercritical carbon dioxide extraction. The first two processes are based on high-pressure extraction followed by separation induced by pressure reduction. Solvent recovery in these two schemes was done either in the liquid or gas phases. The third alternative is based on an isobaric scheme, i.e. the oil extracted at high pressure, is separated by changing the temperature in the separator. *Reverchon* and *Osseo* states that operating costs fall in the same range as those for conventional hexane extraction plants. The quality of a soyabean oil is determined by (i) residual white flake fat; (ii) oil yield; (iii) miscella concentration; and (iv) spent flake hexane extraction. Among the many variables that can affect the quality attributes of the solvent extraction process are: flake structure, thickness, temperature, moisture, percentage of fines and expanded flakes, cracked beans in flakes, solvent ratio, temperature, purity, distribution and retention time in the extractor.

Oil Storage

With limited access to air, crude soyabean oil can be stored in larger tanks for a long time. It is recommended that the oil be cooled to near ambient temperature before being pumped to storage. Crude oil in storage should be clean and low in moisture in order to prevent hydrolysis. Crude oils high in phosphatides deposit a layer of gummy material upon long standing in tanks or during long shipments in tank cars.

EDIBLE OIL PROCESSING

The extracted crude oil need further treatment to convert it to a bland, stable and nutritious product because crude soyabean oil contains small and variable amounts of non glyceride impurities impacting on the quality of the oil. These impurities also have an effect on the processing behaviour and yield of the finished product.

The impurities are of two general types: oil-insoluble and oil-soluble. The insoluble impurities consist of seed fragments, excess moisture and a waxy fraction that appears in refrigerated oil, thus making it appear turbid or cloudy. Some of these impurities such as seed fragments and meal fines are normally removed by filtration.

The oil-soluble impurities are more difficult to remove. They include free fatty acids, phosphatides, gummy or mucilaginous substances, colour bodies, protein or protein fractions, tocopherols, sterols, hydrocarbons, ketones and aldehydes.

Some of the nonglyceride constituents are useful. The tocopherols, for example, help protect the oil from oxidation. The sterols—being odourless, tasteless, colourless and heat stable—also are not detrimental to oil quality. The other impurities, however, should be removed because they adversely affect the flavour, odour, colour heat stability and storage stability of the oil.

Basic Processing Operations and Principal Edible Oil Products

Following steps are followed for basic oil processing; (i) crude oil storage; (ii) degumming and/or caustic refining; (iii) bleaching; (iv) hydrogenation and/or winterisation; (v) deodourisation; and (vi) finished oil storage. Each step removes specific components from the oil.

The specific components removed from the oil during these steps may be classified as follows:

1. Naturally occurring: gums, phospholipids, prooxidant metals, pigments, colour bodies, tocopherols and free fatty acids.
2. New compounds formed during processing: soaps, oxidation products, hydroperoxides, polymers and their decomposition products; colour bodies, isomers and high-melting triglycerides.
3. Processing adjuncts: hydrogenation catalysts, bleaching clay and metal scavengers that precipitate from finished oils, i.e. citric acid.
4. Contaminants introduced by processing: moisture, trace metals, carbonaceous and oil-insoluble materials.

All soyabean oil processing steps consists of independent operations that are usually not connected with each other by continuous flow. Thus, crude soyabean oil is converted into a quality edible oil by a series of refining operations, usually in the sequence as shown in Fig. 11.6.

Oil Degumming and Soyabean Lecithin

The degumming step removes the phosphatides and mucilaginous gums that when hydrated become insoluble in the oil. Free fatty acids, colour bodies and metallic prooxidants are removed to a varying degree in the alkali-refining step. Although the crude soyabean oil may or may not be degummed prior to caustic refining, the removal of nearly all of the phospholipids is essential to produce a finished oil of optimum colour, flavour and oxidative stability. The presence of substantial amount of phospholipids can lead to dark-coloured oils and can also serve as precursors of off-flavours.

Crude soyabean oil, when processed from undamaged beans, contains 500–700 ppm phosphorus (1.5–2.1 per cent phospholipid). Degumming of the oil from undamaged soyabeans removes 79–98 per cent of the phosphorus. Phosphorus content of the oil is lowered from 500–900 ppm in the crude to 12–170 ppm in the degummed oil. Water-degummed oil from damaged beans may have an abnormally high phosphorus level. The nonhydratable phosphatides are believed to be calcium and magnesium salts and may not all be removed by alkali refining either.

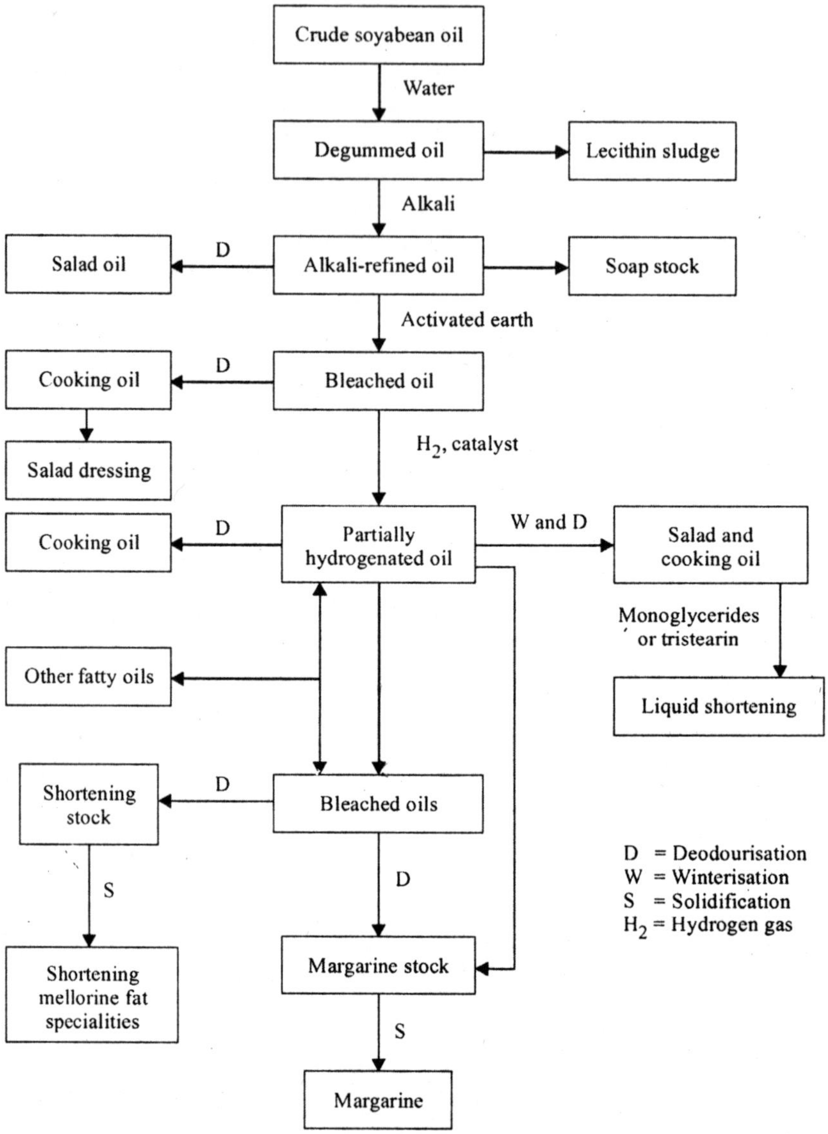

Fig. 11.6. Schematic diagram for manufacture of edible soyabean oil products.

In a commercial operation, crude soyabean oil is mixed with 1–3 per cent of water, based on the oil volume and the mixture agitated mechanically—care being taken to avoid air entrainment—for 30–60 minute at about 70°C (158°F) to hydrate the phosphatides and other colloidal impurities. The hydrated phosphatides and gums can be separated from the oil by settling, filtering, or centrifuging—centrifuging being the more prevalent practice. In some extraction plants the gums are allowed to hydrate from steam condensed in the last stage of steam stripping for removal of solvent from the miscella; in others a small portion of water is mixed into the dry oil.

A combined treatment with citric acid, sodium silicate and silica adsorbents has been described for processing crude and water-degummed oils to overcome difficult separation problems normally requiring a larger number of centrifuges to achieve good separation of the hydrated phospholipids from the oil. Using this treatment, oils of good quality are obtained by providing enhanced phospholipid removal without the generation of an aqueous effluent.

In another development, a crude soyabean oil was degummed in a reactor by countercurrently contacting the oil with supercritical CO_2 at 55 MPa at 70°C (158°F). The phosphorus content of the crude oil was reduced from 620 ppm to less than 5 ppm. The degummed feedstocks can be fed (without further processing, i.e. bleaching) directly to a batch physical refining step consisting of simultaneous deacidification/deodourisation with and without citric acid. Supercritical CO_2 processed oil had equivalent flavour scores, both initially and after 60°C(140°F) ageing and light exposure, when compared with the control soyabean oil. Results also showed that bleaching with absorbent clays may be eliminated by the supercritical CO_2 countercurrent processing step because considerable heat bleaching occurs during deacidification/deodourisation.

Ultrasonic degumming has also been explored and shown to be effective as a pretreatment to physical refining. Odour and flavour evaluation indicated that the salad oil produced by the process of ultrasonic degumming/deodourising/deacidifying was equivalent in quality and stability to a conventionally processed salad oil.

Refining

Almost all edible soyabean oil can be refined with alkali, usually caustic soda, in a continuous processing system. As discussed later, steam refining or physical refining, which is similar to steam deodourisation, is another method of refining for oils where the impurities are primarily free fatty acids. The following discussion covers the key points of refining soyabean oil.

Chemical (caustic) refining

Either a batch or a continuous process can be used to refine soyabean oil. Almost all soyabean oil is now refined by the continuous process in the United States. In a commercial operation the soyabean oil from the storage tank—or directly from the oil dryer in the solvent extraction plant (and in case of degummed oil, from the degumming centrifuge)—is pumped through a heat exchanger [to adjust the temperature to 38°C (110°F) into a 'day' tank] (Fig. 11.7). The oil is thoroughly agitated, a sample is taken and analyses are made for free fatty acids and neutral oil contents (to determine the amount of caustic required for the refining process). The oil is then pumped to a line mixer where it is continuously mixed with a proportioned steam of dilute caustic soda solution.

Where the crude oil is refined directly without degumming, it is common practice to pretreat the oil for 4 hr. with 300–1000 ppm (75 per cent concentration) of phosphoric acid to facilitate removal of the last traces of phosphatides. Degummed oil usually does not require pretreatment. The caustic concentration can be varied to compensate for changes in the oil feed characteristics.

Most soyabean oils are refined with 0.10–0.13 per cent excess dry basis. Caustic strength of 17–18 Bé (12–13 per cent sodium hydroxide content) are used with soyabean oil.

Pressure- or hermetic-type centrifuges are used for both degumming and refining of the oil. Changes in the position of the zone of separation can be achieved by adjusting the back pressure applied to the light-phase discharge. Complete separation of the two phases is difficult to achieve, however, with either system.

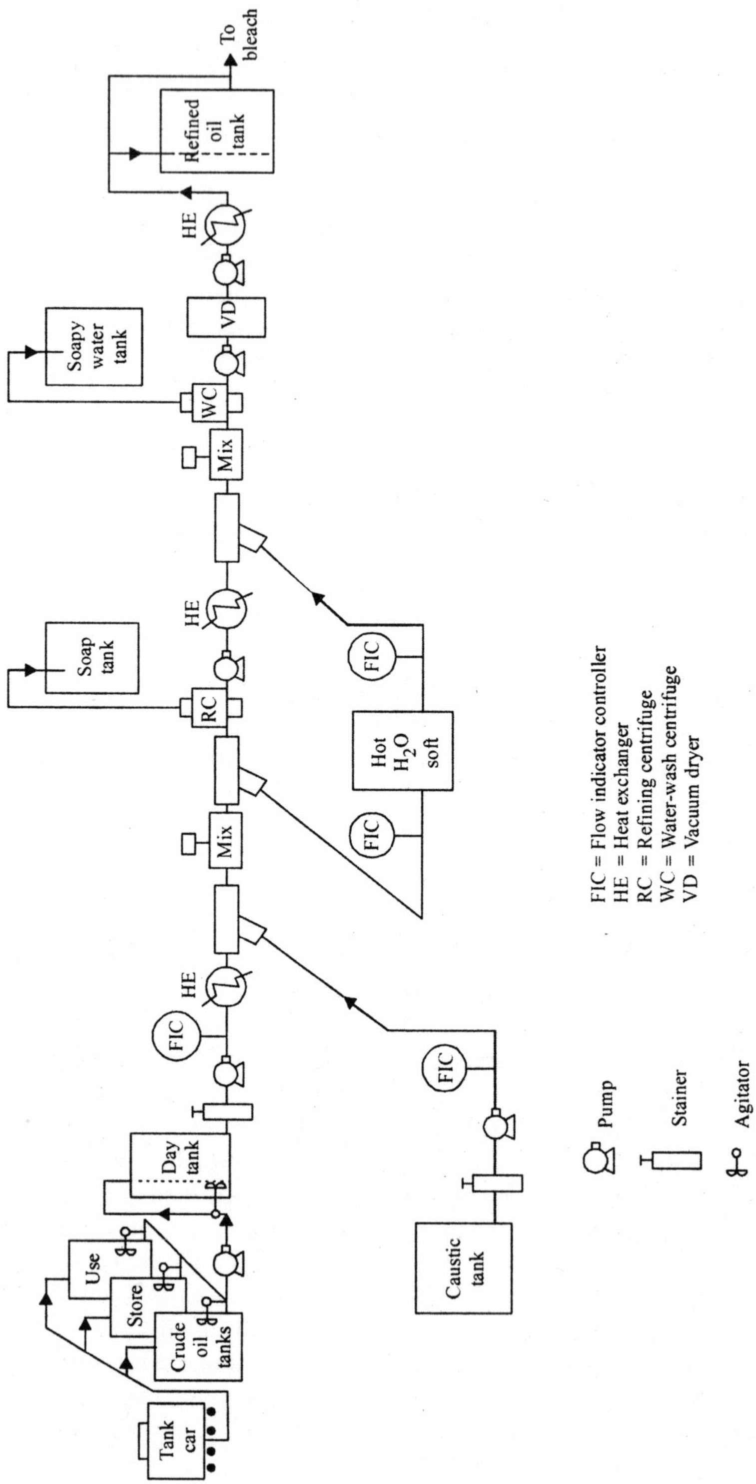

Fig. 11.7. Caustic refining process.

The subsequent steps in the refining process as follows:

The caustic refined oil discharge from the centrifuge is heated to 190°F (88°C) and is mixed with 10–12 per cent by weight of soft water that has been heated to 200°F (93°C). The water-oil mixture then passes through a high-speed shear mixer to obtain intimate contact for maximum soap transfer from the oil to the water phase. The water-oil mixture next passes through a second centrifuge to remove residual soaps; the water-washed oil is discharged as the light phase and the soapy water as the heavy phase. The water-washed, refined oil is then sprayed into a continuous vacuum dryer operated at 28 inch (58 mm) Hg pressure where the moisture content is reduced to less than 0.1 per cent. The refined soyabean oil is next transferred to the bleaching process, or alternatively is continuously cooled to 120°F (49°C) and shipped as once-refined soyabean oil, or it is pumped to storage to await further processing into edible products.

Soap-containing oils must contain less than 0.2 per cent moisture before silica adsorbents are added to ensure complete soap removal. The major by-products of refining soyabean oil are the fatty acids. After the free acids have been saponified with caustic, the soapstock is separated from the natural oil in a centrifuge. Both batch and continuous methods of acidulation are used. In either process, the diluted soapstock is treated with an inorganic acid, normally sulphuric, to convert the soap into fatty acids. The soapstock from the refining centrifuge and the wash water phase from the wash centrifuge are combined and sent to storage prior to acidulation.

From the acidulator the material flows by gravity into the settling tank where two distinct layers are formed. These layers are then separated by centrifugation in the continuous system or by gravity settling in the batch process. The lower aqueous layer, having a very low pH, is discharged to the sewer after neutralisation with caustic; the upper layer, which is a highly acidified fatty acid solution (acid oil), is then dried by settling and drawing off the water or it is pumped through a vacuum chamber to storage. This material is sold to fatty acid producers, soap makers and some direct users (e.g. foundries). Much of the acidulated soapstock goes to the animal feed industry, where it serves as a high-energy feed ingredient.

In most recent developments in oil refining, the distinction between degumming and alkali refining is becoming more and more blurred. With one type of oil and quality, the most economical refining method can be achieved just by degumming alone; this would favour an acid/water degumming method. On the other hand, another oil type or quality may favour a combined degumming/alkali refining treatment. Water degumming, as such, remaining important in operations that want to produce lecithin. Multipurpose degumming/alkali refining processes are now advocated that can be used either to degum in a variety of ways or to do complete alkali refining, depending on oil type and quality.

In recent years, there is also great interest in increasing energy recovery in edible oil refining.

Physical (steam) refining

The change from water degumming to acid–water degumming has made it possible to eliminate alkali refining entirely in favour of what is called physical refining. This eliminates one significant source of water usage and of waste-water; it also requires less capital investment in process equipment. One negative aspect is that it somewhat restricts the types of oils that can be processed at a plant.

In the stream refining process the phosphatide-free oil from a pretreatment process is deodourised by operations conducted under high vacuum at high temperature. To achieve the full potential of the steam-refining deodouriser, the process is designed with the objectives such as: (i) reduction of the free fatty acids from 5.0 to 0.03 per cent or less; (ii) production of a fully deodourised product; (iii) operation without substantially greater utilities consumption than a standard deodouriser; and (iv) recovery of the fatty acids from the sparge steam.

In a modified physical refining (MPR) process, soaps are generated by adding caustic to crude or degummed oil to facilitate adsorption removal of phospholids on an amorphus silica hydrogel.

In physical refining, total adsorbent usage can be reduced while improving product quality. Also the primary and water wash centrifuge steps and the associated by-product waste streams, which are needed in causting refining, can be eliminated. Acid treatment, however, is required when nonhydratable phospholipids are present in the feedstock.

Other novel refining processes

A total degumming process for removing essentially all the phosphatides from soyabean oil, using first an acid and then an alkali and two centrifuges, has shown higher yields than conventional refining. This process, however, does not remove prooxidant metals efficiently and, for this reason, has not found commercial acceptance in the United States.

Dijkstra has described a novel process wherein the washing water is recycled to the oil feed and used to dilute concentrated alkali. This process does not generate an aqueous effluent and can be used for both acid and alkali refining, thus allowing refiners to change gradually from alkali refining to physical refining.

Neutralisation of soyabean oil in miscella with alkali solutions assures elimination of free fatty acids without notable change in the phosphatides content. The phosphatidic concentrate obtained from an oil previously neutralised in the miscella was of higher quality than the phosphatidic concentrate obtained from the oil of the starting miscella.

Aqueous ammonia has the advantage of being safe for the environment because the deacidification agent can be repeatedly reused. Oils, especially soyabean oil with low degree of oxidation, can be fully deacidified only with the help of ammonia. The same effect can frequently be achieved by a preliminary desliming with 5 per cent formic or citric acid. Deodourisation at 210°C of oils that have been deacidified with ammonia and washed with water yields bland and pale edible oils having good storability.

Effects of refining on oil quality

List and *Erickson* state that of all the unit processing operations, refining has the most significant effect on oil quality as measured by colour, oxidative stability and storage properties. If soyabean oil is not properly refined, subsequent processing operations such as bleaching, hydrogenation and deodourisation will be impaired so that finished products will fail to meet quality standards. Also, poor refining will reduce the yield of natural oil, thereby lowering manufacturing profits.

Colour reduction during refining entrails the removal of three types of pigments and colour bodies; (i) those largely removed during refining that may be associated with iron, phosphatides, or both; (ii) those remaining after refining that are absorbable by bleaching clay; and (iii) those that change to a colourless form or are destroyed upon heating to a high temperature during deodourisation. The amount of colour at all stages of processing is somewhat related to phosphorus content; but the influence of phosphorus-associated pigments on the colour or refined oil disappears below a phosphorus content of 25–30 ppm, leaving only the influence of the other two types.

Caustic refining removes free fatty acids to a 0.01–0.03 per cent level and removes virtually all the phosphatides. Crude soyabean oil contains trace amounts (several part per million) of prooxidant trace metals such as iron and copper. Caustic refining usually removes 90–95 per cent of these metals. However, it should be emphasised that even though caustic refining reduces metallic contamination to low levels, residual iron and copper still remain strong prooxidants in refined oils and must be taken into account during storage and handling.

At a constant percentage of water, the total amount of caustic used influences colour removal, i.e. the more caustic used, the lower the colour of refined oil. However, part of the colour removal can be associated with type 1 pigments that are removed by water but not caustic.

Beal and co-workers—using the active oxygen method (AOM) test—have concluded that soyabean oil should be refined to a phosphorus content of no less than 2.0 ppm but no more than 20 ppm. Above or below these values, oxidative stability was impaired. Study has shown that phosphorus removal during refining is a direct function of the excess amount of caustic used and the lye strength and for a given lot of crude oil with a known phosphorus content, refining conditions leading to a optimum phosphorus content can be approximated.

List and *Erickston* report that—plots of residual iron versus residual phosphorus contents of deodourised oils showed that iron increases at phosphorus content below about 1 ppm, reaches a constant value at about 2–20 ppm phosphorus, then begins to increases. Thus, the decreased oxidative stability at phosphorus content above 20 ppm can be explained by the sufficiently high iron content (i.e. greater than 0.2 ppm) which exerts a strong prooxidant effect. Similarly, decreased stability at phosphorus content below 2 ppm can also be explained because of the increased iron content.

At the same time, it should also be pointed out that the traditional method for calculating the amount of refining lye is based on the free fatty acid content and therefore, gives no indication of conditions leading to optimum phosphorus removal. Phosphatide content generally exceeds that free fatty acids in crude soyabean oil by a factor of about 6. In refinery process control, crude oil is usually evaluated for refining loss by the American Oil Chemist's Society (AOCS) chromatographic method.

Bleaching

Treatment of the refined oil with a bleaching adsorbent removes additional pigments, metallic compounds, residual soaps and trace amounts of prooxidants that were not removed by caustic refining and improves the oil's taste. These minute quantities of materials, if no removed, can cause problems in subsequent processing. Prooxidant metal ions present in the form of metallic soaps have a detrimental effect on the hydrogenation process as well as on the flavour and stability of the deodourised oil. Bleaching adsorbents also remove sulphur compounds, decompose peroxides and adsorb the aldehydes and ketones resulting from decomposition of peroxides.

When fatty oils are bleached chemically, colour reduction occurs because of oxidising reactions that have an undesirable effect on the flavour and oxidative stability of the oil. For this reason, chemical bleaching is not used for edible oils. Exposure of soyabean oil to light, particularly ultraviolet rays, has a bleaching effect, but light also results in a decrease in the oil's flavour stability. For this reason, it is recommended to keep light exposure of soyabean oil to a minimum.

Acidic earths are generally used for bleaching soyabean oil to obtain good chlorophyll removal, without which the finished oil may have a greenish cast. The degree of bleaching varies somewhat with the type of end product desired. Usually, a highly bleached oil is required to prepare the typical white-appearing shortening, but more colour–a tinge of brilliant yellow–is permissible for the oil used in margarines, salad oils, salad dressings and cooking oils.

In addition to colour removal, bleaching also almost always improves the initial taste, aged flavour and oxidative stability of the deodourised soyabean oil (Table 11.13).

Table 11.13. Comparison of bleached and unbleached soyabean oils[a].

	Flavour score (Peroxide value)	
Factor	*Bleached*	*Unbleached*
Initial flavour[b]	7.9 (0.8)	6.7 (0.5)
Aged flavour	6.4 (1.8)	5.2 (4.0)
8-hr. AOM peroxide value	2.2	20.0

[a] Both oils were refined and deodourised in a commercial plant with bleaching step being omitted for the one sample. Both oils were protected during processing to minimise oxidation and by addition of a metal scavenger to the deodourised oil.

[b] Based on flavour scale of 1–10 (highest quality). Low peroxide values are desired; the lower these values, generally the more stable the oil. Peroxide values expressed as milli-equivalent per kilogram of fat.

Some heat bleaching of the carotenoids also occurs at hydrogenation temperatures and more so at the deodourisation temperatures. Thus, the need for maximum colour removal in the bleaching step is reduced somewhat if the oil later undergoes hydrogenation and/or deodourisation, particularly so when deodourisation temperature exceeds 204°C (400°F). On the other hand, deodourisation of an unbleached soyabean oil can often result in a product with an unacceptable greenish colour.

Overbleaching may have an adverse effect on oil quality (such as colour reversion) and it is more costly. Bleaching conditions in the plant should be judiciously selected so that the bleaching effect and impurities removal in other processing steps will produce a high-quality finished oil.

Refined, bleached and deodourised oils going into shortenings consistently measure below 1.0 Lovibond red (5¼ inch cell). The finished oil used in most other edible products usually have a Lovibond red reading of 2.5 or below, except for certain grades of salad or cooking oils where darker colours are acceptable. Degradation of the protein in the soyabean can result in brownish and dark colours that also need to be removed.

Goebel gives the following Lovibond red colour values for finished oils: margarine basestock 4–5, first-grade frying fat 2.0–2.5, cooking fat 2.0, and white shortening 1.5 maximum.

Removal of the yellow and red pigments from soyabean oil is less difficult than removal of some forms of chlorophyll. If liquid soyabean oil has a pronounced greenish cast, shortening containing substantial amounts of the hydrogenated oil will have a grayish-white rather than a pleasing-white surface appearance and the shortening's interior colour will be a distinct green.

Pritchett and *Jung* found that chlorophyll content in normal crude soyabean oil (1000–1500 μg/l) could be reduced by about 25 per cent by alkali refining. Subsequent bleaching with acid-activated earth can further reduce chlorophyll content to as low as 15 μg/l. For this reason, many processors alkali treat green oils for removal of impurities, followed by bleaching to remove the green colour in the oil. Hydrogenation further reduces the residual chlorophyll content by about two-thirds. Deodourisation of the hydrogenated oil, however, had little influence on the chlorophyll content. *Beal* on the other hand, found that hydrogenation of soyabean oil with a copper chromite catalyst did remove almost all the green colour.

Since subsequent hydrogenation and/or deodourisation does reduce colour, it is customary to lower the Lovibond red reading from 4.0–9.0 units for the refined oil to 1.5–2.5 units for the refined and bleached oil. The combination of hydrogenation and deodourisation removes more of the red and yellow pigments than does deodourisation alone. The greenish tint is more noticeable at low levels of red and yellow pigmentation; therefore, when a partially hydrogenated oil is to be used in shortenings, it will be necessary to remove more of the chlorophylls by bleaching.

Specifications used for commercial trading stipulate that the Lovibond red colour of a refined and bleached oil prepared from a prime rude soyabean oil shall not exceed 6.0 units and that a discount shall apply whenever the colour exceeds 3.5 units. The specifications also call for green colour of the prime crude oil, as determined by a visual match, to be less than that of standard. A No. 2 grade crude soyabean oil has a darker colour and also is subject to a larger discount on the basis of its red colour.

Bleaching of soyabean oil is generally carried out according to one of three basic processes; batch atmospheric, batch vacuum and continuous vacuum.

Batch atmospheric bleaching

In this process, the heated oil [71.1°C (160°F)] is pumped into the tank equipped with steam coils or a jacket and a paddle agitator. Bleaching clay is added at the top of the tank with the agitator running and the oil is heated to the desired temperature [typically 100–104°C (212–220°F) or less]. The clay–oil slurry is then recirculated through a filter press and back to the bleach tank until it is clear; the clear oil is then pumped to storage or fed into the deodouriser.

Batch vacuum bleaching

Following are advantages of vacuum bleaching over atmospheric bleaching: (i) better protection of the oil against the harmful effects of oxidation; (ii) contact or bleach time can be shortened, thus further reducing the degree of oxidation; (iii) lower bleaching earth requirement (resulting in lower oil loss in the spent earth); (iv) lower colour in the bleached oil than obtained by atmospheric bleaching; and (v) when acidic earths are used, lower soap content in the bleached oil, resulting in less hydrogenation catalyst requirement.

In a typically batch vacuum process, bleaching clay is added to a portion of heated oil [71.1°C (160°F)] in an open slurry tank and then pumped to a vacuum bleach tank containing the balance of the oil. Vacuum bleach tanks are usually under 686–711 mm Hg (27–28 inch) of vacuum and are equipped with steam coils or jacket and an agitator. After 15–20 minute agitation at the bleach temperature, the oil is cooled to approximately 71.1°C (160°F), the vacuum is broken and the clay–oil slurry is filtered and cooled again, to 38–66°C (100–150°F) before exposure to air.

Continuous–vacuum bleaching

Countercurrent operation is more feasible in continuous bleaching within a closed system and two-stage systems are actually in commercial use. In a continuous-vacuum bleaching process, the degummed, refined oil is pumped in a properly proportioned steam at about 54°C (130°F) into the slurry tank at a rate regulated by a flow controller. Bleach clay and filter aid are also continuously fed into the slurry tank at a controlled rate that is adjusted as required by the characteristics of the oil being bleached. The slurry of oil, filter aid and bleaching clay is thoroughly mixed and then pumped or sprayed at a constant rate into the deaerating and dehydrating (upper) section of a vacuum bleacher that is maintained at 381 mm Hg (15 inch) absolute pressure. In the bleacher, the slurry is first deaerated by spraying through nozzles and then—after a retention period of about 7 minutes—is pumped or sprayed into the bleaching (lower) section of the tower through an external heat exchanger to be heated, under automatic temperature control, to a bleaching temperature of 104–116°C (220–240°F). The slurry is mixed thoroughly to provide for maximum contact between the oil and clay and is retained in the bleacher for a definite period of time. The bleached oil–clay slurry is then filtered in a closed press and cooled and the vacuum is broken. The bleached oil is then either pumped to storage or sent to a deodouriser.

After the press cake has reached its maximum thickness, or whenever the flow is diverted to another press, the press is blown with air or nitrogen to displace oil held up in the press chamber and then with dry steam (at 15–45 psig for 30–45 minute) to recover the entrained oil in the press cake. Oil content of the steamed cake can vary considerably, but typical values are in the 25–35 per cent range.

Types of adsorbents used

The most frequently used adsorbents used for bleaching soyabean oil are natural earth, activated earth and activated carbon. They differ in their performance and in a number of characteristics that affect the processor's choice—ranging from natural, unactivated earth that adsorb soap but have little other function to the expensive, highly activated earths. For each type of adsorbent, more than one grade is available.

Activated earths have a higher bleaching efficiency than natural earths and therefore, are preferred to bleach soyabean oil, particularly when the oil is dark in colour or high in chlorophyll content.

Tests by *Shaw* showed that highly adsorbent products give reduced or similar deterioration of the stored or aged oils. By taking advantages of reduced addition levels, products treated with high-efficiency adsorbents can produce oil loss and 'spent' bleaching earth disposal.

Effects of filtration bleaching on peroxide value (PV), thiobarbituric acid value (TAV) and carbonyl value (CV) of autoxidised soyabean oil were investigated by using 23 kinds of activated carbons in order to improve oil quality. It was suggested that hydroperoxides, aldehydes and ketones were adsorbed on the acids sites distributed over the surface or within the pores of the activated carbons while the autoxidised soyabean oil flowed through the packed column. The decrease in oxidative stability of treated soyabean oil seemed to be caused by the elimination of α-, β- and γ-tocopherols. δ-Tocopherol was chemically more stable than α-, β- and γ-tocopherol in autoxidised soyabean oil.

Bleaching of cottonseed and soyabean oils has been effected by hydrogen generated *in situ* by the action of aqueous copper sulphate solution on zinc dust. Yellow colour bodies are bleached more slowly by this process than the red bodies. Colour reduction up to a level of 70–74 per cent is attainable by this method.

It has been shown that lutein competes for adsorption sites on silica with other oils components and added solvents. Added alcohols (e.g. isopropanol) effectively inhibit lutein binding. According to *Mingyu* and *Proctor*, the effect of functional groups on lutein inhibition was in the order to alcohol > ketone > aldehyde > acetone.

Measurement of oil colour

The popular Wesson method uses colour matching with yellow-tinted and red-tinted Lovibond glasses. Visual comparative methods, however, have definite shortcomings in that an exact match for oil containing appreciable amounts of green or brown pigments is difficult to obtain. Nevertheless, much of the data reported in the literature, as well as at least one set of specification by the National Soyabean Processors' Association (NSPA) for commercial trading of soyabean oil, are based on Lovibond colour readings.

Absorption spectra obtained by spectrophotometric measurement of oil colours is a more precise determination and it is also better suited to identify and quantify the green and brown pigments in oils. Absorption spectra can also indicate whether the oils have been mishandled because oxidised and polymerised products have altered spectra from that of fresh oils.

New developments

In bleaching, the change from batch operations to continuous processing under vacuum has helped preserve the oxidative stability of oils much better than had been achieved previously. This also meant more efficient use of clay. Many processes installed in recent years have included the ability to give oils and acid pretreatment. Recent developments have been in the area of adsorbents, particularly synthetic materials based on silica. According to *Mag*, 'considerable research effort is focused on these materials and they are being used much more frequently. This may change the bleaching process significantly in the future'.

Hydrogenation Practices

A large portion of the soyabean oil produced is hydrogenated to produce partially hydrogenated and winterised soyabean oil, margarine oil, bakery and confectionery fats and shortening.

Partial hydrogenation of edible oil is carried out to obtain two important objectives: the shifting of the melting range to higher temperatures and the improvement of the stability of the products to oxidation and flavour deterioration. Both oil composition and physical characteristics of the oil are modified by hydrogenation. According to *Mounts*, the free fatty acid content of crude soyabean oil will affect the catalytic hydrogenation process, so it is necessary to refine the oil before hydrogenation. In addition to effecting a reduction of unsaturation, partial hydrogenation also causes geometric and positional isomers to be formed, resulting in a physical hardening of the oil, as manifested in an increase of its congeal point. Lightly hydrogenated oils are useful as dual-purpose cooking/salad oils because they remain liquid but, at the same time, exhibit improved stability during high-temperature applications. Hydrogenation will add, however, a certain amount of disagreeable flavour and odour; the oil usually has to be deodourised later.

According to *Mounts*, hydrogenation of soyabean oil is normally accomplished with a nickel catalyst (that has been chemically reduced with hydrogen to an active state and dispersed in a protective medium of hardened edible oil or tallow at a level of 25 per cent nickel). Since conditions and selectivity can be controlled within certain limits (including the end point at which hydrogenation is stopped), this reaction has great versatility for yielding a great variety of hydrogenated products from different starting oils.

Selectivity

Mounts states that 'during contact with the hydrogenation catalyst, some of the double bonds may be shifted to a new position and also twisted to the higher melting *trans* form. All of the hydrogenation reactions occur simultaneously; the relative rates at which they occur are the selectivities'. The linoleic selectivity is simply the ratio of the hydrogenation rate of linoleic acid compared to the hydrogenation rate of oleic acid.

Since the flavour problems of soyabean oil are usually associated with its linolenic acid content, the hydrogenation of linolenic acid without the hydrogenation of linoleic acid is very desirable.

Conditions

While there are many options for hydrogenating soyabean oil, according to *Mounts* there are basically two main approaches with nickel catalyst: 'nonselective hydrogenation under conditions such as 50 psig, 0.05 per cent catalyst at 250°F (121°C) in which the relative rates of hydrogenation of oleate, linoleate and linolenate are 1, 7.5 and 12.5 and selective hydrogenation under conditions such as 5–14 psig, 0.05 per cent catalyst at 350°C (177°C), in which the relative rates of hydrogenation are oleate 1, linoleate 50 and linolenate 100. In general, the effect of increased temperature or catalyst concentration will results in an increased selectivity for linoleate over oleate. Soyabean stearine or hard fat for bleaching in plastic or liquid shortenings is prepared by hydrogenation of the mono- and polyunsaturates to an IV of 5 at 20–50 psig or higher at 204–218°C (400–424°F).

Hydrogenation with nickel catalysts has some limitations, however, in that the selectivity for linoleate is not absolute, there is little selectivity for linolenic acid over linoleic acid and selectivity and isomerisation are coupled. Choice of conditions and equipment, therefore, require compromises, mainly based on the exothermic heat of reaction and the desired selectively. Practical reaction conditions are usually at temperatures of 120–200°C (250–390°F) and at pressures from 15–90 psig.

Catalytic transfer hydrogenation of soyabean oil by various hydrogen donors and solvents with a palladium-on-carbon catalyst was investigated. The choice of reaction conditions, donor and catalyst in this process allows the manufacture of partially hydrogenated oils or semisolid fats with controlled fatty acid contents, iodine value, melting point and solid content index. The best results were obtained in aqueous solution with sodium formate as hydrogen donor at 60°C (140°F).

Soyabean oil also has been hydrogenated electrocatalytically at a moderate temperature without an external supply of pressurised H_2 gas. In the electrocatalytic reaction scheme, atomic hydrogen is produced on an active Raney nickel powder cathode surface by the electrochemical reduction of water molecules from the electrolytic solution. Adsorbed hydrogen then reacts with an oil's triglyceride to form a hydrogenated product. Electrohydrogenated products are characterised by a high-stearic-acid content and a low percentage of total *trans* isomers, as compared to that produced in a traditional hydrogenation process.

The most common hydrogenation control test are congeal points and the iodine value as determined by refractive index. Refractive index alone is a satisfactory control for hydrogenation reactions under normal plant conditions for soyabean oil products above about 95 IV; when iodine values are less than 90–95, refractive index alone is not sufficient to control the reaction because the increased amounts of *trans* isomers results in a harder oil than the refractive index would indicate. A congeal point determination is necessary for controlling oils hydrogenated to less than 90 IV.

Hodgson has shown that the correlation of refractive index measurements with the temperature in the reactor permits a continuous monitoring of refractive index at a standard temperatures during the reaction. Tests are consistent up to 232°C (450°F).

For margarine oil or shortening, the solid fat index (SFI), determined by dilatometry, is now the most appropriate method for judging the consistency of hydrogenated fats. Content and type of glyceride solid of a shortening at any temperature determines the workability and creaming ability at that temperature. SFI is a good indicator of the plastic range possibilities of a fat or fat blend [e.g. the firmness of a margarine oil at 10°C (50°F) and its liquefaction at mouth temperature].

Hydrogenation process

Hydrogenation of soyabean oil may be carried out either in a batch or a continuous process; in the United States it is usually a batch operation. In the batch hydrogenator as a cylindrical pressure vessel having a capacity of 5–20 tonnes [Section A, Fig. 11.8]. The oil is pumped into a converter equipped with agitator, heating and cooling coils and vent system. The height-to-diameter ratio for the converter is approximately 2, whereas for the oil charge it is about 1.5. Two or more agitators (the turbine type is preferred) are fitted to a vertical axial shaft. The top stirrer is not immersed too deeply in the charge, so that hydrogen gas can be sucked from the head-space and dispersed into the oil as fine bubbles.

List and *Erickson* describe the process from this point on as follows:

A vacuum is pulled on the headspace and the vessel is heated. Meanwhile, the catalyst is weighed into a mix tank and slurried with a small amount of oil; as soon as the desired temperature is reached, the catalyst is pumped into the converter. Hydrogen is then added to the desired pressure. Then the reaction starts and the temperature is increased to the operating temperature and maintained there by cooling reaction mass.

Cooling may be done by circulating water through the cooling coil to counteract the heat of reaction and, when the desired degree of hydrogenation has been reached, cooling the charge to filtration temperature. Cooling the charge can also be achieved by draining it into a drop tank (Section B, Fig. 11.8), which is equipped with an agitator and heating and cooling coils. *Mounts* refers to this type of a hydrogenator as a 'dead-end converter, i.e. the hydrogen gas is allowed to accumulate in the head-space above the oil and is recirculated by the vortex action of the agitator'. As stated previously, the hydrogenation process is usually monitored by measuring refractive index of the oil.

According to *Allen*, 'after the reaction is complete, the oil is pumped through a filter to remove the bulk of the catalyst, then to a bleach tank where final traces or metallic impurities are removed by adsorption, then to a post-bleach filter to remove the spent bleaching clay. At this point, the hydrogenated oil is free of spent clay and metallic impurities.

Before pumping to storage or to the deodourisation step, the oil is given a polish filtration and passed through an oil cooled to prevent oxidation. Some processors have been able to eliminate the precoat step by the use of a fine mash wire cloth (24 × 110 Dutch weave) as the filter medium.

Practice varies in the handling of the catalyst press cake. 'A small catalyst mix tank is usually provided for receiving the used catalyst and resuspending it in oil for reuse, or for melting and reslurrying it with additional fresh catalyst'.

Effect of process conditions

Temperature, pressure, agitation and catalyst concentration have the biggest impact on product characteristics in a hydrogenation process. *Allen* lists the possible changes in a single-process conditions, with others held constant, that have effects on selectivities, the formation of *trans* acids and the rates of reaction (for nickel-catalysed hydrogenations as follows:

1. Temperatures increased: Selectivities increase, *trans* formation rate increases, hydrogenation rate increases.
2. Pressure increased: Selectivities decrease, *trans* formation rate decreases, hydrogenation rate increases.
3. Agitation increased: Selectivities decrease, *trans* formation rate decreases, hydrogenation rate increases.
4. Catalyst concentration increases: Selectivities increases, *trans* formation rate increases, hydrogenation rate increases.

Since free fatty acids, soap and water can all act as catalyst poisons that reduce catalyst activity and selectivity, the oil to be hydrogenated must be first refined, bleached, free of soap (under 25 ppm) and dry. The hydrogen also must be dry and of as high a purity as possible (i.e. > 98 per cent). Hydrogen can be produced by the steam-hydrocarbon or hydrocarbon-reforming process and pure liquid hydrogen is commercially available.

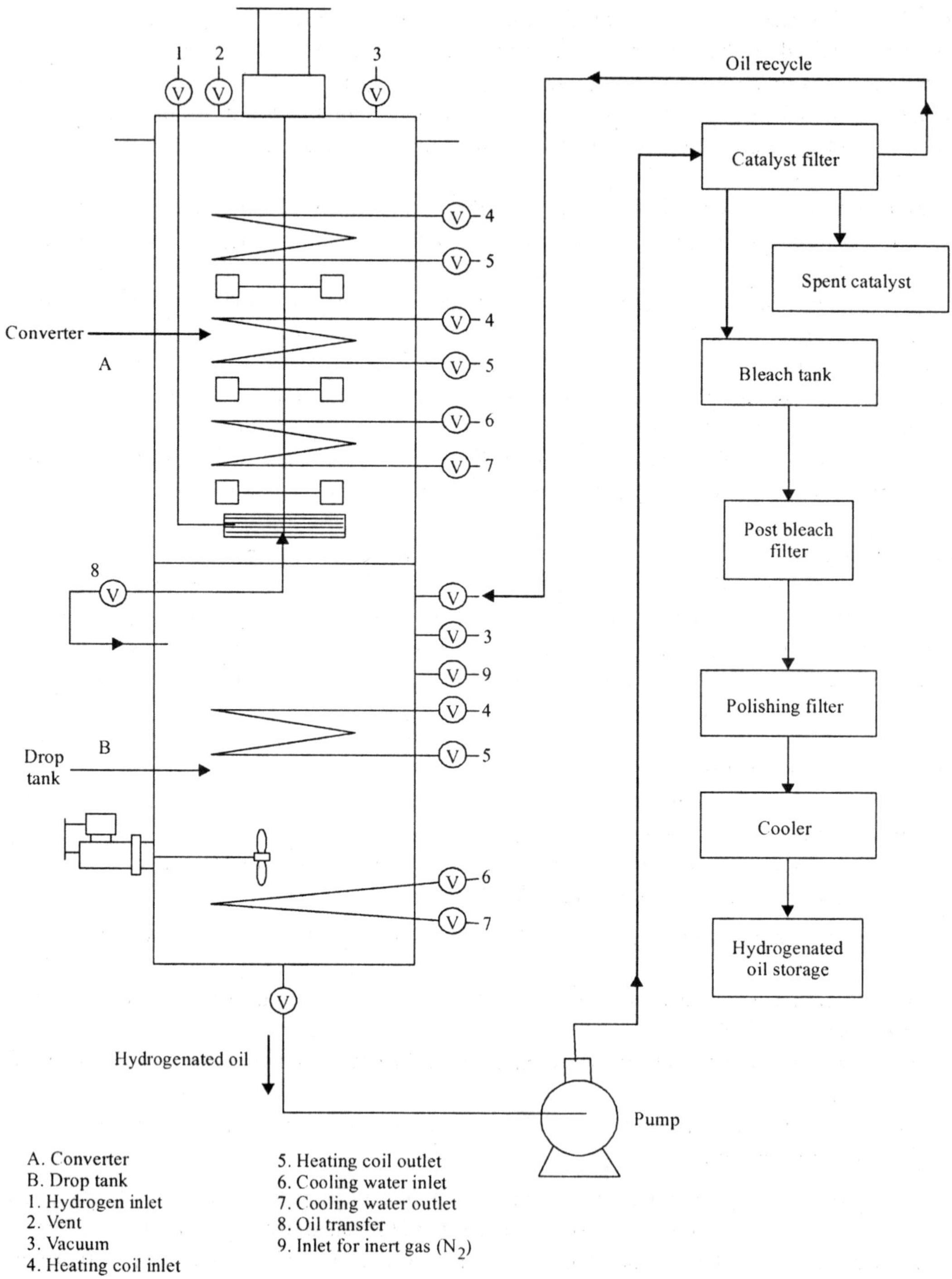

Fig. 11.8. Batch hydrogenation apparatus.

Catalyst activity

Poisons such as sulphur compounds, fatty acids and phosphatides in the oil may decrease catalyst activity. It is possible to reuse a good catalyst, but one has to be prepared for a change in selectivity. The quantity of the catalyst, however, may need to be increased slightly for each reuse to compensate for the decrease in activity. Most manufacturers specify new catalysts only for fully hydrogenated material or hard stock where selectivity is not important.

The poisoning effect of minor compounds in soyabean oil on the activity of nickel-based catalysts during hydrogenation has been investigated. Soyabean oil prepared by combining neutralisation with degumming and then followed by bleaching leads to a slower hydrogenation rate than that from an oil prepared by sequential degumming, neutralisation and bleaching with activated clay. The selection of bleaching earth used in the bleaching process also affects the hydrogenation rate. Higher amounts of phosphorus compounds, oxidation products, β-carotene and iron in these oils accounts for a slower hydrogenation rate. Storage of refined and bleached soyabean oil also greatly affects the hydrogenation rate. An increase in the oxidation products of refined-bleached soyabean oil during storage was the major reasons for the decrease in the hydrogenation rate.

The activities of several commercial nickel catalysts have also been determined by measuring their activation energies. The less active catalysts increase the linolenate selectivity of soyabean oil during hydrogenation. The yields of both *trans* isomers and winterised oil were found higher for the more selectively hydrogenated oil catalysed by the less active catalysts. In sensory evaluation, the fractionated solid fat that contained more *trans* isomers had lower flavour scores than the fractionated liquid oil after hydrogenation and winterisation of soyabean oil.

Hydrogenation-winterisation of damaged beans

An investigation on the effects of hydrogenation and winterisation on the quality of oil from damaged soyabeans showed that the hydrogenated and hydrogenated–winterised oils were of good quality initially when composed to an undamaged commercially processed oil, but they had inferior storage properties when subjected to accelerated storage tests.

When refined and bleached soyabean oil is stored at various temperatures for various lengths of time and then hydrogenated to 70 iodine value, *Covey* and *Wan* have found that peroxides affected the fatty acid composition and the rate of hydrogenation. Peroxides also affected the selectivity of hydrogenation and slope of the solids curve in hydrogenated products.

Alternative processes

Albright indicated that the key advantages of such systems are heat economy and better utilisation of equipment. These goals are achieved by providing part of the heat needed for preheating the oil to reaction temperature by heat exchange with the exiting hot oil.

A series of stirred tank reactors may serve as continuous flow reactors. In such a configuration, the oil, catalyst and hydrogen flow concurretly in a cascade from one stage to the next. A vertical column containing several reactor chamber would also fulfill this purposes. The oil, catalyst and hydrogen are introduced at the bottom of the column and flow concurrently upward. In another system a countercurrent flow of the oil-catalyst slurry and hydrogen gas is used in a unit having trays with vertical baffles in a column to minimise back-mixing.

Concurrent flow of the reactants in a horizontal pipe reactor is also used for continuous hydrogenation. The basic features of this reactor are: stoichiometric consumption of hydrogen, an essentially liquid-filled reactor and agitation provided by jet addition of hydrogen along the reactor length. Kinetics and mechanisms of hydrogenation of soyabean oil in such open-tube reactors have been reported for both nickel-and copper-catalysed reactions.

A semicontinuous hydrogenation system incorporating features of the 'dead-end' reactor with the gas recirculation concept has also been developed. The system provides two alternating dead-end converters employing high-shear mechanical agitation aided by a moderate amount of hydrogen recycled from the headspace back into the oil. While a batch of oil is being hydrogenated in one converter, the batch in the other converter is being filtered. It is claimed that the recycle of hydrogen back into the oil increases the hydrogen available at the catalyst surface, thus increasing the rate of reaction.

There have been few recent changes in hydrogenation. New types of reaction equipment are being tried, some of which demonstrate advantages in catalyst usage and more importantly, in lower *trans* isomerisation rates with some oils. This latter development is of interest because of nutritional concerns.

One very significant change affecting hydrogenation has been the introduction of gas chromatography as an analytical tool. The most recent development in this respect is the combined capillary gas chromatography/infrared spectroscopy method to quantify the various isomers that occur in hydrogenation. This makes it possible to make accurate label declarations as to the content of the nutritionally significant fatty acid isomers in hydrogenated fat products. Also, considerable interest has been generated in the development of new continuous-flow reactor systems. Various design feature for more efficient hydro-heat recovery that could be employed in a hydrogenation plant has been reviewed by *Nobel*. With catalyst particle size ranging from 1 to 100 μm the catalyst does not pass through ceramic membranes; their use could reduce labour expenses amount of oil lost in catalyst recovery and an increase in the use of catalyst.

Alternative catalysts

It has been found that copper-containing catalysts posses a greater selectively for the removal of linolenic acid than do the conventional nickel catalysts used commercially. Linolenic aid is almost completely reduced at iodine values of approximately 110–115 by copper catalysts, whereas oils processed with conventional nickel catalyst still retain about 3–4 per cent at this same iodine value. Copper-hydrogenated oils also retain more of linoleic acid (nutritionally important) than do the nickel-hydrogenated oils. The use of copper catalysts made it possible to hydrogenate soyabean oil to a product that is still liquid at room temperature and has less than 1 per cent linolenic acid content. On the other hand, copper catalysts are less active than nickel catalyst and they are also much more sensitive to poisoning.

Other catalysts that have high linolenate selectivity are: nickel–silver, nickel catalysts surface covered with sulphur and a catalyst system composed of divalent palladium supported on ion-exchange resins.

Nickel–sulphur catalysts can yield hydrogenated products with 90 per cent *trans* configuration of the remaining double bonds. Such high-*trans* products are used in the manufacture of margarines to enhance palatability and physical properties.

Deodourisation

In the conventional edible oil refinery, deodourisaion is the last in a series of process steps used to improve the taste, odour, colour and oxidative stability of the oils by the removal of undesirable substances.

There is still natural odour, flavour and colour of soyabean (i.e. 'beany' odour and flavour and greenish cast) left in a refined soyabean oil. Also, after bleaching, the oil still can have a musty or earthy odour. Furthermore, hydrogenation itself also introduces a typical odour and flavour that has to be removed by deodourisation in order to convert the oil into a bland-tasting, clear and brilliant liquid.

'Deodourisation is basically a high-temperature, high-vacuum, steam distillation process that removes residual constituents left in the oil after refining, such as free fatty acids, flavour and odour bearing compounds (e.g. aldehydes, ketones, alcohols), pigments, sterols, hydrocarbons and other compounds formed by the heat decomposition of hydroperoxides and destroys any peroxides that may be present'. Deodourisation, when properly done, results in a high-quality salad oil, packaged shortening, margarine and other specific products. The concentration of individual constituents in the undeodourised oil is generally no greater than about 0.1 per cent unless the oil has been abused. Normally, for an oil with an initial FFA content of 0.10 per cent –assuming a zero peroxide value–the noticeable flavours and odours will have essentially disappeared when the FFA content has been lowered to 0.01–0.03 per cent.

The three common methods of deodourisation currently in commercial use are: (i) batch; (ii) semicontinuous; and (iii) continuous.

The conditions used commercially for deodourisation of fatty oils vary appreciably with the type of equipment used, the type of oil being deodoursed and the degree to which the oil needs to be deodourised (Table 11.14) Commercial deodourisers also vary considerably with respect to operating temperature and oil residence time.

Table 11.14. Commercial deodourisation conditions.

Factor	*Level*
Absolute pressure	1–6 mmHg
Deodourisation temperature	210–274°C (410–525°F)
Holding time at deodourisation temperature	
Batch deodourisers	3–8 hr.
Continuous and semicontinuous deodourisers	15–120 minute
Stripping steam as wt. % of oil	
Batch deodourisers	5–15%
Continuous and semicontinuous deodourisers	1–5%
Free fatty acid contents	
Feed, including steam refining	0.05–6%
Deodourised oil	0.02–5%

Continuous and semicontinuous deodourisers minimise the time the oil is exposed to high temperatures. For Example, holding time in one commercial batch deodouriser may range from 3 to 8 hr. at temperatures of 232–246°C (450–475°F) with an oil residence time of only 20 minute, whereas another may operate at about the same temperature but with a residence time of 54 minute. It is difficult to extrapolate results obtained in a batch deodouriser with those expected in a semicontinuous or continuous commercial operation.

Deodourisation of soyabean oil

It is important to start with good-quality soyabeans and product quality should be protected in each processing step prior to deodourisation. For example, the proper strength and amount of caustic must be used to ensure adequate removal of the phosphatides during caustic refining and the oil must be washed and bleached properly for removal of soaps. Also, the removal of certain trace thermal decomposition constituents is sometimes necessary for producing a good-quality soyabean oil. Immediately after deodourisation , the addition of citric acid or antioxidants and nitrogen to the oil is recommended.

Batch deodourisation can produce a good-quality soyabean oil, but semicontinuous and continuous deodourisers have a number of advantages. Some of these advantages are inherent in the process while others are the results of equipment design. Removal of odour and flavour-bearing compounds from edible oils by deodourisation in batch systems is dependent on both time and temperature. To obtain the best balance between both initial quality and keeping properties of the oil, deodourisation times should be short when temperatures are high and longer when temperatures are low. Deodourisation times and temperatures producing the most stable soyabean oils—as measured by organoleptic tests and oxidative stability under Schall oven conditions.

Typical commercial conditions for the deodourisation of unhydrogenated soyabean oil in a semicontinuous deodouriser as follows: 'top temperature 475 to 500°F (246–260°C), 15 to 40 minute at the deodourisation temperature with the shorter time used at the higher temperature and higher stripping steam rate, 3–6 mm Hg absolute pressure, 3–8 per cent stripping and sparge steam based on oil throughout and oil cooled to 150°F (66°C) in the deodouriser. Then 0.01 per cent citric acid and antioxidants as desired are added to the cooled oil either in the deodouriser or after removal'.

The preferred conditions for the EMI continuous deodouriser are: 'top temperature 510°F (266°C), 90 minutes total time with 30 minutes at the top temperature, 3 mm Hg absolute pressure and 2.6 per cent stripping steam. Additives such as citric acid are added as for the semicontinuous deodouriser'.

Effects of Deodourisation on oil quality

A freshly and properly deodourised soyabean oil has a bland or nearly bland taste and even after momentary heating to 177°C (350°F), has almost no odour. The deodourised oil ultimately should have less than 0.03 per cent FFA, a peroxide value of zero, Lovibond colours of 10 yellow and 0.7 red and a bland taste.

The linolenic acid of soyabean oil is able to undergo a partial geometric isomerisation under the influence of temperature and heating. The more drastic deodourisation conditions are, the higher the isomerisation rate; this also depends on the kind of oil, its quality, the apparatus and process. Because of the low-linolenic-acid content in soyabean oil, even at high levels of isomerisation, very little *trans* acids are generated. This type of isomerisation also does not affect significantly the formation of polymers and cyclic acids during the frying operation.

Deodourisation is also the most important processing step affecting the initial flavour of salad oils from damaged soyabeans. By increasing the deodourisation temperature from 210 to 260°C (410–500°F), the initial flavour score increased from about 5.4 to 6.8. The undesirable rubbery, rancid and grassy flavours predominant in the 410°F (210°C) oil can be replaced by the more desirable buttery, beany and nutty flavours at the 500°F (260°C) deodourisation temperature. Evaluations of damaged oils deodourised for 1 and 3 hr. showed that longer deodourisation times are required to produce products of higher initial quality from damaged beans.

New developments

Mag states that 'the changes from batch deodourisation to semicontinuous or even continuous operation has improved oil product quality because of the much more efficient oil/steam contact that can be achieved'. Further improvements have dealt mostly with oil/steam contacting arrangements, heat recovery systems, energy saving in vacuum generation and water pollution abatement. Presently, new deodouriser systems seem to be proliferating. Continuous systems based on this-film technology compete with new versions of batch and semicontinuous systems. Many jurisdictions now require use of high-pressure steam for heating purposes to avoid any chance of edible oil coming in contact with petroleum-based heat transfer fluids.

Partially Hydrogenated–Winterised Soyabean Oil

Soyabean oil intended for use as a salad-cooking oil is partially hydrogenated to iodine values of 110–115 to improve odour, flavour and oxidative stability during high-temperature use. This treatment not only saturates the naturally occurring *cis* double bonds but also isomerises them to the higher melting *trans* forms. In addition, some fully or partially saturated triglycerides may be formed. As a consequence of these reactions, partially hydrogenated soyabean oil (e.g. salad oil) will not solidify or cloud at temperatures of 4.5–10°C (40–50°F). This is a serious problem when partially hydrogenated soyabean oil (IV 110–115) is stored at low temperatures. When storing these oils under low-temperature conditions one encounters two undesirable phenomena: (i) clouding caused by *trans* isomers; and (ii) solidification of higher melting disaturated glycerides. To prevent either phenomenon, partially hydrogenated soyabean oil, is therefore, winterised by removing the materials responsible for clouding and solidification.

Winterisation will achieve the following: (i) removes waxes and other non-triglyceride constituents; (ii) removes naturally occurring high-melting triglycerides; and (iii) removes high-melting triglycerides formed during partial hydrogenation.

The types of crystals formed during the manufacture of winterised oils are amenable to separation from the uncrystallised oil and at the low temperatures required for crystallisation, the oil is also quite viscous. In order to overcome this problem, the oil has to be cooled slowly to produce easily filtered crystals; this separation of solid and liquid phases, however, is time consuming, difficult and inefficient even under ideal conditions.

List and *Mounts* describe the winterisation process for soyabean oil as follows:

Winterisation of soyabean oil begins with dry, hydrogenated, bleached oil (iodine value 110–115) being pumped into a large insulated tank fitted with cooling coils at a temperature of 86–95°F (30–35°C) after which the temperature controller is started. The latter regulates the amount of brine or propylene glycol that is circulated through the cooling coils. In order that the oil not be chilled too rapidly, a small temperature differential between the oil and coolant is maintained. Initially, a satisfactory differential is about 23°F (13°C) but it decreases to about 9°F (5°C) when the oil cools to its final temperature.

The temperature of the oil is slowly brought from about 35 to 6°C (95–43°F) over a period of about 24 hr. According to *Neumunz* 'toward the end of the chilling cycle, which normally takes 24 hr., the crystallisation rate is sufficiently rapid to effect a slight rise in temperature [usually 1.8–3.6°F (1–2°C)], which then drops. At this point, external cooling is stopped and the winterised oil is allowed to stand for 6–8 hr., at a temperature of 42.8°F (6°C). The mixture is then ready for filtration'. The filtration separates the winterised oil from the solid or stearine portion.

It is important that the oil is not chilled too rapidly to prevent the formation of small crystals, which are difficult to filter. Even with slow cooling, agitation is required for proper crystal structure. Agitation is usually accomplished with slow stirring by mechanical (e.g. gate type) agitators or by bubbling cold dry air through the oil.

The influence of soyabean lecithins on the spontaneous solidification of different model fats has been studied in the presence and absence of water. Lecithins added to dry fat do not affect crystallisation but, in the presence of water, clearly delay it. Not only does the crystallisation begin late but the surfusion phenomena are also exaggerated.

Preparation of high-stability soyabean oils by hydrogenation–winterisation

Partial hydrogenation and winterisation are employed for making a high-stability soyabean oil for use as liquid cooling–frying fats. Soyabean oil is usually hydrogenated to an IV of 81–84 for these applications.

Dual-stage hydrogenation–winterisation process for producing high-stability liquid frying oils is given below:

Soyabean oil (IV 136.2) is hydrogenated with a nickel catalyst (104°–107°C, 220°–225°F) (25 psi) to an IV of 105.3. After filtering off the catalyst, the oil is transferred to a tank fitted with cooling coils. The temperature is then slowly lowered over a period of 40 hr. to 16.4°C (61.5°F). At the end of 72 hr., the mixture is filtered. The liquid oil portion (79.5 per cent yield) has an IV of 106.0. The hydrogenated-winterised soyabean oil is then again hydrogenated at 118–174°C (245–255°F) with 0.2 per cent nickel to IV of 75–85.

Winterisation of a twice-hydrogenated soyabean oil for use in shortenings and cooking oils is usually accomplished by cooling the oil in two steps. 'First to 24.4°C (76°F) over a period of 12 hr. in a tank fitted with cooling coils, then to 18°C (65°F) over 24 hr. At this point, the liquid oil-crystal mixture is completely fluid and the crystallised portion could be readily separated by filtration under air pressure'. Products formed by this process reportedly exhibit exceptionally good stability at deep-fat frying temperatures.

Crystal inhibitors

Among the many crystal inhibitors used in the winterisation process, the most important are: aluminium tristearate, paraffin-alkylated hydroxy aromatic compounds, fatty sugar esters, aromatic resins, polymerised vinyl esters, phosphatides, polymerised glycerides (oxystearin) and polyglycerol esters of mixed fatty acids.

Oxystearin and polyglycerol esters have Food and Drug Administration (FDA) approval for use in edible oils. Oxystearin is permitted up to a maximum weight of 0.125 per cent of the combined weight of the soyabean oil plus additive. Polyglycerol esters are permitted in vegetable and salad oils—when use is not precluded by food standards—at levels necessary to obtain the desired technical effect.

Shortenings and Margarines: Base Stock Preparation and Formulation

There is a variety of high-quality shortenings and margarines made with soyabean oil serving as the only fat source except for a small amount of hard fat (e.g. fully hydrogenated cottonseed or palm oil) in some cases. Soyabean constitutes about 64, 90 and 80 per cent of fat used in the manufacture of shortenings, margarines and salad and cooking oils, respectively.

Brekke lists the following basic steps in shortenings manufacture: (i) preparation of the individual base stock and hard fats; (ii) formulation of the fat blend and other ingredients; (iii) solidifying and plasticising the fat blend; (iv) packaging; and (v) tempering the shortening, when necessary (e.g. plastic shortenings and margarines used for baking).

Semisolid plastic shortenings are prepared from a base stock consisting of a partially hydrogenated fat and a hard fat or stearine fraction. The base stock is quickly chilled in a scraped surface heat exchanger and 'worked' in a manner such that the shortening has the proper crystal structure and performs satisfactorily when used.

Regular (i.e. plastic) margarines is an emulsion consisting of a base stock, a so-called 'milk' phase, salt, emulsifiers, preservatives, colouring and flavouring ingredients. The base stock generally is a mixture of two or three lots of oils, each partially hydrogenated to the proper degree. The emulsion is quickly chilled and worked in equipment quite similar to that used for shortenings.

Formulations for preparing shortenings

By blending a partially hydrogenated base stock with a hard fat of IV ca. 1–8, one can produce general-purpose shortenings. The hard fat is prepared by fully hydrogenating palm or cottonseed oil that crystallises in the β′ form. Since partial hydrogenation of soyabean oil increases the oxidative stability of the shortening, the amount of hard fat used in the formulation can be decreased accordingly. When hard fats are added to shortenings, they extend both the plastic range (and thereby improve the heat tolerance) and provide crystal stability). Table 11.15 shows the conditions that can be used for hydrogenated soyabean oil to produce seven base stocks for shortening and margarine formulations. Typically, base stocks, nos. 1 and 1A in Table 11.15, are used for general purpose shortenings.

Table 11.15. Conditions for hydrogenation of soyabean oil to produce shortening and margarine-base stocks.

	Primarily shortening bases			*Primarily margarine bases*			*Shortening margarine base*
Base stock no[a].	*1*	*1A*	*2*	*3*	*4*	*5*	*5a*
Hydrogenation conditions							
Initial temperature °C (°F)	148.8 (300)	148.8 (300)	148.8 (300)	148.8–162.7 (300–325)	148.8–162.7 (300–325)	148.8 (300)	140.5 (200)
Hydrogenation temperature, °C (°F)	165.5 (330)	165.5 (330)	165.5 (330)	218.3 (425)	218.3 (425)	176.6 (350)	140.5 (200)
Pressure (psig)	15	15	15	5	10	15	40
Catalyst concentration (% nickel)[b]	0.02	0.02	0.02	0.05[c]	0.02[c]	0.02	0.02
Final iodine value[d]	83–86	80–82	70–72	64–68	73–76	106–108	104–106
Final congeal point, °C (°F)	–	–	25.5–26 (77.9–78.8)	33–33.5 (91.4–92.3)	24–24.5 (75.2–76.1)	–	–
Solid fat index:							
10°C (50°F)	16–18	19–21	40–43	58–61	36–38	4 max.	4 max.
21.1°C (70°F)	7–9	11–13	27–29	42–46	19–21	2 max.	2 max.
33.3°C (92°F)	–	–	9–11	21 max.	2.0 max.	–	–

[a] The soyabean oil must have been properly refined and then bleached to remove all soaps and peroxides prior to hydrogenation.
[b] Based on weight of oil.
[c] Very selective catalyst is needed.
[d] Final iodine value is only approximate.

General-purpose emulsified shortenings (i.e. those used for a variety of products vs. those formulated for icings, toppings, cakes, etc.) can be prepared directly from their nonemulsified counterparts by the addition of mono- and diglycerides. To compensate for the hardness of the added mono- and diglycerides, consistency of the shortening may be adjusted by the addition or deletion of small amounts of hard fat in the formula. It should be possible to produce shortenings for any application by blending one or more of the base stocks with the desired additives (e.g. plasticisers, emulsifiers, antioxidants, metal scavengers, antifoam agents and flavours). Hard fats prepared by fully hydrogenating cottonseed oil or palm oil are considered to be good plasticiser additives.

Formulation of margarine oils

The melting point, fatty acid composition and structure of the individual triglycerides used in margarine influence the structural, nutritional and lubricity aspects of the finished product.

The oil formulation, the manner and degree of supercooling the emulsion and the extent of mechanically 'working' the supercooled emulsion during the crystallising stage are all used to control margarine plasticity. Margarine oils are a blend either of one hydrogenated and one unhydrogenated oil or of two or more hydrogenated base oils. According to *Mounts*, sticks margarines usually are made from a blend of two or three intermediate fats or of a low-IV fat and a liquid oil. Soft margarines contain a high proportion of liquid oil, with fluid margarines having the highest proportions. Since a single hydrogenated base stock cannot be converted into a finished margarine, producers of commercial margarine oils must blend base stocks to meet the specifications of buyers.

The solid triglycerides in a margarine or shortening must have the proper crystal structure. The β form usually produces a grainy or waxy texture in a margarine or plastic shortening. The β′ form, on the other hand, gives a smooth texture that is preferred in margarines and plastics shortening, especially when they are used in cakes and icings.

When the β′ form mixtures are chilled without any hard fat, the composition must contain at least 20 per cent of a fat that crystallises in the β′ form and have a melting point higher than that of the other fat present. In commercial practice, it has been customery to use 20 per cent of partially hydrogenated cottonseed oil or palm oil in some margarines. It is also common to add a third fat of intermediate hardness to a margarine formulation for controlling the crystal structure and obtaining improved product consistency and greater flexibility in formulation of the oil blend. This development has made it possible to produce margarine based entirely on soyabean oil.

The formulations for four margarine oils are given in Table 11.16. With these formulations, the manufacturer has some choice in selecting specific base stocks and the quantify of each required to produce a margarine having the desired SFI in the range shown for oil nos. 2, 3 and 4 in Table 11.16. Oil no. 1 in the example will give a softer margarine than the other three.

Table 11.16. Four margarine oil formulations[a].

	Required SFI for the blend				*Formulation*	
Margarine oil no.	*10.0°C (50°F)*	*21.1°C (70°F)*	*33.3°C (92°F)*	*40°C (104°F)*	*Base stock*	*Amount (%)*
1	20–26	13–17	1.5–3.0	0	Margarine base no. 3	50
					Unhydrogenated soyabean oil	50
2	30.0 max.	17.5 min.	3.5 max.	–	Margarine base no. 3	38
					Margarine base no. 4	20
					Margarine base no 5 (or 5A)	42
3	28–30.5	16–18	2.0–3.0	–	Shortening base no. 1	50
					Margarine base no. 3	25
					Margarine base no. 4	25
4	28–31	16–18	1.5–2.5	–	Shortening base no. 1	60
					Margarine base no. 3	15
					Margarine base no. 4	25

[a] These formulations are developed entirely for base stocks prepared from soyabean oil. It is fairly common practice to put as much as hydrogenated cottonseed oil or other β′ fats in margarine oil to provide crystal stability and thereby prevent later development of a sandy grainy texture in the margarine. This is not absolutely necessary since some manufacturers produce margarine successfully from 100% soyabean oil. When specified by the customer, cottonseed base stock similar to any one of the soyabean base stocks can be prepared and substitute the formulation. Corn oil (100%) margarines contain only corn oil as required to conform with label declarations.

SOYABEAN OIL QUALITY

Broadly speaking, quality of soyabean oil is affected by the following factors:

1. Damage to soyabeans.
2. Excessive moisture.
3. Excessively high temperature in storage and processing.
4. Exposure to air (oxygen) especially at high temperatures.
5. Exposure to prooxidant metals.
6. Exposure to light.

Oxidative Stability of Soyabean Oil

A number of reviews have documented the various factors involved in improving soyabean oil oxidative stability. 'Although processed soyabean oil is bland when used in margarines, shortening and salad oils, it may deteriorated organoleptically when repeatedly heated at cooking and frying temperatures and when exposed to light. Unless properly processed, soyabean oil develops a very complex type of lipid deterioration usually but inappropriately called 'flavour reversion'.

Flavour and colour reversion and oxidation rancidity

Flavour reversion as being 'beany and grassy' at the early stages and as 'fishy or painty' at the more advanced stages. This flavour is characteristics of soyabean oil and other linolenate-containing oils and develops at lower levels of oxidation (i.e. the peroxide value, if measurable, is 10 or below). Since the degree of oxidation is so low, it often cannot be measured chemically but it is detected organoleptically. In contrast, oxidative rancidity is observed at higher levels of oxidation (peroxide value of 10 or above) and is a flavour deterioration common to all unsaturated fats.

Colour reversion

Colour reversion occurs under certain conditions after refining, bleaching and deodourising soyabean oil to a pale light-colour that then becomes dark upon storage. The degree of colour change is dependent on the nature of the raw soyabeans, refining, deodourisation and storage conditions, temperature, exposure to light, and contact with air during storage. The tendency to undergo colour reversion is variable; requirements for reversion vary from a few hours to several months. In all cases, the colour becomes paler after it reaches a peak.

The moisture content of the soyabeans going to the extractor has been implicated in promoting colour reversion in fully refined oils, mainly by affecting the tocopherol content of extracted oil. If the moisture content of the beans is within normal range (7–12 per cent), the tocopherol content of crude soyabean oil also remains normal; in wet beans, however, the tocopherol content is markedly reduced.

Effects of 0–1.0 per cent levels of caprylic, capric, lauric, myristic, palmitic or stearic acids on the oxidative stability of purified soyabean oil were investigated under microwave heating conditions. A prooxidative effect of the fatty acids introduced into the systems was established. The extent of this effect depended on the acyl chain levels of fatty acids. These findings indicate that particular attention should be paid to the free fatty acids content of purified soyabean oil.

The effects of soyabean cultivar and tempering procedures on crude soyabean oil volatiles were investigated by gas chromatography/mass spectrometry (GC/MS). Trends in volatile contents due to tempering were noted primarily in oils stored for 16 days at 60°C (140°F). Differences in volatiles in response to tempering were not apparent in oils that had not been stored. Equilibration of beans with water resulted in oils with higher volatile contents than those from untempered beans. Steaming lowered volatiles from equilibrated values and pressure steaming had an even more dramatic lowering effect.

Control measure/stabilisation

In spite of much progress in the control of flavour and colour reversion, problems caused by light and thermal instability still await resolution when using processed soyabean oils. According to *Frankel*, the greatest improvements so far have been achieved through the use of metal inactivators, better processing, selective hydrogenation and packaging. Other approaches investigated with only limited success include selective removal of linolenate triglycerides, prevention of oxidation by antioxidants and breeding soyabeans to produce oil low in linolenic acid.

The use of minimum temperatures during processing and handling under vacuum during processing, the use of inert gas blanketing during storage and the avoidance of light and metallic prooxidants are preventative in nature. Precautions that need to be followed when preparing and handling soyabean oil can be summarised as follows.

Limiting atmospheric oxidation by minimising contact of the oil with air at ambient temperatures and avoiding all contact at very high temperatures, e.g. in the deodourisation operation. Keeping the process time at elevated temperatures as short as possible. Avoiding, removing and inactivating trace metals (e.g. copper, copper alloys and at high temperatures, common iron) having a prooxidant effect. Avoiding exposure of the oil to light. Practices to be avoided in transporting or handling the oil include cascading of the oil in air and sucking, whipping or bubbling air through the oil.

Specifically, there are certain critical steps that should be carried out during refining, bleaching, deodourisation and hydrogenation in order to obtain a good-quality soyabean oil for edible uses. For example, heating of soyabean oil after alkali refining is an important step in the inactivation of trace metals. 'Improved flavour stability can be expected if bleaching can be made more selective in order to increase the ratio of carotenoids over chlorophylls. To reduce the problem caused by certain oxidative polymers, deodourisation of soyabean oil should only be carried out either after ensuring that the oil is free of peroxides or after their removal. Finally, hydrogenation needs to be more selective toward the 15-double bond of linoenate'.

When continuous processing is used for degumming, refining, bleaching and deodourising the oils, the reduction in process time lessens the chance for an increased rate of oxidation. A secondary benefit of continuous processing from the standpoint of both lessened oxidation and potential energy savings is that when two or more process steps are integrated, the hot oil can be moved directly from one operation to the next without the need for intermediate cooling, storage and reheating.

At ambient and at relatively low process temperatures, carbon steel is quite satisfactory for processing equipment, provided the surfaces exposed to the oil have been properly conditioned. However, at deodourisation temperatures, deodouriser should have stainless steel (usually type 304) or nickel surfaces in contact with the hot oil.

The effectiveness of metal inactivators in stabilising soyabean oil against oxidation have been studied. The investigators have found that the most commonly used metal inactivators are citric and phosphoric acids. Both additives are very effective for improving storage stability at normal temperatures and their effectiveness in soyabean oil is increased by a synergistic action with the naturally occurring tocopherols. Tocopherols act as free radical chain breaking antioxidants and the metal inactivators function as oxidation preventive agents. They suppress both initiation and propagation of free radicals.

Metal inactivators may be used in early stages of processing to protect the oil against oxidation, but they are more effective when added after the oil is heated. The addition of a metal-inactivating or chelating agent such as citric acid or isopropyl citrate at the deodourisation stage, preferably while cooling the oil, helps overcome the prooxidant effect of remaining trances of metallic contaminants. This treatments also improves the oxidative and flavour stabilities of the finished oil.

The use of antioxidants is another approach for stabilising vegetable oils. 'The value of antioxidants in stabilising soyabean oil against flavour deterioration appears to be in situations where the oil is stored in containers that permit transmittance of sunlight or fluorescent light; for example, salad oil packaged in clear flint glass bottles.

As little as 10 ppm citric acid may be effective in stabilising soyabean oil against flavour and oxidative deterioration. The amount of phosphoric acid required in commercial operations is less understood. Excess phosphoric acid may result in extraneous off-flavours in deodourising oils.

The following antioxidants are approved for use in edible fats and oils: butylated hydroxyanisole (BHA), butylated hydroxytoluene (BHT), propyl gallate (PG), tertiary butyl quinone (TBHQ), tocopherols, lecithin, gum or resins guaiac, trihydrobutyrophenone (THBP), 4-hydroxymethyl-2, 6-di-tert-butylphenol, thiodipropinonic acid and dilauryl esters and glycine.

To protect the deodourised oil, antioxidants, usually BHA, BHT, TBHQ or PG, singly but more likely in some combination, may be added to oil leaving the deodouriser. When soyabean oil is stabilised by use of citric acid, the addition of BHA and BHT improves the flavour stability of hydrogenated oil but not of unhydrogenated oil exposed to light. Flavour stability was especially improved when the hydrogenated oil containing 0.01 per cent each BHA, BHT and citric acid has been exposed to the effects of strong fluorescent light.

The addition of the total purified unsaponifiable matter extracted from crude soyabean oil to linoleic acid raised its stability toward oxidative rancidity. An antifoaming agent, e.g. a methyl silicone, is often also added to frying fats.

In order to maintain the proper flow rates, temperature, liquid levels and vacuum during extraction and oil processing, instrumentation is playing an increasing role. 'Instrumentation has the addition benefits of achieving higher product quality, maximum process yields, eliminating losses due to accidental operator error and keeping overall producing costs at a competitive level'.

Evaluation of Finished Oil Quality

The official specifications for soyabean oil and its products as discussed earlier are only minimum standards. The ultimate measures of quality are the odours and flavours of the products. Sensory evaluation, therefore, with or without instrumental analysis, is necessary for both product development and quality control.

Although chemical and physical tests are more reproducible and less time consuming than the sensory evaluations, rancidity determinations must always be done by an organoleptic observation. Nevertheless, establishing correlations between organoleptic scores and various chemical tests are possible and frequently used in commercial practice. For example, the spectrometric characteristics of raw oils can predict the sensorial acceptability of the refined/deodourised final product.

Peroxide values have shown a particularly good correlation with organoleptic flavour scores of soyabean oil and therefore, the peroxide determination is the most widely used chemical test for the determination of fat quality. Since aldehydes having long been identified as oxidation and breakdown products of fats, their determination has also been of great interest in the quality control laboratory.

Total phospholipids, especially phosphatidylethanolamine (PE), were shown to be the major contributors to the development of warmed-over flavour (WOF) in cooked meat when studied in model systems from beef and chicken dark and light meats.

Recently, high-pressure liquid chromatography (HPLC), supercritical, fluid chromatography (SFC) and supercritical fluid extraction (SFE), among other modern analytical methods, have become increasingly available to meet the widespread challenges faced by the fats and oils chemistry in identifying many compounds of interest/concern.

Determination of glyceride composition characteristics by high-temperature gas chromatography or by HPLC gives a more precise indication of composition modifications resulting from hydrogenation and fractionation treatments. Methods of determining density of refractive index, where the response time is usually in inverse proportion, to precision, are apparently not very specific and furnish only relative indications that correspond mainly to manufacturing objectives. However, 'methods intended to determine the solid glyceride content or the glyceride composition, whose indications correspond to more absolute measurements, relate mainly to a *posteriori* verifications'.

The simplest and the most commonly used accelerated storage methods for determining stability of fats or fat products is the oven or Schaal test. Researchers at the USDA Northern Regional Research Centre have been using this method routinely for evaluating soyabean oil stability. Four days of accelerated storage at 60°C (140°F) is approximately equivalent to 3 months storage at ambient temperature. An accelerated method for light stability of soyabean oil has also been developed.

Storage and Handling

Soyabean oil may undergo oxidative, enzymatic and microbiological deterioration in storage or during and after processing. Production of the highest quality soyabean oil product requires close control from harvesting to consumption of finished products.

In relation to handling and storage factors that contribute to oxidative deterioration of soyabean oil, *List* and *Erikson* state that heat accelerates the reaction of atmospheric oxygen with edible oils and therefore, localised overheating is detrimental to fat quality. To avoid this, all storage tanks with heating devices should be equipped with a mechanical agitator.

Going has reported on the oxidative stability of soyabean oil at various processing stages and related the results to commercial storage conditions. According to this study the oxidation rate doubles for every 11.1°C (20°F) increase in temperature. The effect of container size on the oxidative deterioration of refined, bleached soyabean oil stored at 48.9°C (120°F) for 5 weeks has also been reported. The results clearly demonstrate that oxidation is not only a function of time and temperature but also of the amount of oxygen available. Oxygen availability is lower if container size increases; this is so because the surface-to-volume ratio decreases. Peroxide development is a linear function of surface-to-volume ratio.

In field storage tanks the oil is also subjected to conditions that cause development of sizeable temperature gradients that can produce considerable internal oil movement. Such movement would be expected to increase the quantity of oil available at the surface and to accelerate oxygen diffusion. The practical conclusion from this observation is that the rate of oxidation in field storage tanks would be even faster than at similar temperatures in the laboratory. Thus, to minimise peroxide formation and the flavour problems that accompany it in storage, soyabean oil should be stored at as low a temperature as practical. Generally, oils should be kept no warmer than necessary to facilitate handling by pumping. For hydrogenated oils, a holding temperature of 5.6°C (10°F) above the melting point is considered sufficient.

Since the refined-bleached (RB) soyabean oil is very susceptible to oxidation, the RB oil should be stored only for brief periods.

Soyabean oil in general should be preferably stored in carbon steel, stainless steel, stainless clad, or fibreglass-reinforced polyester tanks. Carbon steel tanks should be coated on the interior with an epoxy or polyurethane lacquer. Nitrogen protection should be considered whenever the oil is to be stored for an extended period or held in the hot, liquid state.

As stated previously, avoiding metallic contamination is also very important. Copper and iron are strong prooxidants for soyabean oil and are capable of lowering the flavour and oxidative stability of soyabean oil at levels of 0.01 and 0.1 ppm, respectively. Copper or copper-containing alloys should never be used for equipment used in handling and storing soyabean oil. Iron contamination can be kept to a minimum by proper treatment and cleaning of black iron equipment.

Since all edible fats and oils deteriorate under the effects of light, light deterioration is an important factor in the storage stability of liquid oils packaged for retail trade.

According to *List* and *Erickson*, unhydrogenated or lightly hydrogenated oils that do not require heat for keeping them liquid have greater resistance to deterioration than shortenings. The latter, however, will keep for 2–3 weeks while in a melted condition but should not be allowed to solidify and then be reheated during use.

While the primary concern during extended storage of crude and once-refined soyabean oil is oxidative deterioration, other factors such as increases in moisture and volatile matter, colour after refining and/or bleaching, peroxide value, free fatty acids, and refining losses can also be of considerable importance. In addition to their detrimental effects on flavour and oxidative stability, they also affect the commercial value of soyabean oil.

Special Processing for Off-Specification Oil

Oils from field-, frost- and storage-damaged beans are expensive and difficult to process and the finished oils often have poor flavour, lowered oxidative stability and dark or green colours.

Fryer reported that oils from these frost- and field-damaged beans would not bleach properly by standard methods and yielded poor-quality products that underwent rapid colour and flavour reversion.

It has been found that oil content and iodine value increase with maturity while free fatty content and refining losses decrease. Also, yields of natural oil from mature frost-damaged beans are higher than those from immature frost-damaged beans.

The frost damage affects oil colour. There is an undesirable green colour in crude oils obtained from frost-damaged beans, resulting from the presence of high amounts of porphyrins that have sharp absorption peaks in the 640–680 μm range. The green colour is caused by compounds related to but not identical to chlorophyll or some of its derivatives.

The green oil colour cannot be removed by treatment with acid, alkali, X-rays, ultraviolet rays, electron beams, or ethylene gas. However, pretreatment with charcoal (0.4–1.0 per cent) at 90°C (194°F) or treatment of a cold hexane–oil mixture with charcoal is effective in partially removing the green pigment. When charcoal pretreatment is combined with additional treatment with sugars and activated bleaching clays, complete removal of green pigments is possible.

When bleaching green soyabean oil, it is useful to know that (i) acid-activated clays are more efficient than neutral clays in removing green colour; and (ii) increased amounts of bleaching clay makes the removal of the green colour more effective.

Another practical way to remove green colour from soyabean oil is by hydrogenation. A visible green refined and bleached oil (IV 132) when hydrogenated to IV 110 in the presence of 1 per cent copper chromate catalyst was no longer green after cooling and filtration. Hydrogenation to the same extent with conventional nickel catalysts removed only slight amounts of the green colour.

By varying extracting conditions, it may be possible to produce high-quality, low-phosphorus and low-free-fatty-acid oil extracted from water or freeze-damaged soyabeans. While freezing caused maximum increase in FFA and phosphorus levels compared to other levels in damaged or undamaged soyabeans, these levels were significant when compared to variables caused by flake extraction. Phosphorus and FFA increased when storage temperatures went from 25 to 45°C (77 to 113°F), extraction temperatures from 25 to 50°C (77 to 122°F) and moisture of the flour from 6 to 10 per cent. However, the storage time of soyabean with initially high moisture (20 per cent) did not have a marked influence of FFA and phosphorous levels. Immediately after grinding, moisture of the flavour elevated or lowered the phosphorous level to a great extent, although it had little influence on the FFA level. Phosphatidic acid and phosphatidylchloine were identified as the main phospholipids present when total phosphorus was low in extracted oil. The time taken for the flour to dry to 6 per cent moisture (after grinding and before it was extracted) was critical. Using finely ground flours rather than flakes moderated the increase in FFA and phospholipids as the result of soyabean damage.

Field damage can occur when mature soyabeans are left exposed to rain or humid weather. They develop a dark brown colour and chalky texture and tend to sprout and decay. Also, if soyabeans are stored at moisture levels greater than 13–14 per cent, spontaneous heating may results in storage damage. Field and weather damage usually results in loss of carbohydrates in the bean and in decomposition of the oil fraction. For all practical purposes, the effects of field and storage damage to the oil are similar and probably are caused by enzymatic and microbial processes.

As drought stress increases, measured by accumulating stress degree days, protein content increased linearly and oil content decreased linearly at each air temperature observed. Drought has little effect on the fatty acid composition of the oil, but high air temperatures reduces the proportion of the polyunsaturated components.

One of the major problems encountered in processing oil from damaged beans is that of high refining losses, attributed to the high free fatty acids content of the crude oil and alteration and degradation of phospholipids.

'High refining, losses may be partially overcome by use of phosphoric acid or acetic anhydride degumming. Colour problems may be alleviated in part by use of acidic bleaching earths, increased dosages of bleaching earths and higher temperatures. Hydrogenation with copper-containing catalysts is effective in removing chlorophyll and related pigments from frost-damaged oils.

Specifications for Soyabean Oil

An acceptable quality oil is a light-coloured, bland, or nearly bland-tasting oil, generally obtained by alkali refining, bleaching and deodourisation. For some uses, the oil may also have been partially hydrogenated and/or winterised.

SOYABEAN OIL FOOD PRODUCTS—THEIR PREPARATION AND USES

Mayonnaise, Prepared Salad Dressings, Salad Oils and Margarine

Soyabean oil is the choice of most manufacturers for mayonnaise, prepared salad dressings, and salad oils by virtue of its improved flavour stability, low costs, and ready availability. An additional advantage of soyabean oil in these applications is its ability to form a stronger emulsion than cottonseed oil.

Practically all mayonnaise produced commercially in the United States is prepared exclusively with soyabean oil [usually the RBD and hydrogenated and winterised (HW) types] as the oil component.

The soyabean oil when used in mayonnaise, prepared salad dressing and salad oils must not solidify, cloud, or deposit a crystalline fraction at refrigerator temperatures; otherwise the emulsions will be broken. *Brekke* claims that both the RBD and HW soyabean oils will remain clear after immersion for 5.5 hr. in an ice bath held at 0°C (32°F), as specified in AOCS Method Cc 11-53. Both oils will also pass when the test is extended to 24 hr. as required by some manufacturers. Partial, selective hydrogenation of soyabean oil to lower its linolenic acid content to about 3 per cent (as in HW soyabean oil) improves considerably the oil's flavour and oxidative stability in salad and cooking oil applications. It is also common practice to blend soyabean oil with an oil containing little or no linolenic acid to lower the linolenic acid content of salad and cooking oils (e.g. soyabean with cottonseed oil).

If a margarine of low-*trans*-acid content is desired, interesterification of a nonhydrogenated soyabean oil with about 20 per cent of a soyabean hard fat can be carried out, followed by continuing with the usual steps of emulsion preparation and solidification. It was found that margarine oils having suitable SFIs and of average nutritional value, based on the ratio of linoleic to total saturated fatty acids, can be prepared from blends of soyabean oil alone when the interesterification process is used. Although the *trans*-acid content of their margarine oils was not given, some *trans*-acids were probably present.

Interesterification of edible fats and oils is important, not only to modify their physical properties but also to change fatty acid composition to give products enhanced nutritional benefits. Prior research has indicated that chemical catalysts promote random or directed interesterification reactions. *Konishi* and co-workers have investigated reaction conditions that provide regioselectivity concerning fatty acid positions in the triaclyglycerol through chemical interesterification. Soyabean oil and methyl stearate were used as reactant. Results indicated that regioselectivity can be obtained at 30°C (86°F) in 12–24 hr.

Zeitoun and co-workers have formulated fat blends by mixing fully hydrogenated soyabean oil with nine different commonly used vegetable oils in a ratio of 1:1 (w/w), which were then subjected to interesterification (i.e. rearrangement or randomisation) with a sodium methoxide catalyst. The difference in the endothermic and exothermic peak temperatures, total heat of fusion and crystallisation (β and β′ crystalline content) and solid fat content among the fat blends clearly showed the effect of the composition of each oil on the physical properties. Oils that contained a considerable amount of palmitic acid has a favourable influence on the crystallisation and polymorphic form of interesterified fat blends.

Shortenings

Soyabean oil is an excellent oil for preparation of shortening base stocks and is used in a number of shortenings. Soyabean hard fat can be used in pourable shortenings as well as in plastic shortenings in amounts up to 50 per cent of the total hard fat content. When used alone, soyabean hard fat causes plastic shortenings to become brittle because the hard fat crystallised in the β form. However, it can be used satisfactorily when blended with a greater or equal amount of a β′ hard fat.

Table 11.17 includes possible uses for each types of shortenings.

Great variety of shortenings having different physical states and performance can now be produced for numerous applications by the proper selection of edible vegetable oils and animal fats, by controlled hydrogenation, appropriate blending, plasticising, tempering and the incorporation of plasticisers, emulsifiers and antifoam agents.

Table 11.17. Types of shortenings that contain soyabean oil and some of their uses.

Group	Type and description	Uses
I	Vegetable oils and salad oil	
	The unhydrogenated oils are bland or very mild in flavour and liquid at room temperature. Salad oils are additionally winterised, when necessary, to prevent the oil from solidifying or clouding at refrigerator temperature.	Salad oils Household baking and cooking, pan and grill frying Deep-fat frying when the turnover rate is sufficiently high and frying odour is not a factor. Wholesale bakery production of bread, buns, rolls and cakes.
II	Hydrogenerated winterised oils	
	These oils are similar to those in Group I except for having been lightly hydrogenated and winterised. They possess greatly improved oxidative and flavour stability. Hydrogenated winterised soyabean oil (HWSBO) is a prime example.	Salad oil Pan and grill frying Deep-fat frying in restaurants and institutions when the turnover rate is moderate. Preparation of some potato chips.
III	Hydrogenated vegetable shortenings	
	(a) Oils specially hydrogenated to render them resistant to oxidative changes at frying temperature and in packaged, fried foods. Such shortenings are fairly firm and brittle at room	Deep-fat frying of some food in restaurants and institutions. Preparation of some fried snack foods
	temperature. Most expensive of the plastic frying shortenings. Referred to as the 'heavy-duty' frying fats. Have narrow plastic range. Example–frying shortening. (b) Oil slightly less hydrogenated than those in (a). These shortenings have a lower melting point but impart a better mouth-feel, a very important factor in snack foods.	Fried snack foods
IV	All-purpose vegetable shortenings	
	Oils hydrogenated to a lesser degree than for Group III. Less stable at frying temperatures. Sufficient hard fat added to improve the shortening's plasticity and creaming properties. Examples: general-purpose shortenings. Table 11.15.	Bakery operations where an inventory of shortenings for special uses is not maintained.
V	Blended vegetable shortenings	
	Blend consists of unhydrogenated vegetable oil(s) with a larger percentage of hard fat than used in the all-purpose shortenings. They posses good creaming power, are only slightly less resistant to oxidation and are less costly than the all-purpose shortenings.	Much the same as group IV
VI	Meat fat and vegetable oil blends	
	A blend of a lard and/or tallow and vegetable oils. Proportions vary from supplier to supplier with related effect on melting point and other properties. Some of these shortenings have excellent creaming properties and their oxidative resistance may equal or exceed that of the all-purpose hydrogenated vegetable oil shortenings.	Generally the same as group IV

(Contd...)

Table 11.17. Types of shortenings that contain soyabean oil and some of their uses. *(Contd.)*

Group	*Type and description*	*Uses*
VII	Hydrogenated meat fat–Vegetable oil blends These are similar to those in group VI but have been hydrogenated. They have good resistance to oxidation, are firmer in consistency than the all-purpose shortenings and therefore do not cream as well when used in baking. Often less costly than group III shortenings.	Same as group III
VIII	Hydrogenated, fractionated vegetable oils This group comprises the liquid fraction obtained by selectively hydrogenating a vegetable oil, such as soyabean for example, typically to 80–85 iodine value and then winterising the hydrogenated oil to remove the solids (stearines) fraction[a]. With proper additives, the product is a fluid shortening nearly comparable to oxidative stability to group III (b) shortenings.	Deep-fat frying when turnover rate is low.

[a] The hydrogenation conditions eliminate substantially all of the linolenic acid and reduce the linoleic acid content to under 15% in preparation of the above product.

Uses of Soyabean Oils in Food Products

The following is a review of specific application areas for soyabean oil in food products. In addition to the applications of mayonnaise, margarines, shortenings, prepared salad dressings and salad/cooking oils–already discussed—this section is focusing on applications for these products when used as ingredients in consumer products or in food industry formulations.

Consumer products

Household shortenings used primarily for frying and, to some extent, for general baking and cooking. Household shortenings almost always contain sufficient emulsifiers in the form of mono- and diglycerides to obtain an α-monoglyceride (the active form) content of about 2 per cent. This will make the shortening well-suited for baking.

Baked goods

In baked goods, shortenings contribute to the finished products in a number of ways: (i) impart tenderness and richness; (ii) enhance the aeration of leavened products; (iii) add to the flavour; (iv) promote a desirable grain and flavour; (v) assist in the development of flakiness in products such as pie crusts, Danish and puffed pastry products; (vi) modify the wheat gluten, particularly in the development of yeast-raised doughs; and (vii) act as emulsifier for the retention of liquids.

The main contributions of plastic shortenings to baked goods, however, arise from two chief functions performed by the shortening: a leavening and creaming action and a lubricating function. Large quantities of plastic and pourable shortenings, many made from soyabean oil, are used for bakery, shop and institutional production of yeast-raised goods (bread, buns, rolls, sweet rolls, yeast-raised doughnuts, coffee cakes and cookies).

Liquid or fluid shortenings, or soyabean oil with emulsifier, are very popular today because of better control of product formulation in automated baking systems, reduction in shortening usage and economy of bulk handling of an unheated liquid in the production of batch-process sponge and dough bread, continuously mixed bread and hamburger and frank further buns. The fluid shortenings usually contain hard fat flakes or emulsifier combinations, or both, either dissolved, dispersed, or suspended in vegetable oils. Such shortenings also perform well in specific types of cakes and prepared mixes.

Although the crystalline fats in plastic shortenings play an important role in the development of a fine-grained, cellular structure in cakes, starch, protein and emulsifiers take over this function when fluid or liquid shortenings are substituted and crystalline fat is largely absent.

A bland shortening with good stability is needed for soda cracker doughs to ensure adequate shelf life. Such a shortening can be produced from processed soyabean oil. Some crackers are also coated with a thin layer of a 'cracker spray oil'. A properly hydrogenated and fractionated soyabean oil high in oleic acid is well suited for this use. Soyabean oil is also excellent for preparing shortenings for icings and cream fillers that have good emulsifying, creaming and stiffening performance properties. These shortenings usually contain hydrophilic emulsifiers along with the normal complement of mono- and diglycerides.

Fried foods

A variety of fats can be used for deep-fat frying. These range from unhydrogenated vegetable oils and meat fats to shortening compounds partially hydrogenated for high oxidative stability. Frying and speciality shortenings generally have a narrow plastic range and consists of partially hydrogenated vegetable oils with little or no added hard fat. For potato chips and other fried snack items, the surface should appear shiny and feel wet when the absorbed fat is liquid at room temperature. The dull appearance of fried foods in the frozen state is also improved when partially hardened soyabean oil used fractionated to remove hard fats.

Soyabean oil shortenings are being used in restaurants (including the fast-food type), institutions and homes for deep-fat frying chicken, meat, seafood, potatoes and various vegetables. Sometimes, hydrogenated–winterised soyabean oil is being used for potato chips. Some processors use slightly hydrogenated soyabean oil to which 15–20 per cent cottonseed oil is added for obtaining better flavour.

A soyabean oil hydrogenated to 65–70 IV is frequently used to prepare fried chicken in fast-food restaurants in a pressure frying operation. In this process, a very high fat turnover rate is needed because chicken fat leaches into the shortening and raises its PUFA content.

There is no need for high stability for the frying fat when fried foods are served hot and consumed immediately after preparation. With such high turnover rates, unhydrogenated soyabean and deodourised, stabilised lard can be used as frying fats. Foodservices frying, however, requires higher stability criteria than any other edible oil applications because it does not have the high turnover rates inherent in food-processor-type frying applications. The latter can generally tolerate more unsaturation.

Unemulsified shortenings can be used for pan frying. Lecithin may be used in these products as an antistick compounds along with heat-stable buttery colours and flavours. Based upon a consumer panel evaluation of the most frequently used vegetable oils for pan frying in the Netherlands, *Theunissen* and co-workers found no meaningful differences in the palatability of the foods tested when fried in these oils; they suggest that using the least expensive oil for pan frying will suffice to obtain market acceptance.

Silicone compounds (e.g. polydimethyl siloxanes) are often used in frying shortenings as antifoam agents to raise the smoke point and to increase the oxidative stability of soyabean oil at frying temperatures. The FDA permits their use of addition levels up to 10 ppm in shortenings.

Canned, packaged and dehydrated foods

Soyabean oil is a useful ingredient in many canned and dehydrated foods. Although canned goods are expected to have long shelf stability at ambient temperatures, it is not necessary to provide special features to safeguard resistance to oxidative rancidity as long as the oils are used in vacuum-packed canned goods where oxygen is largely absent.

The same is true for dehydrated foods. For example, when dried egg mixes were prepared commercially with 15 per cent of hydrogenated–winterised soyabean oil and refined, bleached and deodourised corn oil or cottonseed salad oil, a trained panel found them equally acceptable both initially and after accelerated storage tests.

Confectionery coatings and candies

Soyabean oils is being used for the commercial preparation of many confectionery coatings and sweet goods, including certain hard butters. The main areas of use are in fudges, caramels and nougats. In other types of candies such as creams and jellies, smaller quantities are required. Soyabean oil competes in these applications with cottonseed, peanut and the lauric acid oils—coconut, palm and babassu.

Efforts have been made to partially replace cocoa butter by ingredients obtained by interesterifying, hydrogenating and winterising or otherwise fractionating cottonseed, palm and soyabean oils and edible tallow. The objective is to produce hard butters that can be used by themselves as the soluble fat in candy products. These products must also be more uniform and perform in a manner that would make the special tempering conditions required for cocoa butter unnecessary.

Imitation dairy products

Vegetable fats, including soyabean oil, find applications in a number of imitation dairy products, e.g. margarine, mellorine (an ice cream substitute), filled and simulated milk, filled and simulated cheese, frozen and chilled desserts, coffee whiteners, whipped toppings, milk shakes and eggnogs. In 'filled' products an edible vegetable fat replaces only the butterfat whereas in an 'imitation' or simulated product there is no milk present. Liquid vegetable oils, a low-melting, lauric-type hard butter or partially hydrogenated domestic oils are the usual fat components in filled milk.

Partially hydrogenated soyabean oils are also found in whipped toppings, filled evaporated milk and imitation cheese. Selectively hardened soyabean oil compares well in performance with the lauric acid-type fats in a number of these products (e.g. toppings and coffee whiteners).

Fat replacers

The actual market for fat substitutes depends on several factors: FDA approval, quality and versatility, strength of consumer demand and marketing strategies. In some aspects, the fats and oils industry is supplying raw material for some fat replacers or is making fat-based products (e.g. lecithin and monglycerides).

Other uses

Examples of soyabean oil usage in applications not covered so far include—oil in sardines sold in tinned containers, a blend of soyabean and rapeseed hard fats (IV of 12 or less) added to peanut butter (1–2 per cent) to promote rapid peanut flavour release and the many applications where soyabean oil–in unhydrogenated and partially hydrogenated states—is shown on product labels.

The following applications areas, while not all-inclusive, are illustrative of the widespread use of soyabean oil in food products: dehydrated soup mixes, prepared pudding mixes, prepared pancake and waffle mixes, macaroni and cheese mixes, prepared spaghetti sauce, mustard, dry breakfast cereal, frozen fried seafood, hamburger meat extenders, pizza and popcorn.

Nutritive value of soyabean oil

The composition and nutritive value of unhydrogenated soyabean oil are similar to those of other polyunsaturated vegetable oils and soyabean oil is an excellent source of the essential linoleic acid. Because of its high (54 per cent) total PUFA content, it is well suited for diets claiming high polyunsaturate-to-saturate (P/S) ratios.

Unhydrogenated soyabean oil has a higher P/S ratio than do the partially hydrogenated and winterised soyabean oils. As the degree of hydrogenation increases, the P/S ratio generally decreases, although carefully chosen selectivity of the hydrogenation process can lessen this effect. The main effect of hydrogenation on fatty acid composition of soyabean oil is the reduced levels of polyunsaturated fatty acids, increased level of saturates and the formation of *cis* and *trans* positional isomers.

The high (0.095 per cent) tocopherol (vitamin E) level in soyabean oil is also a positive health factor, as is the carotene content, which is precursor for vitamin A.

Metabolic studies

Clandinin states that dietary fatty acid balance is important to membrane composition, membrane-bound enzyme activity, and hormone-mediated functions in a variety of physiological states. The nutritional effect and metabolism of isomeric fatty acids have been investigated and reviewed.

The usual dietary fats and oils are readily absorbed in normal individuals (e.g. soyabean oil and partially hydrogenated soyabean oil are over 96 per cent absorbed). Studies indicate that the main factor affecting fat absorption is its melting point. If the melting point of a fat exceeds approximately 50°C (122°F), the fat is gradually less well absorbed as the melting point increases.

Commercially hydrogenated soyabean oil used in margarines contains varying amounts of *trans* isomers and usually about 15 positional monoenoic isomers in amounts greater than 1 per cent. Double-bond distribution in the *cis* and *trans* monounsaturated fatty acids generally range from 7 to the 14 positions. In addition, *trans-trans, cis-trans* and *trans-cis* diunsaturated fatty acids isomers have been detected and positional diunsaturated isomers are also present.

The distribution pattern of linolenic acid geometrical isomers does not depend on the degree of isomerisation. *Trans* isomers of linoleic acid account for only 1 per cent (soyabean oils) of total fatty acids and the probability of isomerisation of linolenic acid is about 13–14 times that of linoleic acid.

The amount of isomerisation varies with catalyst, degree of hydrogenation and conditions used (i.e. time, temperature and pressure). Hydrogenation of any highly polyunsaturated oil, such as that of peanut, corn, safflower, cottonseed, or sunflower, forms fatty acid isomers similar to those found in partially hydrogenated soyabean oil'.

This means that low levels of the *cis–trans* and *trans–cis* isomers of linoleic acids are found in margarines, along with almost insignificant amounts of the all-*trans* isomers. The possible effect of *trans* isomer levels in margarines and spreads (particularly in the stick products) continues to occupy the attention of the scientific community and the public. Two groups have found an effect, while two other have found none.

Long-term rate feeding studies, with partially hydrogenated soyabean oil, did not have an effect on growth, longevity, or reproduction. Histological examination of various organs also showed no effect. Actually, lightly hydrogenated soyabean oil was effective in lowering both total and low-density lipoprotein cholesterol, but less so than the more unsaturated triglycerides.

Nutritive value of heated and oxidised oils

Lipid oxidation is now recognised to be a critically important reaction in physiological and toxicological processes as well as in food products. Redox-active metals are the major factors catalysing lipid oxidation in biological systems.

Many oxidation products resulting from heating pure triglycerides have been identified. Some of these compounds formed by oxidation and/or pyrolysis of triglycerides are given below:

Aliphatic hydrocarbon	Cyclic hydrocarbons	Ketones
Alcohols	Cyclic monomeric acids	Keto hydroxy fatty esters
Aldehydes	Dibasic acids	Keto fatty esters
Aliphatic acids	Epoxides	Lactones
Cyclic dimeric acids	Hydroxy fatty acids	Polymeric fatty acids
Cyclic ethers	Hydroxy fatty esters	

Nonfood Uses for Soyabean Oil

Industrial (nonfood, nonfeed) uses for soyabean comprise only about 0.5 per cent of the protein and 2.6 per cent of the oil produced from soyabeans grown in the United States. While soyabean oil provides over 50 per cent of all the fats and oils consumed, soyabean oil accounts for less than 10 per cent of the industrial needs for fats and oils in the United States. The major nonfood markets for soyabean oil include drying oil products, paints and varnishes, resins and plastics, printing inks, and foots (as a source of soyabean fatty acids). Epoxidised oils (resins and plastics) and coating vehicles (paints and varnishes) comprise 50 per cent of the industrial market for soyabean oil.

Because of environmental concerns, there is a recent interest in biodiesel fuels and soya-based inks. *Murayama* has reviewed biological derived fuels, their importance, production and present consumption, as well as their performance and costs. Overall conclusion was that, it is possible with these fuels to obtain clean, efficient combustion with esterised fuels or with blends with diesel fuels; this can be achieved in boilers or engines without redesign or other changes being necessary.

Some vegetable oils offer excellent lubricating properties (e.g. mild antiwear/extreme pressure performance and inherently high-viscosity index).

There are, however, some serious performance concerns: poor low-temperature fluidity and poor oxidation resistance at high temperatures. Development of different physical characteristics will be needed before these oils can have an impact on their industrial uses in the low and very high temperature ranges.

Other more recent applications include brominated soyabean oil in fruit-based beverages to improve the cloud stability; as a component in many oleochemicals (e.g. adhesive tapes, shaving compounds, textile water repellents, carbon paper and typewriter ribbons); as fatty acids fractions in candles, crayons, cosmetics, polishes, buffing compounds, and mould lubricants; as a dust supressant in grain elevators and in livestock feed; as soyabean soapstock in similar dust control applications; as a practical carrier in herbicide and insecticide applications; as an antifoam agent in aerated fermentation such as production of antibodies and as a spray to delay the onset of blooms on fruit trees and to reduce susceptibility to frost damage.

The traditional industrial markets for soyabean oil such as coatings, plasticisers, dimer acid synthesis, surfactants, factice production, core binders and numerous minor markets, are stable and nongrowth ones at present, but they have been maintained in the face of past economic pressure from pertochemical derivatives because of properties and advantages not always obtainable from petrochemicals.

In recent years, more concern about environmental pollution by synthetic chemicals may help natural fat- and oil-based products to become more competitive in some markets. Recent commercial interest in biodiesel fluids and soya-based inks are good examples for this new trend.

Soyabeans, as with other oilseeds, are renewable resources that can be produced in abundant supplies. For soyabean products to be adopted, however, the performance characteristics of new soyabean products will have to meet the quality standards of ingredients currently being used, regardless of price and environmental benefits.

SOYA—A HEALTHY COMPLEX OF PHYTONUTRIENTS

Soyabeans have been a valuable resource for mankind by providing excellent protein and other nutrients. Evidence is increasing that the consumption of soya foods can lower the risk for various series diseases such as cardiovascular disease, osteoporosis and certain cancers. In addition to the well studied cholesterol lowering effects of soya protein, other general health improving effects are attributed to several minor components in soya. These minor components are by natural concentrated in the soya hypocotyl or germ the part of the soyabean, which can grow into a new plant upon germination.

An overview of the biological relevance for each component is presented below:

Protein

Soya protein is highly nutritious as it is an excellent source of required amino acids. It has a protein digestibility corrected amino acid (PDCCA) value of 1. PDCAA indicates to what level the protein composition meets the protein needs of adults when consumed as the sole source of protein at a recommended level of intake (0.8 g/kg BW). Soya protein's ability to lower serum cholesterol levels is well recognised. The most commonly quoted research is the meta-analysis by *Anderson* correlating the consumption of soya protein with the risk on Cardivascular disease (CVD). By combining the results of 38 clinical studies examining the effects of soyabean protein on serum lipids profiles, it was concluded that a minimum of 25 gram of soya protein daily substantially reduces total serum cholesterol, LDL cholesterol and triglycerides. Results from additional studies were sound enough for the FDA to approve a health claim for soya protein.

Essential Fatty Acids

The role of dietary fatty acids in CVD is becoming well established. Diets high in unsaturated fatty acids have been shown to improve plasma lipid profiles. More recently, research has focused further on specific unsaturated fatty acids: the ω-3s and the ω-6s. Omega 3 fatty acids are typically found in fish oils, though a few plants including soya contain them. Plant sources of ω-3 fatty acids differ from fish sources but our bodies can covert a small percentage to the fish-oil-type. Typically in soyabean oil the ω-6/ω-3 ratio is 7.5. The oil fraction present in 'soyalife complex' is even more unsaturated and does have an improved ω-6/ω-3 ratio of 3.

Isoflavones

In addition to soya protein, the role of soya isoflavones in reducing the CVD risk factors is under investigation. Since isoflavones act as antioxidants, it can be expected they have a function in preventing or inhibiting the oxidation of LDL cholesterol, thereby improving the LDL/HDL ratio. What's more, it has been shown that isoflavone products do lower blood pressure, further reducing CVD risk. Isoflavones fall within a group of compounds named phytoestrogens. This name indicates their character as estrogen like components from plant origin. Phytoestrogens can be found throughout the plant kingdom, but soya has been found to be an excellent source. The isoflavone concentration in the soya hypocotyl is 10–20 times higher compared to the whole soyabean.

The isoflavone composition of hypocotyl is different from whole soyabeans (% of total):

Isoflavone composition

Description	*Soyabean*	*Soyagerm*
Genistein	50–60	10–20
Daidzein	30–40	50–60
Glycitein	10–20	30–40

Isoflavones have been shown to reduce menopausal complaints in menopausal women, which is considered to be a real estrogenic effect. Further, isoflavones are being studied for their potential role in prevention and inhibition of the development of osteoporosis, generally showing positive effects on bone mass concentration and bone mass density. The exact mechanisms remain unsolved, though it has been suggested that daidzein in particular plays an important role, with respect to the number of bone types sensitive to isoflavone supplementation. *Vitro* studies suggest that genistein might reduce the Ca-resorption from bone by osteoclasts, as a consequence of its tyrosine kinase inhibition activity. This enzyme is also involved in the development of cancers. The inhibition of tyrosine kinases is one of the mechanisms by which isoflavones interfere with cancer development. Amongst other mechanisms under investigation, including the binding to the estrogen receptor itself or by the inhibition of angiogenesis leading to starvation of the tumour. *Vitro* studies further suggest an immune stimulating role. The isoflavone being researched in this area is daidzein, which stimulates both phagocytic activity and proliferation of lymphocytes.

Oligosaccharides

The levels of oligosaccharides in soya may cause flatulence in individuals. This inconvenience is caused by a prebiotic effect or stimulation of the gut microflora and will disappear in time as the microflora adjust to these fibrous products. The positive side of this initial inconvenience is that stimulation of the intestinal microflora results in overall improvements of absorption and metabolism: including the metabolism of many phytochemicals such as the isoflavones. Oligosaccharides have also been shown to lower blood pressures.

Fibre

Fibres can be defined as undigestible polymeric substances. Many of them show a high capacity to bind water resulting in the so-called fecal bulking effect. Although fibres are indigestible, they stimulate the gut microflora in a positive way. It has been known for some time that fibre lowers cholesterol levels due to its capacity to bind intestinal bile acids resulting in excreting these components in the faeces.

Tocopherols

Vitamin E is a fat soluble antioxidant, which protects lipid structures from oxidation. Prevention of cancer and heart disease is associated with the consumption of sufficient amounts of vitamin E. Tocopherols are also shown to improve immune function. 2 gram of 'soyalife complex' adds 5–14 per cent of the recommended daily intake for α-tocopherols. The α-tocopherols, which are even more prevalent in 'soyalife', are gaining attention as well, being potent inhibitors of specific types of oxidative damage of membrane lipids.

Saponins

Saponins are a large group of glycosilated steroid complex compounds. Soya saponins are classified into three groups A, B and E, identified by having the same α-glucone basic structure. Group A saponins are confined to be 'soya hypocotyl', whereas the other groups are present throughout the soyabean, including the germ. 'Soya saponin AB' (major representative of the group A saponins) was shown to be a potent inhibitor of hydrogen peroxide oxidation compared to the other tested soya saponins, proving its antioxidant activity.

Saponins, which are concentrated in the soya hypocotyl, play a part in reducing cholesterol. Similar to fibre, saponins have roles as carrier for bile acids, causing them to be excreted in the faeces. Removal of the bile acids stimulates a need for their production. As a results, cholesterol is pulled from the blood and the liver to form replacement bile. Saponins are gaining attention in research, since they have recently been shown to have potential anti-carcinogenic activities in *in vitro* experiments. Further studies are required to confirm these interesting findings.

Phytosterols

Phytosterols can also assist in reducing plasma cholesterol levels by reducing particularly LDL cholesterol. Although phytosterols are characterised by a limited absorption, they do interfere with intestinal cholesterol absorption. A second hypothesis is that phytosterols increase bile acid production, although they have not been found to interfere with cholesterol synthesis as such.

Lecithin

Lecithin is a well-known antioxidant and emulsifier and major component of all cell membranes. Soya lecithin has been shown to lower total plasma cholesterol and non-HDL-C in monkeys and hamsters, indicating its potency for human cholesterol control. Immune modulatory effects have been described, as a result of increased phagocytosis upon bacterial and fungal infections.

Inositol-6-Phosphate (Phytate)

Phytate has antioxidant effects and has been shown to reduce cholesterol. However, it is known as a chelating agent as well, binding bivalent cations, often suspected to reduce the bio-availability of these ions. The *in vivo* detrimental effect is in most cases hard to estimate since our diets do contain more components, which show interactions as well. Fibre is well known as another chelating agent, although its beneficial effects on intestinal functioning are considered to be more important. The same counts for phytates; our diets are so sufficient in delivering the metallic ions for the general healthy public, that our bodies would hardly notice any limitation in ion absorption. Phytic acid is increasingly investigated for its anti-carcinogenic activities. Both preventive and therapeutic effects have been found for a selection of cancers (colon and mammary cancer), by anti-proliferative and differentiation promoting mechanisms.

In conclusion, 'soyalife complex' is found to be more then just a source of excellent protein and isoflavones, as it is produced from the soya hypocotyl. It contains many more constituents, as discussed above, which also play their own important roles in modifying risk factors for various diseases and are beneficial to general health.

SOYA PROTEINS AND HEART DISEASE

Cardiovascular Disease

Cardiovascular disease (CVD) is used as collective term used to describe a wide variety of conditions related to the heart and blood circulation. CVD is currently the number 1 killer of Americans and has been since 1900. The USA is not an exception with CVD death rates of 37 per cent (1997) for males between 35–74 years old. Australia for instance experienced death rates of 35 per cent in 1995. The European numbers in general are similar (American Heart Association, 2000).

There are many different types of cardiovascular disease like angina, atherosclerosis, hypertension, heart failure and arrhythmia. These diseases often have progressed quite far before there are any symptoms. Contributing risk factors for cardiovascular disease include family history, gender, age, race, smoking, stress, obesity, diabetes, high blood pressure and high serum cholesterol levels. The latter parameters are often used in research to investigate the condition of an individual. A person is considered to have high blood pressure when he or she has a systolic pressure of at least 140 mm Hg and/or a diastolic pressure of at least 90 mm Hg.

Desirable total cholesterol levels are below 200 mg/dl with levels above 240 mg/dl considered high blood cholesterol. In addition to total cholesterol it is useful to determine its distribution in the different lipoprotein fractions, of which VLDL (very low density lipoproteins), LDL (low density lipoproteins) and HDL (high density lipoproteins) are the most important. The LDL fraction is responsible for the transfer of most of the cholesterol from the liver to tissues where it is taken up by LDL receptors. Cholesterol accumulation may occur if there is excess LDL circulating in the blood. The HDL fraction is a scavenger lipoprotein, responsible for the removal of excess cholesterol from tissues. Low LDL levels (below 130 mg/dl) and high HDL levels (about 35 mg/dl) are preferred. Thus the LDL/HDL ratio is a method often used to assess CVD risk. A ratio below 5:1 is desired; the optimum ratio is 3.5:1.

A high fasting plasma triglyceride is often associated with high plasma cholesterol, but is not by itself a risk factor for heart disease. Hypertriglyceridaemia is suggested at levels over 400 mg/dl.

Effect of Soya on Risk Factors

Soya protein's ability to lower serum cholesterol levels is almost old news. The most commonly quoted research is the *meta*-analysis by *Anderson*. Correlating the consumption of soya protein with the risk on CVD. By combining the results of 38 clinical studies examining the effect of soya protein on serum lipid profiles, it was concluded that a minimum of 25 gram of soya protein daily reduces total serum cholesterol, LDL cholesterol and triglycerides substantially. Results from additional studies were sound enough for the FDA to approve the health claim for soya protein. In addition to soya protein the role of soya isoflavones in reducing the CVD risk factors is under investigation. Since isoflavones act as antioxidants it can be expected they have a function in preventing or inhibiting the oxidation of LDL cholesterol, thereby improving the LDL/HDL ratio. What's more, it has been shown that isoflavone products do lower blood pressure, further reducing CVD risk.

A study of postmenopausal women given an aqueous ethanol extract of soya hypocotyl ('soyalife extra') containing 115 mg isoflavone-glucosides is summarised in Fig. 11.9. The fasting blood lipid profiles and blood pressures were established at baseline and after six weeks of treatment.

Evidence is mounting for the soya complex story: soya protein, either alone or in combination with soya isoflavones, is essential for its role in decreasing the risk of CVD. Although the actual mechanisms is not fully understood, it is likely that both soya protein and soya isoflavones play complex roles. But, there is more to the soya complex story.

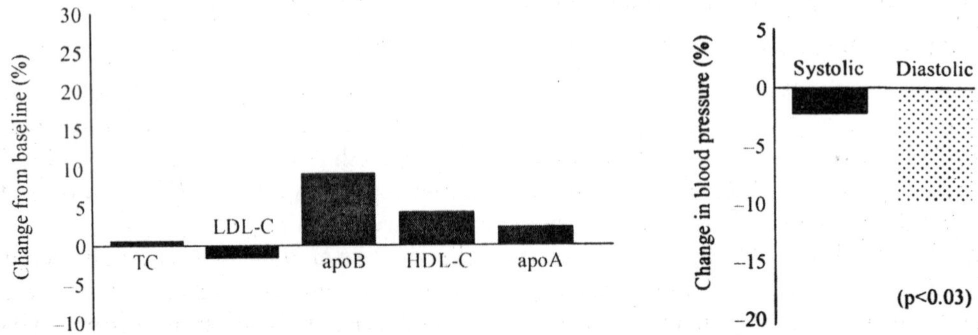

Fig. 11.9. Per cent change in CVD risk factors in normo-cholesterolemic postmenopausal women after 6 weeks isoflavone treatment (115 mg/day), (study was presented at the Washington Soya Symposium, 1999).

CVD risk factors can be mediated by many components, in nature laid down in soya and in some cases concentrated in the soya hypocotyl. The role of soya protein and isoflavones has been discussed above. Other soya components comprise a healthy fatty acid profile, fibre and oligosaccharides and saponins, tocopherols, lecithins, phytosterols and inositol-6-phosphate (minor components). They all add to the health benefits of 'soyalife complex' with respect to reducing CVD risk factors.

Tocopherols

Vitamins E, being a fat soluble antioxidant, protecting lipid structures for oxidation, needs no further explanation. 2 gram of 'soyalife complex' adds 5–14 per cent of the recommended daily intake for α-tocopherols. The γ-tocopherols, which are even more prevalent in 'soyalife', are gaining attention as well, being potent inhibitors specific types of oxidative damage of membrane lipids.

Essential Fatty Acids

The role of dietary fatty acids in CVD is becoming well established. Diets high in unsaturated fatty acids have been shown to improve plasma lipid profiles. More recently, research has focused further on specific unsaturated fatty acids: the ω–3s and the ω-6s. Omega 3 fatty acids are typically found in fish oils, though a few plants including soya contain them. Plant sources of ω-3 fatty acids differ from fish sources but our bodies can convert a small percentage to the fish-oil type. Typically in soyabean oil the ω-6/ω-3 ratio is 7.5. The oil fraction present in 'soyalife complex' is even more unsaturated and does have an improved ω-6/ω-3 ratio of 3.

Oligosaccharides, Fibre and Others

The levels of oligosaccharides in 'soyalife' may cause flatulence in individuals. This inconvenience is caused by a prebiotic effect or stimulation of the gut microflora and will disappear in time as the microflora adjust to these fibrous products. The positive side of this initial inconvenience is that stimulation of the intestinal microflora results in overall improvements of absorption and metabolism; improving the metabolisms of many phytochemicals including the isoflavones. Oligosaccharides have also been shown to lower blood pressure.

It has been known for some time that fibre lowers cholesterol levels, in addition to prebiotic and fecal bulking effect. Saponins, which are concentrated in the soya hypocotyl, also play a part in reducing cholesterol. Both fibre and saponins have roles as carrier for bile acids, causing them to be excreted in the faeces. Removal of the bile acids stimulates a need for their production. As a result, cholesterol is pulled from the blood and the liver to form replacement bile. Saponins may further acts as antioxidants. Phytosterols can assist in correcting plasma cholesterol levels as well, reducing particularly LDL cholesterol. Although phytosterols are characterised by a limited absorption, they do interfere with intestinal cholesterol absorption. A second hypothesis is that phytosterols increases bile acid production, although they have not been found to interfere with cholesterol synthesis as such.

Two other minor components do have cholesterol reducing activity as well. One of them is lecithin, a well known antioxidant and emulsifier, a major component of all cell membranes. Soya lecithins has been shown to lower plasma total cholesterol and non-HDL-C in monkeys and hamsters indicating its potency for human cholesterol control. Phytate has antioxidant effects and has been shown to reduce cholesterol. However, it is known as a chelating agent as well, binding bivalent cations, often suspected to reduce the bio-availability of these ions. The *in vivo* detrimental effect is in most cases hard to estimate since our diets do contain more components, which show interactions as well. Fibre is a well known other chelating agent, although its beneficial effects on intestinal functioning are considered to be more important. The same counts for phytates; our diets are so sufficient in delivering the metallic ions, for the general healthy public, that our bodies would hardly notice any change in ion absorption.

In conclusion, 'soyalife complex' is found to be more than just a source of excellent protein and isoflavones, as it is produced from the soya hypocotyl. It contains many more constituents, as discussed above, which also play their own important roles in modifying risk factors for CVD.

Chapter 12

Sunflower Oil

INTRODUCTION

Sunflower (*Helianthus annuus* L., 2n = 34) belongs to the family of only '*Compositae asteracees*' and to the genus Helianthus. Its original habitat is North America where it was traditionally cultivated by native Indians. It was first cultivated in Europe towards the middle of the 16th century when it was grown there essentially as an ornamental plant.

The achene of sunflower, incorrectly called seed, is formed by a kernel and a pericarp: the hull (Fig. 12.1). Different types of achene can be observed on a flower-head such as (i) sterile achenes, originating either from a ligulate flower or from a malformation at the time of floral initiation; (ii) empty achenes, resulting from improper seed-set or malformation of the fruit; and (iii) full achenes, containing a kernel of variable size filling more or less the hull; there are about 1500 achenes per flower-head.

The size of these seeds varies depending on their position: bigger at the periphery, they are also richer in oil. Depending on the variety, the seed is 7–25 mm long and 4–13 mm wide. The weight of 1000 seeds of today's hybrids varies between 30 and 80 g.

Both thickness of the hull and the hull:kernel ratio also vary, depending on the variety and environmental conditions. A full achene contains:

1. A pericarp (Fig. 12.2) which accounts for 18–35 per cent of the total weight. It contains a cellular layer forming the epidermis and is covered by a light cuticle. Presence of hair can also be noted on the surface.
   As the pericarp develops further, one observes the appearance of several bigger cellular layers (phelloderm). At the bottom end, three to five rows of polygonal cellular layers develop. The inner walls of these layers are sclerified on ripening. Xylem vessels are found in this layer.
   On coming in contact with the kernel, the cellulose fibres ensure, on ripening, the hardness of the hull. An internal epidermis separates the pericarp from the kernel and often comes loose to form an ovarian cavity.
2. The kernel is formed by a coat (three layers of parenchyma, the central one being spongy), an endosperm comprising a single layer of cells rich in protein seeds, closely tied to the coat and finally an embryo containing two cotyledons.

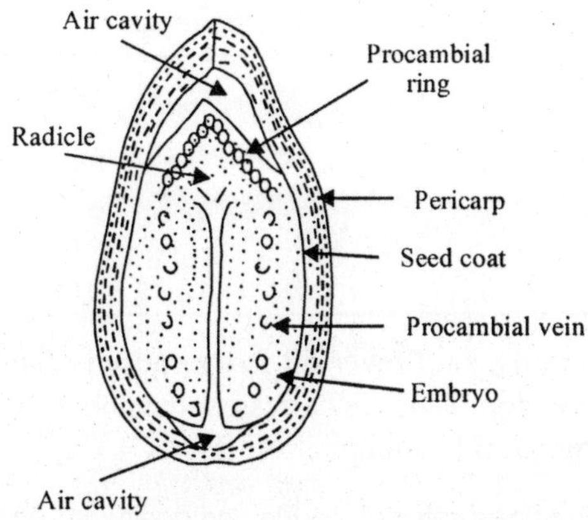

Fig. 12.1. Longitudinal section of a sunflower achene.

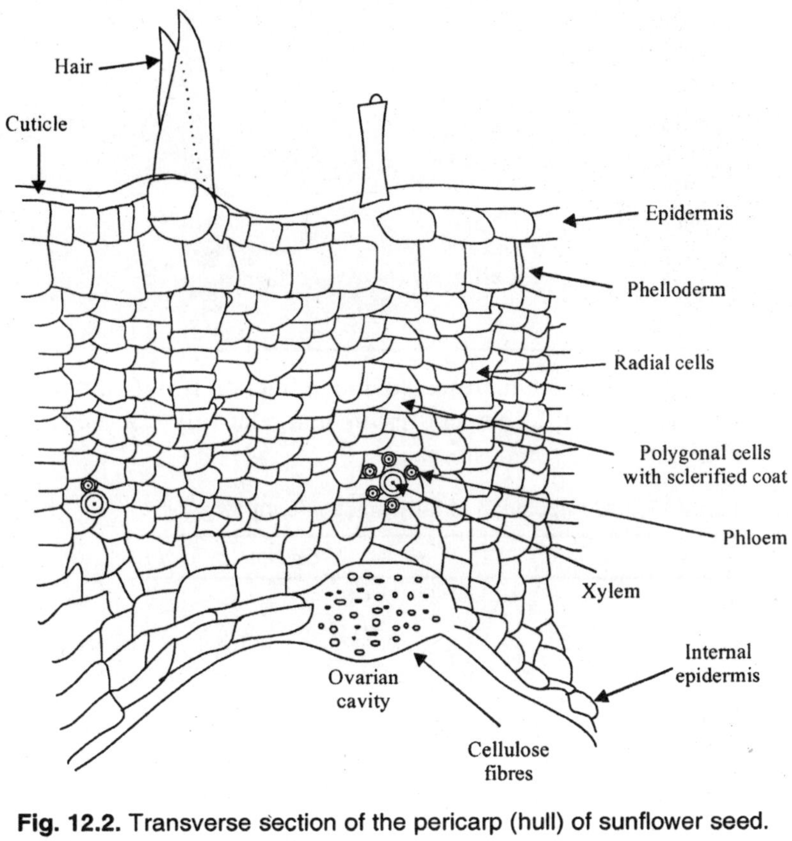

Fig. 12.2. Transverse section of the pericarp (hull) of sunflower seed.

OIL COMPOSITION

Table 12.1 shows the main physical and chemical characteristics.

Table 12.1. Main physical and chemical characteristics of sunflower oil.

D_{20}	0.920–0.925
V_{20} (cp)	51–57
n_D^{20}	1.474–1.476
Solidification point	–
Iodine value	120–134
Saponification value	188–193

Like the majority of vegetable oils, sunflower oil is composed mainly of triglycerides (98–99 per cent) and a very low proportion of various substances designated under the common term 'unsaponifiable fraction. Table 12.2 shows the triglyceride composition (HPLC).

Table 12.2. Sunflower oil triglyceride composition (HPLC).

Nature of TG	*% of total TG*
LLL	36.3
OLL	29.1
PLL	11.3
OOL	6.5
SLL	7.5
POL	4.0
PPL	0.5
OOO	0.6
SOL	2.1
POO	0.4
PPO	0.7

About one-third of triglycerides are represented by the fraction LLL (trilinoleine). Combination of this with a high oleo-dilinaleine (OLL) content results in a relatively low melting point of sunflower oil. Table 12.3 shows the fatty acid composition.

Table 12.3. Composition of fatty acids in sunflower oil.

Nature	*Content, % total FA*
C16:0	5–7
C16:1	$\leqslant$ 0.4
C17:0	$\leqslant$ 0.1
C18:0	4–6
C18:1	15–25
C18:2	62–70
C18:3	$\leqslant$ 0.2
C20:0	< 1
C20:1	< 0.5
C22:0	< 1

The total fatty acids in percent, as presented in the preceding table, shows the important role of the unsaturated fatty acids (87 per cent) with respect to saturated fatty acids (13 per cent). We should first of all point out the importance of linoleic acid (C18:2) and the very low content of linolenic acid (C18:3).

Table 12.4 shows the composition of the unsaponifiable fraction.

Table 12.4. Composition of unsaponifiable fraction of sunflower oil.

Unsaponifiable fraction: 0.5 to 1.5%			
Sterol content (mg/100 g)	325–515	Tocopherol content (mg/100 g)	44–120
Composition of sterols: (% total sterols)		Composition of tocopherols: (% total tocopherols)	
Cholesterol	< 0.4	α-Tocopherol	91–97
Brassicasterol	0	β-Tocopherol	3–6
Campesterol	8–11	γ-Tocopherol	⩽ 2
Stigmasterol	7–10	δ-Tocopherol	–
β-5-sisterol	58–64	Tocotrienol	–
Δ-7-Avenasterol	2–7		
Δ-7-Stigmasterol	9–14		
Δ-7-Avenasterol	4–6		
Isofucosterol	0.4–1		
Fucosterol	2–3		
Hydrocarbon content (mg/100 g), squalen	15–20	Aliphatic alcohol content (mg/100 g)	100

Oleic Sunflower

Sunflower is rich in linoleic acid and the appearance of mutants rich in monounsaturated (oleic acid).

Selection rapidly resulted in the finding of genetic material whose seed contains more then 50 per cent of oleic acid, rich in vitamin E and in good equilibrium with polyunsaturated fatty acids.

Today, the oleic sunflower hybrids (still sometimes called 'Oleisols') are recommended for cultivation. The main differences between the two types of sunflower are shown below:

Description	*Standard sunflower (% total FA)*	*Oleic sunflower (% total FA)*
Palmitic acid, C16:0	7	3
Stearic acid, C18:0	4	5
Linoleic acid, C18:2	70	9
Oleic acid, C18:1	16	83
Linolenic acid, C18:3	Traces	Traces

Here, the most striking characteristic is the inversion in the contents of linoleic acid and oleic acid between the two types of sunflower.

It has already been pointed out that the conditions at the end of the growth cycle (and specially dry conditions often associated with high temperatures) lead to an overall reduction in the content of oil. Likewise, the equilibrium of fatty acids is subjected to fluctuations. In the case of oleic sunflower, the main reasons for these fluctuations are:

1. *Genetic*: Any pollution of the fragment of oleic sunflower by the pollen of standard sunflower brings about a reduction in the oleic acid content of the harvested seed. Thus there is a need to maintain an adequate separation between the sunflower fragments of the two types (minimum 500 m).
2. *Environmental*: In this case, the temperature effect is the main one. If conditions of low temperature occur during ripening, a decrease in the oleic acid content is observed to the advantage of linoleic acid content (Fig. 12.3). The deduction is that oleic sunflower should preferably be cultivated in regions where the autumn is warm.

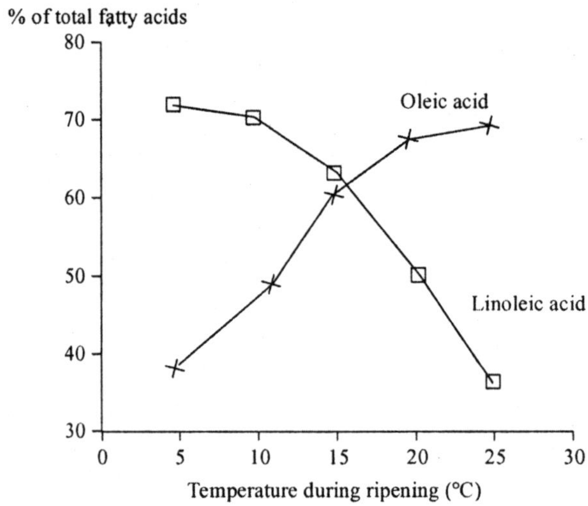

Fig. 12.3. Influence of temperature on the oleic acid and linoleic acid contents of sunflower oil.

SUNFLOWER OIL

Sunflower oil is an excellent vegetable oil both in terms of its healthy characteristics and its wide variety of uses. It has a very clean, bland taste—an appealing trait since it allows the natural flavour of foods to come through. The oil is also appealing visually, with its light yellow—almost clear—colour and clarity. These characteristics make sunflower a superior oil for salads, baking and frying. Its smoke point of approximately 232.2°C (450°F) (higher than any other major vegetable oil) is another benefit for frying operations.

Sunflower oil is among the healthiest vegetable oils available. It ranks high in the percentage of polyunsaturated fat, which can be of benefit in lowering cholesterol level. Sunflower oil is an excellent source of linoleic acid—one of the essential fatty acids required by the human body. And at 45 mg per 100 g, the vitamin E (α-tocopherol) content of sunflower oil is the highest of all leading vegetable oils.

Linolenic is another type of fatty acid. Because it oxidises quickly, vegetable oils containing linolenic acid can develop an off-flavour over a period of time. Also, oils containing more than 2 per cent linolenic acid must be partially hydrogenated during processing. Because sunflower oil contains just 0.5 per cent linolenic acid, it requires less hydrogenation and possesses a longer shelf life than oils with higher linolenic acid contents. A third type of fatty acid in vegetable oils is oleic. High oleic sunflower oil has a much lower percentage of polyunsaturated fat and a large amount of monounsaturated fat. Like high linoleic sunflower oil, it contains a small percentage of saturated fat.

The processing of sunflower oil from a crude state to a consumer-ready quality level is shown in Fig. 12.4. Sunflower oil is an uncomplicated oil to process, resulting in a high yield of refined oil.

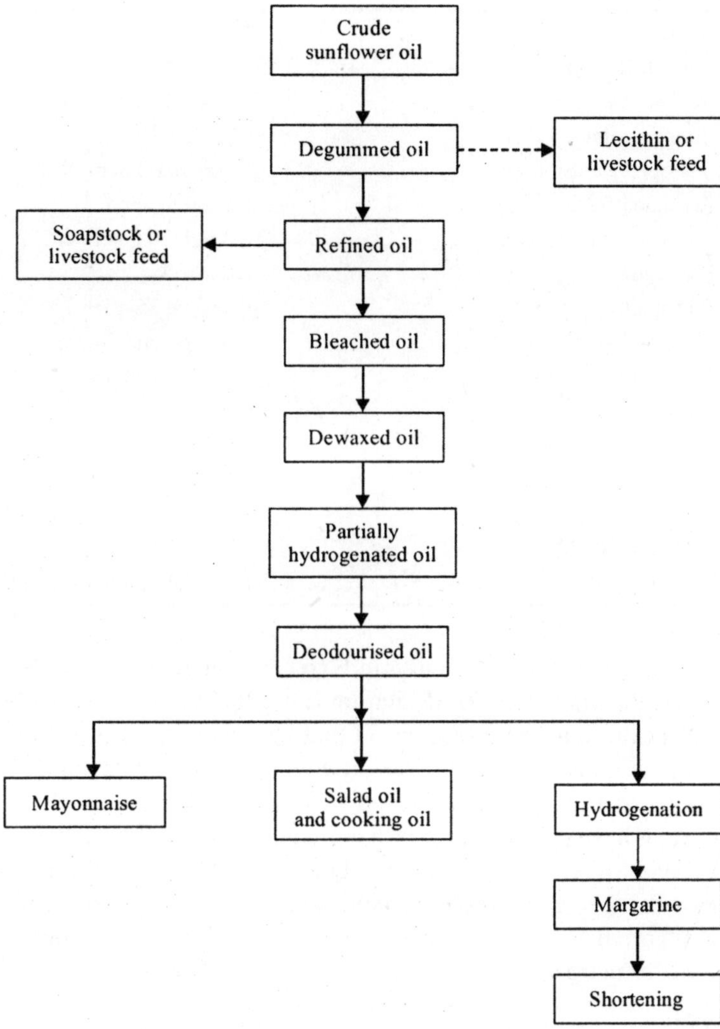

Fig. 12.4. Sunflower oil refining flow chart.

An average quality crude oil, having a free fatty acid content of 1 per cent should produce a refined, bleached and dewaxed oil having a yield of 96.5–97.0 per cent. The specifications for refined, bleached, deodourised and dewaxed sunflower oil are shown in Table 12.5.

Its health benefits, high smoke point, bland flavour and light clear consistency make it an excellent choice for baking, frying and salad dressing. Sunflower oil is accepted universally for use in margarine. Due to these special qualities, along with the outstanding health advantages offered by sunflower oil, an increasing number of food processing companies are recognising the consumer preference for this quality vegetable oil.

Table 12.5. Specifications for refined, bleached, deodourised and dewaxed sunflower oil.

Item	*Specification*
Iodine value	130–144
Saponification value	188–194
Refractive index at 25°C	1.4740–1.4745
Smoke point	252–254°C
Phosphates	1 ppm maximum
Unsaponifiables	1.5% maximum
Lovibond colour	2.0 red, 20.0 yellow
Peroxide	2.0 meq/kg
Fat stability by AOM	Peroxide 35 after 8 hr.
Chlorophyll	0.03 ppm maximum
Moisture and volatiles	0.10% maximum
Free fatty acids	0.05% maximum
Soap (sodium oleate)	0.003% maximum
Chill test:	
0°C (32°F)	48 hr.
4.4°C (40°F)	120 hr.
Flavour and odour	Not rancid, bitter or sour

Sunflower oil is an excellent source of vitamin E (α-Tocopherol), it naturally occurrs at 45 mg per 100 g. Sunflower oil has the highest level of vitamin E of all the major vegetable oils in use.

It is a desirable for health reasons to use an oil that has a low level of saturated fat. Sunflower oil contains a low level of saturated fat. Sunflower oil contains a monounsaturated fatty acid, oleic acid and a high percentage of the polyunsaturated acid linoleic acid. Both monounsaturated and polyunsaturated fatty acids have been found to be beneficial for health. Vegetable oils containing linolenic acid develop an off-flavour over a period of time, because linolenic acid oxidises quickly. Sunflower oil contains less than 0.5 per cent linolenic acid and thus does not usually need to be partially hydrogenated. A light, partial hydrogenation is somctimcs cmployed for sunflower oils used as salad oil. Sunflower oil used for margarine or shortening stocks is hydrogenated only to the desired level.

Crude Sunflower Oil

Sunflower oil is obtained from oil-type sunflower seed. The oil is extracted from the seed either mechanically (i.e. pressing or expelling), with a solvent, or with a combination of mechanical and solvent extractions as shown in Fig. 12.5.

The general fatty acid composition for the oil has been shown in Table 12.3.

DEGUMMING

Purpose of Degumming

Degumming is an optional process and is used to remove some of the lecithin and foreign materials that are present in the crude sunflower oil. Certain oil impurities such as phosphatide, protein, protein fragments and gummy or mucilaginous substances are soluble in oil only in the anhydrous form. The impurities can be precipitated and removed by hydrating them.

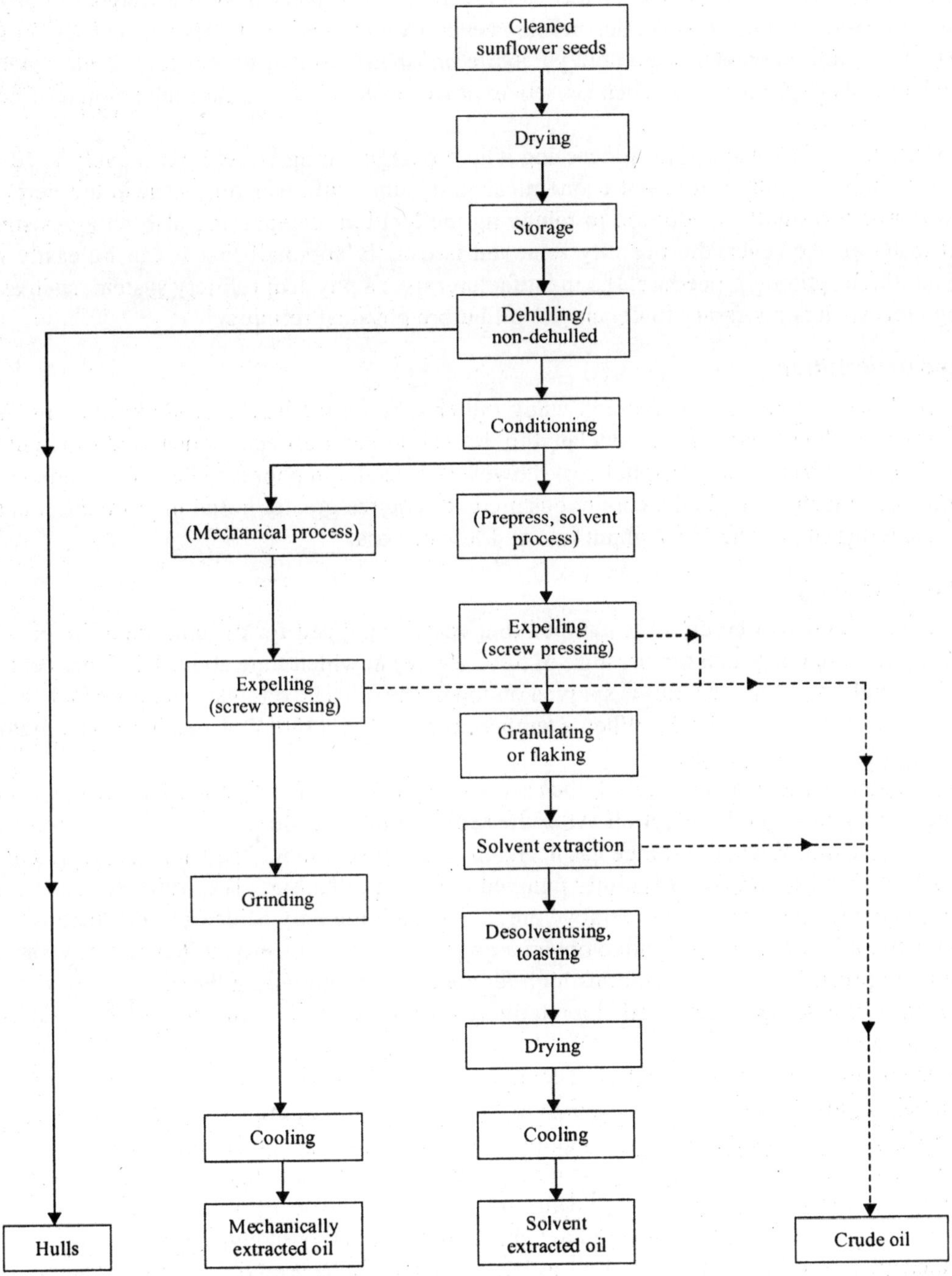

Fig. 12.5. Sunflower oilseed processing flow chart.

Phosphatide and the mucilaginous substances are emulsifying agents and, in excessive quantity, they may increase the loss of neutral oil in caustic refining of the oil. If the oil is going to be physically or steam refined, the contents of phosphatide and metallic prooxidants must be lowered by a degumming process before refining the oil. Water degumming generally removes about 90 per cent of the phospholipid material. There are, however, some nonhydratable phosphatides that do not react with water and are removed during caustic refining. Their presence, however, tends to increase the amount of neutral oil lost in the soapstock.

Contrary to crude soyabean oil processing wherein degumming is used extensively to remove the high percentage of lecithin, it is not economical to degum sunflower oil, because the percentage of lecithin in it is too small. In addition to removing the lecithin, degumming also removes some other foreign material. However, the quantity removed is usually so small that it can be easily removed during the alkali refining procedure. If a manufacturer uses a physical refinery system, such as a steam refining process, it is necessary to degum the oil before physical refining.

Process Description

Crude sunflower oil can be degummed using either a batch method or a continuous method. The economics of each process can be calculated to determine the most economical and efficient method. The batch method has the least capital cost; however, it results in poorer yields of degummed oil than the continuous method. While the continuous method requires greater initial investment, it is the most efficient in terms of a higher throughput rate and less oil loss.

Batch Degumming

The required quantity of crude oil is pumped to a vessel equipped for degumming. The oil is heated, with the agitator running, to a temperature of 32°C (90°F) at which time about 1.5–3 per cent of water (based on weight of oil in the tank) is sprayed on top of the oil. The oil–water mixture is then heated to 76–82°C (170–180°F) in 1–1.5 hr. When a temperature of 82°C (180°F) is reached, the agitator is shut off and the oil is allowed to rest.

The hydrated material and water will then settle to the bottom of the vessel. The settling time should be about 3 hr. to get a good separation. After the settling time, the sludge is slowly drawn off from the bottom of the vessel. When the sludge has been removed, the oil should be vacuum dried to maintain a good quality. The degummed oil is either pumped to a storage tank or alkali refined.

The sludge may be acidulated to remove the oil and split the fatty acids from the lecithin; however, it does require a lot of sulphuric acid to obtain a good split; consequently, it may not be economical. It may be more profitable to use or sell this high fat material for upgrading the meal or feed.

Caution: if this sludge is not used in a relatively short time, it will start to spoil and sour due to the presence of the water.

The equipment used for batch method of degumming is a mild-steel tank having a steep, conical bottom (about a 60° angle) and steam coils sufficient to heat the quantity of oil to be degummed to 76–82°C (170–180°F) in about 1 hr. It also must be equipped with an agitator properly sized for the tank and turning at a rate of about 60 rpm to provide a good dispersion of the water into the oil. Some means of spraying water onto the top surface of the oil should also be included in the design of the vessel.

It is desirable, but not mandatory, to have this tank equipped with a vacuum system for drying the degummed oil. If another tank or system is available that could be used for drying the degummed oil, it would be satisfactory. If the oil is going to be either physically or alkaline refined, it may be possible to bypass the vacuum drying step. Pumps are required to transfer the oil to and from the processing vessels (Fig. 12.6).

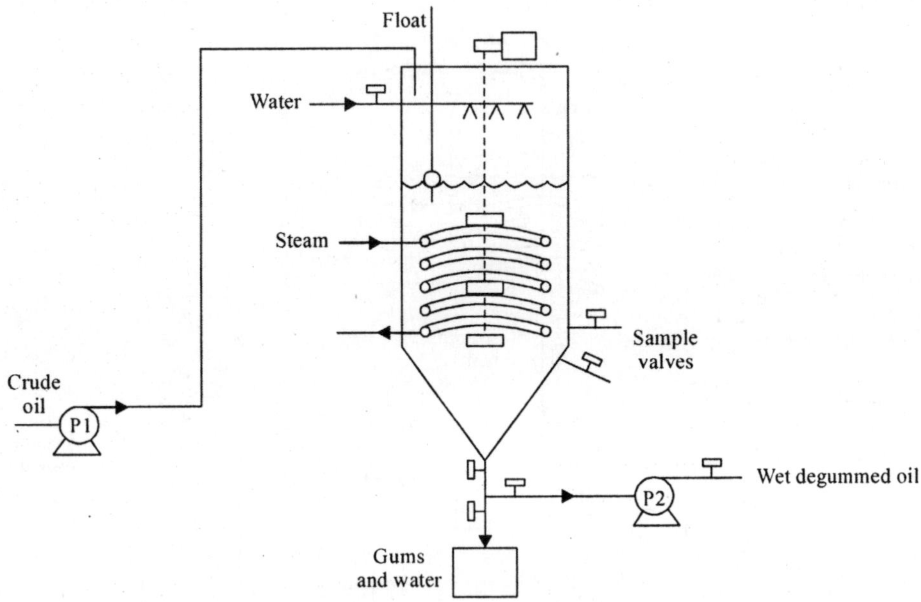

Fig. 12.6. Batch degumming.

Continuous Degumming

Using a ratio controller, the crude sunflower oil and the desired quantity of water are pumped together either through a mixing device such as a static mixer or directly to a holding tank equipped with heating coils (steam) and an agitator. The mixture of oil and water is heated to 76–82°C (169–180°F) and should be allowed to mix for about 30 minutes. This can be accomplished by having a holding tank capacity about equal to the capacity of the hourly rate of feed through the centrifuge. For example, if the centrifuge or centrifuges have a capacity of 30,000 kg/hr., the holding tank should have a capacity of 30,000 kg.

From the holding tank, the oil is pumped through a degumming-type centrifuge to remove the gums. The oil is then either vacuum dried or pumped to a feed tank for alkali refining. The gums are discharged to a tank that can be used for acidulation to recover as much oil as possible or the material (lecithin and gums) may be used or sold as a product to upgrade meal or cattle feed. The same precautions for storing the emulsion before use as given in the batch operation should be observed.

Continuous degumming requires a mild-steel conical bottom tank. A slight cone angle is acceptable for this retention tank. A ratio controller is required to control the quantity of water being added to the oil, which is being metered into the process. A means of heating the oil and/or the oil–water mixture being pumped to the tank is desirable. If not, the tank must be equipped with a steam coil for heating and maintaining the oil–water mixture at 76–82°C (169–180°F). A temperature controller should be used to control the quantity of steam entering the coils for maintaining the desired temperature. An agitator must be there in the tank to move continually the oil–water mixture and keep the hydrated phosphatide and other foreign material from precipitating in the oil. A centrifuge is required to remove the hydrated material from the oil. The centrifuge should be able to handle the heavy materials (such as lecithin) and to run for a long period of time before requiring cleaning. After centrifuging, the oil is vacuum dried and either transferred to storage or is further refined. All the necessary pumps to transfer the oil from one step to the next must be included (Fig. 12.7).

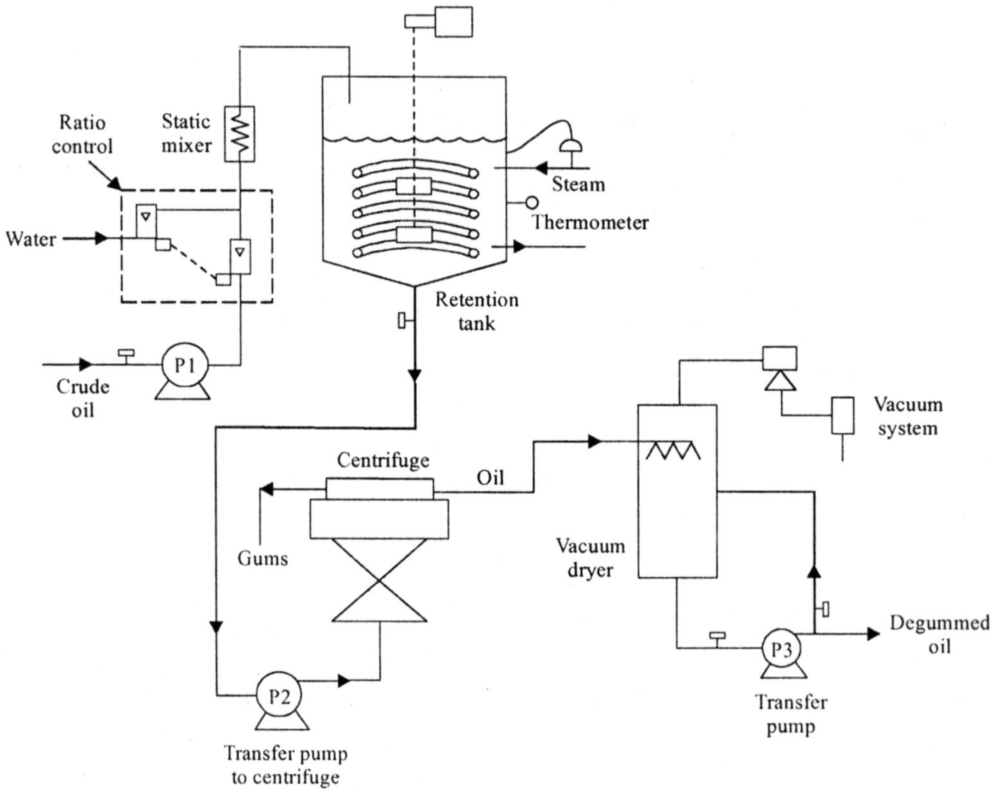

Fig. 12.7. Continuous degumming.

NEUTRALISATION OR REFINING

There are basically two types of refining. One is a chemical, or caustic, refining method; the other is physical, or steam, refining method. The purpose of refining is to remove nonglyceride impurities that consist principally of free fatty acids and significant quantities of mucilaginous substances, phosphatide, chlorophyll and colour bodies. The refining process removes the objectionable impurities in the oil with the least damage to either the glycerides, the tocopherols or other natural antioxidants.

Chemical or Caustic Refining

This method of refining converts the free fatty acids to oil-insoluble soaps, removes all substances that become insoluble on hydration and removes some of the colour bodies present in the oil. Reactions occur at a relatively low temperature, allowing for the refining equipment to be fabricated essentially from iron or preferably, from mild steel.

The technology of caustic refining critically depends on the proper choice of (i) the alkali to be used; (ii) the quantity of alkali strength; and (iii) the refining technique, to achieve the desired purification without causing excessive saponification of the neutral oil. The method used for efficient separation of the refined oil from the soapstock is also an important criterion.

Disadvantage of the method

The major disadvantage of this method of refining is the loss of a small portion of neutral oil in addition to the free fatty acid as a result of saponification. This loss can be minimised, however, by utilising a better technology.

Sunflower oil can be alkali refined using either a batch method or a continuous method of refining. The batch method requires less capital investment, but (i) the yield of refined oil is considerably less than via the continuous method; (ii) the throughput rate is less; and (iii) there is a higher labour cost.

The correct amount of caustic soda (NaOH) used is important to obtain good yields and a quality finished product. If the feed to the refinery is in volume (litres) the quantity must be converted to mass by using the specific gravity of the oil being refined. Also, the volume of caustic (litre) should be adjusted based on the specific gravity of the solution being used.

The percent of excess caustic is important because a large excess will contribute to a high loss of refined oil. This is particularly true if the refining is being accomplished at a temperature of 76°C (169°F) or higher. Some excess caustic soda is required to obtain a good soap formation, removal of nonhydratable phosphatides and a minimum carry-over to the wash water centrifuge and into the oil. If the parts per million of soap in the washed and dried oil is high, it may be necessary to increase the quantity of excess caustic soda slightly to lower the soap content. The oil sample for the soap determination should be obtained from the vacuum dryer. A start with an excess of 0.10–0.15 per cent is recommended if the per cent free fatty acid of the crude oil being refined is in the range of 0.7–1.2 per cent. The quantity of the excess caustic soda should then be adjusted, if necessary, after the process has reached an equilibrium. If the per cent free fatty acids in the crude oil is lower than 0.7 per cent, it may be possible to use an excess of 0.05–0.10 per cent. If the free fatty acids content is > 1.2 per cent, the excess must be increased to between 0.15 and 0.20 per cent.

An excess of 0.25 per cent represents an extremely high quantity; it may, however, be necessary for some kinds of refining. It is suggested that a 16° Bé caustic solution be used for refining sunflower oil. The concentration is not so high as to result in a significant amount of saponification of the neutral oil if the excess quantity, were greater than the optimum.

It is better to use a caustic solution of 14.5–15.5° Bé if the crude oil contains an excessive amount of gums and lecithin (which are generally present when refining tank bottom material). The additional quantity of water in the solution will assist in the hydration of the gums and lecithin.

Use of Phosphoric Acid

Phosphoric acid removes the calcium and magnesium content of the oils as insoluble phosphates. The treatment quantity varies from 0.02 to 1.0 per cent, usually about 0.2 per cent of 85 per cent phosphoric acid. Temperature during the treatment varies from 70 to 90°C (158 to 194°F).

Phosphoric acid may also be used to remove some phosphatides that were not removed by water degumming. These can be removed before alkali refining by treating the degummed oil with phosphoric acid. This phosphoric acid treatment of the oil is necessary if the oil is going to be either physically refined or steam refined. Lecithin cannot be used for human consumption if it has been treated with phosphoric acid.

Experience with sunflower oil indicates that extraneous flavours develop when 0.01 per cent phosphoric acid is added on the cooling side of deodourisation; use of excessive amounts of phosphoric acid can lead to the development of melon, fruity or cucumber off-flavours for the aged oil.

Additional caustic is required to neutralise the excess phosphoric acid if phosphoric acid is added before caustic refining; this increases the cost of producing alkali refined oil.

Phosphoric Acid Use in Alkali Refining

Phosphoric acid treatment is usually employed to assist in removing the nonhydratable phosphatides from the oil. The quantity of phosphatides that contain calcium and magnesium is relatively low in sunflower oil and, consequently, the phosphoric acid treatment is not ordinarily required when using the long-mix refining system. 'In phosphoric acid treatment, the acid is added in the form of an aqueous solution'. It is recommended that the oil–acid solution be agitated for approximately 30 minutes (i.e. a long-mix system) before adding the caustic solution for neutralising the acid and free fatty acids of the oil.

The phosphoric acid treatment is quite often used if a multimix refining process is available. Due to the short contact period of this treatment, it is important that the correct quantity of caustic solution be used to obtain a good refined oil.

The long-mix refining system is usually recommended for refining sunflower oil. The longer retention time (3–6 minutes) in the long-mix system removes more of the colour bodies and permits the use of a milder caustic solution and a lower refining temperature. This system produces a good yield and a high quality oil.

Batch Refining

Batch refining is conducted in a mild- steel, steep conical bottom tank (about a 60° cone angle) equipped with an agitator (slow speed) and paddle-type blades that are pitched to lift the product from the bottom of the tank to the top of the liquid level. It should also have internal steam coils and a method of spraying the caustic solution and water on the surface of the oil.

A quantity of crude oil is pumped into the refining tank without the agitator running. The quantity pumped into the refining tank should be about 70 per cent of the capacity of the vessel being used for refining. The reason for this is that room must be allowed for expansion during the heat-up period as well as for the liquids added during the refining procedure.

After the oil has been pumped into the refining vessel, the agitator is turned on and the steam is allowed to flow into the coils. When the temperature of the oil is at 32°C (90°F) the desired quantity of caustic solution is sprayed on the surface of the oil. Heating is continued during this neutralisation step, at a rate such that the oil reaches 71°C (160°F) in 1 hr. When that temperature is reached, the steam and the agitator are turned off. The contents of the refining tank are allowed to come to a rest and the soapstock and nonoily constituents are allowed to settle to the bottom of the tank.A settling time of 30 minutes should be allowed for a good refining.

A sample drawn at a valve [located either at a point at 15 cm (6 inches) above or below the straight side of the tank] will indicate if the soaps have settled or are still settling. When the oil has been correctly refined, oil drawn from the lower sample valve will be relatively free of soapstocks. Also, inspect the surface of the oil in the refining tank, as occasionally the soap will float, which is caused by the air being entrained in the oil and combining with the soap. If the soapstock floats, the oil usually requires an additional water washing (Fig. 12.8).

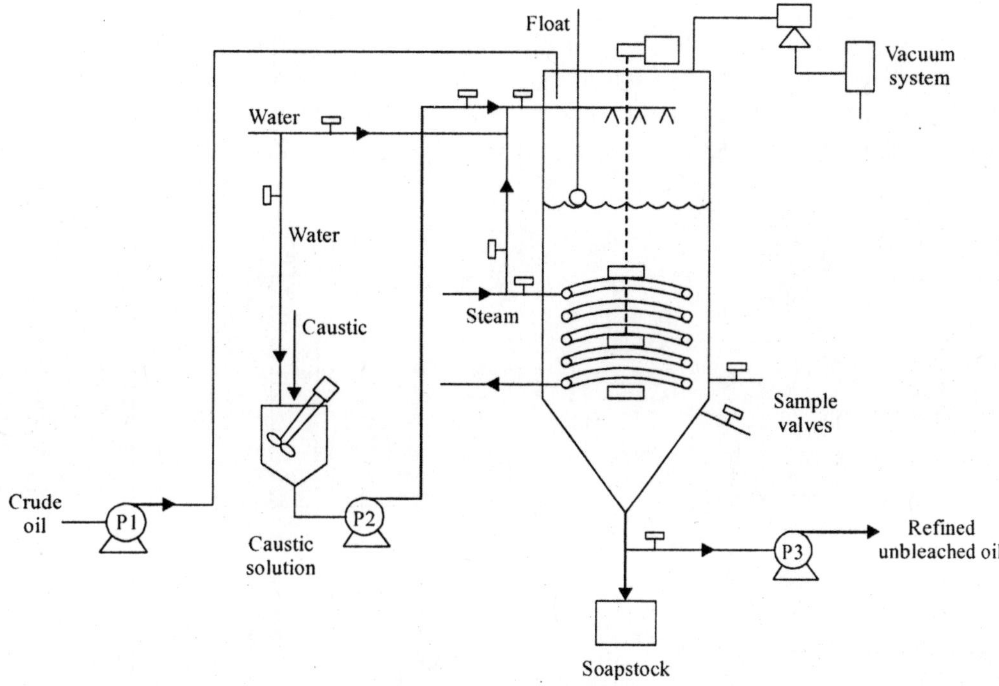

Fig. 12.8. Batch refining.

Start washing the oil if it appears relatively free of soapstock at the bottom side valve. The wash is done by spraying on a 5 per cent hot water (82–88°C; 180–190°F) wash on top of the oil. The percentage of water is based on the weight of the oil in the refining tank. This first hot water wash is then allowed to settle for 30 minutes; then another 5 per cent hot water wash is applied. After the second wash, the batch is allowed to settle for about 2 hr.

The soapstock and water are then drawn off slowly and are pumped to an acidulation tank. After the soapstock has been drawn off, a 10 per cent hot water wash is applied. This is allowed to settle for 1 hr. and then is drawn off. If a lot of heavy soapstock is present, it is pumped to the acidulation tank. If it is just soapy water, it is handled and disposed of in the proper manner.

After drawing off this first 10 per cent of water wash, at least three more 10 per cent hot water washes are given to the neutral oil.

It is desirable in the last 10 per cent hot water wash to try to remove most of the soap by running the agitator. If a high percentage of soap is still present in the oil, additional washes must be given to get the quantity of residual soap down to a desirable level. After the last water wash, the water is drawn off and the neutral oil is vacuum dried, either in the refining tank or in a vessel that can also be used as a bleaching tank for decolouring the oil.

Continuous Refining

A continuous method of refining requires a considerable investment in equipment; however, it is the most efficient and economical method. There are a number of excellent centrifuges available for this type of refining (Fig. 12.9).

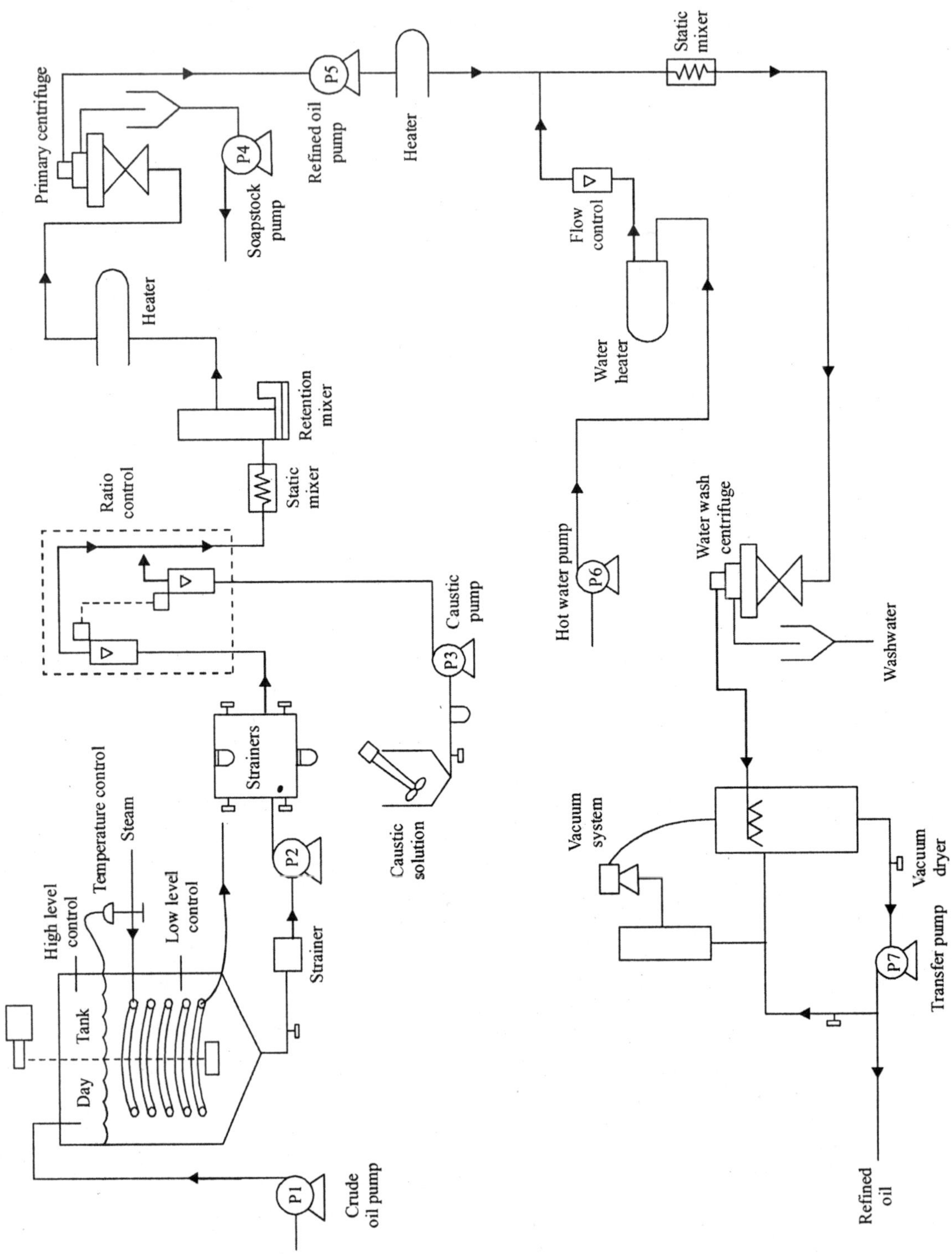

Fig. 12.9. Continuous refining.

The system that is generally used consists of a day tank, which should hold enough crude oil for a minimum of 2 hr. based on the maximum hourly rate of the centrifuges. For example, if there are two centrifuges rated at 15,000 kg/hr, the day tank should have a capacity of at least 60,000 kg. The day tank should be equipped with an agitator and have a control system for maintaining a desired level in the tank. The day tank should have a heating system so as to maintain a uniform temperature. The oil is pumped from the day tank through a strainer and them through a ratio controller, where the desired quantity of caustic solution is added to the oil. From there the oil flows through a mixing device (usually a static mixer) to a system for retention to get a thorough mix).

From the retention system, the oil flows through a means of heating the mixture to the desired refining temperature. The heater used for this operation is generally a fin tube-type heater. From the heater, the mixture is fed to the primary centrifuges to remove the heavy phase, which consists of the soaps, excess caustic solution and nonoily constituents. The light phase, which is the neutral oil, is also discharged from the primary centrifuge and is ready for the water washing. The oil is pumped from the primary centrifuge through a continuous fin tube heater, where it is heated to about 82.2°C (180°F).

After it has passed through the heater, a controlled quantity of hot water is introduced to the oil. This is usually about a 15 per cent hot water (82–87.7°C) wash. After the water has been mixed with the oil, the oil–water mixture is passed through a static mixer to obtain a good mix. From there, the oil–water mixture goes to a centrifuge for the removal of the water phase from the oil. The water-washed neutral oil is then vacuum dried using a continuous vacuum dryer. The dried oil is pumped to storage or is further processed. In some instances the oil may be given a second water wash. It is done after the first water wash and before the oil is vacuum dried. The process would be the same as used for the first water wash.

The quantity of caustic soda solution to be used is calculated as previously shown for the batch method. The excess and degrees Baumé for continuous refining are similar to those for batch refining. A method available for determining the excess alkali should be used occasionally during the refining to be sure that the process is in control. It is desirable to determine the soap content after the water washing and vacuum drying. This helps in making slight changes to the refining conditions so as to minimise the soap level before further processing the oil.

If the oil is going to be deodourised or further processed without decolourisation with an acid-activated clay, the slight addition of a weak acid solution (such as citric acid) before vacuum drying may be desirable. However, if the oil is bleached using a bleaching clay, the acidity in the clay will usually neutralise the slight alkalinity of the oil. Using a continuous method of refining, one can vary the refining temperature considerably to get a good separation of the heavy phase using the centrifuge. A temperature as low as 38°C (100°F) has been used to produce a good quality refined oil. It has been found that a neutralisation temperature of 57°C (135°F) gives both good quality and good yield.

Other Neutralisation Processes

There are different types of caustic refining processes offered by the equipment manufacturers: the long-mix method and the various multimix method. The multimix methods, with some modifications, are referred to as intermix, short-mix and ultra-short-mix. They were developed to process poor and unstable feedstocks. One of the major differences among these methods is retention time. The disadvantage of the multimix method is that the contact and retention times with the caustic solution is so short that the colour bodies of the oil are not reduced or changed by the caustic solution. The removal of impurities such as gums, protein, oxidising materials and colour bodies is as important as the removal of fatty acids.

In the multimix methods, the oil is heated to the processing temperature of 85–95°C (185–203°F). The oil is then mixed with 0.05–0.15 per cent phosphoric acid in a high shear knife mixer for 30 seconds and then mixed with the caustic solution in a neutralisation mixer.

Intermix method

The intermix method is reported to be suitable for most feedstocks with a high nonhydratable phosphatide content or oils requiring rerefining. In this case 30–60 per cent excess caustic above that which is consumed by the phosphoric acid is used at a strength of 20–28° Bé. The mixing is done in a high shear knife mixer for about 30 seconds.

Short-mix method

When processing oils that have a tendency to emulsify, or when the neutralised oil is to be rerefined, the short-mix method may be preferable. In this system, 10–40 per cent of caustic is used at a strength of 12–20° Bé for about 3 seconds using a disk mixer. This type of mixer is designed to provide intensive contact with the oil and caustic without setting off emulsions.

Ultra-short-mix methods

The ultra-short-mix method may be more suitable for processing an oil with a high FFA, which reacts easily. When this method is used, the oil and the caustic solution are mixed directly before the inlet to the separator in an in-line mixer, using 10–30 per cent excess caustic having a 12–28° Bé strength. The retention time for this system is 0.3 second.

Refining Temperature

The refining temperature is critical to obtaining a good yield. A compromise must be made between removing more of the colour bodies and obtaining a low yield by using a higher refining temperature versus obtaining an improved yield using a lower temperature and achieving the desired colour by decolourising with an acid-activated clay.

Yields from continuous refining of relatively good grades of crude sunflower oil (no tank bottoms) at various temperatures indicate that refining at 43°C (110°F) would produce a refining loss approximately 1.42 times the percent of the free fatty acid. At 57°C (135°F) the loss factor is 1.62, while at 71°C (160°F), the loss factor is 1.84. Thus, assuming an FFA content of 1.1 per cent, refining at 43°C versus 71°C saves 0.46 per cent of oil.

Sunflower oil has been refined successfully at 57°C (135°F), using a continuous refining method. A high quality refined sunflower oil also has been produced at a temperature as low as 43°C (110°F). The chlorophyll content of the oil refined at 43°C (110°F) was somewhat higher than oil refined at 57°C (135°F). Chlorophyll was removed by decolourising with an acid-activated clay.

Caustic refining of other oils is often done at temperatures higher than those recommended for refining sunflower oil. A temperature as high as 82°C (180°F) has been used to refine some of the other oils. Refining at such a high temperature removes more of the colour bodies if done in conjuction with a high excess-caustic level. But this results in a significantly greater oil loss due to saponification of neutral oil. Though it is possible to refine sunflower oil at the high temperature used for other oils, that is not recommended. If a temperature higher than 57°C (135°F) is used in refining sunflower oil, the level of caustic employed must be cautiously controlled.

Physical and Steam Refining

This method of refining is used primarily on lower molecular weight oils such as palm, coconut, palm kernel, animal fats and to some extent on soyabean oil. This method has been used in Europe to lower the high free fatty acid content of an oil or fat to a level of 0.5–0.8 per cent. The oil is then caustic refined, bleached and then steam deodourised. Short chain fatty acids, such as those found in palm or coconut oil, are removed very easily by physical refining. It is possible to lower high free fatty acid content of those oils to as low as 0.02–0.03 per cent.

The oil loss could be about 1 per cent greater than the percentage of free fatty acids in the starting oil. For example, if one is refining an oil with 1.25 per cent, free fatty acid, the total oil loss would be about 2.25 per cent. There is, however, a slight tendency for the oil to hydrolyse during the process, which produces more free fatty acids and increases the loss.

The lower the concentration of free fatty acids in the oil, the more difficult they are to remove by distillation. This is not a universally acceped method of refining, as some acids darken at high temperatures and in some oils the colour 'sets' at high temperatures rather than bleach. Glycerides are not appreciably damaged by long heating in the absence of oxygen. It may be possible, however, to strip out some tocopherols and related substances if the oil is subjected to distillation for a long period of time. The distillation process operates like a single-shell deodouriser.

The objectives in physical or steam refining are to reduce the free fatty acids content, fully deodourise the oil, operate without substantially greater utilities consumption than standard or regular deodourisation and recover the free fatty acid from the distillate. The principal disadvantages of the steam refining method are that the equipment must be made of 316 stainless steel to withstand the acidic conditions and corrosion occurs at the high temperatures. The process is limited by the kind of oils which can be refined. Furthermore, oils must be thoroughly degummed and bleached before physical refining can be started. In addition, the per cent of bleaching clay required for decolourising is greater than when the oil is caustic or chemically refined. If sunflower oil is physically refined, the oil will have to be dewaxed after steam refining and then perhaps deodourised again to produce an acceptable product. This process may not lend itself to oils that contain a relatively high percentage of wax.

Pretreatment

Pretreatment is a phosphoric acid degumming method (or treatment of the water degummed oil) before bleaching and steam distillation or refining. The degummed oil is pumped to a mixing tank where it is agitated and maintained at 95°C (203°F). Phosphoric acid in an amount of 0.1–0.5 per cent (by weight) is added to the oil. In a continuous method, the phosphoric acid is added to the oil, using either a proportionating metering pump or a ratio controller, prior to being pumped into the mixing tank.

The oil–acid mixture is agitated at 95°C (203°F) for about 15 minutes. The mixture is pumped to a slurry tank where a quantity of bleaching clay and, if necessary, filter aid is added. The quantity of bleaching clay required depends on the desired colour of the oil before steam refining and deodourisation. After bleaching clay has been added, the oil, acid and clay mixture is transferred to a vacuum bleaching tank where it is held for a period of time to obtain the desired colour. It is then pumped through filters to remove the bleaching clay and into a receiving tank.

Steam Refining

From the receiving tank, the oil is pumped through spray nozzles in a deaerator tank, which is under vacuum, to remove the dissolved air. From the deaerator, the oil is pumped into the refining deodouriser where it is refined. From the refining deodouriser, the oil is cooled to the desired discharge temperature and is pumped through a polishing filter to the desired tank for further processing or storage. If antioxidants are added, it should be done as the oil is being pumped from the polishing filter. Also, the oil should be handled or stored under a nitrogen blanket after the deodourisation.

BLEACHING

Purpose of Bleaching

Bleaching by absorption is used to improve the colour of the oil, remove impurities such as soap and catalyst poisons, and to remove metallic compounds and prooxidants. Bleaching improves the colour of the oil by removing the pigments that are either dissolved in the oil or present in the form of colloidally dispersed particles. If the chlorophyll is not removed, the finished oil may have a greenish cast. Acidic earths are used to obtain good chlorophyll removal.

Bleaching also removes impurities such as traces of soap and catalyst poisons that may be left in the oil after refining. Soap and catalyst poisons interfere with hydrogenation of the oil. The refined oil (neutral oil) can be bleached either continuously or via a batch method. Regardless of the method used, it should be done under vacuum.

Batch Bleaching

In batch bleaching, a quantity of neutral oil is pumped to a tank equipped with an agitator, steam coils and a vacuum system. The oil is heated to 71–77°C (160–170°F) with the agitator running. Vacuum is applied and the temperature is maintained at 71–77°C until 50–75 mm (27–28 inches) of absolute vacuum are obtained. The desired quantity of bleaching clay (0.03–1.5 per cent) is then pulled into the oil through a pipe that discharges below the oil level. This prevents the clay from discharging into the head space of the tank and being lost into the vacuum system. After all of the bleaching clay has been added to the oil, the vacuum and the temperature should be maintained for about 1 hr. (Fig. 12.10).

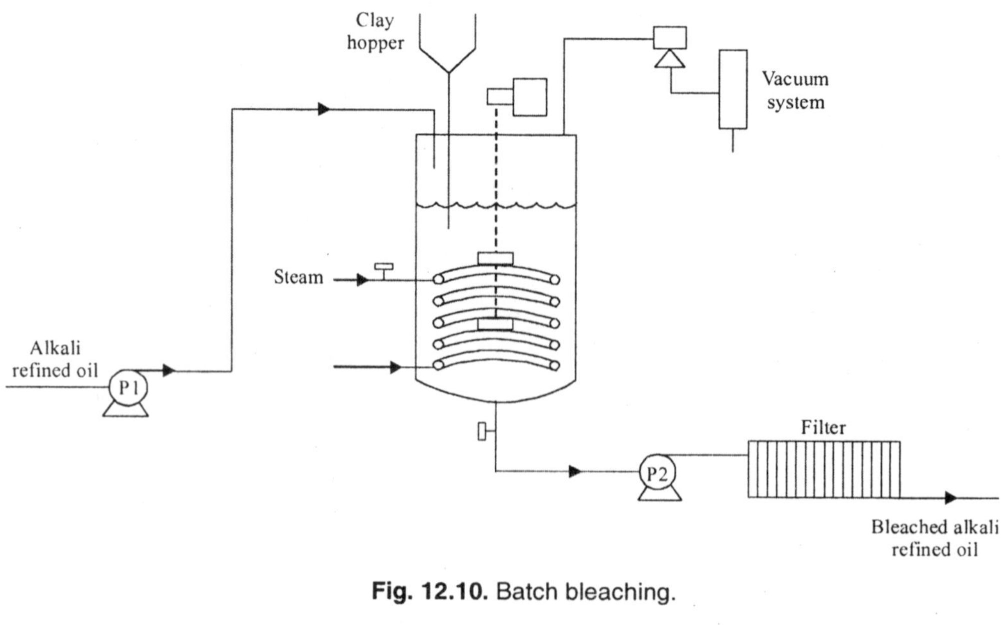

Fig. 12.10. Batch bleaching.

After the oil has been subjected to this bleaching process for 1 hr. and the desired colour is obtained, the vacuum is released and the oil–clay mixture is pumped to a filter where the clay is removed and the decolourised or bleached oil is either pumped to storage or further processed. If the oil is pumped to a storage tank, it should first be cooled, preferably to 38°C (100°F).

Note that (i) the oil must be dry before adding the bleaching clay, because the FFA increases if the oil contains some moisture and (ii) if the oil is pumped to storage hot, it may pick up moisture, which was caused by condensation, in the storage tank. Also, the oil is subject to some oxidation when exposed to the air at 71°C (160°F) for a prolonged time.

Dressing the Filter Press

The recommended dressing of the press, when a plate-and-frame filter press is used for removing the bleaching clay, is as follows. A canvas material (100 per cent cotton duck-type canvas, weighing 0.339 kg/m^2 or 10 oz/yd^2) should be placed over the plate. Filter paper, if available, should then be placed over the canvas.

If the feed or discharge system of the press requires holes in the cloth, cut the holes in the canvas before the canvas is placed on the plate. Any holes needed in the paper should be made after the paper has been placed over the canvas.

Continuous Bleaching

There are various continuous bleaching processes available. Continuous bleaching methods have been used, such as using a single large vacuum tank, a single large atmospheric tank and multiple tanks where the oil is 'cascaded' or pumped from one to another. Regardless of the system used, the tank capacity must be large enough for the residence time in the bleaching operation to be at least 1 hr.

As it is desirable to decolourise the oil under vacuum, a simple continuous vacuum process is reviewed here. The oil is pumped from the vacuum dryer to a slurry tank in the bleaching system. The desired quantity of bleaching clay is added to the oil in the slurry tank. The quantity of bleaching clay can be added in various ways. The simplest method could be a screw-type conveyor set to discharge a given weight of clay per hour based on the rate of oil coming from the vacuum drier.

The slurry tank is equipped with an agitator and a level control. The oil–clay mixture is then pumped to a vacuum tank that is large enough to hold the oil–clay mixture for about 1 hr. The tank must be equipped with an agitator and steam coils to maintain a constant temperature and keep the clay in contact with the oil. A level control should also be on this tank to maintain the desired levels. A pump from this tank transfers the oil–clay mixture at a uniform rate to a filter feed tank and then through a filter press to remove the bleaching clay.

If the oil is going to storage, it should be cooled by means of a heat exchanger before being discharged to the storage tank. This heat exchanger could be used to heat the incoming oil to the slurry tank, provided the incoming oil is cold enough to assist in reducing the temperature of the bleached oil that is being pumped from the system (Fig. 12.11).

DEWAXING

The dewaxing process removes the waxy type of nonoily constituents that are not removed either by alkali refining or by degumming. Various methods can be used for dewaxing the oils, such as a batch method or various types of continuous methods.

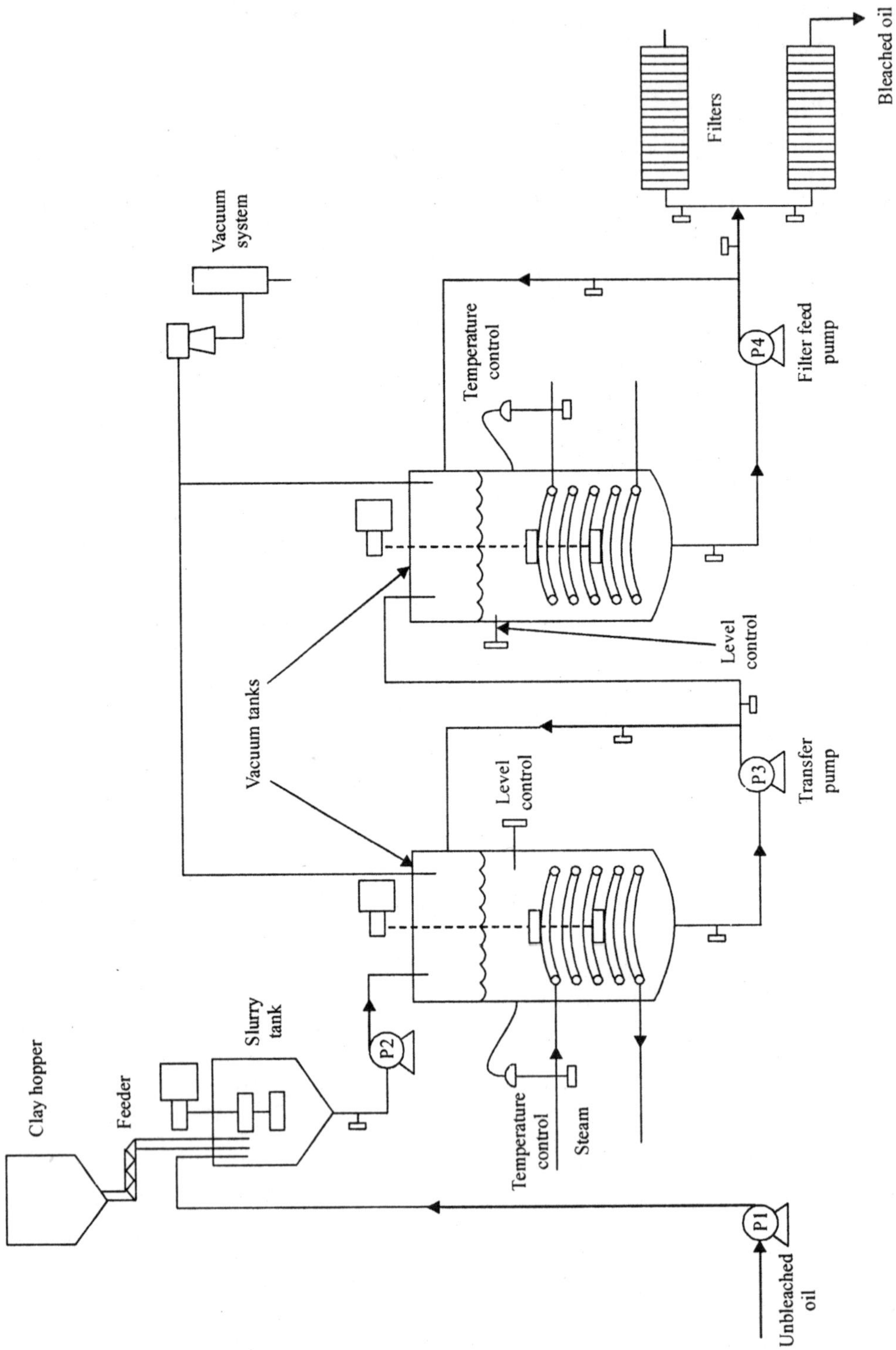

Fig. 12.11. Continuous bleaching.

Batch Dewaxing Method

The batch method is relatively simple. It consists of a well-insulated tank of a given capacity, equipped with some means of agitation. A relatively slow speed mechanical agitator is preferred; however, it is possible to use a pump to circulate the oil from the bottom to the top oil layer. Under no circumstances should air be used for agitation of oil. Air bubbles in the oil inhibit the heat transfer, causing an extended processing time and a higher peroxide value of the oil. The tank used for cooling the oil should also be equipped with a large number of relatively small diameter coils to give a large cooling surface. It is preferable to have a vertical-type tank rather than horizontal, though either is acceptable. However, the agitation should be such that the oil moves rather slowly across the cooling coils, yet the rate of movement is enough to keep the waxes from settling out of the oil.

Various cooling mediums can be used such as an ammonia expansion system or a cold glycol–water system. Whichever system is used for cooling, the oil should be allowed to remain in the tank subject to the cooling process until the temperature of the oil reaches the desired level to produce a satisfactory oil after filtration. The oil level in the cooling tank must be high enough to cover the cooling coils (or jacket, if it is a jacketed vessel) to prevent condensation of moisture in the tank during and after refrigeration.

The oil is cooled from 33 to 0°C (91 to 32°F), which takes about 4 hr. It may be necessary to refrigerate the oil to 0°C (before pumping to a filter feed tank for filtering), if the finished, deodourised oil is going to be stored in a refrigerator in the market or by the customer. If it is not necessary to refrigerate at 0°C, refrigerate to such a temperature that the filtered oil has the desired chill test properties. Following cooling, the refrigerated oil is pumped to a filter feed tank equipped with a slow speed agitator where about 0.25–0.50 per cent of filter aid, based on the weight of oil (having a wax content of 100–400 ppm), is added to the oil. When the filter aid is thoroughly mixed with the oil, the mixture is pumped through a filter.

Various filters may be used, but a simple plate-and-frame filter press, dressed with flannel, muslin and a shark-skin-type paper, is recommended. If the press is so dressed, the loss of refined oil during dewaxing is approximately equal to the quantity of filter aid used in the filtration (Fig. 12.12).

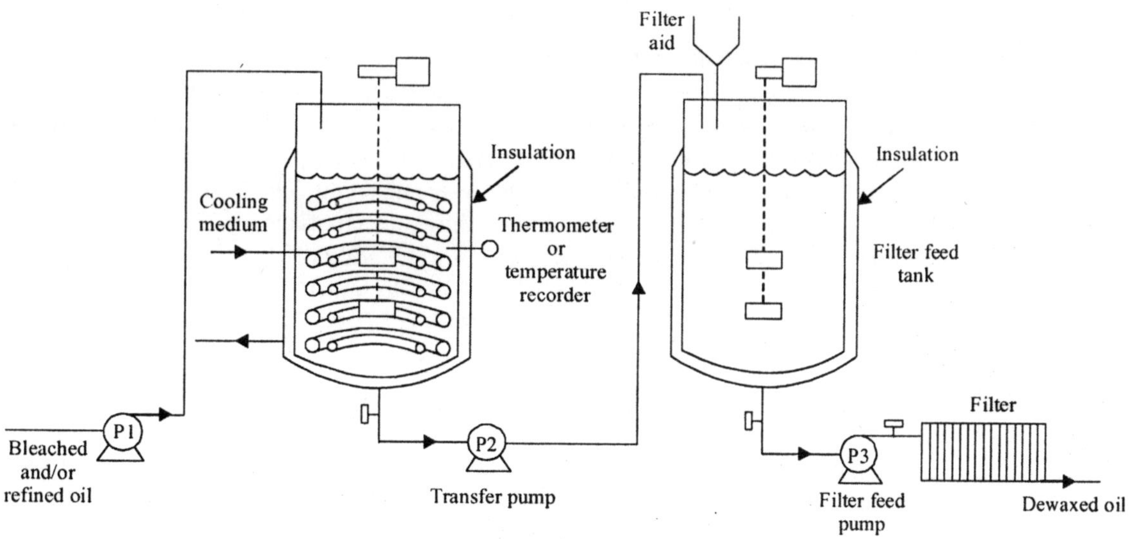

Fig. 12.12. Batch dewaxing.

Continuous Dewaxing Method

A continuous system may use multiple tanks or a continuous chiller. The tanks should be insulated. The oil cools progressively as it is transferred from one tank to the next. The residence time in the system is long enough to permit reducing the temperature of the oil to the desired temperature before it is pumped from the final tank.

The chilled oil is then pumped to a filter feed tank, where the filter aid is added and then pumped through a filter to remove the filter aid and waxes. Detailed information about agitation and filtration of the refrigerated oil is already given in batch dewaxing method.

Alternatively, a continuous process involving a continuous chiller may be used. With this process, the oil being chilled passes through a continuous pipe that has a cooling medium on the outside of a jacketed pipe. The flow rate is such that the oil is at the desired temperature when discharged from the end of the continuous pipe. When this system is used, it is desirable to discharge the refrigerated oil to an insulated tank equipped for agitating the oil. The residence time in the tank permits agglomeration of the waxy material. Agglomeration of the material permits easier filtering.

Filter aid may be added to this residence tank. After a thorough mixing, the oil is pumped through a filter press to remove the waxy material and the filter aid. The residence time should be about 1 hr before filtration.

Continuous Dewaxing Using a Centrifuge

Some caustic refining systems may incorporate a dewaxing process. The basic equipment consists of heat exchangers, crystallisation tanks, a cooling or refrigeration system, pumps and a centrifuge.

The continuous dewaxing process might be more advantageous than the lower temperature dewaxing of the refined oil if the cold test requirement is not stringent [i.e. 0°C (32°F) for 48 hr. and 4.5°C (40°F) for 120 hr. minimum]. Low temperature (0°C, 32°F) crystallisation and filtering of refined oil produce a better cold test oil and meet the stringent requirements as stated above (Fig. 12.13).

Use of a Surfactant

Some suggest incorporating a surfactant to improve the dewaxing process. The method generally used for this type of dewaxing diverts the oil from the normal alkali refining process right after the oil has passed through the primary centrifuge (or soapstock separator). The oil is then pumped through a heat exchanger to a crystallisation tank or tanks. The soap content of the refined, unwashed oil may be in the range of 1000–2000 ppm, which could be sufficient to act as a surfactant. The moisture content of this oil–soap mixture should be 3–4 per cent to facilitate the binding and satisfactorily discharge the waxes. If the moisture content is not this high, additional water should be added.

The unwashed oil should be held in the crystallisation tanks until the oil reaches the desired temperature for dewaxing. The temperature should be 5–7°C (41–45°F). It is also recommended that the residence time in the crystalliser be about 4 hr. to permit a good crystal growth. Note: a quick chill will result in small wax crystals, while a longer residence time will produce larger wax crystals.

After the crystallisation step, the oil, wax and water mixture is heated by means of a heat exchanger to 12–15°C (54–59°F). Then the mixture is centrifuged to remove the wax crystals along with some water and soaps. After the oil has been centrifuged, it is heated and washed using hot water to reduce the soap content; it is then vacuum dried. The soapy water, wax and oil stream recovered from the centrifuge is then handled as a by-product.

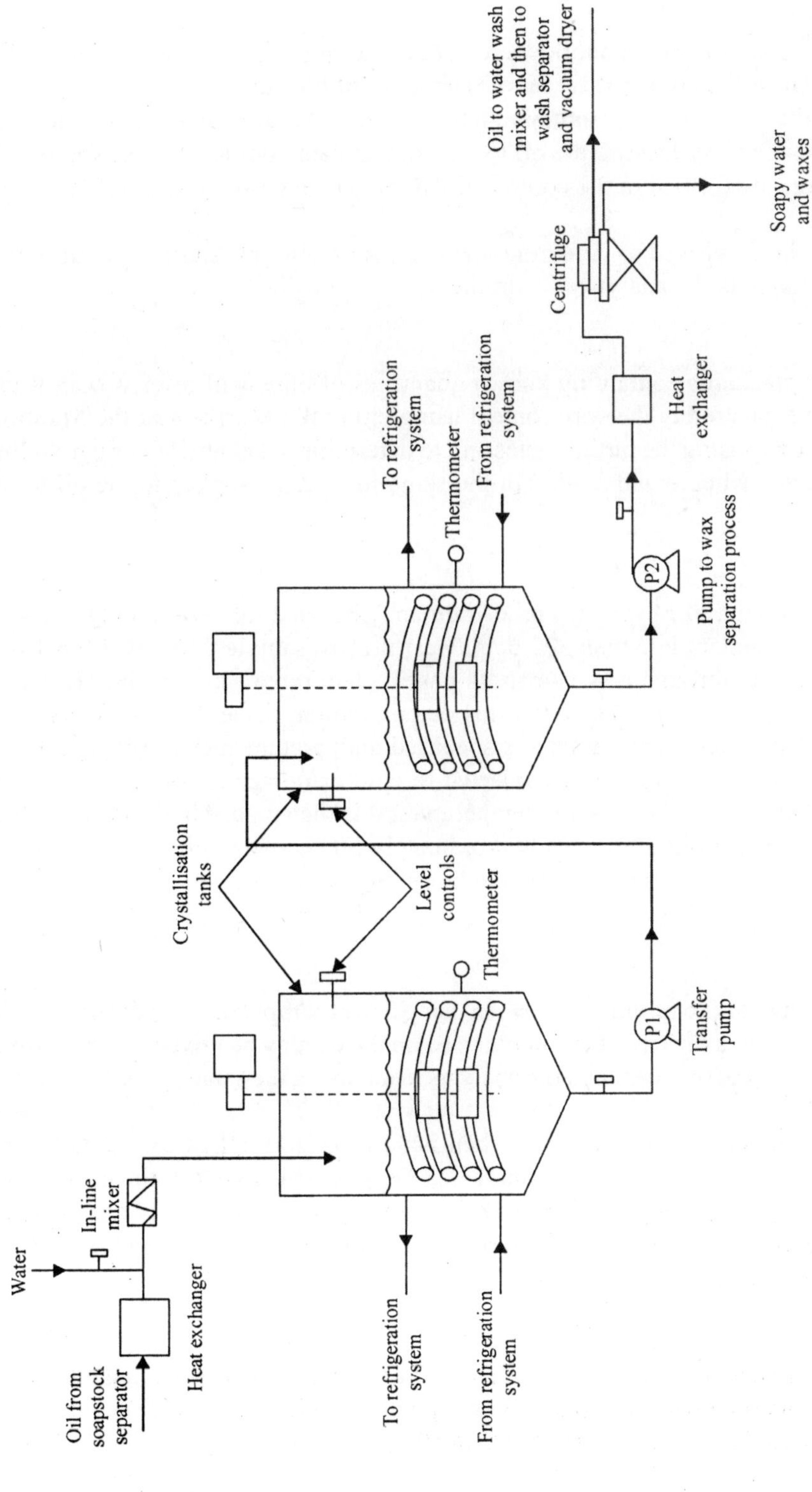

Fig. 12.13. Continuous dewaxing using a centrifuge.

Testing for Wax

There are two procedures for testing of wax in sunflower oil. The traditional qualitative method is the 'cold test', AOCS Cc 11–53. The alternative test is via measurement of turbidity.

The measurement of turbidity method is quantitative and objective. A rapid estimate of the wax content in sunflower oil can be made by heating the oil to 130°C and then cooling it in ice for 10–15 minutes. Microcrystalline waxes are formed in the cooled oil that may be measured by a turbidimetric technique.

These two procedures may be employed to determine the wax content by measuring the turbidity. Both procedures use a turbidimeter as the analytical instrument.

Preparation of Standards

It will be necessary to prepare standards containing known quantities of pure sunflower wax in wax-free sunflower oils to determine accurately the wax content using either the Morrison or the Moulton method. Such standards must be run using the turbidimeter unit to determine the nephelometric turbidity units (NTUs) and must be heated, filtered and cooled in the same manner as applies to the oil to be tested.

Refined Oil Wax Testing

The following method, a modification of Morrison's procedure, may be used for determining the wax content of refined oil with a wax content less than 300 ppm. Heat the test sample to 130°C (266°F) to remove any moisture and then filter through a Whatman #4 paper. Then run a sample while it is still hot, using the turbidimeter to determine the NTU. After this determination, place the sample cell in a constant temperature bath for 1 hr. Such a bath is set at a specified temperature such as 0°C (32°F) or 4.4°C (40°F) or higher, if the oil is not going to be subjected to cold or refrigeration. After 1 hr. the sample is allowed to settle for several minutes at room temperature; it is then placed in the turbidimeter to determine the NTU. The reading should be taken approximately 30 seconds after the sample has been placed in the unit. A flow chart for this method would be useful to understand how it determines the wax content of an unknown sample.

Crude Oil Wax Testing

The readings may exceed the range of the turbidimeter when using a neat sample of oil to determine the wax content of crude oil. When such is the case, the sample to be analysed may be mixed with wax-free sunflower oil to a level such as 25 or 50 per cent and the analysis made using the same procedure as has been previously outlined.

The results obtained from the chart or curve would then have to be multiplied by the factor of dilution. For example, if the sample being analysed contains 25 per cent unknown oil, the quantity of wax as determined from the previously prepared chart would have to be multiplied by four to arrive at the correct wax content. Using this method does not require the use of either acetone or an oil–acetone mixture (WS solvent) for the analysis.

Cold Test

Table 12.6 shows the relationship between the wax content and the temperature at which the oil passes a 24-hr. cold test. For example, an oil having 14 ppm wax is expected to pass a 0°C (32°F) cold test. A refined oil with 50 ppm wax should pass a cold test of 6°C (42°F).

Table 12.6. Suggested maximum ppm wax content for 24-hr. cold test at various temperatures.

Temperature, °C (°F)	*ppm*
0 (32)	14
3 (37)	40
6 (42)	50
9 (48)	59
12 (53)	69
15 (59)	78
18 (64)	86
21 (69)	96
24 (75)	104
27 (80)	119
30 (86)	137
33 (91)	165
36 (96)	215

Filter Aid

Filter aid is used either as a precoat for the filter screens or a filter cloth or to assist in the removal of any foreign material from the product being filtered. The quantity used varies, depending on whether it is used to precoat or to remove foreign material.

The quantity of filter aid used to precoat the filter screen is generally stated by the manufacturer of the filter. If filter aid is used to assist in the removal of foreign material in the product, the percentage used varies from 0.125 to 1.75 per cent (based on the weight of the product being filtered). If the product being filtered contains a substantial quantity of foreign material, or if the foreign material is a slimy or waxy type substance, it may be necessary to use up to 0.5 per cent of filter aid to filter a reasonable quantity of oil. Using less than the correct amount of filter aid causes the press to blind or the filtration rate of slow down, necessitating a cleaning of the filter.

The filtration time during the dewaxing process of the refined oil using a 32-inch plate-and-frame press with 30 plates averages 1.5 hr./10,000 kg (22,000 lb). This calculates to a rate of about 165 kg/m^2 (35 ft^2). The oil loss during the dewaxing of the refined oil is approximately equal to the quantity of filter aid used in the filtration (Table 12.7).

Table 12.7. Percent filter aid (diatomaceous earth) to filter wax from refined sunflower oil.

ppm wax	*Percent filter aid*
50	0.125
100	0.250
200	0.250
300	0.375
400	0.500

(Contd...)

Table 12.7. Percent filter aid (diatomaceous earth) to filter wax from refined sunflower oil. *(Contd.)*

ppm wax	*Percent filter aid*
500	0.625
600	0.750
700	0.830
800	0.920
900	1.000
1000	1.080
1100	1.170
1200	1.250
1300	1.330
1400	1.420
1500	1.500
1600	1.580
1700	1.670
1800	1.750

FILTRATION AND CLARIFICATION SYSTEMS

Filtration and clarification are used to remove the solid type of foreign material from the oil. Various types of equipment are available to accomplish these processes.

Plate-and-Frame Presses

The plates of a plate-and-frame press form the support for the filter fabric and the frames act as spacers and provide space for the cake. This type of press may be obtained in both an open and closed delivery system.

Filter Precoating

A precoating may be used to protect the filter cloth from getting plugged (or fouled) with waxes if flannel, muslin and paper are not available. The precoat consists of a mixture of filter aid and clean, dewaxed oil. The mixture is agitated so that the filter aid is well dispersed. This mixture is then circulated by means of a rather high capacity pump through the press to deposit a thin layer of precoat on the leaves. The thickness of the precoat should be 0.125–0.25 inch. After the precoat has been deposited on the filter cloth, the unfiltered dewaxed oil containing the recommended quantity of filter aid can be started through the filter press. The waxes and filter aid are then removed from the oil and the dewaxed oil is either pumped to storage or further processed. The refined, dewaxed oil obtained by using this procedure has a chill test of more than 36 hr. at 0°C and more than 120 hr. at 4.5°C.

Vertical Tank-type Pressure Filters

The tank-type pressure filter is a more modern type of filter. It usually consists of vertical leaves covered by screen, which are manifolded to a discharge pipe from the filter. This type of press may or may not contain a heel leaf. The heel leaf is a screen located on the bottom of the cylindrical tank that permits the contents of the tank to be discharged as a clarified, finished product instead of an unfiltered one that requires additional processing.

This type of filter generally does not have a cloth cover on the screens, which means that the screens must be precoated to build up a thin filter cake, which removes the foreign material, before a satisfactory product can be produced. There are, however, some tailor-made cloth covers for this type of filter screen, which eliminate the precoating. For dewaxing, however, precoating with dewaxed oil is recommended to obtain better flow rates and less blinding (plugging) of the fabric.

Advantages

1. It is relatively easy to clean because the leaves move out of the tank and the cake can usually be removed by some shaking of each leaf.
2. It is a closed discharge system.

Disadvantages

1. If the flow of oil is not sufficient to fill the tank rapidly, the precoating will not be uniform.
2. If there is an interruption in the flow during filtration, it is possible that some of the cake can slough off the screen and allow unfiltered oil to enter the filtered oil.
3. The cost for this type of filter is usually higher than for a plate-and-frame press.

Horizontal Leaf-type Filter

A horizontal leaf-type filter is usually dressed with paper and used as a polishing filter because the cake capacity is rather limited. This filter consists of a set of horizontal, hollow plates that have a hole in the centre. The plates are stacked one on the other so that the holes in the centre match. The tank is charged with oil in the outside peripheral area. After passing through the filter paper, it is discharged into the centre hole and then passes out of the filter. The disadvantages of this type of filter are:

1. It has a rather limited use because it does not have a large cake capacity.
2. It is more difficult to clean because the plates must be removed vertically for cleaning.
3. It is somewhat more difficult to reassemble without resultant leakage. (It is possible to have one extra complete set of leaves that are cleaned and dressed and inserted as a cartridge. This reduces the downtime required for filter cleaning.

Continuous or Vacuum Filters

Basically, continuous filters are rotating horizontal cylinders partially submerged in an agitated horizontal tank. This type of filter has various segments or compartments that permit the oil and waxes to be drawn onto the fabric. As the cylinder rotates, the vacuum pulls in air through the cake, thus drying the cake. There are usually low and high levels of vacuum. The first section, with vacuum, picks up the oil and filter aid and removes a large portion of the oil. The oil is then pumped to storage. As the cylinder or drum continues to rotate, the high vacuum step is applied to dry the cake and remove much of the residual oil. The oil entrained in the vacuum system is removed by a separator. The dried cake is then removed from the filter cloth with a doctor blade and is usually discharged into a screw conveyor for removal.

Centrifugal Filters

The centrifugal filter is used rarely, if at all, in the vegetable oil industry. It consists of a rotating perforated basket inside a housing, which serves as a filtrate collector. In this system, the inside of the basket is lined with a fine screen or a filter medium. The machine is brought to operating speed, the feed slurry is introduced and feeding is continued until a full filter cake builds up on the filter medium or screen. The feed is then stopped and the cake is spun until dry.

After all of the possible filtrate has been removed from the cake, a plough is advanced into the cake while the basket rotates, discharging the cake through the bottom of the basket or through a chute. These units are either horizontal or vertical. They can be instrumented for complete and automatic cycling and speeds can be varied within the cycle. These units can also be built to include washing cycle and built with vapour tight housings. They are used in various chemical manufacturing facilities and are expensive.

Bag-type Filters

Bag-type units are rather inexpensive and can be manifolded to give a high flow rate. This type of unit consists of a metal vertical cylindrical housing into which a fabric bag is placed inside a perforated or wire basket, which fits inside the housing. Units are available having multiple bags inside one housing. The feed enters the top of the housing and passes through the bag and basket, and the filtrate is discharged from the bottom of the unit.

The advantages of this type of unit are its low initial costs easy cleaning and low maintenance costs. Disadvantages are the rather low solids capacity of the units and relatively high cost of the fabric filter bags. Oil loss is relatively high, because it is difficult to blow these filters effectively (it is necessary to drain the bag manually, leaving a wet residue) and the disposal of the bag may be a problem.

Cartridge-type Filters

The initial cost of a cartridge-type unit is rather low; as with the bag-type filters, they can be manifolded to give a high flow rate. These units consist of a metal vertical steel housing in which various numbers of cartridges may be installed. The cartridges are cylindrical and have a hole through the length. A retainer is placed in the housing so that all the centre holes are in line vertically. The top of the housing seals the top of the holes in the cartridge and the feed is pumped into the vessel on the outside of the cartridge. The flow passes through the cartridge and discharges to the centre hole and then through the bottom of the housing. The cartridges are available in various micron sizes, different fabrics, filter media and lengths. One housing is capable of holding multiple cartridges. The housings also are available in various lengths. Cartridge-type filters are used primarily as polishing filters.

The advantages of this type of filter are the relatively low initial costs, easy cleaning and low maintenance costs. The disadvantages are the high costs of the filter cartridges, low solids capacity and relatively high oil loss because the filter cannot be effectively blown free of oil. The disposal of the cartridge may present a problem.

DEODOURISATION

Deodourisation is the last of the processing steps to produce an edible oil. It is steam distillation under high temperature and high vacuum. When properly deodourised, sunflower oil has a nearly bland taste and aroma and even after heating to 177°C (350°F), it has almost no odour.

Deodourisation reduces free fatty acid content, removes undesirable volatile flavour and odour compounds, removes some sterols and pigments and destroys peroxide. The flavour and odour compounds are aldehydes, ketones, alcohols and hydrocarbons, which even at low concentrations are detectable by taste or smell. The amount of these undesirable components is usually 0.1 per cent of the oil. However, many flavour and odour compounds can be detected by taste and smell at extremely low concentrations that are difficult to detect chemically. The previous refining processes may impart characteristic odours such as an earthy odour from bleaching or the typical hydrogenation odour described as tinny or metallic.

A good quality deodourised oil cannot be produced unless the deodouriser feedstock is of good quality. Cutting corners in the neutralisation and bleaching processes results in less than acceptable stability, odour and flavour of the deordourised oil, which means the oil has to be rebleached and redeodourised.

Operating Parameters for Deodourisation

The operating parameters for efficient removal of undesirable materials are as follows:

1. The vapour pressure of the materials to be removed.
2. The deodouriser operating conditions of temperature, holding time and absolute pressure.
3. The quantity of stripping steam per quantity of oil.
4. The efficiency of the equipment to mix the stripping steam and oil intimately.
5. The quality of the steam.

Deodourisation Operation

Deodourisation is the process of physically removing the flavour, odour compounds and free fatty acids from the oil. It is essentially the steam distillation of volatile compounds from the oil under high temperature and high vacuum conditions.

First, the oil is deaerated at the rather low temperature of 49–54°C (120–129°F) at the same vacuum as in the deodouriser. The deaeration vessel can be external to the deodouriser or included in the design of the deodouriser, such as in the continuous and semicontinuous designs.

A liquid level control is used to regulate the flow of the oil to the deodouriser. The oil is held under the deodourisation conditions until it meets the quality criteria. The oil is then cooled to 38–66°C (100–150°F), at which time a metal inactivator such as citric acid and, if desired, an antioxidant may be added. The oil should then be protected with nitrogen until it is packaged, to reduce the chance of oxidation, which could have a detrimental effect on the oil.

Deodourisation Equipment

There are three basic types of deodourisers: the batch type, the continuous deodouriser and the semicontinuous deodouriser. Continuous units are of two types: single-shell units and double-shell units. The continuous unit offers the greatest operating economy because it involves various heat exchanges. The semicontinuous unit is reasonably efficient, as it is possible to incorporate a heat transfer equipment in this unit. The batch system has rather limited provisions to incorporate heat transfer systems to produce the efficiency found in the continuous or semicontinuous deodouriser.

Batch Deodourisation

The conventional batch deodouriser is a vertical stainless steel cylindrical vessel that has dished heads capable of withstanding a full vacuum. It should be a welded unit and well insulated. The steam is emitted into the bottom of the vessel through a sparge line or a spider, which consists of perforated pipes radiating from the central steam delivery line. The holes in the pipe should be rather small in diameter—not greater than 3.2 mm (0.125 inch).

There should be some means of controlling and determining the volume of steam entering the vessel, using either an orifice plate and/or a valve. The vapour line leading to the entrainment separator or booster should be as short as possible to avoid excessive refluxing. The vessel should be equipped with a steam ejector system, thermometer and pressure gauge to indicate accurately the low pressure in the deodouriser independently of the existing barometric pressure. It should also have the necessary deaerator and pumps to transfer the oil to and from the system.

The operating conditions dictate the quantity of steam to be used in the deodourisation. The total quantity of stripping steam varies between 10 and 50 kg per 100 kg oil, based on a processing time of 3–8 hr. with the average being about 25 kg. If the vacuum is at 12 mm absolute, instead of 6 mm, twice the quantity of steam is needed; at 24 mm, four times steam is needed. The temperature of deodourisation of sunflower oil should be in the range of 250–255°C (482–491°F). Most commercial deodourisers operate within 245–275°C (473–527°F) (Fig. 12.14).

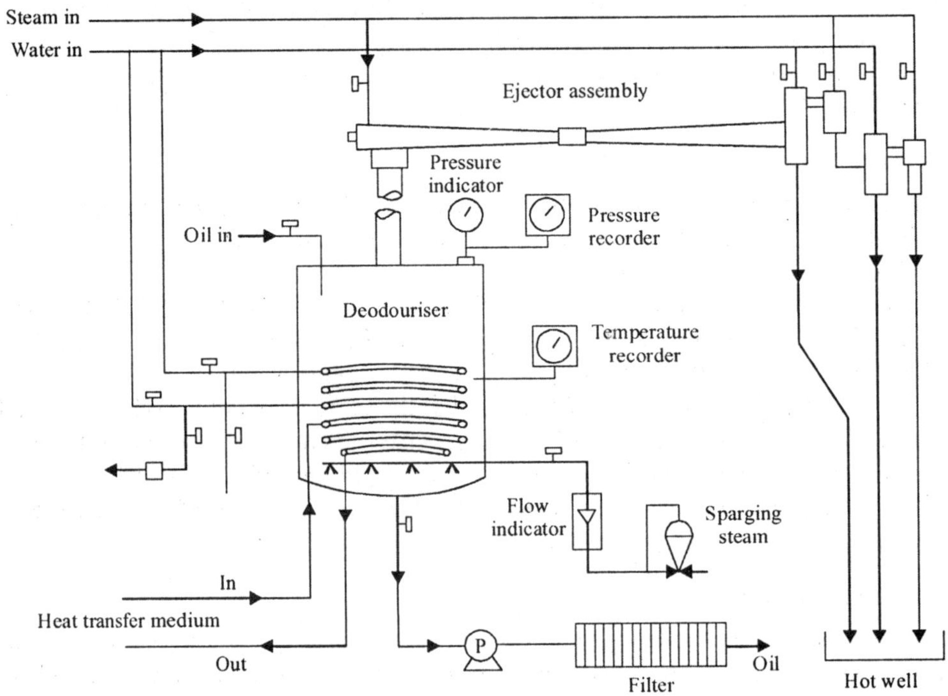

Fig. 12.14. Batch deodouriser.

Continuous Deodourisation

The continuous deodouriser operates so that deodourisation is conducted in a series of five to seven vessels or trays that are mounted within a single type 304 stainless-steel tower. The vapour lines from the tower are manifolded and are made of carbon steel. The oil flows continuously through the deaerating, heating, holding and cooling phases of deodourisation with the retention times being controlled by overflow weirs and pipes. The carbon-steel vapour pipe connections to the stainless-steel tower are located so that the steam used in sparging and stripping, along with the volatile impurities, pass from each section directly into the vapour take-off pipe and are removed through a single connection to the vacuum system. Wire mesh entrainment separators are provided at the top of each section.

Semicontinuous Deodourisation

This system consists principally of a tall cylindrical shell, which is constructed of carbon steel, in which five type 304 stainless steel trays are placed.

A semicontinuous unit operates on the basis of handling finite batches of oil, which move batchwise in a timed sequence through the various steps of deaeration, heating, holding and steam-stripping and cooling in such a manner that each quantity of oil is completely subjected to each condition before proceeding to the next step. The semicontinuous system ensures identical treatment for all portions of the oil. Plants of this type are normally automated and controlled from a central panel; a time cycle controller and interlocks allow the sequence steps to be interrupted when there is an insufficient batch size or an improper drop-valve opening or closing. The sequence can also be interrupted if the oil does not reach the heating or cooling temperature in the allotted time. The design of this type of deodouriser permits frequent changes of feedstock with little intermingling of the successive changes. It also reduces the wide variations in the utility loads encountered with a batch deodouriser. The heat recovery for a semicontinuous system is 40–45 per cent, which is not as good as the continuous system. However, it is much greater than could be achieved with the batch system.

Protection and Handling of Deodourised Oil

The edible oil processor should protect the oil from oxidation by the following measures:

1. Exclusion of air during processing.
2. Cooling the deodourised oil to proper temperature before exposure to the atmosphere.
3. Protection of deodourised oil from the atmosphere by nitrogen blanketing.
4. Addition of chemicals that serve as metal scavengers and antioxidants.

Also, a full vacuum should be maintained in the deodouriser even when not in use to avoid oxidation of the adhering film of residual oil, which could then contaminate the next lot.

Protection from air

Oil must be scrupulously protected from air throughout the deodourising operation. At deodourisation temperatures, the oil reacts rapidly with oxygen, which has a decidedly bad effect on the oil's flavour and oxidative stability. Periodic preventive maintenance is needed to prevent air leaks from occurring in deodouriser fittings below the oil level and in external pumps, heaters and coolers that contact the oil. Welded construction of the deodouriser shell greatly reduces the possibility of leakage. The double-shell feature of semicontinuous or continuous deodourisers eliminates the possibility of the hot oil being exposed to air entering through leaks in the outer shell. The stripping steam also must be oxygen free and, therefore, generated from deaerated water.

Oil cooling

Unhydrogenated oil should be cooled to 38–49°C (100–120°F) before exposure to the atmosphere. Partially hydrogenated oil that melts above the room temperature is somewhat more resistant to oxidation and should be cooled to within 15°C (59°F) of the melting point.

During cooling within the deodouriser, the sparge steam flow rate should be sufficient to provide the agitation needed for a good heat transfer. At a deodouriser pressure of 6 mm (depending on the oil depth), the oil can be cooled only to 54–66°C (129–151°F) before steam condensation occurs, resulting in a wet oil. After removal of dry oil from the deodouriser, the oil should be further cooled to the appropriate temperature in a heat exchanger following the polishing filter.

Additives

A common practice is to add citric acid (0.0005–0.1 per cent, based on weight of the oil), preferably at the cooling stage in the deodouriser, to protect the oil against oxidation. The acid inactivates traces of metal, particularly iron and copper, which appear in the feedstock oil both naturally and as a result of processing. Processors should note that citric acid decomposes rapidly at temperatures above 150°C (302°F) and may be ineffective as a metal scavenger when added before deodourisation. Thus citric acid is added at the cooling stage of deodourisation. Also, complexes formed by trace metals with oxygen and hydroperoxides presumably are decomposed during deodourisation, which permits metal inactivators added after deodourisation to be more effective. It is desirable to add the acid as an aqueous solution. Up to 50 ppm of citric acid can be dissolved in the oil 130°C (266°F) or less at the cooling stage. This amount will remain dissolved or dispersed during the storage of the oil. As little as 10 ppm of citric acid will protect the oxidative stability and flavour stability of the oil.

Filtration

After cooling to 66°C (151°F) or less, the deodourised oil is filtered using a polishing filter, usually through paper, to remove any solids that appear in the oil. There are several reasons for the presence of solids. Small carbonaceous particles may be present because the oil was subjected to high temperatures and localised overheating in the deodourisation process. The oil may have got contaminated with traces of bleaching earth from the feedstock storage tanks. Also, not all of the citric acid added during cooling of the oil may have dissolved.

The filter medium should be capable of removing 30 μm (0.0012 inch) particles for normal use oils and 10 μm particles for bottled oils. The filter should be easy to clean, be simple to assemble correctly and have low volume holdup to reduce the risk of contamination.

Nitrogen blanketing

Nitrogen blanketing is often used in storage, packaging and shipping to protect the deodourised oil. The storage tank is first filled with nitrogen, then oil is pumped in through a pipe. The pipe should stretch down to the tank bottom to avoid splashing of the incoming oil.

Stabilisers

Stabilisers other than metal-inactivating agents are often added to the oil (i) while it is being cooled in the deodouriser; (ii) while it is in storage; and (iii) before it is shipped. Because these additives break down under heat, they are added at oil temperatures < 82°C (< 180°F). Sufficient agitation and mixing time are required to obtain uniform dispersion and true solution.

Propyl gallate, tertiary-butylhydroquinone (TBHQ), *t*-butylhydroxytoluene (BHT) and *t*-butylhydroxyanisole (BHA) are used as antioxidants. The latter two are used to improve the shelf life of the oil and to provide carryover protection into baked and cooked foods. Propyl gallate and TBHQ increase the stability of the oil. The code of federal regulations permits use of these antioxidants in the United States of up to 0.01 per cent, level if used singly, or at a maximum concentration of 0.02 per cent if used in combination of two or more.

HYDROGENATION

Hydrogenation is employed to convert liquid vegetable oils into semisolid plastic fats, which are used in a variety of fatty products such as shortenings, margarine, frying fats and a host of speciality fats (icing fats, peanut butter stabilisers and emulsifiers). The composition of the hydrogenated or partially hydrogenated fat is affected by the fatty acid composition of the original oil as well as the hydrogenation conditions and the extent to which the process is carried out.

Hydrogenation Process

Hydrogenation is the process by which hydrogen is added directly to points of unsaturation (the double bonds in fatty acid chains) in the fatty acids. Hydrogenation of fats has developed as the result of the need to convert liquid oils to a semisolid form for greater utility in certain food applications and to increase the stability of a fat or oil to oxidative rancidity. Hydrogenation, more than any other known process, has contributed to the present high degree of interchangeability among a wide variety of fats and oils.

The degree of hydrogenation is directly related to iodine value (IV). The IV of an oil expresses the amount of iodine with which an oil combines as a percentage of its own weight. The decrease in the IV when an oil is hydrogenated divided by 127, represents the percentage hydrogen gain by weight. It is important to calculate the volume of dry hydrogen, measured at certain set conditions, to drop the iodine value by one unit. At 0°C and 760 mm, 0.9277 m^3 of hydrogen is required to drop 1000 kg of oil by 1 IV. At 15°C and 760 mm, 0.9785 m^3 is required.

The required amount of hydrogen, including a hydrogen loss of 5 per cent, represents the hydrogen factor of 1.05. In normal circumstances, the hydrogen factor for a hardening plant lies between 1.05 and 1.10. A value of 1 m^3 of hydrogen for a drop of 1 IV unit for 1000 kg of oil is useful as it corresponds with a factor of 1.07 at 15°C and 760 mm.

A number of factors influence the course of the hydrogenation reaction: glyceride structure, temperature, pressure, rate of agitation, batch size, the quality of the hydrogen gas and, most important, the kind and amount of catalyst. The hydrogenation of fats and oils is complicated by the mixed nature of the glyceride and the fact that many fatty acids contain more than one double bond, which means that hydrogen can be added at more than one point along the fatty acid chain.

The course that hydrogenation may follow varies. Hydrogenation is ordinarily selective in the sense that hydrogen tends to bond first with the most unsaturated acids. However, selectivity is never complete, resulting in a rather large mixture of various fatty acid components. Another complication that appears in the hydrogenation process is that the reaction may occur at either α (outside) or β (inside) position of the molecule. Yet another complicating factor is the tendency of isomerisation during the hydrogenation reaction. Isomerisation is affected largely by temperature, pressure and the kind and amount of the catalyst.

Catalytic Hydrogenation

Catalytic hydrogenation, like other chemical reactions, is accelerated by an increase in temperature. Under most conditions, within a temperature range of 50–150°C, the time for hydrogenation to a given iodine value could be reduced by about 5 per cent by raising the temperature by 2–7°C. These figures are only approximations, as the effect of temperature is different for different oils. At for > 175–180°C, the effect of increasing the temperature is much less, with the maximum reaction rate attained at about 200–220°C. The temperature at which the hydrogenation reaction is carried out markedly affects isomerisation: the higher the temperature, the greater the degree of isomerisation.

Because gaseous hydrogen must first become dissolved in the oil before it can reach the catalyst–oil interface (site of hydrogenation), any factor influencing the rate of solution will have a corresponding effect on the rate of hydrogenation. The factors specifically affecting the rate of solution are hydrogen gas pressure and the degree of agitation of the catalyst–oil charge. Within the range of ordinary commercial hydrogenation, pressures of 1–6 atmospheres are inclined to increase the hydrogenation rate proportionately.

Nature of the Catalyst

The nature of the catalyst has less influence on the composition of hydrogenated oils. Most of the catalysts used in commercial practice yield approximately equivalent products. A variety of metallic catalysts is available; the most commonly employed catalysts are nickel derivatives. They are used at levels on the order of 0.04–0.1 per cent (nickel basis). Sometimes, the usage level is reduced to about 0.02 per cent (nickel basis) for the hydrogenation of shortening base stocks. Higher catalyst concentrations tend to promote isomerisation. A commonly used technique for the production of margarine base stocks is to hydrogenate with a nickel catalyst that has been used in previous hydrogenations. In this case, catalysts that have been used 3–5 times are often employed at a concentration of 0.2–0.3 per cent (nickel basis). The used catalyst has a greater tendency to promote isomerisation, which is desirable for margarine preparation. Nickel catalysts that have been treated, or poisoned, with sulphur have been used in the production of confectioners' fats (cocoa butter substitutes) because of their tendency to promote isomerisation.

Selectivity in Hydrogenation

Selectivity in hydrogenation means a preference for hydrogenating one class of unsaturate over another. An example of this is the conversion of linolenic acid to linoleic acid, then to oleic acid and finally to stearic acid. This preference in selective hydrogenation remains unaltered until the concentration of the preferred unsaturate is reduced to the desired level.

Effect of Feedstock Impurities on Hydrogenation

Impurities can be divided into three categories: catalyst poisons (sulphur, nitrogen, phosphorus, chlorine and arsenic), catalyst inhibitors (decomposition products, gums, alkali soaps and oxidised fatty acids) and catalyst deactivators (acids, water and carbon monoxide). Catalyst poisons have the most detrimental effect. These can be generally classified as the compounds that contain elements of groups V, VI, VII of the Periodic Table of elements that possess lone electron pairs, such as sulphur and phosphorus. These poisons make catalysts irreversibly inactive.

Catalyst inhibitors are impurities such as decomposition products, gums, oxidised fatty acids and alkali soaps that are absorbed at the entrance of the pores of the catalyst surface. The physical blocking of the active centres of the catalyst inhibits the effect of hydrogenation. However, the influence of these impurities is not as pronounced as those observed for the catalyst poisons.

Catalyst deactivators are those impurities that decrease the hydrogenation reaction rate by chemically reacting with the nickel crystallites.

Carbon monoxide

Carbon monoxide was shown to be a powerful catalyst deactivator below 160°C. Water and carboxyl groups were also identified as possible catalyst deactivators.

Nickel crystallites

Small and active nickel crystallites readily react with water to form nickel oxide, which in turn reacts with free fatty acids to form nickel soaps. For the most part, the reduction of nickel surface of the catalyst results in nickel crystallite increase. Hence, the effect of water is complex and depends on the overall hydrogenation conditions.

Sulphur

Sulphur is a powerful nickel catalyst poison. For example, *Ottessen* determined that 1 ppm sulphur poisons 0.004 per cent (4000 ppm) of nickel catalyst. During hydrogenation, hydrogen interacts with sulphur-containing compounds at the catalyst surface. As a result, sulphur is released, which may react chemically with nickel at the surface of the catalyst. This reaction inhibits the capacity of nickel to absorb and dissociate hydrogen, thereby reducing the total activity of the catalyst.

Sulphur of about 5 ppm has a significantly detrimental effect on the hydrogenation efficiency. As the sulphur content increases to 25 ppm, end points become proportionately higher until virtually no hydrogenation is obtained at a level of 50 ppm. Sulphur dioxide deactivates a catalyst according to the sulphur content and is practically independent of the compounds.

Phosphorus and phosphatides

The poisoning effect of the phosphorus of lecithin and phosphatides is somewhat less than that of sulphur. With respect to hydrogenation, phosphorus in the form of phosphatide mainly affects the catalyst selectivity. As has been previously described, sulphur primarily affects the activity of the hydrogenation catalyst.

Periodic fluctuations in the conventional caustic refining, washing and bleaching processes can pass on to hydrogenation feedstocks that have higher than desired phosphatide. This problem can accentuate in the superdegumming, bleaching and physical refining processes. Feedstocks with phosphatide content higher than 1–2 ppm may have problems during hydrogenation and later with shelf life. The various effects of the catalyst poisons, deactivators and inhibitors may result in significant increases in the hydrogenation operating costs as well as variability in the final hydrogenated product.

Colour bodies

The physical colour of the final product is important to the consumer. Feedstocks used in hydrogenation can contain significant quantities of red and yellow pigments, usually carotenoids. During hydrogenation, the high temperature treatment plus the hydrogenation of double bonds tend to remove the colour effect from carotenoids. Green colour can then become quite apparent. Usually, the green pigments (pheophytin derivatives) can be removed by a bleaching step following hydrogenation.

Hydrogenation Equipment

There are two basic hydrogenation processes: batch and continuous. Most commercial operations employ batch processing. Batch processing has the distinct advantage of product flexibility whereas continuous processing has the advantages of requiring less floor space and lower operating costs. Continuous processing lends itself to the production of large volume products. However, changing from one base stock to another is more difficult with a continuous operation.

The basic process for both batch and continuous processing calls for adding hydrogen to the oil under pressure and agitating the oil in a heated, closed, previously evacuated vessel. The catalyst, usually activated nickel metal, is suspended in the oil in a proportion of 0.02–0.1 per cent (nickel basis). The nickel catalyst consists of approximately 25 per cent nickel suspended in a 75 per cent oil carrier for easy dispersion in the oil to be hydrogenated. Hydrogenation requires that the agitator disperse the gas in the form of fine bubbles.

Three modifications of the batch method are in commercial operation today: gas recirculation, continuous purge and dead end. The latter two systems are employed more often.

The heart of the process is the reactor. This vessel is often referred to as the hydrogenator, the converter and sometimes as the hardening machine. The reactor is a closed cylindrical vessel of about 5,000–27,000 kg capacity. The converter should be so built as to be able to withstand a full vacuum and 7–10 bars pressure, for which carbon steel is a suitable material of construction. In the continuous purge or dead-end system, the converter should be equipped with a properly designed agitator to give efficient hydrogen gas dispersion. Coils should be included for the initial heating, exothermic reaction control and final cooling. A drop tank can be made a part of the process so that the hot, hydrogenated oil can be transferred for further cooling and filtration. This technique increases batch turnover and improves the overall efficiency of the process (Fig. 12.15).

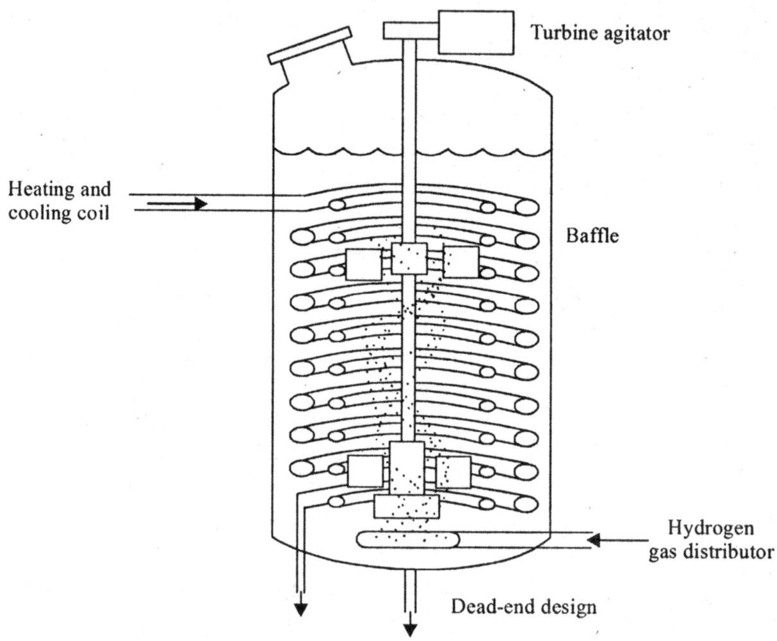

Fig. 12.15. Hydrogenation converter.

Ancillary Systems

Two important ancillary systems include a steam ejector system capable of maintaining a vacuum of about 50 mm Hg absolute and a filtration system to remove the spent catalyst. These filters are commonly referred to as black presses. Any residual nickel catalyst is removed in the subsequent post–bleaching operation. Residual nickel may also be removed by centrifugal washing with a citric acid solution.

Due to the large volume of hydrogen consumed in the process, most plants have their own hydrogen-generating facility. Hydrogen gas can be produced by a number of ways: electrolytic, steam iron, water gas-catalytic and hydrocarbon reforming. The electrolytic process is preferred where electrical power cost is reasonable because the purity of hydrogen produced by this process is better as compared to other processess.

CONTINUOUS HYDROGENATION

In the continuous hydrogenation process, the oil and catalyst are premixed and sent to a reactor that consists of a column divided into a number of reaction chambers created by a series of plates. Only enough hydrogen is metered into the reactor to obtain the desired end point. Stationary catalysts, or fixed-bed reactors, have the advantage of a simplified processing (no catalyst filtration).

Buss Loop Reactor

The buss loop reactor design is suitable for use in a continuous hydrogenation system. Such a system can be set up with two or more loops placed in series in a cascade fashion. The oil feedstock is pumped through an economiser into the hydrogenation reactor, which is under vacuum. This, in effect, deaerates and dries the oil before hydrogenation. The catalyst slurry is drawn into the reactor by the vacuum during the heating phase. As soon as the desired hydrogenation temperature is attained, the vacuum is broken with hydrogen and the reaction commences in the mixing nozzle. Because the consumed hydrogen is continually replaced, the system pressure remains constant. The heat of the hydrogenation reaction dissipates in the economiser, which now acts as a cooler. This cooling effect is brought about by vapourising enough of the condensate collected in the condensate collection tank during the previous heating period to maintain a constant temperature in the hydrogenation system. The product and catalyst are transferred to a holding tank by the residual hydrogen pressure in the reaction loop, thus freeing the reactor to receive the next batch without delay.

The second batch is hardened in the same manner as the first batch. However, during the second hydrogenation cycle the heat recovery system is started up. The feedstock for the second batch is pumped through the second heat exchanger unit and into the last holding tank. Simultaneously, the transfer pump countercurrently pumps the product out of the middle holding tank through the second exchanger into the last holding tank. Thus, the feedstock of the second batch is heated to the reaction temperature, degassed and dried, while the product of the first batch is cooled from the reaction temperature to the filtration temperature. The cycle of operation is continued and as long as the cycle is maintained, there is no further heating or cooling. To complete the process, the product is filtered from the last holding tank to storage.

Use of Hydrogenated Sunflower Oil

Hydrogenated sunflower oil can be used in a variety of fats and oil products. A blend of two or more, partially hydrogenated base stocks can be used to prepare either stick or soft (tub) margarine. As an example, a blend of approximately 35 per cent 65 IV partially hydrogenated sunflower oil and 35 per cent 98 IV partially hydrogenated sunflower oil should provide the proper consistency for stick margarine. In making a soft margarine, as much as 50 per cent liquid (unhydrogenated) sunflower oil can be incorporated into the formula by using a like amount of 75–80 IV partially hydrogenated sunflower oil.

An all-hydrogenated, household-type shortening can be prepared by using about 92 per cent 86 IV partially hydrogenated sunflower oil and 8 per cent completely hydrogenated cottonseed or palm oil. Cottonseed and palm oils are preferred hardstocks, as they possess a β' crystalline habit. Hardstocks made from sunflower or soyabean oil are β phase inclined. Transformation from the β' to the β crystalline form causes a shortening to have undesirable textural characteristics (unduly soft, grainy and ripply). It is possible also to incorporate at least 10 per cent partially hydrogenated cottonseed oil (60 IV) or palm oil into a margarine oil formula when using sunflower oil. This minimises the chance of the margarine developing a sandy, grainy texture.

PROCESSING AND USE OF SUNFLOWER OIL REFINING BY-PRODUCTS

Purpose of Processing the By-products

The processing of vegetable oil produces some by-products. The by-products, with the exception of soapstock, are usually not treated at the refining plants, because their relatively small quantity and low market value make their processing uneconomical. Instead they are normally (i) sold to salvage operators; (ii) sent to land fills (solids); and (iii) emptied into the sewer (liquids).

The increasingly stringent environmental protection rules and the ever-present need to minimise costs are forcing many plants to seek ways to process the by-products to comply with environmental protection rules as well as recover a part of the costs. Most municipal sanitary districts refuse to accept effluent with fatty material concentration over 50–100 ppm. The fatty materials are readily biodegradable, but high concentrations interfere with lagoon digestion. Some of the by-products can be processed into saleable products and some can be regenerated for further use in the plant. This section presents some methods of handling by-products obtained from the alkali refining, decolourisation and dewaxing of sunflower oil. The by-products obtained from these processes are as follows (Fig. 12.16).

1. Soapstock from alkali refining.
2. Clay from decolourisation.
3. Filter aid from dewaxing.
4. Soaps and wax from continuous dewaxing.

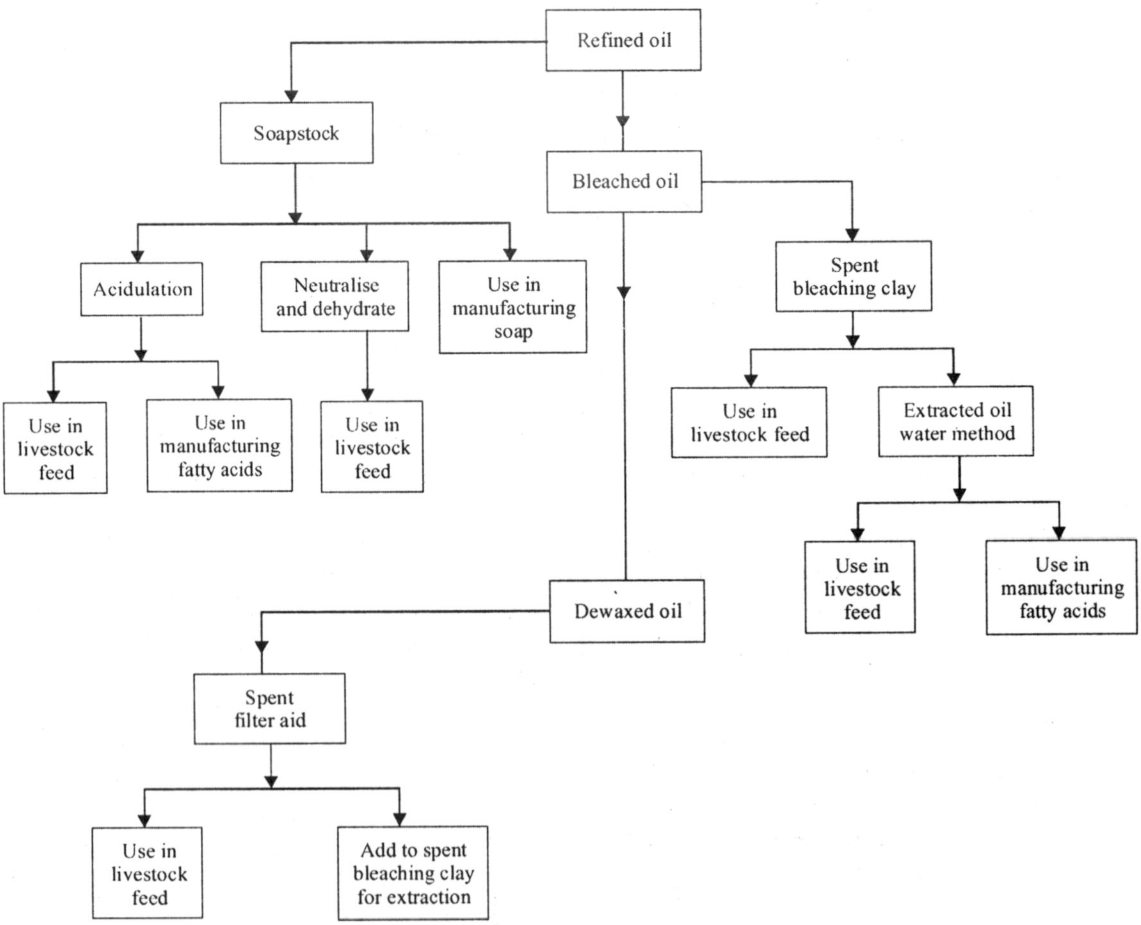

Fig. 12.16. Flow chart showing refining of by-products.

Soapstock from Alkali Refining

The sodium soapstock obtained from the alkali refining process may be used in three different ways: (i) used as an additive to a feed or meal; (ii) further processed into the manufacture of soap; and (iii) acidulated to obtain the fatty acids and neutral oil that were removed during the refining.

Soapstock as an Additive to Meal or Feed

Adding soapstock to a feed or a meal is usually acceptable. It may be done by adding the rich or heavy soapstock to the meal or feed before the final grinding or blending for shipping. Alternatively, soapstock may be added in the desolventiser. The condition of the meal or feed must be checked and monitored if the soapstock is added before the meal is shipped. The additional moisture could cause heating when stored for a longer period of time. The disposal of the soapstock by-product in meal or feed may be prohibited by government regulations and/or the quality specifications for the feed product.

Soapstock Dehydration

Soapstock contains excessive moisture, which may cause the degradation of the meal or feed and limit the quantity of soapstock that can be added. Consequently, it may be necessary to first dehydrate the soapstock before adding it to the meal. Dehydration would also be necessary if the meal were to be stored.

Votator (Eaglin) has developed a process to reduce the alkalinity of the soapstock and dehydrate the neutral soap to a moisture content of about 2 per cent. This is an appropriate method if a high moisture content of the soapstock is a problem. This may be done by a batch method or a continuous system.

Batch Method

The batch process consists of a neutralisation tank that is equipped with an agitator, a transfer pump, a thin film evaporator and a condenser along with a vacuum system. Using the batch method, the soapstock is discharged to an agitated tank, where the soapstock is neutralised (pH 7) using sulphuric acid. The neutralised soapstock is then heated to about 68°C (155°F) and pumped through a thin film evaporator that has a jacket temperature of approximately 166°C (330°F). The dehydrated soapstock is then pumped to a storage tank (Fig. 12.17).

Continuous Method

The continuous method consists of a soapstock holding tank that has an agitator, acid tank and pump, in-line mixer, in -line pH meter and recycle tank, feed pump, preheater, thin film evaporator and a vacuum system along with the required pumps and storage tank (Fig. 12.18).

Using a system to neutralise and dehydrate the soapstock may allow a greater quantity of this material to be added to the meal or feed without causing excessive moisture problems. It is not recommended that either the unneutralised or the neutralised soapstock be added to the meal or feed unless the government regulations permit such an addition and unless the analysis of the final product meets the buyer's quality specifications.

Acidulation of the Soapstock

An almost universally used method for the processing of soapstock is the acidulation method. The acidulation process may be either a batch or a continuous method.

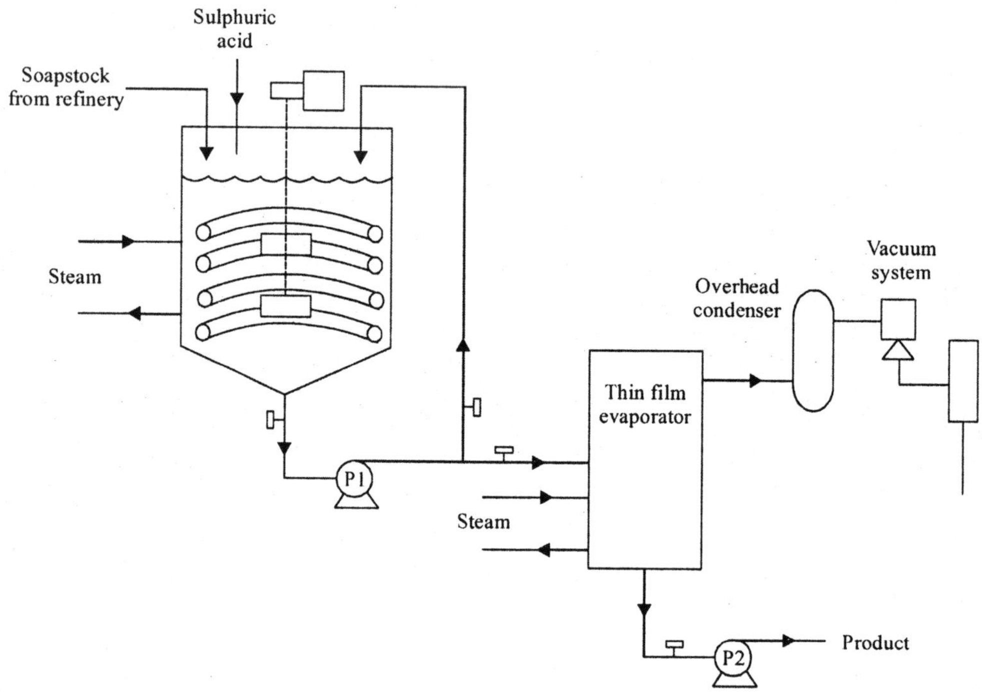

Fig. 12.17. Neutralisation and dehydration of soapstock: batch process.

Batch method

The batch method for acidulation requires good agitation of the soapstock mixture while the sulphuric acid is being slowly added to the soapstock. After the required quantity of acid has been added (about 3.3 per cent) based on the weight of the soapstock—usually enough to bring the pH down to 1–2, the mixture is carefully heated to boiling and allowed to continue boiling until the soapstock is split and the oil (which is almost black) can be seen in the tank. When this occurs, the heating and agitation are shut off and the oil is allowed to rise to the surface of the acid–water phase. After the oil is separated, either it can be pumped off the acid–water phase or the acid–water phase may be drained from the tank until the oil phase is reached, which can then be water washed to remove traces of the mineral acid. After the water washing, the oil can be stored for further processing or dried by means of atmospheric heating or vacuum drying before pumping to storage. The oil can be used to increase the fat content of meal or feed, sold as a feedstock oil for manufacture of free acids, or used in the manufacture of soap.

Batch acidulation system

A batch system consists of an acid-resistant processing tank equipped with a method of heating and agitating the mixture of soapstock and acid. A simple method for heating and agitating that has been used successfully consists of the introduction of live steam into the mixture using a multiple eductor system. The eductors mix the steam with the soapstock–acid mixture and also provide good agitation. Some tanks may be equipped with acid-resistant coils for heating and an acid-resistant agitator, which work just as well as the eductor system (Figs. 12.19 and 12.20).

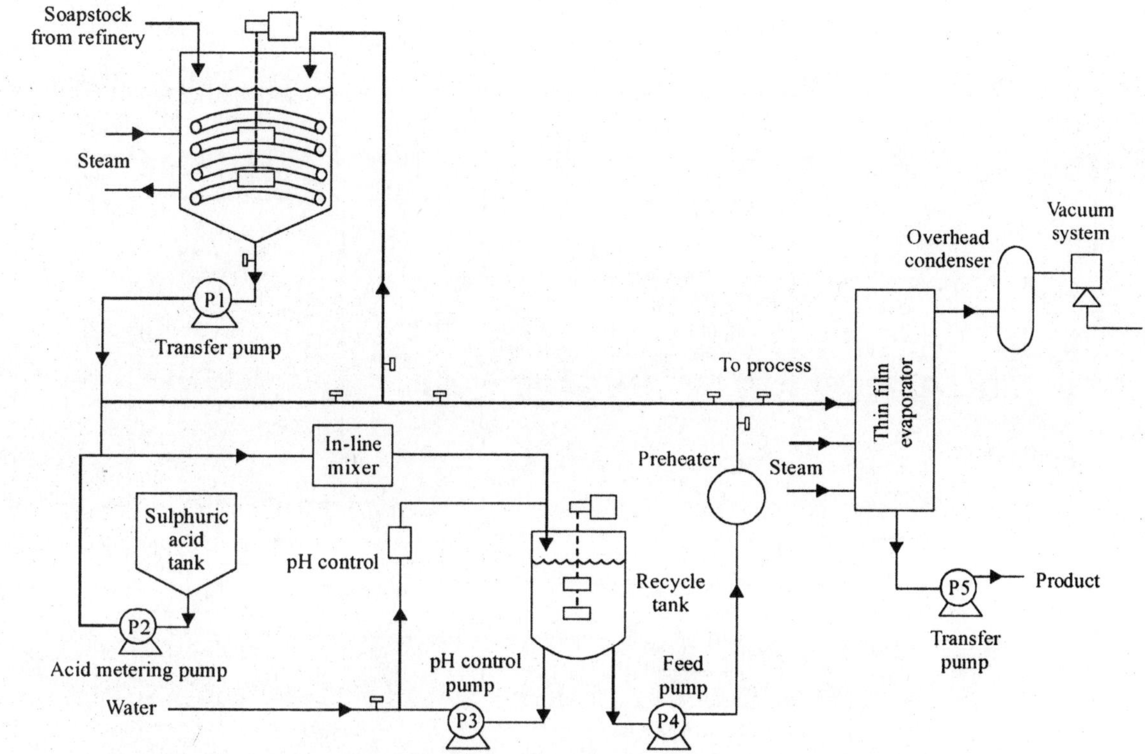

Fig. 12.18. Neutralisation and dehydration of soapstock: continuous process.

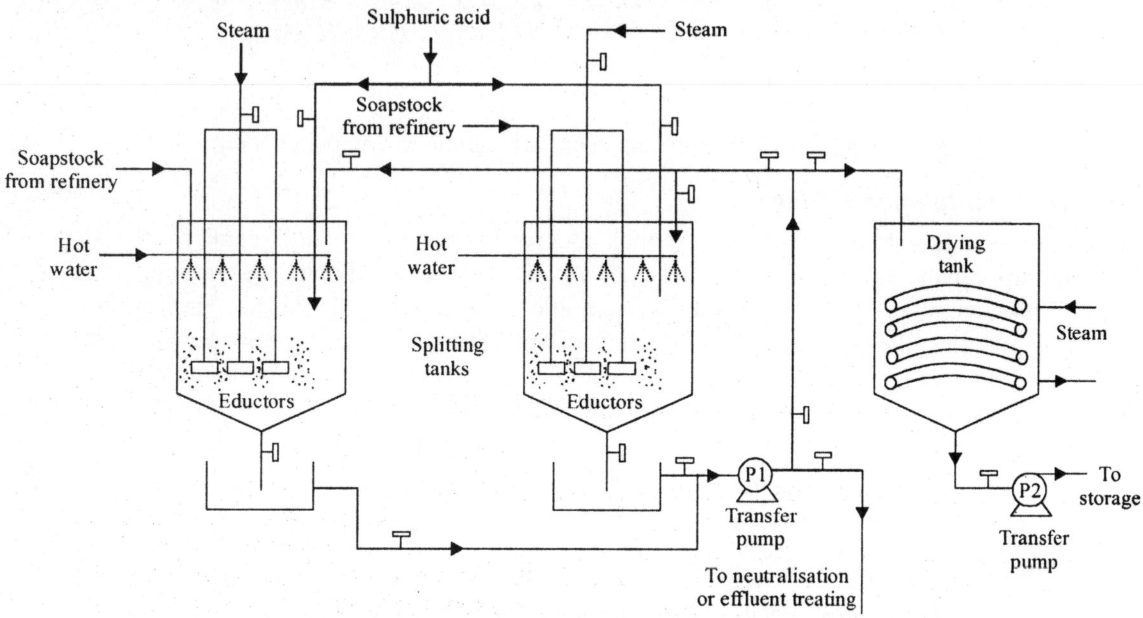

Fig. 12.19. Acidulation of soapstock: batch process (atmospheric drying).

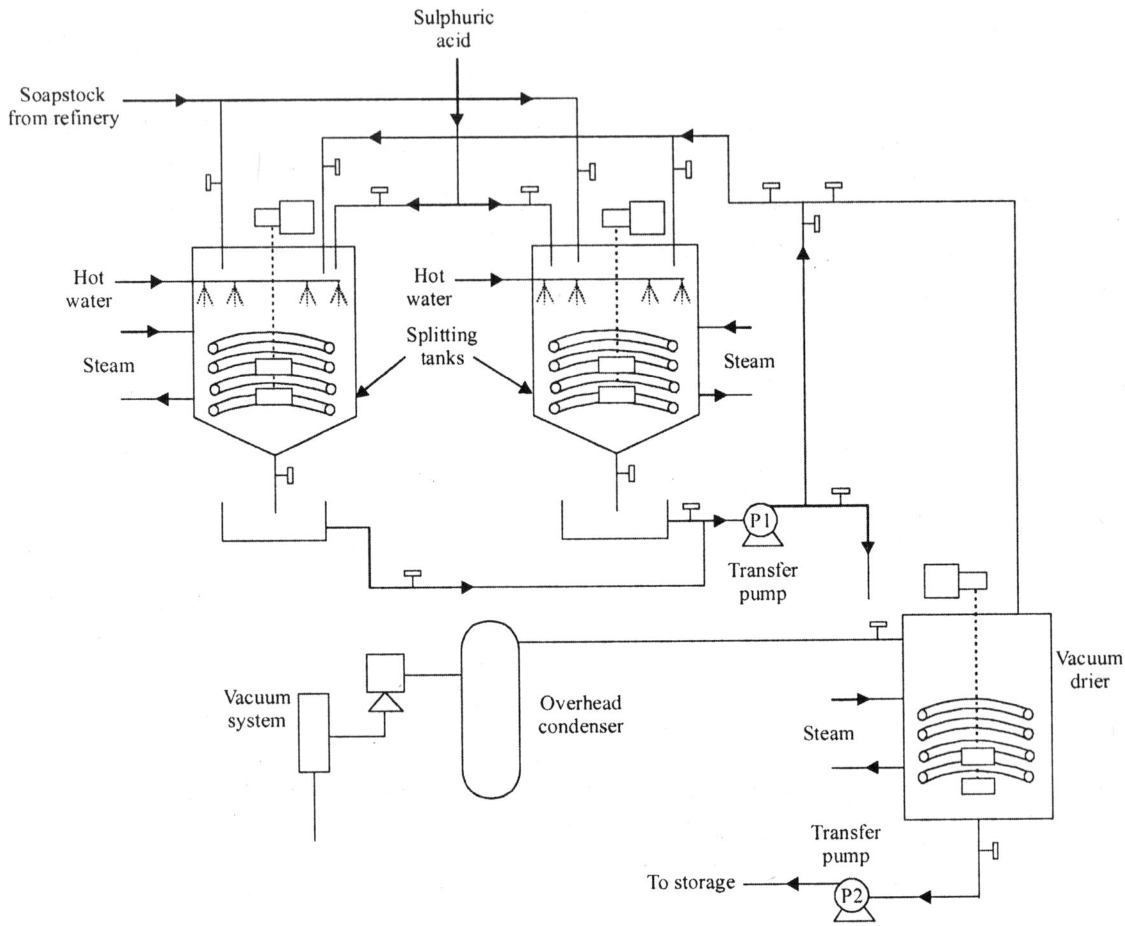

Fig. 12.20. Acidulation of soapstock: batch process (vacuum drying).

Continuous acidulation system

The continuous acidulation system for soapstock consists basically of a holding tank for the soapstock, a holding tank for the sulphuric acid, a pH meter, a mixer, a heater, a splitting column, an oil–water separator, a water washing and separation system and a vacuum drying system. Briefly, the process consists of pumping the soapstock from a holding tank through a mixing device, where the acid is proportioned into the soapstock and then via a pH meter and heater to the splitting column. From that point, the oil and water phases are separated (centrifuges could be used).

The oil is then water washed to remove the residual mineral acid, after which the oil and water phases are separated. Using a centrifuge, the oil is then vacuum dried and pumped to storage (Fig. 12.21).

Clays from Decolourisation

Another by-product consists of the activated clay that was used in decolourisation and also some of the decolourised oil. Again, it may be possible to add this product at low levels directly to meal or feed (following government regulations). The clay by-product should be added at a point in the system that ensures that the filter clay is well mixed and dispersed in the product.

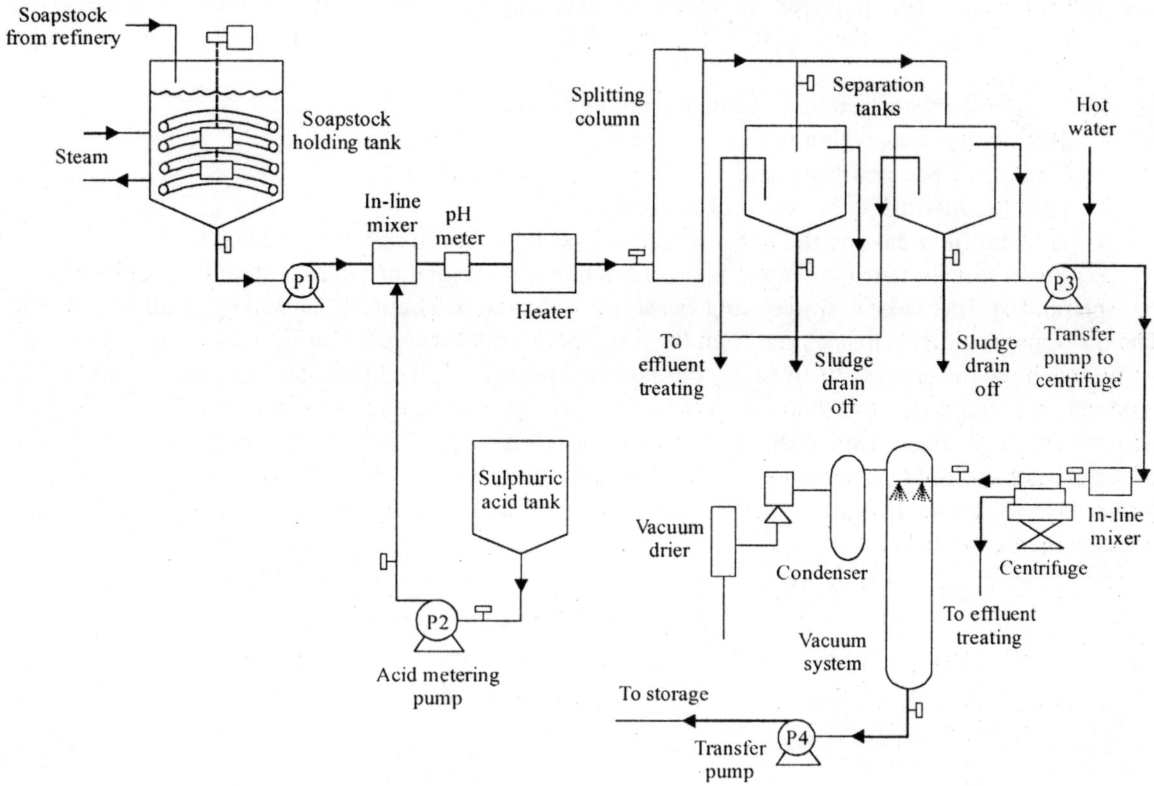

Fig. 12.21. Acidulation of soapstock: continuous process.

Solvent extraction

Solvent extraction of the clay by-product has been used for many years to remove the oil and colour bodies from the decolourising clay. The solvent used was hexane, which required a completely explosion-proof building and equipment. The clay was mixed with the hexane; this mixture was then pumped through a closed delivery press to remove the clay from the oil–solvent mixture. When the press was full of clay, it was blown with steam to remove the residual solvent and oil from the filter cake. The miscella (oil and solvent) was then pumped from a storage tank to a still to remove the hexane. The oil is dark and can be sold or incorporated into meal or feed to increase the fat content. The recovered solvent is reused. The disadvantages of a solvent extraction plant for this operation are that it is expensive and hazardous. Other solvents (e.g. kerosene) have been used to remove the oil from the decolourising clay. If they are used, extreme care should be exercised to prevent fires. Air pressure must not be used to blow the filter press if a flammable solvent is being used; either steam or nitrogen should be used.

Water extraction

Water extraction of oil from the decolourising clay has been used successfully and it does not require the expensive equipment that a flammable solvent extraction requires. This system requires a mild steel mix tank equipped with steam coils and a good agitator. A tank-type pressure leaf filter is preferred; it may be possible, however, to use a plate-and-frame filter press, provided the fabric used in the press is not affected by water. A receiving tank from the filter and the pumps necessary to transfer the material from the mix tank through the filter and from the receiving tank either to storage or to further processing, are required.

The steps in the water extraction process are as follows:

1. Fill the mix tank with a predetermined quantity of water.
2. Add 0.5–1 per cent soda ash.
3. Heat the mixture to 82–88°C (180–198°F).
4. Add the filter cake to the hot soda ash and water solution and agitate for 2 hr.
5. Pump (diaphragm-type pump is recommended) through a pressure leaf tank-type filter.

The ratio of filter cake to water–soda ash solution should be about 910–7,750 kg (2,000–17,045 lb). The oil–water mixture is discharged from the filter into a tank for separation. When the batch of material is filtered, blow the press with air to dry the cake and remove all the oil–water mixture. Allow the oil to separate from the water and then pump the recovered oil to a tank for its drying or, if possible, use it without drying (in meal, feed, soap manufacture) or sell it (Fig. 12.22). The temperature of the mixture of clay, water and soda ash is important to obtaining a good extraction and should be maintained during mixing and filtration. The water that is separated from the oil through filtration may be reused for the next batch of filter cake.

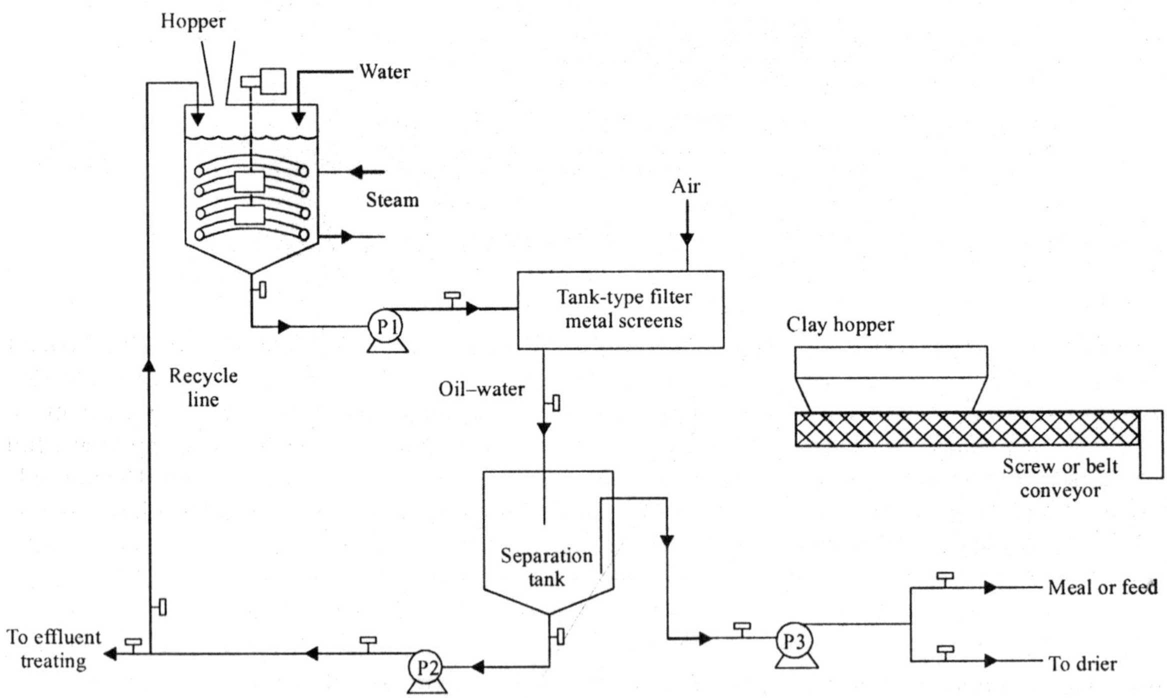

Fig. 12.22. Water extraction of oil from clay and filter aid.

The clay that is removed from the filter after the water extraction may be added back to the meal or feed or added to sandy soil to improve its moisture retention property. It is not recommended that the oil obtained from either solvent extraction or water extraction be added to the crude oil because this oil contains a high percentage of colour bodies. Addition of these colour bodies will in effect increase the quantity of decolourising clay required for subsequent batches.

Filter Aid and Dewaxing

The by-product recovered from the dewaxing process contains the filter aid, waxes and oil. The disposal and recovery of this material and the recovery of oil and wax, may be achieved using the previously discussed methods. Either the solvent extraction system or the water extraction system can be used. It is recommended, however, that the filter aid be mixed with the clay from decolourisation before the water extraction or solvent extraction, as the mixture of these two products filters better than if extracted separately.

Continuous Dewaxing By-products

The continuous water dewaxing process produces a wet soapy by-product consisting of wax and oil. The wax and oil may be recovered by collecting the effluent from the dewaxing centrifuge. This by-product can be used as such in the meal or feed to increase the fat content of such products.

The soapy wax and oil effluent may also be neutralised and dehydrated to remove the water and the wax-and-oil portion may be used in the manufacture of margarine. Various methods may be used to dehydrate this type of a by-product. The use of a thin film evaporator is a good method.

Regeneration of Clay, Filter Aid and Catalyst

It is technically feasible to regenerate bleaching clay and filter aid by physical and chemical means so that it can be reused. However, this is not normally done due to economic reasons. The spent earth can be used as soil conditioners for sandy soil or as a fill. As a precautionary note, vegetable oils adsorbed on such things as bleaching clay or filter aid are extremely susceptible to spontaneous combustion and are considered a fire hazard. Recovery of nickel in the metallic form from spent catalyst used in the hydrogenation of sunflower oil is advisable when nickel prices rise to a favourable economic level or the cost of disposal becomes prohibitive due to environmental mandates.

Commercial Uses for Fatty Materials

Soapstock and phosphatides can be used in animal feed. The fatty materials have a high calorific value and are a good source of phosphorus from the phospholipids. However, utilisation of soapstock is limited to local animal feeding because of two reasons: its transportation costs are high and the product ferments easily. The soapstock can be dehydrated to facilitate transportation, storage and incorporation into the feed. There is a market for soapstock in the form of the acidulated product (acid oil). Acid oil can be stored and shipped in the same fashion as crude oil. It can be used for road dust control with good results. Free fatty acids can be used in the production of soap. Sunflower wax has been successfully used as an ingredient in cattle feed. The fatty materials are normally used along with other ingredients such as meal, grains, silage, etc.

Miscellaneous uses of by-products include their use as starting point, normally after molecular distillation, for a number of chemical syntheses. The materials can be used in the encapsulation of fertiliser and pesticide formulations. This process reduces the application hazard of some of the more dangerous materials and gives a slower release of the active ingredients. Hot well oil from deodourisation can be disposed of in the same fashion as the other fatty materials. At one time a considerable premium was placed on this product as a source of vitamin E (an isomer of tocopherol).

Chapter 13

Safflower Oil

INTRODUCTION

Safflower was used primarily as a source of dye, in paints, varnishes, alkyd resins, a food colouring, a cosmetic, or for medicinal purposes. It faced the competition of other oils, (soyabean, castor, linseed, tall oil). Because of the increasing nutritional interest in polyunsaturated oils, safflowers has gained the market of edible oils, especially for frying and seasoning.

Dried safflower florets are commonly used as an adulterant or substitute for colourful saffron, *Crocus sativus* L., a much more costly spice. *Pliny* pointed out that it could be used as a substitute for castor oil for nonedible purposes. While it had become known as an edible oil during pre-Christian times in Mesopotamia, it was only in more recent times that it began to be used in India as an edible oil and of course, it was not until the middle of this century that it began to enter world commerce, first as an industrial oil and then as an edible product.

Because safflower was introduced to many lands, it is known by a number of different names, some of which are azafrancillo, bastard saffron, benihana, cartamo, cnikos, false saffron, ghurtom, hung hua, kafsha, kahil, kajireh, kardi, khardam, kusumba, onickus, saffron, thistle saffron and ssuff.

Safflower, *Carthamus tinctorius L,* of the compositae family. Other members of this family are the artichoke, chrysanthemum, niger and sunflower. There are at least 25 species of the *Carthamus* genus that grow in the wild, but only *C. tinctorius,* which we call safflower, has been domesticated; some quantities of *C. oxyacantha* have been gathered and used as oil or food sources in India and Pakistan.

The safflower plant as we known it resembles the Scottish thistle but has yellow, orange, or red florets rather than the purple bloom of the thistle. However, the commercial species of safflower, *C. tinctorius*, does not become a weed.

It responds well to moisture and nitrogen. Its seed has the ability to germinate almost immediately if exposed to moisture at the proper temperature, unlike a sunflower seed, which must go through a period of dormancy before germination. The deep root and the many fine laterals that extend from it have the ability to seek out water and nutrients deep in the soil. These properties, while they allow safflower to survive in periods of moisture shortage also limit the areas of the world where safflower can be cropped successfully.

TECHNOLOGY

The seed (humidity on storage < 8 per cent) is pressed to produce oil which can be directly deodourised, if poor in free fatty acids (0.15–0.3 per cent) for seasoning oil uses. The oily meal from pressing rich by about 15 per cent of oil is extracted with solvent (hexane). After extraction, its oil content is 1–1.5 per cent. Crude oils coming from pressing and extraction generally undergo a tradition refining. Table 13.1 lists the typical fatty acids composition of linoleic and oleic safflower oils.

Table 13.1. Typical fatty acid composition of linoleic and oleic safflower oils (%).

Fatty acid	*Normal*	*Oleic*
Palmitic	5.25	4.5
Stearic	1.50	1.5
Oleic	15.0	77.0
Linoleic	77.0	15.0
Others	1.25	2.00

PHYSICAL AND CHEMICAL PROPERTIES

Safflower Seed

Safflower seed (technically an achene) consists of tough fibrous hull that protects a kernel comprised of two cotyledons and an embryo. Hulls make up 18–59 per cent of the seed weight. *Weiss* characterised normal hulled seeds as 38–49 per cent and *Lowes* and co-workers noted percentages of 25–87.5 per cent. This diversity also shows up in seed weight per 1000 seeds (14–105 gram), oil content (11.48–47.45 per cent), and fatty acid distribution (linoleic acid, 11.13–85.6 per cent; oleic acid, 6.74–81.84 per cent, stearic acid, 0.01–4.88 per cent and palmitic acid, 2.1–29.03 per cent.

Safflower seeds are normally cream to white, but since 1960, breeding has resulted in great variation in colour, ranging from normal hull to thin hull (which tend to show part of the underlying coloured layers) to types with gray, purple, or brown-striped hulls. Most of this research has been aimed at creating a thinner hull to increase oil content (Table 13.2). Although reduction of the hull fraction increases valuable oil and protein percentages, too much reduction can produce other problems. Brown-striped seeds have a distinctly musty odour. Thin hull types can create harvesting, storage, handling, and processing problems, although the more gentle combine harvesters in use for the last decade can probably deal with the harvesting worry. Brown striped seeds also were shown to contain colourless precursors in the hull and kernel that could create dark extracted oil.

Three phosphatides have been identified that are responsible for colour formation from oil extracted from the kernel of safflower seed: phosphatidyl ethanolamine, phosphatidyl myoinositol, and phosphatidyl choline. Oleic types of safflower are produced primarily in the United States and to a minor degree in Mexico. The commonly available types exhibit oleic fatty acid levels in the 76–81 per cent range. Linoleic level decreases proportionally as oleic level increases. Safflower seeds in the Northern Hemisphere tend to be higher in linoleic acid at more Northern latitudes because cooler temperatures are usually experienced at and after the time of flowering. Similarly, oleic levels increase in safflower exposed to higher temperatures. Oil contents of commercially available oleic types tend to be slightly lower than linoleic types, but field yields are equal or superior to linoleic types. Table 13.3 lists some of the characteristics of safflower seeds and Table 13.4 highlights the identification and distribution of sugars in safflower hull and kernel.

Table 13.2. Analysis of US safflower seed[a].

Type	*Oil*	*Protein*	*Fibre*
Analyses of whole seed			
Gila	38.1	16.7	22.3
U-5	38.5	17.2	21.1
US-10	36.8	19.4	22.3
Frio	40.1	15.4	20.8
Thick-hull hybrid	37.8	17.3	21.5
Brown striped	47.7	20.3	11.7
Pigmentless brown stripped	42.8	22.5	13.6
Thin hull	47.2	21.1	11.2
Analyses of hull			
Gila	3.2	4.3	57.1
U-5	2.2	5.0	58.4
US-10	1.4	3.6	60.0
Frio	2.7	4.1	60.4
Thick-hull hybrid	2.2	4.1	63.9
Brown striped	5.7	8.4	46.9
Pigmentless brown striped	5.6	8.6	46.2
Thin hull	5.1	10.0	45.3
Analyses of kernel			
Gila	60.9	24.9	1.5
U-5	61.8	25.4	1.5
US-10	59.0	29.4	1.5
Frio	64.0	23.0	1.0
Thick-hull hybrid	58.1	24.7	2.8
Brown striped	52.7	24.8	0.9
Pigmentless brown striped	55.9	27.4	2.7
Thin hull	62.6	25.5	0.9

[a] All analyses are percentages on a moisture-free basis.

Table 13.3. Characteristics of commercial safflower production.

Country of region	*Oil content*	*Moisture (%)*	*Protein (% in solvent extracted meal)*	*Linoleic (% in total fatty acids)*
United States	39.5–44	4–5	25	75–78
Canada	32–35	5–9	24	76–81
Mexico	30–38	5–12	24	60–70
Argentina	35–36	6–12	23–24	70–76
India	32	7–8	21–24	72–78
China	28–32	7–8	25–28	76–82
Australia	35–38	5–9	24	70–76

Table 13.4. Identification and distribution of sugars in safflower hull and kernel.

Safflower component	*Sugars present*	*Percent distribution of sugars*	*Percent sugars*	*Percent sugar on defatted basis (calculated)*
Kernel	Uronic sugar glycosides	14.3	0.43	–
	Raffinose	35.8	1.08	–
	Sucrose	46.9	1.42	–
	Galactinol	3.0	0.09	–
	Total	–	3.02	7.74
Hull	Uronic sugar glycerosides	38.9	0.37	–
	Raffinose	6.8	0.06	–
	Sucrose	22.6	0.21	–
	Galactinol	2.7	0.025	–
	D-Gluose	14.6	0.14	–
	D-Fructose	14.2	0.13	–
	Total	–	0.94	0.79

Safflower Oil

Physico-chemical characteristics

Table 13.5 gives the principal characteristics of this fluid yellow-reddish oil.

Table 13.5. Main physical and chemical characteristics of safflower oils.

	Linoleic	*Oleic*
D_{20}	0.922–0.927	0.910–0.916
V_{20} (cp)	52–54	70–75
n_D^{20}	1.474–1.478	1.470–1.474
Solidification point (°C)	– 15	–
Iodine value	135–150	91–95
Saponification value	186–198	186–198

Fatty acid composition

Table 13.6 gives the compositions of fatty acid of safflower oils. Depending on the area of production, linoleic acid content of the linoleic variety (LS) can vary from 70 per cent (Australia) to 80 per cent (China).

The range for the oleic varieties (OS) is 73 to 80 per cent of oleic acid.

For certain oleic and linoleic varieties of Italian origin, *Conte* gave some unusual values for the main fatty acids: oleic safflower, (average) C16:0, 6.7 per cent; C18:1, 59 per cent; C18:2, 33 per cent and linoleic safflower (average) C16:0, 13.7 per cent; C18:1, 25 per cent; C18:2, 58 per cent. *Raie* also gave for the Pakistani varieties of linoleic safflower the following values: C14:0, 0.9–3.1 per cent; C16:0, 9.4–12.0 per cent; C18:0, 2.3–5.5 per cent; C18:1, 14.0–15.7 per cent; C18:2, 66–73 per cent.

Table 13.6. Fatty acids composition of safflower oils.

Nature	% Total FA Oleic	% Total FA Linoleic
C14:0	< 0.1	< 0.1
C16:0	6–7	6–7
C16:1	< 0.2	< 0.5
C17:tot.	–	–
C18:0	< 2	2–3
C18:1	73–80	10–20
C18:2	12–20	68–83
C18:3	< 0.2	< 0.2
C20:0	< 0.3	< 0.5
C20:1	< 0.2	< 0.2
C22:0	< 0.2	< 0.2

Composition of unsaponifiable fraction

Table 13.7 provides composition on the unsaponifiable fraction of safflower oil. The composition of the sterolic fraction given in the table are from the specifications of Italian Standardisation Organisation.

Table 13.7. Composition of unsaponifiable fraction of safflower oil.

Unsaponifiable fraction: 0.5–1.5%			
Sterols (mg/100 g)	240–380	Hydrocarbons (mg/100 g)	126–162
		Tocopherols (mg/100 g)	30–60
Sterol composition (% total sterols)		Tocopherol composition (% total tocopherols)	
Cholesterol	< 0.2	α-Tocopherol	83–96
Brassicasterol	–	β-Tocopherol	2–3
Campesterol	10–16	γ-Tocopherol	7–17
Stigmasterol	8–15	δ-Tocopherol	1–4
β-Sitosterol	52–60	Tocotrienols	–
Δ5-Avenasterol	–		
Δ-7-Stigmasterol	15–22		
Δ-7-Avenasterol	–	Triterpenic alcohols (mg/100 g)	60–66

Conte gives for two varieties of oleic safflower 65 per cent β-sitosterol and 12 per cent Δ7-stigmasterol. From the contents indicated by *Horstmann* (26 seed or oil samples of Australian and Californian safflower, with only one oleic oil sample), one deduces a different average composition (11 per cent campesterol, 6 per cent stigmasterol, 41 per cent β-sitosterol, 26 per cent Δ7-stigmasterol, 5 per cent Δ7-avenasterol).

It should be noted that Δ7-stigmasterol content is correlated positively to the linoleic acid content.

For example, from the contents given by *Hortsmann*, one deduce a ratio of 3.3, but *Kiosseoglou* gives a ratio of 1.1 to 1.5 (on two oil samples).

Finally *Hortsmann* gives for triterpenic alcohols, an average content of 27 mg/100 gram oil, which is much lower than the values given by *Itoh*.

Normal safflower oil is pale yellow to golden and has a slightly nutty flavour. Table 13.8 summarises the important chemical and physical characteristics of US safflower oil. Safflower oil exhibits the highest level of linoleic fatty acid of any commercially available oil. This high level, combined with an absence of linolenic fatty acid, is what has made safflower oil attractive to consumers, initially as a quick-drying oil that could produce films that would not yellow with age and, more recently, as an edible oil with the highest available level of polyunsaturation. As an edible oil, the high level of

Table 13.8. Physical and chemical characteristics of US safflower oil.

Characteristic	*Usual range of California oil*	*Minimum*[a]	*Maximum*[a]
Physical			
Colour (Gardner)	8–10	–	11[b]
Colour after heat bleaching, 315.5° (600°F)	2–3+[b]	–	4[b]
Colour, refined, bleached, deodourised[c]	0.5–1.0 red[d]	–	15 yellow and 1.5 red[d]
Specific gravity , 25/25°C	0.919–0.924	–	–
Refractive index, n_p 25°C	1.473–1.476	–	–
Titer, °C	15–17	–	–
Flash point, °C (°F)	148.8 + (300 +)	121.1(250)	–
Chemical			
Free fatty acids, % as oleic	0.15–0.6	–	2
	0.03–0.05[b]	–	0.05[d]
Saponification value	186–194	–	–
Iodine value (Wijs)	141–147	140	155
Unsaponifiable, %	0.3–0.6	–	1.5
Peroxide value (at time of shipment)	0–1.0[d]	–	1.0[d]
Moisture and volatile, %[e]	0.03–0.1	–	0.8
Insoluble impurities, %[f]	0.01–0.1	–	0.3
Moisture and impurities, %	0.05–0.1[b]	–	0.1[b]
Principal fatty acids, % TFA			
Palmitic	4–6	–	–
Stearic	1–2	–	–
Oleic	16–12	–	–
Linoleic	75–79	72	–
Linolenic	Nil	–	–

[a] Per NIOP trading rule; [b] Nonbreak grade, NIOP; [c] AOCS method Cc 13b-45; [d] Edible grade, NIOP; [e] AOCS method, Ca 2d-25; [f] AOCS method Ca 3–46.

unsaturation also creates problems. Home consumers using safflower oil for frying must be careful to clean pans quickly after use or a tough varnish film results, which is difficult to remove. Fresh safflower salad-grade oil has excellent flavour and odour characteristics, and because it lacks linolenic fatty acid, it does not display the fishy or beany odours sometimes associated with poorly refined soyabean oil. Unfortunately, it does have a relatively short shelf life (typically 9–12 hrs. AOM), which means the oil should be kept cool after the bottle is opened to maintain freshness.

Oleic safflower oil displays most of the same characteristics as the linoleic type, except for its fatty acid structure (Table 13.1). It has been noted that a blend of linoleic and oleic edible oils would improve the dietary value of commercial safflower oil. Blends of this type are used in japan and appear to be achieving good acceptance by the public.

A few studies have reported on the location of the fatty acids on the triglyceride in varying ways. One study used argentation TLC with lipase hydrolysis on a sample of safflower oil from Kenya that contained 10 per cent total saturated fatty acids. It was found that 2 mol per cent was configured with two saturated acyl chains (S) and one unsaturated acyl chain (U), 26 mol per cent had a SU_2 configuration and 72 mol per cent had a U_3 configuration. It was also reported that 3 mol per cent had two double bonds attached, 3 mol per cent had three double bonds 23 mol per cent had four double bonds, 19 mol per cent had five double bonds, and 47 mol per cent had six or seven double bonds. Another study measured the position of linoleic acid on the triglyceride in a study on hydrogenation. It was found that 84.6 mol per cent of linoleic acid was located in the 2-position in a safflower oil containing 76.4 per cent linoleic acid.

Safflower Meal

The by-product of the extraction of safflower oil is a grayish tan to brown cake or meal that exhibits flakes or shreds of whitish safflower hulls. Table 13.9 presents typical analysis for safflower meal.

Table 13.9. Typical analyses for US safflower meal.

Characteristic	*A*[a]	*B*[b]	*C*[c]	*D*[d]
Crude protein, %	21.03	20.0	42.0	25.4
Crude fat, %	6.6	0.5	1.3	1.5
Moisture, %	9.0	10.0	9.2	8.0
Crude fibre, %	32.2	37.0	15.1	32.5
Ash	3.7	5.0	7.8	5.9
Calcium, %	0.23	0.24	0.4	0.37
Total phosphorus, %	0.61	0.24	0.4	0.8
NFE, %	–	–	–	40.0
TDN, %	–	–	–	57.0

[a] Expeller pressing of safflower seed without decortication.

[b] The low end fraction of meal that resulted from prepress–solvent extraction of safflower seed followed by two fraction tail-end decortication.

[c] The high end fraction of meal that resulted from prepress–solvent extraction of safflower seed followed by two-fraction tail-end decortication.

[d] Prepress–solvent extraction of safflower seed without decortication.

Most meal produced in the united states is of a solvent-extracted type. The amino acid and mineral contents of meal are shown in Table 13.10.

Table 13.10. Amino acids and minerals in safflower meal[a].

Factor	*A*	*B*	C
Methionine	0.4	0.33	0.69
Cystine	0.5	0.35	0.7
Lysine	0.7	0.7	1.3
Tryptophane	0.3	0.26	0.6
Threonine	0.47	0.5	1.35
Isoleucine	0.28	0.27	1.7
Histidine	0.48	0.5	1.0
Valine	1.0	1.0	2.3
Leucine	1.1	1.2	2.5
Arginine	1.2	1.9	3.7
Phenylalinine	1.0	1.0	1.85
Glycine	1.1	1.1	2.4
Calcium	0.28	0.37	0.44
Phosphorus	0.78	0.80	1.41
Potassium	0.79	0.79	1.33
Magnesium	0.36	0.37	1.33

[a] Table 13.9 for explanation of A, B and C. Numbers are percents.

PROCESSING

Steps involved in processing of safflower oil are discussed below:

Extraction

Most of the safflower processed in India in the past was crushed by a mortar and –pestle-like device called a *ghani*. Seed was cleaned by hand and then introduced into a *chakki*. This machine, which consisted of two horizontal stone wheels, one of which was turned by a blindfolded bullock, partially dehulled the cleaned seed passing between the stones. Hand winnowing and sieving next removed the hulls from the seed kernels. The meats were pressed into balls after the addition of about 6 per cent water. About 15 kg of the 'balled' kernels were introduced into the *ghani*, an inverted conical mortar into which a heavy pole was placed. The pole was held to the side of the mortar by heavy weights and dragged around the perimeter by a team of oxen. A small amount of heated oil was added, and crushing then proceeded for 45 minute, after which the oil was allowed to drain out through a small hole. A ghani could process about 100–120 kg of seed per day.

More recently, *ghanis* capable of processing 150–175 kg per day were sometimes motor driven. Animal-powered ghanis could obtain 11–16 per cent residual oil in the extracted meats, while motor driven models could extract 10–12 per cent. Today, some cast-iron *ghanis*, expellers, and solvent-extraction plants are used, in addition to the older stone devices. The oil extracted by a ghani is clarified by settling and decanting or by water washing. The oil is placed in tins for local sale.

Most safflower was first processed in the United States by continuous screw press expellers. Some processors attempted decortication, by the nature of safflower seed acts against successful decortication. To prevent the oil from scorching, water-cooled shafts were recommended. Oil so treated could easily be heat bleached to a level below 4 Gardner colour. Expellers such as the 'Anderson super duo duplex' could process about 15 tonne of safflower seed per day, leaving 7–8 per cent residual oil in the remaining cake. However, the principal problem encountered in processing safflower seed through expellers was the propensity of expeller-processed safflower meal to burn in storage. The combination of a reactive polyunsaturated residual and the fibrous texture of safflower meal created many fires in the past. Once safflower processing shifted to prepress-solvent extraction, which brought residual oil contents down below 1.5 per cent, most storage problems were eliminated.

These same expellers, if employed in a prepressing mode wherein 15–17 per cent residual oil remains in the cake that is sent to the solvent extraction unit, can process 45–50 tonne of seed per day. Prepressing of safflower produced under California conditions (or the equivalent) requires no cooking, flaking or cracking of the seed before extraction and results in oil capable of being heat bleached to 1–3 Gardner colour.

During early period Pacific Vegetable Oil (PVO) produced an air gun device that decorticated safflower seed satisfactorily, but the idea was abandoned because it required too much energy and was extremely noisy. A PVO researcher developed a method for decorticating safflower meal after extraction, which employed the principle that the fine particles produced in grinding safflower cake are high in protein and the coarser particles are more fibrous. This method, using vertical hammer mills to grind the cake and a combination of air classification and screening was employed by several California mills to produce safflower meal of 42 per cent protein, in addition to an 18–20 per cent protein middle fraction and 6 per cent protein hull fraction. More recently, most mills have returned to only producing 'as is' meal of approximately 25 per cent protein content, because the high amount of energy consumed by the tail-end process cost more than the additional return gained from the high protein fraction.

The high cost of energy encouraged some mills in the 1980s and 1990s to replace or supplement prepress expellers with caged expander-extruders, which are capable of removing approximately 66 per cent of the available oil through the caged portion of the extruder and to produce collets that are ideal for efficient solvent fraction. Extruders require much less horsepower per tonne of seed processed than expellers and cost less to maintain.

Horizontal basket or moving bed solvent extractors are preferred over vertical tower extractors in processing safflower cake. The fibrous nature of safflower provides a natural channel through which the solvent can move, and the bed acts as a natural filter medium. Tower extractors generally have problem extracting safflower seed because the hulls tend to float, sometimes carry over in the top of the extractor and cause excessive wear in a tower's rotary seal.

Refining, Bleaching and Deodourising

Safflower oil that is extracted from seed in good condition is easy to refine because it is low in FFA and contains few gums or impurities. Conventional caustic refining systems work well. This most important factor in handling safflower oil, destined for edible use, is to limit exposure to air throughout the extraction, refining, bleaching, deodourising and packaging cycle. Nitrogen blanketing should be employed if deodourised oil is to be stored for more than few hours. Generally speaking, safflower oil processed by expeller processing will contain just enough free fatty acids and impurities to require refining before deodourisation; in most cases.

Safflower seed that is produced in areas with late summer rains or cool weather cycles that interfere with maturation can produce dark-coloured or greenish oils that are often higher in FFA as well. If the seed has sprouted before or during harvest or has been attacked by *Alternaria, Pseudomonas*, or other head-rot diseases, the resulting oil can be quite difficult to refine and extremely difficult to bleach.

While safflower oil may, on occasion, display minute traces of a fine, lacy wax, most US refiners neither winterise nor dewax safflower oil, feeling that a brilliant oil can be delivered without it. Japanese refiners generally insist on refining and bleaching safflower oil to under 1.0 red colour, followed by winterisation to avoid problems with minute amounts of wax that may appear in the oil in the winter months in the north.

Production of Margarine and Mayonnaise

If proper steps are not taken, physical crystal changes (polymorphism) can take place in the production of safflower margarines, resulting in a product with a sandy texture. The β-crystalline form that results consists of large crystals instead of the smooth, uniform mixture desired in a margarine; safflower's uniform triglycerides structure encourages production of β-crystals. This problem can be solved by incorporation of small amount of more saturated oil into the margarine mix.

Soft safflower margarines, wherein a highly hydrogenated safflower lattice was employed to encapsulate a larger portion of liquid safflower oil, have been successfully produced by several companies. The methods employed to produce these types of margarine structures have been reviewed. It has been shown that $Cr(CO)_3$ catalysts can be used to selectively hydrogenate safflower oil and retain a 90–95 per cent *cis* configuration. Several studies have reported on safflower oil's good taste, appearance, odour and texture in mayonnaise and frozen salad dressings, where it exhibits excellent qualities in repeated freeze-thaw cycles.

Industrial Processing

Although this Chapter is concerned with oils that are used in edible products, it is to remember that safflower oil's recognition in modern times occurred because of interest in its excellent properties as semidrying oil. Safflower oil's light colour, ability to heat bleach to near water whiteness, low level of free fatty acids and impurities and lack of linolenic acid make it an ideal vehicle for white house paints and varnishes and for the production of alkyd resins. It is easy to polymerise via kettle bodying without the need for vacuum equipment; capable of producing excellent blown, limed, or maleated oils; and acts as a good source for conjugation or methyl esters.

Safflower Oil

In today's marketplace safflower occupies a unique position. It is the oil with the highest level of linoleic acid available commercially. It continues to enjoy a favourable reputation in the mind of consumers. Most safflower oil produced today reaches consumers as a refined, deodourised and bleache salad oil; as a principal ingredient in margarine; and in several forms of mayonnaise and salad dressings. A small percentages of the total oil produced, primarily prepress oil, is bottled and sold to consumers without any further refining, bleaching, or treatment of any kind other than filtration. In the United States and Europe, a small segment of the market wants an oil that has not been exposed to chemicals (in this case hexane). Bottled prepress oil generally has a short shelf life, perhaps of less than two weeks and once opened the oil needs to be refrigerated so that it does not develop strong odours.

In Mexico, safflower oil occupied a preferred status for many years in grocery stores catering to the affluent. When first produced in Mexico, a sizable portion of the safflower oil produced was used as an adulterant in sesame oil. Over time, safflower became the premier oil in the marketplace and *puro cartamo* would command a substantial premium. Safflower oil itself soon began to be adulterated with sunflower and other oils and eventually consumers became aware of this and switched loyalty to branded oils that were cheaper.

Recently the most successful programme has been operated by Saffola Grocery, product co., which markets so-called Grown without 'Pesticides Safflower Oil'. Saffola has chosen to use this method, because it wishes to establish its own definition for the 'purity' of its product, in contrast to organic oils, which are usually defined by government edict, subject to periodic change. The Grown without Pesticides (GWP) regime requires that the safflower is planted on land that has had no chemicals applied to it for at least 6 months and that shows no residue levels. The farmer is allowed to apply fertiliser but no planting seed fungicides (allowed in organic farming) or other chemicals. A thorough auditing scheme that includes inspection of the crop throughout the growth cycle and inspection of harvesters, trucks and storage facilities for cleanliness and lack of chemical sprays is employed. This is more rigorous that the standards employed by the organic industry, which works primarily on the honour system.

In India much oil is still sold by small mills that simply filter oil from the press and supply the product in small tins or even in the consumer's own vessel. Safflower production is by and larger a neighbourhood affair in India.

The European market consists of three areas. First, safflower oil is an ingredient in sunflower-based margarines, helping to maintain a guaranteed level of polyunsaturation. This market area may be in decline, because some manufactures lowered their polyunsaturated guarantee levels in early 1994, opting perhaps to feature low saturation or higher monounsaturation attributes in the future. Safflower oil also finds a small but dedicated audience because of its high level of unsaturation. A portion of this market prefers to use either unrefined process oil or a form that has been gently deodourised. The third market area is for safflower and particularly oleic safflower, oil that is used for blending with other oils. When safflower oil became more expensive than other oils, this market area virtually vanished. Much more rigorous and sophisticated control measures by government authorities have also restricted attempts at blending.

Safflower Meal

Safflower meal, the by-product of the production of safflower oil, contains all of the hull. The high fibre content of the hull limits its value. In California, safflower meal is employed primarily as an ingredient in dairy feeds; it is also used in beef cattle feed and to a limited extent in poultry mixes.

Safflower hulls find their best market when incorporated in safflower meal and none has been produced separately for many years in the United States, because most mills produced only two fraction when decorticating, 20 and 42 per cent portions. Today, the energy consumed in separating safflower meal fractions exceeds the premium that can be gained from the high protein fraction, so most mills confine themselves to offering as is meal of 25 per cent protein.

Numerous studies have shown safflower oil to be a good feed product for beef cattle, dairy cattle, poultry and lamb and is generally available at price levels that are similar to be lowest prices for alfalfa hay, gain screenings almond hulls and other low protein feeds.

Promising experiments have been done to produce protein flour or protein isolates from safflower meal. The USDA compared safflower protein isolate with isolate from soya and found the safflower product to be quite useful. The study also outline the cost of investment and production for the process envisioned.

QUALITY ASSESSMENT

Although most of the standard tests for measuring physical and chemical characteristics of a product work well for safflower seed and its products, some unique problems have arisen over the years.

Safflower Seed

During its growth cycle, a safflower head will respond to the amount of moisture available. If there is plenty of moisture, many of the individual seeds that have begun to form in the head will fill completely. If moisture is restricted or if a sudden trauma such as diseases or removal of water occurs, some of the seeds that have begun to fill will stop filling and others will not even begin. This results in a mixture in each safflower head of some seeds that are plump and completely filled and others that appear to be the same but that, on inspection, are empty or only partially filled. For the laboratories performing dockage tests on the thousands of samples representing each truckload delivered, it can be a daunting task to find which seeds are empty or partially filled. Originally, the dockage analysis method adopted by the state of California employed a series of hand screens, followed by winnowing through a Bates aspirator and hand picking of the resultant sample to find empty seeds that escaped the aspirator. This method was too slow and when used to measure samples containing high amount of empty hulls, as it often encountered in Great Plains safflower, up to 30 passes through an aspirator were required to find all empties.

Determining the oil content of safflower seed in the laboratory by solvent extraction is also more difficult than for other oilseeds because of the vast difference in texture of the hull compared to the kernel within. The hull must be cracked or all of the oil will not be extracted. But in cracking the seed, the kernel tends to mash as well as small amounts of oil can be lost in the process, a small amount is important when the sample contains only 5 grams of seeds. Since many people expressed dissatisfaction in safflower oil content analyses, PVO's control laboratory worked for along time to develop a better method than the standard AOCS procedure. This method of analysis is now part of the NIOP rules for safflower.

The NIOP also conducted extensive tests to develop methods for better sampling of safflower seed. Field run safflower seed is fairly difficult to sample. Although pure safflower seed is relatively smooth flowing, the seed delivered by a farmer can contain portions of stalks and stems; parts of the head that held seed; and leaves and other foreign material. Safflower seed, which has traveled over bumpy roads for 50–200 km inside a truck, for 3000 km in a railroad hopper car, or for 10000 km in the hold of a heavy grain carrier on its way to Japan, tends to stratify and unless the sampling device reaches all levels of the product, the sample is not representative. Japanese buyers, who were receiving 5000–15000 tonnes of safflower at a time, found that the oil content and dockage analyses performed at time of shipment did not reflect what the oil mills obtained as a final outturn in the milling of the same seeds. The NIOP adopted standard sampling and dividing procedures aimed at reducing variation in results and these procedures now are incorporated in their rules.

It is particularly important to remember that the sample used in analysing safflower oil contents must be first cleaned of all dockage (including empty hulls), unlike the common method of measuring sunflower oil contents, which is performed on seed containing admixture. This puts a premium on good sampling, good cleaning, good division of the sample and consistent performance of the analysis itself.

Safflower seed oil content can also be determined by the use of nuclear magnetic resonance (NMR) and today most plant breeders employ NMR techniques to measure their new lines. NMR techniques can be performed on only one half of a seed, so the other half can be planted if the results of the analysis are promising. In its earlier versions, processors tended to feel that NMR analysis produced oil content results that were slightly higher than found by standard solvent extraction analysis or than what was actually obtained at the oil mill. This has been disproven in the case of safflower seed and the industry has adopted NMR analyses in larger part to speed up paperwork. Because of the relatively small amount of safflower seed being measured for oil content annually, no one has taken the time to prove that present-day NMR procedures should be used to substitute for the standard AOCS procedure.

The first practical procedure for quickly determining if a truckload of seed is a linoleic or oleic variety. It involves squeezing a few seeds in a small hand-powdered press to obtain a few drops of oil. A drop of oil is placed on the glass prism cell of a hand-held refractometer. The refractive index has a straight-line relationship with the iodine value or fatty acid distribution of the oil, hence it is easy to determine if the seed in question meets an oleic standard or not, so long as a temperature correction is applied. Recently, it has become simpler to compare the unknown sample to a known oil standard, eliminating the need to apply a temperature correction. Temperatures corrections are difficult to measure accurately in the field under the time pressure of harvest.

Safflower Oil

Measurement of safflower oil's various chemical and physical characteristics is quite straight forward and only minor changes have occurred over 60 years in the rules governing the safflower trade. In 1990, the requirement for certification that safflower oil demonstrate a negative halphen test was dropped. The emergence of better and better GLC technology eliminated the need for a colour test of cottonseed oil adulteration.

Oxidation reactions of safflower oil and methods for following heat-generated changes in composition during deep-fat frying were studied in depth. *Peoria* analysed the head-space volatiles of safflower methyl esters and safflower oil, respectively, subjected to accelerated oxidation and found them to be the most reactive of all oils tested. One study showed vinyl-*n*-amylkelone to be the compound responsible for the generation of metallic off-flavours in oxidised safflower oil.

It has been found that tocopherols in linoleic safflower oil were more stable than tocopherols in oleic safflower oil. The odour studies showed that oleic safflower did well compared with all other oils used in the study. A broad study was conducted of the effects of various substances on the oxidation of safflower oil in deep frying of high temperature reactions in the presence of amino acids and of the effect of amino acids on emulsions, dried emulsions and chemical and organoleptic properties.

Safflower Meal

As already mentioned, safflower meal tends to stratify in storage so the principal problem in quality control is making sure a truly representative sample is obtained. A survey of the world collections for seeds high in lysine was undertaken and this work has been continued for both lysine and methionine. The work included studies of safflower steroids. Another study demonstrated how to remove deleterious glucosides from safflower meal and then demonstrated possibilities for removal of these and other negative factors to make safflower meal a more useful product. Others have isolated three conjugated serotin factors and their related phenolic factors.

STORAGE AND TRANSPORTATION

Safflower Seed

The most important element in the storage of safflower seed is anticipation of problems. If safflower seed buyers maintain contact with the suppliers and inspect the fields, most problems can be solved before they escalate. Safflower seed that is below 8 per cent moisture; is free of green weed, seeds, or trash; and has been brought to room temperature gradually is quite stable and can be stored indefinitely with no problem. Arranging for outside cleaning and/or drying before delivery to the oil mill and possible rejection, if the grower is unable to cope with weeds in the field, is much better than handling such problems on an emergency basis at time of delivery.

Because oleic and linoleic safflower seeds are virtually identical in appearance, extreme care is necessary to prevent inadvertent mixing. If a positive paper trial can be established for identifying fields of linoleic and oleic safflower from time of planting until delivery to the oil mill or storage point, much more confidence is possible when the sample are taken and the seed is checked for refractive index to verify positively the type delivered. Each load can be properly directed to its appropriate discharge point. If safflower seed is put into storage free of included green weeds or other plant material, with moisture level that has been brought to equilibrium at 8 per cent or under, it can be stored indefinitely.

At the time of the year that safflower is harvested, air temperatures are often in excess of 38°C. If safflower seed is being harvested from a particularly weedy field, the farmer must be careful to monitor the temperature of the seed in the truck or trailer into which the harvest is loaded. If a truck is forced to wait overnight in such conditions, the seed can begain heating to a dangerous level. Similarly, if seed is brought to a warehouse on a hot day, it is prudent for the warehouse to pull air through the seed pile or silo until the temperature of the entire mass reaches equilibrium. Monitoring the temperature within the seed mass by means of thermocouples is mandatory. If temperatures start to rise, air circulation can be started again until equilibrium is restored. If a hot spot cannot be controlled, it is prudent to reach that area as soon as possible by turning part of the seed column (in a tank or silo) or by digging into the side of the pile (if stored in a flat warehouse). This is not a time to move slowly; tear off the side of the building if that is what it takes to reach a hot spot. Safflower seed seems to be quite hygroscopic, absorbing moisture from wet material adjacent to it, which in turn causes germination to start and hence heating. Safflower seed that has heated will char to a dark brown to black mass and will develop strong odours that can permeate the storage unit if not separated quickly. This is not to portray safflower seed as a problem seed. Most seed is received in fine condition and stored without problems.

If a problem load is received, it is usually best to process it through a simple grain cleaner to remove the problem of foreign material or, if that is not possible, to pile the seed in a thin layer on a concrete slab for a few days until it stabilises. Safflower seed can be dried in forced air grain dryers. It is wise, however, to avoid letting temperatures in the dryer exceed 82°C to prevent safflower oil from scorching. If necessary, two passes through a dryer are to be preferred to one pass at high temperatures.

Safflower seed is normally not attacked by grain weevils, but these pests are often attracted to the forgein material contained in seed stored longer than 5 months. Again, careful monitoring is necessary if seed is destined for export. Treatment with approved fumigants in a timely fashion will avoid having the shipment graded 'weevily', which in turn makes the safflower seed samples grade and subject to rejection by the buyer.

Safflower seed is quite stable under carriage in ocean vessels. Although natural separation within the vessel's interior are preferred, safflower seed can be successfully separated from safflower bulk cargoes in the hold by a temporary separation built from layers of plywood, plastic and burlap.

Safflower Oil

Crude safflower oil can be stored and conveyed in normal black-iron vessels or pipelines without problems. The most important factor is to avoid exposure to air.

Safflower salad oil, because of its reactivity, should be stored and shipped in stainless equipment. It was found that safflower was more light stable than soyabean or cottonseed oils but still recommended packaging in brown bottles. It was found that α-tocopherol does little to inhibit photoxidation and it has been recommended that brown bottling be used to inhibit photoxidation and to reduce formation of free radicals, which generate objectional off-flavour. In today's world of environmental concerns, the only antioxidant used is the addition of citric acid during deodourisation. If safflower salad oil is to be stored, nitrogen sparging and blanketing is recommended.

Safflower Meal

As noted, expeller safflower meal is dangerous to store under any conditions. Solvent-extracted safflower meal is much more stable and can be stored safely so long as moisture levels are kept low in storage. Safflower meal can hold 10 per cent moisture under California weather conditions, but it is preferable to maintain it at 5–7 per cent moisture content if storage is for an extended period. Long term storage in small diameter tanks should be avoided, because safflower meal tends to bridge under such condition and become difficult to remove.

UNIQUE USES

Bird Feed

Safflower seed is employed in many wild bird feeding mixes because of its high oil content; it is also used to feed caged birds, particularly parakeets and parrots. Safflower seed is used in the feeding of tame ducks in China and Taiwan and for tame pigeons in Europe.

Ornamental

Safflower seeds and flowers were used in Egypt to make ornaments, wreaths, and jewellery. Horticultural use is limited because of safflower's prickly foliage, but is employed in some places for this reason as a protection for other plants against children and dogs. The spineless varieties are often used by Spanish and Portuguese farmers as garden flowers and as sources of food colouring. More recently, the red or orange flowers of spineless safflower have been harvested and then either dried or sold as a fresh display by florists. If dried carefully, safflower plants and flowers retain their colour and can be used for years in a static display.

Food Colouring

Safflower has been employed as a source of colour and flavour in cooking in every country where it has been grown. In the areas of Spain and Portugal where saffron is grown, fields of safflower exist that the local farmers do not like to acknowledge. Almost surely it was being used as an adulterant to the much more expensive saffron.

Dye

Another ancient use for safflower is to make dyes. The principal dye, carthamin, is a bright red colourant that is extracted from red-flowered plants after the yellow dye has been leached with water. Carthamin ($C_{43}H_{42}O_{22}$) imparts a scarlet red colour to silk and cotton. Fine examples of its durability from ancient times can be found in museums in Egypt, China and Japan. Safflower yellow pigment ($C_{16}H_{20}O_{11}$) must be removed to allow the red dye to be extracted; in earlier times the yellow was discarded.

Medicinal

Safflower seed, pollen, florets and oil have been used for medicinal purposes almost since cultivation began. Safflower oil, called oleum cnicium, was used as a milder substitute for castor oil and Pedanius Dioscorides, in *De Materia Medica.*

Several Arab texts dating to 1000–1500 mention safflower as an antidote for poisons and as an agent to induce sweating as a fever cure. Known as Kusumbha oil, safflower was regarded as a purgative in ancient Africa and India. Charred safflower oil is used in India as a treatment for sores and rheumatism and safflower seeds are employed as diuretics and tonics. China has a long history of use of all part of the safflower plant in medicinals combined with many herbal products. A tea is manufactured in Beijing and a polyamino acid nutrient is manufactured from safflower protein isolates.

USP safflower oil has been used as a carrier for penicillin. Data on use of safflower oil as a vehicle for injection of androgens have been published and it has been demonstrated that safflower oil could be used as dipersant for solutions of intravenous feedings of dextrose.

Cosmetic

Safflower florets were used to colour ceremonial ointments in Egyptian tombs. The Journal of American College of Toxicology published a report on the safety of safflower oil. The article concluded that safflower oil was safe as a cosmetic ingredient in the current practices of use. Oleic safflower is considered equally safe and safflower oil was found to be nonallergenic.

Powdered safflower florets have been employed since ancient times as a rouge in Egypt, China and Japan, where it was known as *beni*. The *Miselnz* of ancient Hebrew literature speaks of safflower's use as a rouge. It was mixed with French chalk and used as a rouge in old England. Soot from charred safflower plants was used until recently as the sources of *kohl*, a cosmetic used to darken the eyelids of Egyptian women. In India, safflower oil is used as an ingredient in soap and the preparation of *Macassar* hair oil.

Chapter 14

Peanut Oil

INTRODUCTION

The peanut (*Arachis hypogaea L.*) belongs to the sub-family of Papilionaceae. This plant is extremely plastic; its optimal temperatures are between 20 and 35°C, growth being inhibited below 10°C and above 45°C. Visible alterations appear only at temperatures below 5°C or above 50°C. As a matter of fact, the temperature affects the speed of growth: the vegetable cycle varies from 85 days (accelerated growth in tropical countries) to 180 days (cold areas). The resistance of the plant to drought is equally remarkable and the best output in rain-fed farming is obtained in the countries where the duration of the rainy season is very close to that of the vegetative cycle (from 400 mm in Northern Senegal to 1,400 mm or more in the forest zone). The physical factors of the soil also play their part particularly owing to the special mode of fructification of the peanut which requires an easy penetration of the gynophores and an easy uprooting of the crop. The most favourable conditions are, therefore, realised in loose, well drained and well structured soil with a rough or fine texture.

The peanut grows throughout the entire intertropical zone and to the north up to approximately 40° latitude, where in spite of the higher latitude the hot summers enable the plant to complete its cycle. The largest diversity of varieties is found in South America and, to a lesser degree, in the Philippines-Malaysia-Indonesia area and West Africa. The peanut is an annual crop. It's main stem and primary ramifications may grow to a length of 20 to 70 cm. Two subspecies are distinguished according to their sequential or alternate ramification.

The leaves of the peanut plant are alternate, with stipules. They have four leaflets. The flowers are aerial, yellow and papilionaceous with five petals. Self-fertilisation is practically guaranteed after which the ovary elongates to produce an organ called the gynophore which grows to a length of 15 cm or more and penetrates the soil tanks to positive geotropism. The fruit develops horizontally at the end of the gynophore at a depth of 2 to 7 cm below the surface of the soil.

The pod includes a cellulosic hull and one to four seeds. The straw-yellow pod is 2 to 8 cm long and the round or slightly elliptic seeds are covered by a seminal tegument (testa) of varying colour (white, pink, red or even purplish).

The seed weighs between 0.2 to 2 gram and is the more important element on account of its richness in oil (about 50 per cent) and proteins (25 per cent).

On account of its composition in fatty acids making it perfectly suitable for frying and seasoning, peanut oil enjoys a higher price compared to other main edible oil (30–40 per cent costlier on an average as compared to sun flower oil or soyabean oil). The protein content and composition of peanut oil meal are much appreciated in cattle feed. The oil meal guaranteed as free from aflatoxine competes with the soyabean oil meal. Edible peanuts are sold at a high price, about two times that of the seeds meant for oil-mills and often at a still higher price for the varieties of best quality.

PEANUT OIL

It is a nondrying vegetable oil [defined as oils whose iodine values (IV) are <125] and has a high proportion of unsaturated fatty acids: oleic acid (50–65 per cent) and linoleic acid (18–30 per cent). Stearic, arachidic, behenic and lignoceric acids account for 10–12 per cent of the total fatty acids. Arachidic acid is the characteristic component of peanut oil.

Peanut oil is used mainly for edible purposes in the preparation of shortening, margarines and mayonnaise, as a cooking and frying oil and as a salad oil. Outside of North America, production of peanut oil is the major use of edible peanuts and the oil may be hydrogenated into vanaspati, an Indian analogue to margarine. Although there is no US standard of identity *per se*, peanut oil must be suitable for human consumption and conform to the identity characteristics defined by the Codex Alimentarius commission. Because of a high smoke point (229.4°C), refined arachis oil is often used in deep-fat frying, but hydrolysis of glycerides into free fatty acids (FFA) during frying leads to a decrease in smoke point. For both frying and as a salad oil, peanut oil is considered to be superior to soyabean oil and develops fewer flavour defects with long-term use. Peanut oil is considered to be superior in the manufacture of pourable dressings because of its ability to hold solids in suspension longer. However, because peanut oil solidifies at 0–3°C, it does not meet the definition for salad oil, which must remain clear after 5.5 hr. of immersion in an ice bath at 0°C. A nonedible use of peanut oil is as a diesel fuel in specifically modified engines, but it is more expensive than conventional diesel fuel and has the added drawbacks of lower heating value, greater surface tension, greater viscosity and greater density.

It is obtained from seeds by pressure or pressure-extraction. The oil mills often procure the peanuts in hulls; 100 kg of peanuts yield an average of 30 kg of shells and 70 kg seeds which in turn yield 34 kg oil (pressure-extraction) and 36 kg oil meal (cake).

The speciality of peanut oil is in its composition which varies widely according to the botanical types of nuts, but also the variety within the same type and the environment for a given variety.

CHEMICAL AND PHYSICAL CHARACTERISTICS

Table 14.1 gives various chemical and physical characteristics for peanut oil.

Acetyl Value

The acetyl value is the number of milligrams of KOH required to neutralise the acetic acid produced by the hydrolysis of 1 gram of acetylated fat and is a measure of the free hydroxyl groups present in the oil. The acetyl number of peanut oil is lower than other vegetable oils but higher than coconut oil, palm oil and the animal fats and oils.

Acid Value

The acid value is the number of milligrams of KOH required to saponify the free fatty acids in 1 gram of the fat or to neutralise 1 gram of fatty acids. For peanut oil, the acid value is approximately the same as for lard or cottonseed, rape and colza oils. The low acetyl number is attributed to sterols and other unsaponifiable components.

Table 14.1. Chemical and physical characteristics of peanut oil.

Characteristic	*Value*
Acetyl value	8.5–9.5
Acid value (maximum)	0.30 mg KOH/g oil
Acid value (nonvirgin oil, maximum)	0.6 mg KOH/g oil
Acid value (virgin oil, maximum)	4 mg KOH/g oil
Calculated gums (phosphatides × 32)	0.35%
Colour (Gardner, maximum)	4
Colour (Lovibond, maximum)	Yellow 16–25; 2.0 red
Colour (visual)	Light yellow
Ester number	188–192
Flavour and odour	Bland
Free fatty acid (as oleic acid, maximum)	0.05 %
Heat of fusion (unhydrogenated)	21.7 cal/g
Heating value	40.3 mJ/kg
Hehner value	95–96
Insoluble impurities	0.43%
Iodine no. (Wijs)	–26
Kinematic viscosity (21.1°C)	70.7 cSt
Melting point	0–3°C
Melting point of the fatty acids	22–30°C
Moisture (% maximum)	0.06
Moisture and volatiles	0.23%
Peroxide value (maximum)	10 meq peroxides oxygen/kg oil
Polenske value	0.5
Refractive index (n_D 40°C)	1.46–1.465
Reichert-Meisl value	0.5
Saponification number	188–195
Smoke point (minimum)	229.4°C
Specific gravity (21°C)	0.915
Specific heat (C_p, liquid oil)	0.4914 + 0.004 T (°C)
Surface tension	35.6 mN/m
Thiocyanogen value	0.5
Titer	26–32°C
Unsaponifiable lipids	0.40%

The acid value (percent free fatty acids) is a measure of the extent of hydrolysis of triglycerides that has occurred in a fat and is generally accompanied by a rancid odour and lowering of the smoke, flash and fire points. Peanuts contain a low level of phosphatides ('gums'), reflected by the low value for calculated gums. In peanuts, the phosphatides have a higher palpitate content than the triglycerides.

Colour

The light yellow colour of peanut oil is due to β-carotene and lutelin and becomes lighter as the peanuts mature. Although oil colour may be used to assess maturity other methods are preferred because many factors influence oil colour. Colour measurement is frequently done by visual comparison under a CIE standard light source. Two measurements are commonly used in the paint and coating industry. The Gardner colour is determined using a Gardner-delta colour comparator, which has a scale between 1 and 18. Three calibrated glasses (red, yellow and blue) are used in Lovibond instruments; the Lovibond colour gives the thickness of these three glasses, which match the colour of an unknown sample under a standard illuminant. Both instruments are widely used to measure colours for materials that transmit light, but they are being supplanted by less subjective automated spectrophotometry methods.

Ester Number

The ester number is the number of milligrams of KOH required to saponify the neutral oil in 1gram of the sample exclusive of that required to neutralise the free fatty acids. Because peanut oil is mainly triglyceride, this value represents the percentage of glycerine yielded on saponification

Crude peanut oil has a nutlike flavour, which is removed by refining American oil chemists' society (AOCS) flavour quality ballots for oil quality have been described and incorporate separate ballots for grading and flavour intensity, which describe the flavour characteristics rather than the suspected cause or process. Standard flavour descriptors for peanuts have been given.

Free Fatty Acids

The percent free fatty acids in peanut oil varies between 0.02 and 0.6 per cent. Lipase hydrolysis of triacylglycerols into free fatty acids and glycerol occurs before germination and during adverse storage. Consequently, high FFA values indicate poor handling, immaturity, mould growth, or other factors that lead to triglyceride hydrolysis.

Heat of Fusion

The heat of fusion, or latent heat, is the quantity of heat required to change 1 gram of solid to a liquid with no temperature change. This latent heat increases with increasing molecular weight. The heating value of heat of combustion is the amount of heat produced by the combustion of 1 kg of oil. The heat of combustion increases with the chain lenght of the fatty acids for both monoglycerides and triglycerides.

Hehner Value

Hehner value expresses the percentage of water insoluble fatty acids plus unsaponifiable matter in an oil or fat. This method is of greatest value in testing butterfat purity. Like most vegetable oils, peanut oil has a higher Hehner value than butterfat. Lipids with soluble fatty acids will have lower Hehner values than those with a greater proportion of high molecular weight fatty acids. The iodine value (IV), or Wijs iodine number, is the number or grams of iodine absorbed under standard conditions by 100 gram of fat and is an official method of the AOCS and AOAC. Peanut oil's (IV) of 82–107 indicates it is more saturated than corn, cottonseed, or linseed oil but is less saturated than coconut, palm, or butter oil.

Kinematic Viscosity

The kinematic viscosity (cSt) is the dynamic viscosity (poise) divided by the fluid density. Peanut oil is a non-Newtonian fluid whose dynamic viscosity is a function of the rate of imposed shear **and** temperature. In contrast to safflower, soyabean, corn and olive oils, at refrigeration temperatures (0–3°C) peanut oil sets to a gel, which is not filterable during winterisation.

Peroxide Value

The peroxide value measures a fat's reactive oxygen content in terms of milliequivalents per 1000 gram fat, following AOCS method Cd 8–53 or AOAC method 965.33. Elevated peoxide values indicate that lipid oxidation has taken place. Free fatty acids can serve as substrates for lipoxygenase and peroxidase, both of which are inactivated during heating. Once the cell structure is disrupted by pressing, lipoxygenase reacts with linoleic, linolenic, or arachidonic acid (either as the free acid, triglyceride, or methyl or ethyl esters) to form hydroperoxides. Hydroperoxides can undergo further decomposition to form pentanal and hexanal, both of which are detectable by headspace analysis. These oxidation products are correlated with reduced flavour scores and cardboard or painty flavour defects. Although the peroxide value is used as an indicator of oil oxidation, the *Kreis* test was found to be a better predictor of oxidation than the peroxide value for peanut oil.

Polenske Value and Reichert-Meissel Values

Polenske value and Reichert-Meissel values are indicators of steamvolatile water-soluble (butyric, caproic and caprylic) or insoluble (capric and lauric) fatty acids, respectively. These tests were designed for detecting low molecular weight fatty acids in oil and adulteration in butterfat. Butterfat has Reichert-Meissel value of 17.0–34.5.

Thiocyanogen Value

The thiocyanogen value (TV) is a measure of the amount of this reagent absorbed by 1 gram of fat. GLC methods have largely displaced this method for determining the content of oleic, linoleic and linolenic acids when IV's are determined.

Melting Point and Titer Value

For soap making, the melting point of the fatty acids (titer value) is an important parameter and is more characteristic than the melting point. The titer value for peanut oil is lower than that for cottonseed oil (30–37°C), cocoa butter and animal fats and oils but is higher than that for corn (14–20°C) and/or linseed oil (19–21°C).

Unsaponifiable Matter

The unsaponifiable matter is largely sterols and methylsterols. Detailed compositional analysis of the unsaponifiable fraction are discussed later in this chapter.

Gas Chromatography

Before the development of gas chromatography and high pressure liquid chromatography, the presence of peanut oil (as an olive oil adulterant) was detected because peanut oil contains about 5 per cent arachidic acid as a glyceride. Arachidic acid is insoluble in cold alcohol unlike stearic and palmitic acids. Methods for the detection of arachidic acid include the Bellier, Evers, Evers-Bellier and Renard tests. Arachidic acid is predominant in the lecithin and cephalic fractions of peanut oil. Detection methods for toxic oils as an adulterant in edible oils such as peanut oil have been reviewed.

Oil Content of Peanuts

Oil content of peanuts ranges from 36–56 per cent and increases with direct fertilisation, but it is generally independent of market type and growth habit. Small-seeded Spanish market types tend to have a higher oil content than the large-seeded Virginia types. Oil content and lipid composition, however, are affected by environment, insecticide, growth regulators and herbicide application; maturity; soil temperature and variety. Oil content peaks between 10 and 12 weeks after pegging and then declines as these lipids are used as a substrate for respiration. High temperature drying (curing) at a moisture content of 42–57 per cent reduces oxidative stability.

Composition of Peanuts

Peanut oil is composed of mixed glycerides of approximately 80 per cent unsaturated and 20 per cent saturated fatty acids. In mature peanuts, the oil is 96 per cent triaclyglycerol with the main fatty acids being palmitic, oleic and linoleic. During germination, these triacylglycerides are hydrolysed by lipases to glycerol and fatty acids, which are oxidised to provide energy for the newly emerging plant. These lipid fractions are usually separated by column chromatography and TLC. A detailed examination of the fatty acid composition of peanut oil is shown in Table 14.2. These values differ from *Osman* and co-workers who did not isolate a phosphatidylserine fraction, possible due to varietal differences. Because a portion of the phosphatides (cephalins, phosphatidylserine and lecithins) is complexed. Storage of peanuts at high moisture content (>9 per cent) reduces oil stability and promotes lipid oxidation. A similar antioxidant effect for phosphatides in the oxidation of fish oil has been noted.

Feeding studies have shown that peanut oil is more atherogenic than its iodine value would indicate. Because chemical treatment to randomise peanut fatty acid triglyceride structure reduces atherogenicity, triglyceride structure is believed to be involved and was found to vary with the production source. Table 14.3 gives triglyceride structures for major peanut varieties.

Table 14.2. Fatty acid composition of triglycerides and phosphatides of cold pressed peanut oil.

	Fatty acid methyl esters (% of total)			
Fatty acid	*Triglycerides*	*Phosphatidylethanolamines (Cephalins)*	*Phosphatidylserines*	*Phosphatidylcholines (Lecithins)*
C16:0	8.1	24.7	33.9	12.9
C16:1	Trace	0.1	0.2	Trace
C17:0	Trace	0.2	–	Trace
C18:0	1.5	2.6	4.7	2.8
C18:1	49.9	39.5	30.9	47.0
C18:2	35.4	28.2	27.5	35.6
C18:3	Trace	0.1	Trace	Trace
C20:0	1.1	0.1	0.2	0.3
C20:1	0.9	1	0.2	0.4
C22:0	2.1	0.5	0.5	0.3
C22:1	–	–	–	0.1
C24:0	1.0	3.0	2.0	0.5
C24:1	–	–	–	0.1

Table 14.3. Stereospecific analysis of triacylglycerols in peanut oil.

Variety		*Fatty acid (mole%)*							
		C16:0	*C18:0*	*C18:1*	*C18:2*	*C20:0*	*C20:1*	*C22:0*	*C24:0*
Wild types									
A. monticola	*sn*–1	23.09	5.43	34.14	35.62		0.59	0.46	0.62
	sn–2	5.23	1.54	36.02	55.27			0.41	0.5
	sn–3	8.22	2.43	43.69	25.43	4.76	3.58	9.02	4.05
A. nambyquarae	*sn*–1	18.7	3.51	48.16	24.64		0.96	2.88	0.66
	sn–2	5.38	1.52	55.22	35.34			0.59	0.75
	sn–3	7.63	0.52	46.12	28.74	3.52	4.12	7	4.64
Hirsuita									
PI 241633	*sn*–1	18.12	3.98	51.72	23.86		0.98	0.79	0.47
	sn–2	6.8	1.87	54.06	35.83			0.52	0.4
	sn–3	4.42	1.17	55.46	18.04	3.74	3.28	10.4	5.57
PI 158850	*sn*–1	19.17	3.92	39.12	35.37		0.72	0.58	0.48
	sn–2	5.82	2.56	39.54	49.32			0.5	0.42
	sn–3	6.54	1.28	48.14	23.27	4.35	3.26	10.97	4.91
Virginia									
Florigiant	*sn*–1	20.1	4.9	50.7	22.6	0.5	0.7	0.4	0.3
	sn–2	2.2	0.7	51.5	45.3	0.1	0.3	0.1	
	sn–3	10.3	3.2	57.2	14	4.8	2	5.3	3
Florunner	*sn*–1	20.7	3.5	49.5	24.4	0.3	0.7	0.5	0.5
	sn–2	2.1	0.6	47.8	48.8	0.1	0.4	0.1	0.1
	sn–3	11.4	2.3	55.5	14.1	4.4	2.3	6.5	3.4
Greg coll. 486	*sn*–1	17.75	4.74	49.84	24.96		0.78	0.46	0.45
	sn–2	4.63	1.34	54.3	38.24			0.39	0.44
	sn–3	7.15	2.85	49.3	19.74	4.96	3.57	10.68	4.96
PI 313946	*sn*–1	18.51	4.7	45.85	27.65		0.94	1.18	1.29
	sn–2	3.52	0.93	45.41	48.19			0.31	0.44
	sn–3	6.94	2.41	52.11	19.32	4.02	3.4	8.54	5.02

(Contd..)

Table 14.3. Stereospecific analysis of triacylglycerols in peanut oil. *(Contd.)*

Variety		Fatty acid (mole%)							
		C16:0	*C18:0*	*C18:1*	*C18:2*	*C20:0*	*C20:1*	*C22:0*	*C24:0*
Spanish									
Argentine	*sn*–1	20.73	5.58	41.22	30.27		0.52	1.12	0.56
	sn–2	3.82	0.90	42.88	50.04			0.30	0.36
	sn–3	10.66	4.48	50.48	16.78	5.14	2.12	9.29	3.49
PI 25744	*sn*–1	19.37	6.11	41.30	31.82		0.58	0.48	0.46
	sn–2	4.50	1.32	45.00	47.68			0.39	0.41
	sn–3	6.05	4.98	44.90	21.20	7.34	3.25	10.21	4.88
Spanhoma	*sn*–1	19.38	4.88	42.96	30.13		0.69	0.89	0.83
	sn–2	4.63	1.32	47.14	44.67			0.37	0.48
	sn–3	10.59	3.72	46.81	21.83	4.66	2.04	9.36	2.98
Starr	*sn*–1	24.20	4.90	40.40	28.40	1.80	0.60	0.70	0.30
	sn–2	2.40	0.80	39.50	56.90	0.40	0.20	0.10	
	sn–3	16.00	4.20	50.00	13.80	0.10	2.40	7.30	1.80
Tennessee Red	*sn*–1	19.68	5.53	44.28	27.50		0.79	1.43	1.08
	sn–2	4.59	1.39	46.06	46.19			0.40	0.36
	sn–3	10.09	3.25	42.67	26.14	4.77	1.87	9.36	3.37
Valencia									
Zaire A65	*sn*–1	22.00	2.00	39.60	33.20	1.30	0.60	0.80	0.50
	sn–2	0.90	0.20	30.30	68.30		0.30		
	sn–3	16.70	5.70	47.50	16.20	4.20	2.40	6.10	1.20
Zaire P43	*sn*–1	17.90	7.70	40.01	28.90	1.60	0.60	2.50	0.70
	sn–2	1.20	0.20	36.50	62.10				
	sn–3	14.50	7.00	44.20	17.10	6.20	2.50	6.90	1.60
Average (± SD)	*sn*–1	19.96 ± 1.90	4.76 ± 1.32	43.93 ± 5.17	28.62 ± 4.13	1.10 ± 0.67	0.72 ± 0.15	1.01 ± 0.75	0.61 ± 0.27
	sn–2	3.85 ± 1.75	1.15 ± 0.63	44.75 ± 7.38	48.81 ± 9.12	0.20 ± 0.17	0.30 ± 0.08	19.96 ± 1.90	0.42 ± 0.15
	sn–3	9.81 ± 3.67	3.30 ± 1.77	48.94 ± 4.54	19.71 ± 1.54	4.46 ± 1.54	2.81 ± 0.71	8.46 ± 1.81	3.66 ± 1.37

Peanut triglycerides tend to contain saturated fatty acids and fatty acids with shorter chain lengths predominantly in the *sn*-1 position. The *sn*-2 position consists almost entirely of unsaturated fatty acids. As part of the peanut breeding programmes efforts were made to screen and identify peanuts whose triglyceride structure was less atherogenic. Although varietal differences were found, location and location by year effects had more pronounced effects on fatty acid composition and breeding programmes were modified to concentrate on developing varieties with a high oleic acid content and lower O:L ratio. In soyabeans, as the bean matures, there is a rapid synthesis of triglycerides, which is accompanied by a increase in linoleic and oleic acid. Similar results have been found for peanuts. Abbreviations of fatty acids used in the discussion of triglycerides are given below.

P, palmitic (C16:0)	Po, palmitoleic (C16:1)
S, stearic (C18:0)	O, oleic (C18:1)
A, arachidic (C20:0)	E, eicosenoic (C20:1)
B, behenic (C22:0)	L, linoleic (C18:2)
Lg, lignoceric (C24:0)	Ln, linolenic (C18:3)
C, cerotic (C26:0)	

TRIGLYCERIDE STRUCTURE OF PEANUT VARIETIES

Table 14.3 gives triglyceride structures for 15 major peanut varieties. Oleic acid is present at high concentrations at all three positions, while linoleic acid is found primarily in the *sn*–2 position. The shorter chain length saturated fatty acids (palmitic and stearic) are mainly located at the *sn*–1 position and less at the *sn*–3 position, while the longer chain length saturated fatty acids (arachidic, behenic and lignoceric) are located on the *sn*–3 position. Eicosenoic acid is also more frequently located at the *sn*–3 position. In a later study, using HPLC with a proprionitrile mobile phase, C18 column and temperature programming, commercially available peanut oil was separated into diglycerides, polar components and 36 triglycerides.

Six triglycerides (OOL, OOO, OLL, PLO, PLL and BLL) comprise more than 75 per cent of the triglycerides in peanuts. Table 14.4 shows the percentage of each of the 36 triglycerides in commercial peanut oil.

Table 14.4. Percentage of triglycerides in commercial peanut oil.

TG	*TG*	*Percent Area*	*TG*	*TG*	*Percent Area*	*TG*	*TG*	*Percent Area*
OOL	–	19.61	LLL	–	2.33	POP	–	0.97
OOO	SLO	15.26	OOB	PLLg	2.31	ELL	–	0.73
OLL	–	13.23	ELO	–	2.08	POLg	SOB	0.69
PLO	–	12.19	SOO	PLA	1.97	POB	–	0.40
						OOC	–	0.40
POO	PLS	7.45	PLB	OOA	1.62	OLC	–	0.34
PLL	–	4.69	OOLg	–	1.26	POC[a]	SOLg[a]	0.26
BLL	–	2.95	OLLg	–	1.14	POA[a]	SOS[a],LLC	0.25
OLB	–	2.60	OLA	–	1.12	POS	–	0.21
PLP	–	2.36	LLLg	–	1.07	–	–	–

[a] POA with SOS and POC with SOLg eluted together.

MAJOR ENVIRONMENTAL FACTORS THAT INFLENCE LIPID CONTENT

Because peanut grown in cooler climates or during periods of cooler temperatures in the latter weeks of the growing season tend to have a greater degree of unsaturation, a lower O:L ratio and a shorter shelf life, the emphasis of peanut breeding in the northern areas of peanut production (Virginia and North Carolina) has been to develop varieties with early maturation, increased storage stability, a higher O:L ratio and low iodine value.

As the peanut matures, there is an increase in lipid content, which is accompanied by a change in fatty acid composition. It has been found that mature seed to contain more stearic and oleic acids and less arachidic, behenic and lignoceric acids than immature seed. The oleic and linoleic acid content and the O:L ratio increase as peanuts mature. A variety of methods are used to assess maturity and peanuts are harvested when there is the greatest proportion of sound mature seeds. However, because of the indeterminate fruiting habit of peanuts, at any given harvest date, peanuts of the same size may differ in maturity. Cooler soil temperatures result in size distributions containing more large peanuts.

Major environmental factors that influence lipid content or metabolism include light, temperature, water stress, soil constituents, atmospheric constituents, herbicides and insecticides, physical damage and pest attack. Significant studies have been conducted on the environmental effects on the fatty acid composition. These studies found that it is possible to obtain oil with increased stability and a high O:L ratio with a low IV by avoiding stress due to cold, drought, or dryland farming and by selecting soils with more basic pH and increased iron while avoiding overfertilisation. Reductions in O:L ratio also generally occur when peanuts are sprayed with growth regulators. There is an alteration in fatty acid composition as well as a decreased eicosenoic acid content, which are accompanied by an increased oil yield. Herbicides have a slight effect on oleic and linoleic acid content.

MODIFICATION OF FATTY ACID COMPOSITION

Modification of fatty acid composition has been a particular goal of breeding programmes because oil quality, fatty acid composition and protein composition are highly heritable. Iodine values may be measured in the field by measuring the oils's refractive index; a linear relationship was found between the specific gravity of peanut kernels and oil content. Methods of peanut improvement through breeding programmes have been discussed in detail.

Although most varieties have 36–67 per cent oleic acid, 15–43 per cent linoleic acid and O:L values between 1.19 and 4.46, two peanut lines with 80 per cent oleic acid and 2 per cent linoleic acid (O:L = 40) have been identified and introduced into breeding programmes. Cultivars having these high O:L ratios do not have significant differences in oil content and are not accompanied by differences in colour, aroma, flavour, or texture. Breeding programmes have also concentrated on drought-resistant cultivars because mid to late season drought is most detrimental to pod yield.

On sterol analysis in vegetable oils, it has been found that peanut oil contains 0.09 to 0.3 per cent sterol (as free and fatty acid esters), which depend on the analytical method. Refining could remove 61.1 per cent of the sterol content. Detailed analysis of Nigerian peanut oil (IV = 90.7) found that peanut oil contained 0.4 per cent unsaponifiable lipids, which were divided into four fractions by TLC. For this unsaponifiable fraction, 60 per cent was sterols and 27 per cent was hydrocarbons, aliphatic alcohols, etc. Two components (β-sitosterol and campesterol) comprised 64 and 15 per cent of this sterol fraction. The major triterpene alcohols included 24–methylenecycloartanol (46 per cent) and cycloartanol (33 per cent). Using TLC, *Gaydou* separated the unsaponifiable lipids of Madagascar peanut oil (1.09 per cent of the oil) into sterol, methylsterol and triperpene alcohols. β-Sitosterol and campesterol composed 72 and 16.6 per cent of the sterol fraction. In the 4–α methylsterol fraction,. four components comprised > 10 per cent of the fraction: citrostadienol (20.5 per cent), obtusifoliol (17.2 per cent), gramstisterol (14.7 per cent) and cycloeucalenol (14.4 per cent). The major components in the triterpene alcohol fraction were 14–methylene-cycloeucalenol (42 per cent), cycloartenol (22.4 per cent), cycloartanol (15 per cent) and lupeol (10.1 per cent). In studies on free and bound sterols in vegetable oils, it was found that campesterol, stigmasterol and β-sitosterol were evenly distributed within the free sterols, sterol esters and sterol glycosides and in sterol-containing lipoproteins.

Tocopherols are natural antioxidant present in peanut oils. Tocopherol content differs according to variety and production location. Tocopherol content was found to vary between 268 and 510 ppm but was not correlated with fatty acid composition, fat content, or kernel weight. Tocopherol content also varied with production location and was generally higher in peanuts grown in the United States and in runner peanuts. After refining, the tocopherol content decreases. Commercial peanut oil has been found to have 21 and 15 ppm α-and γ-tocopherols Composition of peanut oil unsaponifiable fraction Africa/ South America is shown in Table 14.5.

Table. 14.5. Composition of peanut oil unsaponifiable fraction.

Description		*Unsaponifiable fraction (Africa); 0.6–1.0 %*	
Sterols (mg/100g)	190–330	Hydrocarbons (mg/100g)	70–110
		Squalene	40–70
		Tocopherols (mg/100 gram)	28–73
Sterol composition (% total sterols)		Tocopherol composition (% total tocopherols)	
Cholesterol	< 0.7	α-Tocopherol	42–59
Campesterol	12–15	β-Tocopherol	2–5
Stigmasterol	8–11	γ-Tocopherol	32–52
β-Sitosterol	58–66	δ-Tocopherol	2–4
Δ 5–Avenasterol	9–14		
Δ 7–Stigmasterol	< 1		
Δ 7–Avenasterol	1–2		
Description		*Unsaponifiable fraction (South America); 0.6–0.9 %*	
Sterols (mg/100g)	260–320	Hydrocarbons (mg/100g)	70–110
		Squalene	40–70
		Tocopherols (mg/100 gram)	30–100
Sterol composition (% total sterols)		Tocopherol composition (% total tocopherols)	
Cholesterol	< 0.4	α-Tocopherol	47–65
Campesterol	16–19	β-Tocopherol	1–3
Stigmasterol	6–9	γ-Tocopherol	30–50
β-Sitosterol	61–67	δ-Tocopherol	2–3
Δ 5–Avenasterol	7–9		
Δ 7–Stigmasterol	< 1		
Δ 7–Avenasterol	< 1		

Other Peanut Products

Oil meal

A by-product of oil milling, it contains 48–50 per cent of protein fractions and is now utilised as cattle feed, it is appreciated in this respect for its composition in amino acids and its light cellulose content. Peanut flour is used for human consumption in some producing countries, particularly in the USA and Asia where it is incorporated in a wide range of preparations.

Peanuts for direct consumption

This includes peanuts which are directly consumed in various forms starting with the complete seed: roasted seeds, peanut butter, candies, etc. Most of the US production is consumed in this form, as in numerous producing countries where only a fraction of peanuts is processed.

Tops

Tops are dried after harvesting and used as animal feed (0.4 fodder unit/kg). The hulls are mainly utilised as fuel for steam and energy production in oil mills.

GENERIC TECHNOLOGY FOR ASSURED BIOCONTROL OF GROUNDNUT INFECTIONS LEADING TO ITS YIELD IMPROVEMENT

Importance of Groundnut Crop

Groundnut *(Arachis hypogaea L.)*, commonly known as peanut, is a member of Leguminosae family. It is an important oil and food crop. Since mature seeds possess a delicate flavour and high calorific value, it is mainly consumed by human beings as salted and roasted seeds. Naturally, it is termed as poor man's *badam.*

The seed is also crushed to produce peanut butter, oil and oil-cake. Its high calorific value is due to 22–32 per cent protein, 38–39 per cent carbohydrates and 42–52 per cent oil content. Out of total groundnut produced in India, over 80 per cent is used for oil extraction, 12 per cent for seed purposes, 6–7 per cent for domestic consumption and about 1–2 per cent in shelled form for export.

Heavy Pathogen Infestation Affecting Groundnut Production

In spite of the above-mentioned economic importance, every year, severe economic losses to groundnut crop are caused due to soil-borne fungal pathogens *viz, Sclerotium rolfsii, Fusarium oxysporum, Aspergillus flavus etc.*

Therefore, its export to the extent of 37–45 per cent two decades before, has reduced to mere 1–2 per cent. To control them, excessive use of chemical fungicides has been found to create food contamination and environmental risks, posing hazards to human health.

Suggested Eco-Friendly Remedy

It involves an application of a formulation of siderophoregenic fluorescent pseudomonads, which are known to produce a number of novel secondary metabolites such as siderophores, pyoluteroin, pyrrolnitrin, pseudomonic acid, oosmycin A, sorbisitin, IAA, cytokinins etc. while:

1. Some of them posses anti-fungal properties against a wide range of pathogens.
2. IAA and cytokinins, lead to improvement in groundnut growth.
3. Siderophores provide biocontrol against a group of pathogens.

In the light of these facts, the suggested siderophore-based formulation, christined as 'sideroplus' derived from siderophoregenic Pseudomonas, controls phytopathogenic fungi of groundnut. This treatment enables healthy growth of plants, leaving no residue on groundnut crop or soil and its foliage which is used as a fodder for cattle by virtue of its biodegradability. Therefore, it is eco-friendly

Cost-effective Biocontrol Technology

Besides promoting vibrant growth of groundnut crop, application of siderophore producing pseudomonads controls phytopathogens most affecting its productivity. This treatment ensures increase in the yield of groundnut upto 15–17 per cent, more nutritious fodder for cattle and fertile soil for the next cropping and thereby qualifies to be cost-effective. These facts have been borne out from laboratory tests and field trials.

Laboratory Demonstration of Anti-fungal Activity

Bioefficacy of sideroplus is measured by bore well method by killing the phytopathogens in 48 hr. Contrary, untreated (control) set permits continued growth of pathogens, certifying the effective biocontrol power of 'sideroplus'.

Bioefficacy Profile at Farm Scale Trials

Bioefficacy studies were undertaken to demonstrate the effect of sideroplus on the growth and productivity of groundnut crop. A 10 per cent jaggery solution (100 ml) and 150 gram formulation was used for of coating (bacterisation) of 10 kg groundnut seeds, which were subsequently air dried and sown. Its effects was observed from the height of groundnut plant, root ramification, foliage vigour, more chlorophyll, healthy pods and higher yield. Several field trials on acre scale ascertained reproducibility in its performance.

Complementary Interaction (s) with Rhizobacteria

Bacteria found in the root zone (Rhizobacteria) of plants maintain soil health. The sideroplus application was not deleterious to Rhizobacteria. It was found to accord complementary interaction with *Bradyrhizobium Japonicum* NCIM 2746, *Rhizobium japonicum* IARI 102 (nitrogen fixers for legumes), *Azotobacter vinelandii* (nitrogen fixer for non-legumes) and *Pseudomonas putida* NCIM 2847 (phosphate solubiliser).

Salient Features of Biocontrol Package

Its salient features are:

1. Anti -fungal activity against phytopathogens without adverse effect on rhizobacterial population.
2. About 150 gram formulation is optimum dose for 10 kg of groundnut seeds for controlling infection.
3. Effective control of phytopathogens during complete duration of crop (90–105 days), as compared to only 10 days protection with chemical fungicides.
4. Eliminates necessity of application of fungicide/pesticide for crop protection.
5. Improved shelf-life of formulation upto 9 months (10^9 cfu/gm).
6. Growth promotion of complementary *Azotobacter* and *Rhizobium* sp.
7. Protects plant from iron chlorosis and reduces application of iron.

8. Improved per cent germination, root ramification, nodulation, height foliage and chlorophyll content.
9. Allows peanut crop to tolerate moisture stress upto 8–10 days.
10. Improved yield in bioefficacy trials on six varieties of groundnut in two seasons and at different locations demonstrated its resilience.
11. Improves yield (15–17 per cent) and oil content (1–2 per cent) of groundnut, irrespective of variety and season.

Thus, being an eco-friendly, cost-effective and simple alternative approach for arresting soil-borne fungal pathogens of groundnut, thereby improving its yield, it is hoped to be acceptable to the farming community.

Chapter 15

Sulphated and Sulphonated Oils

INTRODUCTION

Sulphated and sulphonated oils are among the direct derivatives of fats, used for non-edible purposes and are strongly hydrophilic. They can be easily emulsified, in the form of water in oil as also oil in water and belongs to the category of anionic compounds.

Sulphonation is the formation of a sulphonic acid, i.e. a compound containing the —SO_2OH group. Common sulphonating agents are: concentrated sulphuric acid, fuming sulphuric acid, sulphur trioxide, alkali disulphates, pyrosulphates, chlorosulphonic acid and a mixture of manganese dioxide and sulphurous acid. Sulphonated oils are vegetable or animal oils which have been treated with sulphuric acid and the excess of acid washed out and the oil neutralised with a small amount of caustic soda and ammonia.

Most of the fats, whose industrial production is sufficient and cost price is compatible with the desired uses, are either sulphated or are sulphonated in order to be of use in various industries (metallurgy, leather, textile, paper industry, maintenance products, cosmetics, biotechnology, chemistry, etc.), not to mention all industries, as manufacturing auxiliaries and/or for waste treatment.

Thus, the market offers sulphated and sulphonated oils of animal (mammals, fish) and plant origin of any nature, as well as some compounds obtained by their hydrolysis: fatty acids and fatty alcohols.

Sulphated and sulphonated oils are remarkable due to the variety of base raw material they provide and the number of substances prepared from the same fat. Since these oils are soluble (emulsifiable) in water and are claimed to be capable of—supplying lubrication for various purposes; emulsifying (dispersing) in water various other materials; plasticising; defoaming; dispersing; penerating; scouring; wetting out; and softening. For some of these operations specially formulated oils may be necessary.

Their various uses as greases, softeners, conditioners, defoaming agents etc. are due to their mass and superficial properties which in turn depend on their constitutive fatty chains.

Sulphated and sulphonated oils commonly used in industries are only partially sulphated or sulphonated; their SO_3 combined which measures the degree of sulphatation or sulphonation, is between 1 and 5 per cent in most of the cases. This, for an oleic glyceride, for example, means a maximum of one-fourth of the molecules carrying sulphonated group. Also they should be considered as oils emulsified by emulgators (sulphates or sulphonates) formed *in situ*.

Sulphated oils, obtained by adding sulphuric acid to fats, were described by *Fremy* who studied this action on olive oil and almond oil and on oleic acid.

MAIN REACTIONS FOR THE ADDITION OF SO_3 GROUP ON FATS

The main reactions for the addition of the SO_3 group on fats are discussed below.

Sulphatation

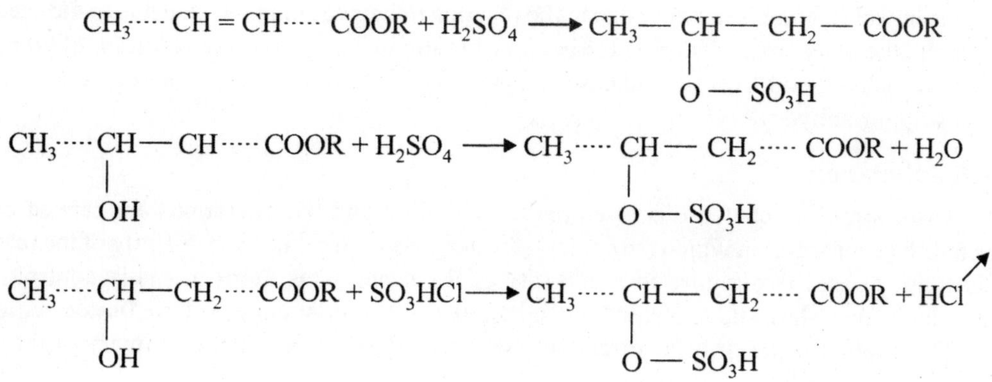

Sulphonation

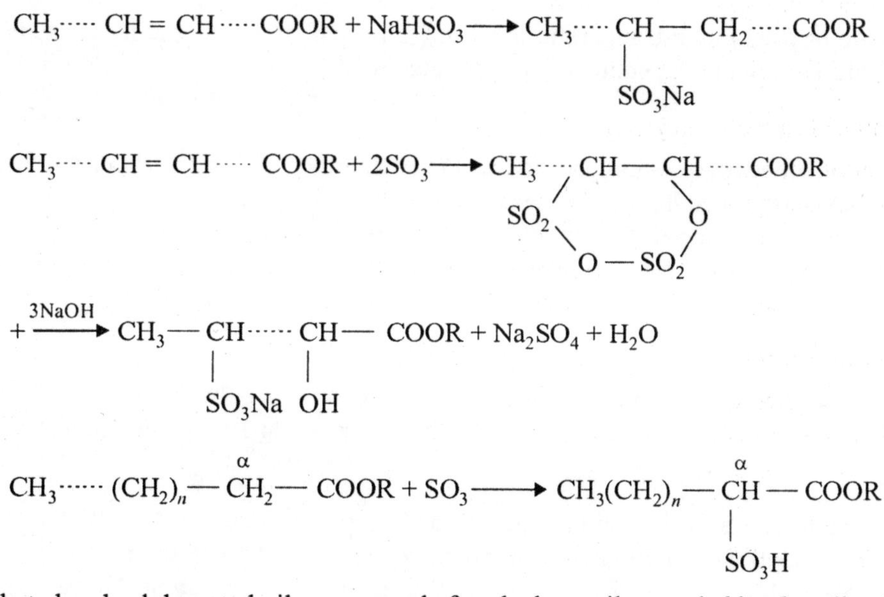

The sulphated and sulphonated oils are named after the base oil preceded by the adjective sulphated or sulphonated e.g. sulphated grape seed oil, sulphated tallow, sulphonated coconut oil. The names of these compounds must, however, conform with their constitution in order to avoid false interpretation by the user and to enable him to choose the appropriate product for the intended use.

SULPHATED OILS

Sulphation of unsaturated fat chains or of hydroxyl groups present on fat chains is quite easy; it results in an exothermic reaction which, if not properly contained can result in carbonising the oil and reducing the sulphuric acid to sulphur anhydride.

This overheating was used for evaluating the degree of reactivity of oils and fats. 'Sulphuric heating' and 'Maumene's degree' are two methods based on this principle.

Sulphuric heating measures the time necessary for obtaining a maximal temperature as a result of the addition of 20 per cent of concentrated sulphuric acid compared to weight of oil to be examined.

An increase in temperature is not only due to the reaction of sulphatation but also due to the secondary reactions which take place, mainly during the action of highly unsaturated fats. Addition of 20 per cent sulphuric acid to fish oil, cod-liver oil and castor oil already leads to reduction reaction of acid with the formation of sulphur anhydride.

Industrial Sulphatation

Industrial preparation of sulphated oils often requires cooling and the operations are carried out in a double layered steel apparatus in which the coolants (water, brine) circulate. The quantity of the reactional medium does not exceed two tonnes in a majority of the cases. This helps in easily controlling the overheating which takes place and to obtain light coloured products containing as little fission compounds as possible. The coolant water or brine circulating in the double hull is often a tributary of the central coolant system which feeds all the equipment of the same plant.

Steel was chosen for constructing the sulphonate equipment as it is resistant to concentrated sulphuric acid: storage tanks and supply pipe lines for intermediary tanks which are used for measuring the amount of acid to be used in these apparatus are also made up of steel. Sometimes steel is replaced by plastics which are immune to the action of concentrated acids.

Main Methods of Fat Sulphatation

Simple addition of concentrated sulphuric acid to the unsaturated and/or hydroxylated fats (castor oil, fish oil, flare, tallow) is the most frequently used method. Sometimes 'monohydrate' acid is also used when higher SO_3 combined are required. The amount of acid normally added varies between 15 and 30 per cent by weight of the oil or fat to be treated. By fixing SO_3H group on triglycerides and unsaturated fatty acids we get different principal and secondary compounds, depending on the sulphating agent used.

Batch sulphatation of fats

This operation is carried out when the quantity of oils involved is between 500 kg and 1 tonne; it is carried out in an apparatus described in Fig. 15.1. This process consists of various phases:

Sulphatation

It is divided into intervals at which acid is added: 1.5–2.5 hr. or more depending on the cooling system available. Time for total sulphatation: 4–5 hr. which varies according to the nature of the fats being treated and the SO_3 combined which has to be fixed. During the entire operation, the reaction temperature should be controlled so that it does not exceed the fixed limits (Table 15.1).

Table 15.1. Industrial sulphatation temperatures of some fats.

Nature of oil	*Temperature before sulphonation*	*Maximum temperature during introduction of the acid*	*Temperature at the end of Sulphatation*
Neat's foot	15°C	30°C	32°C
Olive	20°C	25–27°C	30–32°C
Corn	15–20°C	25–27°C	30–32°C

(Contd...)

Table 15.1. Industrial sulphatation temperatures of some fats. *(Contd.)*

Nature of oil	*Temperature before sulphonation*	*Maximum temperature during introduction of the acid*	*Temperature at the end of Sulphatation*
Grape seed	15–20°C	25–27°C	30–32°C
Rapeseed	15–20°C	30°C	30–32°C
Peanut	20°C	30°C	32–35°C
Fish	15–20°C	25°C	27°C
Cod	15–20°C	25°C	27°C
Castor	20°C	30°C	35°C

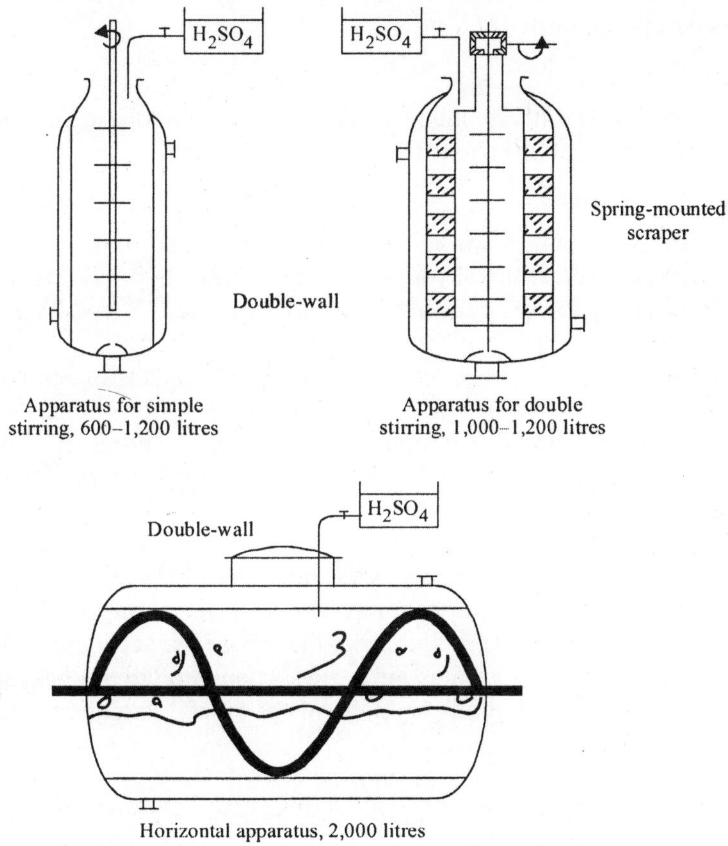

Fig. 15.1. Equipment for batch sulphatation.

Washing

It is done in a conical tank made up of lead or covered with plastic containing brine with 10 per cent sodium chloride or whose volume represents ranging Glauber's salt from 30 to 35°C double the weight of the sulphated oil. The oil is poured into it in a few minutes. The entire solution is stirred with the help of a mechanism with blades rotating at slow speed for a short while in order to avoid the formation of an emulsion which cannot be easily separated.

First decantation

First decantation after a pause of 1–2 hr. of the washed mass, sulphated oil floating on the surface is separated from the decanted acid brine.

First neutralisation

First neutralisation of the mass floating on the surface is then neutralised with concentrated soda, while agitating for 30–45 minutes until the pH reaches around 4 (until the Congo paper turns red). Temperature of the mixture which rises slowly should not exceed 50–55°C at the end of the operation in order to avoid hydrolysis of the molecules.

If conducted properly, these operations influence the content of SO_3 combined of the product and generally it has been found that the higher the final neutralisation temperature and greater the time taken, greater is the loss of combined SO_3.

Second decantation

The neutralised medium is left to settle for many hours, sometimes for one night; then brine is removed. The floating oily product thus contains about 12–18 per cent water.

Second neutralisation and adjusting

After this operation, the oil is homogenised and neutralised until a pH of 6.5–7 or more is obtained by using a solution of concentrated potassium hydroxide or ammonium hydroxide or any other appropriate agent. It is then adjusted to the desired concentration often between 70 and 72 per cent fat, which in commercial terms is known as 100 per cent concentrated.

At this stage co-surfactant agents are added to give the product its commercial form or to make it appropriate for use. The commercial product is generally in the form of a water-oil micro-emulsion.

Sulphated oils look like oily liquids, transparent with colour ranging from light yellow to dark brown depending on the origin of the fats and the manufacturing methods; they are sometimes bleached with strong oxidiser such as hydrogen peroxide or chlorites.

Sulphated oils give milk or opalescent emulsions in water or other solutions depending on the degree of sulphatation and the nature of the co-surfactants used during the adjusting stage. It is also possible to obtain compounds of bicontinuous micro-emulsion, i.e. compounds which do not change their look when they are diluted with water. It is from these base products that auxiliary products that are formulated and which are used by various industries. Though relatively less sulphated, these oils are excellent emulgators of fats, mineral oils (soluble oils or others), organic solvents, etc.

Fast sulphatation

Batch sulphatation can be accelerated by adding sulphuric acid at very short intervals, thus leading to a noticeable change in temperature and a subsequent fixing of the combined SO_3 (Table 15.2).

The mass of sulphated oil is then directed to the washing tank containing sodium chloride or Glauber's salt as in the usual process. It is noteworthy that the use of sodium chloride forms products which do not crystalise much during storage when the outside temperature is low.

On the other hand it is possible to recycle the brine of sodium chloride in order to avoid saline pollution; sodium sulphate formed during the first neutralisation phase is recovered by crystallisation and the brine is filtered and recycled after adjusting the required density. Temperatures and sulphatation, washing, neutralisation and decantation times are variable depending on the methods followed at each plant. The methods are adapted to the material available, the cooling possibilities, the agitating speeds, filling of tanks, but they also vary from country to country and the seasons, as sulphatations and decantations operate under different conditions taking into account the outside temperature.

Table 15.2. Batch sulphatation methods for fats.

Conventional methods

Nature of oil	*% H_2SO_4*	*Sulphatation time*		*Temperature at end of obtained reaction*	*% combined SO_3*
		Acid addition	*Total*		
Cod-liver	25	2 – 3 hr.	4 – 5 hr.	27°C	3.5
Neat's foot	20	2 – 2 hr. 30	3 hr. 50 – 5 hr.	32°C	3.8
Fish	22	2 hr. 30 – 3 hr. 30	4 – 6 hr.	30°C	3.5
Castor	25	2 hr. 30 – 3 hr.	5 – 6 hr.	35°C	6
Seal	25	2 – 3 hr.	4 – 5 hr.	30°C	3.5

Fast methods

Nature of oil	*% H_2SO_4 (concentration, 93%)*	*Sulphatation time*		*Temperature of oil*	*Reaction temperature*	
		Acid addition	*Total*		*Max*	*Final*
Cod-liver	22.5	12′	1 hr.	17°C	36°C	35°C
Olive/wheat	22.5	15′	1 hr.	26°C	56°C	50°C
Wheat	22.5	15′	1 hr.	23°C	40°C	32°C
Oleic acid	22.5	10′	1 hr.	26°C	52°C	50°C

(Quanitity of oil per reaction: 362 kg).

Instant sulphatation

If the temperature attained is not very high and the quantity of sulphatation agents used is relatively low, then an increase in the contact time of sulphuric acid with the fat to be sulphated does not result in much sulphatation; thus it is better to reduce to the minimum the period of action of acid on the fat molecules, more so as the colour of the products obtained is greatly improved.

Temperature of oil before sulphatation

The kinetics of fixing the -O-SO_3H group is influenced by temperature, but the decomposition reaction of the sulphonating agents into SO_2 restricts the possibilities of working at higher temperatures, particularly with easily oxidisable fats. The amount of combined SO_3 fixed thus passes through a maximum depending on the quantity of sulphating agent and the reaction temperatures.

Amount of sulphating agent used

During instantaneous sulphatation, for a given initial oil temperature and a sulphuric acid with known concentration, the SO_3 combined increases with the quantity of acid until the reaction temperature reaches the stage in which secondary reactions take place.

Concentration of sulphuric acid

The concentration of sulphuric acid used noticeably influences the degree of combined SO_3 fixed on the final product (Table 15.3).

Table 15.3. Influence of the concentration of acid used during the process of instantaneous sulphatation.

Nature of oil	Temperature	% of acid compared to oil	Concentration of sulphuric acid 92%	97%	100%
			% of SO_3 combined from the sulphated product		
		14			3.8
Neat's foot	20°C	14.9		3.6	
		17.8	4.1		
		16.5			4
Oleic esters	20°C	20		4.1	
		20	1.7		
		14.3			3.5
Fish	20°C	16.2		3.5	
		19	3.6		
Seal	20°C	15.7	3.6		
		17.6		4.1	
		15.1			3.1
Cod-liver	20°C	17.8		2.9	
		20	2		

Following the very short contact time during instant sulphatation, this influence is all the more important here than in the conventional batch methods.

Continuous sulphatation

In order to obtain interesting results during continuous sulphatation, the oil is heated beforehand. Its flow rate can vary between 300 and 600 kg/hr. and even more. Different factors which influence the quality and the degree of sulphatation of sulphated oils produced are:

1. Temperature of oil before sulphatation.
2. Temperature and concentration of acid used.
3. Proportion of sulphating agents used.
4. Temperature of brine used for washing the sulphated oil.
5. Proportion of brine used for washing as compared to the sulphated oil.

The various methods of continuous sulphatation: inclined or vertical, linear current, counter current or by pulverising acid on a film of oil or even counter-pulverisation are mostly based on the process of cooling the compounds obtained in order to avoid secondary reactions.

Sulphatation can also be carried out without cooling in very short time periods. Under such conditions continuous sulphatation of oils forms products with characteristics similar to compounds obtained by batch sulphatation and even closer to those produced by fast sulphatation. Continuous sulphatation has other advantages such as: reducing the quantity of the sulphatation agent, brine for washing, neutralisers: reducing the washing time, neutralisation time and time taken for acid hydrolysis, improving the colour of the sulphated compound, an easier and more effective control of the manufacturing process.

In discontinuous sulphatation, manufacturers are obliged to homogenise various procedures in order to obtain the desired characteristics. But in case of discontinuous sulphatation, depending on the nature of an oil and its main characteristics it is possible to determine in advance, with the help of simple methods, the reactional temperatures, flow rate of oil, acid and brine so as to obtain products with practically constant characteristics.

Some Sulphated Fats

Obtained from unsaturated or hydroxylated fats, sulphated fats are of various types: of animal origin, (flare, Neat's-foot, horse oil, tallow), of plant origin (castor, olive, rapeseed, graspe seeds, peanut, wheat, soyabean, sunflower, cotton) and fish oils (herring, cod-liver menhaden). Mention must also be made of methyl oleate, oleyl oleate or very short chain alcohols, saturated or unsaturated, butyl ricinoleate, propylene glycol dioleate, glycerol trioleate, partial glycerides which in certain applications are replaced by Neat's foot oils (glycerol trioleate) or spermaceti (glycol propylene dioleate), etc.

Sulphated oils and fats are a mixture of sulphated molecules, different secondary compounds and glycerides or cerides which have not been affected by sulphatation and emulsified by the sulphated molecules. Due to this they are classified according to the amount of combined SO_3 which expresses the number of emulgating molecules that are contained. Normal sulphated oils are considered to be strongly sulphated when the amount of combined SO_3 in a 100 per cent commercial product exceeds 3 per cent; it is medium-sulphated when the content is 2–3 per cent; and has low sulphate content when it is less than 2 per cent.

Calculating the amount of combined SO_3 in a sulphated oil

Procedure of calculating the amount of combined SO_3 in a sulphated oil is given as under:

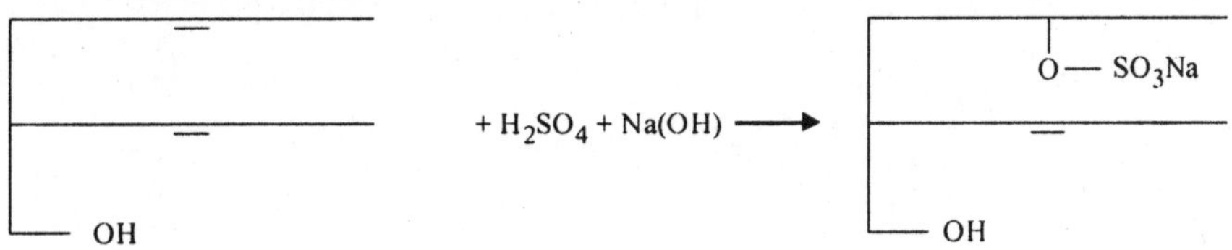

Glycerol dioleate — Monosulphated glycerol dioleate (at the ethylene bond)

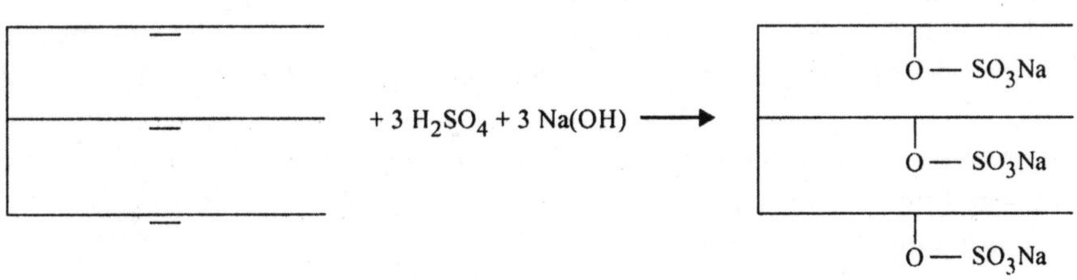

Glycerol trioleate PM = 885.4 — Sodic trisulphate of trioleate glycerol PM = 1245.4

Or the amount of combined SO_3 for sulphated glycerol trioleate:

$$\frac{240 \times 100}{1245.4} = 19.2\%$$

For a 3 per cent SO_3 combined, the amount of emulsifier present is equal to:

$$\frac{100 \times 3}{19.2} = 15.6\% \text{ of the active matter}$$

Strongly sulphated oils produce micro-emulsions in water when the content of sulphated molecules in the product equals about 20 per cent.

SULPHONATED OILS

Sulphonated oils are obtained from unsaturated chains by the action of Na bisulphite in concentrated aqueous solution. Studies already done in this field have lead to the preparation of wetting agents such as dioctyl/sulphosuccinate obtained by esterification of maleic anhydride on two molecules of octylic alcohol, then sulphatation by a concentrated solution of bisulphite.

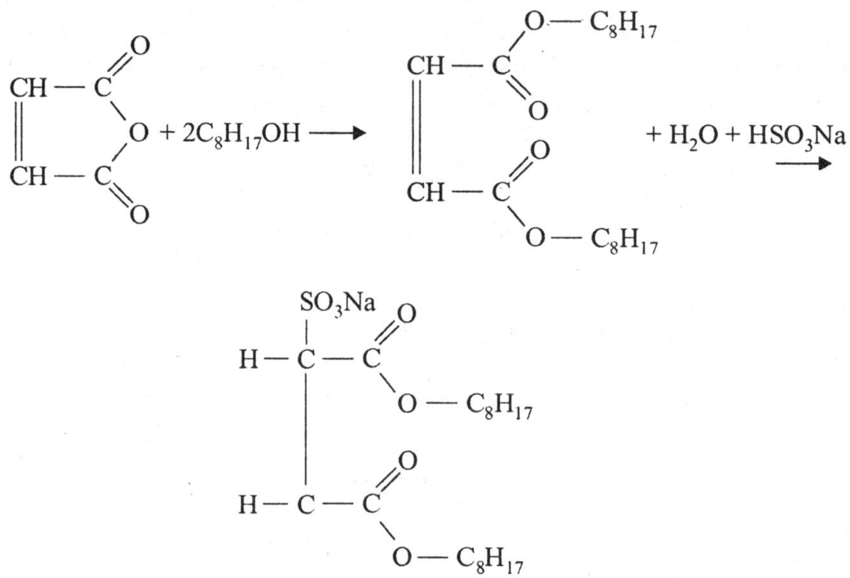

Numerous variants of maleic esters or unsaturated diacids with more or less long chain have been used in the preparation of washing and cleaning agents, but these are particularly wetting properties which are exploited for these types of products in numerous industries: such as rubber, metallurgical and mechanical industries, decantation of ores, paints and inks, pigmentary sizing, cleaning, phytosanitary products, etc.

The results obtained on unsaturated diacids and esters of unsaturated mono-acids have inspired processing of unsaturated fats by the same methods.

Two types of sulphonating agents can be applied:

1. Concentrated sodium bisulphite solution.
2. Metabisulphite solution whose conversion rate in bisulphite is incomplete, thereby not allowing higher levels of combined SO_3.

Sulphonation of Unsaturated Fats by Action of Sodium Bisulphite Solutions

This operation depends on several factors which must all be present in order to produce conclusive results. This highlights the importance of:

1. Partial oxidation of unsaturated fats in the beginning, in view of activating one of the carbons of conjugated ethlyene bonds.
2. Violent mixing, indispensable on contact with oily and aqueous phases.

3. Presence of a water emulsifying agent in oil, generally an aliquot part of a previous operation.
4. Temperature range between 60 and 100°C.
5. Time. The reaction being very slow–between 3 and 6 hrs. sometimes longer–it can be activated by adding a suitable co-surfactant before the operation of each type of fat to be processed and by using catalysts which effectively reduce reaction time. In these conditions, compounds are obtained which produce bicontinuous water-oil micro-emulsions, stable in cold and hot conditions.
6. Preliminary oxidation: Activation of conjugated ethlyene bonds allows the acceleration of the fixation of sodium bisulphite, which is taken in the water-oil heterogeneous medium. Formation of oxirane bridges during oxidation equally contributes, in giving a certain hydrophile to fat molecules at interpenetration phases.

   (a) Oxidised chain:

   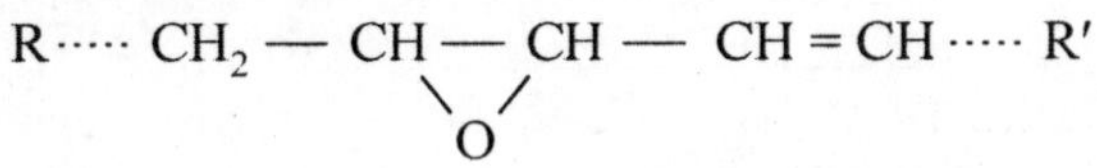

   (b) Oxidised chain, sulphonated by bisulphite: HSO_3Na

   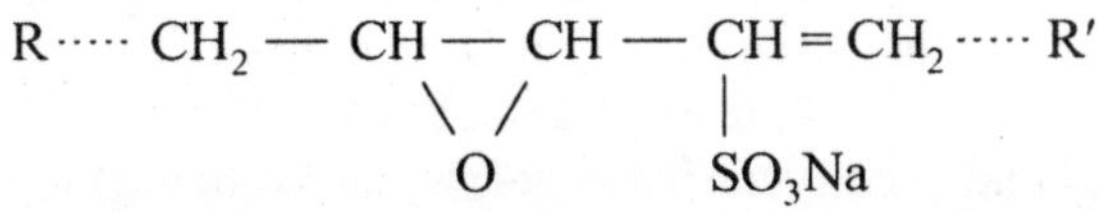

   Oxidation must be carried out in a moderate manner so that the saturation curve does not interfere with oxirane formation and that it blends with it, if possible, without secondary transformations appearing due to esterification or hydrolysis (Fig. 15.2).

   (c) Esterification of oxirane bridges.

   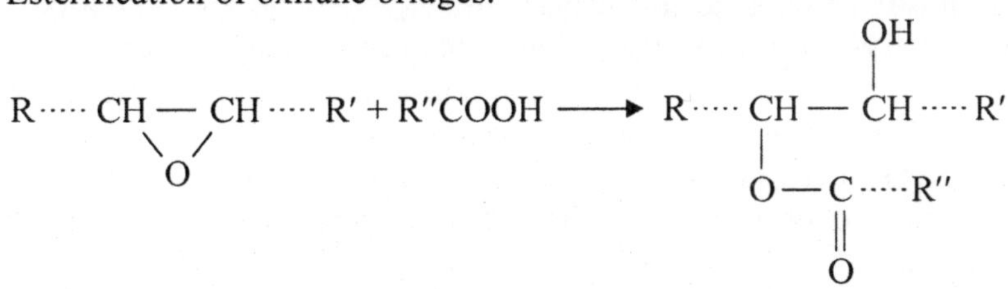

   (Hence the importance of the acidity of oils used. Neutral oils sulphonate better than acid oils).

   (d) Hydrolysis of oxirane bridges.

   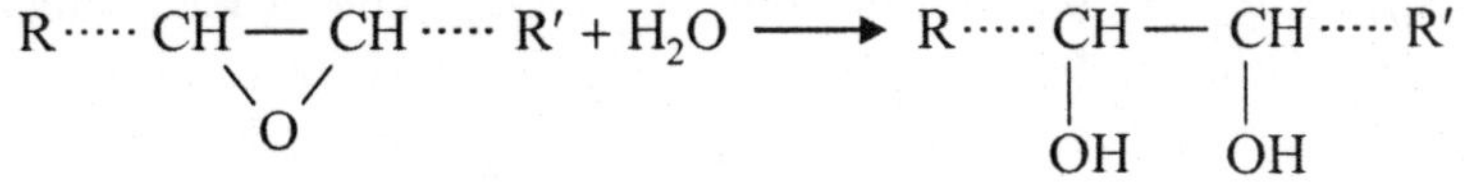

These modifications reduce the effectiveness of operation and, at the same time the speed and yield of the subsequent sulphitation reaction. This is why the methods of prior oxidation of fats are applied mostly at moderate temperatures.

Their evolution is generally followed by determination of the iodine value, peroxide value which gives the quantity of peroxide formed at the time of oxidation, and also by oxirane rate obtained.

This latter method is based on property which is possessed by oxirane bridges to fix hydracids; therefore, it is thus an acidimeteric method which in return permits these determinations to be carried out.

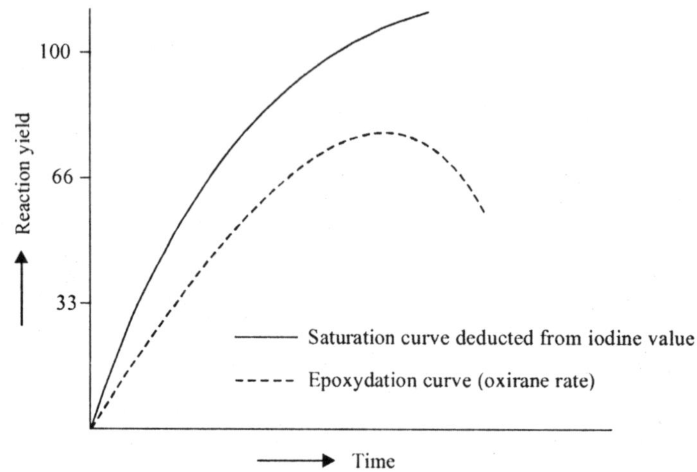

Fig. 15.2. Saturation curve.

Given the possible reactions, it is observed that the forms of activation of molecules to be selected will differ depending on the type of fat chain to be reacted on the bisulphite. Epoxidation is preferred in case of polyunsaturated molecules containing conjugated ethlyene bonds, peroxidation being more favourable to monounsaturated molecules where the formation of oxirane would block the sulphonation reaction. This is the case when processing glycerides and oleic esters.

When oxidation and subsequent sulphitation operations are carried out in suitable conditions, at the end of reaction one finds the presence of a transparent compound, water-oil micro-emulsion whose concentration in water is mostly between 20 and 30 per cent.

While gradually diluting the product, an oil dispersion is obtained in opalescent water, even transparent at times, throughout dilution, if care is taken in adding a small quantity of co-surfactant to oil such as alcohol, glycol or a compound of any other type, such as an amino acid.

Crude sulphitation products often result in milky dispersions, at times coarse, and the presence of small quantities of suitable co-surfactant provokes formation of a transparent water-oil interface when these two phases come into contact.

The most effective product, for a given type of oil, must meet certain hydrophyllic and lipophylic criteria. In this regard, it is necessary to recall that as a general rule, the length and the branching of co-surfactant chains are at least as important as the nature of hydrophyllic classification to which they are linked. Great stability of such emulsions ensures very safe use of this method and particularly a constancy of results during intermediate storage time of pre-emulsions, or by more or less significant dilutions, which modify the stability and the fineness of dispersions according to the zone of their diagram of phases.

Sulphonated oils (also called 'sulphited') are distinguished by their degree of unsaturation and overlapping of their molecular chains. The two extremes are illustrated by oils of sulphited fishes. Very often used as cerid synthesis like cetyl oleate or esters such as sulphited glycol propylene dioleate. In the first case, one has to deal with a molecule of relatively higher molecular weight and significant overlapping. In the second case, the molecules generally linear, show a molecular weight lesser by one third and hence more penetrating in the fibrous mattress (paper, textile, leather).

The choice of applications thus varies considerably, especially as the former retain a strong oxidability, which is distinctly reduced for the latter.

α-SULPHONATED FATS AND OILS

If it is commonly agreed that usual sulphating or sulphonating agents like sulphuric acid do not act on saturated fatty acids and their esters, the use of more active sulphonating agents favour fixation of sulphonic classifications in the α position, under certain conditions, in relation to the carboxylic classification of fatty chains. Preliminary works on α-sulphonates of saturated fatty acids are very old and concern compounds with short chains, study of the properties of their salts and esters such as sulphoacetic, sulphomalonic, sulphopropionic and sulphobutyric acids, etc. Dutch researchers and, more recently, American researchers developed an interest in the preparation of long-chain α-sulphonates.

Preparation of Sulphonic Acids from Saturated Fats

Preparation of α-sulphonic acid from saturated fatty acids, their salts and their esters have been vastly described. Three main methods have been retained:

1. By direct sulphonation of saturated fatty acids or their esters.
2. By substitution of a sulphonic group by a halogen.
3. By sulphitation of a monounsaturated fatty acid.

The first and the last are the most common methods.

Direct sulphonation of saturated fats

The study of direct sulphonation of saturated acids and their esters, by sulphur trioxide or by sulphuric chlorohydrin, shows that fixation of sulphonic classification is done well in α position, as much as on fatty acids with short chain as well as on long chain.

Industrial preparation of α-sulphonates is carried out by direct addition of sulphonating agents to fats brought at a temperature slightly higher than their melting point. The simplest industrial method is to add oleums to fats, mostly oleum with 10 per cent of SO_3.

Addition of stabilised sulphur trioxide can also be done but it requires dissolving of fats in a solvent such as tetrachlor-ethylene (4 to 5 times the quantity of fats); an excess of sulphur trioxide of 60 per cent is necessary for a good yield. The reaction is highly exothermic after the addition of sulphur trioxide is over; the mixture is heated, solvent is distilled under reduced pressure and the fat residue washed. α-Sulpholauric acids, myristic, palmitic, stearic acids can be obtained in this manner.

The neutral salts of α-sulphonic acids are obtained by sulphonation of saturated fatty acids obtained by direct neutralisation in aqueous medium with the help of various neutralising agents. Their low solubility facilitates their separation from brine, formed by neutralisation of excess sulphonating agents.

Acid salts can be obtained by partial neutralisation of α-sulphonic acids. Simple addition of the quantity of neutralising agent for neutralisation of sulphonic classification releases carboxylic acidity; industrially, the first neutralisation is verified on pH meter or more simply on Congo paper.

Properties of α-sulphonic acids

Pure α-sulphopropionic acids are in the form of hygroscopic crystals with high melting point, even for the first terms. α-sulphopropionic acid has a melting point of 84–86°C; just like sulphoacetic acid it would crystallise with a molecule of water. The melting point of α-sulphopalmitic acid is 90–91°C.

Properties of mono-and disodium α-sulphonates

Study of the solubility of mono and disodium salts permits the statement that monosodium salts are less soluble than disodium salts. The disodium α-sulpholaurate and disodium α-sulphomyristate are distinctly more soluble than α-sulphopalmitate and stearate.

Surface activity of α-sulpholaurate and myristate is low; surface tension decreases with the increase in molecular weight. Monosodium salts have lower surface tensions. Measurements show that α-sulpholaurate disodium does not have wetting properties, than α-sulphomyristate has very few of them, but on the contrary, α-sulphopalmitate disodium is a good wetting agent.

Compared to sodium oleyl sulphate and sodium alkylaryl sulphonate, α-sulphonates of saturated fatty acids have lower detergent power.

Properties of esters of α-sulphonates of saturated fatty acids

The melting points of sodium α-sulphonates of saturated fatty acid esters are much higher than those of fatty acids and their esters. It is to be noted that when mapping different melting points of these compounds, depending on the length of the esterifying alcohol chain, the maximum reduction of the melting point is between 4 and 5 carbons for an alcohol chain.

Surface and interface tensions in 0.1 per cent solution at 25°C reveal a considerable decrease depending on the increasing length of the esterifying alcohol chain (Table 15.4).

Table 15.4. Surface and interface tensions of some sodium salts of alkyl α-sulphopalmitate and stearate.

Esters	*Surface tension (milli N/m)*	*Interface tension (milli N/m)*
Na Me α-sulphopalmitate	40.2	9.7
Na Me α-sulphostearate	39	8.4
Na Et α-sulphostearate	38.5	7.8
Na Isopropyl α-sulphostearate	37.6	7.1
Na Propyl α-sulphopalmitate	37	7.2
Na Butyl α-sulphopalmitate	35.3	6.4
Na Amyl α-sulphopalmitate	33.1	4.9
Na Amyl α-sulphostearate	33.6	4.5

Foaming power measured on a Ross and Miles apparatus is much more developed for esters than for monosodium and disodium salts of the α-sulphonic acid of saturated fatty acids (Table 15.5).

Table 15.5. Foaming properties of some α-sulphonated esters in distilled water.

	Foam, mm Solution 0.25%-temperature 60°C	
Esters	*Immediate*	*After 5 min*
Na Me α- sulphopalmitate	216	129
Na Me α-sulphostearate	216	179
Na Propyl α-sulphopalmitate	164	134
Na Isopropyl α-sulphostearate	160	142
Na Butyl α-sulphopalmitate	202	133

On hydrolysis, the stability of sodium salts of the α-sulphonate esters of saturated fatty acids is remarkable, as much in acid medium as in acid and saline media.

Esters of α-sulphonic acids of saturated fatty acids have a good stability in presence of most alkaline-earth and other salts, except aluminium salt.

Preparation of Partial α-Sulphonates from Saturated Fats

It is important to know that most marketed α-sulphonated fats are the product of partial sulphonation of saturated fats, as are the sulphate derivatives obtained by sulphatation of unsaturated fats.

The rate of combined SO_3 fixed on the main α-sulphonated oils such as coconut oil, palm kernel oil, babassu oil, etc. is generally between 2 and 6 per cent. Not all fatty acid molecules are therefore carriers of a sulphonic group, quite to the contrary. Industrial uses of this type of product are plastic deformation of metals, textiles or paper actually do not require this characteristic.

Preparation of partially sulphonated saturated oils

This is carried out with the help of oleum by introducing the sulphonating agent in the oil or vice-versa, at a temperature of 30°C (for lauric type oils). Mixing time is several hours, until the desired SO_3 combined is obtained.

After sulphonation, excess sulphonating agent is removed from the sulphonated product by washing with the help of concentrated brine, then the medium is neutralised by concentrated soda. Washing and neutralisation are carried out at temperatures varying from 40 to 55°C.

This method leads to the production of partial sulphonates whose properties are quite different from sulphated oils, in particular better stability of their emulsions or micro-emulsions in acid and saline medium, and complete absence of oxidability–desirable characteristics for a number of applications.

The degree of sulphonation expressed in combined SO_3 of partial α-sulphonates obtained from sulphonated coconut, palm kernel and babassu oils is variable: between 1/10th and 1/5th of the theoretical quantity of SO_3 likely to be fixed on all the fatty acids present in the molecules.

INDUSTRIAL USES OF SULPHATED AND SULPHONATED OILS AND FATS

The industrial uses of sulphated and sulphonated oils and fats depend on their special properties, on their origin and the nature of solubilising classification which has been fixed on them (sulphate or sulphonate) which is explained by:

1. Stability in an acid medium more or less good.
2. Stability in presence of metallic ions more or less large.
3. Resistance on hydrolysis more or less important.
4. An oxidability more or less strong.
5. An ionic force more or less marked (Table 15.6).

Table 15.6. Comparative ionic force of sulphated and sulphonated oils.

Sulphonic acids and sulfovinic acids have a decreasing ionic force according to the following classification:

Hydroxy-sulphonic acid, obtained by action of sulphuric trioxide on unsaturated fats, is easily dissociable. This dissociation is due to negative inductor effect of the hydroxyl group which tends to favour an ethylene bond and the release of proton H^+.

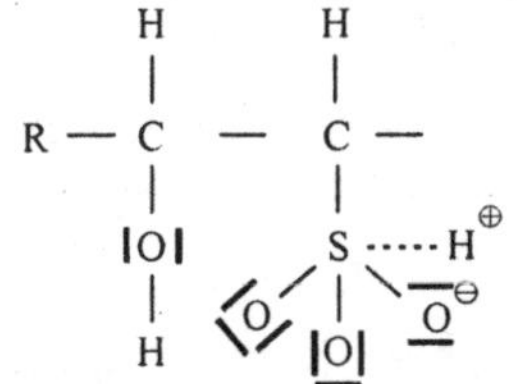

(Contd...)

Table 15.6. Comparative ionic force of sulphated and sulphonated oils. *(Contd.)*

The α-sulphonic acid of a saturated fatty acid is easily dissociable due to the proximity of the carboxylic group. This is a strong acid.

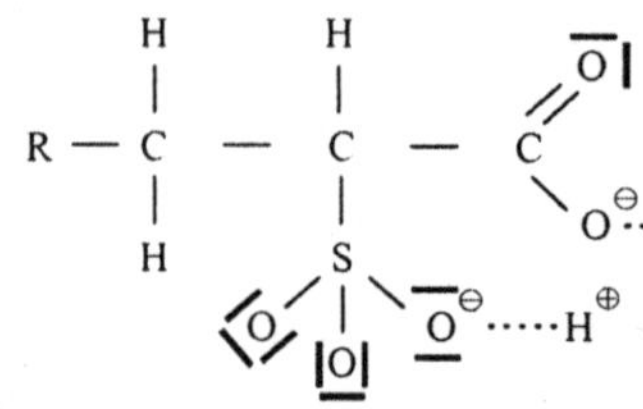

The sulphonic acid corresponding to sodium salt, obtained by the action of sodium bisulphite on ethlyene bonds of unsaturated fats, is also a strong acid.

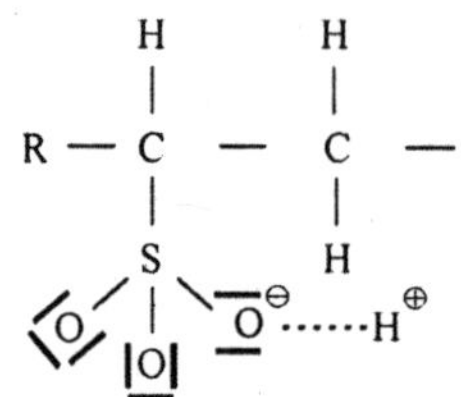

Sulphovinic acid, obtained by the action of sulphuric acid on the ethylene bonds of unsaturated fats or on the hydroxyl groups of some fat chains, is less dissociable due to the presence of an oxygen atom between the sulphur and the carbon of the fatty chain, which tends to reject the bond doublet, linking the proton to the oxygen, thus opposing its release by reducing the acidity.

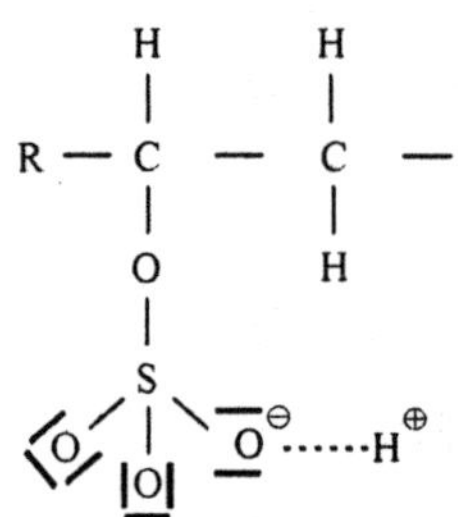

Moreover, the molecular weight of fatty acids, their branching, the presence of a hydroxyl group, the unsaturation of chains are responsible for appreciable secondary qualities: lubricating power, spreadability, extreme pressure behaviour, unoxidisability, etc. characteristics which determine the choice of applications.

The sulphatation and sulphonation oils and fats render them emulsifiable or soluble in water; residual emulsifying power towards various fats and hydrocarbons, direct their usage towards preparation of auxiliary products, right from the beginning, first of all for the replacement of soaps in textile industry, simply on account of their better stability in hard waters, then in other industries such as the paper industry, tannery, ore mining and extraction, biotechnology, etc. The part of fats unaffected by sulphatation is used as a vector for some specific additives for given applications.

Metallurgy

Fats partially sulphonated or sulphated enter in the composition of some soluble oils used for plastic deformation of metals in humid phase or for removal by cutting tool.

Drawing, stretching and pressing operations of steel, copper, brass, aluminium use composition of this type. In these greases, sulphated or sulphonated derivatives act as emulgator: they show the advantage of good stability in calcareous medium which, in drawing of steel bars for example, allows direct processing of scrapped bars after neutralisation with lime. Mechanical drive of lime wash must not cause premature rupture of emulsions.

Textile

In this industry, sulphated and sulphonated oils are used in as it is or in strong proportions in the preparation of auxiliary products meant for:

1. Oiling of natural, artificial and synthetic fibres where the lubricating qualities of vegetable oils are used at the time of combing or carding of material.
2. Oiling of hard fibres: mixing of strings, ropes, etc. of good quality and very clear tints, requires use of inoxidable products such as α-sulphonated oils for example.
3. Reeling of synthetic fibres such as polyamides; some compositions used for this operation contain coconut oils sulphonated in mixture with paraffinic mineral oils. The antistatic properties of these products are probably related to their hygroscopic power due to the presence of sulphonic group.
4. Degreasing and deoiling of rayon creeps: Partial α-sulphonates such as sulphonated coconut oil proposed in this regard act as softening agent at the same time and can also be used for degreasing of wool, cotton and rayon in a dye bath.
5. Finishing and softening: Partial α-sulphonates are used instead of sulphated castor oil, because of their unoxidisability, plasticising power and reasonable price.

Chapter 16

Miscellaneous Oils

INTRODUCTION

The word oil is applied to a wide range of substances that are quite different in chemical nature. Oils derived from animals or from plant seeds or nuts are chemically identical with fats, the only difference being one of the consistency at room temperature. This chapter describes the various miscellaneous oils such as—grapeseed oil, camelina oil, poppy seed oil, almond, hazelnut, walnut oil, wheat germ oil, conjugated oils and avocado oil etc.

GRAPESEED OIL

Like wheat and olive, grape is one of the most ancient, traditional crops. *Vitis vinifera*, semi-arborescent creeper plant of the Vitaceae family, is grown in the temperate regions. All the wine-growing countries, mainly of Mediterranean coast (Italy, France, Spain, Turkey, Romania, Greece, Yugoslavia) have benefited from by-products of wine. The pear-shaped seeds are composed of an outer epidermis constituted of big porous cells, an inner epidermis and an endosperm containing, among other, compounds, proteins, oil and tannins. Grapeseeds may contain from 5 to 20 per cent oil depending on the climatic conditions, the varieties and the ripening of fruits. More generally, the oil content remains between 12 to 13 per cent.

Dry seeds (apparent density: 0.48–0.52) coming from distilleries are sent for pressing in order to get small pellets. These pellets are then sent to a unit for solvent extraction (hexane). Crude oil thus obtained is very coloured and is sent for refining. The presence of very fine waxes requires a winterisation stage which takes place in two phases: degummed, neutralised oil (phosphoric acid, soda) is subjected to an initial cooling at 5°C and then brought to 12°C for centrifugation. A second tempering at room temperature precedes washing and bleaching, followed by post-winterisation. The oil is then filtered before deodourisation. Refined grapeseed oil is used for frying and seasoning. Its richness in linoleic acid is responsible for its nutritional image (dietetics, food). It is also used for non-edible purposes in cosmetology, in the soap industry (liquid soaps) and in lipochemistry for the production of fatty acids.

Oil meals from extraction, poor in proteins and rich in tannins, are used as combustible, fertilisers and even as fodder for rabbits and ruminants. It should also be pointed out that procyanidines (polyphenolic oligomers; content: ≈ 1 per cent) are extracted from deoiled seeds which are used by the pharmaceutical industry (improvement of blood capillary flimsiness) and by the cosmetic industry (skin ageing slackening due to antioxidant properties of these molecules).

Physico-chemical Characteristics

Table 16.1 shows the major chemical indices and physical characteristics of oil.

Table 16.1. Main physical and chemical characteristics of grapeseed oil.

D_{20}	0.923–0.926
V_{20} (cp)	53–58
n_D^{20}	1.473–1.476
Solidification point (°C)	–11 to –17
Iodine value	130–138
Saponification value	188–194

Triglyceride Composition

Table 16.2 shows the triglyceride composition of Tempranillo grapeseed oil.

Table 16.2. Composition of triglycerides of grapeseed oil (HPLC).

Nature	*% of total TG*
LLL	35.7
OLL	21.0
PLL	17.0
SLL + LOP	15.9
LOO	10.3

Table 16.3 shows the glyceridic structure corresponding to three different species of Egyptian grapeseed oil. These varieties have contents close to 60 per cent linoleic acid, 19–24 per cent oleic acid, 4.5–6 per cent stearic acid and 10–13 per cent palmitic acid.

Table 16.3. Fatty acid distribution in 1, 2, 3 positions of grapeseed oil triglycerides.

Nature of fatty acid	*% of total FA in 2*	*% of total FA in 1 + 3*
C16:0	–	14–19
C18:0	–	9
C18:1	22–28	15–25
C18:2	71–77	53–56

Fatty Acid Composition

Table 16.4 gives the composition of fatty acids. For certain varieties, not always stated clearly, some values can deviate from the indicated ranges. Thus, values of up to 13 per cent for palmitic acid, 2 per cent for stearic acid, 35 per cent for oleic acid and values of less than 50 per cent for linoleic acid have been quoted.

Table 16.4. Fatty acid composition of grapeseed oil.

	% of total FA	
Nature	*Usual values*	*Codex*
C14:0	< 0.3	< 0.3
C16:0	7–10	5.5–11
C16:1	< 0.5	< 1.2
C17: tot.	–	–
C18:1	3–6	3.0–6.0
C18:2	14–22	12–28
	65–73	58–78
C18:3	< 0.5	< 1.0
C20:0	< 0.3	< 1.0
C20:1	< 0.2	–
C22:0	< 0.5	< 0.3

This oil belongs to the group of linoleic oils. Its saturated fatty acid content rarely exceed 15 per cent. Highly unsaturated but containing very little linolenic acid, the polyunsaturated saturated ratio is greater than 5 and monounsaturated saturated ratio is close to 1.3.

Composition of unsaponifiable fraction

Table 16.5 summarises the composition of unsaponifiable fraction of grapeseed oil. *Frega* observed that erythrodiol and uvaol are concentrated in pulp and skin but not detectable in the oil. He also mentioned traces of Δ7-stigmasterol, contrary to data given in the Table 16.5 (up to 2 per cent). Finally, he mentions that squalene represents 55 per cent of the hydrocarbon fraction (without any indication on the amount of this fraction).

Table 16.5. Composition of unsaponifiable fraction of grapeseed oil.

Constituents	*Unsaponifiable fraction 0.8–1.5%*	*Constituents*	*Unsaponifiable fraction 0.8–1.5%*
Sterols (mg/100 g)	259–418	Hydrocarbons (mg/100 g)	
		Tocopherols (mg/100 g)	32–52
Sterol composition (% total sterols)		Tocopherol composition (% total tocopherols)	
Cholesterol	< 0.5	α-Tocopherol	26–39
Brassicasterol	–	β-Tocopherol	2–14
Campesterol	11–15	γ-Tocopherol	10–34
Stigmasterol	8–12	δ-Tocopherol	< 2
β-Sitosterol	66–73	α-Tocotrienol	10–26
Δ5-Avenasterol	2–4	γ-Tocotrienols	14–34
Δ7-Stigmasterol	< 3		
Δ7-Avenasterol	1–3	Triterpenic alcohols (mg/100 g)	
Ergosterol	< 2	Erythrodiol	> 2
Fucosterol	< 2		

CAMELINA OIL

Camelina sativa, also called 'false flax' or 'German sesame' is an annual plant of the family of Cruciferae (Glabrata variety). Camelina has a short vegetation cycle and does not require very high quality soil, though this factor influences the yield which can vary from 1 to 2.5 T/ha. Recent studies have shown that notable improvements have to be made in seed quality and yields before camelina oil can become competitive. Oil content of camelina seeds varies from 33 to 41 per cent.

The possible uses for this oil are essentially industrial.

Camelina oil meal is not of major interest as animal feed. The glucosinolate content of the fat-free meal lies between 17 and 67 μmoles/gram.

Physico-chemical Characteristics

Camelina oil is a fluid oil. Crude oil is golden yellow coloured and develops a characteristics cabbage odour. These characteristics disappear after refining. Table 16.6 gives main physical and chemical characteristics of camelina oil.

Table 16.6. Main physical and chemical characteristics of camelina oil.

D_{20}	0.917–0.924
V_{20} (cp)	
n_D^{20}	1.475–1.476
Solidification point (°C)	–11 to – 18
Iodine value	130–145
Saponification value	188–194

Fatty Acid Composition

This oil is remarkable for its high linolenic and icosenoic (gondoic, C20:1) acid content associated with a low erucic acid content which, however, remains greater than that of new rapeseed varieties. Table 16.7 shows the fatty acid composition of camelina oil.

Table 16.7. Fatty acid composition of camelina oil.

Nature	*% of total FA*
C16:0	4–7
C16:1	–
C17:0	–
C18:0	2–3
C18:1	9–24
C18:2	12–19
C18:3	33–40
C20:0	< 2
C20:1	12–18
C22:0	1–2
C22:1	2–5

Unsaponifiable fraction

The unsaponifiable content varies between 0.8 and 1.2 per cent. Tocopherol content of oil varies between 54 and 78 mg/100 gram.

POPPY SEED OIL

Papaver somniferum, is a variety of poppy (Papaveraceae family). Its fruit is a capsule containing many seeds which contain 36 to 50 per cent of oil. Originally grown in Asia, its use as a crop has spread particularly in Europe and in the north of France. The sap of capsules and stems is rich in opium, from which morphine and other alkaloids are extracted for the pharmaceutical industry.

This oil can be used in the preparation of foods. Virgin oil from first 'cold' pressing is marketed as a diet product, but the main uses are the fatty acid and the paints and varnish industries. The oil has been registered in the French Pharmacopoeia as 'iodised poppy oil' or 'ethyl esters of iodised fatty acids of seed oil poppy'.

Oil meals have a protein content of about 35 per cent, which makes them interesting for animal feed. Their development in this sector is limited (appetence and toxicity), however, with a rate of incorporation of < 10 per cent.

Physico-chemical Characteristics

Table 16.8 shows main physical and chemical characteristics of poppy seed oil.

Table 16.8. Main physical and chemical characteristics of poppy seed oil.

D_{20}	0.925–0.928
V_{20} (cp)	51–62
n_D^{20}	1.475–1.476
Solidification point (°C)	−15 to − 20
Iodine value	130–143
Saponification value	188–195

Triglyceride Composition

Table 16.9 shows the triglyceride composition. This composition was deduced by calculation from the fatty acid composition and by determining fatty acids in position 2 (enzymatic method). The result of this study is the following overall distribution between unsaturated (U) and saturated (S): 13.60 per cent, SU2, 30 per cent, S2U, 3.3 per cent.

Table 16.9. Poppy seed oil triglyceride composition (HPLC).

Nature	*% acids of total TG*
LLLn	2.7
LLL	28.4
PLL	14.4
OLL	21.4
POL	7.4
SLL	3.3
LOO	5.4

(Contd...)

Table 16.9. Poppy seed oil triglyceride composition (HPLC). *(Contd.)*

Nature	*% acids of total TG*
PPL	1.8
POO	0.9
PSL	0.8
SOL	0.8

Fatty Acid Composition

Table 16.10 shows the fatty acid composition of poppy seed oil.

Table 16.10. Fatty acid composition of poppy seed oil.

Nature	*% of total FA*
C16:0	9–11
C16:1	< 0.2
C17:1	< 0.1
C18:0	1–2
C18:1	13–18
C18:2	69–77
C18:3	< 3.5
C20:0	< 0.1
C20:1	< 0.1
C22:0	–

Unsaponifiable fraction

Table 16.11 shows the composition of unsaponifiable fraction of poppy seed oil. *Beringer* reported a tocopherol content of 14.7 mg/100 g oil with composition limited to α-tocopherol (17 per cent) and γ-tocopherol (83 per cent).

Table 16.11. Composition of unsaponifiable fraction of poppy seed oil.

Constituents	*Unsaponifiable fraction 0.5–1.5%*	*Constituents*	*Unsaponifiable fraction 0.5–1.5%*
Sterols (mg/100 g)	275	Hydrocarbons (mg/100 g)	160
		Tocopherols (mg/100 g)	14–15
Sterol composition (% total sterols)		Tocopherol composition (% total tocopherols)	
Cholesterol	< 0.1	α-Tocopherol	17
Brassicasterol	< 0.1	β-Tocopherol	–
Campesterol	22	γ-Tocopherol	83
Stigmasterol	3	δ-Tocopherol	–
β-Sitosterol	68	Tocotrienols	–
Δ5-Avenasterol	2		
Δ7-Stigmasterol	2		
Δ7-Avenasterol	–	Triterpenic alcohols (mg/100 g)	35

OLEAGINOUS FLAX

Flax is a dicotyledonous plant of the *Linum usitatissimum* species, genus *Protolinum*, sub-group *Eulinum,* of the linaceae family. Its geographical origin is believed to be Asia in view of the presence of a very large number of allied species in this region. These species, however, have never been developed so far since they are not of major economic interest. Today, flax grows in all temperate zones located in the latitudes of roughly 30° to 60°. The plant has a single stem which bears lanceolate sessile leaves and some branches at the tip. It grows to a height of 0.4 to 1.2 m. According to their qualities, the varieties of the species are divided into oleaginous flax (linseed) and textile flax (fibre flax). The varieties are differentiated by the yield in grain or fibre, height of stem, size of seeds, quality of fibre, etc. The conditions of farming and the effects of the climate further accentuate the differences between the two types of flax.

When mature, the seed most often has a brown colour. With a smooth and glossy surface, it is oblong and flat, with a more or less curved beak. It develops in a capsule with 5 loculi with 2 compartments each and one seed in each compartment. The seed is made up of an embryo, covered with an oily albumen and protected by teguments.

The latter have five cell layers. From the outermost to the innermost layer:

1. A transparent mucilagenous layer.
2. A greenish layer of parenchymatous aspect.
3. A fibrous layer with thick yellow membranes.
4. A thin and colourless, irregular layer of broken cells.
5. A monocellular red layer which gives the seed its colour; if it is lacking, the seed is greenish or yellowish and if it is atrophied, the seed is dark brown in colour.

Full-grown linseeds weight 6 to 14 g per 1000 seeds for the flax varieties, whereas for the fibre variety the same number of seeds weight 4 to 7 gram.

Oil is contained in the cytoplasma of embryo and albumen cells. Oil content varies from 35 to 45 per cent dry weight. For the fibrous flax, it is generally less than 40 per cent. Oil content depends of course on the size and shape of the seed which influence the relative size of the teguments. Compared to the weight of the deoiled dry fraction, protein content is in the range of 36 to 40 per cent.

Oil Compositions

The main physical and chemical characteristics of linseed oil are shown in Table 16.12.

Table 16.12. Main physical and chemical characteristics of linseed oil.

D_{20}	0.928–0.933
V_{20} (cp)	42–47
n_D^{20}	1.479–1.484
Solidification point (°C)	
Iodine value	170–204
Saponification value	189–196

The iodine value (IV), which indicates the oil's drying property, can be calculated by the following formula:

$$1.16 \text{ IV} = 3.04 \text{ Ln} + 2.02 \text{ L} + \text{O}$$

The oil consists essentially of five fatty acids, the percentages of which are indicated in Table 16.13. Traces of other saturated (La, M, etc.) and unsaturated (Po) fatty acids are occasionally found.

Table 16.13. Fatty acid composition in linseed oil.

Nature	*% of total FA*			
	Europe	*Canada*	*Argentina*	*India*
C16:0	4–6	5–6	4–5	9–10
C16:1	< 0.5	< 0.3	< 0.3	< 0.3
C17:0	–	–	–	–
C18:0	2–3	3–4	5–6	7–8
C18:1	10–22	19–20	19–21	10–21
C18:2	12–18	14–16	15–24	13–15
C18:3	56–71	54–61	45–53	50–61
C20:0	< 0.5	< 0.5	< 0.5	< 0.5
C20:1	< 0.6	< 0.6	< 0.6	< 0.6
C22:0	–	–	–	–

The area of farming seems to have a bearing on the composition of fatty acids; the flaxes produced in Canada or in Europe are richer in linolenic acid than those produced in Argentina and as such have a higher iodine index. This point is important for the users in the paint and varnish industry. Table 16.14 shows the relevant composition of the unsaponifiable fraction.

Table 16.14. Composition of unsaponifiable fraction of linseed oil.

Constituents	*Unsaponifiable fraction 0.5–1.3%*	*Constituents*	*Unsaponifiable fraction 0.5–1.3%*
Sterols (mg/100 g)	200–410	Hydrocarbons (mg/100 g) of which squalene	80–100 20–30
		Tocopherols (mg/100 g)	110–130
Sterol composition (% of total sterols)		Tocopherol composition (% of total tocopherols)	
Cholesterol	1–2	α-Tocopherol	
Brassicasterol	0–1	β-Tocopherol	
Campesterol	28–29	γ-Tocopherol	
Stigmasterol	9–10	δ-Tocopherol	
β-Sitosterol	44–53	Tocotrienols	
Δ5-Avenasterol	10–13		
Δ7-Stigmasterol	1–4		
Δ7-Avenasterol	< 1	Triterpenic alcohols (mg/100 g)	150

ALMOND, HAZELNUT AND WALNUT OIL

Almond Oil

Almond tree is cultivated first and foremost for nut production. The crop is largely disseminated in all countries having a temperate or mildly warm climate. The almond tree (*Prunus amygdalus*) belongs to the Rosaceous family. Its fruit contains one or even two seeds or almonds. There are two types of almonds: the sweet almond and the bitter almond. The latter contains cyanogenetic glucoside (amygdaloside) which remains in the meal after oil extraction. The oil content of almonds varies from 50 to 65 per cent.

Almond oils are used for seasoning ('cold'-pressed oils) but also in cosmetic and pharmaceutical formulations. They are labeled as 'sweet almond oil' to avoid confusion with the essential oil (or essence) of bitter almonds extracted from meals of this variety and used for perfumery. Sweet almond oil is listed in the French Pharmacopoeia.

Physico-chemical characteristics

Table 16.15 gives the main physical and chemical characteristics of almond oil.

Table 16.15. Main physical and chemical characteristics of almond oil.

D_{20}	0.911–0.917
V_{20} (cp)	66–76
n_D^{20}	1.471–1.472
Solidification point (°C)	– 10 to – 20
Iodine value	92–106
Saponification value	189–196

Composition of triglycerides

Table 16.16 gives the triglycerides composition of almond oil.

Table 16.16. Triglyceride composition of almond oil (HPLC).

Nature	*% of total TG*
LLL	1–4
OLL	12–20
PLL	1–4
LOO	21–28
POL	7–11
OOO	23–40
POO	7–10
PPO	< 0.4
SOO	1–5

Fatty acid composition

Table 16.17 gives the fatty acid composition of almond oil.

Table 16.17. Fatty acid composition of almond oil.

Nature	*% of total FA*
C14:0	–
C16:0	6–8
C16:1	< 1
C17:tot.	–
C18:0	1–2
C18:1	64–82
C18:2	8–28
C18:3	< 0.2
C20:0	< 0.1
C20:1	< 0.1
C22:0	< 0.2

Composition of unsaponifiable fraction

Table 16.18 gives the composition of unsaponifiable fraction in almond oil. The values indicated for fractions of hydrocarbons and triterpenic alcohols are those given by *Itoh*.

Table 16.18. Composition of unsaponifiable fraction in almond oil.

Constituents	*Unsaponifiable fraction 0.3–1.2%*	*Constituents*	*Unsaponifiable fraction 0.3–1.2%*
Sterols (mg/100 g)	120–400	Hydrocarbons (mg/100 g)	470
		of which squalene	30–50
		Tocopherols (mg/100 g)	25–56
Sterol composition (% total sterols)		Tocopherol composition (% of total tocopherols)	
Cholesterol	< 1	α-Tocopherol	92–99
Brassicasterol	–	β-Tocopherol	< 1
Campesterol	2–3	γ-Tocopherol	1–7
Stigmasterol	0–1	δ-Tocopherol	–
β-Sitosterol	74–87	Tocotrienols	–
Δ5-Avenasterol	7–20		
Δ7-Stigmasterol	0–5		
Δ7-Avenasterol	0–4	Triterpenic alcohols (mg/100 g)	40–70
Fucosterol	1–2		

Itoh, has mentioned a percentage of 230 mg/100 gram for hydrocarbons and esterified sterols fraction in a sample of bitter almond (*Prunus communis*) oil.

As for tocopherol composition, *Salvo*, gives for the bitter almond oil 74 per cent of α-tocopherol, 24 per cent of γ-tocopherol with traces of β-tocopherol, whereas sweet almond is said to contain only α-tocopherol and trace of γ-tocopherol (Table 16.18).

Hazelnut Oil

The hazelnut tree (*Corylus avellana*), which belongs to the Betulaceae family, produces a fruit with a high oil content: 50–65 per cent, sometimes even 65–70 per cent. In Europe, hazelnut oil comes mainly from Italy, Spain and Turkey. It is used in seasoning but mostly in the cosmetic industry, in personal care products as well as by the pharmaceutical industry.

Hazelnut oil meals may contain up to 30 per cent oil. Further solvent extraction is therefore necessary.

Physico-chemical characteristics

Table 16.19 shows the main physical and chemical characteristics of hazelnut oil.

Table 16.19. Main physical and chemical characteristics of hazelnut oil.

D_{20}	0.912–0.915
V_{20} (cp)	66–76
n_D^{20}	1.470–1.471
Solidification point (°C)	– 18 to – 20
Iodine value	85–95
Saponification value	190–195

Triglyceride composition

Table 16.20 shows the triglyceride composition of hazelnut oil.

Table 16.20. Triglyceride composition of hazelnut oil (HPLC).

Nature	*% of total TG*
LLL	< 2
OLL	2–9
OOLn	< 0.5
PLL	< 1
OOL	14–27
POL	1–5
OOO	44–60
POO	5–13
PPO	< 0.3
SOO	2–6
PSO	< 0.1

Fatty acid composition

Table 16.21 shows the fatty acid composition of hazelnut oil. Belonging to the 'oleic oil' group, the saturated fatty acid content of hazelnut oil, is around 8–9 per cent; the range is much wider for polyunsaturated fatty acids (linoleic acid). The sub-committee on vegetable oils of the Italian Technical Commission on Oils and Fats provides the following specifications:

C16:0, 4.5–7.5;; C18:0, 1.8–3.2; C18:1, 77–84; C18:2, 6.0–14.

Table 16.21. Composition of fatty acids in hazelnut oil.

Nature	*% of total fatty acids*
C14:0	< 0.1
C16:0	5–9
C16:1	< 0.3
C17:tot.	< 0.1
C18:0	1–4
C18:1	66–83
C18:2	8–25
C18:3	< 0.6
C20:0	< 0.3
C20:1	–
C22:0	–

Composition of unsaponifiable fraction

Table 16.22 shows the content and composition of the major components. *Colombini* indicates the unsaponifiable fraction content as 1.4–1.6 per cent in an hazelnut oil extracted by solvent in the laboratory. Among sterols, the Δ7-avenasterol content varies largely from one laboratory to another.

Table 16.22. Composition of unsaponifiable fraction of hazelnut oil.

Constituents	*Unsaponifiable fraction 0.3–0.7%*	*Constituents*	*Unsaponifiable fraction 0.3–0.7%*
Sterols (mg/100 g)	75–195	Hydrocarbons (mg/100 g)	–
		Tocopherols (mg/100 g)	30–55
Sterol composition (% total sterols)		Tocopherol composition (% total tocopherol)	
Cholesterol	< 1	α-Tocopherol	79–98
Brassicasterol	< 0.5	β-Tocopherol	2–4
Campesterol	4–7	γ-Tocopherol	2–19
Stigmasterol	< 2	δ-Tocopherol	1–6

(Contd...)

Table 16.22. Composition of unsaponifiable fraction of hazelnut oil. *(Contd.)*

Constituents	*Unsaponifiable fraction 0.3–0.7%*	*Constituents*	*Unsaponifiable fraction 0.3–0.7%*
β-Sitosterol	82–88	Tocotrienols	–
Δ5-Avenasterol	2–5		
Δ7-Stigmasterol	1–3		
Δ7-Avenasterol	1–6	Triterpenic alcohol (mg/100 g)	15–17

Leone has indicated the contents of methylsterols (10.3–12.5 mg/100 g oil) and dimethyl sterols (triterpenic alcohols) along-with the composition of these fractions. No data has been found on the hydrocarbon fractions. Tocopherol composition varies considerably, particularly for δ-tocopherol; irrespective of varieties, the methods of analysis used are probably responsible for this disparity. In all cases, the major component is α-tocopherol (80–98 per cent).

Walnut Oil

The common walnut tree, *Juglans regia*, belongs to the Juglandaceae family. Originating from Persia and the Caucasus, its crop spread to Europe and mainly in France (Périgord and Dauphiné) where it is cultivated essentially for its tasty nuts. These nuts are very rich in oil, containing 50–60, sometimes even 70 per cent. The seasonal production of walnut oil (mid-December to the end of March) is, however, rather small. The oil obtained by the first 'cold' pressing is much appreciated for seasoning because of its special taste. It also has a dietetic value owing to the presence of polyunsaturated fatty acids (C18:2 and C18:3). This high degree of unsaturation calls for special precautions during the oil processing and storage.

A virgin oil is produced by 'cold' pressing of shelled kernels (about 45 per cent of the nut total weight): this is a craft industry, supplemented sometimes by large-scale industrial production.

The damaged shelled kernels, unfit for consumption, are sent to oil mills where they are crushed to produce a paste which is heated and then pressed hydraulically. The temperature affects the yield and also the taste of oil. The average yield of oil is 0.5 1/1 kg nutmeat, i.e. 11 oil from 4.5 kg of whole nuts.

The meal can be used as animal feed, but it quickly turns rancid. It is also used as fertiliser.

Physico-chemical characteristics

Table 16.23 shows the physical and chemical characteristics of walnut oil.

Table 16.23. Main physical and chemical characteristics of walnut oil.

D_{20}	0.924–0.926
V_{20} (cp)	
n^{20}_D	1.475–1.476
Solidification point (°C)	– 15 to – 25
Iodine value	135–151
Saponification value	189–198

Fatty acid composition

Table 16.24 summarises the fatty acid composition of walnut oil. Walnut oil belongs to the 'linoleic' groups and contains a significant amount of linolenic acid (≈10 per cent). Saturated fatty acids constitute about 10 per cent of the total fatty acids and the approximately 90 per cent unsaturated fatty acids are mainly of the polyunsaturated type (about 72 per cent).

Table 16.24. Fatty acid composition of walnut oil.

Nature	*% of total FA*
C14:0	< 0.1
C16:0	6–8
C16:1	< 0.2
C17:0	< 0.1
C18:0	1–3
C18:1	14–21
C18:2	54–65
C18:3	9–15
C20:0	< 0.3
C20:1	< 0.3
C22:0	< 0.2

Composition of unsaponifiable fraction

Table 16.25 summarises the composition of unsaponifiable fraction in walnut oil.

Table 16.25. Composition of unsaponifiable fraction of walnut oil.

Constituents	*Unsaponifiable fraction 0.5–1.0%*	*Constituents*	*Unsaponifiable fraction 0.5–1.0%*
Sterols (mg/100 g)	170–245	Hydrocarbons (mg/100 g)	270–392
		e.g. squalene (mg/100 g)	1–3
		Tocopherols (mg/100 g)	55–60
Sterol composition (% total sterols)		Tocopherol composition (% total tocopherols)	
Cholesterol	< 0.5	α-Tocopherol	1–5
Brassicasterol	< 0.4	β-Tocopherol	< 2
Campesterol	4–7	γ-Tocopherol	82–95
Stigmasterol	< 2	δ-Tocopherol	4–15
β-Sitosterol	82–90	Tocotrienols	–
Δ5-Avenasterol	4–7		
Δ7-Stigmasterol	< 4		
Δ7-Avenasterol	< 1	Triterpenic alcohols (mg/100 g)	50–180
Ergosterol	< 0.3		
Fucosterol	< 1		

Grace reports a content of 0.1 per cent for the unsaponifiable fraction in refined oil and 0.6 per cent in crude or virgin oil. The sterol content of refined oil is reported as 54 mg/100 g oil, which seems low. Sterol composition shows small quantities of cholesterol. Some authors report the presence of iso-fucosterol (0.7–1.5 per cent). The chemical nomenclature of this sterol (24-Z-ethyl idenecholestene-5, 3β-ol) leads to the conclusion that its structure is identical to that of Δ5-avenesterol. The improper use of common names in the case of phytosterol may be a source of confusions and inaccuracies.

EVENING PRIMROSE, BORAGE AND BLACKCURRANT OILS

It has been shown that the needs of the body for essential fatty acids (EFA) vary according to age, physiological and pathological conditions. The use of oils rich in linoleic acid (sunflower, corn, safflower, grapeseeds, etc.) and linoleic and α-linolenic acids (soyabean, rapeseed) is generally sufficient for meeting these needs especially if the subjects are in good health and able to metabolise the intake of the essential fatty acids.

However, the metabolism of the essential fatty acids (EFA) the regulation of the synthesis process (Δ6 and Δ5 desaturases), the local conditions of cellular exchanges and absorption of fatty acids may be disturbed and hence may lead to deficiencies not always having an visible clinical manifestation. The use of oils with a high γ-linolenic 18:3 (*n*-6) content makes it possible to overcome restrictive stages of the biosynthesis of arachidonic acid. That is why nutritionists took recourse to some rare cultivated plants whose seeds contain γ-linolenic acid, namely primrose, borage and blackcurrant seeds.

These are rare species of flowering plants capable of producing γ-linolenic (γ-Ln) acid, giving thus the evidence of presence of a group of genes which code for the Δ6 desaturase enzyme in this genome. Three plants have given rise to industrial and commercial developments: primrose (*Œnothera biennis*) borage (*Borago officinalis*) and blackcurrant (*Ribes nigrum*).

Evening Primrose Oil

The evening primrose pertains to the order of Myrtles because the flower is made up of a green calyx and clearly visible corolla, both growing within a tube called hypanthium which prolongs the ovary (the ovary is inferior). Within the Myrtles, the evening primrose belongs to the Onagraceae family as its flower follows the pattern of four (4 sepals, 4 petals, 4 + 4 stamens and 4 joined carpels). Amongst the Onagraceae, the evening primrose of the Enothera genus is characterised by its yellow or white flower producing a fruit with a capsule providing four compartments each and containing about one hundred seeds. The seed of the evening primrose is prismatic and angular with a length of 1.2–2.2 mm and a breadth of 0.5–1.0 mm according to the species. It contains only the embryo (exalbuminous seed); the lipid reserves are therefore those of the cotyledon cells. In the 25 species which have been analysed, the qualitative composition of oil has been found to be very homogeneous (Table 16.26). All the evening primroses produce γ-Ln, a property confirmed by the studies carried out on seeds of the species belonging to other sections and sub-sections of the evening primrose.

On the other hand, the content of fatty acids varies widely from one specie to another. For agronomic purposes, we are considering the content of oil in the seed and the content of γ-Ln in the oil: the former varies between 12 and 30 per cent and the latter between 4 and 12 per cent (two species, *O. blandina* and *O. purpurata*, however, have a γ-Ln content of less than 1 per cent).

Table 16.26. Intra and interspecific variations of content of oil and γ-Ln in the seeds of evening primroses (8 species of the enothera section, sub-section enothera).

Species	N^+	*(1)* *% oil in seed*	*(2)* *% γ-Ln in oil*	*(3)* *% γ-Ln in seed*	*(4)* r^{++}
O. hookeri	42	24.0 ± 3.2	10.0 ± 1.6	2.4 ± 0.4	–0.3 ns
O. cokerelli	51	28.6 ± 1.6	10.6 ± 0.6	3.0 ± 0.1	–0.6 ***
O. biennis, var. biennis	45	24.0 ± 4.0	10.7 ± 1.6	2.5 ± 0.3	–0.8 ***
O. biennis, var. cruciata	46	24.7 ± 2.5	10.1 ± 1.2	2.5 ± 0.2	–0.9 ***
O. rubricaulis	36	23.5 ± 1.7	8.5 ± 1.1	2.0 ± 0.3	–0.5 ***
O. issleri	27	33.1 ± 0.1	9.4 ± 0.9	3.1 ± 0.3	–0.5 ***
O. rubricuspis	24	33.1 ± 2.6	11.1 ± 1.2	3.7 ± 0.3	–0.8 ***
O. silesiaca	15	24.5 ± 3.0	12.1 ± 1.4	2.9 ± 0.2	–0.8 ***

Columns (1) to (3) give the average ± standard variation (× ± σ) by species.
Variable (3) is calculated from variables (1) and (2).
N^+ is the number of plants which supplied a sample of seeds weighing 3 g, produced by controlled self-fertilisation.
r^{++} is the coefficient of linear correlation between the variables (1) and (2).
ns: not significant.
***: significant with a risk of 0.001.

Upgrading of the evening primrose for production of γ-Ln

Upgrading of this plant consists in measuring the variation in the content of oil and γ-Ln in individual species having more than 8 per cent γ-Ln in oil; amongst the 25 species mentioned above, only 8 fell in this category and their data are recorded in Table 16.26. The variation in the content of oil is very high, its value depending on the particular species. The content of γ-Ln in oil and in the seed vary, however, very narrowly since the content of oil in the seed and that of γ-Ln in oil have a significant negative correlation, we must also take into account their capacity to produce γ-Ln in the seed, for assessment of the different species.

Four species stand out on the basis of these two criteria:

1. *O. rubricuspis.*
2. *U. silesiaca.*
3. *O. cockerelli.*
4. *O. issleri.*

The same species were tested in the field. The data on the basis of 2 years of cultivation shows that the content of oil was slightly less and that of γ-Ln remains unchanged with regard to individual data. The output of seeds varies widely from one cultivation to another. The average for all the test cases comes to 18 T/ha. The best production of oil is 6 T/ha., the average being 4.8 T/ha and the best production of γ-Ln is 71 kg/ha, the average being 47.7 kg/ha.

However, even if the production of γ-Ln could be upgraded by selection of the species, the problems arising during the cultivations of the evening primrose are very important. They are three in number: there is no herbicide which is specific for the evening primrose, the harvesting of capsules on swath leads to loss of seeds which may go up to 10 g/ha and the method of extraction of oil from seed is not yet satisfactory. These defects cost heavily on conditions of contract for cultivation of the evening primrose.

The evening primrose oil is generally obtained by mechanical pressure followed by extraction by hexane. It can also be obtained by pressure applied when cold.

Table 16.27 indicates the distribution of fatty acids on position 1, 2 and 3 of the triglyceride and Table 16.28 gives the composition of the fatty acids.

Table 16.27. Fatty acid distribution in 1, 2, 3 positions of evening primrose oil triglycerides.

Nature of fatty acids (FA)	*% of total FA on 2*	*% of total FA on 1*	*% of total FA on 3*
C16:0	1.8	9.6	6.8
C18:0	1.5	4.4	0.3
C18:1	7.6	8.0	7.1
C18:2	78.9	73.3	74.2
C18:3	–	–	–
C18:3 (*n*-6)	10.7	3.6	13.5

Table 16.28. Fatty acid composition of evening primrose oil.

Nature	*% of total FA*
C14:0	<-0.1
C16:0	5–9
C16:1	< 0.1
C18:0	1–2
C18:1	8–12
C18:2	70–79
C18:3	< 0.1
C18:3 (*n*-6)	8–13
C20:0	< 0.3
C20:1	< 0.1
C22:0	< 0.1

Borage

Borago officinalis is a herbaceous plant found in the western Mediterranean region. The Borago genus includes three species, including *Borago officinalis*, commonly known as borage.

The plant was initially used as a medicinal plant before the exceptional composition of the oil of *Borago officinalis* was noticed. The French name 'bourrache' is spelt either with a single 'r' or with two 'r's. The botanical nomenclature uses only one 'r' for the name of the family or of the genus as it is believed that the etymology of the term is derived from 'Abu Rach' meaning 'Father of sweat' an expression linked to the reputation of its flower being sudorific.

Other properties have been ascribed to different parts of the plant. The extract from the stem, the foliage and the flowers has some interesting chemical properties; the flower contains allantoine. The stem is rich in potassium nitrate which could explain the diuretic and sudorific properties of borage. We also find in it pyrrolizidinic alkaloids which do not evince the mutagenous and hepatotoxic alkaloids often found in this type of substances. Lastly, the plant is rich in borensitol and contains a mucilage which releases glucose, galactose and arabinose by hydrolysis.

Seeds are slightly ovoid in shape, the seed measures, about 6 mm in length and 3 mm in breadth. It weights about 15–20 mg. The contents of oil in the seeds of borage varies according to the degree of their maturity. There is a problem with harvesting, as all seeds do not ripen at the same time. The percentage of oil extracted varies in different samples between 28–33 per cent for ripe seeds and 13–18 per cent for seeds which are unripe. The use of herbicides gave negative results.

The most delicate stage of cultivation is the harvesting. It could be undertaken as soon as the first seeds (of pepper-grain size) start falling. Two methods were used for the sake of comparison:

1. Defoliant treatment with diquat, followed after eight days by the use of a harvester-thresher.
2. Cutting with a swash power at a certain height (10–15 cm) followed by the use of harvester-thresher after six days.

The two methods produced comparable results. In case of extensive cultivation, the output varies between 300 and 500 kg/ha whereas the real potential can vary from 800–1000 kg or more/ha. The various trials (already carried out or in process) aim at devising either varieties with a lesser period of flowering or a harvesting system adapted to this progressive maturity.

There are different methods for extraction of borage crude oil:

1. Application of hot pressure.
2. Extraction by solvents (particularly by hexane).
3. Application with cold pressure.
4. Extraction of super-critical CO_2 (pilot industrial stage).

The cost, the output of oil and the quality of the crude oil vary according to the process used. The average composition of fatty acids in borage oil is given in Table 16.30. The distribution of fatty acids in triglyceride positions 1, 2, 3 is given in Table 16.29.

The borage oils found at present in the French and European markets are either of 'virgin' quality, i.e. obtained by cold pressure, or of 'refined' quality specifying the operations and the approved products.

Only 'specific' refining can guarantee the stability of the oil in terms of timing, organoleptic qualities and the total absence of contaminants (pesticides, defoliants). By 'specific' means taking particular care during the process of deodourisation (at low temperature so as to avoid the formation of *trans* isomers and loss of tocopherols) as well as saturation of nitrogen before, during and after deodourisation in order to obtain a percentage of dissolved oxygen below 0.5 ppm. Thus, refined borage oil will have a very low level of free acidity (in terms of oleic acid), a peroxide value PV < 5 meq O_2/kg with total absence of phytosanitary products.

Table 16.29. Fatty acid distribution in 1, 2, 3 of borage oil triglycerides.

Nature of FA	*% of total FA in 2*	*% of total FA in 1*	*% of total FA in 3*
C16:0	2.3	16.9	13.1
C18:0	1.3	5.3	3.1
C18:1	13.7	13.8	18.7
C18:2	42.5	54.3	17.6
C18:3 (*n*-6)	40.4	4.0	30.1
C20:1	0.8	3.7	7.7
C22:1	0.4	0.3	5.4

On the other hand' 'virgin' borage oil, suitably cold pressed, should essentially come from fresh seeds, pure and contamination-free. Otherwise, the pollution, if any, existing in the seed will inevitably pass onto the oil. 'Virgin' oil has a good standing in dietetics (in capsule form).

The studies performed by *Carnat*, quoted above, provide the following results regarding the stability of the non-refined, solvent-extracted oil:

Table 16.30. Fatty acid composition in borage oil.

Nature	*% of total FA*
C16:0	9–15
C16:1	< 0.4
C18:0	3–7
C18:1	15–19
C18:2	32–38
C18:3	< 1
C18:3 (*n*-6)	18–25
C20:0	< 0.4
C20:1	2–4
C22:1	2–4

1. Oil preserved for one year at 4°C in the dark.
   PV in mmol/kg 95
2. Oil preserved for one year at 22–23°C in daylight
   PV in mmol/kg 835

Original oil had a PV of 6. Refined oil, when stored in dark for a year, does not show any appreciable change in PV, in acid value and in the content of γ-linolenic acid.

Main uses of borage oil

The main interest of borage oil lies in the occurrence of a high percentage of γ-linolenic acid (20–25 per cent). The pharmacological properties of γ-Ln are a result of its structure and its special position in the metabolic chain of the *n*-6 family. These properties are multiple and diversified. The main uses are in hospital dietetics, dispensary pharmacy and cosmetology. They are comparable to those of the evening primrose which has about double the content of γ-Ln.

Blackcurrant Oil

The blackcurrant (*Ribes nigrum*) is a round, dull black berry. The pulp is rather sour whereas the skin is aromatic. Because of its taste and smell, the blackcurrant is not a very popular fruit. Like the two congeneric fruits, the gooseberry bush and the red currant bush, the blackcurrant is cultivated in Europe and in Asia. It grows naturally in France in the humid forests and valleys of Auvergne and Mont-Cenis, in Piémont and in Central Europe.

Blackcurrant seed oil is extracted by the use of hexane (ratio 1:3) on washed seeds. Cleaning includes special processing in alcohol at 70°C, followed by vacuum drying at 90°C.

The extracted oil is then refined by traditional alkaline refining techniques and then stabilised by addition of 200 ppm ascorbyl palmitate.

Specifications for blackcurrant seed oil are given below:

1. Acid value max. 0.6.
2. Peroxide value 5 milliequivalents O_2/kg oil.
3. Added antioxidant Ascorbyl palmitate, 200 ppm.

The composition of fatty acids given in Table 16.31.

Table 16.31. Fatty acid composition in blackcurrant seed oil.

Nature	*% of total FA*
C16:0	6–8
C16:1	< 0.2
C18:0	1–2
C18:2	9–13
C18:3 (*n*-3)	44–51
C18:3 (*n*-6)	12–14
C18:4 (*n*-3)	15–19
C20:1	2–4
C22:0	< 0.5

The peculiarity of the blackcurrant seed oil, as compared to evening primrose and borage oils, is the occurrence of α-linolenic acid mainly stearidonic acid and this fatty acid is situated after the limiting stage of δ-6-desaturase. For the *n*-3 series, it is the equivalent, from the metabolic point of view, to γ-Ln in the *n*-6 series. The glyceride structure and the composition of the insaponifiable fraction are given in Tables 16.32, 16.33, 16.34 and 16.35.

Table 16.32. Triglyceride composition of blackcurrant seed oil (GC).

Nature	*% of total TG*
C48	–
C50	0.6
C52	19.3
C54	77.0
C53	3.1

Table 16.33. Triglyceride composition of blackcurrant seed oil (HPLC).

Nature	*% of total TG*	*Nature*	*% of total TG*
Ln Ln-γLn Ln-γLn-γLn γLn-γLn-γLn	5.5	O L Ln O L γLn P L Ln	3.7 5.1 2.4

(Contd...)

Table 16.33. Triglyceride composition of blackcurrant seed oil (HPLC). *(Contd.)*

Nature	*% of total TG*	*Nature*	*% of total TG*
S_4 Ln L ⎤		P L γLn	3.2
S_4 γLn L ⎥		O L L	6.8
S_4 S_4 P ⎦		P L L	6.3
Ln-γLn L	7.9	P O Ln ⎤	
L L Ln	10.2	P P γLn ⎦	2.1
O Ln Ln ⎤			
O-γLn-γLn ⎥	6.0	O O L	2.0
O L S_4 ⎥			
P Ln-γLn ⎥		S L L	1.7
L L L ⎦	12.9		

S_4: Stearidonic

Table 16.34. Fatty acid distribution in positions 1, 2, 3 of triglycerides of blackcurrant oil.

Nature of fatty acid	*% of total FA in 2*	*% of total FA in 1*	*% of total FA in 3*
C16:0	2.0	14.2	4.4
C18:0	1.5	4.9	–
C18:1	14.1	12.6	5.1
C18:2	53.1	42.7	43.8
C18:3 (*n*-3)	8.1	17.2	14.1
C18:3 (*n*-6)	17.4	4.1	25.8
C18:4 (*n*-3)	2.6	0.7	–

Table 16.35. Composition of unsaponifiable fraction in blackcurrant oil.

Constituents	*Unsaponifiable fraction 1.0–1.2%*	*Constituents*	*Unsaponifiable fraction 1.0–1.2%*
Sterols (mg/100 g)	400	Hydrocarbons (mg/100 g),	100
		of which squalene	35
		Tocopherols (mg/100 g)	150
Sterol composition (% total sterols)		Tocopherol composition (% total tocopherols)	
Cholesterol	0.7	α-Tocopherol	6.0
Brassicasterol	–	β-Tocopherol	–
Campesterol	7.2	γ-Tocopherol	82.5
Stigmasterol	0.2	δ-Tocopherol	7.9
β-Sitosterol	85.4	Tocotrienols	–
Δ5-Avenasterol	3.0		
Δ7-Stigmasterol	0.4		
Δ7-Avenasterol	0.4	Triterpenic alcohols (mg/100 g)	105

Main uses of blackcurrant seed oil

The main uses of blackcurrant seed oil as in the cases of evening primrose and borage, are generally based on claims which concern the pharmacological properties of γ-Ln.

Competing biotechnologies

Scientific research has confirmed the interest of γ-linolenic acid in cosmetic, dietetic and pharmaceutical products. The oldest and best known source of γ-Ln is the evening primrose oil. This recent development reflects the active role played by γ-Ln on the market and the renewed interest of industrialists in this product, which was already known for a long time. Besides the vegetable oils already mentioned, it will be appropriate to examine the biotechnological sources which have been developed specially in Japan. The micro-organisms producing 'particular' fats may constitute an alternative to the agricultural products. The algae of the Cyanophyceae family or certain moulds (Mucorals) generally produce fatty acids of γ-Ln type.

Two types of cultivating methods are used for optimising mould cultures: cultivation of free cells and cells immobilised on a solid support or retained in a gel.

Within the reactors, important parameters of cultivation (pH, temperature, dissolved oxygen, carbon dioxide) are controlled with the help of probes and a computerised system of management.

In future, the use of reactors with immobilised cells or of fermenters may lead to a definite improvement of output not only of biomass but mostly of γ-Ln.

In France, the Fat Chemistry Department of IRHO in Montpellier, in particular, has distinguished itself during the last decade thanks to γ-Ln research in different vegetable species, utilisation of evening primrose, microbiological production of γ-Ln-rich triglycerides and production of β-monoglycerides by borage oil lipolysis.

As regards industrial production, besides methods of utilisation of algae and bacteria, those using filament mushroom are also applied in different countries of the world. However, the use of lipids produced by biotechnology for medical or dietetic purposes should compulsorily be supported by a thorough toxicological experimentation (for detection of an unidentified elements in the unsaponifiable fraction). Use in cosmetics could be one of the very first utilisations for γ-Ln obtained by fermentation.

A 'good' fermentation medium, a good strain and an adequate 'know-how' are the three fundamental elements of biotechnological success. As for economic feasibility it will depend essentially on the prices that the market can afford, marketing policy and production cost.

WHEAT GERM OIL

Wheat germ oil is used in dietetics and cosmetics. Main physical and chemical characteristics of wheat germ oil are given in Table 16.36 and fatty acids composition is given in Table 16.37.

Table 16.36. Main physical and chemical characteristics of wheat germ oil.

D_{20}	0.924–0.925
V_{30} (cp)	45–47
n_D^{20}	1.474–1.478
Iodine value	118–128
Saponification value	190–196

Table 16.37. Fatty acid composition of wheat germ oil.

Nature of fatty acid	*Content in % of total FA*
C16:0	13–20
C16:1	< 0.1
C17:0	–
C18:0	< 2
C18:1	13–21
C18:2	55–60
C18:3	4–10
C20:0	< 0.2
C20:1	–
C22:0	–
C22:1	–
C24:0	–

Wheat germ oil is famous for its richness in tocopherols (Table 16.38). The main isomers of wheat germ oil are: α (vitamine E) and β-tocopherol. *Piironen* found different contents for β and γ tocopherols: respectively 13 and 22 per cent of total tocopherols; he detected no β-tocotrienol but some traces of β-tocotrienol. Some commercial wheat germ oils may contain up to 200 mg α-tocopherol/100 gram oil.

Table 16.38. Composition of the unsaponifiable fraction of wheat germ oil.

Constituents	*Unsaponifiable fraction 3–4%*	*Constituents*	*Unsaponifiable fraction 3–4%*
Sterols content (mg/100 g oil)	500–900	Tocopherol content (mg/100 g oil)	190–240
Sterol composition (% total sterols)		Tocopherol composition (% total tocopherols)	
Cholesterol	–	α-Tocopherol	60–65
Brassicasterol	–	β-Tocopherol	20–25
Campesterol	20–25	γ-Tocopherol	5–9
Stigmasterol	2–7	δ-Tocopherol	2
β-Sitosterol	65–67	α-Tocotrienol	2
Δ5-Avenasterol	4–5	β-Tocotrienol	2
Δ7-Stigmasterol	–	γ-Tocotrienol	tr
Δ7-Avenasterol	< 2	δ-Tocotrienol	–

CONJUGATED OILS

Tables 16.39 and 16.40 give the main physical and chemical characteristics and fatty acid composition of tung, oiticia and ongokea oils. Tung oil is obtained from the kernel of the tung tree (*Aleurites fordii*) of the Euphorbiaceae family, which is a native of China and Indo-China. Abrasin oil is obtained from *Aleurites montana*. For commercial purposes, these two oils having very similar compositions are mixed together. The kernels contain 50–60 per cent oil.

Table 16.39. Main physical and chemical characteristics.

Description	*Tung*	*Oiticica*	*Ongokea*
D_{20}	0.934–0.943	0.966–0.977	0.975–0.995
V (cp)	–	–	–
n_D^{20}	1.518–1.525	1.511–1.517	1.505–1.515
Titre	36–40°	42–47°	5° approx.
Iodine value	166–170	140–180	180–205
Saponification value	188–197	187–193	188–195

Table 16.40. Fatty acid composition.

Nature	*% Total FA Tung*	*% Total FA Oiticica*	*% Total FA Ongokea*
C16:0	3–5	6–8	2–4
C16:1	–	–	–
C17	–	–	–
C18:0	1–2	4–6	1–2
C18:1	4–9	5–7	7–17
C18:2	4–10	5–7	4–5
C18:3	1–3	–	–
C18:3 conj.	77–78	–	–
C18:3 conj. keto	–	48–78	–
C18:1 diynoic	–	–	70–80

Ongokea (or isano) oil is obtained from the seeds of *Ongokea Gore* (containing 60–78 per cent) which is a native of Equatorial Africa. Oiticica oil is obtained from the nuts of *Licania rigida* (oil content: 40–65 per cent) which is a native of Brazil.

The main component of tung oil is α-eleostearic acid (9, 11, 13 C18:3 eventually accompanied by its all *trans* isomer β-eleostearic acid). The main component of oiticica oil is licanic acid (4-keto-eleostearic acid). Ongokea oil is characterised by the presence of acetylenic acids: isanic (octadec-17-ene-9, 11-diynoic), bolekic (conjugated octadecenediynoic), isanolic (hydroxy-octadec-17-ene-9,11-diynoic).

These oils are mainly used in the manufacture of paints and varnishes. The polymerisation of ongokea oil is a very strong reaction which must be carried out with adequate precautions.

OTHER OILS

The data on oils which, because of their peculiar fatty acid composition, are interesting for oleochemistry. The plants from which they are extracted are presently the subject of agricultural research.

Crambe Oil

This oil is characterised by its high content of erucic acid (Tables 16.41 and 16.42).

Table 16.41. Main physical and chemical characteristics of crambe oil.

D_{20}	0.908
V_{25} (cp)	85
n_D^{20}	1.472
Melting point (°C)	6
Iodine value	93
Saponification value	160–175
Unsaponifiable fraction	0.5%

Table 16.42. Fatty acid composition of crambe oil.

Nature of fatty acids	*% of total FA*
C14:0	–
C16:0	2
C16:1	–
C17:0	–
C18:0	–
C18:1	15
C18:2	10
C18:3	7
C20:0	–
C20:1	3
C22:0	2
C22:1	55–60
C24:0	–

Vernonia Oil

This oil is obtained mainly from the seeds of *Vernonia galamensis*. It is a naturally epoxidised oil having a lower viscosity than industrially epoxidised soyabean and linseed oil (Tables 16.43 and 16.44).

Table 16.43. Main physical and chemical characteristics of Vernonia.

D_{25}	0.905–0.973
V_{20} (cp)	–
n_D^{25}	1.474–1.486
Melting point (°C)	–
Iodine value	104–108
Saponification value	165–210
Unsaponifiable fraction	1–8%

It is characterised by its high content of vernolic acid (*cis* 12–13 epoxyoleic acid) which may be as high as 80 per cent.

Table 16.44. Fatty acid composition of vernonia oil.

Nature of fatty acids	*% of total FA*
C14:0	–
C16:0	2–3
C16:1	–
C17:0	–
C18:0	2–4
C18:1	3–6
C18:2	12–14
C18:3	< 0.3
C20:0	< 0.5
C20:1	< 0.4
C22:0	–
C22:1	–
C24:0	–
cis 12–13 epoxyoleic	72–78

Cuphea Oil

This oil is obtained mainly from four species of cuphea selected for cultivation: *Cuphea lutea, laminuligera, lanceolata, C. viscosissima* (Table 16.45).

Table 16.45. Fatty acid composition of cuphea oil.

Nature of fatty acids	*Proportion, % of total FA*	
	Capric variety	*Lauric variety*
C8:0	< 1	–
C10:0	26	5
C12:0	59	81
C14:0	4	5
C16:0	< 1.5	2
C16:1	–	–
C17:0	–	–
C18:0	< 0.5	< 0.5
C18:1	3	3
C18:2	6	4
C18:3	< 0.5	< 0.5
C20:0	–	–
C20:1	–	–

This oil is a subject of agricultural research. It is characterised by its high content of medium chain triglycerides, which are a source of capric acid (C10:0) and lauric acid (C12:0) for oleochemistry.

Avocado Oil

Avocado (*Persea gratissima*) belongs to Lauraceae family. It is cultivated widely in all tropical and sub-tropical regions. More than 200 varieties have been identified. The pulp (60–75 per cent of the fruit) has a remarkably high oil content. Depending on the varieties, it varies between 5 and 30 per cent. The avocado kernel, on the other hand, has very little oil (≤ 2 per cent). This oil can be used in seasoning, but considering its high cost, it is usually not used for such purposes in Europe. Part of this oil is used in the soap industry but most of it is used by the pharmaceutical and cosmetic industries.

Extraction of avocado oil is difficult because the oil is found in a very fine emulsion in the fruit cells. Two processes exit: either processing fresh fruit or dried fruit. Treatment of fresh fruit is not economically advantageous and produces an oil loaded with water, making the storage of the oil difficult.

For these reasons, the process on dried fruit is preferred: the fruits are cut into slices and dried (80°C, 24–48 hours); after dehydration (residual humidity: 5–9 per cent) the pulp is pressed and/or solvent extracted. Crude oil which is viscous, strongly coloured, 'winterised' at 5°C has to be refined (giving a yellow or yellowish green coloured oil) for its main applications.

The residue obtained after oil extraction (deoiled pulp) has very little nutritive value. A product with strong antibiotic properties has been obtained from the avocado's kernel. Since, avocado oil has a particularly high unsaponifiable content, many industrial processes have been developed to extract this fraction (concentration by molecular distillation, then purification by saponification and counter-current solvent extraction). The unsaponifaible matter of avocado is used in many cosmetic formulations.

Physico-chemical properties

Table 16.46 shows the essential characteristics of avocado oil. *Jacobsberg* gives a saponification value of 220 for unrefined oil. *Swischer* reports a free fatty acid content of 1–3 per cent in crude oil and ≤ 0.5 per cent in refined oil. *Rancurel* notes an oleic acid acidity range from 0.6 to 2.4.

Table 16.46. Main physical and chemical characteristics in avocado oil.

D_{25}	0.910–0.920
V (cp)	
n_D^{20}	1.465–1.474
Solidification point (°C)	7–9
Iodine value	63–95
Saponification value	177–198

Chemical composition

Triglyceride composition

Table 16.47 shows the triglyceride composition of avocado oil. These values can vary considerably according to variety, area of production and fruit maturity.

Table 16.47. Triglyceride composition of avocado oil (HPLC).

Nature	*% of total TG*
OLL	5–8
PLL + PoOL	2–7

(Contd...)

Table 16.47. Triglyceride composition of avocado oil (HPLC). *(Contd.)*

Nature	*% of total TG*
OOL	10–18
POL + PoOO	12–27
PPL + PPoO	0–3
OOO	12–30
POO	19–25
PPO	3–11
PPoO	< 0.2

Fatty acid composition

Table 16.48 shows the fatty acid composition of avocado oil. Avocado oil is noted for its high palmitoleic acid content (C16:1, 2–13 per cent according to the varieties and extraction process chosen). Besides, this oil belongs to the 'oleic oils' group. Large ranges are observed in oleic and linoleic acid composition due to the numerous existing varieties.

Table 16.48. Fatty acid composition of avocado oil.

Nature	*% Total fatty acids*
C14:0	< 0.3
C16:0	7–32
C16:1	2–13
C17: tot.	< 0.3
C18:0	< 1.5
C18:1	36–80
C18:2	6–18
C18:3	0–5
C20:0	< 1
C20:1	–
C22:1	< 0.7

Composition of the unsaponifiable fraction

Table 16.49 shows the composition of unsaponifiable fraction of avocado oil. In addition to the different varieties, the extraction process (extraction in a laboratory or industrial process) influence the results considerably.

The unsaponifiable content is high: often it ranges between 1 and 5 per cent. Sterols represent 35 to 70 per cent of the unsaponifiable matter. *Rancurel* gives a different range, varying between 4 and 20 per cent. *Itoh*, reports the sterol content of avocado oil extracted from different Camaroonian varieties as 560 mg/100 g (34 per cent of the unsaponifiable fraction representing 1.6 per cent of the oil).

The sterolic fraction contain mainly β-sitosterol (60–92 per cent), compesterol (5–12 per cent), Δ5-avenasterol (1.8–10 per cent). Table 16.49 gives wider noticeable ranges. Differences are reduced in the case of peculiar varieties.

Table 16.49. Composition of unsaponifiable fraction of avocado oil.

Constituents	*Unsaponifiable fraction 1.0 to 12%*	*Constituents*	*Unsaponifiable fraction 1.0 to 12%*
Sterols (mg/100 g)	350–560	Hydrocarbon (mg/100 g),	310
		of which squalene	34–37
		Tocopherols (mg/100 g)	8–15
Sterol composition (% total sterols)		Tocopherol composition (% total tocopherols)	
Cholesterol	1–2	α-Tocopherol	77–100
Brassicasterol	–	β-Tocopherol	–
Campesterol	5–12	γ-Tocopherol	0–23
Stigmasterol	1–10	δ-Tocopherol	–
β-Sitosterol	60–92	Tocotrienols	–
Δ5-Avenasterol	2–10		
Δ7-Stigmasterol	1–2		
Δ7-Avenasterol	1–2	Triterpenic alcohols (mg/100 g)	15–40

Hydrocarbon content varies between 5 and 20 per cent of the unsaponifiable fraction; *Itoh* reports a content of 310 mg/100 g oil (19 per cent of the oil unsaponifiable fraction for Camaroonian varieties). Squalene ranges from 0.3–0.7 per cent (34 to 37 mg/100 g oil).

The triterpenic alcohol content varies between 6 per cent—approximately 100 mg/100 g oil and 15 to 40 per cent of the unsaponifiable fraction.

Numerous constituents of avocado unsaponifiable still remain unidentified.

The tocopherol content depends largely on the oil extraction process, without dwelling on the analytical method used for its determination (colourimetry, GC, HPLC, UV or fluorescent detection). Thus, depending on the technology used, *Turatti* reports a range of 2 to 22 mg/100 g oil (Brazilian variety). Concentrations reported in the literature vary between 8 and 15 mg/100 g, the tocopherol fraction being ≤ 0.1 per cent of the unsaponifiable. Only α and γ-tocopherols are found.

Tall Oil

Origin, production, applications

Tall oil is a by-product of the wood-pulp industry (sulphate pulping process Kraft). It is one of the major fatty acid sources for industrial applications. The composition of tall oil depends on tree species (mixed conifers—firs and spruces or mixed hardwood/pinewood furnishes), their origin and the industrial pulping process. Crude tall-oil contains mainly free fatty acids, rosin acids (colophony) and 'unsaponifiable neutrals'. The characteristics of good quality tall oil are shown in Table 16.50.

Table 16.50. Characteristics of crude tall oil (origin: South East USA).

Acid value	132–165
Saponification value	142–172
Rosin acids (%)	23–40
Fatty acids (%)	52–57
Unsaponifiable fraction (%)	8–20

Technology

Wood chips are cooked in an alkaline solution of soda and sodium sulphide called 'white liquor' (Kraft pulping process); during this process the triglycerides are saponified and sodium soaps of fatty and rosin acids dissolve in the cooking fluid which is now changed into a 'black liquor'. After separation from the wood pulp, this black liquor is cooked again and when the content of dry substances reaches 18–22 per cent, alkaline soaps are precipitated and in the presence of air, rise to the surface as a tall oil froth called 'skimmings'. They are skimmed off and treated with sulphuric acid to regenerate organic acids; the rosin acids crystallise during cooling. Crude tall-oil is then washed, dehydrated (under vacuum at 260°C) and then distilled.

The pitch obtained from first distillation contains sterols and sterol esters (neutral or unsaponifiable) which, can be recycled. Distillation processes have been improved: it is now possible to obtain a fraction of 98 per cent pure fatty acids and a rosin acid fraction containing less than 3 per cent fatty acids; such results require a good quality crude tall-oil (acid value: 150–160).

An analysis of crude tall-oil reveals the following average quantity for the different fractions.

Rosin	30 per cent
Fatty acids	28 per cent
Distilled tall-oil	7 per cent
Distillation heads	11 per cent
Distillation residues	19 per cent
Losses	5 per cent

Chemical composition

Fatty acid composition

It should be remembered that the rosin fraction contains abietic, dehydroabietic, neoabietic, palustric, pimaric and isopimaric acids.

The fatty acid fraction is divided into three commercial grades: Type 1 (98 per cent), Type II (96 per cent), Type III (90 per cent) (ASTM, 1976).

Table 16.51 gives the composition of crude tall-oil and tall-oil fatty acids. Tall-oil contains a substantial quantity of conjugated linoleic acid isomers produced during the pulping process.

Table 16.51. Composition of tall-oil fatty acids.

	% of Total FA	
Nature	*Crude tall-oil*	*Tall-oil fatty acids*
C16:0	3–4	< 1
C16:1		< 1
C17:0	–	–
C18:0		1–2
C18:1	22–26	45–55
C18:2	19–22	35–40
C18:2 conjugated	2–6	5–10
C20:2		1–2
Resin acids	40–45	< 1

Composition of unsaponifiable fraction

The unsaponifiable fraction of crude tall-oil varies between 8 and 20 per cent. β-Sitosterol is the principal component of the sterol fraction. Aliphatic alcohols are represented mainly by lignoceric alcohol (C24). *Conner* reports the composition of 'neutrals' (unsaponifiable fraction) in the foam of black pine liquor: they contain mainly diterpenes (41 per cent), a fraction of 'steroids', i.e. sterols, methylsterols, triterpenic alcohols and ketonic derivatives (32 per cent), fatty alcohols (6 per cent), stilbenes (6 per cent) and other constituents (15 per cent) such as monoterpene, polyprenols and minor constituents including triterpenes (0.6 per cent of neutral substances in foam), with squalene representing 33 per cent of particular fraction. Table 16.52 shows the composition of the sterolic fraction.

Table 16.52. 'Neutral' substances in the foam of pine tall-oil.

Steroid fraction: 32.4% Sterols: 30% Composition (%)	
Cholesterol	< 0.2%
Campesterol	8.4%
Campestanol	1.0%
β-Sitosterol	84%
Stigmastanol	6.4%
Triterpenic alcohols: 1.3%	
Methyl sterols and ketonic derivatives: 1.1%	

References

Brydson, J.A., *Encyclopedia of Food Science and Technology,* Elsevier Applied Science, London.

Brydson, J.A., *Chemistry and Technology of Edible Oils,* Applied Science, London.

Bunn, C.W., *Bailey's Industrial Oil and Fat Products,* Butterworth Scientific, London

Treloar, L.R., *Fatty Acids in Industry,* Clarendon Press, London.

Gardon, J.L., *Introduction to Fats and Oil Technology,* Interscience, New York.

Bateman, L., *Encyclopedia of Chemical Technology,* Applied Science, London.

Stannett, V., *Vitamins and Minerals,* Academic Press, London.

Wall, L.A., *Antioxidants,* Interscience, New York.

Noll, W., *Chemistry and Technology of Fats and Oil,* Academic Press, New York.

Blow, C.M., *Soyabean Utilisation Alternatives,* Butterworth Scientific, London.

Hofmann, W., *Essential Oils and Aromatic Chemicals,* Applied Science Publishers, London.

Smith, C.K., *Spices as Antioxidants,* John Wiley and Sons, New York.

Holden, G., *Chemistry and Technology of Lipids,* Reinhold, New York.

Lewis, F.W., *Soap Manufacture,* John Wiley and Sons, New York.

Brown, R.P., *Food Sciences,* Applied Science Publishers, London

Morton, M., *Dairy Technology and Engineering,* Van Nostrand Reinhold, New York.

Craig, A.S., *Fatty Acids and their Derivatives,* John Wiley and Sons, New York.

Blackley, D.C., *Basic Concepts of Food Technology,* Applied Science Publishers, London.

Saunders, J.H., *Chemistry of Animal and Vegetable Oils,* Interscience, New York.

Wright, P., *Edible Fats and Oil Processing,* Maclaren, London.

Alliger, G., *Chemistry and Biochemistry of Natural Waxes,* Reinhold, New York.

Scott, G., *Cosmetics and Toiletries,* Elsevier, Amsterdam.

Bruins, P.F., *Sunflower Science and Technology,* Reinhold, New York.

Kraus, G., *Olive Oil Technology,* Interscience, New York.

Donnet, J.B., *Elements of Food Engineering,* Marcel Dekker, New York.

Nourry, A., *Chemistry of Toxicity*, Maclaren, London.

Allen, P.W., *Cottonseed Oil,* Maclaren, London.

Anon, W., *Frying Technology,* Applied Science, London.

Collyer, H.J., *Dairy Technology,* Interscience, New York.

Khromov, M.K., *Nutrition and Dietics,* Maclaren, London.

Mooney, M., *Chemistry of Fatty Acids,* Academic Press, New York.

Norman, R.H., *Coconut Oil,* Maclaren, London.

Penn, W.S., *Castor Oil,* Maclaren, London.

Scott, J.R., *Chemistry of Emulsifiers,* Maclaren, London.

Stamberger, P., *Rendering, the Invisible Industry,* Oxford University Press, Oxford.

Stern, H.J., *Flavour and Fragrances,* Maclaren, London.

Thompson, D.C., *Encyclopaedia of Chemical Technology*, John Wiley and Sons, New York.

Sprague, G.R., *Handbook of Lipids,* Vanderbilt, New York.

Wheelans, M.A., *Bakery Products,* Maclaren, London.

White, J.L., *Science and Technology of Surfactants,* Academic Press, New York.

Nauntron, W.J., *Handbook of Vegetables Oils,* Arnold, London.

Kuryla, W.C., *Soyabean Utilisation Alternatives,* Maclaren, London.

Killeffer, D.H., *Sunflower Science and Technology,* Palmerton, New York.

Cooper, W., *Oleochemicals and Allied Products,* Applied Science Publishers, London.

Brennan, J.J., *Toxicity and Safe Handling of Oils and Fats,* Academic Press, New York.

Allen, N.S., *Chemistry of Lipids,* Elsevier, USA.

Bark, L.S., and **Allen, N.S.,** *Analysis of Oils and Fats*, Elsevier, USA.

Alfrey, T., *Chemistry of Oil and Fats,* Interscience, New York, USA.

Bernhardt, E.C., *Chemistry of Lipids,* Pergamon, UK.

Bernhardt, E.C., *Vegetable Protein Utilisation in Human Foods and Animals Feedstuffs,* Van Nostrand Reinhold, USA.

Bruins, P.F., *Canola Meal Feed Industry Guide,* Wiley Interscience, New York.

Ceresa, R.J., *Corn Chemistry and Technology,* Wiley Interscience, New York.

Ferry, J.D., *Handbook of Soap Technology,* Wiley Interscience, New York.

Harper, C.A., *Degumming, Refining, Washing and Drying,* McGraw-Hill, New York.

Iwen, E.D., *Antioxidants: The Ultimate Answer of Oxidation,* Elsevier, USA.

MacDermott, C., *Linseed Refining,* Marcle Dekker, Reinhold, USA.

Matthews, F.G., *Deodourisation—A Basic Understanding,* Van Nostrand Reinhold, USA.

McKelvey, J.M., *Spices, Herbs and Edible Fungi,* Wiley Interscience, New York.

Cherry, J.P., *Lecithins: Sources, Manufacture and Uses,* American Oil Chemists Society.

Alliger, G., *Chemistry and Technology of Lips,* Arnold, London.

Saunders, J.H., *Food Sciences,* Pergamon, UK.

Campbell, R., *Nutrition and Dietics,* Arnold, London.

Index

Accelerated storage tests, 99
Acid value/Percent free fatty acid, 12
Acidulation of the soapstock, 631
Active oxygen method, 96
Additives, 623
Adsorptive bleaching, 69
Advantages of methyl esters, 31
Alkali refining, 109
Alkanolamides, 48
Almond oil, 692
Ambient storage tests, 99
Analysis of olive oils, 221
Analysis of rapeseed lecithin, 484
Ancillary systems, 628
Anti-nutritional factor, 399
Antioxidants, 86
Application of castor oil and its derivatives in lubricant industry, 393
Applications of castor oil in biodegradable and environmental friendly lubricants, 395
Enzyme catalysed hydrolysis, 387
α-Sulphonated fats and oils, 679
Avocado oil, 710
Azelaic acid, 355

Batch degumming, 600
Batch deodourisation, 621
Batch neutralisation, 8
Batch refining, 604
Batch sulphatation of fats, 670
Biogas from palm oil mill effluent, 289
Biological processes and storage of cottonseed, 129
Blackcurrant oil, 702
Bleachability of palm oil, 243
Bleaching, ,9 10, 75, 610
Blown castor oil and other products, 350
Borage, 700

Camelina oil, 687
Canola oil extraction and processing, 61
Canola oil: physical and chemical properties, 76
Carbohydrates, 523
Cardiovascular disease, 589
Cardiovascular disease in Indians, 410
Castor cake, 391
Castor hulls, 391
Castor oil derivatives in grease industry, 395
Castor oil derivatives in textile industry as lubricants, 394
Castor oil emerging as vital chemical industry input, 397
Castor seed meal, 399
Castor stalks, 393
Catalytic hydrogenation, 625
Caustic soda-ash process for refining crude cottonseed-oil, 160
Changes in seeds with maturation, 121
Characteristics of palm oil refinery effluent, 302
Chemical or caustic refining, 602
Chemical properties of palm oil, 317
Chemistry of castor oil, 393
Chemistry of esterification and transesterification, 32
Chemistry of fat splitting, 25
Chemistry of hydrogenation 36
Chlorophylls, 220
Cholesterol and other blood fats, 411
Cholesterol and triglycerides as lipids, 412
Cholesterol and triglycerides circulate in blood as lipoproteins, 412
Clays for decolourisation, 634

Coconut cooking oil, 17
Coconut health oil, 17
Coconut oil as a body oil, 18
Coconut oil as hair oil, 19
Coconut oil in cosmetics, 18
Coconut oil in dental cleaning products, 20
Coconut oil in perfumes and flavours, 20
Coconut oil in shampoo, 19
Coconut palm, 1
Coconut water, 3
Commerical potential of enzymatic fat splitting, 388
Composition and characteristics of olive oil, 205
Composition of coconut oil, 10
Composition of palm oil and palm kernel oil, 236
Composition of peanuts, 659
Composition of soyabean oil, 540
Conjugated dienes and trienes, 94
Conjugated oils, 706
Continuous caustic soda refining method, 158
Continuous degumming, 601
Continuous deodourisation, 622
Continuous dewaxing by-products, 637
Continuous hydrogenation, 628
Continuous neutralisation (short mix), 9
Continuous sulphatation, 674
Copra, 4
Copra meal, 7
Cottonseed kernel, 120
Cottonseed-cake or meal, 181
Crambe oil, 707
Crude sunflower oil, 598
Crystallisation, packaging and tempering, 74
Cuphea oil, 709
Cyclic dihydroxy compounds, 220
Cyclic monohydroxy compounds, 220

Daily fat intake, 432
Deep frying fat, 305
Degumming, 10, 75, 598
Dehulling, 140
Dehydrated castor oil (DCO), 339
Dehydrated castor oil (DCO) fatty acids, 344
Delinting, 138
Deodourisation, 9, 74, 463
Deodourisation equipment, 621
Deodourisation of soyabean oil, 567
Deodourisation operation, 621
Dewaxing, 73, 611
Dietary fat recommendations, 414
Differentiation between olive oil and
olive-pomace oil, 223
Direct injection, 95
Direct sulphonation of saturated fats, 679
Disadvantage of the method, 603
Dynamic headspace, 95

Edible cottonseed flour for human consumption, 186
Edible oil processing, 550
Edible products, 15
Edible products from rapeseed oil, 465
Edible uses of rice bran oil, 436
Effect of feedstock impurities on hydrogenation, 626
Effect of palm oil on blood lipids, 326
Effect of rapeseed oil on blood cholesterol, 498
Effect of rapeseed oil on longevity, 501
Effect of rapeseed oil on the liver, 497
Effect of rapeseed oil on the thyroid gland, 497
Effect of rapeseed oil on vitamin requirement, 499
Energy potential from oil palm by-products, 288
Enzymatic splitting, 385
Enzyme catalysed hydrolysis
Enzyme inactivation, 507
Epoxidised castor oil, 348
Erucic acid and its cleavage products, 480
Erucyl amide, 480
Ethoxylated castor oil, 348
Evening primrose oil, 698
Extraction of olive oil, 212

Fat development in oilseeds, 121
Fate of dietary cholesterol and trigycerides, 412
Fatty acid composition, 12, 695
Fatty acid distillation and fractionation operations, 28
Fatty acid fractionation, 29
Fatty acid profile of castor oil, 399
Fatty acids, 25, 80
Fatty acids and derivatives from coconut oil, 20
Fatty alcohols, 35
Fatty alcohols from natural fats and oils, 36

Filter aid and dewaxing, 637
Filtration and clarification systems, 618
Flavour quality of canola oil, 92
Flavour stability of canola oil, 96
Food beans and oil beans, 536
Formulation of margarine oils, 571
Fractionation of fatty alcohols, 41

Gas chromatography, 658
Glucosinolates, 504
Glycerine, 41
Glycerine from fats and oils, 41
Glycerine recovery and bleaching, 163
Glycerol monoricinoleate, 354
Grapeseed oil, 684

Handling and practical storage of cottonseed, 134
Hazelnut oil, 694
Health and soyafoods, 538
Health benefits of rice bran oil (RBO), 434
Heavy pathogen infestation affecting groundnut production, 665
Heptaldehyde, 363
Heptanoic acid, 364
High pressure hydrogenation processes, 38
Hydrogenated canola oil in edible oil products, 72
Hydrogenated castor oil, 345
Hydrogenation, 71, 75, 624
Hydrogenation process, 625
Hydrogenation equipment, 627
Hydrolysis of castor oil, 377

Immune system, 115
Impurities in cottonseed, 135
Industrial uses of sulphated and sulphonated oils and fats, 681
Insoluble carbohydrates, 538
Instant sulphatation, 673
Integrated technology for refining rice bran oil and by-products recovery, 419
Interesterification, 73
Intermix method, 608
Iodine value, 13, 86
Isoflavones, 586
Isolation of the protein, 510

Kernel, 3
Kinematic viscosity, 657
Kinetics of enzyme catalysed hydrolysis, 387

Lauric, 305
Lecithin, 417
Lipids, 518
Lipolysis, 132
Lurgi fatty acid hydrogenation method, 40

Main methods of fat sulphatation, 670
Main uses of blackcurrant seed oil, 705
Major environmental factors that inflence lipid content 663
Margarine, 304, 472
Mayonnaise, prepared salad dressings, salad oils and margarine 578
Mechanical extraction 5
Medical and infant food formulations 15
Metabolism of erucic acid 500
Methyl 12-hydroxy stearate 355
Methyl esters 30
Methyl ricinoleate 353
Modified castor oils, 339
Modified soda-ash process for refining cottonseed-oil, 160
Monoalkyl phosphates, 47
Monounsaturated fatty acids, 88
Mycotoxin removal, 111

Nickel crystallites, 626
Nonfood uses for soyabean oil, 585
Non-mechanical traditional methods of extraction of oil and kernels, 249
Non-nutritional uses of rapeseed oil and rapeseed fatty acids, 476
Nucellus, 120
Nutritional aspects of cottonseed oil, 180
Nutritional effects of palm oil, 326
Nutritional properties of minor components in palm oil, 327

Nutritional quality of glandless cottonseed-meal, 185
Nutritional significance of canola oil composition, 87
Nutritional significance of rice bran oil, 406
Nutritional value and use of rapeseed meal, 501
Nutritional value of rapeseed oil, 486
Nutritive value of cottonseed flour, 187
Nutritive value of heated and oxidised oils, 585
Nutritive value of soyabean oil, 584
Nylon-11, 358

Oil content of peanuts, 659
Oil degumming and soyabean lecithin, 551
Oil palm by-products, 286
Oil palm products, their formation and characteristics, 230
Oils of Elaeis oleifera, 239
Oleaginous flax, 690
Oleic sunflower, 595
Oleochemical derivatives from coconut oil and their general applications, 24
Oligosaccharides, 587
Olive oil components, 219
Origin analysis and technology developments of castor oil and its derivatives, 400
Oryzanol, 416
Oxidative stability of soyabean oil, 574
Oxygen absorption tests, 96

Palm kernel cake, 239
Palm kernel oil, 239
Palm kernel production control, 235
Palm kernels, 233
Palm kernels, palm kernel oil and palm kernel cake, 245
Palm oil, 236, 240
Palm oil delays ageing, 329
Palm oil mill effluent (POME), 287
Palm oil refinery effluent treatment, 299
p-Anisidine values, 94
Pathological changes induced by rapeseed oil, 499
Peanut oil, 655
Peroxide value/stability test, 14
Pesticide removal, 111
Phosphoric acid use in alkali refining, 604
Phosphorus and phosphatides, 627
Physical and steam refining 609
Physical properties of soyabean oil, 543
Physico-chemical properties, 710
Physico-chemical properties of castor oil, 398
Pigments of cottonseeds, 128
Platelet aggregation, 115
Polar lipids, 83
Polenske value and Reichert-Meissel Values, 658
Polyunsaturated fatty acids, 88
Poppy seed oil, 688
Post-expeller treatment, 6
Preparation of edible cottonseed flour and its utilisation, 187
Preparation of high-stability soyabean oils by hydrogenation–winterisation, 569
Preparation of partial α-sulphonates from saturated fats, 681
Preparation of partially sulphonated saturated oils, 681
Preparation of sulphonic acids from saturated fats, 679
Pre-treatment and degumming, 155
Principles and methods in the manufacture of oleochemicals, 25
Process control of alkali refining, 75
Processing and utilisation of cottonseed oil, 154
Processing of cottonseed, 135
Processing of crude palm oil, 298
Production and utilisation of cottonseed and cottonseed products, 118
Products of ozonolysis, 480
Properties of castor oil, 332
Purpose of bleaching, 610
Purpose of processing the by-products, 629

Quality assurance, 75
Quality of palm oil and kernels, 240
Quality standards, 245
Quantity controls, 235

Rapeseed fatty acids, 479
Rapeseed lecithin, 481
Rapeseed oil, 476
Recovery of by-products from rice bran oil (RBO) refining, 420
Recovery of oil from soyabeans, 547

Refined cottonseed oil, 200
Refining of olive oils, 217
Refining of pomace oil, 218
Refining, bleaching, dewaxing and deodourisation, 109
Regeneration of clay, filter aid and catalyst, 637
Reichert-meissl value, 13
Resistance to oxidative rancidity, 96
Respiration and lipolysis, 132
Rice bran oil—Heart-friendly cooking oil, 431
Rice wax—utilisation, 428
Ricinoleic acid, 352
Ricinoleyl alcohol, 354
Role of fat in human body, 432
Role of fatty acids and micro-nutrients on cardiovascular disease with special reference to rice bran oil, 410

Safflower meal, 644
Safflower oil, 641, 647
Safflower seed, 639
Salad and cooking oils, 471
Saponins, 588
Saturated fatty acids, 88
Sebacic acid, 364
Sedimentation phenomenon in canola oil, 86
Selectivity in hydrogenation, 626
Semicontinuous deodourisation, 622
Sensory evaluation, 92
Sesame cake and meal, 529
Sesame seed and other by-products, 532
Shell and husk, 4
Shortening, 304, 579
Shortenings or compound cooking fats, 473
Soaps, 353
Soapstock as an additive to meal or feed, 631
Soapstock dehydration, 631
Soapstock from alkali refining, 631
Solvent extraction, 6
Sources of cholesterol, 412
Sources of triglycerides, 412
Soya chemistry, 536
Soya proteins and heart disease, 589
Soya—A healthy complex of phytonutrients, 586
Soyabean oil food products—their preparation and uses, 578
Soyabean oil quality, 573
Speciality fats, 305
Stability to photochemical oxidation, 99
Steam stripping and deodourisation, 10
Steps involved in processing rapeseed oil, 438
Storage of coconut oil, 18
Studies on cottonseed-cake, 181
Sulphatation, 669
Sulphated castor oil, 346
Sulphated oils, 669
Sulphonated oils, 676
Sulphonation, 669
Sulphonation of fatty acid methyl esters, 57
Sunflower oil, 596
Surfactants, 51
Suspension process, 39
Synthetic esters from castor oil, 394

Tall oil, 712
Tertiary amines, 58
Testa, 3
Testing for wax, 616
Titer, 15
Tocopherols, 587, 590
Tocopherols and sterols, 84
Treatment of palm oil mill effluent, 267
Triacylglycerols, 80
Triglyceride structure of peanut varieties, 662
Triglycerides: Another important blood fat, 411
Triterpenic acids, 220
Twitchell process, 384
Types of fatty acids and derivatives from coconut oil and their general applications, 21
Types of lipoprotein particles, 412

Undecanedioic acid, 363
Undecyl alcohol, 362
Undecylene aldehyde, 362
Undecylenic acid, 356
Unsaponifiable matter, 13
Uses of coconut oil, 15
Uses of cottonseed protein flour, 189
Uses of soyabean oils in food products, 581
Utilisation of cottonseed oil, 174
Utilisation of rapeseed lecithin, 485

Vanaspati, 304
Vernonia oil, 708
Versatility of palm products, 330
Virgin coconut oil, 16
Volatile analysis by gas chromatography, 94
Walnut oil, 696
Western soya foods, 539
Wheat germ oil, 705